Genetics

SECOND EDITION

John B. Jenkins
SWARTHMORE COLLEGE

HOUGHTON MIFFLIN COMPANY   BOSTON

Dallas
Geneva, Illinois
Hopewell, New Jersey
Palo Alto
London

To Fereshteh, John, and Soraya

The *Drosophila* shown on the cover is a mosaic mutant—that is,
one that has two genetically distinct classes of tissue. It
is a very rare event under normal circumstances, but it can
be induced by mutagenic agents. Tissue in this animal is either
heterozygous for the white-eye mutation, in which case the
eye is red, or it is lacking the wild-type gene, in which case
it is white.

PRINTED IN THE U.S.A.

Library of Congress Catalog Card Number: 78-69608
ISBN: 0-395-26502-9

Contents

crossing-over   The mechanism of crossing-over
Cytological verification of crossing-over   The significance of
crossing-over   Mitotic crossing-over

Overview *100*

Days

Daughters of Time, the hypocritic Days,
Muffled and dumb like barefoot dervishes,
And marching single in an endless file,
Bring diadems and fagots in their hands.
To each they offer gifts after his will,
Bread, kingdoms, stars, and sky that holds them all.
I, in my pleached garden, watched the pomp,
Forgot my morning wishes, hastily
Took a few herbs and apples, and the Day
Turned and departed silent. I, too late,
Under her solemn fillet saw the scorn.

RALPH WALDO EMERSON
1851

Preface

This book is intended for use in a one-semester genetics course. It presupposes some background in basic biology and chemistry as they might be taught in freshman college courses or perhaps some of the more advanced high school courses. However, a deficiency in chemistry or biology will not seriously impede the student's progress through the book.

This book develops the principles of genetics within a historical framework designed to enhance the student's understanding of key ideas. I have tried to unravel the fabric of modern genetic concepts enough to expose the threads of thought that hold that fabric together. First we explore the ideas and experiments that have contributed to our understanding of transmission genetics in higher organisms. Then we analyze the research leading to the identification of the genetic material and to the elucidation of its structure and replication. Next we examine the function of the genetic material, then analyze transmission genetics in viruses and bacteria. After this, we consider the regulation of genes; then we study population genetics and the process of speciation. I explain the chapter-by-chapter sequence of topics, as well as the major changes made in this second edition, in greater detail in the Prologue addressed to readers of this book.

My motivation for writing this text has its origins in my graduate training with Eldon J. Gardner at Utah State University and Elof A. Carlson at UCLA, and it has been strengthened by the many students I have had the pleasure of teaching at Swarthmore College. While studying with Professors Gardner and Carlson, I acquired a deep appreciation of the historical forces that led to the emergence of important genetic concepts. From my students has come the encouragement and stimulation to carry out the approach I have chosen to use.

It is, of course, impossible to complete a project of this magnitude without the assistance of many people. The extremely talented staff at Houghton Mifflin provided much needed support and assistance at various points along the way. Many of my friends and colleagues read all or parts of the manuscript as it was taking shape; their advice was crucial to the development of this second edition. I would like especially to thank my good friend Harry O. Corwin at the University of Pittsburgh for critically reading the entire manuscript and offering many helpful suggestions. My gratitude is also extended to Benjamin W. Snyder of Swarthmore College, Gareth R. Babbel of the University of South Florida, Eugene Katz of the State University of New York at Stony Brook, Henry J. Wehman of King's College, Pennsylvania, and Charles H. Ellis, Jr., of Northeastern University, for their perceptive advice as the manuscript took shape. To these individuals, to users of the first edition who commented on the book, and to others too numerous to mention, I want to say that this second edition has greatly benefited by their input. The final content and accuracy of this book is, of course, my own responsibility.

I could not have even begun this second edition without the love and support of my family. Writing is an intensely personal and private occupation that requires long hours of solitude. Fereshteh and I sacrificed a great deal of togetherness while this book was developing. Making up for that lost time together is a task we both look forward to. John and Soraya have been absolutely marvelous during the writing of this book. Their enthusiastic love and understanding has been a constant source of joy for me. They will now see more of their father.

John B. Jenkins
Swarthmore, Pennsylvania

Prologue

Of all the biological sciences, none has had a more exciting or rapid course of development than genetics. In 1865 an obscure Austrian monk, Gregor Johann Mendel, suggested that every cell contained various pairs of "factors" and that each pair determined a specific trait. The members of each pair segregated from each other in the process of sex-cell formation; and the segregation by each pair—and its resulting trait—was unaffected by the other pairs.

Profound as this deceptively simple observation was, it was not recognized as such for 35 years. At least part of the reason for the delay was the absence of a firm cytological basis for understanding what Mendel had observed. By 1900, though, this was no longer the case. In that year Mendel's work was once more brought to light, this time with a solid cytological framework. We can mark 1900, then, as the birth year of modern genetics.

The spectacular unfolding of genetic concepts has taken about three-quarters of a century. We have moved from units, or factors, segregating and assorting independently in the nucleus of a cell, to the identification of DNA as the genetic material, to the incredible yet beautifully organized sequence of events by which genes (DNA) produce their products.

This unfolding of genetic concepts is really much more than the erection of a structure within which we can interpret various aspects of heredity. It is a model of scientific methodology, of human genius, and, above all, of human potential. The cast of characters that have contributed to the effort reflect human qualities that we dare not ignore. We witness individuals who discover or observe, but fail to interpret correctly; others who integrate diverse observations and arrive at new and significant findings that have a major influence on people's thinking; and still others who, though incorrect in their thinking, serve as catalysts for the thinking of others. Hence genetics is more than just a conceptual science; it is a vital and dynamic science that touches all facets of our being.

The chapters that follow discuss not only current research, recent trends, and basic concepts of genetics, but also the activities and ideas that have brought us to our present state of advancement. Therefore, whenever the opportunity presents itself, we present the sequence of ideas and experiments leading to the elucidation of major principles along with the principles themselves.

The organization of the chapters follows a logical progression of ideas. In Chapter 1 we consider the events and influences leading to the work of Mendel and its subsequent fate. We develop the chromosome theory of inheritance and show that Mendel's "factors" are indeed chromosomes. In Chapter 2 we pursue this theory and show that the behavior of chromosomes during meiotic division parallels the behavior and distribution of genes; that is, that there is a cytological basis for Mendelian patterns of inheritance. Chapters 3 and 4 extend Mendelian principles into other areas. They specifically look at quantitative inheritance, environmental effects, gene–gene interaction, and cytoplasmic inheritance (a type of inheritance that does not obey Mendelian principles because the traits are not chromosomally associated). Chapter 5 explores variation in chromosome number and structure.

What the first five chapters do, then, is lead us from pre-Mendelian concepts of inheritance, to Mendel and Mendelian principles, and to the ramifications of Mendelian principles.

Chapter 6 introduces us to the intensely exciting and sometimes frustrating developments leading to the identification of nucleic acids (DNA and RNA) as the carriers of genetic information. The pivotal point in this chapter is the determination of the structure of DNA, because once this is known, models for replication, mutation, and information storage become apparent. Chapter 7 explores replication, one of the logical consequences of the DNA model and an important aid to understanding subsequent chapters.

In Chapters 8 and 9, we examine the process by which genes (DNA) produce gene products (usually proteins). Chapter 8 sets forth the events leading to the discovery that genes produce proteins and to an understanding of the actual process of protein assembly in living systems. Chapter 9 discusses the fascinating unraveling of the genetic code—how an alphabet with four letters (the four bases in DNA) translates into one with twenty (the twenty amino acids in protein), and how such a nucleic acid-protein association could have evolved.

Mutation processes are examined in Chapter 10 because, in order to study gene function, we need to know when a gene's function has changed and what caused that change. Moreover, the study of mutation leads to an understanding of gene structure, thus allowing us to correlate gene structure and gene function.

Chapters 11 and 12, dealing with the genetics of bacteria and viruses, are conceptually similar to the first four chapters in that they concern the transmission of genetic information from one generation to the next. The similarities cease here, however, because bacteria and viruses have evolved unique methods of transmitting and recombining genetic material. These methods, along with the events leading to their discovery, are discussed in depth.

It is known that not all genes in a cell are functioning at the same time. Chapter 13 views the phenomenon of gene regulation by studying first the experiments pointing to differential gene activity and second the various factors that regulate gene activity in prokaryotes and eukaryotes.

Chapters 14 and 15 explore the integration of genetics with Darwinian evolution. First we apply Mendelian principles of inheritance to populations of sexually reproducing organisms, then we examine the process of speciation.

The last chapter is a rather personal look at how genetics has been used and misused in the past and present. We also look into the future and assess some of the areas of human existence in which genetics may play a key role.

In this edition, as in the first, you will find that we often couple historical perspectives with thorough examinations of the experiments and ideas that led to major genetic concepts. But this edition also contains some important advances. I have greatly elaborated on the material concerned with the expansion of Mendelism, so that it now spans two chapters (Chapters 3 and 4) instead of one. The replication of the genetic material now has a chapter of its own (Chapter 7). The regulation of gene activity in eukaryotes has been expanded in a major way (Chapter 13). And population genetics has been developed to the point at which there are now two chapters: one dealing with genes in populations (Chapter 14) and one dealing with the process of speciation (Chapter 15). Chapter 16, Genetics: Past, Present, and Future, is entirely new.

In addition to these structural alterations, each chapter has been rewritten and updated. More human examples have been provided where appropriate. A large number of questions and problems have been added to the ends of the chapters, and a new section entitled Problems for Review has been added at the end of the book. It contains additional questions and problems to help the student review basic material. To offer students and instructors the greatest flexibility in using the questions and problems in the text, I have provided answers to selected end-of-chapter problems, plus answers to all the Problems for Review, at the back of the book. I have also put in a glossary to help students with some of the vocabulary of genetics.

Answers and solutions to *all* end-of-chapter questions and problems are contained in the Instructor's Manual that accompanies this text. Instructors who wish to provide their students with all the answers (rather than just those appearing in the book) are hereby given permission to reproduce pages from the Instructor's Manual for distribution to their classes. The Instructor's Manual also contains additional resources, including lists of films and laboratory manuals for genetics, suggestions for term-paper topics, and a chart correlating chapters of the textbook with the readings in several collections of fundamentally important papers in genetics.

One

The establishment of Mendelian principles of inheritance

It has long been known that although offspring are never identical to parents, kinship is often visible and clear. Throughout most of human history, however, the mechanism of heredity that accounts for the similarities and differences among parents and progeny was unknown. Theories, of course, abounded. The subservient position of women in many societies, both past and present, may reflect the belief that the female simply housed the hereditary material provided by the male. And we find in medieval Germanic literature the idea that the uncle, not the father, sponsors the growing youth, which may denote confusion regarding the male's role in reproduction and inheritance. Indeed, for decades a debate raged as to whether it was the male or female that contributed a miniature, preformed individual to the next generation.

In the midst of such confusion, it is not surprising that, in an effort to explain the origin of variations, people sought answers in terms of magic, folklore, gods, and spirits. Perhaps the most reasonable interpretation of the phenomenon of inheritance came from the ancient Greeks. A writer of the Hippocratic school of medicine, which flourished around 400 B.C., suggested a mechanism for variation remarkably like the one Darwin and Lamarck proposed some 2200 years later. He surmised that the environment could direct variation along a specific pathway—a belief later referred to as the "theory of acquired characteristics." Noting the practice of a people who bound children's soft skulls to elongate and point them, the writer deduced that in the course of time the elongated and pointed head became an inherited characteristic, because the trait would somehow be transmitted to the semen and in turn be passed on to the next generation. Not long afterward, Aristotle (384–322 B.C.), somewhat more cautiously, also suggested that acquired traits could be inherited. He pointed out, however, that although deformed parents occasionally produced deformed offspring, the offspring of cripples were usually of sound body. He added that children sometimes resemble their grandparents rather than their parents. Thus, without any idea of the mechanism or even the concept, Aristotle (Figure 1.1) grasped the common genetic phenomenon of recessiveness. In one form or another, observations and ideas such as these were important forces in biological thought well into the nineteenth century.

If we take 1900—the year in which Mendel's work was discovered after 35 years of obscurity—as the birthdate of modern genetics, then we can refer to the whole preceding era of biological thought as the *pre-Mendelian era*. Like most great scientific theories, Mendel's theories did not spring up from nowhere. They were the logical outgrowth of observations, speculations, and experiments that had been carried on at an accelerating pace for over a century. Some of his predecessors had discovered many of the things he found, but they were less thorough and less discerning. Their work makes up the immediate pre-Mendelian era, and to trace it can be instructive.

Figure 1.1
Aristotle

Linnaeus and the fixity-of-species doctrine

Figure 1.2
Carl von Linné

The Bettmann Archive, Inc.

Among the views dominating biological thought during the 100 years before Mendel described his findings in 1865, none were more influential than those of the Swedish taxonomist Carl von Linné (often Latinized as Carolus Linnaeus, 1707–1778). Today he is best remembered for his work on the systematic classification of plants and animals. In keeping with his view of the scientist as simply a recorder of observations, Linnaeus (Figure 1.2) held strong antiexperimentalist views, and his prestige stimulated the kind of research that dealt with the collection and classification of data rather than original investigation.

One notable idea—believed to be a scientific principle in the seventeenth century and rooted in the distant past—owed much of its nineteenth-century prominence to the support of Linnaeus and his followers. This is the *fixity-of-species* doctrine, which stemmed from the religious belief in *special creation*. According to this view, animal and plant species were created by God and remained unchanged and unchangeable. For over 100 years, this concept tended to warp or retard scientific thinking. It is a tribute to the insight and strength of scientists like Lamarck and Darwin that they were able to withstand the pressures of such a doctrine, proposing instead that organisms change over time and that new species evolve from preexisting species.

The doctrine of fixed species imposed on its adherents the idea that nature is static. Thus it excluded the possibility of evolution except for minor intraspecific variations, such as those found within the dog family. The doctrine also influenced the interpretation of experiments in plant hybridization conducted over nearly a century. Before Mendel, plant hybridizers interpreted their experiments within the constraining notion that species had been fixed by special creation. Thus prevented from seeing their hybrids as a means of analyzing the laws that govern the inheritance of variation, they failed to appreciate the full implications of their own work and ignored or rejected many of their own findings.

***Naturphilosophie* and its effect on genetic thought**

A second major influence on biological investigation, especially as it related to inheritance, was the German school of thought known as *Naturphilosophie*. Led by such influential figures as the philosopher G. W. F. Hegel (1770–1831), the poet-philosopher J. W. Goethe (1749–1832), and the physician A. W. Henschel (1790–1856), this group held that nature was unity, that the individual organism was the expression of its interacting constituents, and that the study of the parts when separate from the whole offered little promise. (This is sometimes referred to as the *holistic* view of living systems.) Unfortunately for the science of genetics, two plant breeders (J. Kölreuter and K. F. von Gärtner) whose work Mendel later showed to be on the right track were adversely affected by the nature-philosophers; progress in genetics was perhaps delayed by 75 years.

Figure 1.3
John Dalton

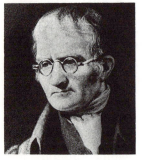

The Bettmann Archive, Inc.

Yet another influence on the study of heredity resulted from atomic theory. In 1808, the British chemist John Dalton (Figure 1.3) published what is now a classic, *A New System of Chemical Philosophy*. It contained much that is relevant to biology, including the idea that all matter is composed of atoms, indivisible and hence the basic building blocks of all matter. This theory was particularly influential in undermining the holistic viewpoint of the nature-philosophers, because it suggested that organisms could be usefully studied in terms of their parts. The view was impressively applied to living things as early as 1839, when Matthew Schleiden and Theodor Schwann proposed that all living things are composed of *cells*, the basic units of life.

The cell theory spread swiftly, and by the time Mendel's ideas were discovered in 1900, much was known about cell structure and function, including the role of the nucleus in fertilization and the existence of chromosomes and their behavior during mitosis and meiosis. Indeed, the analytical view was rapidly replacing the holistic view of the nature-philosophers by the time Mendel was performing his experiments.

Darwin and the
origin of variation

Figure 1.4
Charles Darwin

The Bettmann Archive, Inc.

Charles Darwin (Figure 1.4) developed some highly novel views of inheritance, views that became prominent in the latter part of the nineteenth century. In 1859, Darwin published *The Origin of Species*. In his five years as naturalist on *H.M.S. Beagle*'s voyage exploring the South American coast and the South Seas, Darwin had an almost unparalleled opportunity to observe the teeming forms of life and their complex interrelationships. He noted that a species tends to reproduce exponentially; in the absence of environmental pressures, however, while under normal conditions, the size of a population remains stable over long periods. He further observed considerable variation within a species, even among individuals of the same subspecific group. He concluded that there must be a competition for survival, or a ''struggle for existence,'' and that some variants are better able to survive than others.

Darwin's book-length statement of the theory of evolution underlined the need for insight into the mechanism of variations in living things. But it was here that Darwin came up against his principal difficulty. He knew of no mechanism that could account for the variations he observed everywhere. He proposed that change occurred gradually, over generations, and that large, sudden changes (*macromutations*) played little part in the mainstream of evolution, because organisms already adjusted to their environment could tolerate only minor changes and still remain viable. What he sought was a mechanism that would allow for the gradual accumulation of these minor changes.

In his search for such a mechanism, Darwin also rejected hybridization—that is, the production of offspring from the mating of individuals belonging to different species or varieties. He reasoned that if crossbreeding were common and if, as demonstrated, it resulted in a blending of characteristics, then it would not result in an accumulation of minor changes but would tend to produce homogeneity among species.

Darwin was also familiar with the work of Jean Baptiste Lamarck. In his *Philosophie Zoologique* (1809), Lamarck had argued that the environment directly influences the characteristics an organism inherits, because the organism responds to the environment by gradually acquiring adaptive characteristics that over generations become inherited. Thus, in time, the giraffe acquired a long neck because of its continuous need to stretch it in order to reach the leaves it feeds on. Put somewhat more broadly, Lamarck believed that the more a structure is used, the more prominence it assumes, and vice versa. This came to be known as the *doctrine of use and disuse*. Darwin was at first reluctant to accept the Lamarckian explanation of variation, but because hybridization and macromutation appeared to be evolutionary dead ends, he felt compelled to accept a theory of variation based on environmental influence.

In 1868 (three years after Mendel announced his discovery of the laws governing inheritance), Darwin published his *Variations of Animals and Plants Under Domestication,* in which he presented what is called a *theory of pangenesis* (Figure 1.5). According to this theory, each part of every

Figure 1.5
Darwin's theory of
pangenesis

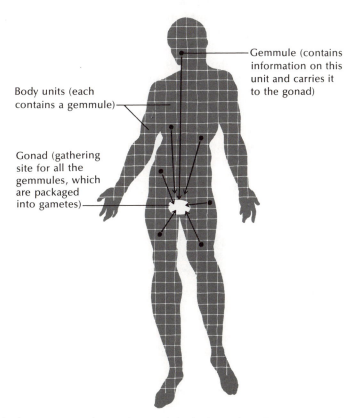

Body units (each
contains a gemmule)

Gemmule (contains
information on this
unit and carries it
to the gonad)

Gonad (gathering
site for all the
gemmules, which
are packaged
into gametes)

This theory assumes that each part of the body produces its own gemmules and that these are transported by the bloodstream to the gonads, where they become part of the semen.

6

*The establishment of
Mendelian principles of
inheritance*

living body contains small units called *gemmules*, which contain the essence of that part. These gemmules filter into the circulatory system and are carried to the gonads, where they are incorporated into the gametes and eventually transmitted to the next generation. The gemmules, Darwin stated, can respond to the environment in specifically directed ways and hence are the basis for the origin of variation.

Darwin's cousin, Francis Galton (1822–1911, Figure 1.6), who was skeptical of the inheritance of acquired characters, tested the theory of pangenesis (Figure 1.7). He transfused blood from a dark rabbit to a white one, reasoning that if there were gemmules in the circulatory system, they would be transferred with the blood. Mating the transfused white rabbit with a true-breeding white should then produce offspring generally darker than the white parents, or offspring with dark patches. Galton saw no such result and therefore rejected pangenesis as Darwin had described it.

Galton's own concept of the inheritance mechanism was somewhat closer to the facts as they are now understood. He proposed that the hereditary units were in the gonads but were not carried there from other parts of the body. The process of cell division was then known in broad outline, and Galton hypothesized that the gemmules were distributed to sex cells by this process. He theorized that a gemmule is replicated prior to cell division so that each daughter cell receives a full complement of hereditary material. Galton thought that a specific number of gemmules were responsible for each inherited trait and that the trait varied with the number of gemmules influencing it. Thus for color—white, black, or a shade between—two gemmules would give three possible combinations: two white gemmules in a cell, one white and one black, or two black. Three gemmules would give four possible combinations and hence four possible shades, and so on. The relationship between number of gem-

Figure 1.7
Galton's test of the pangenesis theory

Transfusion
of blood

Offspring of transfusion
recipient when mated to
another white rabbit;
no black appears

Galton tested Darwin's theory of pangenesis by transfusing blood from a dark-coated rabbit to a white-coated one. The latter was then mated to another white-coated rabbit; but since the offspring showed no black pigmentation, Darwin's view of pangenesis was discredited.

mules and number of possible expressions of a trait followed a predictable pattern expressed in what is known as *Pascal's triangle:*

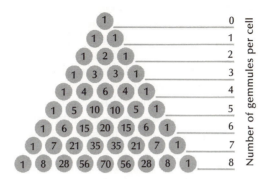

Thus six gemmules for color would give seven classes: white, black, and five shades of gray. The ratios of these classes would be 1:6:15:20:15:6:1. To explain the origin of variation, Galton suggested that gemmules could exist in different stable chemical states. He used as an analogy a polygon whose chemical effect might vary with the side it rested on (Figure 1.8). Although this interpretation would give discrete classes, Galton later rejected the idea of discontinuous variation (discrete classes) in favor of a blending pattern of inheritance (continuous variation), which would better account for the gradual changes Darwin had noted. Yet Galton's work resembled Mendel's in two important ways: It envisioned hereditary units that segregate at random, and it used statistical methods.

Figure 1.8
Galton's analogy between gemmules and polygons

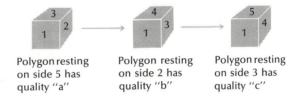

Polygon resting on side 5 has quality "a"

Polygon resting on side 2 has quality "b"

Polygon resting on side 3 has quality "c"

Mendel's predecessors

It is unfortunate that during much of the pre-Mendelian period, scientists had little communication with practicing gardeners and farmers who, by various techniques of hybridization and selection, were developing new varieties of plants and animals and improving breeds already in existence. On the continent, the Hegelian nature-philosophers were at least partly responsible for this gap in communication, because they discredited observation and experiment and elevated only intellect and intuition. The work of hybridizers studying sexuality and heredity in plants also suffered somewhat from this attitude.

8

*The establishment of
Mendelian principles of
inheritance*

Kölreuter   One of the earliest of the plant hybridizers with a direct link to Mendel was Josef Gottlieb Kölreuter (1733–1806, Figure 1.9), who in 1760 published his *Preliminary Reports of Experiments and Observations Concerning Some Aspects of the Sexuality of Plants.* Although largely overlooked by the scientific community of its time, the book is a landmark, describing over 500 plant hybridization experiments. In one series, Kölreuter used tobacco plants, and in a cross between *Nicotiana rustica* and *Nicotiana paniculata,* he secured an intermediate hybrid, vigorous and healthy but sterile when self-fertilized. (Kölreuter was the first to follow crosses systematically through more than one generation and the first to observe and record the segregation of characteristics, though he did not fully appreciate what he saw.) He believed in the doctrines of special creation and fixity of species, however, so the vigor of some of the first-generation hybrids disquieted him; he was reassured to find them sterile. When the crossing of a red with a white carnation produced pink offspring that were self-fertile and produced white, pink, and red plants in the second generation—a phenomenon called *segregation*—he suggested that in such cases, the hybrid was unstable because it was unnatural and that it tended to revert to the original parental types. Again he was able to reconcile his findings with the doctrine of fixity of species.

Presumably at the instigation of Goethe, Kölreuter's work was attacked by the Hegelian nature-philosophers in the person of the physician A. W. Henschel. Henschel opposed hybridization experiments. Pollen, he argued, could not transmit the entire nature of one plant to another, because it was only part of the whole. Perhaps because of the intellectual climate of the time—and perhaps because of Kölreuter's own ambivalence—the attack was successful. His work went overlooked, and he died a frustrated and bitter man.

Knight, Seton, and Goss   In the 1820s, three British botanists independently experimented with, among other things, the garden pea, the plant Mendel would successfully study 30 years later. Curious about how peas of two different colors could grow in the same pod, they crossed yellow-seeded plants with green-seeded plants and observed that the first-generation offspring had yellow seeds. After planting and self-fertilization, these seeds gave rise to second-generation plants in which some seeds were yellow and some green. Thus the green trait had not disappeared in the first generation; rather it had been "masked" by the yellow trait. Another way of putting it is that yellow was the dominant form of the trait and green the recessive. The plants grown from the second generation of green seeds, when self-fertilized, produced only green-seeded offspring. In addition to showing that yellow masked, or was dominant to, green, these experiments also demonstrated that yellow and green do not blend. Instead, they retain their integrity and segregate from each other in the next generation. And finally, they showed that green-seeded plants always breed true.

Figure 1.9
Josef Gottlieb Kölreuter

Drawing by William Carroll

Figure 1.10
Karl Friedrich von Gärtner

Drawing by William Carroll

Figure 1.11
Charles Naudin

Drawing by William Carroll

T. A. Knight, A. Seton, and J. Goss were the first to record such observations (although they did not interpret them) and were thus instrumental in demonstrating the phenomena that later came to be known as *segregation, dominance,* and *recessiveness.*

Gärtner The findings of Kölreuter, Knight, Seton, and Goss appear to have been known to Karl Friedrich von Gärtner (1772–1850), another German plant hybridizer, whose work Mendel discussed along with Kölreuter's. Gärtner (Figure 1.10) carried out a number of crosses among garden peas and noted the disappearance of the green seed color in the first generation of crosses between yellow- and green-seeded plants, an observation Mendel would later analyze in detail. More important were Gärtner's corn and tobacco experiments, in which he quantified his data and demonstrated ratios similar to Mendel's. Gärtner crossed yellow-kernel corn plants with others having red kernels and produced hybrids all of the yellow type. When these were self-fertilized, the second generation approximated three yellow to one red—essentially the 3:1 ratio Mendel was to find in garden peas.

In his experiments with tobacco, Gärtner observed that the hybrids were not always intermediate, which disagreed with the findings of some of his predecessors. He noted two other inheritance patterns possible in the first generation: a hybrid identical to one of the parent forms and a hybrid resembling one parent more than the other.

But Gärtner too was limited by his preconceptions. He had attended a German school influenced by the nature-philosophers. Accepting the doctrines of special creation and fixity of species, he believed that hybrid forms were unnatural and that they reverted to a parental type. Perhaps because of his holistic viewpoint, he did not concentrate on single traits, as Mendel was to do, and so failed to perceive the full significance of his findings.

Naudin Finally, the French botanist Charles Naudin (1815–1899) anticipated what Mendel was later to call hereditary "factors" by proposing that "essences" in the gametes controlled the characteristics of the progeny and that these essences segregated from each other during the formation of gametes. Working with various plants, Naudin (Figure 1.11) traced several generations of crosses, but he was not interested in the ratios Mendel was to develop and stopped short of mathematical analysis.

It is perhaps remarkable that none of these investigators fully appreciated the larger implications of his own work. All reported dominance and recessiveness (though not in those terms) and the segregation of inherited characteristics. Some even obtained results close to Mendel's 3:1 ratio of dominant to recessive. Why did they see no further? The doctrine of fixity of species, lent prestige by the support of Linnaeus, was one reason. A second was probably the influence of the nature-philosophers. Moreover, the nineteenth century was a period of observation, fact-gathering, and broad generalizations. Naturalists were making long exploratory voyages, collecting, and classifying. The stay-at-home experimentalist was not yet in the mainstream of biological thought. It was in this environment that the theory of evolution, presented jointly by Wallace* and Darwin in 1858, sought to explain the whole spectrum of observed variation.

One stimulus toward viewing living systems more analytically came later from cytology, which was growing rapidly during the period. But it did not have this effect until near the close of the century. Gärtner, for example, rejected the work of the cytologists on the ground that they destroyed the integrity of the organism by examining its parts. At any rate, the achievements of the plant hybridizers, including Mendel, were long neglected. How, then, was Mendel able to shake free of the intellectual restraints that hampered so many investigators of his time? How was he able not only to perceive but also to interpret patterns of inheritance? An examination of his life will give us some direction in answering these perplexing questions.

Gregor Mendel

Figure 1.12
Gregor Johann Mendel

The Bettmann Archive, Inc.

Born in 1822 to a peasant family in the Czech village of Heinzendorf, then part of Austria, Gregor Johann Mendel (Figure 1.12) was early recognized for his keen, disciplined mind and was sent to good schools. After graduating from the gymnasium (high school) at 18, he enrolled in college in Olmütz, where he tried without success to get work as a tutor. Perhaps because of this failure, perhaps yielding to financial pressures, he left school and spent the next year on the family farm. When he returned to college, he did find work as a tutor, but the pressures of studying and teaching, coupled with money problems, were more than he could manage, and he left school at 21 to enter the Augustinian monastery of St. Thomas at Brno, in what is now Czechoslovakia. There Mendel found both tranquility and intellectual stimulation. He was made a substitute teacher and liked the work. Ironically, his failure to pass the examination for a regular teaching certificate brought him some first-rate education in science. One of his examination readers reported to the monastery that, although he lacked formal training, he had a high potential, and the reader suggested that the University of Vienna could give him the background he needed.

From 1851 to 1853, Mendel attended the university. There he was influenced by two scientists, a plant physiologist named Franz Unger and Christian Doppler, world-famous physicist and discoverer of the Doppler

---

* Alfred Russel Wallace (1823–1913) was a naturalist who arrived at basically the same theory of evolution as Darwin, but offered much less supporting evidence.

effect. Unger was one of the few men of his time who rejected the fixity-of-species doctrine; he believed that speciation could occur through crossbreeding. Unger was also impressed by Matthew Schleiden's work on the cellular composition of plants. Included on the reading list for Unger's course were the works of Kölreuter and Gärtner, so Mendel may have learned of their hybridization experiments at this time.

Doppler's chief influence on Mendel was probably to sharpen his mathematical awareness. Unlike biology, nineteenth-century physics was highly mathematical, and Mendel's firm grasp of statistical thinking and procedures probably owes much to Doppler. Biologists then approached research in what might be called a Baconian fashion. They collected data and observed extensively, then sought an underlying principle to give shape and meaning to their observations. Physicists, on the other hand, generally worked the way most scientists do today—making preliminary observations and framing hypotheses, constructing and carrying out experiments, and accepting or rejecting their hypotheses on the basis of their results. It appears that when Mendel later began the experiments that led to his discovery of the laws of heredity, he approached his problem in the style of a physicist. Some say he had answers in mind before he actually started the experiments, which he then carried out to confirm his hypotheses.

Mendel's experiments

Mendel began his experiments with garden peas in 1856, three years before the publication of Darwin's *Origin of Species* (Figure 1.13). Unlike most of his predecessors, Mendel was interested in observing the inheritance pattern for specific characteristics, which he selected only after careful observation. He conducted his experiments over a period of eight years, carried them through the fourth generation, recorded his findings with painstaking care, and was the first to analyze such experiments statistically and to assess the significance of the ratios he discovered. Except for Naudin (who discussed essences), he was the first to postulate the existence of "factors," the heredity determiners now known as genes and chromosomes.

Mendel studied seven carefully selected pairs of contrasting characteristics in the garden pea: round versus wrinkled seeds, yellow versus green seeds, gray versus white seed coats, full versus constricted seed pods, yellow versus green seed pods, tall versus dwarf plants, and axial versus terminal flower positions (Figure 1.14). Because these characteristics took one form or the other and did not blend to produce intermediate types, he was able to trace them through successive generations without ambiguity or confusion. Moreover, these factors were inherited independently of each other, so each could be studied without interference from the others. The structure of his experiments was thus remarkably clear.

Mendel's selection of the garden pea as his experimental organism was indeed fortunate. In addition to having well-defined traits, the flower

Figure 1.13
Mendel's garden

12

*The establishment of
Mendelian principles of
inheritance*

Figure 1.14
The seven contrasting
characteristics of the
garden pea used by
Mendel

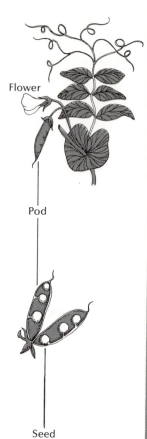

Flower

Pod

Seed

Contrasting traits

	Form 1	Form 2
Stem		
1. Axial pods and flowers along stem		Terminal pods and flowers on top of stem
2. Tall (6–7 ft)		Dwarf (3/4–1 ft)
Pods		
3. Full		Constricted
4. Yellow		Green
Seeds		
5. Round		Wrinkled
6. Yellow cotyledons		Green cotyledons
7. White coat (white flowers)		Gray coat (violet flowers)

Figure 1.15
Garden pea flower, showing male and female reproductive parts

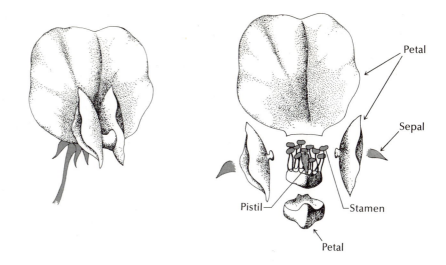

Petal

Sepal

Pistil

Stamen

Petal

(Figure 1.15) contains both male parts (*stamens*) and female parts (*pistils*) and is normally self-fertilizing. Cross-pollination in the garden pea is uncommon except when an experimenter intervenes. Self-fertilization is an important feature of garden-pea reproduction, because after several generations it results in the development of *pure lines* (strains that are comparatively homogeneous, genetically). In carrying out his experiments, Mendel ensured against cross-fertilization by covering flowers with bags to prevent foreign pollen from fertilizing the ovule. When he wanted cross-pollination, he would remove the stamens from the flower, dust the pistil of that flower with pollen from another flower, then cover the stamenless flower with a bag to prevent further pollination. Thus, in all Mendel's experiments with the garden pea, he was able to dictate the parentage of the offspring.

Mendelian principles of inheritance

Working with the garden pea, Mendel demonstrated that the seven carefully selected, contrasting, inherited traits behaved in mathematically precise and predictable ways. The pattern of inheritance exhibited by these traits featured a *dominant* and *recessive* relationship between alternative forms of each trait. For example, round seed shape was dominant over wrinkled shape, tallness over dwarfness (Figure 1.16), and yellow

Figure 1.16
Pattern of inheritance of pea plant height: tallness versus dwarfness

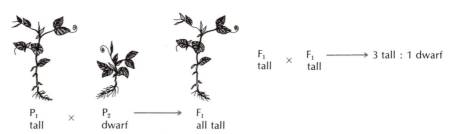

P_1
tall

×

P_2
dwarf

F_1
all tall

F_1
tall

×

F_1
tall

⟶ 3 tall : 1 dwarf

13

14

*The establishment of
Mendelian principles of
inheritance*

seed-pod color over green. Mendel also deduced that the factors (genes) responsible for each form of a trait *segregated* from each other into different gametes and, further, that each pair of factors influencing a given trait *assorted independently* of the other pairs of factors.

The principle of segregation   The segregation of members of a pair of factors was demonstrated by Mendel in a cross of a tall pea plant with a dwarf pea plant. When self-fertilized, tall plants produced tall offspring only. Dwarf plants, when self-fertilized, produced only dwarf offspring. This held true no matter how many generations of peas were followed, demonstrating that these strains were *pure lines*. However, when cross-pollination occurred between a tall plant and a dwarf plant, the first-generation plants (F_1) were all tall, illustrating that tallness was *dominant* over dwarfness, which was *recessive*. When these tall hybrid F_1 plants were self-fertilized, the second (F_2) generation produced was composed of both tall and dwarf plants in a 3:1 ratio. (The actual figures were 787 tall plants and 277 dwarfs. For a perfect 3:1 ratio, the figures would have been 798 and 266.) This type of cross, involving one trait in alternative forms, is termed a *monohybrid* cross. A cross involving two traits in alternative forms is a *dihybrid* cross, and a cross involving three traits is a *trihybrid* cross.

To prove that this 3:1 ratio was indeed the result of a pair of segregating genes,* with the gene for tallness being dominant over the gene for dwarfness, Mendel self-fertilized the F_2 generation. Because dwarfness was recessive, he predicted that the F_2 dwarf plants would produce only dwarf F_3 progeny. He also predicted that one-third of the F_2 tall plants would be *homozygous* for the dominant tall gene—that is, both members of the gene pair were the same, both determining tallness. Therefore, one-third of the F_2 tall plants would, when self-fertilized, produce only tall F_3 progeny. The remaining two-thirds of the F_2 tall plants, Mendel predicted, would be *heterozygous*—that is, one member of the gene pair would specify tallness and the other dwarfness. These heterozygous plants would have progeny in a tall-to-dwarf ratio of 3:1. Mendel's predictions were borne out by his experimental results.

The basis for Mendel's prediction was his thesis that a pair of genes determined each trait, one member of the pair being dominant over the other and the members segregating from each other into different gametes. A diagram of this cross will illustrate the principle of segregation (Figure 1.17) and the basis of Mendel's predictions. The symbolism used to describe genetic crosses is usually based on the abnormal or *mutant* trait. In the tall-by-dwarf cross, dwarf is mutant and recessive, while tall is normal and dominant. Therefore we assign *D* to denote the gene for tallness and *d* to denote the gene for dwarfness. The *D* and *d* genes are *alleles*, or contrasting forms of a gene pair. The original, pure-line tall and dwarf plants were *homozygous*, because the alleles in each parent plant

* We shall change our terminology at this point from "factor" to "gene," although the two terms are not entirely equivalent. Mendel, along with some of the early twentieth-century geneticists, used "factor," but ambiguities surrounding its usage led to the emergence of "gene" as the most appropriate term for the basic unit of inheritance.

Figure 1.17
Monohybrid cross between a pure-line tall pea plant and a pure-line dwarf pea plant (a summary of the phenotypic and genotypic ratios is included)

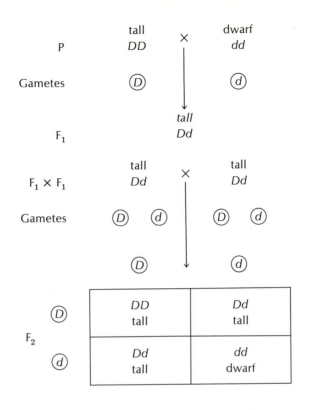

Phenotypes	Genotypes	Genotype frequency	Phenotypic ratio
tall	DD	1	3
	Dd	2	
dwarf	dd	1	1

were the same—*DD* or *dd*. The gametes of both parent plants contained only one member of the allelic pair, the tall plant producing all *D* gametes and the dwarf plant all *d* gametes. The F_1 generation produced by these gametes was *heterozygous* and tall, because its pair of genes differed (*Dd*) and because *D* is dominant over *d*.

Half the gametes produced by the F_1 hybrid tall plants contained the *D* gene, and half contained *d*. Assuming random fertilization, the probability of a *D* gamete meeting another *D* gamete is $\frac{1}{2} \times \frac{1}{2}$, or $\frac{1}{4}$. The same holds true for a *d* gamete meeting another *d* gamete: $\frac{1}{2} \times \frac{1}{2}$, or $\frac{1}{4}$. However, the probability of a *D* gamete meeting a *d* gamete is $\frac{1}{2}$, because it can happen in two ways, either $D \times d$ or $d \times D$. Since the probability of each separate occurrence is $\frac{1}{4}$, we have $\frac{1}{4} + \frac{1}{4} = \frac{1}{2}$. The *genotypic ratio* of the F_2 generation is thus $\frac{1}{4}$ *DD*, $\frac{1}{2}$ *Dd*, and $\frac{1}{4}$ *dd*. But *D* is dominant over *d*, so *DD* and *Dd* would have the same appearance, or *phenotype*. Therefore our F_2 *phenotypic ratio* would be $\frac{3}{4}$ tall plants to $\frac{1}{4}$ dwarf plants. The introduction of

Figure 1.18
Testcross between a
heterozygous F_1 tall plant
and a homozygous
recessive dwarf plant,
showing the 1:1
phenotypic and genotypic
ratios

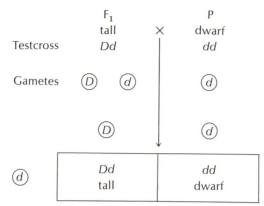

Summary: ½ tall, ½ dwarf

two terms here helps us to distinguish the genetic constitution of an organism from its actual appearance. *Phenotype* refers to the appearance of an organism as it might be detected visually or with the aid of special instruments, such as chemical analyzers. The actual genetic constitution of an organism is its *genotype*. In the foregoing discussion, tallness and dwarfness are phenotypes, but *DD*, *Dd*, and *dd* are genotypes.

The F_2 generation has a 3:1 phenotypic ratio but a 1:2:1 genotypic ratio. Mendel correctly predicted that the F_2 tall plants would be composed of two genotypic classes, *DD* and *Dd*, in a 1:2 ratio. Self-fertilization of the *DD* tall plants would give rise to all tall F_3 progeny, but self-fertilization of the *Dd* tall plants would give rise to tall and dwarf F_3 progeny in a 3:1 ratio.

Figure 1.19
Albinism, caused by a
recessive gene that blocks
pigment formation (Photo
by United Press
International)

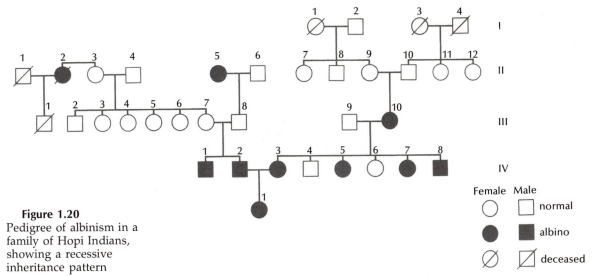

Figure 1.20
Pedigree of albinism in a family of Hopi Indians, showing a recessive inheritance pattern

	Female	Male	
	○	☐	normal
	●	■	albino
	⊘	⊠	deceased

C. M. Woolf and R. B. Grant, *Amer. J. Hum. Genetics* 14:391 (1962). Used by permission of The University of Chicago Press. Copyright 1962 by Grune and Stratton, Inc., for the American Society of Human Genetics.

A different type of cross, the *testcross*, crosses the F_1 heterozygote to a homozygous recessive individual. The testcross confirms the pattern of inheritance outlined by Mendel (Figure 1.18). The F_1 heterozygote from the preceding cross, when mated with a homozygous recessive plant (*Dd* × *dd*), produced the expected 1:1 ratio of tall to dwarf offspring.

The seven traits studied by Mendel all exhibited the same patterns of inheritance. The members of a pair of genes did not blend, but rather *segregated* cleanly from each other during the formation of gametes, thus establishing the *principle of segregation*. This principle can be demonstrated in a wide range of organisms, including humans.

The principle of segregation as seen in humans One form of albinism, a condition affecting humans, is caused by a recessive gene that results in the failure of melanin (a dark pigment) to form in the body (Figure 1.19). The inheritance of this gene follows a Mendelian pattern, as shown in Figure 1.20, a pedigree of a family of Hopi Indians. Parents II9 and II10 and III7 and III8 are phenotypically normal, but they produce albino offspring. Therefore all four parents must have been carriers for the recessive gene (*a*) causing albinism:

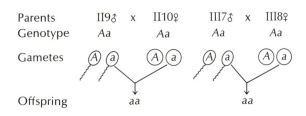

17

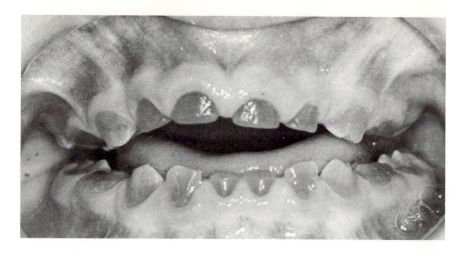

A mutant or abnormal trait can also be a consequence of a dominant gene. Consider the inheritance of the condition known as *dentinogenesis imperfecta* (Figure 1.21), a rare, genetically determined tooth defect. The teeth of affected individuals have an unusual, opalescent brown color due to abnormal dentine formation, and their crowns wear down readily. Figure 1.22 shows a pedigree of a family with a long history of the condition. There is no skipping of generations, such as we saw in the case of the recessive albinism. Note too that roughly half of an affected person's children have had the trait transmitted to them, pointing to the heterozygous genotype of the affected individual. Individuals who do not express the trait do not transmit it to any of their descendants. These are all criteria for dominant inheritance.

Examining the pedigree shown in Figure 1.22 should convince you that this is indeed a dominant inheritance pattern. In order for the trait to be recessive, we would have to assume that all of the spouses (I1, II5, III1,

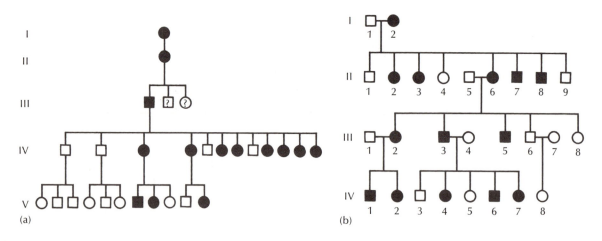

(a)

(b)

and III4) were carriers of the mutant gene and that all affected individuals were homozygous:

Normal, carrier 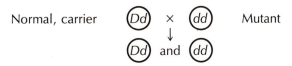 Mutant

This assumption granted, we could explain the rough 1:1 ratio of mutant to normal observed in the pedigree. But the assumption that four unrelated people are carriers for such a rare trait (incidence of 1/8000) is not reasonable. It is much more reasonable to suggest that the trait is the result of a dominant gene:

Mutant 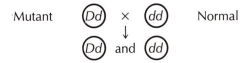 Normal

It is important to note here that Mendelian principles of segregation apply not just to pea plants. They apply to a vast array of species, including humans.

Another trait that owes its existence to a rare dominant gene is the fatal illness called the "Joseph Family Disease." Known to occur only in the descendants of a nineteenth-century Portuguese whaler named Anton Joseph, the illness is a late-onset hereditary disease that affects the nervous system. Its expression usually begins around age 30 with problems in walking and balance. The affected individual gradually loses the ability to walk, speech becomes thick and slurred, muscles in the arms and legs stiffen and lose their coordination, swallowing becomes difficult, the person can no longer control body secretions, and pneumonia and death usually follow. The biochemical basis of the disease is unknown.

What is it about the inheritance pattern of this disease that leads us to suggest that the "Joseph Family Disease" is a dominant trait? Anton Joseph came to the United States in 1845, settling in the San Francisco area. His father died of the disease. Three of Anton Joseph's six children died of the disease, and Anton Joseph himself died of the disease at the age of 45. All of the descendants of his normal children have remained free of the disease. But the descendants of his 3 affected children continue to express the disease. Anton Joseph's first daughter, Mary, bore 14 children. Two of them died in early childhood of unrelated causes, but 8 of the remaining 12 developed the disease and died. The same pattern continues today: The individual marries, has children, begins expressing the disease in his or her early thirties, dies in his or her forties, leaving the surviving spouse and children to live in fear, knowing that about half of the children will die of the disease. This tragic illness is a classic example of a dominant inheritance pattern.

Figure 1.23
Dihybrid cross between a pure-line tall pea plant with smooth seed coat and a dwarf plant with wrinkled seed coat (a summary of the phenotypic and genotypic ratios is included)

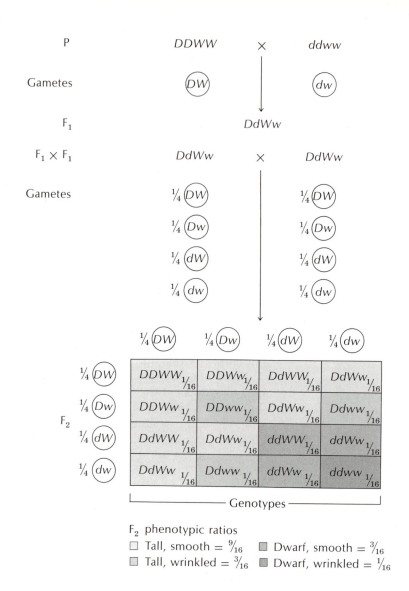

The principle of independent assortment   In addition to studying the inheritance of alternative forms of a single trait, such as plant size (tall versus dwarf), Mendel also crossed plants differing in two or more pairs of alleles. For example, a dihybrid cross between a tall plant with smooth seed coats and a dwarf plant with wrinkled seed coats produced F₁ progeny that were tall with smooth seed coats (Figure 1.23). This established tall (D) as dominant over dwarf (d) and smooth (W) as dominant over wrinkled (w). Self-fertilizing the F₁ hybrids produced an F₂ with the phenotypic ratio of 9:3:3:1—tall and smooth, tall and wrinkled, dwarf and smooth, and dwarf and wrinkled, respectively. Mendel recognized that the 9:3:3:1 ratio of this dihybrid cross resulted from the *independent assortment* of two monohybrid crosses. The segregation of one pair of alleles is

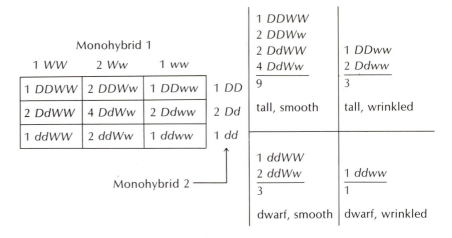

Monohybrid 1

	1 *WW*	2 *Ww*	1 *ww*	
	1 *DDWW*	2 *DDWw*	1 *DDww*	1 *DD*
	2 *DdWW*	4 *DdWw*	2 *Ddww*	2 *Dd*
	1 *ddWW*	2 *ddWw*	1 *ddww*	1 *dd*

Monohybrid 2

1 *DDWW*	
2 *DDWw*	
2 *DdWW*	1 *DDww*
4 *DdWw*	2 *Ddww*
9	3
tall, smooth	tall, wrinkled
1 *ddWW*	
2 *ddWw*	1 *ddww*
3	1
dwarf, smooth	dwarf, wrinkled

independent of the segregation of other pairs of alleles, as indicated in the above diagram.

The F_1 hybrid of a dihybrid cross would produce four kinds of gametes: *DW, Dw, dW, dw*. Again assuming random fertilization, the probability of any pair of gametes meeting is $\frac{1}{4} \times \frac{1}{4}$, or $\frac{1}{16}$. Each square of the checkerboard in Figure 1.23 (called a *Punnett square*) represents a probability of $\frac{1}{16}$, with all possibilities totaling 1. Dominance makes some of the genotypes phenotypically identical, even though they may differ genotypically. The final 9:3:3:1 phenotypic ratio results from two pairs of genes segregating independently of each other, with one member of each pair dominant over the other.

A testcross of an F_1 hybrid from a dihybrid cross gives the 1:1:1:1 phenotypic ratio shown in Figure 1.24. A trihybrid cross (Figure 1.25)

Figure 1.24
Testcross between an F_1 pea plant heterozygous for tallness and smooth seed coat and a homozygous recessive dwarf pea plant with wrinkled seed coat, showing the 1:1:1:1 phenotypic and genotypic ratio

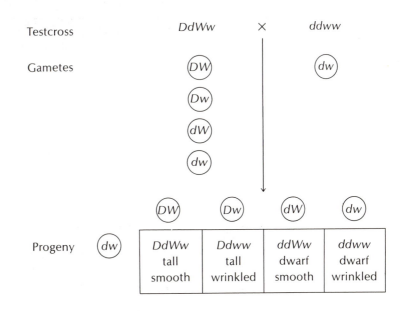

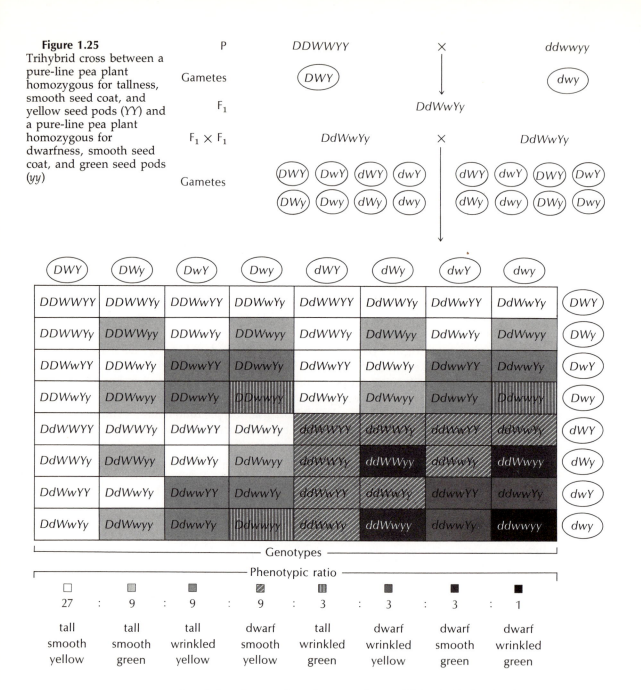

Figure 1.25
Trihybrid cross between a pure-line pea plant homozygous for tallness, smooth seed coat, and yellow seed pods (YY) and a pure-line pea plant homozygous for dwarfness, smooth seed coat, and green seed pods (yy)

involving three pairs of genes gives a 27:9:9:9:3:3:3:1 F_2 phenotypic ratio, based on the same principles of independent assortment described for the dihybrid cross.

When dominance is incomplete (for example, when the heterozygote is phenotypically intermediate between the two parents), the same principles of segregation and independent assortment hold true, but the

Figure 1.26
Monohybrid cross
exhibiting incomplete
dominance in carnations

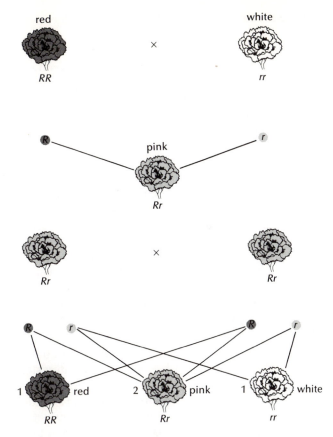

phenotypic ratios change (Figure 1.26). Returning to Kölreuter's experiments, a red-flowered carnation plant (*RR*) mated to a white-flowered plant (*rr*) produced pink F₁ plants (*Rr*). Self-fertilization of the F₁ pink hybrids produced an F₂ phenotypic ratio of 1 red to 2 pink to 1 white. We must bear in mind that, although phenotypic ratios may differ from Mendel's because of different dominance and recessive relationships, these ratios not only do not contradict Mendelian principles but in fact add force to them.

Why Mendel's work went unrecognized

Mendel concluded the bulk of his research on the garden pea in 1863. He presented his findings before the horticultural society of Brno in 1865 and published them in the society's proceedings the following year. A reasonably well-knit scientific community existed at that time, and the paper was moderately well distributed. Certainly, it found its way into all the major Western countries then interested in genetics, including the United States, Great Britain, France, Germany, and Austria, as well as other Central European countries. But there was no response from any quarter: no discussion, no criticism, not so much as a word of recognition.

Puzzled and concerned, Mendel began corresponding with a man he thought could help—Karl von Nägeli, a noted botanist whose work

24

*The establishment of
Mendelian principles of
inheritance*

Unger had introduced him to. Nägeli was interested in Mendel's work but doubted its significance. He suggested that Mendel experiment with hawkweed, Nägeli's special interest at the time, and Mendel did so. But neither of them knew that hawkweed, unlike the garden pea, reproduces without fertilization and hence could not possibly give comparable results. (Mendel's ratios can be obtained only with organisms that reproduce sexually.) Although the correspondence between the two men covered a number of years, Nägeli was never able to grasp the full import of Mendel's work and continued to urge that he experiment with hawkweed.

Mendel became Prelat, or abbot, of Königskloster in 1868 and gave up systematic research in about 1871, when he became involved in a tax dispute between the monastery and the state. He died in 1884, his findings on heredity ignored. Indeed, it was not until 1900 that his work was discovered independently by the Dutch hybridist Hugo de Vries, the German botanist Carl Correns, and the Austrian botanist Erich von Tschermak. There has been much speculation about the reasons for the long interim, but no wholly satisfying explanation has emerged. It has been argued that Mendel's paper might have met with a happier fate had it been published in a more influential journal. But the Brno journal reached many countries, and Mendel also had 40 reprints of the article, though only two recipients of them have been identified. It has also been suggested that biologists were interested in little else but Darwinism in that period, especially during the 1860s, and there is probably some truth to this. It has even been said that Mendel himself contributed to his own neglect. Twice a failure in the examination that would have qualified him to become a regular teacher, rejected by Nägeli, unable to corroborate his data in work on hawkweed and some other plant species, he may have lacked the force and confidence necessary to gain a hearing.

But it is more likely that the scientific community was for many of those years simply unready to appreciate Mendel's work. By 1900, however, the role of the nucleus in fertilization was understood, the behavior of chromosomes in cell division had been observed, and there was a body of cytological knowledge to which the behavior of Mendel's factors could be linked. In short, two lines of biological thought, Darwinism and cytology, had reached the point at which they could make use of the legacy of the obscure Czechoslovakian monk.

The cytological basis of inheritance

Cells, or "microscopical pores," in cork and wood were drawn and described by Robert Hooke as early as 1665, but the concept of the cell as the building block of life did not develop until the second third of the nineteenth century. In 1831, the naturalist Robert Brown described dark bodies of unknown function that he found in plant cells. These were the nuclei. Five years later, animal cells were compared structurally with the cells in plants. In 1838, Schleiden published a paper on cells in plants and, more importantly, described them to Schwann, who at once recognized their similarity to cells in animal nerve tissue. The following year, Schwann published his paper "On the Correspondence in the Structure and Growth of Plants and Animals" and gave currency to what has been called one of the greatest generalizations in the history of biology, the *cell theory*. The cell theory states that all living tissue is composed of units (cells), that each unit has a complex nucleus surrounded by a differentiated substance called *cytoplasm*, and that the units are capable of self-duplication.

The association of inheritance with chromosomes

Knowledge of internal cell structure was greatly enhanced by improvements in the microscope and by the development of staining in the 1840s. Because different parts of the cell took stains differently, it became possible to see structures not visible before. It was soon discovered that the nucleus was a permanent part of the cell and that it contained a substance that stained darkly and was hence called *chromatin*. But the nucleus was not thought to play a role in inheritance until the late 1870s. At this time, O. Hertwig, H. Fol, and E. Strasburger studied fertilization in various organisms and noted that it involved the fusion of nuclei from sperm and ova.

Studying unfertilized sea urchin eggs that had been exposed to sperm suspensions, Hertwig (1876) observed that the unfertilized egg contained one nucleus, but that on contact with sperm, two nuclei were visible. He suggested that this second nucleus was contributed by the sperm, although he did not see the actual contribution occur. He also observed the fusion of the two nuclei and concluded that this was the primary event of fertilization. A year later, Fol was able to detect the actual transfer of the nucleus from sperm to egg. He also made the important observation that only one sperm was required to fertilize an egg, contradicting an earlier belief that several sperm were necessary for fertilization.

Strasburger studied plant fertilization, noting that it involved only the fusion of pollen and ovule nuclei and that pollen cytoplasm had no part in embryogenesis. These three investigators correctly inferred that the nucleus was the transmitter of hereditary information. But the way had been prepared for this observation nearly a century earlier by Kölreuter.

Kölreuter demonstrated that in the inheritance of alternative forms of a particular trait, such as red or white flower color, the sex of the parents did not influence the outcome of the cross. Red-flowered males mated to white-flowered females and white-flowered males mated to red-flowered females produced the same type of offspring. But it was not until the nineteenth century that Kölreuter's findings were appreciated for their true significance: If the quantity of cytoplasm in an egg is several times

26

*The establishment of
Mendelian principles of
inheritance*

that in a sperm and if the nuclear volume is the same in both, it stands to reason that the nucleus must be the major factor in fertilization and inheritance. If the cytoplasm were the key to fertilization and inheritance, Kölreuter's reciprocal crosses would have given different results.

Before fertilization and inheritance could be more firmly linked, there had to be additional observations about chromosomes, the nucleus, and inheritance. In 1884 and 1885, Hertwig, Strasburger, A. Weismann, R. A. Kölliker, and Nägeli made a series of definitive statements about the role of the nucleus in fertilization and inheritance. Nägeli, having studied Kölreuter's work, was deeply impressed by the disproportionate sizes of the sperm and egg. He marveled that, although the egg supplies as much as a thousand times the cytoplasm contributed by the sperm, it supplies no greater proportion of the hereditary properties. Yet Nägeli failed to suggest that the key to understanding heredity was to be found in the nucleus. Hertwig had familiarized himself with the views of Nägeli and Strasburger and gathered impressive evidence to support the theory that the nucleus is the significant structure in the transmission of inherited characteristics. He even referred to "nuclein," a substance isolated in 1871 and later shown to be largely deoxyribonucleic acid (DNA), the genetic material (see page 238). That nuclein was found in the nucleus was significant to Hertwig, who obliquely suggested a connection between it and inherited properties. Weismann postulated the existence of a hereditary substance found in the nuclei of all cells and passed on from generation to generation through the germ (sex) cells, but he did not mention nuclein.

The association of chromosomes, nuclei, and inheritance began with the work of A. Schneider, who in 1873 described the behavior of the nucleus during cell division. Earlier observers had thought that the nucleus in germ cells disappeared and that a new nucleus was formed after cell division. Schneider noted that the nuclear contents did not disappear, but condensed and were distributed to the two new nuclei. This finding strongly supported the idea that the nucleus and its contents were continuous and passed from one generation to the next. Two years later, E. van Beneden described the division of the nucleus in detail.

Mitosis and meiosis   The next few years saw the processes of mitosis and meiosis fully described. Walther Flemming performed the definitive study of mitosis in 1878, using epithelial cells and red blood cells of the salamander. He noted that when the chromosomes arrived at the metaphase plate, each was composed of two subunits and hence appeared double. During anaphase, the chromosomes split longitudinally and moved to opposite poles of the cell. Thus all the chromosomes in a nucleus doubled, then were exactly and equally partitioned between the two daughter nuclei. (Flemming himself used the term "mitosis"; the term "chromosome" was not used until 1888, when it was coined by W. Waldeyer. The process of mitosis as we understand it today is summarized in Figure 1.27.)

Figure 1.27
The process of mitosis

1

Interphase

The chromosomes are uncoiled and not visible as such. There are four (two pairs) in the nucleus. Chromosome replication has occurred.

Centrosome
Chromatin
Nucleolus
Nuclear membrane
Cell membrane

2

Early prophase

The chromosomes have begun to condense and are now visible. Each chromosome consists of two sister chromatids joined at the centromere. The centrosome has divided to form centrioles, thus spindle formation has begun. The nucleolus has not yet dispersed.

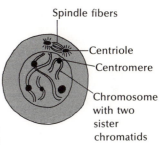

Spindle fibers
Centriole
Centromere
Chromosome with two sister chromatids

3

Middle prophase

The chromosomes continue to coil and condense.

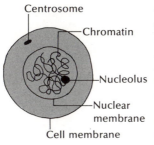

4

Late prophase

The chromosomes continue to coil and condense.

5

Metaphase

The nuclear membrane has broken down, and the nucleolus has dispersed. The chromosomes are short and thick and aligned at the equatorial region, attached to the spindle fibers by their centromeres.

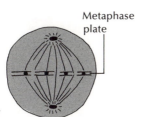

Metaphase plate

6

Early anaphase

The centromeres divide and the sister chromatids, now called daughter chromosomes, move toward opposite poles, the centromere leading the way.

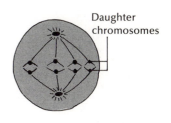

Daughter chromosomes

7

Late anaphase

Daughter chromosomes continue to move toward opposite poles.

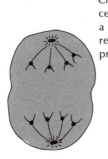

8

Telophase

Chromosomes have ceased moving and a nuclear membrane reforms. The cell prepares to divide.

Cleavage furrow

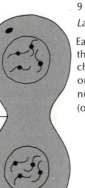

9

Late telophase with cytokinesis

Each daughter cell has the same number of chromosomes as the original interphase nucleus, four (or two pairs).

Daughter cells

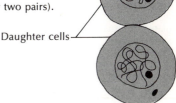

Figure 1.28
The process of meiosis

Interphase

The chromosomes are uncoiled and not visible as such. There are four (two pairs) in the nucleus. Chromosome replication has occurred.

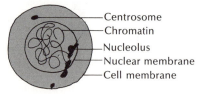

— Centrosome
— Chromatin
— Nucleolus
— Nuclear membrane
— Cell membrane

Early prophase₁

Chromosomes coil and condense, becoming more visible.

— Spindle fibers
— Centrioles
— Chromosome with two sister chromatids

— Homologous pair of chromosomes

Middle prophase₁

Chromosomes pair up and coil around each other, a process called *synapsis*. They continue to condense.

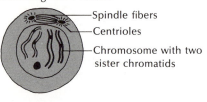

Late prophase₁

Chromosomes have exchanged segments with each other. The chromatids remain attached at x-shaped regions called *chiasmata*.

Metaphase₁

Chromosomes line up so that the centromeres of each pair lie on either side of the metaphase plate. The nuclear membrane and nucleolus have dissipated.

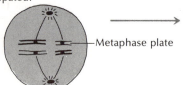

— Metaphase plate

Anaphase₁

The chromosomes move to opposite poles of the cell, one member of each pair to each pole. The centromeres remain undivided.

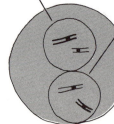

Telophase₂

Chromosome movement has ceased; there are four haploid nuclei.

Anaphase₂

The centromeres divide and the daughter chromosomes move to opposite poles of the cell.

Metaphase₂

Chromosomes line up along the metaphase plate.

Prophase₂

The chromosomes coil and condense.

Interphase

The chromosomes become diffuse, but there is no replication.

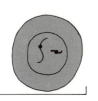

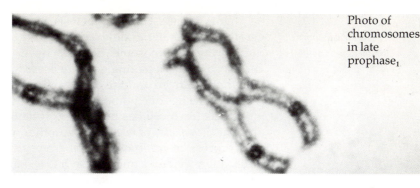

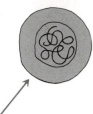

Telophase₁

Chromosome movement has ceased and a nucleus forms, yielding two haploid nuclei, but each chromosome consists of two chromatids.

In the "resting," or *interphase* cell, the chromosomes are in an uncoiled condition and thus difficult to see, but it is during interphase that each chromosome is replicated. The first stage of mitosis, *prophase*, begins when the chromosomes coil and condense and a spindle apparatus forms. By mid-prophase, the chromosomes are easily visible, each longitudinally divided into identical sister *chromatids* that are held together by a *centromere*. The nucleolus disperses throughout the nucleus during the mitotic prophase and thus becomes undetectable with the light microscope.

The next stage, *metaphase*, begins when the nuclear membrane breaks down and the nucleolus is no longer visible. The chromosomes continue to coil and condense, and are aligned at the equator or *metaphase plate*. *Spindle fibers* extend from the *centrioles* to the centromere of each chromosome. These fibers contract and move the chromosomes to opposite poles of the cell. Plant cells do not usually have centrioles, but they do have the spindle structure.

The next stage, *anaphase*, begins when the centromeres divide and the sister chromatids of each chromosome separate and move toward opposite poles of the cell. After separation, the sister chromatids are referred to as *daughter chromosomes*.

The last mitotic phase, *telophase*, occurs when chromosome movement is completed. Nuclear membranes now form; the spindle apparatus disappears; *cytokinesis*, or cell division, occurs; and two genetically identical daughter cells exist where one did before.

In 1883, van Beneden revealed that gametes contain only half as many chromosomes as the somatic, or nongerm, cells and that the characteristic somatic number is reestablished at fertilization. He accurately described the process of meiosis, whereby the chromosome number is reduced by one-half. But he erroneously concluded that all maternal chromosomes (all those chromosomes contributed to the individual by the mother) go to one nucleus and all paternal chromosomes to the other. (In fact, maternal and paternal chromosomes are randomly distributed to daughter cells during meiosis.) The behavior of the chromosomes during mitosis and meiosis was thus recorded, but their role in heredity was still not even guessed at. The process of meiosis as we understand it today is summarized in Figure 1.28.

In its broadest sense, meiosis is a lengthy process in which the chromosomal material doubles once and the cell divides twice, thus reducing the chromosome number by one-half. Before meiosis, each chromosome has a pairing mate, or *homolog*. For each pair, one member is maternally derived and the other paternally derived. A cell in which each chromosome has one homologous mate is called *diploid*. At the conclusion of meiosis, each cell has but one representative of each homologous pair and thus is *haploid*, or *monoploid*. Another way of characterizing meiosis is as a process of *reduction division* from a diploid to a haploid state. At fertilization, haploid nuclei fuse to reestablish the diploid condition, in which each chromosome has a homologous mate.

Meiosis has two successive divisions. The first, the reduction division, is the more complex. During this division, homologous chromosomes pair, then separate and move to different nuclei. The second meiotic division, which is less involved than the first, consists of the splitting of the centromere and the separation of sister chromatids into different nuclei. Thus four haploid nuclei result from the meiotic division of one diploid nucleus.

The first stage of meiosis is a lengthy and complex *prophase I* made up of five substages. The first substage that can be recognized as distinct from interphase is *leptotene*. During this substage, the chromosomes first appear as distinct, threadlike nuclear objects. The chromosomes have replicated by this substage, but by all appearances they seem to be single structures. During the next substage of prophase, *zygotene*, homologous chromosomes pair tightly (a process called *synapsis*) and continue to condense. They still appear to be single structures.

The third substage in the prophase of meiosis I is called *pachytene*. For the first time, the chromosomes are often seen to be double—that is, each chromosome consists of two chromatids. The chromosomes continue to shorten and thicken during pachytene. Nonsister chromatids within each homologous pair may have exchanged segments of genetic material with each other, a process called *crossing-over*. The synapsed, homologous chromosome pair with its four chromatids is called a *tetrad*, or *bivalent*. (The genetic significance of crossing-over will be explored in more detail in the next chapter.)

During the fourth substage of prophase I, *diplotene,* the homologous chromosomes seem to repel each other, but remain attached at points called *chiasmata* (sing., *chiasma*). These chiasmata are X-shaped areas and may represent regions in the chromosome pair in which exchange of chromosomal material has occurred. As we have said, we refer to this exchange process as *crossing-over*, a process that has far-reaching genetic consequences.

The diplotene stage in many animals is extremely prolonged. In human females, for example, the diplotene stage of meiosis is completed in the ovary during fetal life. About 400,000 *oocytes* (immature eggs) reach this stage and then stop. They remain in this suspended state until the onset of puberty. After puberty, one egg a month goes on to complete meiosis in the Fallopian tube during the menstrual cycle [but only after fertilization is the polar body II formed (see Figure 1.32) and meiosis completed]. Some of the eggs may remain in this stage of arrested meiosis for as long as 50 years!

During the last substage, *diakinesis*, the bivalents continue to thicken and move toward the nuclear membrane, distributing themselves evenly. The nucleolus generally is dispersed, the nuclear membrane dissolves, and the spindle fibers attach to the centromeres. This marks the conclusion of prophase I and the beginning of metaphase I.

At *metaphase I*, the members of the bivalents are highly condensed and arranged on opposite sides of the equatorial plate, each centromere connected to the spindle fiber apparatus. The chiasmata have moved toward the ends of each chromosome, a process called *terminalization.*

During *anaphase I*, the members of a homologous chromosome pair move toward opposite poles. The centromeres have not divided, so each chromosome still consists of two chromatids (a *dyad*).

Telophase I occurs when the chromosomes reach the opposite poles. Often this is accompanied by the formation of nuclear membranes around each haploid chromosomal complement and then by a brief *interphase.* Sometimes, though, membrane formation and interphase do not take place, and the cells move immediately into the second meiotic division.

At *prophase II*, the dyads condense and move to the equatorial region. The centromere divides at *metaphase II*, and the former chromatids, now daughter chromosomes, move toward the opposite poles in *anaphase II.* When the chromosomes stop moving (*telophase II*), nuclear membranes form around each haploid complement, and *cytokinesis* follows.

To summarize the differences between meiosis and mitosis: (1) The pairing of homologous chromosomes takes place in meiosis, but not in mitosis. (2) The segregation of chromosomes *without* centromere division occurs in meiosis, but not in mitosis. (3) The second meiotic division is not preceded by chromosome duplication, unlike the situation in mitosis, in which duplication *has* occurred.

The significance of mitotic cell division was given theoretical consideration in 1883 by Wilhelm Roux, a German embryologist. Fascinated by the precision with which a cell distributes its nuclear contents to its daughter cells, he considered such a mechanism to be essential to maintaining constancy of "nuclear qualities" that all cells must possess. Roux recognized that these "qualities" must be contained in the chromosomes and thus saw chromosomes as probable carriers of the units of heredity.

Between 1887 and 1891, Weismann developed his theory of the continuity of the "germ plasm," which he identified with the substance of the chromosomes as carriers of the units of heredity. He believed that the hereditary material of the nucleus had to remain constant from one generation to the next and that it did so by reducing the number of chromosomes in the gamete by half and restoring the original number by fertilization. The continuity of the germ plasm proved an idea of lasting value.

In 1900, Mendelian principles were rediscovered, and in 1902 Walter

Sutton, a student of E. B. Wilson at Columbia, and Theodor Boveri, a well-known German cytologist, united the fields of cytology and heredity. Sutton and Boveri independently published papers that described the continuity of chromosomes through cell division and announced the important discovery that there are morphological differences among chromosomes. The most important relationship they developed, however, was the parallel between meiotic chromosome behavior and Mendel's principles. They pointed out that the segregation of chromosomes during meiosis could account for Mendel's principle of segregation. They further reasoned that the random alignment of chromosomes along the metaphase plate could support Mendel's principle of independent assortment. In other words, one could postulate a mechanism for Mendel's principles by assuming that Mendel's "factors" were positioned on chromosomes—or indeed were themselves chromosomes.

Meiosis and its relationship to Mendelian principles

With the merger of cytology and heredity into a unified discipline of genetics, a reinterpretation of Mendelian principles became possible. If we substitute the term "chromosome" for Mendel's "factor," segregation and independent assortment describe perfectly the behavior of chromosomes during meiosis. In meiosis, homologous pairs of chromosomes synapse and then *segregate* into different nuclei. The pairs also align randomly along the metaphase plate, thus providing a cytological mechanism for independent assortment (Figure 1.29). These parallels form a strong argument in support of the *chromosome theory of inheritance,* which says that the genetic determinants are located on chromosomes.

It may now be instructive to apply our knowledge of Mendelian principles and meiosis to some genetic analysis. Mendel's experiments, data, and ratios were based on the *sampling* of meiotic products. That is, gametes are produced by the millions, but only a random sample is actually analyzed. This type of genetic analysis is called a *strand analysis,* and because it is a statistical type of analysis, its accuracy depends to a large extent on the size of the sample analyzed—the larger the sample, the more accurate the results. When using the garden pea, humans, *Drosophila* (the fruit fly), or any organism containing two sets of chromosomes (*diploid*), strand analysis is usually the only type of analysis possi-

Figure 1.29
Diagram of two pairs of homologous chromosomes aligning independently along the metaphase plate in meiosis I (note that each chromosome consists of two sister chromatids joined at the centromere)

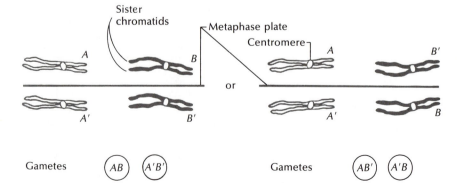

ble, because all the products of any single meiotic event can seldom be individually assessed. In some organisms, however, *all* the products of a single meiotic event are held in a common structure as tetrads. The study of tetrads is *tetrad analysis.*

To understand tetrad analysis, which explicitly strengthens the tie between Mendelian principles and the meiotic process, it is helpful to examine the life cycles of two common haploid organisms—*Neurospora* and *Chlamydomonas*—and then compare them with the life cycle of a typical diploid organism, in this case *Drosophila.*

Neurospora (Figure 1.30), the common bread mold, is composed of a

Figure 1.30
Life cycle of the common bread mold *Neurospora*

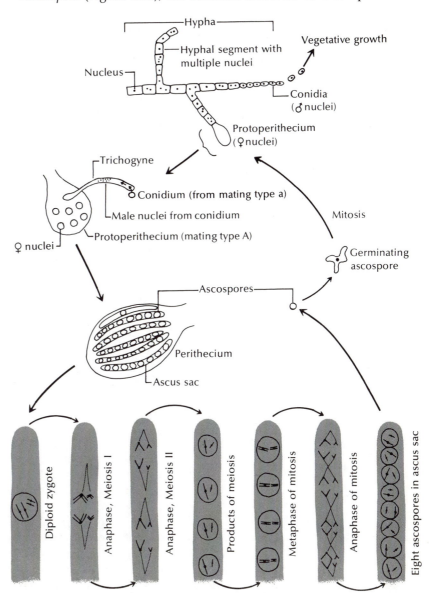

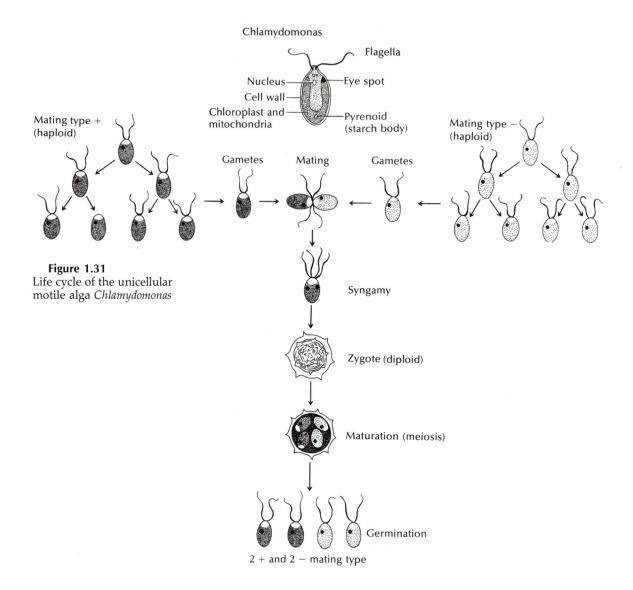

Figure 1.31
Life cycle of the unicellular motile alga *Chlamydomonas*

mass of intertwined filaments, or *hyphae,* that form a spongy pad called a *mycelium.* The individual hyphae are segmented and usually contain more than one haploid nucleus per segment. *Neurospora* can reproduce asexually in two ways: by mitotic replication of the nuclei followed by subsequent fragmentation and growth of the hyphae, or by the formation of special haploid spores called *conidia* that, on germination, form new hyphae. It is possible for hyphae from different strains to fuse. When this happens, nuclei of different strains and different genotypes occupy the same cytoplasm (a condition referred to as *heterokaryon* formation).

Neurospora has a sexual cycle too. The hyphal segments can differ-

Figure 1.32
Life cycle of the fruit fly
Drosophila

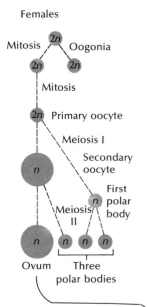

Females

Mitosis — 2n — Oogonia
2n   2n

Mitosis

2n   Primary oocyte

Meiosis I

Secondary
oocyte

First
polar
body

Meiosis
II

n   Ovum

n   n   n
Three
polar bodies

Males

2n   Mitosis
2n   2n

Spermatogonia

Mitosis

Primary
spermatocyte

2n

Meiosis
I

Secondary
spermatocyte

n   n

Spermatids

Meiosis
II

n   n   n   n

Maturation

Spermatozoa

entiate into immature fruiting bodies called *protoperithecia* that contain haploid maternal nuclei. The hyphal segments, as we mentioned, can also differentiate into conidia, the male gametes. *Neurospora* exists in two genetically determined mating types, both of which produce conidia and protoperithecia. When a conidium of one mating type comes in contact with a special mating filament (*trichogyne*) on the protoperithecium of an opposite mating type, the nucleus of the conidium moves into the fruiting body, dividing mitotically, and fertilizes several female nuclei. Each diploid zygote is contained in an *ascus sac*, and a mature fruiting body (*perithecium*) may contain as many as 300 such sacs. After fertilization, each diploid zygote undergoes meiosis to form four haploid cells, which in turn divide mitotically once to form eight haploid *ascospores*. The arrangement or sequence of the ascospores in the ascus sac is very specific and reflects the temporal sequence of meiotic events. Knowing the spatial sequence of ascospores in an ascus sac can tell us when and where crossover occurred, for example, or whether a pair of alleles segregated from each other in meiosis I or meiosis II. (This matter will be explored shortly.)

Chlamydomonas (Figure 1.31), a single-celled green motile alga, also has two distinct genetically determined mating types. A *Chlamydomonas* can differentiate into a mature gamete under the proper environmental circumstances, and a gamete of one mating type can fuse with a gamete of the opposite mating type to form a diploid zygote. The zygote undergoes meiosis to form four haploid spores that, on release from the spore case, form mature haploid organisms.

The life cycle of both *Neurospora* and *Chlamydomonas* should be compared with the life cycle of a diploid organism such as *Drosophila* (Figure 1.32). In the *Drosophila* male gonad (*testis*), *spermatogonia* divide mitotically

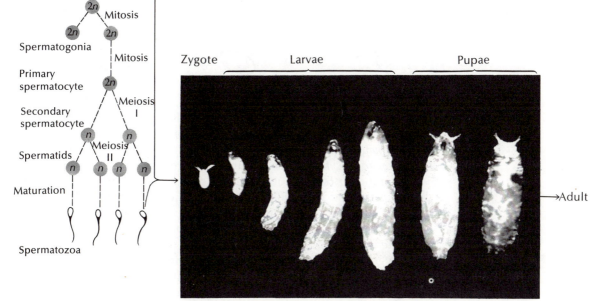

Zygote   Larvae   Pupae

→Adult

Photo from M. W. Strickberger, *Experiments in Genetics with Drosophila,* New York: John Wiley, 1962

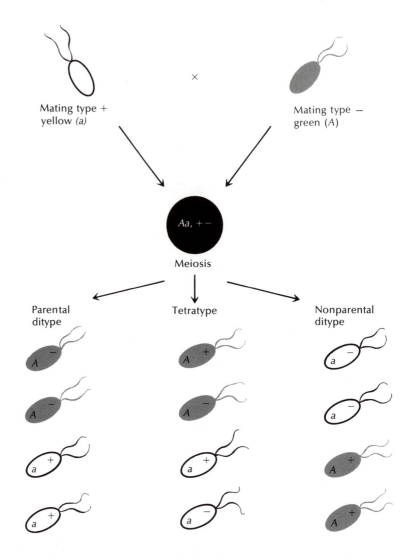

Figure 1.33
Dihybrid cross in *Chlamydomonas,* illustrating the tetrad analysis

Mating type +
yellow *(a)*

×

Mating type –
green *(A)*

Aa, + –

Meiosis

Parental
ditype

Tetratype

Nonparental
ditype

to produce *primary spermatocytes,* which undergo meiosis I to form *secondary spermatocytes* and meiosis II to form *spermatids.* The spermatids mature into *spermatozoa.* In the female gonad *(ovary), oogonia* divide mitotically to form *primary oocytes.* These in turn undergo meiosis I to form *secondary oocytes* and *polar bodies* (nonfunctional sex cells) and meiosis II to form a mature *ovum* and two additional polar bodies. A zygote forms from the union of a spermatozoon and an ovum.

The principles of segregation and independent assortment are clearly illustrated in haploid organisms. Problems of dominance, recessiveness, homozygotes, and heterozygotes do not arise, because each allele is uninfluenced by the presence of an allelic mate. A cross in *Chlamydomonas* involving, for example, the two mating types (mt^+ and mt^-) and a yellow (*a*) and green (*A*) strain would result in three different types of tetrads (Figure 1.33): a *parental ditype* (PD), a *nonparental ditype* (NPD), and a *tetratype* (T). Given that the genes for these traits are on different chromosomes, PD and NPD result from the segregation and independent assortment of the chromosomes at meiosis. The last tetrad class, the tetratype, arises from a *crossing-over* event in the first meiotic division. In the first meiotic prophase, the paired chromosomes—each composed of two chromatids—coil tightly around each other and can exchange segments with each other. The crossover event shown in Figure 1.34 results in a tetrad composed of all four possible combinations: *A mt⁺*, *A mt⁻*, *a mt⁺*, *a mt⁻*. Thus meiosis and Mendel's principles of segregation and independent assortment are clearly associated in this mating system—and even more so in *Neurospora* crosses.

Figure 1.34
Generation of a tetratype by meiotic crossing-over

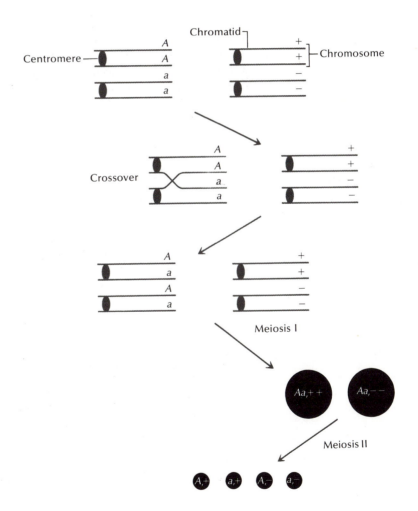

38

*The establishment of
Mendelian principles of
inheritance*

Neurospora has a unique advantage over *Chlamydomonas* in genetic studies—the tetrads are ordered. This offers the possibility of learning at what stage in the meiotic process genes segregate, when crossing-over occurs, and which chromatids engage in crossing-over. A more thorough analysis of crossing-over, as it relates to gene mapping, will be presented in the next chapter, but a brief example here will illustrate the advantages of using *Neurospora* for studying Mendelian principles. A strain requiring the amino acid *tyrosine* as a supplement in its medium (*tyr⁻*), mating type + (*mt⁺*), is crossed with a normal tyrosine-independent strain (*tyr⁺*), mating type − (*mt⁻*). Genes for both traits are on different chromosomes. As in *Chlamydomonas*, three types of tetrads are observed: PD, NPD, and T (Figures 1.35, 1.36).

The PD and NPD tetrads result from independent assortment and segregation. The T tetrad results from crossing-over between chromatids numbered 2 and 3 in the first meiotic division in an area between the centromere and the *tyr* locus, giving the sequence *tyr⁻ mt⁻*, *tyr⁺ mt⁻*, *tyr⁻ mt⁺*, *tyr⁺ mt⁺*. An important observation with respect to the crosses of *Chlamydomonas* and *Neurospora* is that, if crossing-over occurs, the genes involved do not segregate until the second meiotic division (see Figure 1.34). If no crossing-over occurs, segregation takes place during the first meiotic division. (Other features of crossing-over in *Neurospora* and *Drosophila* will be discussed later.)

Figure 1.35
Dihybrid cross in
Neurospora, illustrating
the tetrad analysis

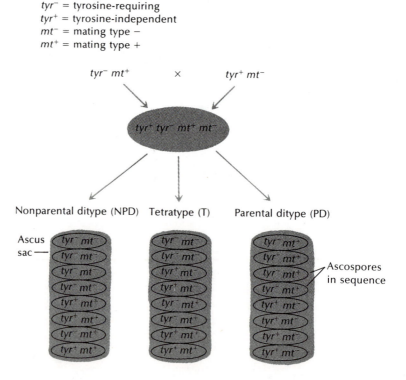

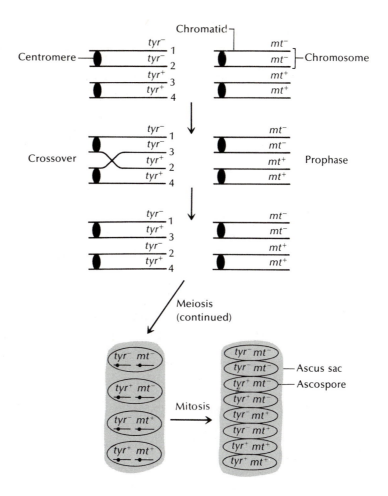

Figure 1.36
Meiotic basis of the ascospore sequence in the tetratype ascus sac shown in Figure 1.35

Overview

Mendel's principles have been substantiated in a variety of experimental systems, and their rediscovery ranks as one of the most momentous events in the history of science. But this is not to say that his work has been free of debate and criticism. For example, two aspects of Mendel's work have been interpreted by some historians as indicative of data tampering. First, the data supporting his principles are almost too perfect, given the sampling techniques he employed. Second, Mendel studied a total of seven contrasting traits in the garden pea and they all showed independent assortment. In order for these traits to be independently assorting, they would have to be determined by genes on different chromosomes (and the garden pea has only seven pairs of chromosomes), or the gene pairs could be on the same chromosome, but so widely separated that crossing-over between them results in their independent as-

40

*The establishment of
Mendelian principles of
inheritance*

sortment. This latter circumstance has in fact proved true (see Blixt among the references listed at the end of this chapter), but the probability of either event occurring randomly is quite small.

Rather than tampering with his data, it is more likely that Mendel knew what to expect before he did his experiments and conducted preliminary experiments to find seven independently assorting traits. As for his data fitting expectations too closely, it may well be that Mendel or his gardening assistants stopped collecting data when the numbers fit his expectations. This is not the most objective procedure to follow. However, because his principles have been verified by numerous investigators since his time, we feel that harsh criticism of Mendel's methodology is not warranted.

When Mendel's work was discovered in 1900, its validity did not escape challenge. In 1905, William Castle observed that traits were rarely as constant as Mendel's work implied. To support this, he showed that when an albino guinea pig was crossed to a black guinea pig, the offspring were all black, implying that black was dominant over albino. When he crossed F_1 heterozygotes, the result was a 3:1 ratio of black to albino. However, the F_2 albino animals had small patches of black color on their coats. Castle interpreted this as suggesting that members of the gene pair that determine the contrasting coat colors did not segregate cleanly at meiosis and contaminated each other (Figure 1.37).

Castle's suggestion of gene contamination was shown to be incorrect when *modifier genes* were discovered. A gene is a modifier if its activity affects the activity of other nonallelic genes. In the case of guinea pig coat color, there are genes located on other chromosomes or elsewhere on the same chromosome whose activity affects these color genes. The genes for coat color were segregating cleanly, but their activity was being affected by other genes.

Another major challenge to Mendel's work came from Galton, who suggested that Mendel's patterns of discontinuous variation were of no help in understanding evolutionary mechanisms. Recall that Galton was seeking a mechanism that would give a continuous spectrum of variation. To Galton's way of thinking, Mendelism was not saying that. Galton and his colleagues saw Mendelism as saying, for example, that plants were *either* tall or dwarf. But they sought a mechanism to explain tall, dwarf, and all sizes in between.

Galton's objections were laid to rest when it was demonstrated that more than one gene pair can contribute to a given trait. Let us say, for example, that three gene pairs are involved in expressing a trait such as plant height and that each gene pair is segregating and assorting independently. Assuming a situation in which the phenotype is determined by the number of dominant alleles present, there would be tall and short plants and five sizes in between. This accords both with Darwin's observations of continuous variation and with Mendelian principles.

Figure 1.37
Schematic representation of Castle's theory of gene contamination. When F_1 guinea pigs heterozygous for blackness and albinism were crossed, the presence of patches of black on the coats of the F_2 albinos suggested to Castle his theory of gene contamination.

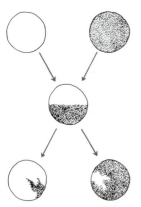

From E. A. Carlson, *The Gene: A Critical History.* Philadelphia: W. B. Saunders Company, 1966

These and other less prominent challenges marked some of the years immediately following the discovery of Mendel's work. But despite all the challenges, Mendel's principles have been upheld and have served as the framework within which other genetic phenomena have been interpreted. To be sure, there are patterns of inheritance that do not follow those set down by Mendel, but these patterns tend to broaden rather than contradict his principles. In the next three chapters, we will examine the expansion of the Mendelian principles of inheritance.

Questions and problems

1.1 Francis Galton carried out a transfusion experiment to test the validity of Darwin's theory of pangenesis. Design another experiment to test this theory, giving predictions for results consistent with or inconsistent with Darwin's proposal.

1.2 Several factors contributed to Mendel's success in discovering the principles of inheritance. List as many of these factors as possible, indicating the significance of each.

1.3 What is the significance of the testcross?

1.4 Reconstruct the threads of thought that led to the idea that chromosomes are the carriers of the genetic material.

1.5 Why is sexual reproduction essential to the demonstration of Mendelian principles? (Be sure to include asexual reproduction in your discussion.)

1.6 Examine Figure 1.34 on page 37. We show a tetratype generated by a crossover between the centromere and the *A* gene.
   (a) What would the tetrad type be if the crossover were between the centromere and the + gene instead?
   (b) What would the tetrad type be if both crossover events occurred?

1.7 Study the diagrams (at left) of dividing cells. Is either of these cells haploid? Explain your answer. What stages of mitosis do these cells represent?

1.8 What meiotic stages are represented by the following cells?

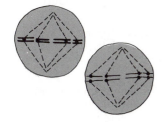

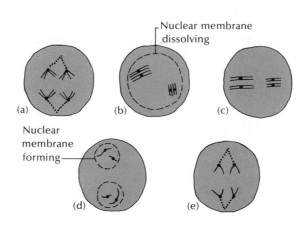

Nuclear membrane dissolving

(a)   (b)   (c)

Nuclear membrane forming

(d)   (e)

1.9 Here is a pedigree for notch ears in Ayrshire cattle. Is this trait due to a dominant or a recessive gene? Defend your answer.

Solid: notch ears
Open: normal

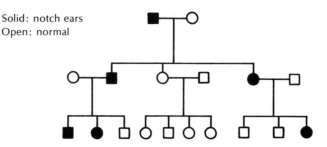

1.10 The following is a pedigree for calluses on the index finger in humans. What conclusions do you draw about the inheritance pattern of this trait?

Solid: calloused index finger
Open: no calloused index finger

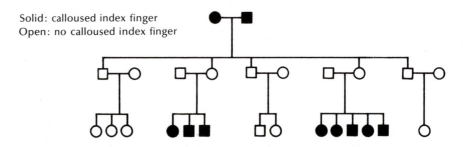

1.11 Black coat color in guinea pigs is the result of the dominant gene W, and white coat color is the result of the recessive gene w.
    (a) A black female guinea pig produces a white offspring. What must her genotype be?
    (b) What genotypes and phenotypes could the male have had?
    (c) What kinds of offspring (phenotypic) would the female produce in a testcross?

1.12 In a certain plant species, it is known that purple color appears only in the presence of a dominant A allele along with a dominant B allele. In a purple $\times$ white cross, an F_1 is produced of which 13 are purple and 17 white. What are the parental genotypes?

1.13 In cocker spaniels, coat color is determined by a pair of independently assorting gene pairs. The presence of a dominant A and a dominant B gene determines black coat color, aa with a dominant B produces a liver-colored coat, bb with a dominant A is red, and $aabb$ is lemon colored. A black cocker is mated to a lemon cocker, and they produce a lemon pup. If this same black cocker is mated to another of his own genotype, what kind of offspring would be expected?

1.14 In summer squash, white fruit color is determined by the dominant allele Y and yellow color by the recessive allele y. Disc-shaped fruit is determined by the dominant allele S and spherical fruit by the recessive allele s. The two gene pairs are on different chromosome pairs. A cross between a white disc-shaped variety and a yellow spheroid variety produces an F_1 that is all white disc-shaped. If these F_1 plants are allowed to self-fertilize, what phenotypic ratio would you expect in the F_2?

1.15 In *Drosophila*, vestigial wings are determined by the recessive allele *vg* and normal wings by the dominant allele *vg*⁺. Ebony body color is determined by the recessive allele *e* and normal body color (tan) by the dominant allele *e*⁺. When a fly with normal body color and normal wings is crossed to another fly, they produce 160 progeny.

> 43 normal body, normal wings
> 35 normal body, vestigial wings
> 40 ebony body, normal wings
> 42 ebony body, vestigial wings

> (a) What is the genotype of the normal parent?
> (b) What are the genotype and the phenotype of the other parent?
> (c) What are the genotypes of the offspring?

1.16 The normal split-hoof condition in swine is determined by the recessive allele *m*, and a mule-footed condition is determined by the dominant allele *M*. The recessive allele *b* determines black coat color, and the dominant allele *B* determines white. In one litter from a white, mule-footed female (sow), the 12 offspring were all white and mule-footed. What is the genotype of the sow if the phenotype of the male (boar) was black and split-hoofed?

1.17 In the cross *AABBCCDDEE* × *aabbccddee*, with all gene pairs on different chromosomes, how many different kinds of gametes can be formed by the F₁?

1.18 In snapdragons, flower color is determined by a pair of alleles: *R* results in red and *r* in white. Leaf shape is determined by another independently assorting pair of alleles: *N* results in narrow leaves and *n* in broad leaves. A true-breeding red, broad-leafed plant is crossed to a true-breeding white, narrow-leafed plant. The F₁ has pink flowers and medium-width leaves. What will the F₂ phenotypic and genotypic ratios be if the F₁ is self-fertilized?

1.19 Cystic fibrosis is a serious genetic disease typified by a defect in protein metabolism. Some of the consequences of the defect are pancreas degeneration, lung destruction, and chronic respiratory infection. A man not exhibiting any symptoms of the disease marries a woman not exhibiting any disease symptoms. They have three children, all of whom have cystic fibrosis. Is cystic fibrosis caused by a dominant or a recessive gene?

1.20 The following is a pedigree for retinitis pigmentosa (an eye disease) and cystinuria (high urinary levels of cystine, usually accompanied by kidney stones).

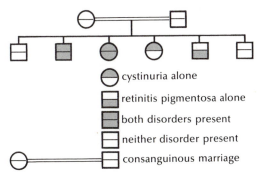

cystinuria alone
retinitis pigmentosa alone
both disorders present
neither disorder present
consanguinous marriage

> (a) Are the traits due to dominant or recessive alleles?
> (b) What are the parental genotypes?

References

Blixt, S. 1975. Why didn't Gregor Mendel find linkage? *Nature* 256:206.

Carlson, E. A. 1966. *The Gene: A Critical History.* Philadelphia, Saunders.

Corwin, H. O., and J. B. Jenkins. 1976. *Conceptual Foundations of Genetics.* Houghton Mifflin, Boston.

Crew, F. A. E. 1966. *The Foundations of Genetics.* New York, Pergamon.

Dunn, L. C. 1965. *A Short History of Genetics.* New York, McGraw-Hill.

Iltis, H. 1932. *The Life of Mendel.* London, Allen and Unwin.

Mendel, G. 1866. Versuche über Pflanzen-Hybriden [Experiments on plant hybrids]. Verh. Naturf. Vei. in Brünn, Abhandlungen, IV:3–47.

Olby, R. C. 1966. *Origins of Mendelism.* London, Constable.

Orel, V. 1973. The scientific milieu in Brno during the era of Mendel's research. *J. Hered.* 64:314–318.

Peters, J. 1959. *Classic Papers in Genetics.* Englewood Cliffs, N.J., Prentice-Hall.

Roberts, H. F. 1929. *Plant Hybridization before Mendel.* Princeton, N.J., Princeton University Press.

Stern, C. (ed.). 1950. The birth of genetics. *Genetics.* 35, supplement.

Stern, C., and E. Sherwood. 1966. *The Origin of Genetics: A Mendel Source Book.* San Francisco, Freeman.

Stubbe, H. 1972. *History of Genetics.* Cambridge, Mass., MIT Press.

Sturtevant, A. H. 1965. *A History of Genetics.* New York, Harper and Row.

Two

Mendelism and the chromosome theory of inheritance

In Chapter 1, several independent lines of research were shown to converge on the emerging science of genetics. The cytological work carried out during the nineteenth century traced inheritance first to the nucleus and then specifically to units in the nucleus called chromosomes. And when Mendel's work was discovered in 1900, it seemed reasonable to associate his particulate factors with chromosomes. In 1903, a penetrating analysis of the relationship between Mendel's observations and the behavior of chromosomes was published by W. Sutton. He perceived a clear parallel between the behavior of chromosomes during meiosis and the behavior of pairs of Mendelian factors. During meiosis, members of a pair of chromosomes segregated into different daughter gametes—just as Mendel's factors did. Also, the segregation of one pair of chromosomes was independent of other pairs—a cytological basis of independent assortment. Thus the parallel between meiosis and Mendel's principles was the first clear association between genes and chromosomes—an association referred to as the *chromosome theory of inheritance*. Other observations consistent with this theory will be explored in this chapter.

The localization of genes to chromosomes prompted investigations into how genes are physically arranged on chromosomes relative to each other. Some of the issues raised by this question will also be dealt with in this chapter.

The chromosome theory of inheritance

The chromosome theory of inheritance postulates that genes are located on chromosomes. The first clear association came about when meiosis was proposed as the mechanism for Mendel's principles of segregation and independent assortment. However, although the parallel between Mendelian principles and meiotic chromosomes was consistent with the chromosomal theory of inheritance, it was by itself insufficient to establish the soundness of the theory. A great advance in the theory took place between 1902 and 1905 when the inheritance of a specific trait, sex, was linked to the inheritance of a specific chromosome. The chromosome theory of inheritance was further strengthened in 1910 when a gene in *Drosophila*, the gene determining white eyes, was shown to be located on the X chromosome, a sex-determining chromosome. This section will explore these avenues of research, which supported the premise that genes and chromosomes are indeed associated.

The inheritance of sex   Although all living organisms reproduce, sexual reproduction is by no means universal. Of those that reproduce sexually, some individuals, called *monoecious* organisms, produce both male and female gametes. (Most flowering plants are monoecious.) In these cases, the individual organisms are not sexually distinct. *Dioecious* organisms are sexually dimorphic individuals, each producing only one type of gamete—male or female. Finally, some species appear to reproduce asexually, though many of these reproductive processes might actually be considered *parasexual,* because genetic information is exchanged between individuals but not through gamete formation or nuclear fusion (for example, bacterial conjugation, discussed in Chapter 12).

Sex is sometimes difficult to pinpoint, because it is one of the less well-defined biological entities. A vague term that reflects our total impression of the differences between male and female, *sex* is commonly used to indicate certain of an individual's physiological, anatomical, and behavioral attributes.

Nearly all dioecious organisms possess the potential for both male and female phenotypes, but normally only one predominates. In most organisms, including humans, both male and female characteristics are developed to varying degrees, resulting in what we might call *intersexual* phenotypes. Generally, one set of sexual characteristics predominates, a result of the genetic constitution and the hormonal environment during development. After discussing the genetic basis of sex determination, we will examine the influence of hormones in the determination of sex.

The chromosomal basis of sex determination   C. E. McClung (1902) was the first to associate an inherited characteristic with a specific chromosome, thus providing the initial *experimental* support for the chromosome theory of inheritance. In so doing, McClung extended the earlier investigations of Henking, who observed that in the hemipteran insect *Pyrrhocoris,* the male possessed an odd number of chromosomes (23) and the female an even number (24). As a result of spermatogenesis, some sperm contained 11 chromosomes and some 12. All female gametes contained 12 chromosomes. Henking made no association between unequal chromosome numbers and the inheritance of sex, but McClung did. Working with grasshoppers, McClung recognized two different types of sperm: One contained an "accessory" chromosome and one did not. He argued that such dimorphic sperm could account for the 1:1 sex ratio observed in dioecious organisms, but he incorrectly concluded that a sperm carrying the "accessory" chromosome was male-determining and that a sperm lacking the "accessory" was female-determining. Actually, the reverse is true.

McClung had no knowledge of these *sex chromosomes,* as they came to be called, occurring in females, so he surmised that only males carried them. In 1905, however, Stevens and Wilson confirmed and extended McClung's observations, suggesting two distinctly different mechanisms of sex determination. In the first mechanism, they asserted that females possess one more chromosome than the male. As an alternative, they proposed that both sexes have the same number of chromosomes, but that in the male, one of the chromosomes (**Y**) is smaller and morphologically distinct from its homolog (**X**). No matter which of these chromosomal compositions a species has, meiosis produces two types of sperm, either **X** and **Y** or **X** and **O** (**O** means no sex chromosome), and one type of egg, which is **X**-containing. We can summarize the Wilson-Stevens proposals for the inheritance of sex in the following manner.

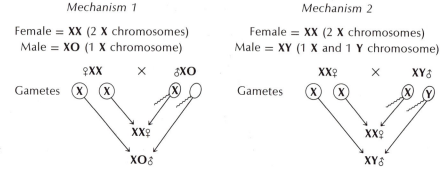

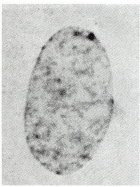

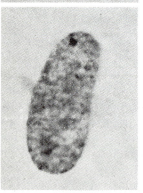

Figure 2.1
Barr bodies in the nuclei of human female cells. [From U. Mittwoch and D. Wilkie, *Brit. J. Exp. Pathol.* 52:186 (1971); reproduced in U. Mittwoch, *Genetics of Sex Differentiation,* Academic Press, New York, 1973, page 62]

The basic error in the McClung model—not repeated in the Wilson-Stevens model—was his suggestion that the sperm carrying the "accessory" chromosome (**X**) determined maleness, whereas in fact it determined femaleness (grasshoppers are **XO** or **XX**). These two most common mechanisms of sex determination are called the **XO-XX** and **XY-XX** mechanisms, respectively. But other mechanisms do exist. In birds, moths, butterflies, and some fishes and amphibians for example, the roles are reversed. The male is *homogametic* (**XX**) and therefore produces one type of gamete, and the female is *heterogametic* (**XY** or **XO**) and therefore produces two types of gametes.

In sexually dimorphic organisms, then, there are two basic chromosome classes: sex chromosomes and non-sex chromosomes. The latter are called *autosomes*.

The presence of two **X** chromosomes in the female and only one in the male creates a problem of no small magnitude. Since the **X** and **Y** chromosomes are for the most part not homologous, females have twice the dose of **X**-chromosome genes found in the male, simply because they have two **X** chromosomes to the male's one. This raises the interesting question of whether a mutant allele in the homozygous state on the female's **X** chromosomes produces the same phenotype produced by that same allele in the hemizygous state (a gene locus present only once in the genotype) on the male's **X** chromosome. Studies indicate that the phenotypes are the same. The resolution of this perplexing problem began in 1949 when a Canadian researcher, Murray Barr, found that nerve cells in female cats contain a dark-staining body in their nuclei, but that no such nuclear body is observed in the neurons of male cats. A similar dark-staining body was subsequently observed in the cells of females of many species, but not in those of males.

Figure 2.1 (cont.)

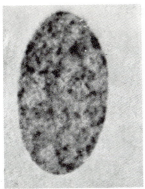

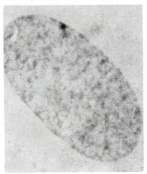

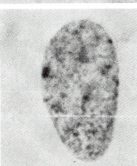

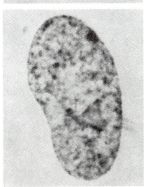

In 1961, Mary Lyon suggested that during mammalian development, one of the female's **X** chromosomes is inactivated and that this inactivated **X** chromosome forms the dark-staining nuclear body in females (or in males, in species wherein the male is **XX**). This is the *Lyon hypothesis.* The inactivated chromosome is referred to as the *Barr body* (Figure 2.1). The inactivation process is random, so that some cells have one of the **X** chromosomes inactivated, and other cells have the other **X** chromosome inactivated. The inactivation of the **X** chromosome results in a phenomenon called *dosage compensation:* the genetic mechanism that compensates for genes that exist in two doses in the homogametic sex, and in one dose in the heterogametic sex.

Many of you have undoubtedly seen the Lyon hypothesis in action. Female cats with calico or tortoise-shell coats (a mosaic of black and yellow hairs) have been known for a long time. Male calicos are extremely rare. Black hairs are the result of a dominant gene (Y), and yellow hairs are the consequence of a recessive allele (y). These alleles are located on the **X** chromosome. A Yy female is calico and not black because of the random inactivation of **X** chromosomes (and therefore of the genes). Cells that inactivate the Y-carrying chromosome produce yellow hairs; cells that inactivate the y-carrying chromosome produce black hairs.

We do not know how dosage compensation works in *Drosophila,* because Barr bodies are not seen in this organism. But it appears that dosage compensation here is a function of a differential gene activity rate in the two sexes.

The behavior of the sex chromosomes during meiosis is quite similar to that of the autosomes, but there are differences. In **XO** males or females, there is no homolog for the **X** chromosome to pair with, so it normally goes to one of the poles in advance of the autosomes. In **XY** organisms, the two chromosomes have limited segments of homology, permitting only limited synapsis during meiosis.

In *Drosophila,* the **Y** chromosome has no effect on sex determination, but it does control male fertility. In general, if the ratio of **X** chromosomes to sets of autosomes is 1:1 or more, the organism is a female. If the ratio is 0.5–1.0:1.0, the organism is an intersex (having both male and female characteristics), and if the ratio is 0.5:1.0 or less, the organism is male. These general statements are reflected in Table 2.1, which summarizes sex determination in *Drosophila.*

In mammals, the **Y** chromosome is usually sex-determining. In humans, for example, **XO** is phenotypically female and **XXY** is male (see page 198). In addition to its sex-determining properties, the human **Y** chromosome—when present twice, as in **XYY** males—is also thought by some to influence height and aggressiveness. But genetic sex-determining mechanisms are not always simple and do not always fit into clean model systems. We know, for example, of **XX** human beings who are phenotypically male. We do not know how genetic females come to express a male phenotype, but the situation has certainly received a lot of attention. One

Table 2.1
The chromosomal basis of sex determination in *Drosophila*

Chromosomal constitution			
Autosomes	Sex chromosomes	Ratio of X chromosomes to sets of autosomes	Sex
Haploid (*n*)	X	1.0	Female (recorded as patch of tissue only; female designation is theoretical)
Diploid (2*n*)	X	0.5	Male (sterile)
Diploid (2*n*)	XX	1.0	Female
Diploid (2*n*)	XY	0.5	Male
Diploid (2*n*)	XXX	1.5	Metafemale (sterile, low viability)
Diploid (2*n*)	XXY	1.0	Female
Diploid (2*n*)	XYY	0.5	Male
Diploid (2*n*)	XXYY	1.0	Female (reduced fertility and viability)
Diploid (2*n*)	XYYY	0.5	Male (usually sterile)
Triploid (3*n*)	X	0.33	Metamale (sterile)
Triploid (3*n*)	XX	0.67	Intersex (sterile)
Triploid (3*n*)	XY	0.33	Metamale (sterile, low viability)
Triploid (3*n*)	XXX	1.0	Female (sterile)
Triploid (3*n*)	XXY	0.67	Intersex (sterile)
Triploid (3*n*)	XXXX	1.33	Metafemale (sterile, reduced viability)
Triploid (3*n*)	XXXY	1.0	Female (sterile)
Tetraploid (4*n*)	XX	0.5	Male
Tetraploid (4*n*)	XXX	0.75	Intersex (sterile)
Tetraploid (4*n*)	XXXX	1.0	Female

Metafemale

Metamale

suggestion is that under normal circumstances, **Y**-chromosome genes "switch on" male-determining genes located on the **X** chromosomes. In these rare **XX** individuals, there is a mimic of the **Y**-chromosome gene(s) that activates the male-determining genes on the **X** chromosome.

But the situation becomes even more confusing when we look at the wood lemming (*Myopus schisticolor*). Fredga and his colleagues recently discovered that female lemmings could be either **XX** or **XY**! Both types of female are normal and indistinguishable from each other. It has been suggested that, in this species, genes on the **X** chromosome inactivate the male-determining effect of the **Y** chromosome.

Both these unusual sex-determining mechanisms seem to suggest that **Y**-chromosome genes activate male-determining **X**-chromosome genes. In one case, the **Y**-chromosome genes are mimicked, and in the other case, the **Y**-chromosome genes are repressed. (Keep in mind that this is speculation.)

There is one case, however, in which we have more information about the basis for genetic sex differing from phenotypic sex. This is the *testicular-feminization syndrome,* wherein the genetic sex is **XY**, or male, and the phenotypic sex is female (often described as voluptuous). These females do not menstruate and have little if any pubic hair. They have testes that remain undescended and normal levels of androgen, the male hormone. The gene that causes this syndrome is **X**-linked, and its effect appears to be to make the males' target tissues unresponsive to the male hormone, androgen. Because the target tissues do not respond to the androgen, the phenotype is female.

The testicular-feminization syndrome raises complicated ethical problems. Consider a happily married, attractive woman with the syndrome. She does not menstruate and is, of course, infertile. Her physician discovers that she has a blind vagina and undescended testes. Since testes in these cases have a tendency to become cancerous, they should be removed. But to remove them, the physician must inform the patient that she is a genetic male complete with testes. One can imagine the shock that this person and her spouse would experience. This is a very complex issue, and it is not easy to determine what course medical ethics dictates to the physician.

Sex determination by sex chromosomes is the common, but by no means the only, genetic sex-determining mechanism. Autosomal genes can also be influential in determining sex, as Sturtevant showed (1945) in his account of a mutant gene in *Drosophila* that transformed genetic females into sterile males. The gene is a recessive, autosomal gene called "transformer" (*tra*). Genetic females (**XX**) that are homozygous for *tra* are transformed into phenotypic males that have the characteristic male sex combs, male-patterned abdomens, and male internal and external genitalia. They have rudimentary testes but produce no sperm.

An **XXY** genetic female that is homozygous for *tra* is also a sterile male, but an **XY** male that carries *tra* in a homozygous state is a normal male. The transformer phenomenon in *Drosophila* brings out an interesting feature of sex determination in this organism: Maleness appears to be determined largely by autosomal genes, femaleness by **X**-chromosome genes.

Hormones and sex determination   Hormones are known to play an important role in sex determination. We have seen, for example, how an **XY** genetic male was phenotypically female because the male hormone target tissue was unable to respond to the male hormone. In the sea worm *Bonnelia*, some of the larvae produced by the female worm swim away and settle on the ocean floor; these develop into females. Others land on the female's proboscis, and these develop into males. The basis for this appears to lie in the production of a hormone by the female that determines maleness. An extract of the female has been shown to make all *Bonnelia* larvae into males, illustrating that hormones can strongly influence sexual differentiation.

It has long been known that when cattle, sheep, pigs, and goats give birth to *dizygotic* (two-egg, or fraternal) twins, of which one is male and one female, the female twin is, in over 90 percent of the births, phenotypically malelike. This genetic female with male features is called a *freemartin* (Figure 2.2). In these cases, it has been shown that the fetal membranes fuse, allowing cross-circulation between the two fetuses. Hormones released by the developing male pass through the placenta to the female and cause a partial sex reversal of her ovaries. This interferes with the development of the female reproductive tract and stimulates the development of male sexual characteristics. If no membrane fusion occurs, no freemartin results.

A more dramatic example of how hormones can affect sexual development occurs in *Medaka*, a small teleost fish that can be made to experience complete sex reversal by the administration of gonadal hormones.

Figure 2.2
A pair of twins in cattle, a male (left) and a female (right), whose blood circulations are united. The female becomes a hormonal intersex, or freemartin

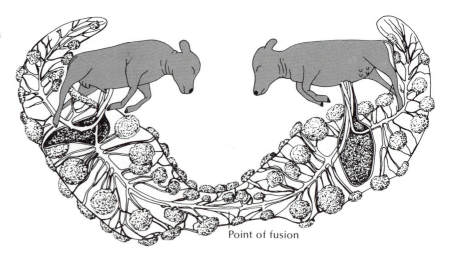

Point of fusion

The male *Medaka* is chromosomally **XY**, the female **XX**. Body color in *Medaka* is determined by a pair of genes (*R* and *r*) found on the sex chromosomes—both **X** and **Y**. *R* is dominant over *r* and results in a red-orange color, whereas *rr* results in white. An X^rY^R red-orange male, when treated with the female hormone estrone, is converted to an X^rY^R red-orange functional female. When crossed to an X^rY^R male, this transformed male gives a phenotypic ratio of three red-orange to one white and a male-to-female ratio of 3:1. One-third of the males (Y^RY^R) produce only male offspring in the F_2. An X^rX^r white female, when treated with the male hormone testosterone, is converted to an X^rX^r white functional male. When crossed to a white female, it produces only white female progeny (Figure 2.3).

Like fish, most amphibians also undergo sex reversal when treated with hormones of the opposite sex. For example, male amphibians treated with estrogen become females, and female amphibians treated with androgen become males.

Some birds undergo dramatic sex reversal. In chickens, only one of the two gonads develops into an ovary. If that ovary is removed from the hen, the remaining gonad frequently develops into a testis and produces male hormones. The hen's comb grows, she crows like a cock, develops

Figure 2.3
Sex reversal, mediated by sex hormones, in the teleost fish *Medaka*. [Data from T. Yamamoto, *J. Exp. Zool.* 137 (1958)]

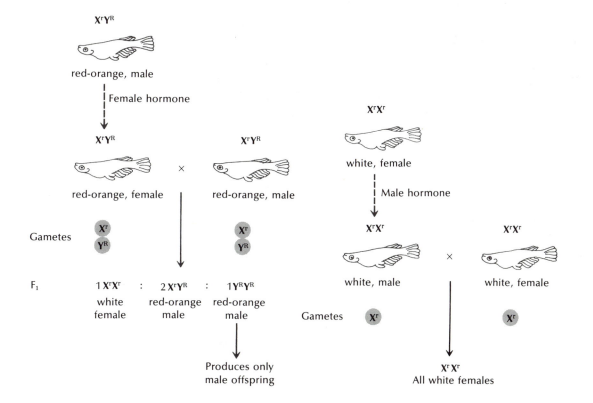

male plumage, and this once-egg-laying hen often becomes a sperm-producing rooster.

Sex linkage We have discussed how one trait, sex, is generally determined by a specific pair of chromosomes, the sex chromosomes. However, "sex" is a complex phenotype. No single gene determines it except in such organisms as *Neurospora* and *Chlamydomonas,* in which mating type is determined by a single gene pair. One of the first tasks undertaken by early twentieth-century geneticists was to associate specific genes with specific chromosomes.

Figure 2.4
Thomas Hunt Morgan

T. H. Morgan (Figure 2.4) was the first to show clearly that a gene was located on a chromosome. Others before him had reported that the inheritance of traits paralleled the inheritance of the sex chromosomes, but it was Morgan who clarified the correlation between sex-chromosome inheritance and sex-linked traits.

About 1909, Morgan began experimenting with *Drosophila* and attempted to induce mutations with ionizing radiation and chemicals. He failed to induce a mutation in his experimental system, but during his control experiments, a white-eyed male appeared (*Drosophila* normally has red eyes). This determiner of white eyes (*w*) was the first gene to be demonstrably associated with a specific chromosome. Crossing the white-eyed male with a normal, red-eyed female produced an F_1 that was all red-eyed. He concluded from this that the gene for red eyes was dominant over the gene for white eyes. When he crossed F_1 red-eyed males with F_1 red-eyed females, he obtained in the F_2 a 3:1 ratio of red-eyed flies to white-eyed flies—just as Mendelian principles might lead one to expect. But the F_2 white-eyed flies were all males! Morgan had expected equal numbers of males and females in both classes. The cross is outlined as follows.

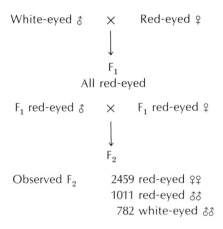

White-eyed ♂ × Red-eyed ♀
↓
F_1
All red-eyed

F_1 red-eyed ♂ × F_1 red-eyed ♀
↓
F_2

Observed F_2 2459 red-eyed ♀♀
1011 red-eyed ♂♂
782 white-eyed ♂♂

Note: The lower than expected number of white-eyed flies in this cross is due to the reduced viability of white-eyed mutants.

From this cross, Morgan proposed that the gene determining white eyes and its normal allele are located on or associated with the **X** chromosome. This predicts that the F_1 females will be heterozygous for white eyes (X^wX^{w+}), where w symbolizes the white-eye gene and w^+ is the normal allele producing red eyes). Testcrossing such a female to a white-eyed male (X^wY) would, therefore, give progeny of one white-eyed to one red-eyed fly, with equal numbers of males and females in each class. His prediction was borne out, as shown in the following cross.

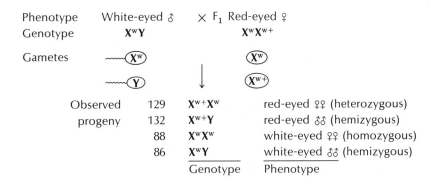

		Genotype	Phenotype
Phenotype	White-eyed ♂	× F_1 Red-eyed ♀	
Genotype	X^wY	X^wX^{w+}	
Gametes	X^w	X^w	
	Y	X^{w+}	
Observed progeny	129	$X^{w+}X^w$	red-eyed ♀♀ (heterozygous)
	132	$X^{w+}Y$	red-eyed ♂♂ (hemizygous)
	88	X^wX^w	white-eyed ♀♀ (homozygous)
	86	X^wY	white-eyed ♂♂ (hemizygous)

This is essentially a 1:1 ratio (white-eyed flies have reduced viability), and it demonstrated to Morgan that the trait is not *sex-limited* (that is, found only in one sex) but rather *sex-linked*. In other words, the trait is linked to the inheritance of sex. (Actually, Morgan spoke of the white-eye "factor" being linked to the sex "factor." He did not use the term "chromosome" until later.) A modern chromosomal interpretation of Morgan's crosses is given in Figure 2.5.

In the latter part of 1910, Morgan found two more mutant genes in *Drosophila*—causing yellow body color (y) and miniature wings (m)—that were also sex-linked. With them, he was able to demonstrate that genes could change linkage associations by means of chromosomal *crossing-over* (Figure 2.6). Morgan crossed a white-eyed male with normal body color $\left(\underrightarrow{w\ y^+} \right)$ to a red-eyed female with yellow body color $\left(\dfrac{w^+\ y}{w^+\ y} \right)$.

The F_1 was composed of red-eyed, normal-bodied females and red-eyed, yellow-bodied males. This demonstrated that both mutant genes were recessive. The F_1 females were testcrossed to white-eyed, yellow-bodied males and the offspring analyzed. Only white-eyed, normal-bodied and red-eyed, yellow-bodied flies were expected, but two additional classes appeared (though in low frequency): red-eyed, normal-bodied flies and white-eyed, yellow-bodied flies.

The latter two categories must have been the result of chromosomal

Figure 2.5
The chromosomal
interpretation of Morgan's
crosses with white-eyed
Drosophila

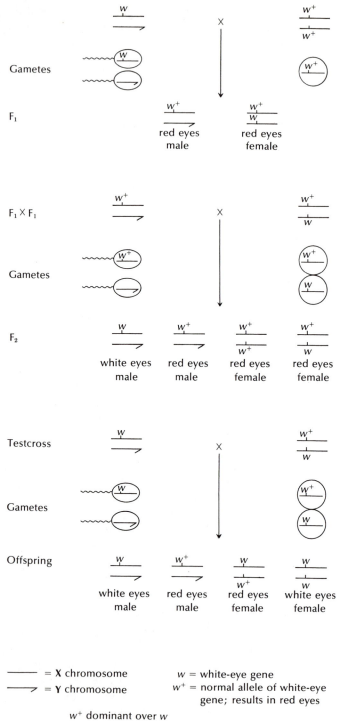

= **X** chromosome w = white-eye gene

= **Y** chromosome w^+ = normal allele of white-eye
                              gene; results in red eyes

w^+ dominant over w

Figure 2.6
Morgan's cross involving two sex-linked genes, demonstrating crossing-over

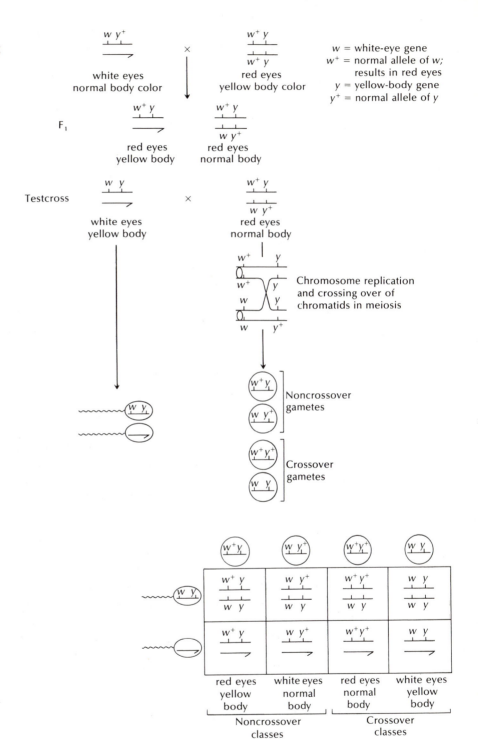

crossing-over in the F_1 female parent, which would have produced $w^+ y^+$ and $w\ y$ gametes.

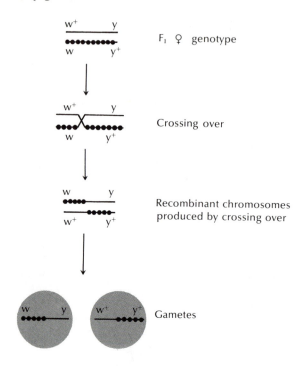

F₁ ♀ genotype

Crossing over

Recombinant chromosomes produced by crossing over

Gametes

Since Morgan's discovery of sex linkage, dozens of genes in *Drosophila* and other organisms have been traced to locations on the **X** chromosome and, though rarely, on the **Y** chromosome. The **Y**-linked genes (*holandric inheritance*), such as the one causing hairy ear pinnae in humans (Figure 2.7) are *sex-limited*, because they can be passed on to only one sex. Any genes found on both the **X** and **Y** segregate the way autosomal genes do.

There are well over 100 known traits in humans that owe their existence to **X**-linked genes. The *dentinogenesis imperfecta* trait ("brown teeth") mentioned on page 18) is due to a sex-linked dominant gene. Other sex-linked traits are certain types of night- and colorblindness; ichthyosis, a scaly skin condition; rickets due to vitamin-D resistance; certain forms of *diabetes insipidus;* the Duchenne type of muscular dystrophy; and many types of enzyme deficiencies. Hemophilia, the "bleeder's disease," is a well-known sex-linked trait. Sufferers from this disease lack a blood-clotting factor, either antihemophilic globulin (hemophilia A) or plasma thromboplastin component (hemophilia B). Figure 2.8 is a pedigree of a family in which the original mother (I1) was a carrier and the father (I2) a hemophiliac. Hemophilia has plagued European royalty since its introduction into the royal line by Queen Victoria.

The significance of Morgan's studies of sex linkage lay in their twofold support of the chromosome theory of inheritance. He showed

Figure 2.7
Hairy ear pinna, caused by
a **Y**-linked gene in humans

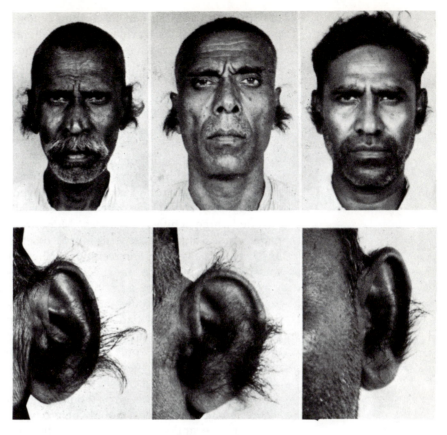

C. Stern, W. R. Centerwall, and S. S. Sarkar, *Am. J. Human Genetics,* University of Chicago Press, Vol. 16, 1964.
Copyright Grune and Stratton, American Society of Human Genetics

Figure 2.8
Inheritance of hemophilia
in a kindred (group of
related people), including
affected females. Black
symbols represent affected
individuals. IV-7 may
have been hemophilic,
since she had some of the
symptoms. (From
Merskey, C. 1951. *Quart. J.
Med.* 20:299–312)

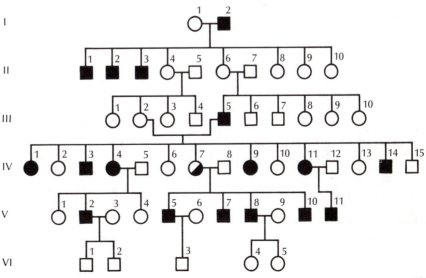

first that genes are linked to chromosomes and second that there is a parallel between meiotic chiasmata and altered linkage associations. That is, chromosomal crossing-over during meiosis could explain Morgan's observed linkage change.

In 1916 C. B. Bridges (Figure 2.9), one of Morgan's graduate students, showed an unexpected pattern of inheritance to be caused by abnormal segregation of sex chromosomes during meiosis. This correlation of a cytological event with a genetic event was the strongest support yet for the chromosome theory of inheritance.

Bridges based his discovery on three flies that developed unexpected phenotypes. Morgan had crossed white-eyed male *Drosophila* with red-eyed females. The F_1 was expected to consist of all red-eyed flies, but it included three white-eyed males. Morgan, in what is now his classic paper on sex linkage, chose to ignore the three white-eyed males, dismissing them as simply the result of further mutation. Bridges, however, was intrigued by these unexpected F_1 white-eyed males (Morgan followed Bateson's lead by admonishing all his students to "treasure your exceptions"). To explain the anomaly, Bridges suggested that perhaps the males with white eyes had received an **X** chromosome from the male parent and *no sex* chromosome from the female parent. Unfortunately, these F_1 males, which would be **XO**, were shown to be sterile and could not be used for further genetic analysis. Unexpected results were again obtained from reciprocal crosses of white-eyed females to red-eyed males. All the F_1 males from this cross would be expected to be white-eyed and all the females red-eyed (though heterozygous; $w^+//w$). Bridges noted, however, the appearance of a few white-eyed females and a similar number of red-eyed males. He hypothesized that the white-eyed females had received two **X** chromosomes from the female parent, and since these exceptional females were fertile, he was able to test and confirm his predictions, as shown in Figure 2.10.

The basis for these results was *chromosomal nondisjunction,* the occasional failure of homologous chromosomes to separate during meiosis. The original parental female normally produces gametes containing a single **X** chromosome, but in rare cases she produces gametes with either no **X** (nullo-**X**) or two **X** chromosomes. The two-**X** gamete, when fertilized by a **Y** sperm, produces a female that is phenotypically identical to the parental female, in this case white-eyed. The exceptional white-eyed female, when crossed to a normal red-eyed male, produces four kinds of gametes: **XX**, **Y**, **XY**, and **X**. The offspring resulting from the fertilization of these four gametes by the two types of male gametes produce six viable offspring (**XXX** and **YY** are seldom viable), two of which are unusual because they show a mother-to-daughter and father-to-son pattern of inheritance. Recall that the normal pattern of inheritance for a sex-linked recessive trait is mother to son and father to grandson.

Nondisjunction can be classified as either *primary* or *secondary*, de-

Figure 2.9
Calvin B. Bridges

Figure 2.10
Bridges' demonstration of chromosomal nondisjunction

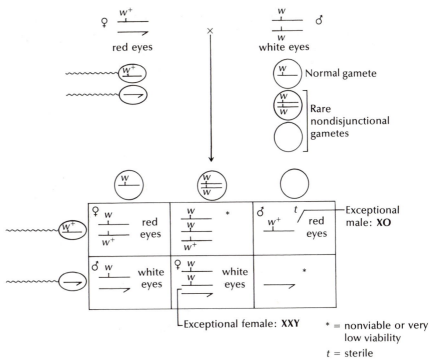

(a) Schematic basis for primary chromosomal nondisjunction in *Drosophila*.

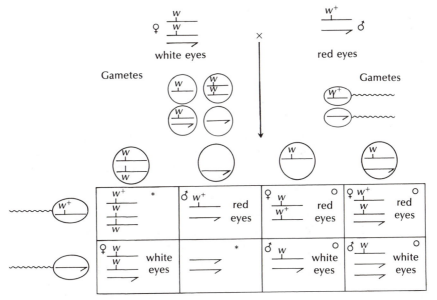

(b) Secondary nondisjunction.

61

pending on the chromosomal constitution of the parent. The original nondisjunctional event in the normal **XX** female is *primary nondisjunction:*

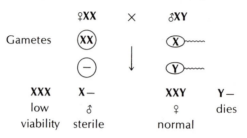

Primary nondisjunction
(occurs in a normal **XX** female)

Secondary nondisjunction involves the production of nondisjunctional gametes by an exceptional **XXY** female:

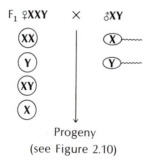

Secondary nondisjunction
(occurs in an exceptional **XXY** female)

Progeny
(see Figure 2.10)

Nondisjunction results from an improper pairing of chromosomes during the meiotic metaphase. Its occurrence is not limited to sex chromosomes and can occur with any chromosome pair. Some consequences of nondisjunction will be explored in Chapter 5, when we discuss chromosome abnormalities and their phenotypic consequences.

Bridges experimented extensively with nondisjunctional flies and made a cytological analysis of their chromosomes. He found that the exceptional female did indeed have an **XXY** constitution, thus establishing a cytological basis for a particular pattern of inheritance.

In 1921, L. V. Morgan (the wife of T. H. Morgan) discovered a strain of flies in which nondisjunction occurred 100 percent of the time. She made this important discovery during an examination of one of her culture bottles. She found a fly whose body was part mutant (yellow) and part normal. In other words, it was a mosaic fly. As she was isolating it for further study, it flipped onto the floor and disappeared. Not one to give up easily, Morgan searched the lab methodically and finally found the object of her search perched on a window ledge. She climbed up and trapped it. L. V. Morgan's investigations of this fly considerably strengthened the chromosome theory of inheritance. In the strain developed from this fly, the females *always* produced female offspring identical to the maternal parent and male offspring identical to the paternal parent. This was in marked contrast to Bridges' stock of **XXY** females, which produced this result in only one-third of the cases (see Figure 2.10).

A white-eyed stock of these exceptional females was developed. When they were mated to red-eyed males, their progeny consisted of exclusively white-eyed females and exclusively red-eyed males. To explain this inheritance pattern, the Morgans proposed that the exceptional females, though they were **XXY**, produced only **XX** and **Y** gametes because the two **X** chromosomes were physically attached. They called this strain *attached-X*, symbolized as **XX**. L. V. Morgan later verified the attached nature of the chromosomes when she demonstrated cytologically that the **X** chromosomes of the exceptional females were physically joined at the centromere. Figure 2.11 outlines a cross involving attached-**X** females.

In summary, the chromosome theory of inheritance gained strong support from demonstrations by McClung, Wilson, and Stevens that the inheritance of sex is tied to a specific chromosome and from T. H. Morgan's demonstration that genes reflecting a Mendelian pattern of inheritance are linked to the **X** chromosome and that certain deviations from

Figure 2.11
Cross between a white-eyed, attached-**X** female (X^wX^wY) and a normal male

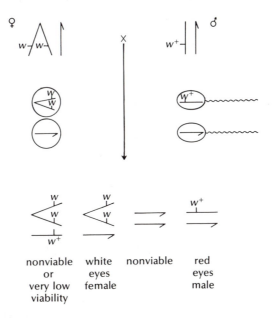

nonviable | white | nonviable | red
or | eyes | | eyes
very low | female | | male
viability | | |

expected Mendelian inheritance patterns can be explained by cytologically verifiable chromosomal nondisjunction. The soundness of the chromosome theory of inheritance was thus established. In the next section, we will discuss the positioning of genes on chromosomes.

Linkage, recombination, and gene mapping

The chromosome theory of inheritance, which proposes that genes are linked to chromosomes, implies that because there are far more genes than chromosomes in an organism, many genes must be located on the same chromosome. A chromosome and its associated genes is called a *linkage group.* Further, meiotic crossing-over suggests that genes can be transferred from one chromosome to its homolog, thus altering the composition of linkage groups. Although the logic of these statements may be clear today, it was not necessarily so in 1908, when the premise that genes are on chromosomes was not always granted.

Complete versus incomplete linkage

W. Bateson and R. C. Punnett were the first to observe that two pairs of genes do not necessarily assort independently of each other. But their interpretation of this observation was strongly influenced by their belief that genes were *not* located on chromosomes. When they crossed purple-flowered (r^+), long-pollen (l^+) pea plants to red-flowered (r), round-pollen (l) plants, they noted a wide divergence from the expected 9:3:3:1 phenotypic ratio in the F_2.

Phenotypic class	Observed numbers	Numbers expected on a 9:3:3:1 basis
Purple-long	296	242
Purple-round	19	80
Red-long	27	80
Red-round	87	27

Figure 2.12
William Bateson (right) shown in his garden with Wilhelm Johannsen (see Chapter 4). (Photograph courtesy American Philosophical Society)

Bateson (Figure 2.12), having temporarily rejected the idea that genes and chromosomes were associated, proposed that independent assortment had occurred but was followed by differential sex-cell reproduction. That is to say that during gamete formation, those cells containing the r^+ l^+ and r l genes underwent mitotic reproduction to a greater extent than the others, thereby giving specific ratios (Figure 2.13). This idea has been termed the *reduplication hypothesis.*

T. H. Morgan's group was by this time gathering large numbers of *Drosophila* mutants and was also noting deviations from expected 9:3:3:1 phenotypic ratios. But Morgan's interpretation differed from Bateson's in that he accepted the chromosome theory of inheritance. However, there was another key to Morgan's interpretation. Morgan had read F. A. Janssens' 1909 cytological study of meiosis in amphibians, in which Janssens observed chromosomes in crosslike configurations at synapsis.

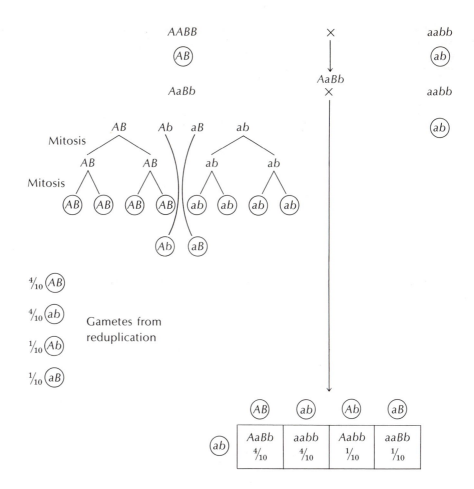

Figure 2.13
Bateson's reduplication hypothesis

He called these configurations *chiasmata* and interpreted them as the result of a fusion at one point between two of the four cromatids in a tetrad, followed by breakage and reunion. The consequence of these events is the exchange of equal and corresponding segments of two of the four chromatids. Morgan proposed that in his dihybrid crosses, the gene pairs he used were in the same linkage group—that is, on the same chromosome—and that the ratios he observed were attributable to *complete linkage* (genes were always inherited together because the chromosomes had not crossed over), to *incomplete linkage* (genes changed linkage groups in some meiotic cells because the chromosomes had crossed over), and to the segregation of linkage groups (see Figure 2.14). Janssens' work was crucial to Morgan's interpretation of the departures from the expected 9:3:3:1 ratio. But, perhaps more importantly, Janssens' work suggested genetic tests of Morgan's interpretation.

Based on cytological studies of meiosis, Morgan suggested that complete and incomplete linkage were simply expressions of the distance between two genes on the same chromosome. If the two genes are very

Figure 2.14
Morgan's interpretation of complete versus incomplete linkage

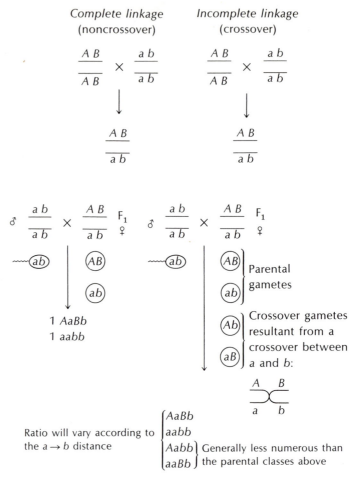

Ratio will vary according to the $a \rightarrow b$ distance

$$\begin{cases} AaBb \\ aabb \\ Aabb \\ aaBb \end{cases}$$ Generally less numerous than the parental classes above

close, crossing-over between them is rare or nonexistent; hence complete linkage is observed. If the two genes are not close together, however, crossing-over can occur between them; hence incomplete linkage is observed.

Evidence supporting the chromosome theory of inheritance eventually disproved Bateson's reduplication hypothesis and upheld Morgan's view. Other observations substantiated Morgan's interpretation of linkage groups. In *Drosophila*, for example, four linkage groups are known and there are four pairs of chromosomes. The garden pea has seven linkage groups and seven pairs of chromosomes, and maize has ten linkage groups and ten chromosome pairs. It was not until 1921, however, that Bateson felt there was sufficient countering evidence for him to abandon his reduplication hypothesis.

Gene mapping in diploids   Morgan suggested that the closer two genes were on a chromosome, the smaller the probability that crossing-over would take place between them. In 1912 A. H. Sturtevant, another of Morgan's students, developed this idea into a model for mapping genes on chromosomes. He proposed that the percentage of crossing-over was a function of the distance separating the genes: The greater the distance, the greater the crossing-over between them. Setting 1 percent crossing-over as equivalent to 1 *map unit* (MU), Sturtevant, using sex-linked genes, made *two-point crosses* (two pairs of linked genes) and *three-point crosses* (three pairs of linked genes). The result was the first genetic mapping of a *Drosophila* chromosome.

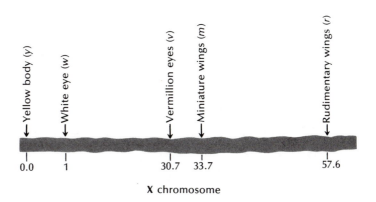

X chromosome

Assignment of a gene to a linkage group   Two-point and three-point crosses will be discussed shortly, but first we should describe how genes are assigned to specific linkage groups. Genes located on the **X** chromosome are easily identified by the patterns of inheritance described earlier, but autosomally linked genes present a more difficult problem. In *Drosophila*, for example, there are three pairs of autosomes, and both sexes are autosomally identical. In one method of linking a gene to a specific autosome, each chromosome carries a different marker gene: curly wings (*Cy*), plum eyes (*Pm*), dichaete wings (*D*), and stubble bristles (*Sb*)—all dominant. Organisms carrying these genes in a homozygous state are usually nonviable. These genes are not sex-linked. Because *Cy*, *Pm*, *D*, and *Sb* segregate from each other and assort independently, we can propose two possibilities. First, they are in different linkage groups. Second, the genes are not on the same chromosome of the linkage group pairs.

The genetic cross that appears in the following diagram shows that different F₂ phenotypic ratios are predicted according to whether the genes are located on the same homologous pair of chromosomes. If, for example, *Cy* and *Pm* are on different chromosome pairs, they will assort independently and give an F₂ with four phenotypic classes. If they are on the same homologous pair, an F₂ composed of only one phenotypic class will result. The latter possibility is in fact observed, thus proving that the two genes are in the same linkage group and on different members of the homologous pair (the same holds for *D* and *Sb*).

1. If *Cy* and *Pm* are located on nonhomologous chromosomes (that is, on different linkage groups).

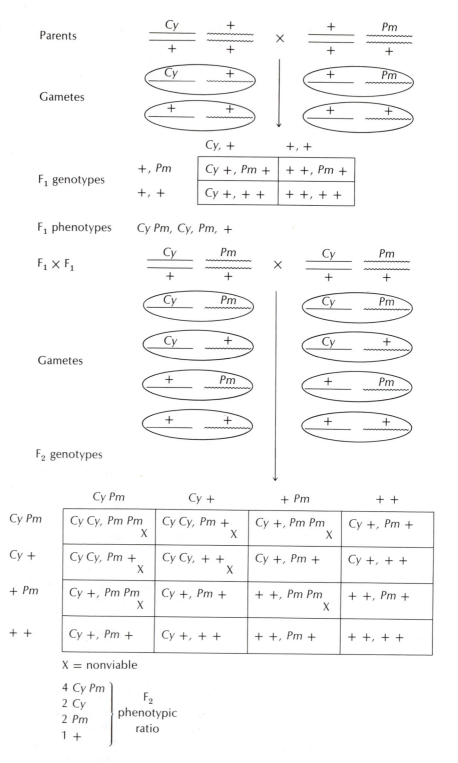

Parents

Gametes

F$_1$ genotypes

	Cy, +	+, +
+, Pm	Cy +, Pm +	+ +, Pm +
+, +	Cy +, + +	+ +, + +

F$_1$ phenotypes Cy Pm, Cy, Pm, +

F$_1$ × F$_1$

Gametes

F$_2$ genotypes

	Cy Pm	Cy +	+ Pm	+ +
Cy Pm	Cy Cy, Pm Pm X	Cy Cy, Pm + X	Cy +, Pm Pm X	Cy +, Pm +
Cy +	Cy Cy, Pm + X	Cy Cy, + + X	Cy +, Pm +	Cy +, + +
+ Pm	Cy +, Pm Pm X	Cy +, Pm +	+ +, Pm Pm X	+ +, Pm +
+ +	Cy +, Pm +	Cy +, + +	+ +, Pm +	+ +, + +

X = nonviable

4 Cy Pm ⎫
2 Cy   ⎬ F$_2$ phenotypic ratio
2 Pm   ⎮
1 +   ⎭

2. If *Cy* and *Pm* are located on homologous chromosomes.

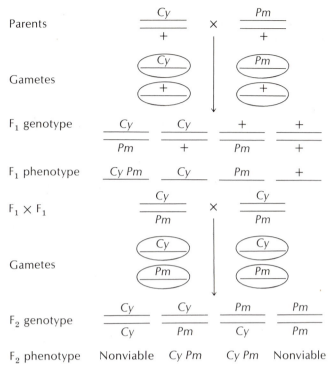

Parents
$$\frac{Cy}{+} \quad \times \quad \frac{Pm}{+}$$

Gametes

F_1 genotype
$$\frac{Cy}{Pm} \qquad \frac{Cy}{+} \qquad \frac{+}{Pm} \qquad \frac{+}{+}$$

F_1 phenotype
Cy Pm   Cy   Pm   +

$F_1 \times F_1$
$$\frac{Cy}{Pm} \quad \times \quad \frac{Cy}{Pm}$$

Gametes

F_2 genotype
$$\frac{Cy}{Cy} \qquad \frac{Cy}{Pm} \qquad \frac{Pm}{Cy} \qquad \frac{Pm}{Pm}$$

F_2 phenotype   Nonviable   Cy Pm   Cy Pm   Nonviable

Note: + = normal (gene or phenotype); homozygous mutants are lethal

Using the *Cy//Pm, D//Sb* stock, a gene can be placed into a specific linkage group—either with *Cy* and *Pm*, with *D* and *Sb*, or perhaps with all of them if the gene is located on the small fourth chromosome pair (the third autosomal pair). The technique for associating a gene with a specific linkage group is shown on page 70. The mutant (*a*) is first crossed to a stock of *Cy//Pm, D//Sb* flies, then F_1 males are testcrossed to the homozygous mutant females.* If the mutant is on chromosome II (chromosome I is the **X** chromosome), the F_2 generation will show *a* appearing with *Sb*. If the mutant is on chromosome III, the F_2 generation will show *a* appearing with *Cy*. If the mutant is on chromosome IV, the F_2 generation will show *a* appearing with *Cy* alone, *Sb* alone, and *Cy Sb* together. Thus, in *Drosophila*, a gene can be assigned to a specific linkage group by mating the unknown to stocks of flies that carry special genetic markers.

The two- and three-point cross   Once it has been determined that a gene is associated with a specific linkage group, the gene's position within that group can be ascertained. Once again, this is accomplished by appropriate genetic crosses. The two-point cross, three-point cross, and testcross are the most commonly used.

* F_1 males are used because in *Drosophila* there is essentially no meiotic crossing-over in males. Crossing-over is restricted to females.

Technique for associating
a gene with a specific
linkage group in *Drosophila*

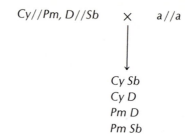

Cy//Pm, D//Sb   ×   a//a

Cy Sb
Cy D
Pm D
Pm Sb

Select any one of these four phenotypic classes for males: *Cy Sb*, for example

(a) If "a" is on chromosome II:

♂Cy//a, Sb//+   ×   ♀a//a, + // +

Cy Sb   a +
Cy +
a Sb
a +

Cy Sb, Cy, a Sb, a

(b) If "a" is on chromosome III:

♂Cy// +, Sb//a   ×   +//+, a//a

Cy Sb   + a
Cy a
+ Sb
+ a

Cy Sb, Cy a, Sb, a

(c) If "a" is on chromosome IV:

♂Cy// +, Sb// +, +//a   ×   +// +, +// +, a//a

Cy Sb+   + Sb +   ++ a
Cy Sb a   + Sb a
Cy + +   +++
Cy + a   + + a

Cy Sb, Cy Sb a, Cy, Cy a,
Sb, Sb a, +, a

(d) If "a" is sex-linked:

♀"a"   ×   wild-type♂

all "a" males
all "wild-type" females

(How would this localization work if "a" were a dominant instead of a recessive?)

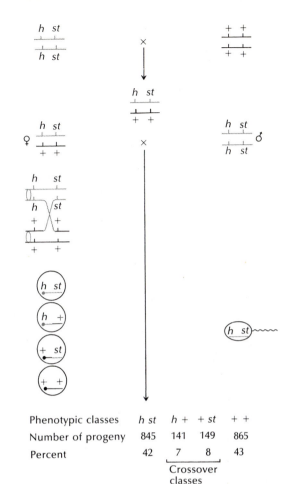

Figure 2.15
Two-point cross for
mapping genes in
Drosophila

Phenotypic classes	h st	h +	+ st	+ +
Number of progeny	845	141	149	865
Percent	42	7	8	43

Crossover
classes

The two-point cross in *Drosophila*—indeed in most diploid organisms—is less useful than the three-point cross, because multiple crossovers between the two genes can be missed. Take, for example, a cross between a hairy-bodied (*h*), scarlet-eyed (*st*) fly and a normal mate (Figure 2.15). These genes are found on chromosome III. The F_2 data show these genes to be about 15 map units apart. But if a *double crossover* had occurred between them, it would have gone undetected (see the diagram at the top of page 72).

Two-point crosses, then, usually result in an *underestimate* of the distance between two genes, because the crossover percentage will reflect only single crossovers or rare triple crossovers. Another weakness of the two-point cross is that it often does not allow sequencing of genes on a chromosome. In the cross described, the gene sequence could have been *h st* or *st h*. The three-point cross overcomes some of these problems.

To ascertain the map distance between two points on a chromosome, it is important to keep the distance as small as possible. If *h* and *st* were linked but separated by a distance of over 50 map units, we would be

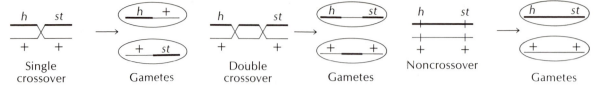

| Single crossover | Gametes | Double crossover | Gametes | Noncrossover | Gametes |

A double crossover may go undetected in a two-point cross.

unable to detect the linkage. They would behave as if they were on independently assorting chromosomes, because about half the time they would show recombination and about half the time they would not. We would find *h st*, *h +*, *+ st*, and *+ +* gametes in essentially a 1:1:1:1 ratio, the ratio for independently assorting alleles in a dihybrid cross.

The three-point cross involves three pairs of linked genes and can give data about the position of these genes relative to each other and, if the genes are relatively close together, the correct distances between the genes. Table 2.2 presents the data of Bridges and Olbrycht (1926), who used the three-point cross and testcross to map three recessive sex-linked genes: scute bristles (*sc*), echinus eyes (*ec*), and crossveinless wings (*cv*).

Table 2.2
Recombination data from a three-point cross involving

```
+   ec   +
+---+---+---+  females and
+---+---+---+
sc   +   cv

sc   ec   cv
+---+---+---+  males
         →
```

Progeny phenotypes	Progeny genotypes		Observed no.	Percent
	Male	Female		
ec	+ ec +	+ ec + / sc ec cv	8576	41.26
sc cv	sc + cv	sc + cv / sc ec cv	8808	42.38
sc ec	sc ec +	sc ec + / sc ec cv	681	3.28
cv	+ + cv	+ + cv / sc ec cv	716	3.44
ec cv	+ ec cv	+ ec cv / sc ec cv	1002	4.82
sc	sc + +	sc + + / sc ec cv	997	4.80
sc ec cv	sc ec cv	sc ec cv / sc ec cv	4	0.02
+	+ + +	+ + + / sc ec cv	1	

Total F$_2$ flies: 20,785

72

The sequencing and spacing of the *sc*, *ec*, and *cv* genes was accomplished by crossing *sc cv* flies with *ec* flies and then testcrossing the heterozygous F_1 females with *sc ec cv* males:

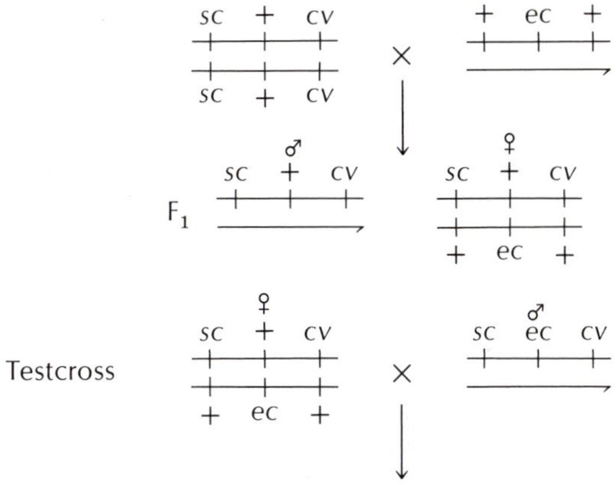

A three-point cross and testcross were used to map three recessive sex-linked genes. (*Note:* + = normal allele)

$$
\begin{array}{ccc}
sc & + & cv \\
\end{array}
\quad\times\quad
\begin{array}{ccc}
+ & ec & + \\
\end{array}
$$

F_1

♂
$$
\begin{array}{ccc}
sc & + & cv \\
\end{array}
$$

♀
$$
\begin{array}{ccc}
sc & + & cv \\
+ & ec & + \\
\end{array}
$$

Testcross

♀
$$
\begin{array}{ccc}
sc & + & cv \\
+ & ec & + \\
\end{array}
\quad\times\quad
♂
\begin{array}{ccc}
sc & ec & cv \\
\end{array}
$$

Progeny (see Table 2.2)

The F_2 was analyzed and the sequence and spacing of the three genes determined. The + *ec* + and *sc* + *cv* classes form 83.64 percent of the total and represent the noncrossover or parental classes; *sc ec* + and + + *cv* (6.72 percent), along with + *ec cv* and *sc* + + (9.62 percent), represent single crossover classes; *sc ec cv* and + + + are extremely rare classes (about 0.02 percent) and are interpreted as representing *double crossovers*.

Using these data, we can position these genes relative to each other:

	F_1♀	F_2
If the sequence were	*Crossovers*[2]	*Phenotypic classes*
1. ec cv sc	SCO	ec cv sc / + + +
	SCO	ec + sc / + cv +
	DCO	ec cv + / + + sc
	NCO	ec + + / + cv sc
2. ec sc cv	SCO	ec sc cv / + + +
	SCO	ec + cv / + sc +

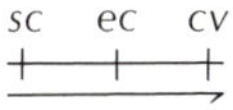

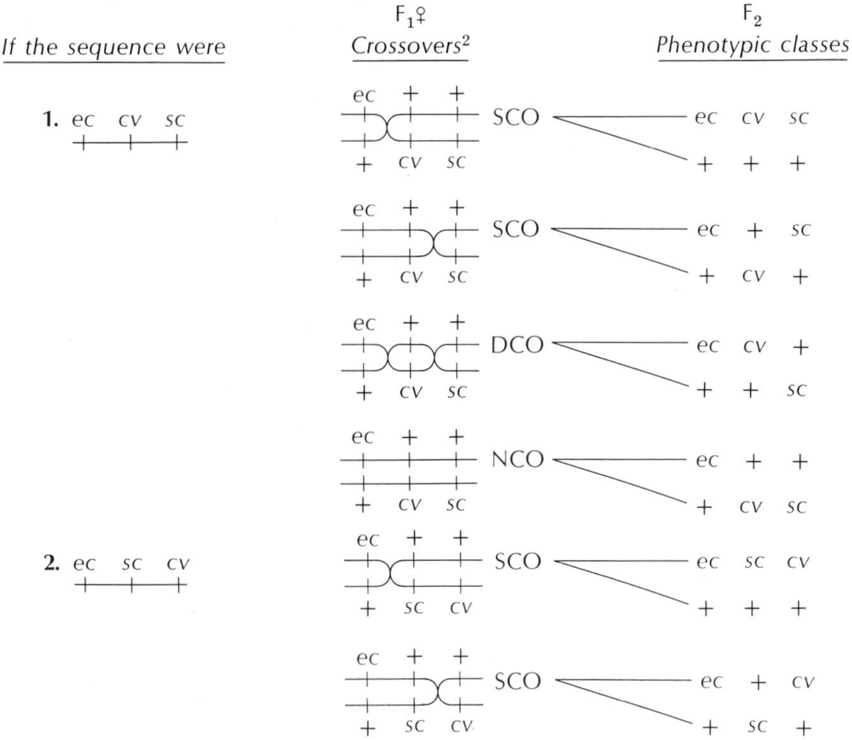

[2]SCO = single crossover, DCO = double crossover, NCO = noncrossover.

74

*Mendelism and the
chromosome theory of
inheritance*

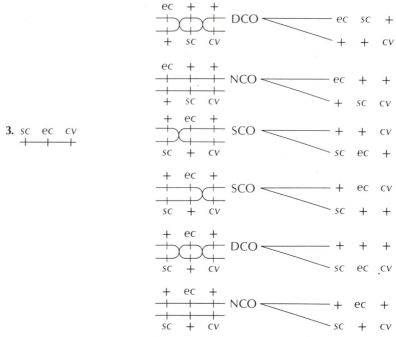

3. sc ec cv

A comparison of these predictions with the data in Table 2.2 tells us that only sequence 3 is compatible with the data. A simpler way of sequencing genes in a three-point cross is to compare the noncrossover class (highest frequency) with the double-crossover class (lowest frequency) and determine which gene has changed places.

Noncrossover: + + ec   and   sc cv +
Double crossover: + + +   and   sc cv ec

By interchanging ec with + in the noncrossovers, we get our double-crossover classes. Therefore ec must be the middle gene.

Having positioned the genes in their proper sequence, we can construct a genetic map by calculating the space or distance between them.

Crossovers between sc and ec   6.72
                                                    +0.02   (DCO class includes a cross-
                                                    ‾‾‾‾   over between sc and ec)
                                                    6.74

Crossovers between ec and cv   9.62
                                                    +0.02
                                                    ‾‾‾‾
                                                    9.64

Therefore   sc 6.74 ec 9.64 cv
                                                   ├──────────┼──────────┤

                                                   ├──────── 16.38 ────────┤

We can use these same procedures derived from *Drosophila* to map genes in other diploid organisms. For example, let us look at an experiment involving gene mapping in corn (the life cycle of corn is diagrammed in Figure 2.16). In this experiment, there are three gene pairs, all located on chromosome 9.

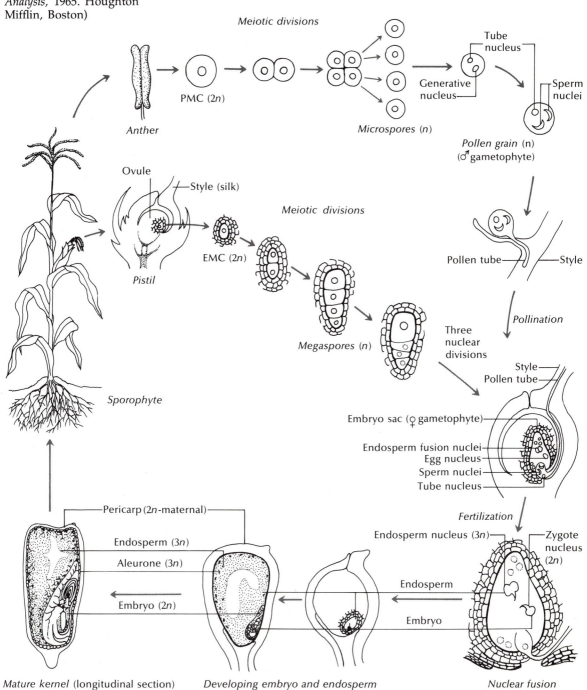

Figure 2.16
Life cycle of *Zea mays*
(After W. K. Baker, *Genetic Analysis*, 1965. Houghton Mifflin, Boston)

Meiotic divisions

PMC (2*n*)

Anther

Microspores (*n*)

Tube nucleus

Generative nucleus

Sperm nuclei

Pollen grain (n)
(♂ gametophyte)

Ovule

Style (silk)

Meiotic divisions

EMC (2*n*)

Pistil

Sporophyte

Pollen tube

Style

Pollination

Megaspores (*n*)

Three nuclear divisions

Style
Pollen tube

Embryo sac (♀ gametophyte)

Endosperm fusion nuclei
Egg nucleus
Sperm nuclei
Tube nucleus

Fertilization

Endosperm nucleus (3*n*)

Zygote nucleus (2*n*)

Pericarp (2*n*-maternal)

Endosperm (3*n*)

Aleurone (3*n*)

Endosperm

Embryo (2*n*)

Embryo

Mature kernel (longitudinal section)

Developing embryo and endosperm

Nuclear fusion

75

76

*Mendelism and the
chromosome theory of
inheritance*

c^+ = colored aleurone
c = white aleurone
sh^+ = full endosperm
sh = shrunken endosperm
wx^+ = starchy endosperm
wx = waxy endosperm

In each instance the "+" allele is dominant. The cross was performed as follows.

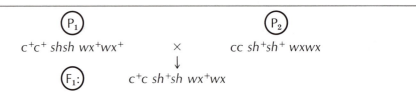

P₁ — $c^+c^+\ shsh\ wx^+wx^+$ × P₂ — $cc\ sh^+sh^+\ wxwx$

F₁: $c^+c\ sh^+sh\ wx^+wx$

Testcross: $c^+c\ sh^+sh\ wx^+wx$ × $cc\ shsh\ wxwx$

Phenotype	Number of progeny	Frequency	
1. White, shrunken, starchy	116	0.0173	SCO
2. Colored, full, starchy	4	0.0006	DCO
3. Colored, shrunken, starchy	2538	0.3784	P
4. Colored, shrunken, waxy	601	0.0896	SCO
5. White, full, starchy	626	0.0933	SCO
6. White, full, waxy	2708	0.4037	P
7. White, shrunken, waxy	2	0.0003	DCO
8. Colored, full, waxy	113	0.0168	SCO
	6708	1.0000	

Data from R. A. Emerson, G. W. Beadle, and A. C. Fraser (1935). *Cornell Univ. Agric. Exp. Stn. Mem.*

What is the gene sequence? Classes 3 and 6 are the parental or non-recombinant classes. Classes 2 and 7, being the rarest in this three-point cross, are the double-crossover classes. Comparing classes 2 and 7 with classes 3 and 6, we see that only *sh* has changed places. Therefore it must be the middle gene. The sequence is

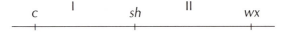

The map distances are

c—sh = Sum of all crossovers in region I
= (8) + (1) + (2) + (7)
= 0.035, or 3.50 map units

sh—wx = Sum of all crossovers in region II
= (4) + (5) + (2) + (7)
= 0.1838, or 18.38 map units

And the final map is

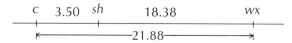

Note that in both these mapping experiments, the double crossovers were important to our calculations. If the experiments were done as two-point crosses (*sc—cv* and *c—wx*), these distances would have been 16.34 and 21.71 map units, respectively, both underestimates.

Interference and the coefficient of coincidence   H. J. Muller, another of Morgan's students, studied double crossovers in *Drosophila* and discovered that double crossovers (DCO) are much less frequent than expected. If, in our *Drosophila* example, the expected frequency of DCO equals the probability of SCO between *sc* and *ec* times the probability of SCO between *ec* and *cv*, then the expected number of DCO is 0.0674 times 0.0964 times 20,785, or 134. But since the observed DCO is 5, Muller inferred that a crossover in one region interfered with crossing-over in an adjacent region. This phenomenon is termed *interference*. The degree of interference is quantified as the complement of the *coefficient of coincidence*, a term expressed by the following ratio:

$$\frac{\text{Observed DCO}}{\text{Expected DCO}}$$

In the foregoing data, the coefficient of coincidence is $\frac{5}{134}$, or 0.037; therefore the interference is 1.00 minus 0.037, or 0.963. In the corn-gene mapping experiments

$$\text{Expected DCO} = 0.0350 \times 0.1838 \times 6708 = 43$$

$$\text{Observed DCO} = 6$$

$$\text{Coefficient of coincidence} = \frac{6}{43}, \text{ or } 0.14$$

$$\text{Interference} = 1.00 - 0.14, \text{ or } 0.86$$

An interference of 1.00 means that an SCO completely inhibits a crossover in the adjacent region. An interference of 0.00 means that the coefficient of coincidence is 1.00—that is, there are as many DCO as expected. In general, the closer three genes are, the greater the interference, and vice versa (Figure 2.17). Although the reason for such interference is not yet clear, it may result from the tension generated by chromosomal coiling during early prophase. This tension could interfere with multiple crossing-over in limited chromosomal regions.

In bacterial viruses, it is not uncommon to find a coefficient of coincidence above 1.00. This phenomenon, called *negative interference*, occurs because a single chromosome can have numerous crossover opportunities. Recombination in bacterial viruses will be discussed in Chapter 11.

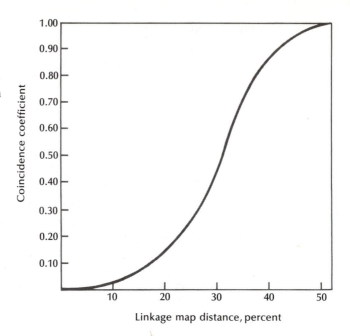

Figure 2.17
The coefficient of coincidence as a function of the distance separating two genes. [Calculated by Weinstein (1958) from data of Bridges (Morgan, Bridges, and Schultz, 1935)]

Negative interference is also observed in some *Neurospora* crosses, but the mechanism for it is not well understood.

These basic techniques of the three-point cross and the testcross have been used to map the chromosomes of *Drosophila* (Figure 2.18) and other organisms (*Neurospora*, mouse, corn). Once a map has been constructed, it is possible to work backward. That is, taking the gene distances and coefficient of coincidence we obtained in Figure 2.17,* we can predict noncrossover, single-crossover, and double-crossover classes. For example, suppose we are given the cross

$$♀ \quad \begin{array}{ccc} a & 10.6 \; b & 13.4 \; c \\ \hline & & \\ \hline + & + & + \end{array} \quad × \quad \begin{array}{ccc} a & b & c \\ \hline & & \\ \hline a & b & c \end{array} \quad ♂$$

We can predict the offspring.

Double crossover: $a + c$ and $+ b +$
     $=$ SCO $(a - b)$ × SCO $(b - c)$ × coefficient of coincidence
     $= 0.106 × 0.134 × 0.25$
     $= 0.0036$, or 4 in 1000 progeny
Single crossover: $a + +$ and $+ b c$
     $= 106 - 4 = 102$[†] in 1000 progeny
Single crossover: $a b +$ and $+ + c$
     $= 134 - 4 = 130$ in 1000 progeny

* These values would not be applicable to such organisms as bacteria, because the curve would be different

† Remember that we must reduce the expected SCO by the DCO.

Noncrossover or parental: a b c   and   + + +
= 1.00 − (0.0036 + 0.102 + 0.130)
= 0.764, or 764 in 1000 progeny

Figure 2.18
Partial linkage map of
Drosophila (data from
Bridges)

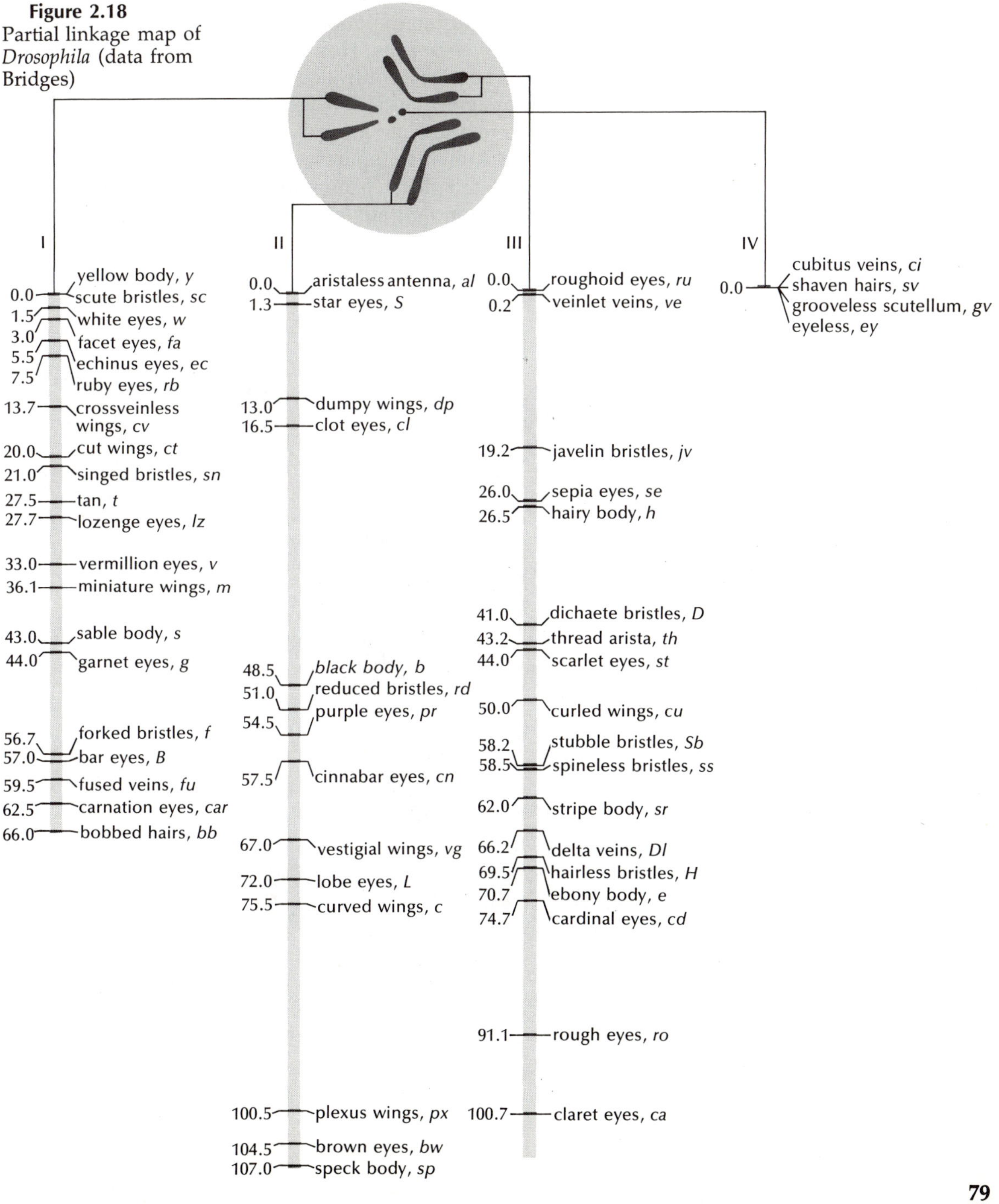

I

0.0	yellow body, *y*
	scute bristles, *sc*
1.5	white eyes, *w*
3.0	facet eyes, *fa*
5.5	echinus eyes, *ec*
7.5	ruby eyes, *rb*
13.7	crossveinless wings, *cv*
20.0	cut wings, *ct*
21.0	singed bristles, *sn*
27.5	tan, *t*
27.7	lozenge eyes, *lz*
33.0	vermillion eyes, *v*
36.1	miniature wings, *m*
43.0	sable body, *s*
44.0	garnet eyes, *g*
56.7	forked bristles, *f*
57.0	bar eyes, *B*
59.5	fused veins, *fu*
62.5	carnation eyes, *car*
66.0	bobbed hairs, *bb*

II

0.0	aristaless antenna, *al*
1.3	star eyes, *S*
13.0	dumpy wings, *dp*
16.5	clot eyes, *cl*
48.5	black body, *b*
51.0	reduced bristles, *rd*
54.5	purple eyes, *pr*
57.5	cinnabar eyes, *cn*
67.0	vestigial wings, *vg*
72.0	lobe eyes, *L*
75.5	curved wings, *c*
100.5	plexus wings, *px*
104.5	brown eyes, *bw*
107.0	speck body, *sp*

III

0.0	roughoid eyes, *ru*
0.2	veinlet veins, *ve*
19.2	javelin bristles, *jv*
26.0	sepia eyes, *se*
26.5	hairy body, *h*
41.0	dichaete bristles, *D*
43.2	thread arista, *th*
44.0	scarlet eyes, *st*
50.0	curled wings, *cu*
58.2	stubble bristles, *Sb*
58.5	spineless bristles, *ss*
62.0	stripe body, *sr*
66.2	delta veins, *Dl*
69.5	hairless bristles, *H*
70.7	ebony body, *e*
74.7	cardinal eyes, *cd*
91.1	rough eyes, *ro*
100.7	claret eyes, *ca*

IV

0.0	cubitus veins, *ci*
	shaven hairs, *sv*
	grooveless scutellum, *gv*
	eyeless, *ey*

79

Gene mapping in humans Humans, like many other animal and plant species, cannot be experimentally bred in the same way that we breed *Drosophila, Neurospora,* or maize. Humans, of course, do not participate as subjects in controlled breeding experiments. But even if this were not an obstacle, the nine-month gestation period and the "litter size" of 1 would make humans unattractive subjects. How do we solve genetic problems occurring in the human species if controlled breeding experiments are not possible? How do we assign genes to linkage groups in humans, and how do we map genes within a linkage group?

Gene mapping in humans requires devising alternative mapping strategies. Though we are still far from constructing a detailed map of human chromosomes, important advances have been made. By far the

Figure 2.19
Schematic representation of Sendai-virus-induced fusion of two mononucleate cells (A and B) from different species into a binucleate heterokaryon, which then divides and gives rise to two mononucleate hybrid cells (synkaryons). These AB hybrids continue to divide, while gradually eliminating most of the chromosomes originating from parental cell B. (From Ringertz and Savage, *Cell Hybrids,* Academic Press,New York, 1976)

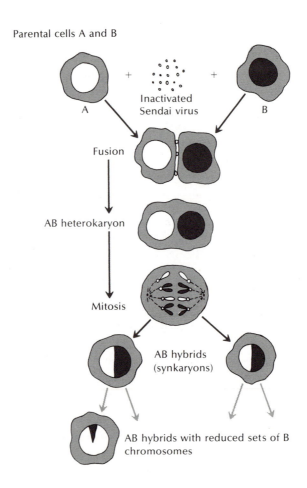

Parental cells A and B

Inactivated
Sendai virus

A   B

Fusion

AB heterokaryon

Mitosis

AB hybrids
(synkaryons)

AB hybrids with reduced sets of B
chromosomes

most important techniques for mapping genes in humans have emerged from studies of somatic cells grown in culture media.

Cell fusion is one such technique (Figure 2.19). Somatic cells from distantly related organisms (such as mice and humans), in the presence of inactivated parainfluenza viruses called *Sendai*, fuse. The resulting cell is called a *heterokaryon*, a single cell with two different nuclei. When the nuclei fuse, we have a hybrid cell called a *synkaryon*. The synkaryons proliferate, and as they do so, an intriguing phenomenon occurs: Human chromosomes are lost and mouse chromosomes retained, setting up different populations of hybrid cells. Different clones stably retain human chromosomes in different populations.

Cell fusion enables us to place human genes into specific linkage groups. Suppose, for example, that a population of human cells has a gene that codes for a specific enzyme—call it *E*. A population of mouse cells is mutant for that gene and thus has none of that enzyme. We fuse the two cell types and examine different populations of the hybrid cells. In each population, many human chromosomes are missing, and in each we can test for the presence of the enzyme. The following data summarize our observations.

	Hybrid cell population					
	①	②	③	④	⑤	⑥
Human chromosomes present	21, 18, 16 7, 5, 2	18, 16 8, 3	21, 18 1	23, 19, 17 15, 14, 11 9, 8, 3, 1	18, 6, 2	21, 5
E activity present	yes	yes	yes	no	yes	no

From these data, we can conclude that the *E* gene is on chromosome 18, because every time 18 is present we have *E*, and every time it is absent we have no *E*. Adding to this basic strategy, we can say that genes that segregate together during chromosome elimination are very probably on the same chromosome.

To map human genes, we draw on two main sources. We employ pedigrees involving two or more linked genes and calculate crossover percentage the same way we did in *Drosophila*. Alternatively, we use somatic cell genetics to study the correlation between changed linkage groups and points of chromosome breakage.

Somatic cell genetics has proved an extremely valuable tool in human genetics. Figure 2.20 is a recent map showing confirmed regional assignments of some human genes, assignments based primarily on somatic cell genetics.

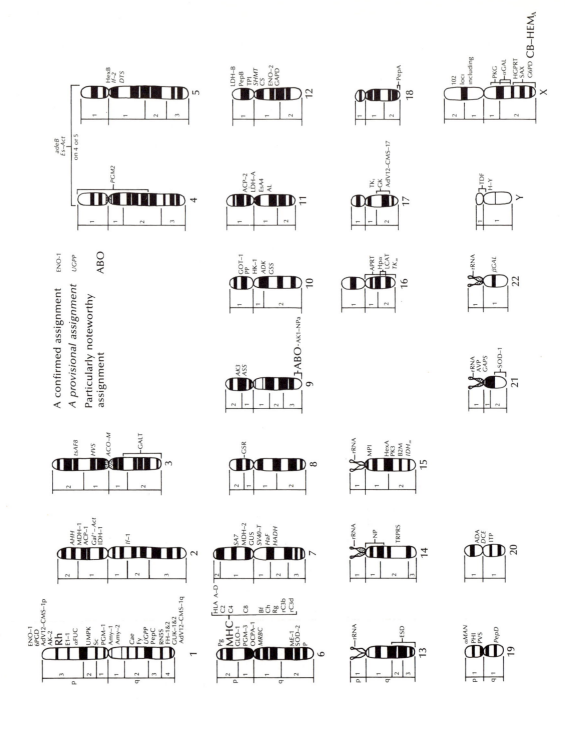

A confirmed assignment

A provisional assignment

Particularly noteworthy assignment

ABO	ABO blood group (chr. 9)
ACO	Aconitase, mitochondrial (chr. 3)
ACO-S	Aconitase, soluble (chr. 9)
ACP-1	Acid phosphatase-1 (chr. 2)
ACP-2	Acid phosphatase-2 (chr. 11)
ADA	Adenosine deaminase (chr. 20)
adeB	FGAR amidotransferase (chr. 4 or 5)
ADK	Adenosine kinase (chr. 10)
AdV12-CMS-1p	Adenovirus-12 chromosome modification site-1p (chr. 1)
AdV12-CMS-1q	Adenovirus-12 chromosome modification site-1q (chr. 1)
AdV12-CMS-17	Adenovirus-12 chromosome modification site-17 (chr. 17)
AHH	Aryl hydrocarbon hydroxylase (chr. 2)
AK-1	Adenylate kinase-1 (chr. 9)
AK-2	Adenylate kinase-2 (chr. 1)
AK-3	Adenylate kinase-3 (chr. 9)
AL	Lethal antigen: 3 loci (a1, a2, a3) (chr. 11)
Amy-1	Amylase, salivary (chr. 1)
Amy-2	Amylase, pancreatic (chr. 1)
ASS	Argininosuccinate synthetase (chr. 9)
APRT	Adenine phosphoribosyltransferase (chr. 16)
AVP	Antiviral protein (chr. 21)
Bf	Properdin factor B (chr. 6)
$\beta2M$	$\beta2$-Microglobulin (chr. 15)
C2	Complement component-2 (chr. 6)
C4	Complement component-4 (chr. 6)
C8	Complement component-8 (chr. 6)
Cae	Cataract, zonular pulverulent (chr. 1)
CB	Colorblindness (deutan & protan) (X chr.)
Ch	Chido blood group (chr. 6)
CS	Citrate synthase, mitochondrial (chr. 12)
DCE	Desmosterol-to-cholesterol enzyme (chr. 20)
DTS	Diphtheria toxin sensitivity (chr. 5)
EI-1	Elliptocytosis-1 (chr. 1)
E11S	Echo 11 sensitivity (chr. 19)
ENO-1	Enolase-1 (chr. 1)
ENO-2	Enolase-2 (chr. 12)
Es-Act	Esterase activator (chr. 4 or 5)
EsA4	Esterase-A4 (chr. 11)
ESD	Esterase D (chr. 13)
FH-1 & 2	Fumarate hydratase-1 and 2 (S & M) (chr. 1)
αFUC	Alpha-L-fucosidase (chr. 1)
Fy	Duffy blood group (chr. 1)
Gal+-Act	Galactose + activator (chr. 2)
αGAL	α-Galactosidase (Fabry disease) (X chr.)
βGAL	β-Galactosidase (chr. 22)
GALT	Galactose-1-phosphate uridyltransferase (chr. 3)
GAPD	Glyceraldehyde-3-phosphate dehydrogenase (chr. 12)
GAPS	Phosphoribosyl glycineamide synthetase (chr. 21)
Gc	Group-specific component (chr. 4)
GK	Galactokinase (chr. 17)
GLO-1	Glyoxylase I (chr. 1)
GOT-1	Glutamate oxaloacetic transaminase-1 (chr. 10)
G6PD	Glucose-6-phosphate dehydrogenase (X chr.)
GSR	Glutathione reductase (chr. 8)
GSS	Glutamate-γ-semialdehyde synthetase (chr. 10)
GUK-1 & 2	Guanylate kinase-1 & 2 (S & M) (chr. 1)
GUS	Beta-glucuronidase (chr. 7)
HADH	Hydroxyacyl-CoA dehydrogenase (chr. 7)
HaF	Hageman factor (chr. 7)
HEM$_A$	Classic hemophilia (X chr.)
Hex A	Hexosaminidase A (chr. 15)
Hex B	Hexosaminidase B (chr. 5)
HGPRT	Hypoxanthine-guanine phosphoribosyltransferase (X chr.)
HK-1	Hexokinase-1 (chr. 10)
HLA	Major histocompatibility complex (chr. 6)
Hpα	Haptoglobin, alpha (chr. 16)
HVS	Herpes virus sensitivity (chr. 3)
H-Y	Y histocompatibility antigen (Y chr.)
If-1	Interferon-1 (chr. 2)
If-2	Interferon-2 (chr. 5)
IDH-1	Isocitrate dehydrogenase-1 (chr. 2)
IDH$_m$	Isocitrate dehydrogenase, mitochondrial (chr. 15)
ITP	Inosine triphosphatase (chr. 20)
LCAT	Lecithin-cholesterol acyltransferase (chr. 16)
LDH-A	Lactate dehydrogenase A (chr. 11)
LDH-B	Lactate dehydrogenase B (chr. 12)
αMAN	Lysosomal α-D-mannosidase (chr. 19)
MDH-1	Malate dehydrogenase-1 (chr. 2)
MDH-2	Malate dehydrogenase, mitochondrial (chr. 7)
ME-1	Malic enzyme-1 (chr. 6)
MHC	Major histocompatibility complex (chr. 6)
MPI	Mannosephosphate isomerase (chr. 15)
MRBC	B-cell receptor for monkey red cells (chr. 6)
NP	Nucleoside phosphorylase (chr. 14)
NPa	Nail-patella syndrome (chr. 9)
OPCA-I	Olivopontocerebellar atrophy I (chr. 6)
P	P blood group (chr. 6)
PepA	Peptidase A (chr. 18)
PepB	Peptidase B (chr. 12)
PepC	Peptidase C (chr. 1)
PepD	Peptidase D (chr. 19)
Pg	Pepsinogen (chr. 6)
PGK	Phosphoglycerate kinase (X chr.)
PGM-1	Phosphoglucomutase-1 (chr. 1)
PGM-2	Phosphoglucomutase-2 (chr. 4)
PGM-3	Phosphoglucomutase-3 (chr. 6)
6PGD	6-Phosphogluconate dehydrogenase (chr. 1)
PHI	Phosphohexose isomerase (chr. 19)
PK3	Pyruvate kinase-3 (chr. 15)
PP	Inorganic pyrophosphatase (chr. 10)
PVS	Polio sensitivity (chr. 19)
Rg	Rodgers blood group (chr. 6)
Rh	Rhesus blood group (chr. 1)
rRNA	Ribosomal RNA (chr. 13, 14, 15, 21, 22)
rC3b	Receptor for C3b (chr. 6)
rC3d	Receptor for C3d (chr. 6)
RN5S	5S RNA gene(s) (chr. 1)
SA7	Species antigen 7 (chr. 7)
SAX	X-linked species (or surface) antigen (X chr.)
Sc	Scianna blood group (chr. 1)
SHMT	Serine hydroxymethyltransferase (chr. 12)
SOD-1	Superoxide dismutase-1 (chr. 21)
SOD-2	Superoxide dismutase-2 (chr. 6)
SV40-T	SV40-T antigen (chr. 7)
TDF	Testis determining factor (Y chr.)
TK$_m$	Thymidine kinase, mitochondrial (chr. 16)
TK$_s$	Thymidine kinase, soluble (chr. 17)
TPI	Triosephosphate isomerase (chr. 12)
TRPRS	Tryptophanyl-tRNA synthetase (chr. 14)
tsAF8	Temperature-sensitive (AF8) complementing (chr. 3)
UGPP	Uridyl diphosphate glucose pyrophosphorylase (chr. 1)
UMPK	Uridine monophosphate kinase (chr. 1)

Figure 2.20 Human chromosome map (From McKusick and Ruddle. 1977. *Science* 196: 390–404)

Gene mapping in haploids

In Chapter 1, we discussed the fact that the principles of segregation and independent assortment are clearly illustrated in haploid organisms such as *Chlamydomonas* and *Neurospora* (see pages 33–38). In this section, we will build on our understanding of tetrad analysis, integrating it with our knowledge of gene mapping, and extending it so that we can examine the process of gene mapping in haploids.

Mapping genes in a haploid organism like *Neurospora* follows the same principles as mapping genes in diploids like *Drosophila*. But there are subtle differences. Recall that in a *Neurospora* cross of $A \times a$, if no crossing-over occurred, the alleles segregated from each other in meiosis I (first-division segregation). If crossing-over occurred, segregation of alleles did not occur until Meiosis II (second-division segregation). Figure 1.36 (page 39) demonstrates this point. First- and second-division segregation are differentiated by the spore sequence in the ascus sac. In *Neurospora*, we can use the proportion of ascus sacs showing second-division segregation in our mapping calculations. Let us examine the way this works in practice.

Figure 2.21 is a drawing of a series of actual asci sacs from a single perithecium from a cross between a wild-type strain of *Neurospora* and a lysine-requiring strain (lys^-). Of the 14 sacs shown, 8 are nonrecombinant (show first-division segregation), and 6 are recombinant (show second-division segregation). The six showing second-division segregation result from crossing-over between the *lys* locus and the centromere. So, in our calculations, we are in fact mapping the distance between the centromere and the *lys* locus. Figure 2.22 illustrates the crossover events for each ascus sac shown.

The distance between the centromere and the *lys* locus is

$$\frac{\text{Second-division segregants}}{\text{Total number of asci}} \times \frac{1}{2} = \frac{6}{14} \times \frac{1}{2} = 21.4 \text{ MU}$$

We count each ascus sac as a recombinant, but only one-half of the chromosomes in each one have undergone crossing-over. Therefore we need to multiply by $\frac{1}{2}$ to calculate the actual crossing-over percentage.

If a crossover occurs between a gene and its centromere, there are four possible ascospore sequences: *AAaaAAaa*, *AAaaaaAA*, *aaAAaaAA*, *aaAAAAaa*.

Two other possible sequences, *AAAAaaaa* and *aaaaAAAA*, may be noncrossover sequences, or they may be the result of a multiple-crossover event between the centromere and the *A* locus.

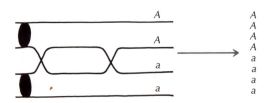

Figure 2.21
Neurospora ascus sacs (From Hayes, *Genetics of Bacteria and their Viruses*, 2d ed., Wiley, New York, 1968)

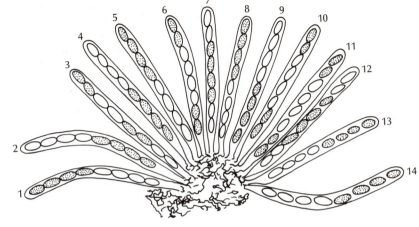

Figure 2.22
Diagram of crossovers producing ascus sacs in Figure 2.21

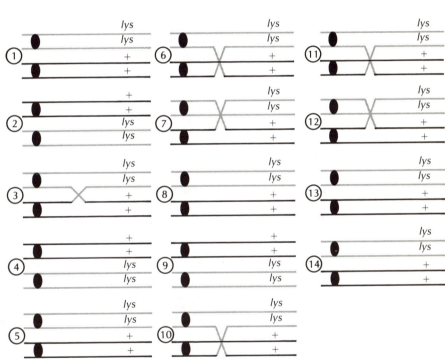

Multiple-crossover events can produce asci that are indistinguishable from noncrossover asci. Without the complication of multiple-crossover events, then, the maximum percentage of recombinant asci is $\frac{4}{6}$ or 67 percent, no matter how great the distance between the centromere and the locus under investigation. The maximum crossover percentage is 33. But because of multiple crossing-over events, the upper limit of 67 percent recombinant asci is only approached.

The two-point cross and centromere mapping Consider a cross in

Neurospora involving two pairs of alleles. We need to determine first whether the two allele pairs are linked. Once we ascertain their linkage relationship, we can determine their positions relative to each other, their positions relative to the centromere, and the map units separating them.

One strain requires thiamin as medium supplement (*thi*), and the other strain grows in buttonlike colonies (*but*).

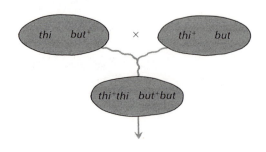

Class	Number of Asci	Spores				
		1 + 2	3 + 4	4 + 5	6 + 7	
A	272	*thi but*⁺	*thi but*⁺	*thi*⁺ *but*	*thi*⁺ *but*	PD*
B	48	*thi but*⁺	*thi*⁺ *but*⁺	*thi but*	*thi*⁺ *but*	T
C	21	*thi but*⁺	*thi but*	*thi*⁺ *but*⁺	*thi*⁺ *but*	T
D	3	*thi but*⁺	*thi*⁺ *but*	*thi but*⁺	*thi*⁺ *but*	PD

* See Chapter 1 (page 37) for a review of these terms.

In analyzing these data, we conclude that the two loci are indeed linked, rather than being on different chromosomes. If two pairs of alleles were on different chromosomes, then they would exhibit independent assortment and produce equal numbers of PD and NPD asci.

There were no NPD asci recovered and 275 PD asci recovered, leading us to the conclusion that the loci are not independently assorting but rather are linked.

Examining classes A and B leads us to the conclusion that the *thi* locus is more distant from the centromere than the *but* locus. How do we conclude this? Class A is a noncrossover class in which both pairs of alleles show first-division segregation. In class B, the second most frequent class, the *but* locus shows first-division segregation, but the *thi* locus shows second-division segregation. Loci that exhibit a high frequency of second-division segregation are generally more distant from the centromere than loci exhibiting a low frequency of second-division segregation. Thus the *thi* locus is further than the *but* locus from the centromere. At this point we can offer a hypothesis.

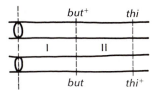

Examining classes C and D, we are forced to reject this hypothesis. In class C, we have first-division segregation for *thi* and second-division segregation for *but*. The tetratype (T) ascus thus generated requires a double crossover.

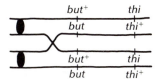

In class D, a single-crossover event would generate that particular PD ascus, since both loci show second-division segregation.

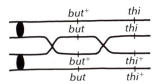

However, as we learned earlier, double-crossover events are rarer than single-crossover events. Assuming our hypothesis, we are saying that the single-crossover class (D) is rarer than the double-crossover class (C). This does not make sense, so we reject the hypothesis and offer another that is consistent with our understanding of genetic mapping. In this case, we suggest that the centromere is between the *but* and *thi* loci, with *thi* more distant than *but*.

From this new hypothesis, then, class A is a noncrossover class; class B is a single crossover between the centromere and *thi*; class C is a single crossover between the centromere and *but*; and class D is a double crossover.

88

Mendelism and the
chromosome theory of
inheritance

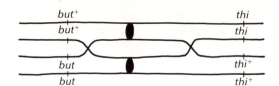

All this is more reasonable, is consistent with known facts, and is therefore more acceptable to us as a viable hypothesis.

To calculate the actual map distances, we use the raw data and convert them into map units.

1. Centromere—*but*

$$\frac{21 + 3}{344} \times \frac{1}{2} = 0.0349 = 3.49 \text{ MU}$$

2. Centromere—*thi*

$$\frac{48 + 3}{344} \times \frac{1}{2} = 0.0741 = 7.41 \text{ MU}$$

3. Final map

but 3.49   7.41   thi

|←——— 10.90 ———→|

Figure 2.23
Tetrad analysis
(a) Tetrad analysis in
Neurospora, showing how
crossover events
correspond to ascospore
sequences.
(b) Types of crossover
events that can lead to
specific ascospore
sequences in *Neurospora.*
(Three different sequences
of genes *A* and *B* are
considered relative to
centromeres.)

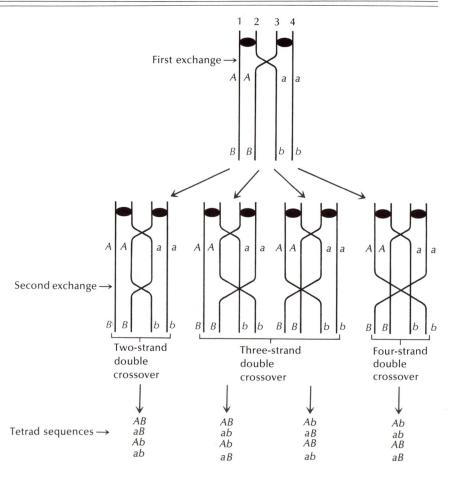

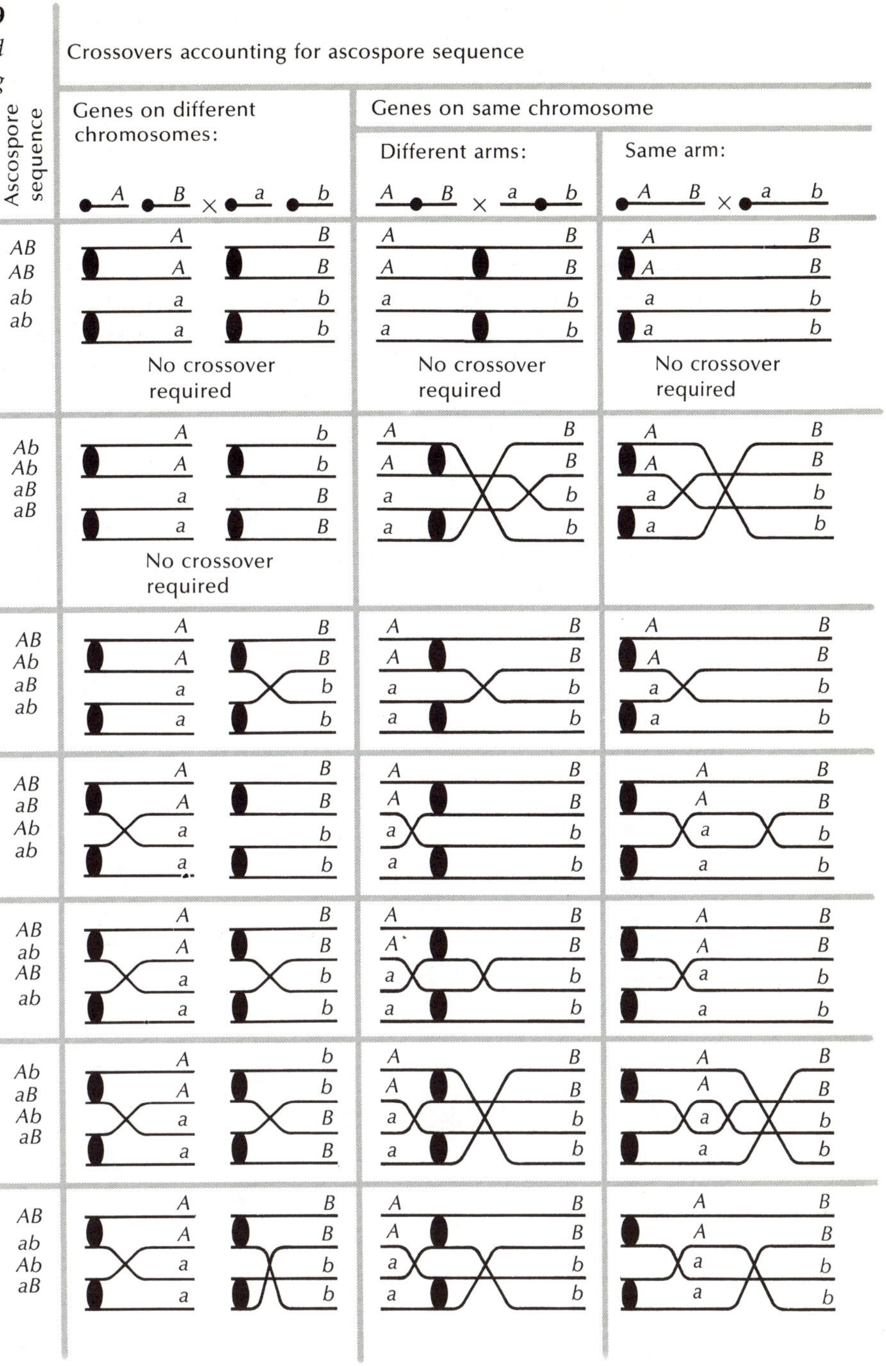

Ascospore sequence

Crossovers accounting for ascospore sequence

Genes on different chromosomes:	Genes on same chromosome	
	Different arms:	Same arm:
$A \bullet B \times a \bullet b$	$A \quad B \times a \bullet b$	$A \quad B \times a \bullet b$

AB / AB / ab / ab

No crossover required — No crossover required — No crossover required

Ab / Ab / aB / aB

No crossover required

AB / Ab / aB / ab

AB / aB / Ab / ab

AB / ab / AB / ab

Ab / aB / Ab / aB

AB / ab / Ab / aB

Tetrad analysis in *Neurospora* reveals some of the complicated four-stranded crossover events and resulting ascospore sequences that can occur (Figure 2.23). For example, with the four chromatids of a pair of chromosomes numbered as shown and also with a single crossover occurring between chromatids 2 and 3 in the area between the centromere and the end of the chromosomes, there are four distinct ways in which a second exchange could occur between nonsister chromatids. (This assumes that the first exchange does not prohibit a second exchange.) The ascospore sequences that would be generated by these types of double crossovers, as well as others, are outlined in part (b) of Figure 2.23 (study this figure carefully before proceeding).

The stage of crossing-over

Crossing-over might occur at one of two stages in the meiotic process, either before chromosome replication or after chromosome replication, when each chromosome consists of two chromatids.

In haploid organisms such as *Neurospora* and *Chlamydomonas*, we have noted that crossing-over appears to take place at the four-strand stage, because tetrad analysis often reveals the presence of four distinct meiotic products, a *tetratype*, in one ascus sac. If crossing-over occurs prior to chromosome replication, only two classes of meiotic products would appear. Exploring these two possibilities in *Drosophila*, E. G. Anderson in 1925 crossed an attached-**X** female heterozygous for the dominant bar-eyed gene (*B*) with a normal male.

Point of attachment

If crossing-over occurred *before* the chromosomes replicate—that is, before the four-strand stage—then crossing-over should not produce any novel genotypes.

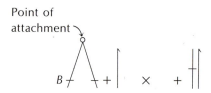

But if crossing-over occurs *after* the chromosomes replicate—the four-strand stage—then the possibility for forming novel genotypes arises.

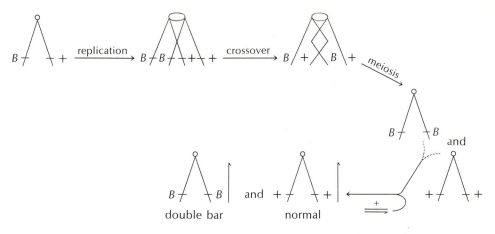

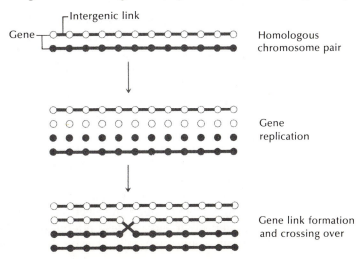

double bar   and   normal

Crossing-over at the four-strand stage can produce *BB* homozygotes, which are phenotypically distinguishable from *B+* heterozygotes, and wild-type females. Anderson found *BB* and + offspring, so crossing-over in *Drosophila* as well occurs at the four-strand stage. We can generalize to say that all known crossing-over occurs at the four-strand stage.

The mechanism of crossing-over   Although a number of factors can influence the frequency of crossing-over —sex, age of the mother, temperature, nutrition, chemicals, radiation— the crossover mechanism itself remains poorly understood. The actual breakage and reunion of chromosomes, which is the strongest theory for recombination in prokaryotes (see Chapter 11), also appears to be the best explanation for recombination in eukaryotes. Breakage and reunion is, however, not the only theory for explaining crossing-over. J. Belling theorized in 1933 that genes duplicate after chromosome pairing but do so in an unconnected state. When replication is complete, the genes become connected. Recombinant chromosomes can form when genes on different homologs are linked together (Figure 2.24). The Belling theory has been

Figure 2.24
Belling's theory of recombination

Intergenic link

Gene

Homologous chromosome pair

Gene replication

Gene link formation and crossing over

Figure 2.25
Darlington's theory of chromosome crossing-over

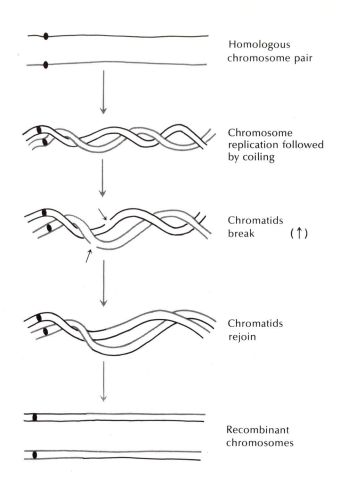

Homologous chromosome pair

Chromosome replication followed by coiling

Chromatids break (↑)

Chromatids rejoin

Recombinant chromosomes

modified over the years as more information on chromosome structure and replication was uncovered, but it is not well supported by experimental data.

Another theory, proposed by C. D. Darlington in 1932, suggests that chromosomes coil around each other so tightly that breaks in chromatids occur and that the breaks may result in the fusion of chromatid segments from homologous chromosomes and the formation of recombinant chromosomes (Figure 2.25). Darlington's theory is more widely supported than Belling's, primarily because of the evidence favoring a breakage-and-reunion type of crossover mechanism, especially in microorganisms. But there are objections to both theories, and no clear-cut choice as to the crossover mechanism seems possible at present.

The key to crossing-over appears to be a structure called the *synaptonemal complex*, or simply SC. First observed in 1956 by M. J. Moses and independently by D. W. Fawcett, this meiotic prophase structure (Figure 2.26), visible through the electron microscope, has been observed almost universally in eukaryotes undergoing meiosis. It is observed in synapsed chromosomes, where it takes on the appearance of a ribbon composed of

Figure 2.26
Synaptinemal complex: a section through synapsed chromosomes in the primary spermatocyte prophase of the rooster, showing boundaries of the bivalent chromosome and the various elements of the synaptinemal complex. Compare the photo with the diagram. (1) Central element: protein and perhaps some DNA. (2) Central space: traversed by protein filaments, at least some of which contain DNA. (3) Inner segments of axial elements: DNA and protein. (4) Outer segments of axial elements: DNA and protein. (5) Main chromosomal mass: DNA and protein.

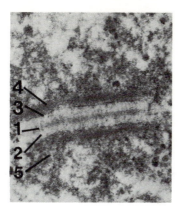
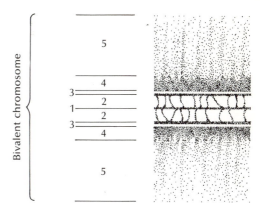

three dense, longitudinal bands, the lateral bands being more dense than the central band. These bands are mainly protein, but DNA, the genetic material, is also present. In the spaces between the bands, there are fine transverse elements, again composed mainly of protein, but containing some suggestion of DNA.

It is currently believed that during early prophase of meiosis, homologous chromosomes first pair up or align themselves, then form the SC. But exactly how this SC leads to crossing-over is not yet clear. A number of observations suggest that it is important in crossing-over. There is no crossing-over in *Drosophila melanogaster* males, and no SC's have been observed in their meiotic cells. We can observe chromosomal pairing in these males, but no SC formation. Also in *Drosophila melanogaster*, there is a third chromosome mutation called c3G, which, when homozygous in females, results in no crossing-over. Cytological examination of these mutants reveals the complete absence of SC.

There are also observations that tend to cast doubt on the importance of the SC in recombination. Though there are many instances in which the absence of genetic crossing-over is accompanied by the absence of the SC, there are cases in which there is genetic crossing-over but no observable SC formation. Studying *Drosophila ananassae*, Rhoda Grell and her colleagues noted that, in males of the species, the crossover frequency is only one-third that found in females. However, even though crossing-over does occur, there is no evidence for the presence of SC.

Keeping the ideas in mind, we can ask about the relationship between chiasma formation and crossing-over. There are two major hypotheses to consider. The *classical theory* (Figure 2.27a) suggests that chiasmata may or may not be manifestations of physical exchange between homologous chromatids. According to this theory, chromosomes pair and coil during meiotic prophase. The tension generated by this coiling may lead to chromatid breaks and subsequent fusion of chromatid segments. But, according to the classical theory, it is clear that the chiasmata do not necessarily indicate a physical exchange of chromatid segments.

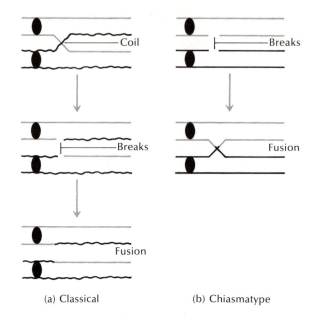

Figure 2.27
Two major hypotheses to account for the relationship between chiasma formation and crossing-over

(a) Classical   (b) Chiasmatype

An alternative theory, the *chiasmatype theory* (Figure 2.27b), suggests that each chiasma observed is a manifestation of a physical exchange of chromatid segments that probably occurred during the pachytene stage of the meiotic prophase. In other words, there is a one-to-one correlation between chiasmata and crossing-over. Most of the evidence today favors the chiasmatype theory, though there are some problems with it. The presence, for example, of chiasmata in *Drosophila* males, in which no crossing-over is occurring, is puzzling. However, we noted earlier that in these males there were no observable synaptonemal complexes. So the chiasmata in *Drosophila* males may be nothing more than superficial entanglements with no bearing at all on crossing-over. The study on the relationship between chiasmata and crossing-over in wheat by Fu and Sears (1973) describes a fine example of the one-to-one correlation between chiasmata and crossing-over.

The actual molecular mechanism of recombination in eukaryotes is—so far at least—not well understood. In bacteria and viruses, our knowledge is more complete, as we shall see in later chapters. However, it is not appropriate at this point to suggest that recombinational mechanisms in bacteria and viruses are the same as in eukaryotes. They appear to be quite different in most respects. But both appear to involve breakage and reunion.

Cytological verification of crossing-over

For the ultimate acceptance of the chromosome theory of inheritance, it was necessary to demonstrate that a recombinational event could be correlated with a cytological event. Prior to 1929, the most important evidence linking genes to chromosomes was the sex-linked work of Bridges and Morgan. Between 1929 and 1931, an important series of papers appeared correlating genetic recombination studies with cytological studies.

The first study (Dobzhansky, 1929) employed x rays to induce breaks in genetically marked *Drosophila* chromosomes. The result of many of these breaks was the union between a piece of one chromosomal linkage group and another linkage group—for example, a segment of chromosome III fusing with chromosome II, or vice versa. Such events, called *translocations,* will be dealt with in Chapter 5. By noting when genes or blocks of genes changed linkage groups, Dobzhansky was able to detect these translocations genetically. Once a translocation was so identified, Dobzhansky made a cytological examination of the flies and identified which chromosomes had exchanged segments. (*Drosophila* chromosomes, which become banded when stained, are easily distinguishable by size and banding patterns.) Thus Dobzhansky was able to correlate a genetic event (a changed linkage group) with a cytological event (a translocation).

In 1931, two additional studies demonstrated that cytological crossing-over occurs in conjunction with the expected types of genetic crossing-over. H. B. Creighton and B. McClintock reported that, in a strain of maize, one of the chromosome pairs (the second smallest) had large, cytologically conspicuous knobs at their ends. In this same strain, the knobbed chromosomes also had a translocation. By cytological means, knobbed chromosomes could be distinguished from knobless, and translocated chromosomes could be distinguished from normal. In addition, the following genetic markers were employed in the study.

$$c^+ = \text{Colored aleurone}$$

$$c = \text{Colorless aleurone}$$

$$wx^+ = \text{Starchy endosperm}$$

$$wx = \text{Waxy endosperm}$$

(Here c^+ is dominant over c and wx^+ is dominant over wx.) The experiment, diagrammed in Figure 2.28, showed that genetic and cytological recombination could be correlated. It was necessary for a plant that was heteromorphic for the knob and translocation and heterozygous for the c and wx genes to be crossed to a plant that had two normal, knobless chromosomes and was homozygous for c but heterozygous for wx. The chromosomal and genetic compositions of the parents employed in this cross are shown in Figure 2.28. From an examination of the types of crossover and noncrossover gametes possible, it can be seen that in each case a genetically determined phenotype can be associated with a cytologically visible chromosomal morphology.

Also in 1931, C. Stern published an account of cytological verification of genetic crossing-over. Stern employed a stock of *Drosophila* carrying an **X** chromosome that was genetically marked with carnation-colored eyes (*car*) and bar eyes (*B*) and cytologically marked either by a piece of **Y** or by existing in two fragments. The experiment, summarized in Figure 2.29, again verified that genetic recombination is correlated with cytological crossing-over.

Figure 2.28
Genetic recombination correlated with cytological crossing-over in maize

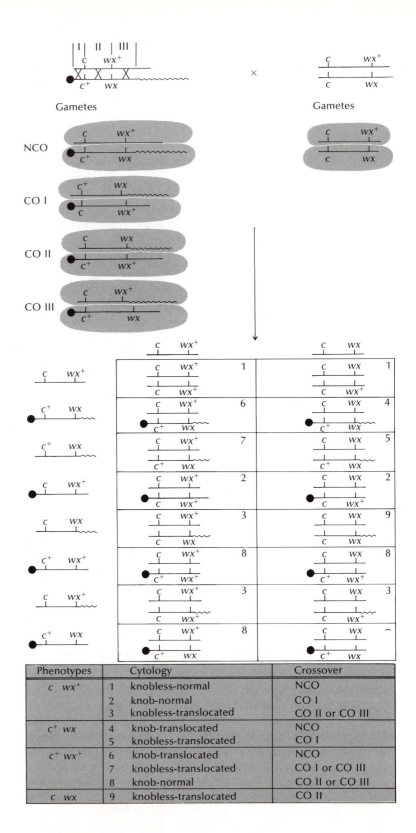

Phenotypes		Cytology	Crossover
$c\ wx^+$	1	knobless-normal	NCO
	2	knob-normal	CO I
	3	knobless-translocated	CO II or CO III
$c^+\ wx$	4	knob-translocated	NCO
	5	knobless-translocated	CO I
$c^+\ wx^+$	6	knob-translocated	NCO
	7	knobless-translocated	CO I or CO III
	8	knob-normal	CO II or CO III
$c\ wx$	9	knobless-translocated	CO II

Figure 2.29
Genetic recombination
correlated with cytological
crossing-over in *Drosophila*

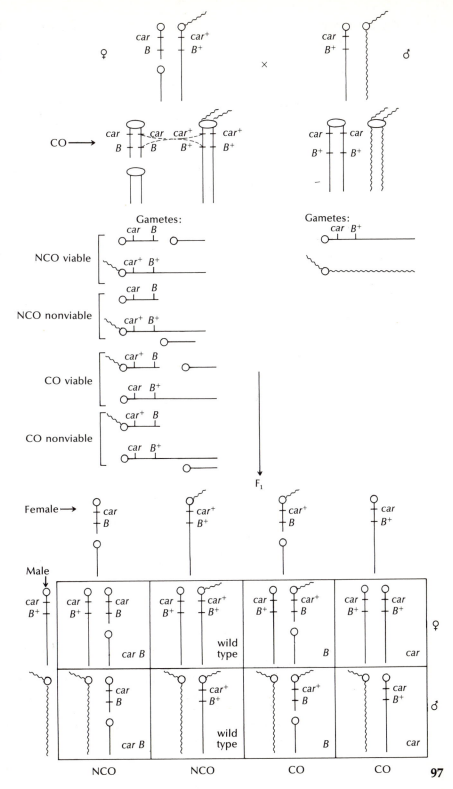

The significance of crossing-over

Crossing-over is known to occur in a variety of living organisms from viruses to humans. That it is so common a process in plants, animals, and microorganisms indicates that it must be evolutionarily advantageous, and indeed it is. Its basic significance is that it increases genetic diversity through the interchanging of blocks of genes. This increases phenotypic diversity, which maintains plasticity in a species through polymorphism. And polymorphism in a species permits adaptation to a wider range of habitats, enhancing the potential for evolutionary success.

Mitotic crossing-over

Crossing-over is not a phenomenon restricted exclusively to meiosis. It has also been found to occur in somatic cells undergoing mitosis. Curt Stern (Figure 2.30), pursuing an observation recorded earlier by Bridges, was the first to demonstrate mitotic crossing-over. Using *Drosophila*, Stern crossed a yellow-bodied fly (y) to a fly with singed bristles (sn). Both mutations are sex-linked and recessive, so we expect F_1 females to be wild type.

Figure 2.30
Curt Stern

Indeed, the F_1 females were wild type. But Stern noted that some flies had yellow spots on their bodies. These spots could have been due to the mutation of y^+ to y in some of the somatic cells, producing clones of cells yellow in phenotype. But the frequency of their occurrence was much greater than the known mutation rate for y^+ to y, so Stern was skeptical of this explanation. The idea was rejected altogether when Stern observed flies with "twin spots": a wild-type female with a patch of yellow tissue bordered with an equal-sized patch of singed tissue (Figure 2.31). The probability of both sn and y mutations occurring together and next to each other was far too small to explain the relatively high frequency of these twin spots.

Stern proposed instead the idea of mitotic crossing-over (Figure 2.32). He interpreted the twin spot and yellow spot as due to single crossover events in somatic cells. The very rare flies with only a patch of sn tissue he attributed to double crossing-over. Since Stern's pioneering efforts in the area of somatic crossing-over in *Drosophila*, the work has continued and expanded. The phenomenon has even been used as a tool for mapping genes in species that are not amenable to the more standard forms of mapping experimentation.

Figure 2.31
Twin spot in *Drosophila*

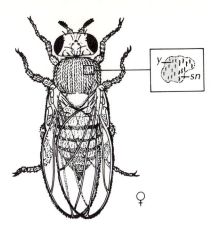

Figure 2.32
The recombinational basis for the generation of mitotic twin spots in *Drosophila*

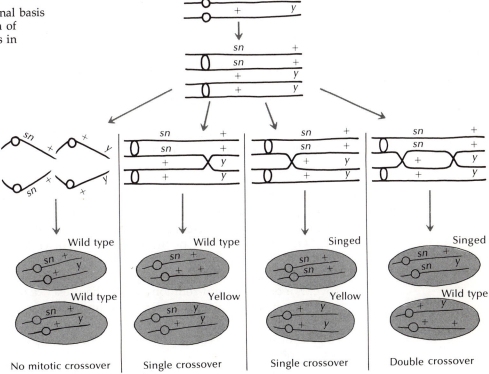

The chromosome theory of inheritance emerged in 1903 when Sutton suggested that Mendelian patterns of inheritance could be accounted for by the behavior of chromosomes during meiosis. The maturation of this very important concept took more than 20 years. The first step occurred between 1902 and 1905, when the inheritance of a specific phenotype, sex, was linked to a specific and morphologically distinct chromosome. But we noted the extreme complexities of the sexual phenotype when we discussed the influence of hormones and autosomes on the unfolding of the sexual phenotype. We showed that the genetic sex and phenotypic sex often do not match.

In 1909, Morgan made the first unambiguous link between a gene and a specific chromosome when he showed that the white-eye gene in *Drosophila* is located on the **X** chromosome. Further studies showed that each chromosome carries many genes, and this led to a most important idea: If a chromosome carries many genes, and if chromosomes cross over as they are known to do during meiosis, then cytological crossing-over should correlate with altered linkage associations. Sturtevant developed this basic theme into a scheme for mapping genes in *Drosophila* and made the first genetic map of a chromosome. Additional correlations were made between genetic and cytological crossing-over by Stern, Creighton, McClintock, and Dobzhansky around 1930.

Additional support for the chromosome theory of inheritance came from one of Morgan's students, Bridges. He showed that an unusual pattern of sex-linked inheritance was the result of chromosomal nondisjunction.

We discussed gene mapping in considerable detail. Though mapping procedures may vary from one species to another, the basic underlying principle remains the same: The farther apart two genes are on a chromosome, the greater the probability of a crossover occurring between them.

Our understanding of the mechanisms of crossing-over is incomplete. We know from studies in *Drosophila* and *Neurospora* that crossing-over occurs at the four-strand stage. We also feel confident that crossing-over is a breakage-and-reunion event. The presence of the synaptonemal complex has enhanced our understanding of chromosome pairing and raised interesting questions about the actual mechanisms of crossing-over. But we do not yet have a clear model that explains crossing-over in eukaryotes.

Crossing-over was long thought to be an event that occurred exclusively in meiosing cells. But Stern showed that it can also occur in mitosing cells.

Genetic recombination—whether through independent assortment of chromosomes or through crossing-over—is extremely important to the evolutionary success of sexually reproducing species.

2.1 The **Y** chromosome in *Drosophila* has no genes. Criticize that statement.

2.2 Differentiate between sex-linked inheritance and sex-limited inheritance.

2.3 Trace the major contributions to the chromosome theory of inheritance (draw on information in Chapters 1 and 2).

2.4 Design an experiment to test Bateson's reduplication hypothesis. Offer predictions based on both the chromosome theory and the reduplication hypothesis.

2.5 Several times in the history of genetic research, the reciprocal cross (a mutant male with a normal female and a normal male with a mutant female) has been a crucial technique. Discuss two or more occasions when the reciprocal cross was employed. In each instance, point out the genetic principle demonstrated.

2.6 What are the advantages of using *Neurospora crassa* for genetic studies as opposed to *Drosophila*?

2.7 A geneticist is unable to demonstrate linked genes in an organism. What explanations could you offer for this failure?

2.8 Why is the maximum frequency of recombinants 50 percent in eukaryotes, even though the length of the chromosome map is often greater than 50 map units?

2.9 Design a cross in which more than 50 percent of the offspring are recombinant.

2.10 Consider the four loci in *Neurospora*: *a*, *b*, *c*, and *d*. Dihybrid matings are performed using various combinations of these four gene pairs, with the following results.

Spore sequence	A		B		C		D		E		F		G	
① ②	+	+	+	−	+	+	+	+	+	+	+	−	+	+
③ ④	+	+	+	−	+	−	−	+	−	−	−	+	−	−
⑤ ⑥	−	−	−	+	−	+	+	−	+	+	+	−	+	−
⑦ ⑧	−	−	−	+	−	−	−	−	−	−	−	+	−	+

Mating			*Offspring (%)*				
	A	B	C	D	E	F	G
$b^+ d^+ \times b^- d^-$	15	18	5	55	3	1	3 (100)
$a^+ d^+ \times a^- d^-$	51	2	12	31	0	2	2 (100)
$a^+ d^- \times a^- d^+$	4	59	10	22	0	4	1 (100)
$c^+ a^- \times c^- a^+$	3	67	15	4	0	11	0 (100)
$c^+ d^+ \times c^- d^-$	71	3	12	11	2	0	1 (100)

Determine the positions of the four loci relative to each other and to the centromere, and their map distances.

102

*Mendelism and the
chromosome theory of
inheritance*

2.11 In *Drosophila*, there are three recessive genes affecting eye color, eye shape, and body color. Kidney-bean-shaped eyes, cardinal-colored eyes, and ebony body color are all on the third chromosome. Their symbols are *k*, *cd*, and *e*, respectively. The F$_1$ from parents homozygous for one, two, three, or none of these recessive genes is testcrossed, and the following progeny are produced.

1. 863 kidney, cardinal
2. 879 ebony
3. 67 kidney, ebony
4. 74 cardinal
5. 53 kidney
6. 47 ebony, cardinal
7.   5 kidney, ebony, cardinal
8.   7 wild type

(a) What were the parental phenotypes?
(b) What is the sequence of the three genes and the map distances separating them?
(c) Calculate the coefficient of coincidence.

2.12 Given the following two *Drosophila* stocks and a genetic map, predict the offspring you would obtain from 2000 progeny of an F$_1$ × testcross (assume all the mutants are recessive).

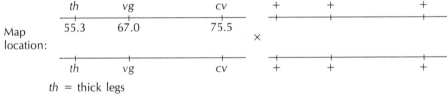

Map location:

th = thick legs
vg = vestigial wings
cv = curved wings

2.13 In some breeds of spotted cattle, the color of the spots can be red or mahogany. A cross between a red male and a mahogany female produces male offspring that are mahogany and female offspring that are red. The F$_1$ red females, when crossed to F$_1$ mahogany males, produce the following F$_2$ ratio.

Red male: 1
Red female: 3
Mahogany male: 3
Mahogany female: 1

How would you interpret this cross?

2.14 Pattern baldness in humans is determined by a pair of alleles (*A* = bald, *a* = nonbald). Two heterozygotes marry and produce the following offspring ratio.

Bald male: 3
Nonbald male: 1
Bald female: 1
Nonbald female: 3

(a) How would you interpret this cross?
(b) What were the parental phenotypes?

2.15 Three linked loci in corn are involved in the following genetic cross.

$$\frac{a\ b\ c}{a\ b\ c} \times \frac{A\ B\ C}{A\ B\ C}$$
$$\downarrow$$
$$\frac{a\ b\ c}{A\ B\ C} \times \frac{a\ b\ c}{a\ b\ c}$$
$$\downarrow$$
F₂ progeny

A B C	291
A B c	68
A b C	111
A b c	21
a B C	15
a B c	119
a b C	72
a b c	288

What are the gene sequence and the map distances between the three points? Calculate the coefficient of coincidence.

2.16 In *Neurospora*, a cross is made between a strain that requires nicotinic acid (*nic*) and a strain that requires adenine (*ad*). The alleles determining these traits are linked.

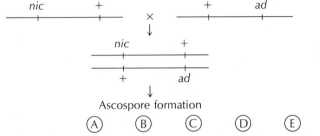

		A	B	C	D	E	F	G
	1	+ ad	+ +	+ +	+ ad	+ ad	+ +	+ +
Spore	2	+ ad	+ +	+ ad	nic ad	nic +	nic ad	nic ad
pair	3	nic +	nic ad	nic +	+ +	+ ad	+ +	+ ad
	4	nic +	nic ad	nic ad	nic +	nic +	nic ad	nic +
		PD	NPD	T	T	PD	NPD	T
		727	1	81	4	80	1	5

Using these data, determine the sequence of the three points (centromere, *nic*, and *ad*) and the distance separating them.

2.17 In a *Neurospora* cross of mating types *A* and *a*, the following data were collected.

$$A \times a$$
$$\downarrow$$
$$Aa$$
$$|$$
meiosis
$$\downarrow$$

		A	B	C	D	E	F
	1	A	a	A	a	A	a
Spore	2	A	a	a	A	a	A
pair	3	a	A	A	a	a	A
	4	a	A	a	A	A	a
		107	112	8	9	9	10

What is the distance between the mating-type locus and the centromere?

Anderson, E. G. 1925. Crossing over in a case of attached X chromosomes in *Drosophila melanogaster. Genetics* 10:403–447.

Baker, B. S., A. T. C. Carpenter, M. S. Esposito, R. E. Esposito, and L. Sandler. 1976. The genetic control of meiosis. *Ann. Rev. Genetics* 10:53–134.

Barratt, R. W., D. Newmeyer, D. D. Perkins, and L. Garnjosst. 1954. Map construction in *Neurospora crassa. Adv. Genetics* 6:1–93.

Bateson, W., E. R. Saunders, and R. C. Punnett. 1905. Experimental studies in the physiology of heredity. *Reports to the Evolution Committee, Royal Society,* Vol. II London, Harrison and Sons.

Belling, J. 1933. Crossing over and gene rearrangement in flowering plants. *Genetics* 18:388–413.

Bridges, C. B. 1914. Direct proof through nondisjunction that the sex-linked genes of *Drosophila* are borne by the X-chromosome. *Science* 40:107–109.

Bridges, C. B. 1916. Nondisjunction as proof of the chromosome theory of heredity. *Genetics* 1:1–52, 107–163.

Bridges, C. B., and T. M. Olbrycht. 1926. The multiple stock "Xple" and its use. *Genetics* 11:41–56.

Carlson, E. A. 1966. *The Gene: A Critical History.* Philadelphia, Saunders.

Cattanach, B. M. 1975. Control of chromosome inactivation. *Ann. Rev. Genetics* 9:1–18.

Chakravartti, M. R. 1968. Hairy pinnae in Indian populations. *Acta Genetica* 18:511–520.

Corwin, H. O., and J. B. Jenkins. 1976. *Conceptual Foundations of Genetics.* Boston, Houghton Mifflin.

Creighton, H. B., and B. McClintock. 1931. A correlation of cytological and genetic crossing over in *Zea mays. Science* 17:492–497.

Darlington, C. D. 1932. *Recent Advances in Cytology.* London, Churchill.

Day, J. W., and R. F. Grell. 1976. Synaptonemal complexes during premeiotic DNA synthesis in oocytes of *Drosophila melanogaster. Genetics* 83:67–79.

Dobzhansky, T. 1929. Genetical and cytological proof of translocations involving the third and fourth chromosomes of *Drosophila melanogaster. Biol. Zentralblatt* 49:408–419.

Dunn, L. C. 1965. *A Short History of Genetics.* New York, McGraw-Hill.

Fincham, J. R. S. 1970. Fungal genetics. *Ann. Rev. Genetics* 4:347–372.

Fogel, S., and R. K. Mortimer. 1971. Recombination in yeast. *Ann. Rev. Genetics* 3:219–236.

Fredga, K, A. Gropp, H. Winking, and F. Frank. 1976. Fertile XX- and XY-type females in the wood lemming, *Myopus schisticolor. Nature* 261:225–227.

Fu, T. K., and E. R. Sears. 1973. The relationship between chiasmata and crossing-over in *Triticum aestivum. Genetics* 75:231–246.

Gillies, C. B. 1975. Synaptonemal complex and chromosome structure. *Ann. Rev. Genetics* 9:91–110.

Grell, R. F., H. Bank, and G. Gassner. 1972. Meiotic exchange without the synaptonemal complex. *Nature New Biology* 240:155–157.

Hastings, P. J. 1975. Some aspects of recombination in eukaryotic organisms. *Ann. Rev. Genetics* 9:129–144.

Henderson, S. A. 1970. The time and place of meiotic crossing over. *Ann. Rev. Genetics* 4:295–324.

Janssens, F. A. 1909. La Théorie de la chiasmatypie, nouvelle interpretation des cinéses de maturation. *La Cellule* 25:387–406.

Käfer, E. 1977. Meiotic and mitotic recombination in *Aspergillus* and its chromosomal aberrations. *Adv. in Genetics* 19:33–131.

Levine, R. P. 1971. *Papers on Genetics.* St. Louis, Mo., Mosby.

McClung, C. E. 1902. The accessory chromosome—sex determinant? *Biol. Bull.* 3:43–84.

McKusick, V. A. 1962. On the X-chromosome of man. *Quart. Rev. Biol.* 37:69–175.

McKusick, V. A. 1973. Human genetics. *Ann. Rev. Genetics* 7:435–473.

Mittwoch, U. 1973. *Genetics of Sex Differentiation.* New York, Academic.

Morgan, L. V. 1922. Non-criss-cross inheritance in *Drosophila melanogaster. Biol. Bull.* 42:267–274.

Morgan, T. H. 1910. Sex-limited inheritance in *Drosophila. Science* 32:120–122.

Morgan, T. H. 1911. Random segregation versus coupling in Mendelian inheritance. *Science* 34:384.

Morgan, T. H., C. B. Bridges, and J. Schultz. 1935. Constitution of the germinal material in relation to heredity. *Carnegie Institution of Washington Year Book* No. 34, pp. 284–291.

Moses, M. J. 1968. Synaptonemal complex. *Ann. Rev. Genetics* 2:363–412.

Muller, H. J. 1916. The mechanism of crossing over. *Amer. Nat.* 50:193–221, 284–305, 350–366, 421–434.

Novitski, E. 1967. Nonrandom disjunction in *Drosophila. Ann. Rev. Genetics* 1:71–86.

Perkins, D. D. and E. G. Barry. 1977. The cytogenetics of *Neurospora. Adv. in Genetics* 19:133–285.

Renwick, J. H. 1971. The mapping of human chromosomes. *Ann. Rev. Genetics* 5:81–120.

Ringertz, N. R., and R. E. Savage, 1976. *Cell Hybrids.* New York, Academic.

Ris, H., and D. F. Kubai. 1970. Chromosome structure. *Ann. Rev. Genetics* 4:263–294.

Ruddle, F. H., and R. P. Creagan. 1975. Parasexual approaches to the genetics of man. *Ann. Rev. Genetics* 9:407–486.

Sears, E. R. 1976. Genetic control of chromosome pairing in wheat. *Ann. Rev. Genetics* 10:31–52.

Stadler, D. R. 1973. The mechanism of intragenic recombination. *Ann. Rev. Genetics* 7:113–127.

Stern, C. 1931. Zytologisch-genetische Untersuchungen ab Beweise für die Morgansche Theorie des Faktorenaustauchs. *Biol. Zentralblatt* 51:547–587.

Stern, C. 1936. Somatic crossing over and segregation in *Drosophila melanogaster. Genetics* 21:625–730.

Stern, C. 1973. *Human Genetics.* San Francisco, Freeman.

Stevens, N. M. 1905. *Studies in Spermatogenesis with Especial Reference to the Accessory Chromosome.* Washington, D.C., Carnegie Institution of Washington, No. 36.

Sturtevant, A. H. 1913. The linear arrangement of six sex-linked factors in *Drosophila,* as shown by their mode of association. *J. Exp. Zool.* 14:43–59.

Sturtevant, A. H. 1945. A gene in *Drosophila* that transforms females into males. *Genetics* 30:297–299.

Sturtevant, A. H. 1965. *A History of Genetics.* New York, Harper and Row.

Sutton, W. 1903. The chromosomes in heredity. *Biol. Bull.* 4:231–251.

Taylor, J. H. 1965. *Papers in Molecular Genetics.* New York, Academic.

The meiotic process, a review. 1976. *Nature* 259:82.

Thomas, C. A. 1971. The genetic organization of chromosomes. *Ann. Rev. Genetics* 5:237–256.

Turner, C. D. 1960. *General Endocrinology.* Philadelphia, Saunders.

Voeller, B. 1968. *The Chromosome Theory of Inheritance.* New York, Appleton-Century-Crofts.

Weinstein, A. 1958. The geometry and mechanics of crossing over. *Cold Spring Harbor Symp. Quant. Biol.* 23:177–196.

Westergaard, W. and D. von Wettstein. 1972. The synaptinemal complex. *Ann. Rev. Genetics* 6:71–110.

Wilson, E. B. 1905. The chromosomes in relation to the determination of sex in insects. *Science* 22:500–502.

Wolff, S. 1977. Sister chromatid exchange. *Ann. Rev. Genetics* 11:183–201.

The expansion of Mendelian principles, I

In every branch of learning, the discovery of major principles is followed by detailed applications and qualifications that fill out the broad picture. Genetics exemplifies this process. We have seen that Mendel's "factors" quickly became genes located on chromosomes. In this chapter, we shall see that more than two allelic forms of a gene can exist. We shall also see that many phenotypes are the result of two or more pairs of alleles mapping at different genetic loci, and that phenotypes are expressions of genes interacting with each other and with the environment.

Multiple alleles and complex genetic loci

As investigators began confirming Mendel's principles, a problem of no small magnitude emerged: How does one explain dominance and recessiveness? In 1906 Bateson, a strong advocate of Mendelism but an opponent of the chromosome theory of inheritance, proposed his *presence-and-absence* hypothesis (Figure 3.1), which stated that dominance is due to the presence of a particular gene and that recessiveness results from the loss of that gene. The theory's advocates visualized each gene as an entity floating free in the cell nucleus, but when it became evident that genes were linked to chromosomes and indeed that several genes were on the same chromosome, the hypothesis ran into its severest criticism. In 1910, Morgan argued that if presence and absence represented a process by which parts of chromosomes were lost—parts corresponding to the presence factors—then chromosomes would be of unequal size in a heterozygote. Pairing between homologs would then be difficult, for how could a "nothing" pair with a "something"? This argument was also applied to the theory of evolution. If presence and absence were causal, then evolution would proceed by the constant loss of genetic material, a difficult argument to sustain when so many of the more recently evolved animals and plants have more chromosomal material than the more primitive ones.

Actually, a damaging argument against the presence-and-absence hypothesis had been available since 1904. This date marks the discovery of *multiple allelism:* more than two allelic forms of a gene that map to the same genetic locus. L. Cuénot (1903) demonstrated that the number of alternatives for a particular gene pair is not necessarily limited to two. He crossed three different strains of mice: gray coat, yellow coat, and black coat. Gray was the normal coat color, and when crossed to mice with black coats, the F_1 was gray, suggesting that gray was dominant over black. A gray-by-yellow cross produced an F_1 that was yellow, suggesting that yellow was dominant over gray. When Cuénot crossed yellow by black, he found that they behaved toward each other like alleles, with yellow dominant over black. In other words, the genes for yellow, gray, and black fur appeared to form a series of alleles, a series we call *multiple alleles.*

Figure 3.1
Bateson's presence-and-absence theory (From E. A. Carlson, *The Gene: A Critical History,* Philadelphia: W. B. Saunders Company, 1966)

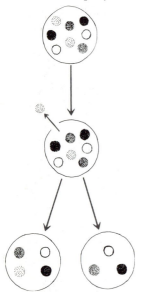

At each meiotic division the four pairs of factors segregate, a factor occasionally being lost. The loss of a factor leads to the formation of a cell lacking that factor. In Bateson's theory the "recessive" condition is the loss of a factor rather than the presence of a mutated form of it.

The original black and yellow strains were derived from a mutant albino strain. If this albino strain was itself the result of the loss of a gene, how could a "nothing" mutate to a "something" twice—yellow in one case and black in the other? This was a most damaging argument against presence and absence.

Perhaps the best-known example of multiple alleles was discovered by Morgan in 1912. Morgan had developed a stock of white-eyed *Drosophila* from his normal red-eyed stock. The presence-and-absence advocates would argue that this white-eye mutant arose through the loss of the red-eye factor. But in the white-eyed stock another mutation arose: an eosin eye-color mutant.

Genetic analysis showed this eosin mutant to be an allele of white, because it mapped at the same locus on the **X** chromosome and because, when it was heterozygous with white (w), it produced a mutant eosin eye color, demonstrating that the two alleles were functionally related. This new mutant was symbolized as w^e and was allelic to w; but how could w^e result from the loss of the mutant gene (w)? That is, how could a "nothing" have lost a "something" to mutate to w^e?

Another argument against the presence-and-absence hypothesis developed in the form of a number of mutations for the coat color of rabbits. Soon after the discovery of the eosin mutant in *Drosophila*, a series of coat-color mutants in rabbits were shown to be allelic.

$$c^+ = \text{Normal coat color}$$
$$c^{ch} = \text{Chinchilla coat color}$$
$$c^h = \text{Himalayan coat color}$$
$$c = \text{Albino}$$

Genotype	Phenotype
$c^+//c^{ch}$ =	Normal
$c^+//c^h$ =	Normal
$c^+//c$ =	Normal
$c^{ch}//c^h$ =	Chinchilla
$c^{ch}//c$ =	Chinchilla
$c^h//c$ =	Himalayan

The same countering argument was employed. Since c^+ is dominant over everything, then c^{ch}, c^h, and c must be "nothing"; but how could there be three varieties of "nothing," and how could one "nothing" be dominant over another "nothing"? The problems with this type of reasoning are apparent.

Eye-color mutations in *Drosophila* and coat-color mutations in rabbits and mice are examples of *multiple alleles*: more than two alternative forms of a gene, all of which are functionally related and map at the same locus on the chromosome. The discovery of multiple allelism laid to rest the presence-and-absence hypothesis as formulated by Bateson to account for dominance and recessiveness. Since 1912, multiple allelic systems have been found at several loci in *Drosophila* and other organisms.

Persisting into the 1940s was a view that multiple allelic series consisted of several functionally related alleles mapping at the same genetic locus with no *intralocus recombination* (crossing-over occurring only between genes and not within them). Before this time, genetic crossing-over had been observed to occur only between different loci and never within a locus, resulting in multiple alleles mapping at the same point on the chromosome. In 1940, however, C. P. Oliver found that crossing-over could occur between the lozenge-eye alleles in *Drosophila* and inferred that alleles could be spatially separate:

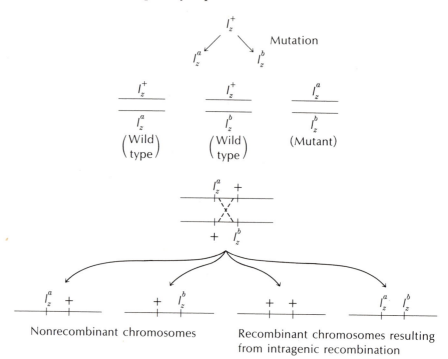

Nonrecombinant chromosomes   Recombinant chromosomes resulting from intragenic recombination

Soon afterward E. B. Lewis demonstrated the same thing at the star-asteroid eye locus in *Drosophila*. The frequency of crossing-over was very low in both cases, because the distance separating the alleles was small indeed. Nevertheless, crossing-over did occur and the mapping of multiple alleles was now possible. Proof of recombination between alleles meant that the concept of the gene had to be reevaluated. No longer could genes be viewed as a series of beads on a string with crossing-over occurring only between beads. The term *pseudoalleles* was coined in 1945 to describe a series of functionally related genes within which recombination could occur.

Since the concept of the gene as a locus within which no recombination could occur was inconsistent with pseudoallelism, the term *complex locus* came into existence to describe any nest of pseudoalleles between which crossing-over can occur. Figure 3.2 presents recent maps of the

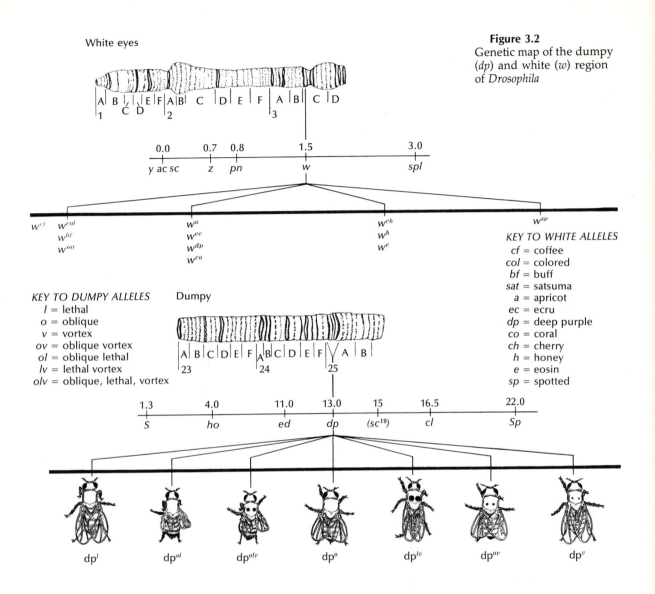

White eyes

Figure 3.2
Genetic map of the dumpy (*dp*) and white (*w*) region of *Drosophila*

0.0	0.7	0.8	1.5	3.0
y ac sc	z	pn	w	spl

w^{cf} w^{col} w^{a} wch w^{sp}
w^{bf} w^{ec} w^{h}
w^{sat} w^{dp} w^{e}
      w^{co}

KEY TO WHITE ALLELES
cf = coffee
col = colored
bf = buff
sat = satsuma
a = apricot
ec = ecru
dp = deep purple
co = coral
ch = cherry
h = honey
e = eosin
sp = spotted

KEY TO DUMPY ALLELES
l = lethal
o = oblique
v = vortex
ov = oblique vortex
ol = oblique lethal
lv = lethal vortex
olv = oblique, lethal, vortex

Dumpy

1.3	4.0	11.0	13.0	15	16.5	22.0
S	ho	ed	dp	(sc¹⁹)	cl	Sp

dp^{l} dp^{ol} dp^{olv} dp^{o} dp^{lv} dp^{ov} dp^{v}

white and dumpy complex loci of *Drosophila* based on current recombinational data. At both loci, there remain several pseudoalleles that have not yet been separated by crossing-over. Either these are mutants at the same *site* (a single DNA base pair), or they are so close together that recombination between them is too rare to have been detected. The A-B-O and Rh blood groups in humans are multiple allelic series that have proved unfruitful to recombinational analysis because humans are not particularly willing or convenient experimental animals. However, should techniques such as somatic cell genetics make recombinational analysis possible, then these loci too will probably turn out to be complex.

Multiple alleles have altered our concept of the genetic locus and the gene. We now visualize a gene, physically, as a segment of DNA, and multiple alleles would represent mutations at different points in the DNA segment (see the diagram below).

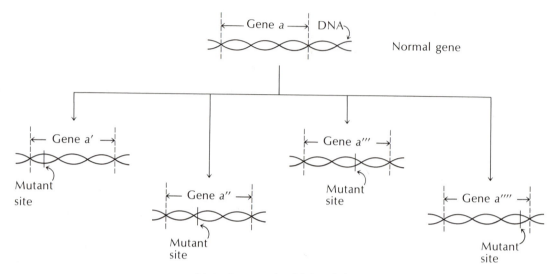

Mutant genes (multiple alleles)

The concept of dominance

The interpretation of dominance and recessiveness within the context of the presence-and-absence hypothesis was shown to be incorrect. But it was still necessary to understand the functional implications of dominance and recessiveness. By examining some of the relationships that can exist between members of an allelic pair, perhaps we can develop a functional interpretation of these terms.

The phenotypic ratios Mendel obtained with garden peas have been observed many times in a wide range of organisms. Such ratios require that one member of an allelic pair be dominant over the other. But in fact this kind of relationship between alleles is quite uncommon. Four principal dominance relations are known. In *complete dominance,* that described by Mendel, the heterozygote has a phenotype identical to that of one of the original parents. In *incomplete dominance,* first clearly described by Kölreuter, the phenotype of the heterozygote is intermediate between the parental phenotypes. In *overdominance,* the phenotype of the heterozygote is more extreme than that of either parent. For example, the amount of fluorescent eye pigment in a heterozygous white-eyed *Drosophila* exceeds that found in either parent. Finally, in *codominance,* the heterozygote ex-

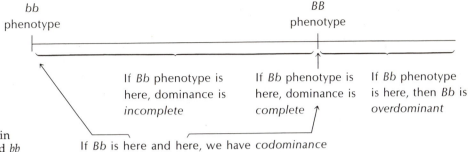

Figure 3.3
Types of dominance in relation to the *BB* and *bb* phenotypes

presses both parental phenotypes. An example of this is the AB blood type in humans, which is the expression of both the I^A and I^B alleles. Figure 3.3 summarizes these relationships.

Classification of alleles as dominant or recessive ultimately depends on the function of the gene products produced and the level of analysis applied to the phenotypes in question. For example, in the analysis of a white-eyed heterozygous *Drosophila* ($w//w^+$), a number of conclusions are possible, depending on how the phenotype is analyzed. Visual inspection of the white-eyed heterozygote would not enable us to distinguish it from a normal red-eyed fly. So on this basis, we would classify the normal allele (w^+) as dominant over the mutant allele (w). But analysis of the pteridine eye pigment of this heterozygote would point to overdominance, because it is in greater concentration than in either parent. Furthermore, analysis of the spermathecae (sperm-storing organs) of the heterozygous female fly would reveal structural abnormalities less extreme than those of the white-eyed homozygote, which, in addition to white eyes, has partially defective reproductive organs. This indicates incomplete dominance. Basically, then, dominance is often a function of the level of analysis employed.

What does it really mean when we say that a gene is dominant or recessive? For many years, genes were not thought of in functional terms. "Dominant" and "recessive" were applied to traits we could see, such as height, color of seed coat, and so on. A gene was considered dominant if it was expressed in the heterozygous condition and recessive if it was masked. The bar-eye gene (*B*) exemplifies what has been called a dominant gene in *Drosophila*, and ebony body (*e*) might be a typical recessive. A heterozygous bar-eyed fly is phenotypically bar, and a heterozygous ebony-bodied fly is phenotypically normal. Dominant genes have traditionally been assigned upper-case letters, and recessives have been assigned lower-case letters. But such designations are often misleading, because they give a false impression of how a pair of genes relate to each other. For example, close examination of the bar-eyed heterozygote reveals that the phenotype is actually intermediate, exhibiting incomplete dominance. Similarly, an ebony-bodied heterozygote is often more deeply pigmented than a normal fly, another suggestion of incomplete dominance. A more useful and perhaps less confusing symbolism is the one that employs a lower-case letter with a superscript plus (a^+) to designate the form of the gene found most often in nature (wild-type allele) and the lower-case letter without a superscript (a) to designate a mutant form. Multiple alleles would be given superscripts such as a^1, a^2, a^3, and so on.

In this notation, nothing is stated or implied about the dominant-recessive nature of either allele, which will in fact vary from one allelic pair to another. It will even vary with the environment. Consider, for example, the inheritance of a temperature-sensitive lethal gene at different temperatures:

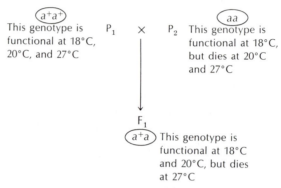

P_1 × P_2

a^+a^+
This genotype is functional at 18°C, 20°C, and 27°C

aa
This genotype is functional at 18°C, but dies at 20°C and 27°C

F_1

a^+a This genotype is functional at 18°C and 20°C, but dies at 27°C

At 18°C, the heterozygote is identical to both parents, and no dominance relations exist. Indeed, at 18°C, we have no evidence that there is a pair of alleles in existence. At 20°C, a^+ is dominant over a. And at 27°C, a is dominant over a^+. The meaninglessness of classifying these genes as simply dominant or recessive is evident.

By regarding alleles from a functional perspective, we gain a clearer understanding of dominance and recessiveness. Let us say that a gene (a^+) produces an enzyme necessary for the normal functioning of the organism. It catalyzes the following reaction.

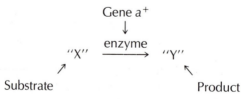

Gene a^+
↓
"X" $\xrightarrow{\text{enzyme}}$ "Y"

Substrate   Product

Suppose a mutation occurs that changes gene a^+ to a. In a heterozygote, the mutation can have any of the following three consequences:

First, gene a may give rise to a nonfunctional enzyme or to none at all. Assuming the organism is diploid, it still has a single a^+ gene and may, therefore, still produce enough of the normal enzyme to function normally. Only when the organism is homozygous for a will the reaction

"X" to "Y" *not occur* and a mutant phenotype result. Such a mutant gene can be classified as recessive, with a^+ completely dominant over a.

Second, gene a may give rise to a nonfunctional enzyme or to no enzyme at all, but this time the single a^+ gene may *not* produce enough of the normal enzyme. The reaction rate may then be sufficiently slowed so that not enough "Y" is produced for a normal phenotype, and an intermediate phenotype is produced instead. The mutant gene a is then incompletely dominant over a^+ (or vice versa).

Third, the mutant gene a may produce an enzyme that has a *greater* affinity for the substrate "X" than does the normal a^+ gene product, but instead of converting "X" to "Y", the enzyme converts "X" to a new product, "Y_2". The presence of "Y_2" causes a mutant phenotype to appear in the heterozygote that is identical to that of the homozygous a phenotype. Therefore, a may be classified as a dominant mutation. Dominance of a over a^+ may also arise when the a gene product inactivates the normal a^+ gene product. Assuming that aa is nonfunctional, then a^+a will also be nonfunctional, and a will be dominant over a^+.

Other similar models could be constructed to present a functional interpretation of dominance and recessiveness, but the point is that dominance and recessiveness have meaning only when discussed in terms of gene products. A diagrammatic summary of possible dominant-recessive relationships appears in Figure 3.4.

Environmental and genetic interactions

The consequence of a dominance relationship is the modification of genetic expression. In other words, given a pair of differing alleles, the expression of one of the alleles is modified by the other in most cases. In a broad sense, this is a form of gene–gene interaction, and in this section we shall explore other forms of such interaction. Genetic expression can also be modified by the environment, which we may regard as everything but the chromosomes, and we shall now discuss some examples of this phenomenon.

**Gene–environment
interactions** Most phenotypes observed are the result of the interaction between the genotype and the environment. Put more simply,

Phenotype = Genotype + Environment

Organisms inherit a genotype, but during the course of development the genotype encounters several environmental situations that may make it subject to modification. External factors such as temperature, light, and nutrition and internal factors such as sex, age, and the availability of substrates can combine to modify the expression of a genotype.

Granted that a genotype's expression can be modified by the envi-

Figure 3.4
Models of dominant, recessive, and incompletely dominant genes

(1) a^+ is completely dominant (a is completely recessive):

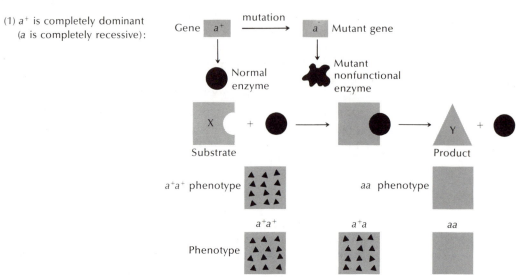

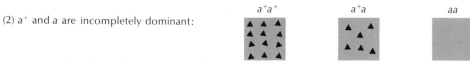

One dose of a^+ produces enough product to give an a^+a phenotype identical to a^+a^+

(2) a^+ and a are incompletely dominant:

One dose of a^+ produces half as much product as a^+, resulting in an intermediate phenotype

(3) The mutant a is dominant over a^+:

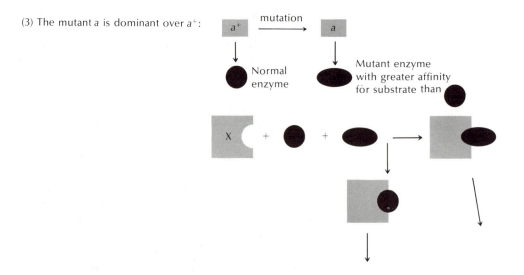

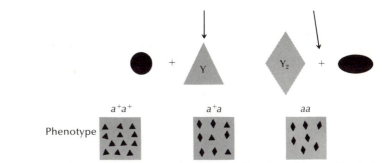

Phenotype

a^+a^+ a^+a aa

The a gene product produces a new phenotype so that a^+a and aa are identical

(4) The mutant a is dominant over a^+:

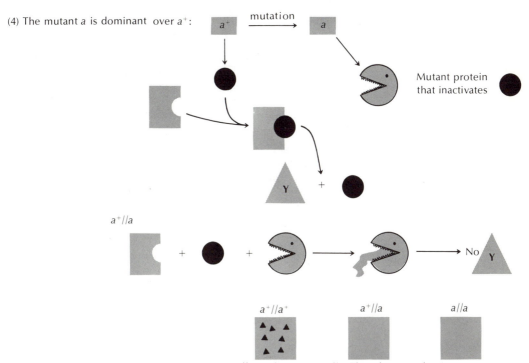

Mutant protein
that inactivates

$a^+//a$

$a^+//a^+$ $a^+//a$ $a//a$

$a^+//a$ is identical to $a//a$. Therefore a is dominant over a^+.

117

ronment and indeed by other genes, it stands to reason that the appearance of an organism is the result of a good deal more than just its genetic constitution. When we compare observable effects of genes with genetic constitution, two terms—*penetrance* and *expressivity*—are quite useful. Penetrance is the proportion of individuals that show an expected mutant phenotype, and expressivity is the degree to which a particular mutant phenotype is expressed by the individual organisms. In the traits Mendel studied, a mutant phenotype was expressed by all individuals carrying the mutant genes in a homozygous condition. Penetrance in these cases was 100 percent. The expression of the mutant genes showed little variation. Thus the mutant genes that Mendel studied were completely penetrant and not subject to extensive environmental modification. In contrast is the gene in humans that determines retinoblastoma (tumor of the eye). This is considered a dominant mutation, because heterozygotes can express a mutant phenotype, but it is a gene with *incomplete penetrance*, because not all the individuals who have the gene express retinoblastoma. Because the severity of the disease varies among the individuals who express it, we say that the expressivity of this gene also varies.

Let us examine how the environment can affect our interpretation of the inheritance of a pair of alleles that follow a Mendelian pattern of segregation. Suppose that a pair of genes have segregated to give a 1:2:1 genotypic ratio. If each genotypic class were phenotypically distinct and also not subject to extensive environmental modification, a 1:2:1 phenotypic ratio would be observed. If, on the other hand, each genotypic class were subject to extensive environmental modification, the phenotypic classes could so overlap that there would be a broad distribution of phenotypes around a central mean, and there would be no way to be certain that two genes were segregating:

1. Phenotypic variability when the environment has little influence on genotype expression.

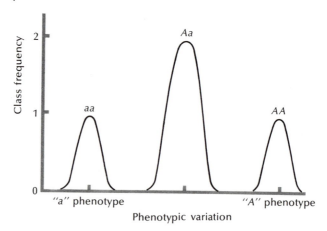

2. Phenotypic variability when the environment can extensively modify genotype expression.

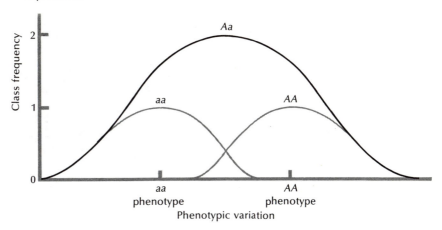

Phenotypic variation

Genotypes vary to the degree that their expression can be modified. That is, if a genotype could be placed in a range of different environments, only a limited range of phenotypic expressions would be observed, and the number would vary from one genotype to another. The *norm of reaction* of a genotype is the entire range of environments within which genotypic expression will vary. This norm can be narrow or quite wide. The A-B-O blood groups in humans, for example, show almost no variability, irrespective of environment. The norm here is narrow. Human intelligence, on the other hand, can vary widely with environment. The norm here is broad.

Dramatic environmental influences   The role of the environment in influencing a phenotype can be subtle or dramatic. Occasionally, a response to the environment is so great that a presumably normal phenotype is altered enough to resemble a known, genetically caused, mutant phenotype. A phenotype produced by the environment that simulates the effects of a particular gene or group of genes, even though the simulated phenotype is not inherited, is called a *phenocopy*. Many examples are known. In humans, for instance, certain drugs can produce mutant phenotypes.

One of the most tragic instances of induced phenocopy occurred in the early 1960s, when deformed children resulted from the administration of the tranquillizing drug *thalidomide* to women in their sixth week of pregnancy. The children, born in Great Britain and West Germany, expressed a phenotype called *phocomelia* (Greek: *phoke*, seal + *melos*, limb), in which one or more of the limbs were reduced to flipperlike appendages

Figure 3.5
Malformed baby with phocomelia, caused by thalidomide, in forelimbs [Data in table from W. Lenz, *Proc. 2nd Int. Conf. Congenital Malformations,* 263–276 (1964).]

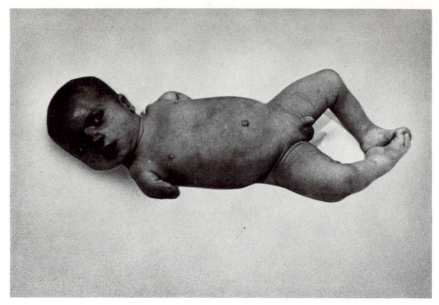

Photo courtesy of Dr. W. Lenz, Munster, West Germany and Dr. H. Taussig, Johns Hopkins Hospital, Baltimore, Md.

	1960	1961	1962 Jan.–July	Aug.–Oct.	Total
Total births	19,052	19,917	13,326	5,542	57,837
Phocomelia-like malformations	28	60	40	2	130
History:					
Thalidomide taken	13	46	33	2	.94
No evidence of intake	0	5	1	0	6
No history available	15	9	6	0	30

(Figure 3.5). The common factor in all of the cases was thalidomide, considered so harmless by the manufacturer that it was often dispensed without prescription. The human fetus during the first six weeks of development is most susceptible to the drug. Although the reasons are not clear, it appears that the drug interferes with key metabolic reactions at the time when limbs are being formed, thus producing phocomelia, a phenocopy of rare dominant mutations in humans.

Another drug that can induce phenocopies is *LSD*. Schizophrenia, a mental disease that is in part genetically controlled, is characterized by severe disintegration of emotional stability. A genotypically normal individual under the influence of LSD may express some symptoms of the schizophrenic phenotype.

Mutant genotypes can also be expressed as normal phenotypes, as in diabetes control and temperature-sensitive mutations. *Diabetes mellitus* is to a large extent an inherited disease. Two theories about its genetic basis are that it is controlled either by several gene pairs or that it is controlled by a recessive gene with low penetrance. Diabetes is characterized by a deficiency or defectiveness of a pancreatic hormone, insulin. Sufferers from the disease cannot utilize glucose, so they must depend on an external source of insulin. Injection of the hormone adjusts the internal environment so that the mutant genotype is not expressed and a normal phenocopy is produced.

In the genetic disease hemophilia A, blood is unable to clot because a specific protein required for normal clotting is defective or present in insufficient quantities. Through injections of this factor, hemophiliacs clot normally—a mutant genotype with a normal phenotype. Unlike insulin, which can be synthesized or obtained from nonhuman animals, the antihemophiliac factor can be obtained only from humans. Thus it is difficult to come by and extremely expensive.

Temperature-sensitive mutations (ts) are mutant genotypes that express mutant phenotypes at one temperature (*restrictive temperature*) and normal phenotypes at another (*permissive temperature*). In bacterial viruses, lethal mutations can occur that affect virus development. If viruses carrying temperature-sensitive lethal mutations are grown in bacterial hosts at the restrictive temperature (42°C), they will develop to a certain point and then stop. But if the viruses are allowed to grow at a permissive temperature (25°C), the lethal mutation is not expressed and development proceeds to completion.

Temperature-sensitive mutations are also known in *Drosophila*, but the techniques for detecting them differ from those used in viruses. Wild-type *Drosophila* are treated with a mutagenic agent. The treated flies are mated to *Cy//Pm* flies (remember that *Cy* and *Pm* are lethal when homozygous), and the F_1 *Cy* flies are raised at the permissive temperature of 25°C and saved for further genetic analysis to see if a dominant temperature-sensitive lethal (DTS-L) mutation was induced (lethal at 29°C in the heterozygous state). The F_1 *Cy* flies are testcrossed to a *Cy//Pm* fly and are allowed to deposit eggs for four days at 25°C before being transferred to a new container, where the eggs laid will be raised at 29°C (restrictive temperature). If the flies raised at 29°C do not produce the *Cy* and *Pm* classes, it is probable that a DTS-L was induced. If all three expected classes are found among the flies raised at 29°C, no DTS-L was induced (Figure 3.6).

These examples of phenocopies point out dramatic alterations of phenotypes by environmental manipulation. Mutant phenotypes can be induced from genetically normal organisms, and normal phenotypes can be induced in genetically mutant organisms.

Subtle environmental influences: Twin studies   The relationship between the genotype and the environment is not always obvious. Traits such as intelligence, certain mental illnesses, and some forms of cancer show

Figure 3.6
Screening system in *Drosophila* for detecting dominant temperature-sensitive lethal mutations (*DTS-L*)

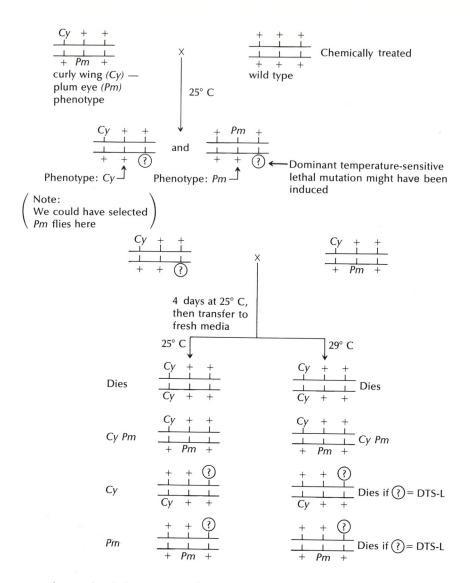

complex and subtle gene–environment interactions. For such traits, the relative influence of genes and environment is difficult to determine. Techniques such as twin studies have been extensively used to measure the contributions of each.

Francis Galton was among the first to recognize the difference between identical and fraternal twins and to use them in genetic studies. He suggested that identical twins are *monozygotic* (from the same fertilized egg) and therefore genotypically identical, and that fraternal twins are *dizygotic* (from two fertilized eggs) and therefore no more alike genetically than any pair of nontwin *siblings* (brothers and sisters).

Using monozygotic (MZ) twins, we can study the effect of different environments on identical genotypes; with dizygotic twins, we can study the effects of different (though related) genotypes in similar environments. Twin studies are one of our most valuable tools in resolving the age-old nature–nurture problem.

Galton ascribed differences between monozygotic twins to the environment. It is now known, however, that data from identical-twin studies must be used cautiously, because it is sometimes difficult to confirm the monozygotic origin of twins. It is also known that the *in utero* environments, although similar, are not necessarily identical for monozygotic twins, so there may be important environmentally induced differences before birth.

In the postnatal development of monozygotic twins, other factors can complicate matters considerably. When the zygote first divides into two separate cells or zygotes, the zygotes are identical because of the mitotic process that generated them. However, all subsequent mitoses may not be so error-free. A somatic cell may actually lose a chromosome and produce a clone of cells all missing that particular chromosome. This sets up a genetic difference in otherwise genetically identical twins. Gene mutations in somatic cells create the same type of problem. We can thus have clones of cells genetically different from cells in the other sibling of a monozygotic pair.

Problems are also inherent in studies with dizygotic twins. Though presumably these twins arise from two different eggs produced by meiosis in two different oocytes, an ovum may divide in two after meiosis by a parthenogenetic process. This yields two genetically identical cells. Each is fertilized by a different sperm, but the maternal components are identical. These dizygotic twins are more different genetically than monozygotic twins, but less different genetically than normal dizygotic twins.

These and other problems suggest the limitations of twin studies. Twin studies are extremely valuable as research tools, but we must be cautious in the reliance we place on data derived from them. We must bear in mind that twin studies may tell us something about the genetic predisposition to develop a particular trait, but they rarely reveal anything about the number of genes involved, their mode of action, or their transmission pattern. Twin studies also make certain unwarranted assumptions about environmental constancy. Monozygotic twins, because of their striking similarities, tend to seek out very similar environments. Dizygotic twins, on the other hand, are more different and perhaps are even of the opposite sex, so they tend to seek out different environments. We do know that as dizygotic twins age, they grow more dissimilar, and as monozygotic twins age, they remain remarkably similar. To cut down somewhat on these inherent problems, one should compare monozygotic twins only with dizygotic twins of the same sex.

Nevertheless, when controlled breeding experiments are not possible (and in human beings they usually are not), twin studies offer the best means of assessing how much of the variability observed between different individuals is due to hereditary differences and how much is due to the environmental differences encountered during the individuals' development.

An elegantly simple though effective way to assess differences between twins is to examine traits that are either present or absent in the members of the twin sets. If twins express (or fail to express) the same phenotype, they are *concordant* for that phenotype. If they differ in the phenotype, they are *discordant* for it. To assess the roles of heredity and environment, we must ascertain the degree of concordance and discordance in both monozygotic and dizygotic twins. Suppose, for example, that 99 percent of all identical twins studied were concordant for eye color (in 99 percent of all identical twins, the eye color in each twin pair was the same), compared with a rate of 28 percent for fraternal or dizygotic twins (72 percent of the twin pairs were dissimilar in eye color). This suggests that eye color is determined largely by the genotype. The incidence of measles, on the other hand, has a concordance rate of 95 percent in identical twins and 87 percent in fraternal twins, suggesting that the environment largely determines the phenotype. Figure 3.7 summarizes concordance and discordance values for some physical and behavioral traits.

Figure 3.7
Concordance and discordance in twins for a variety of traits (After Stern, *Principles of Human Genetics*, Freeman, San Francisco, 1975; and other sources)

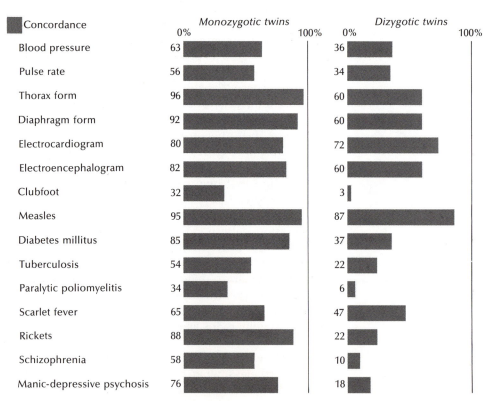

Subtle environmental influences: Human intelligence Intelligence is a complex trait determined by the genotype and the environment. It is a trait composed of a constellation of related abilities, such as the ability to

1. Define and understand words
2. Think of words rapidly
3. Analyze mathematical relationships
4. Analyze spatial relationships
5. Memorize and recall
6. Perceive differences and similarities among objects
7. Formulate rules, principles, or concepts for solving problems or understanding situations

Galton suggested that intelligence was determined almost entirely by the genotype, but this view has not been widely accepted. Although it may be granted that there is an innate general ability akin to our concept of general reasoning ability, other distinct attributes—musical ability, verbal ability, or quantitative ability—may be unrelated to that ability or its components. Intelligence tests, like those devised by Binet and revised by Termin and Merrill, employ a single number, the intelligence quotient or IQ, and are intended to measure a person's innate intelligence. Whether they actually do so has been widely debated.

In 1965, Cattell suggested that there are two components of general ability. The first, called *fluid intelligence,* is primarily genetic and a function of brain development; this reaches its peak at about the age of 14, when the brain reaches its full growth. Fluid intelligence can be measured by tests free of cultural and environmental bias. *Crystallized intelligence* is a function of training and education and can change after the completion of brain growth.

Attempts to quantify the relative contributions of genotype and environment to intelligence are complicated by disagreement in defining intelligence and skepticism of the tests designed to measure it. Nevertheless, it is generally agreed that intelligence is determined by several gene pairs interacting with the environment. Twin studies reveal that identical twins raised together are more concordant for IQ than identical twins raised separately or fraternal twins raised together, which emphasizes the influence of the genotype and the environment on the trait. A study of identical twins separated at birth showed that the greater the environmental differences between them, the greater the IQ differences—up to a maximum of 24 points—thus suggesting a limitation on environmental influence. Conversely, separated identical twins raised in similar environments showed IQ differences of only 6 points. In subsequent studies, the IQ of adopted children was compared with that of the biological mother and the adoptive mother. It was found that the IQ correlated much more with that of the biological mother than with that of the adoptive mother.

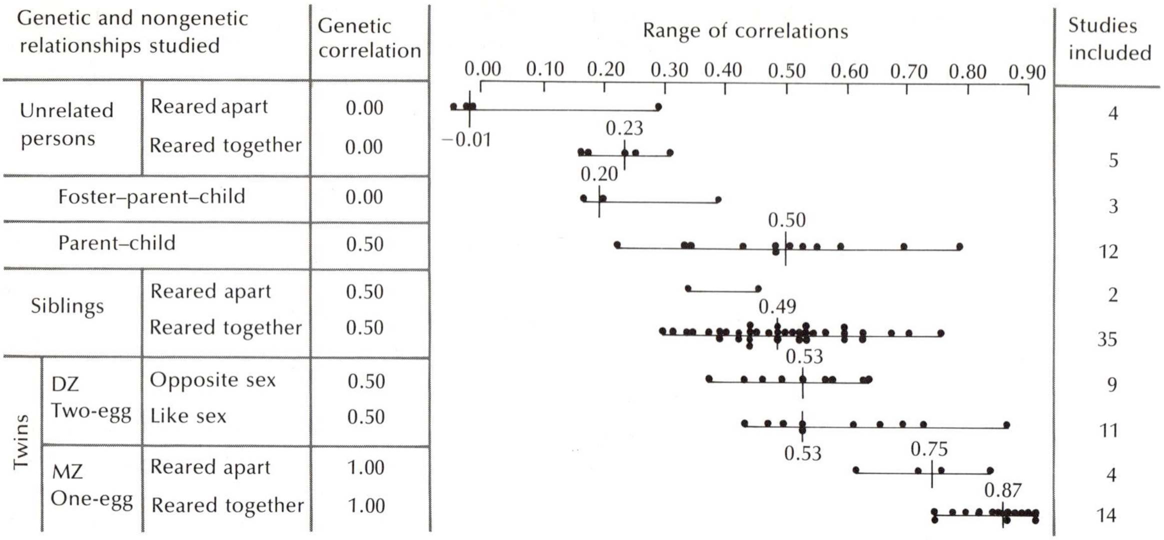

Genetic and nongenetic relationships studied		Genetic correlation	Range of correlations (average)	Studies included
Unrelated persons	Reared apart	0.00	−0.01	4
	Reared together	0.00	0.23	5
Foster–parent–child		0.00	0.20	3
Parent–child		0.50	0.50	12
Siblings	Reared apart	0.50		2
	Reared together	0.50	0.49	35
Twins — DZ Two-egg	Opposite sex	0.50	0.53	9
	Like sex	0.50	0.53	11
Twins — MZ One-egg	Reared apart	1.00	0.75	4
	Reared together	1.00	0.87	14

Figure 3.8
Correlations of intelligence-test scores for pairs of people with different degrees of hereditary and environmental similarity. The lengths of the lines indicate the ranges of the correlations obtained in the various studies. The dots show the correlations in each study. The short vertical slashes indicate the average correlation in each type of study. (After Erlenmeyer-Kimling and Jarvik, 1963. *Science* 142: 1477–1479)

Although we lack precise quantitative estimations of genotypic and environmental contributions to intelligence, current estimates are that from half to three-fourths of intelligence is genetically based, the remainder environmentally controlled. Twin studies afford the basis for such estimates. Study Figure 3.8 carefully. It is a comparison of IQ scores for people with different degrees of genetic relationship to each other. The range goes from totally unrelated people to identical twins. To assess how much of the IQ variability observed between different individuals is due to genetic differences, we use *correlation coefficients*. These values are expressions of the degree to which pairs of individuals obtain similar IQ scores. A correlation coefficient of 1.00 tells us that there is perfect agreement between pairs of individuals. A value of 0.00 indicates no agreement, and a value of −1.00 is a negative correlation, meaning that an increase in one score is accompanied by a decrease in the other.

In Figure 3.8 we see the results of correlational studies done on many people with a diverse array of genetic relatedness. In 14 different studies of IQ testing among monozygotic twins raised together, we have a correlation coefficient range of 0.76 to 0.92, with a mean of 0.87. Presumably these are genetically identical individuals raised under similar conditions. Note that, as the amount of genetic similarity decreases, with the environment held relatively constant, the correlation also decreases. Hence unrelated people living apart have a negative correlation coefficient. This is certainly a strong argument for the importance of the genotype in the trait we call human intelligence.

The strongest evidence favoring a high genetic contribution to human intelligence comes from the comparison of correlation coefficients of monozygotic twins raised together or apart with foster parents and their children. We have values of about 0.8 contrasted with a value of about 0.2. This is a powerful hereditarian argument, but it has been challenged by those supporting the environmentalist perspective. They point out that relatively few studies have been done, certainly not enough to warrant sweeping generalizations. They also criticize the method of data collection and the testing procedures. Though the debate goes on about the relative importance of genotype and environment to human intelligence, we cannot avoid the conclusion that human intelligence is strongly determined by the genotype. Perhaps as much as 75 percent of our intelligence is genetically determined. But at the same time, studies show that the environment exerts a major influence in the development of that intelligence.

Our discussion of intelligence and its genetic and environmental roots is pertinent to a controversial study published by Arthur Jensen in 1969. Jensen's thesis is that there are inherent differences in intelligence among different racial groups. At the core of his argument is the syllogism that if races differ genetically, and if intelligence is a genetically determined trait, then perhaps we can expect to observe differences in intelligence between racial groups.

Jensen reports that blacks as a group score about 15 points lower than whites on IQ tests. He contends that this difference is due to genetic factors. He argues further that about 80 percent of intelligence is determined by the genotype and that there *might* be black–white group differences.

Jensen's thesis has been hotly debated. Critics point out the difficulties in defining intelligence, measuring it, and defining a human race. Many argue that the IQ differences observed by Jensen could be explained by environmental differences. And still others suggest that Jensen's methodology is open to criticism (his sampling procedures, for example). The Jensen report has generated an enormous amount of discussion, and for good reason. It is an explosive issue. If accepted, it could dramatically alter our social and educational programs, and it could lead to an insidious racism. There are many who question the appropriateness of conducting studies such as Jensen's. Differences of opinion will probably come to light if you engage in discussions of this most important topic and its ramifications. Some of the references at the end of this chapter can serve as a starting point.

Subtle environmental influences: Schizophrenia   Another behavioral trait whose environmental and genetic components are difficult to quantify is schizophrenia. Over half the patients in mental hospitals today are classified as schizophrenics, even though the disease is a difficult one to define. There is, however, a characteristic syndrome associated with schizophrenia that includes gradual withdrawal from reality, disordered and bizarre thinking, emotional deterioration, and disturbed impulses or conduct. Twin studies of schizophrenia suggest a strong genetic component, but how does one quantify the contributions of genotype and environment to the schizophrenia phenotype? Table 3.1 summarizes twin studies of schizophrenia.

Earlier studies of schizophrenia indicated a concordance rate as high as 86 percent, but more recent investigations show considerably lower rates—one as low as 6 percent. Pollin recently published figures showing

Table 3.1
A summary of twin studies of schizophrenia

Investigator	No. of MZ pairs	Percent MZ concordance	No. of DZ pairs	Percent DZ concordance	MZ/DZ ratio	Sampling	Diagnosis by
Luxenberger							
1928	17	60–76	33	0	—	R,C[a]	Author
1930	21	67	37	—	—	—	
1934	27	33	—	—	—	—	
Rosanoff							
1934	41	61	101	10	6.1	R	Hospital
Essen-Moller							
1941	11	55–64	27	15	4.3	C	Author
Kallmann							
1946	174	69–86	517	10–15	5.7	R,C	Author
Slater							
1953	41	68–76	115	11–14	5.4	R,C	Author
Inouye							
1961	55	76	17	12–22	3.5	R	Author
Tienari							
1963	16	6–31	21	5	1.2	B	Author
Harvald & Hauge							
1965	9	44	62	10	4.4	B	Hospital
Gottesman & Shields							
1966	24	42–65	33	9–17	3.8	C	Hospital
Kringlen							
1967	55	25–38	172	4–10	3.0	B	Author
Pollin and associates							
1969	80	14–16	146	4.4	3.5	B	Hospital

[a]R, resident hospital population; C, consecutive hospital admissions; B, birth register.

From W. Pollin et al., "Psychopathology in 15009 pairs of veteran twins: evidence for a genetic factor in the pathogenesis of schizophrenia and its relative absence in psychoneurosis," *Amer. J. Psychiatry* 126:43–56 (1969). Copyright © 1969 by the American Psychiatric Association.

that 85 percent of monozygotic twins studied were discordant for schizophrenia. He concluded that the genetic contribution to the disease was minimal. Yet an average of all twin studies preceding Pollin's gives a concordance for schizophrenia of 61 percent for monozygotic twins and 12 percent for dizygotic twins, suggesting that genetic factors are largely responsible for schizophrenia, although their presence is not always sufficient for the disease to occur. Hard data on the inheritance of schizophrenia are lacking, but it appears that, as with high intelligence, there is an inherited potential for the trait that can come to full expression if the environment encourages it.

A model for viewing the interaction between the genotype and the environment in the production of the schizophrenic phenotype is shown in Figure 3.9. The genotype produces enzymes, which in turn produce metabolites that can induce schizophrenic behavior. In our model, we suggest that environmental stress increases steroid production and that

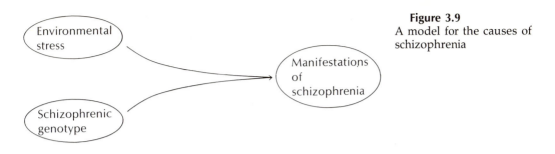

Figure 3.9
A model for the causes of schizophrenia

Schizophrenia is the product of gene–environment interaction

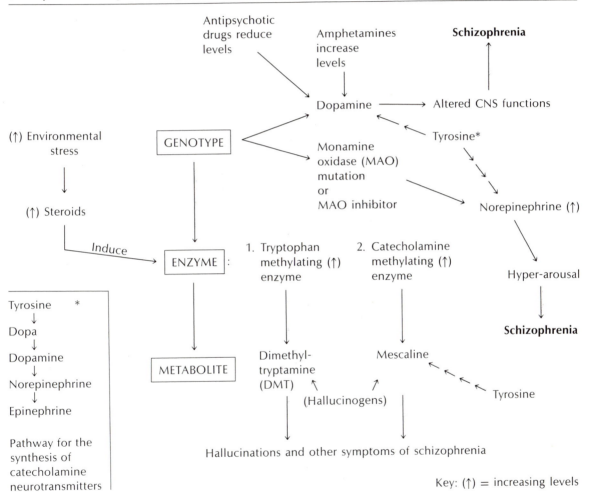

Antipsychotic drugs reduce levels

Amphetamines increase levels

Schizophrenia

Dopamine ⟶ Altered CNS functions

(↑) Environmental stress

GENOTYPE

Tyrosine*

Monamine oxidase (MAO) mutation or MAO inhibitor

(↑) Steroids

Norepinephrine (↑)

Induce

ENZYME :

1. Tryptophan methylating (↑) enzyme

2. Catecholamine methylating (↑) enzyme

Hyper-arousal

Tyrosine   *
↓
Dopa
↓
Dopamine
↓
Norepinephrine
↓
Epinephrine

Schizophrenia

METABOLITE

Dimethyl-tryptamine (DMT) ↑

Mescaline

Tyrosine

(Hallucinogens)

Pathway for the synthesis of catecholamine neurotransmitters

Hallucinations and other symptoms of schizophrenia

Key: (↑) = increasing levels

these steroids induce the synthesis of certain enzymes. The schizophrenic genotype may be more susceptible to stress, or certain genes may be more susceptible to the inducing action of the steroids. Increased levels of methylating enzymes raise the levels of methylated metabolites with hallucinogenic properties, such as mescaline and DMT, and these hallucino-gens can induce schizophrenic symptoms. Or the genotype may produce increased levels of the catecholamine neurotransmitter, dopamine, and this alters the function of the central nervous system to induce a hyperaroused, schizophrenic state. MAO, which normally inactivates the catecholamines, may be mutant. In this case, increased levels of another neurotransmitter (norepinephrine) occur, accompanied by hyperarousal and schizophrenia.

All this suggests that there may be more than one cause and more than one form of schizophrenia and that the environmental input may vary, depending on the nature of the genetic defect. This is an intriguing possibility and may well account for the wide array of phenotypic expressions and the variances in the data. However intriguing the possibilities may be, the genetic basis for schizophrenia is unknown. Some speculate that the disease is caused by a single, autosomal, dominant gene with 25 percent penetrance and variable expressivity. Another suggestion is that schizophrenia is due to a single, autosomal pair of genes exhibiting incomplete dominance, so that the disease is expressed by all homozygous mutants and some heterozygotes. A third theory is that several pairs of genes interact to cause the disease.

Manic-depressive psychosis is characterized generally by cyclical patterns of mania, or exaggerated elation, followed by depression. Available evidence (see Figure 3.7) suggests that the genotype is very important in the disturbance, but we do not yet know how the genotype acts to produce the phenotype.

Subtle environmental influences: Cancer   Cancer is characterized by abnormal cell division and tissue growth. Although cancer in general has no genetic basis, there are specific types that do, and these are usually restricted to specific tissues. It appears that certain combinations of genes create a potential for expressing a specific type of cancer but that the potential may be realized only if the environment is conducive to the expression of the genotype. This means that an individual may be phenotypically healthy yet genetically defective.

Breast cancer and stomach cancer typify cancers that apparently have no strong genetic basis, but the cancer of the colon that is accompanied by intestinal polyposis (numerous polyps, or lumps of tissue projecting into the intestine) appears to be the result of a single dominant mutant gene with almost complete penetrance—that is, the role of the environment is minimal. The pedigree in Figure 3.10 shows that eight members of one family died of cancer of the colon (VII-7, VIII-1, VIII-2, VIII-3, IX-1, IX-8, IX-9, IX-15). Since one of the individuals examined (IX-12) had polyposis but no cancer, it was suspected that perhaps the intestinal polyposis was the predisposing factor for the cancer. Of 51 family members, 6 had polyposis (IX-2, IX-12, IX-14, X-7, X-9, X-29), 39 were normal, and 6 were not examined. Later examination of the 6 expressing polyposis revealed that 2 had developed malignant polyps.

Based on family histories and the dominant nature of this gene, pre-

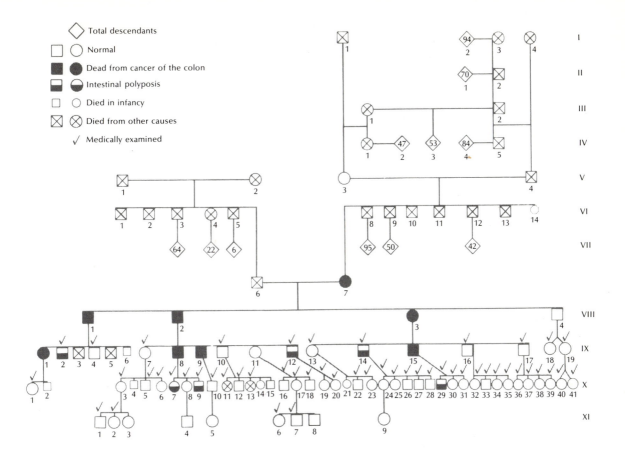

Figure 3.10
Pedigree chart showing cases of intestinal polyposis and cancer of the colon (From Gardner, *Genetics of Cancer*, Utah State University Press, Logan)

dictions were made about the expected inheritance of intestinal polyposis. Every child that expressed polyposis had a parent who had died of cancer of the colon. Half the children who had one parent with cancer of the colon had intestinal polyposis. Normal parents produced all normal offspring. When individuals with polyposis married normal individuals and produced offspring, half the offspring also had polyposis. All this argues for a dominant gene inheritance.

Individual   *Pp*   ×   *pp*   Normal
with polyposis   ↓   individual

Pp and *pp*

P = Dominant gene for polyposis
p = Normal allele

Additional studies of intestinal cancer suggest that over 90 percent of the incidence of this particular form of the disease derives from the genotype and less than 10 percent from the environment. Apparently, the chief role of the environment is in the conversion of nonmalignant polyps to cancerous polyps.

Other cancers also appear to be the result of inherited genes. Xeroderma pigmentosum is a disease caused by a recessive gene, and one of the characteristics of the disease is the appearance of skin tumors. Retinoblastoma, inherited as a simple dominant gene, is a disease characterized by retinal tumors. And the gene for the Zollinger–Werner syndrome causes tumors in various endocrine glands. In all these cases, the environment modifies the expression of the genotype so that perhaps the tumors may not appear, or the severity of the disease may be reduced.

But while the genotype and the environment interact to produce some forms of cancer, other cancers are increasingly attributed to the effects of certain viruses (see Chapter 11). As a matter of fact, the genetically based cancers we have just discussed may well be explained by mutations that result in the failure of specific cells to protect themselves against the debilitating effects of cancer-inducing viruses. It may also be that certain viruses with the potential for inducing cancer are incorporated into a cell's chromosomes where they become new "genes", waiting for the right environmental cue to transform the host cells into unregulatable cancerous cells.

In summary, the environment of a gene is the sum total of all physical and chemical factors external to it. Since most genes induce the production of proteins and since most of these proteins are enzymes whose activity depends on the environment, it seems reasonable to infer that a change in the environment may alter the activity of enzymes so as to produce a mutant phenotype or a variation of one.

Gene–gene interactions

Strictly speaking, not genes but gene products interact. Usually these products are, at least indirectly, proteins whose activity can be modified by the environment. The products of nonallelic genes form part of the general environment and hence can be expected to interact and result in modified Mendelian ratios. Here we shall focus our attention on cases in which the products of one gene pair inhibit, enhance, or otherwise alter the function of products of pairs of nonallelic genes. The interaction between alleles in various dominance relationships has already been discussed, and in the next chapter we shall expand on our theme of gene–gene interaction when we examine the inheritance of quantitative traits.

Bateson and Punnett were among the first to record an example of gene–gene interaction. In 1905, they observed that the inheritance of comb shape in chickens followed typical Mendelian dihybrid ratios, but that two of the F_2 phenotypic classes were novel; that is, they were not found in the parental generation. The combs on the heads of chickens have four possible shapes—pea, rose, single, and walnut (Figure 3.11). When two pure-line single-comb chickens are crossed, all the offspring have single combs. Likewise, pure-line pea-comb chickens, pure-line

Figure 3.11
Comb shapes in poultry,
produced by different
combinations of alleles in
two pairs of genes

Rose

Walnut

Pea

Single

Courtesy of Ralph G. Somes

rose-comb chickens, and pure-line walnut-comb chickens breed true. But a pure-line pea-comb mated with a pure-line single-comb produces all pea-comb F_1 offspring, indicating that the single comb is produced by recessive genes. And a pure-line rose-comb mated with a pure-line single-comb produces all rose-comb F_1 offspring, confirming the recessive nature of the genes responsible for the single comb.

The cross that demonstrates gene–gene interaction is that between a pure-line rose-comb and a pure-line pea-comb, because all the F_1 offspring express a novel phenotype, the walnut comb. When these F_1 walnuts were mated, the F_2 contained another novel phenotype, the single comb. The following data are taken from the original paper by Bateson and Punnett and summarize the crosses discussed so far (s = single, w = walnut, p = pea, r = rose).

133

Female comb type[a]	Male comb type[a]	Offspring comb types			
		s	w	p	r
s	s	131			
p	p			25	
r	r				26
r	s				36
p	s			37	
p	r		36		
F_1 w[b]	F_1 w[b]	17	94	39	27

[a] Homozygous strains unless otherwise indicated.
[b] F_1's from the p x r cross.

Except for the crosses p × r and F_1 w × F_1 w, this experiment suggests a multiple allelic series analogous to that of rabbit coat color discussed earlier. The F_1 w × F_1 w cross is the key, however. Calculation of the phenotypic ratios gives us

	Observed numbers	Expected numbers based on a 9:3:3:1 ratio
Walnut	94	100
Rose	27	33
Pea	39	33
Single	17	11

A statistical test of these data can be made to determine whether they fit the dihybrid hypothesis predicting a 9:3:3:1 phenotypic ratio. One such test is the *chi-square test,* a statistical procedure enabling the investigator to determine how closely an experimentally obtained set of values fits a given theoretical expectation. The techniques for obtaining a chi-square (χ^2) value and a probability are described in Tables 3.2 and 3.3. The χ^2 value of 5.81 corresponds to a probability of 13 percent, which means that on chance alone, we would expect such deviation (departure from expected) in this experiment 13 percent of the time. This is well above an acceptable lower limit of 5 percent, so we conclude that the data of Bateson and Punnett conform to this expected ratio, and we accept the hypothesis of a 9:3:3:1 ratio. The walnut-comb trait is the product of the interaction of two pairs of independently assorting nonallelic genes (pea and rose).

$$\text{(pea) } AAbb \quad \times \quad aaBB \text{ (rose)}$$
$$\downarrow$$
$$F_1 \quad AaBb$$
$$\text{(walnut)}$$
$$AaBb \quad \times \quad AaBb \quad (F_1 \times F_1)$$
$$\downarrow$$

F_2 genotypes
$\begin{cases} 1\ AABB & 1\ AAbb & 1\ aaBB & 1\ aabb \\ 2\ AABb & 2\ Aabb & 2\ aaBb \\ 2\ AaBB \\ 4\ AaBb \end{cases}$

F_2 phenotypes   9 walnut   3 pea   3 rose   1 single

This interaction described involves two gene pairs, both influencing the same trait and assorting independently of each other.

Another trait that demonstrates similar interactions but gives a different ratio is flower color in the sweet pea. As with comb shape, both gene pairs exhibit complete dominance, but when one pair is homozygous recessive, it masks the effects of the other. We say that the homozygous recessive pair is *epistatic* (masking) to the other. Gene pair A produces either purple pigment or no pigment, and gene pair B determines whether the purple pigment will be expressed. Bateson and Punnett discovered this form of epistasis when they crossed two different inbred strains of sweet peas, both with white flowers, and obtained a purple-flowered F_1. Self-fertilizing the F_1 purple-flowered plants produced an F_2 composed of purple and white flowered plants in a 9:7 ratio. Since the common denominator here is 16, the findings suggest a Mendelian pattern for two independently assorting pairs of genes. Homozygous recessiveness for either pair results in a white flower.

P_1 — White $AAbb$ × White $aaBB$ ↓

F_1 — Purple $AaBb$

F_1 $AaBb$ × F_1 $AaBb$ ↓

1	AABB	1	AAbb	1	aaBB	1	aabb
2	AaBB	2	Aabb	2	aaBb		
2	AABb						
4	AaBb						
9	purple			7	white		

A = purple
a = white
B = presence of color
b = absence of color

Table 3.2
A chi-square analysis of Bateson–Punnett data on the inheritance of comb shape in chickens. [The data obtained by Bateson and Punnett fit a 9:3:3:1 ratio. If we were to repeat the experiment, 5 percent or more of the time we would expect to find as great or greater differences.]

Class	Observed (o)	Expected (e)	$(o - e)$	$(o - e)^2$	$\dfrac{(o - e)^2}{e}$
w	94	100	−6	36	0.36
r	27	33	−6	36	1.09
p	39	33	6	36	1.09
s	17	11	6	36	3.27
	177	177	0		5.81

$$\chi^2 = \sum \frac{(o - e)^2}{e} = 5.81$$

Degrees of freedom = 3 = maximum number of variables that can be freely assigned before the rest of the variables are completely determined (when three of the classes are fixed, the fourth is automatically fixed by the population size).

From Table 3.3, for 3 degrees of freedom, our χ^2 value of 5.81 lies between 0.20 and 0.10. Interpolation gives us 0.13. This means that we would expect such deviation from the expected results 13 percent of the time, and since we set our lower limit at 5 percent, the data fit the hypothesis.

Table 3.3
Probability values correlated with chi-square values and degrees of freedom (A = probability for the values of χ^2 given below, df = degrees of freedom)

df	A = 0.99	A = 0.98	A = 0.95	A = 0.90	A = 0.80	A = 0.70	A = 0.50
1	0.00016	0.00063	0.0039	0.016	0.064	0.15	0.46
2	0.02	0.04	0.10	0.21	0.45	0.71	1.39
3	0.12	0.18	0.35	0.58	1.00	1.42	2.37
4	0.30	0.43	0.71	1.06	1.65	2.20	3.36
5	0.55	0.75	1.14	1.61	2.34	3.00	4.35
6	0.87	1.13	1.64	2.20	3.07	3.83	5.35
7	1.24	1.56	2.17	2.83	3.82	4.67	6.35
8	1.65	2.03	2.73	3.49	4.59	5.53	7.34
9	2.09	2.53	3.32	4.17	5.38	6.39	8.34
10	2.56	3.06	3.94	4.86	6.18	7.27	9.34
11	3.05	3.61	4.58	5.58	6.99	8.15	10.34
12	3.57	4.18	5.23	6.30	7.81	9.03	11.34
13	4.11	4.76	5.89	7.04	8.63	9.93	12.34
14	4.66	5.37	6.57	7.79	9.47	10.82	13.34
15	5.23	5.98	7.26	8.55	10.31	11.72	14.34
16	5.81	6.61	7.96	9.31	11.15	12.62	15.34
17	6.41	7.26	8.67	10.08	12.00	13.53	16.34
18	7.02	7.91	9.39	10.86	12.86	14.44	17.34
19	7.63	8.57	10.12	11.65	13.72	15.35	18.34
20	8.26	9.24	10.85	12.44	14.58	16.27	19.34

df	A = 0.30	A = 0.20	A = 0.10	A = 0.05	A = 0.02	A = 0.01	A = 0.001
1	1.07	1.64	2.71	3.84	5.41	6.64	10.83
2	2.41	3.22	4.60	5.99	7.82	9.21	13.82
3	3.66	4.64	6.25	7.82	9.84	11.34	16.27
4	4.88	5.99	7.78	9.49	11.67	13.28	18.46
5	6.06	7.29	9.24	11.07	13.39	15.09	20.52
6	7.23	8.56	10.64	12.59	15.03	16.81	22.46
7	8.38	9.80	12.02	14.07	16.62	18.48	24.32
8	9.52	11.03	13.36	15.51	18.17	20.09	26.12
9	10.66	12.24	14.68	16.92	19.68	21.67	27.88
10	11.78	13.44	15.99	18.31	21.16	23.21	29.59
11	12.90	14.63	17.28	19.68	22.62	24.72	31.26
12	14.01	15.81	18.55	21.03	24.05	26.22	32.91
13	15.12	16.98	19.81	22.36	25.47	27.69	34.53
14	16.22	18.15	21.06	23.68	26.87	29.14	36.12
15	17.32	19.31	22.31	25.00	28.26	30.58	37.70
16	18.42	20.46	23.54	26.30	29.63	32.00	39.25
17	19.51	21.62	24.77	27.59	31.00	33.41	40.79
18	20.60	22.76	25.99	28.87	32.35	34.80	42.31
19	21.69	23.90	27.20	30.14	33.69	36.19	43.82
20	22.78	25.04	28.41	31.41	35.02	37.57	45.32

Abridged from R. A. Fisher and F. Yates, *Statistical Tables for Biological, Agricultural and Medical Research*, 6th edition, Harlow, Essex: Longman, 1974, by permission of Longman Group Limited.

There are many other examples of two different independently assorting genetic loci being involved with a common phenotype. For example, we observe the 9:7 ratio among the F_2 of *Drosophila* flies with scarlet eyes (*st*) crossed to flies with vermilion eyes (*v*). Both scarlet and vermilion eyes are bright red, as opposed to the more orangeish-red of wild types.

Genotype	$stst\ v^+v^+$	×	$st^+st^+\ vv$
Phenotype	(scarlet)	↓	(vermilion)
F_1 genotype		$st^+st\ v^+v$	
F_1 phenotype		(wild type)	

$$F_1 \times F_1$$
$$\downarrow$$

9	wild type	$st^+(st\ \text{or}\ st^+)\ v^+(v^+\ \text{or}\ v)$	= 9
		$\begin{cases} stst\ v^+(v\ \text{or}\ v^+) \\ st^+(st^+\ \text{or}\ st)\ vv \\ stst\ vv \end{cases}$	$\begin{matrix} = 3 \\ = 3 \\ = 1 \end{matrix}$
7	bright red		

Both genetic loci are involved in different parts of the pathway for the synthesis of the wild-type eye pigments, so a mutation at either locus blocks the synthesis of the pigments required for wild-type eye color.

Another interesting example of gene–gene interaction involves coat color in mice. An albino strain (*cc*) is crossed to a strain with black fur (*aa*), and the F_1 expresses a new phenotype called agouti (Figure 3.12). In the F_1 × F_1 cross, we get an F_2 ratio of 9 agouti, 3 black, and 4 albino. Here, too, we see that there is independent assortment between two genetic loci, along with the interaction of these loci in the production of a specific phenotype. Shortly we shall examine some models that might help us interpret some of these instances of gene–gene interaction.

Figure 3.12
Pigment patterns in agouti, albino, and black fur

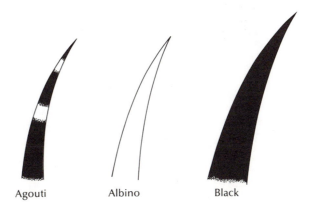

Agouti   Albino   Black

Figure 3.13
Photographs showing
different ways in which
the tumorous-head
phenotype is expressed.
Arrows indicate abnormal
growths. [From E. J.
Gardner, *Adv. Genetics*
15:116–146 (1970)]

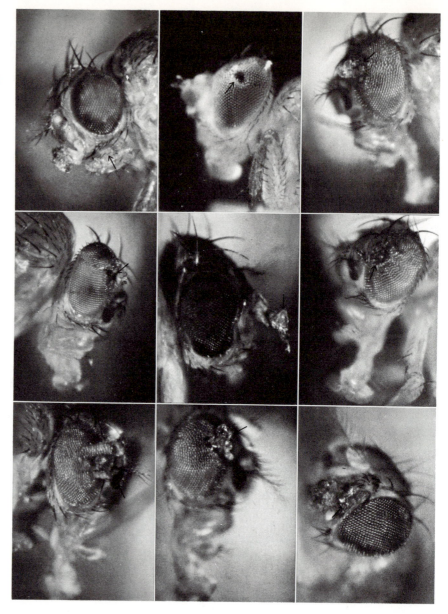

An interesting example of both gene–gene and gene–environment
interaction is the inheritance of a tumorous-head growth in *Drosophila*
(Figure 3.13). The tumorous-head (*tu-h*) phenotype is a function of the
interaction of two major gene pairs:

tu-1 on the **X** chromosome (*tu-1*$^+$ = normal allele)

tu-3 on the third chromosome (*tu-3*$^+$ = normal allele)

In a highly inbred *tu-h* stock (homozygous for *tu-1* and *tu-3*), the expression of the *tu-h* phenotype (the position and size of the tumor) may vary, but over 90 percent of the flies will express it in some form. Penetrance is thus over 90 percent with variable expressivity. The inheritance pattern of *tu-h* is seen in the following reciprocal cross.

1.

90-percent penetrance $\dfrac{tu\text{-}1}{tu\text{-}1} \quad \dfrac{tu\text{-}3}{tu\text{-}3}$ $\female \times \male$ $\dfrac{tu\text{-}1^+}{\quad} \quad \dfrac{tu\text{-}3^+}{tu\text{-}3^+} \longrightarrow$ 0-percent penetrance

$\downarrow$

F_1 $\dfrac{tu\text{-}1}{tu\text{-}1^+} \quad \dfrac{tu\text{-}3}{tu\text{-}3^+}$ $\female$ and $\male$ $\dfrac{tu\text{-}1}{\quad} \quad \dfrac{tu\text{-}3}{tu\text{-}3^+} \longrightarrow$ 40-percent penetrance expressed in both males and females

2.

0-percent penetrance $\dfrac{tu\text{-}1^+}{tu\text{-}1^+} \quad \dfrac{tu\text{-}3^+}{tu\text{-}3^+}$ $\female \times \male$ $\dfrac{tu\text{-}1}{\quad} \quad \dfrac{tu\text{-}3}{tu\text{-}3} \longrightarrow$ 90-percent penetrance

$\downarrow$

F_1 $\dfrac{tu\text{-}1}{tu\text{-}1^+} \quad \dfrac{tu\text{-}3}{tu\text{-}3^+}$ $\female$ and $\male$ $\dfrac{tu\text{-}1^+}{\quad} \quad \dfrac{tu\text{-}3}{tu\text{-}3^+} \longrightarrow$ 0 to 2-percent penetrance expressed in both males and females

These reciprocal crosses of *tu-h* with normal flies point out the following:

1. For the maximum expression of *tu-h* in the F_1, the maternal parent must be homozygous for *tu-1* and *tu-3*. However, *tu-3* alone also expresses a *tu-h* phenotype when homozygous (not shown) or when heterozygous, though at a very low frequency. Studies of *tu-h* suggest that *tu-3* is an incompletely dominant major gene and that *tu-1* is either a minor gene or a special kind of modifier gene whose activity enhances the expression of the *tu-3* gene. The optimum environment restricts the maximum expression of *tu-h* to just over 90 percent.

2. The unequal results from the reciprocal crosses suggest a cytoplasmic gene product in the female egg cytoplasm that is controlled by *tu-1*. The maternal effect is an environmental one that is contributed by the female parent and maximizes the percentage of *tu-3* homozygotes and heterozygotes expressing the *tu-h* phenotype.

Possible mechanisms of gene–gene interaction   This discussion of gene–gene interaction has been oriented to specific phenotypes and the genes governing their expression. But genes do not produce phenotypes directly; most often, they indirectly produce proteins that have structural or enzymatic functions. Interpreting gene–gene interaction is possible only within the molecular context of gene-product interactions. From this perspective, we can speculate on the type of gene-product interaction that might result in such phenotypes as comb shape and flower color in sweet peas.

We can hypothesize that comb shape is determined by a protein composed of two different chains of amino acids, the basic building blocks of proteins. We can then hypothesize a summary of the activity of each gene involved in the phenotype as follows.

Gene $A \longrightarrow$ Amino acid α-chain ——————
Gene $a \longrightarrow$ No functional protein
Gene $B \longrightarrow$ Amino acid β-chain –––––––––
Gene $b \longrightarrow$ No functional protein

AABB, AaBB, AABb, AaBb ======== = Walnut
aaBB, aaBb ======== = Rose
Aabb, AAbb ========= = Pea
aabb no protein = Single

There is no evidence to support the validity of a model of this type to the shape of the comb, but the model helps us to visualize how gene products might interact.

The inheritance of flower color in sweet peas can be visualized, again speculatively, as involving two gene products—one lays a foundation on which pigment produced by the other is deposited (or a protein that the pigment must interact with).

Gene $A \longrightarrow$ An enzyme that synthesizes purple pigment
Gene $a \longrightarrow$ No enzyme produced, therefore no pigment
Gene $B \longrightarrow$ A protein that forms a foundation for the
deposition of pigment
Gene $b \longrightarrow$ No protein, therefore no foundation

The results could then be

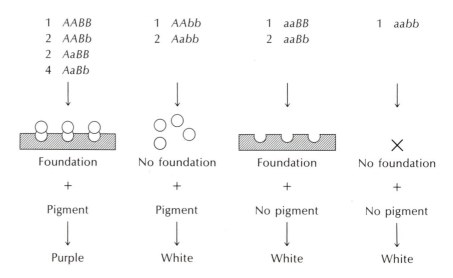

1 AABB	1 AAbb	1 aaBB	1 aabb
2 AABb	2 Aabb	2 aaBb	
2 AaBB			
4 AaBb			

Foundation | No foundation | Foundation | No foundation
+ | + | + | +
Pigment | Pigment | No pigment | No pigment

Purple | White | White | White

An alternative proposal that might explain the interaction of two gene pairs in forming the purple pigment (or the *Drosophila* eye pigment) involves a sequence of reactions. Substrate X is converted to Y by enzyme A, and Y is converted to Z by enzyme B. Enzyme A is coded for by gene *A*, with gene *a* producing a nonfunctional enzyme. Enzyme B is coded for by gene *B*, with gene *b* producing a nonfunctional enzyme. Homozygous recessiveness at either gene locus stops the sequence and causes white color:

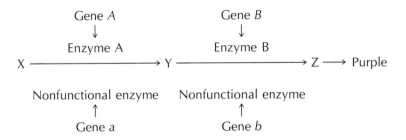

We can similarly construct a reasonable model to explain the inheritance of coat color in mice. The albino (*c*) mutation represents a defect in the biosynthetic pathway for melanin pigment. The black (*b*) mutation specifies that pigment be dispersed throughout the length of the hair. And the dominant allele (*b⁺*) specifies the agouti pattern.

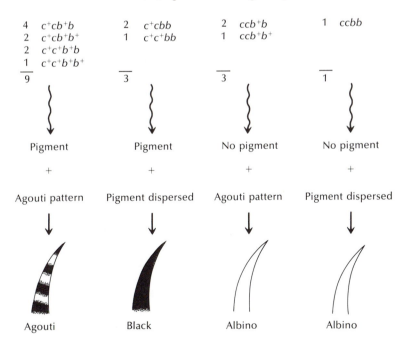

The inheritance of tumorous head is more complicated than the preceding examples. An explanatory model is more difficult, but not impos-

Figure 3.14
Model for the interaction
of two gene pairs to
produce the
tumorous-head
phenotype

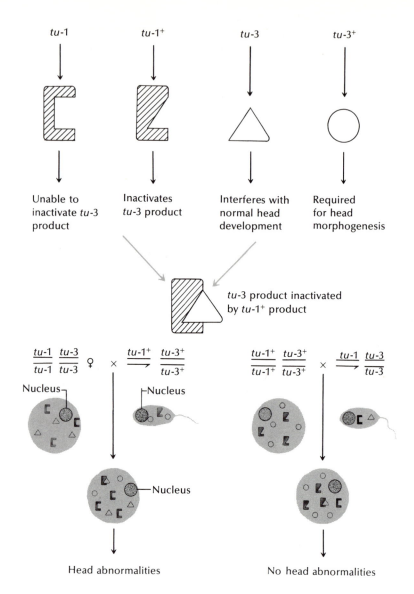

sible, to construct (Figure 3.14). To begin with, we can suggest that during
early embryogenesis, the mutant *tu-3* gene produces a product that interferes with normal development in the head of *Drosophila*. Even a little of
this product can occasionally produce a *tu-h* phenotype, as in the case of
the *tu-3/tu-3+* heterozygote. The *tu-1+* gene produces a product that
blocks the action of the *tu-3* gene product, but the *tu-1+* gene is active *only*
during the formation of gametes. Since the egg contains hundreds of
times as much cytoplasm as the sperm, it can store more *tu-1+* gene
product. Sperm cannot store it.

When *tu-1*, *tu-3* homozygous females are mated to normal males, the
absence of *tu-1+* gene product in egg cytoplasm, or even of the minuscule

amount found in sperm cytoplasm, can result in the expression of the *tu-3* gene. Hence a *tu-h* phenotype is expressed in about 40 percent of the F_1 flies. The reciprocal cross, between a normal female and a *tu-1*, *tu-3* homozygous male, produces essentially no *tu-h* F_1 offspring, because the *tu-1*$^+$ gene product in the egg cytoplasm neutralizes the activity of the *tu-3* gene product.

In summary, before gene function was understood, the term "gene–gene interaction" was quite proper in making observations like those of Bateson and Punnett involving comb shape in chickens and sweet pea flower color. But when it became known that a gene commonly codes for a polypeptide (a chain of amino acids), "gene–gene interaction" had to be reinterpreted in terms of the interaction of gene products or functions.

Many enzymes are known to contain two or more different polypeptide chains; in other words, two or more genes contributed to the final enzyme structure. That enzyme may be specific for only one reaction, but the reaction (a chemical phenotype, if you wish) is clearly dependent on more than one gene pair.

Continuing on this same line, the discovery of gene–gene interaction dispelled the naive notion that each gene produced a single, nonoverlapping individual effect and that all these effects fitted together like a mosaic to produce the organism. This interpretation of gene action might be summarized as "one gene, one character." The contemporary view is that the development of an organism results from the interaction of gene products—with each other and with the environment—which thereby produce the final phenotype. The former viewpoint can be called *preformationism;* the latter, *epigenesis.* Figure 3.15 summarizes the forces that interact to form a phenotype.

Figure 3.15
Environmental and genetic influences on the phenotype of offspring

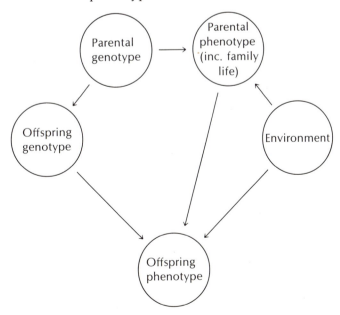

Alterations of the basic Mendelian 9:3:3:1 ratio occur through a variety of means. We have already discussed the way the environment interacts with the genotype in the production of a phenotype. We also discussed the interaction of different genetic loci in the emergence of a phenotype. We also want to mention certain other factors that can modify a Mendelian ratio.

Lethal genes can modify a Mendelian ratio. Consider the case of yellow body color in mice, first reported by Cuenot in 1905. Cuenot found that yellow body color is determined by a dominant gene (Y), but the gene could not be made homozygous. In all *yellow* × *yellow* crosses, a ratio of 2 yellow mice to 1 agouti (normal) results. It was later discovered that yellow homozygotes are indeed formed, but they die early in embryonic development. The Y gene is thus dominant for yellow coat color but recessive for lethality.

$$Yy \quad \times \quad Yy$$
$$\downarrow$$
$$1\ YY \qquad 2\ Yy \qquad 1\ yy$$

$$\text{(dies)} \quad \text{(yellow)} \quad \text{(agouti)}$$

The genes Cy, Pm, D, and Sb, discussed earlier, are similarly described. The DTS-L genes, discussed in connection with dramatic environmental influences on a genotype, are quite different from Y, because they are dominant lethals as opposed to recessive lethals. Since Cuenot's initial discovery, lethal genes have been discovered throughout the prokaryotic and eukaryotic worlds.

Mendelian ratios can also be seriously modified when one or more classes of gametes are specifically inhibited from fully participating in the formation of zygotes. We call this phenomenon *meiotic drive*, a term coined by L. Sandler and E. Novitski. In *Drosophila*, there is a gene called segregation distorter (SD), a dominant second-chromosome gene. Gametes carrying the SD allele (along with eye-color marker genes) fertilize more eggs than gametes carrying the SD^+ allele:

$$bw = \text{Brown eyes}$$
$$bw^+ = \text{Normal}$$
$$cn = \text{Cinnabar eyes}$$
$$cn^+ = \text{Normal}$$

The *bw cn* homozygote has white eyes (an example of gene–gene interaction).

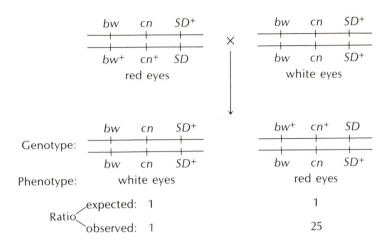

In addition to the effect that *SD* has on the recovery of *SD*-carrying second chromosomes, it also has a marked effect on the recovery of sex chromosomes, as demonstrated by Denell, Judd, and Richardson in 1969.

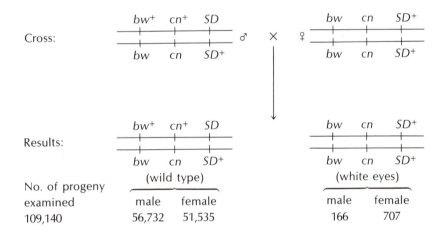

No. of progeny examined	male	female	male	female
109,140	56,732	51,535	166	707

Two striking features are evident in these data: first, the preponderance of wild-type individuals (99.2 percent); and second, the distorted sex ratio of the white-eyed flies (81 percent female).

These results can be interpreted in the following fashion. At the first meiotic division, the sex chromosomes segregate at random with respect to the other chromosomes, but because of the action of *SD*, some or all of the sperm that carry SD^+ do not function in fertilization. The degree to which the SD^+-bearing sperm are rendered nonfunctional is affected by the sex chromosome carried by that gamete. For the *SD*-bearing sperm, the sex chromosome constitution has no effect on the functionality of sperm:

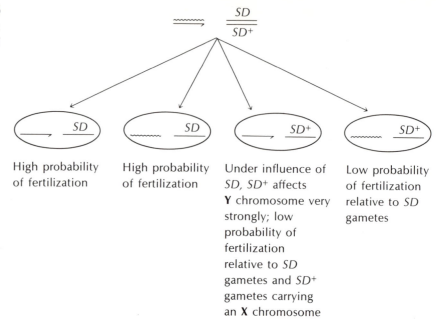

High probability of fertilization	High probability of fertilization	Under influence of *SD*, *SD+* affects **Y** chromosome very strongly; low probability of fertilization relative to *SD* gametes and *SD+* gametes carrying an **X** chromosome	Low probability of fertilization relative to *SD* gametes

The molecular basis for this interaction is not yet known, but the interaction's effect on sperm functioning is readily apparent.

Segregation distortion is also evident in male mice carrying certain of the tailless (*t*) alleles. A male mouse heterozygous for *t* sometimes transmits the *t* allele to as many as 99 percent of the offspring. Evidently, some of the *t* alleles enhance the capacity of those sperm to effect fertilization, compared with those carrying the *T* allele.

Segregation distortion could be a significant force in evolution since even the most adaptively useful genes would be eliminated if the gamete carrying them were made nonfunctional by the influence of segregation distortion. However, we lack experimental evidence showing that segregation distortion is a major evolutionary force in natural populations.

Overview

This chapter developed some of the many ideas that grew out of the fertile ground prepared by Mendel. We first examined the important phenomenon of dominance. Bateson attempted to characterize dominant–recessive relationships in terms of the presence or absence of "factors." But the discovery of multiple alleles led to the rejection of Bateson's idea and cleared the way for a new understanding of the genetic locus: the complex locus concept. As attention focused on gene function, the concept of dominance was interpreted in functional terms. We discussed ways of interpreting dominant, recessive, incompletely dominant, and codominant relationships, and we did so within the context of allelic interaction models.

In a strict sense, allelic interaction leading to various dominance relationships is a modifying type of interaction. In other words, one allele can modify the effects of the other. We continued with this basic idea by examining the plasticity of the genotype as it interacts with the environment. The environment can have a dramatic effect on the genotype, as is the case with temperature-sensitive lethals and phenocopies. Or the environment can operate in subtle ways with the genotype in the production of such complex phenotypes as intelligence, schizophrenia, and cancer. For the analysis of these subtle gene–environment interactions in humans, twin studies have been exceptionally valuable. However, as we pointed out, twin studies must be used with caution.

Another factor influencing gene expression is other genetic loci. Gene–gene interaction (we could more properly call it gene product–gene product interaction) was first reported by Bateson when he was studying the inheritance of comb shape in poultry. We saw how one genetic locus could modify the expression of another genetic locus, an occurrence that frequently leads to altered Mendelian ratios. We developed some models for interpreting this situation.

In subsequent chapters, we will discuss the modification of genotype expression again, though from a different perspective. We shall examine the inheritance of quantitative traits, cytoplasmic influences, mutations, and gene regulation.

Questions and problems

3.1 In this chapter, we developed models for interpreting dominance and recessiveness in functional terms. Develop a reasonable model for the interpretation of codominance (use the A, B, and AB blood groups as your example).

3.2 Distinguish between the terms *gene* and *pseudoallele*.

3.3 A true-breeding strain of brick-red mink, when crossed to a true-breeding strain of white mink, produced an all white F_1. The total F_2 progeny from several $F_1 \times F_1$ matings was: 237 white, 63 black, and 21 brick-red. Propose a reasonable explanation for this observation.

3.4 How would you distinguish between F_2 phenotypic ratios of 9:7 and 9:6:1 when the homozygous double recessive of the 9:6:1 is lethal, thus resulting in a 9:6 ratio?

3.5 A rare disease has been hypothesized as caused by a dominant gene with low penetrance. How would you validate this hypothesis, as opposed to a hypothesis of a recessive gene with very high penetrance?

3.6 Distinguish between the terms *epistasis* and *dominance*.

3.7 There are three different homozygous strains of moth: one lays red eggs, one lays green eggs, and one lays white eggs. Red and white are determined by genes at different loci. White is epistatic to red. The following cross is performed.

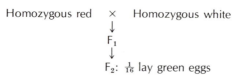

How would you explain the results? What are the F_1 and F_2 phenotypes and genotypes?

3.8 The normal *Drosophila* body color is tan, but there is a sex-linked recessive gene (y) that causes yellow body color. The same phenotype can be induced by feeding the *Drosophila* larvae silver nitrate. What does this say about the influence of the genotype and the environment on the body-color phenotype?

3.9 How would you resolve the paradoxical situation of monozygotic twins consisting of one male and one female?

3.10 How did the discovery of multiple alleles contribute to our interpretation of the gene?

3.11 In chickens, feather color is determined by two different genetic loci. Two different homozygous strains of chickens with white feathers are crossed, and they produce an F_2 composed of 125 white-feathered birds and 35 brown-feathered birds. Use chi-square analysis to test your hypothesis of a modified Mendelian ratio. Explain the pattern of inheritance, the F_1 and F_2 genotypes, and the F_1 phenotype.

3.12 In summer squash, spheroid fruit genes are dominant over genes for long fruits. Two different homozygous spheroid-fruited varieties are crossed and produce the following F_2.

89 disc
62 spheroid
11 long

Use chi-square analysis to ascertain the fit of this data to a reasonable modified dihybrid ratio. Explain the pattern of inheritance, the F_1 genotype and phenotype, and the F_2 genotypes.

3.13 The direction of hair growth in guinea pigs is determined by two different genetic loci. A homozygous rough-coated animal is crossed to a homozygous smooth-coated animal. The F_2 consists of 93 smooth coated, 96 intermediate rough-coated, and 52 rough-coated animals. Use chi-square analysis to ascertain the fit of this data to a reasonable modified-dihybrid ratio. Explain the pattern of inheritance, the F_1 genotype and phenotype, and the F_2 genotypes.

3.14 In swordtails, a popular aquarium fish, the "Montezuma" variety is bright orange-red with black spots. The wild-type variety is olive-green with small patches of black and yellow. In a Montezuma × Montezuma cross, about $\frac{1}{3}$ of the offspring are wild type and the remaining $\frac{2}{3}$ Montezuma. The Montezuma strain cannot be made to breed true. How do you explain this?

3.15 A brilliant mathematician is pregnant. Wanting her child to be proficient in the mathematical sciences, she spends several hours a day solving math problems. A few years after birth, the child displays a striking talent for solving relatively complex math problems. Does this prove that the mother's activities during her pregnancy influenced the development of the child's mathematical skills? What if the story were changed so that the mother was a musician and the child developed a striking talent for music?

3.16 You are given an *Aa* heterozygote. By using examples, discuss the possibility of one allele affecting the penetrance and expressivity of the other allele.

3.17 In any cross involving heterozygotes for multiple alleles, what will the phenotypic ratio(s) be?

3.18 An albino dog is crossed to a brown dog. The F_1 puppies are all black. $F_1 \times F_1$ crosses produce 19 black puppies, 8 white puppies, and 7 brown puppies. From these data, determine the genetic basis of coat color in this breed of dogs.

3.19 Two different inbred strains of evening primrose (*Primula*) lack a compound called malvidin. When these two strains are crossed, the F_1 does not have malvidin. The $F_1 \times F_1$ cross produces 65 plants that lack the compound and 14 that have it. Offer an interpretation of the genetic basis of malvidin production.

3.20 A green-striped eagle is crossed to a solid yellow-colored eagle. The progeny eagles are solid green or solid yellow in a 1:1 ratio. When the F_1 solid-green eagles are crossed to each other, solid-green, solid-yellow, green-striped and yellow-striped eaglets are produced in a 6:3:2:1 ratio. How do you explain this?

References

Bateson, W. 1906. The progress of genetics since the rediscovery of Mendel's papers. In *Progressus rei Botanicae, Association Internationale des Botanites*, J. P. Lotsy (ed.). Jena, East Germany, G. Fisher.

Carlson, E. A. 1959. The comparative genetics of complex loci. *Quart. Rev. Biol.* 34:33–67.

Carlson, E. A. 1966. *The Gene: A Critical History*. Philadelphia, Saunders.

Cattell, R. B. 1965. *The Scientific Analysis of Personality*. London, Penguin.

Corwin, H. O., and J. B. Jenkins. 1976. *Conceptual Foundations of Genetics*. Boston, Houghton Mifflin.

Cuénot, L. 1903. L'Hérédité de la pigmentation chez les souris. *Arch. Zool. Exper. et Gen.* 1 (4th S.):33–41.

DeFries, J. C., S. G. Vandenberg, and G. E. McClearn. 1976. Genetics of specific cognitive abilities. *Ann. Rev. Genetics* 10:179–208.

Denell, R. E., B. H. Judd, and R. H. Richardson. 1969. Distorted sex ratios due to segregation distorter in *Drosophila melanogaster*. *Genetics* 61:129–139.

Galton, F. 1869. Hereditary Genius: An Inquiry into Its Laws and Consequences. New York, Macmillan.

Gottesman, I. I., and J. Shields. 1972. *Schizophrenia and Genetics. A Twin Study Vantage Point*. New York, Academic.

Heston, L. L. 1970. The genetics of schizophrenic and schizoid disease. *Science* 167:249–256.

Jensen, A. R., 1969. How much can we boost I.Q. and scholastic achievement? *Harvard Educ. Rev.* 39:1–123.

Kagan, J. S., J. M. Hunt, J. F. Crow, C. Bereiter, D. Elkind, L. J. Cronbach, and W. F. Brazziel. 1969. How much can we boost I.Q. and scholastic achievement? A discussion. *Harvard Educ. Rev.* 39:273–356.

Kallman, F. J. 1953. *Heredity in Health and Mental Disorder.* New York, Norton.

Levine, L. 1971. *Papers on Genetics, A Book of Readings.* St. Louis, Mo., Mosby.

Levitan, M., and A. Montagu. 1977. *Textbook of Human Genetics.* New York, Oxford University Press.

Lewis, E. B. 1942. The star and asteroid loci in *Drosophila melanogaster. Genetics* 27:153–154.

Lewis, E. B. 1948. Pseudoallelism in *Drosophila melanogaster. Genetics* 33:113.

Lewontin, R. C. 1975. Genetic aspects of intelligence. *Ann. Rev. Genetics* 9:387–406.

McClearn, G. E., and J. C. DeFries. 1973. *Introduction to Behavioral Genetics.* Freeman, San Francisco.

Moody, P. A. 1975. *Genetics of Man.* New York, Norton.

Oliver, C. P. 1940. A reversion to wild-type associated with crossing over in *Drosophila melanogaster. Proc. Nat. Acad. Sci. U.S.* 26:452–454.

Peacock, W. J., and G. L. G. Miklos. 1973. Meiotic drive in Drosophila: New interpretations of the segregation distorter and sex chromosome systems. *Adv. Genetics* 17:361–410.

Peters, J. A. 1959. *Classic Papers in Genetics.* Englewood Cliffs, N.J., Prentice-Hall.

Punnett, R. C. 1911. *Mendelism.* New York, Macmillan.

Simpson, N. E. 1968. Diabetes in the family of diabetics. *Canad. Med. Assoc. J.* 98:427–432.

Snyder, S. H., S. P. Banerjee, H. I. Yamamura, and D. Greenberg. 1974. Drugs, neurotransmitters, and schizophrenia. *Science* 184:1243–1253.

Stern, C. 1973. *Principles of Human Genetics.* San Francisco, Freeman.

Sturtevant, A. H. 1913. The Himalayan rabbit case, with some considerations on multiple allelomorphs. *Amer. Nat.* 47:234–238.

Sutton, H. E. 1975. *An Introduction to Human Genetics.* New York, Holt.

Suzuki, D. T., and D. Procunier. 1969. Temperature sensitive mutations in *Drosophila melanogaster:* III. Dominant lethals and semilethals on chromosome 2. *Proc. Natl. Acad. Sci. U.S.* 62:369–376.

Thompson, J. S., and M. W. Thompson. 1973. *Genetics in Medicine.* Philadelphia, Saunders.

Woolf, C. M., and R. M. Woolf. 1970. A genetic study of polydactyly in Utah. *Amer. J. Hum. Genetics* 22:75–88.

Zimmerling, S., L. Sandler, and B. Nicoletti. 1970. Mechanisms of meiotic drive. *Ann. Rev. Genetics* 4:409–436.

Four

The expansion of Mendelian principles, II

Exceptions to Mendelian principles of inheritance are not necessarily contradictions. Indeed, interpreted within the framework of the chromosome theory of inheritance, the exceptions serve to strengthen Mendelism. Linkage, recombination, gene mapping, multiple alleles, and environmental and genetic interactions are examples of concepts that emerged from Mendelism as exceptions, only to strengthen and broaden Mendelian principles.

In this chapter, we shall examine two other concepts that developed as exceptions to Mendelism. The first is *quantitative inheritance*. Mendel's experiments produced the principles of segregation and independent assortment through studies with traits that existed in contrasting forms (tall versus dwarf, smooth versus wrinkled). But evolutionary changes were not so discontinuous. By contrast, evolutionary changes were gradual and continuous (see Chapter 1). Thus, in its formative years, Mendelism was opposed by such biometricians as Galton, W. F. R. Weldon, and Karl Pearson, who argued for blending models of inheritance that would produce the continuous inheritance patterns so characteristic of the evolutionary process. The work of Johannsen, Nilsson-Ehle, and East showed that Mendelism could explain continuous variation, and the great rift between the biometricians and the Mendelians was resolved to almost everyone's satisfaction. The two fields thereafter progressed together.

The second concept that emerged from noted exceptions to Mendelism is called *extranuclear* or *cytoplasmic inheritance*. This concept developed to explain inherited traits that did not follow Mendelian patterns.

Quantitative inheritance

The seven traits Mendel analyzed in garden peas were all *discontinuous*. That is, all the plants he produced were tall *or* short, had seeds that were smooth *or* wrinkled, yellow *or* green, and so on. But it had long been known that not all traits behave this way. A century earlier, studying height in the tobacco plant, Kölreuter found that a cross of pure-line tall and dwarf plants produced an F_1 that was intermediate in height and an F_2 (from self-pollinated F_1 plants) that ranged from tall to dwarf, with most plants intermediate. In peas, by contrast, Mendel reported a discontinuous F_1, all dominant in phenotype, and a discontinuous F_2, three tall to one dwarf.

Galton (1869) and Karl Pearson (1904) also studied the inheritance of characters exhibiting continuous variation. They showed, for example, that human stature is to a large extent genetically controlled. Galton suggested that those traits showing a continuous variation were ascribable to a pattern of blending inheritance. That is, a child receives one-half of its inheritance from each of its two parents, one-fourth from each of its four grandparents, and so on. Continuing backward indefinitely, this

Figure 4.1
Law of ancestral
inheritance

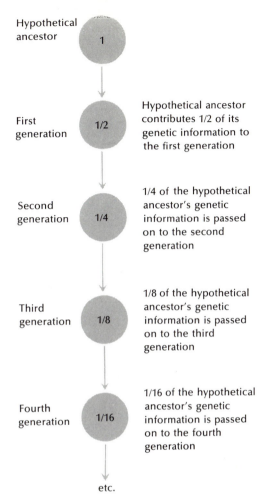

Hypothetical ancestor	**1**	
First generation	**1/2**	Hypothetical ancestor contributes 1/2 of its genetic information to the first generation
Second generation	**1/4**	1/4 of the hypothetical ancestor's genetic information is passed on to the second generation
Third generation	**1/8**	1/8 of the hypothetical ancestor's genetic information is passed on to the third generation
Fourth generation	**1/16**	1/16 of the hypothetical ancestor's genetic information is passed on to the fourth generation

etc.

Each generation, no matter how far removed from a hypothetical ancestor, passes on a specific proportion of that ancestor's genetic information.

would account for the total inheritance of the child. Galton called this the *Law of Ancestral Inheritance* (Figure 4.1).

The advent of Mendelism gave rise to two interpretations of the inheritance of characters showing continuous variation. On first glance, they might appear to be contradictory.

The multiple-gene hypothesis How was the difference in inheritance patterns for the same trait in different organisms to be explained? The biometricians, including Galton and Pearson, held that all differences consisted in series of small—but continuous and measurable—steps brought about by a blending type of

153

Figure 4.2
Johannsen's analysis of
Phaseolus (From E. A.
Carlson, *The Gene: A
Critical History,*
Philadelphia, W. B.
Saunders Company, 1966)

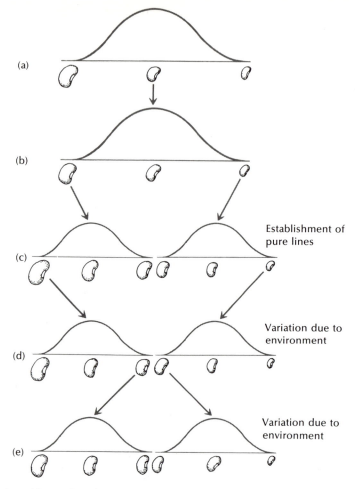

(a)

(b)

Establishment of
pure lines

(c)

Variation due to
environment

(d)

Variation due to
environment

(e)

An unselected bean of medium size (a) repeats the normal distribution, as shown in (b).
If either large or small beans are selected and inbred for several generations, pure lines
are established (c). When the largest bean from the small line or the smallest bean from
the large line is selected (d), the distribution remains normal (e).

inheritance. They pointed out that not all Mendel's tall and short pea
plants were equally tall or equally dwarf. Instead, they ranged along a
continuum from tall–short to short–short and tall–tall to short–tall. These
men also noted variations in the degree of smoothness or wrinkling in the
seed coats of the peas. They therefore suggested that Mendel's so-called
contrasting traits could be interpreted as either quantitative or continuous.

Another group of geneticists, led by Bateson, argued that all inherited traits were interpretable within the framework of Mendel's princi-

ples, but that continuous traits might be determined by a large number of Mendelian genes. This is called the *multiple-gene hypothesis*. Bateson suggested that the variation observed in Mendel's contrasting traits could be attributable to subtle environmental differences.

Until 1909, there was much confusion about the general applicability of Mendel's principles to continuous patterns of inheritance. In that year, W. Johannsen (pictured next to Bateson in Figure 2.12) drew a clear distinction between genotype and phenotype, and H. Nilsson-Ehle obtained evidence supporting the multiple-gene hypothesis as an explanation of the inheritance of quantitative traits.

Johannsen, motivated largely by Galton's theory of heredity, demonstrated that in a pure line of the broad bean (*Phaseolus*), variations in bean size could be attibuted exclusively to environmental influences (Figure 4.2). Under normal field conditions, the beans of *Phaseolus* hybrids vary in weight from 150 to 750 mg, ranging along a quantitative continuum. If either large or small beans are selected and inbred for several generations, pure lines are established. Yet these pure lines still express variability, and further selection from within them fails to alter the phenotypic curve, whether the largest bean from the small line is selected or the smallest bean from the large line (Figure 4.2). Thus it was shown that the environment can, within specific limits, influence the expression of a trait. In its broader application, Johannsen's work led to the realization that a phenotype has two components—one genotypic, the other environmental. In fact, Johannsen coined the terms *phenotype, genotype,* and *gene* as a result of his pure-line studies.

If Johannsen's views helped clarify the relations between heredity and environment, Richard Woltereck's studies of variations and inheritance in small freshwater plankton crustaceans called daphnia clarified matters even further. Woltereck argued that the amount of variation that could occur within pure lines was limited by the genotype as it interacted with the environment. This limited range of variation is the *norm of reaction* (see Chapter 3), and it does not change unless a genetic mutation occurs. If this idea had been more generally appreciated, the growth of evolutionary biology might have been more rapid. The studies of Johannsen and Woltereck were important in the further analysis of quantitative inheritance. They focused attention on the inheritance of quantitative traits and developed biometrical methods for analyzing quantitative traits.

Confirmation of the multiple-gene hypothesis

That more than one pair of genes could influence a given trait was shown independently by H. Nilsson-Ehle in 1909 and E. M. East in 1910, further strengthening Bateson's position of multiple genes (or polygenes) and opening the way to a Mendelian interpretation of continuous variation. Nilsson-Ehle proposed that three pairs of genes influenced grain color in

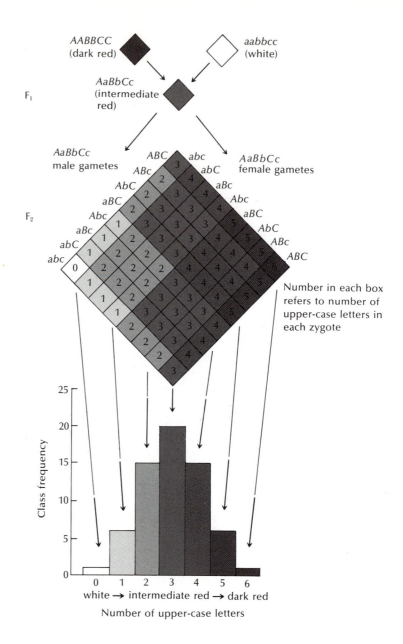

Figure 4.3
Nilsson-Ehle's analysis of the inheritance of grain color in wheat, demonstrating polygenic inheritance

wheat. When a white-grained wheat was crossed with a dark-red-grained variety, the F_1 was intermediate (Figure 4.3). Self-fertilization of the F_1 produced an F_2 phenotypic ratio of 1 dark red:6 moderate red:15 red:20 intermediate red:15 light red:6 light light red:1 white. These ratios resembled Galton's inheritance scheme, discussed in Chapter 1. But we can interpret the ratio to be the result of a Mendelian trihybrid cross among three independently assorting gene pairs, all incompletely dominant.

This work led to the concept of *multiple-gene,* or *polygenic, inheritance.* In multiple-gene inheritance, several gene pairs contribute to one character. The genes involved may or may not assort independently, but they influence one phenotype in a cumulative and, perhaps, seemingly continuous manner.

Nilsson-Ehle studied a trait that showed Mendelian inheritance to confirm the multiple-gene hypothesis, but East (1910) confirmed it for a trait that did not demonstrate Mendelian ratios and was subject to extensive environmental modification (Figure 4.4). East crossed two strains of tobacco that differed in flower length—41 cm versus 93 cm. Each strain was homozygous (a pure line) at the start of the experiment, and although each showed minor variations in flower length—owing, presumably, to the environment—the large difference between the two strains was without doubt genetically based. The F_1 generation was intermediate and again showed minor variations. Since the F_1 was presumed to be genetically uniform, the observed variations were judged to be environmentally caused.

Figure 4.4
Inheritance of corolla length in *Nicotiana longiflora.* The figure shows the percentage frequencies with which individuals fall into various corolla-length classes. The F_1 and F_2 means are approximately intermediate between those of the parents. The means of the four F_3 families are correlated with the corolla length of the F_2 plants from which they were obtained by selfing, as indicated by the arrows. (After Mather, *Biometrical Genetics,* Methuen, 1949; East, *Genetics* 1:164–176, 1916)

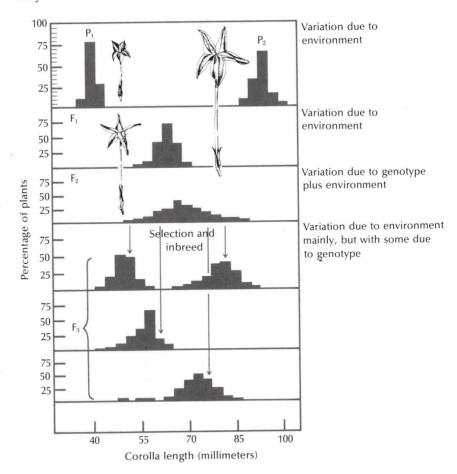

Inbreeding the F_1 produced an F_2 with wide phenotypic variation caused by the segregation and independent assortment of genes coupled with the environmental modification of the different genotypes. In the F_3, produced by inbreeding different F_2 classes, there is less variation among the different F_3 families. This is because there are fewer different gene pairs segregating. Every F_3 peak corresponds closely to the value of the trait in the F_2 individual that was self-fertilized. This is a strong argument for the genetic basis of the trait and suggests that F_3 differences between families are genetic. Within a family, the differences are genetic and environmental.

An estimate of the possible number of gene pairs contributing to this trait can be made by determining the proportion of F_2 individuals that fall into the parental classes. For example, if a quantitative trait in the F_2 shows a parental extreme to be $\frac{1}{16}$ of the total, it may be that two pairs of polygenes are assorting independently. A parental extreme comprising $\frac{1}{64}$ of the total suggests three gene pairs, $\frac{1}{256}$ suggests four gene pairs, $\frac{1}{1024}$ suggests five gene pairs, and so on. East raised 444 F_2 plants without finding a parental phenotype. One could conclude, therefore, that more than four gene pairs are involved in the determination of flower length in tobacco. It is now estimated that nine pairs of genes are responsible for corolla length in this cross. This estimate must be used cautiously, however, because it assumes that all genes produce equivalent effects, that there is no linkage, and that environmental effects are negligible. As a preliminary approximation, though, this method may be useful.

Figure 4.5
Skin-color distributions in blacks and whites. Skin color is measured by the skin reflectance for light of 685-mμ wavelength. (Based on research by G. A. Harrison and J. J. T. Owen, *Annals of Human Genetics*, Vol. 28, pp. 27–37, and graphically interpreted by Bodmer and Cavalli-Sforza.)

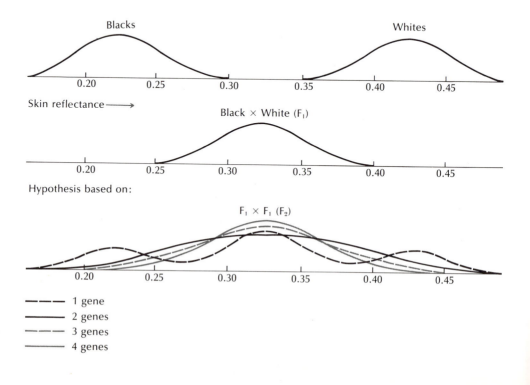

In the human species, we do not have such conveniences as pure lines to study quantitative traits, but the inheritance of skin color comes the closest to meeting our standards for analysis of these traits. It is clear from all studies of skin color that it is a quantitative character influenced by polygenes having at least a partially additive effect. In 1913, C. B. Davenport postulated 5 skin-color classes. In a black × white cross, the F_1 was intermediate. In $F_1 \times F_1$ crosses (not brother–sister), the F_2 consisted of the 5 classes in approximately a 1:4:6:4:1 ratio. This ratio is expected if we are dealing with two independently assorting gene pairs showing incomplete dominance and additivity (note Galton's ratio here, as discussed in Chapter 1). Indeed, this was Davenport's model.

But Davenport's hypothesis has not held up well under closer scrutiny. For one thing, his classification scheme has been successfully challenged. It was too arbitrary. Modern techniques of using a fixed wavelength of light and measuring skin reflections show a continuous distribution of skin color, not a series of discrete classes.

Most hypotheses today propose a system with four or five gene pairs and additive effects. We derive this by comparing theoretical F_2 curves with observed F_2 curves (Figure 4.5).

Transgressive variation

Quantitative inheritance sometimes produces F_2 phenotypes that exceed the parental extremes. This is called *transgressive variation*, and it generally occurs when the parents do not represent the extreme classes. Punnett, in studying the inheritance of body size in two varieties of chickens, crossed a large variety with a small variety and obtained an F_1 of intermediate size. Mating F_1 with F_1 produced an F_2 generation about 1 percent of which contained chickens both larger and smaller than the original parents. Punnett inferred the presence of four pairs of independently assorting genes that produced cumulative rather than discontinuous effects. The parents presumably did not represent the extreme classes:

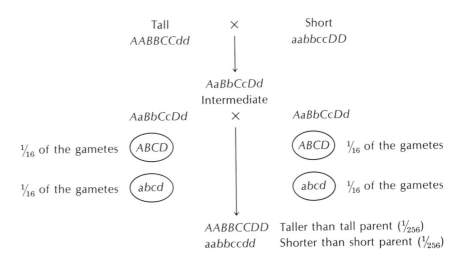

Transgressive variation is less common than the contrasting phenomenon of *regressive variation,* in which offspring tend to approach the mean of the population rather than to exceed the parental extremes. For example, the children of tall parents tend to be shorter than their parents. That is, they tend to approach the mean height of the population. Conversely, the children of short parents tend to be taller than their parents, again tending to approach the mean of the population.

A force that tends to counteract regression is *assortative mating,* or the mating of similar types. Tall men tend to marry tall women, for example, and short women tend to marry short men. The offspring of these parents will resemble them more than if the parents had mated at random.

In summary, quantitative inheritance involves polygenic systems, and although the ratios may be more complex than Mendel's, they do not contradict his principles. The continuous patterns of inheritance are explained by polygenes that can undergo segregation and independent assortment. The fusion of Darwinian evolution and Mendelism occurred in 1909–1910 with the work of Johannsen, Woltereck, Nilsson-Ehle, and East, all of whom demonstrated polygenic inheritance.

Careful analysis of inheritance patterns can often point up the complicating influences of environment and polygenic systems. Classic Mendelian traits, presumably determined by a single gene pair, are often far more complex than once believed. Sometimes a regular Mendelian trait is obscured because of a continuous distribution pattern. More often, however, a continuous distribution pattern is the result of a polygenic system that is for all intents and purposes impossible to organize into phenotypic classes that correspond to genotypic classes. A discussion of some examples and models may help in the analysis of quantitative inheritance patterns.

It has often been pointed out that the ability to taste phenylthiocarbamide (PTC) is determined by a single dominant gene (*T*), and the inability to taste PTC by the homozygous recessive condition:

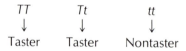

However, when tasters are examined more carefully, it is clear that they exhibit considerable variation in response to different PTC concentrations. For example, in one study of PTC-tasting ability, a series of PTC solutions was made up and administered to people. The results are shown in Figure 4.6. Solution 1 is the strongest, solution 13 the weakest. In this study, people were tested until they reached a solution they could taste (threshold solution). Then they stopped. Solution 5 seems to be the break point for this study. People stopping at solution 4 and above are classed as "nontasters," and those stopping at solution 6 and below are classed as

Figure 4.6
Distributions of taste
thresholds for
phenylthiocarbamide
(PTC) in an English
population (From
Barnicot, *Ann. of Eugenics*,
15, 1950)

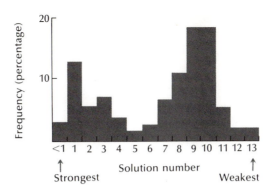

"tasters." Note, however, that there is overlap and that so-called nontasters at one concentration are tasters at another. It has also been shown that different populations have different thresholds and different distributions. PTC tasting, though it is determined by a single gene pair, is evidently influenced by other genes.

Acid phosphatase, an enzyme found in red blood cells, is produced by a multiple allelic series at one locus: P^A, P^B, and P^C. Each allele produces an enzyme form with a different activity; P^C is the most active and P^A the least. The six possible genotypes (AA, AB, AC, BB, BC, CC) are distinguishable under controlled conditions, but there is so much overlap among the functions of the six genotypes that they are usually indistinguishable (Figure 4.7). The overall activity distribution shows a continuous variation. Thus a regular Mendelian trait is obscured by a continuous distribution pattern.

Many traits are determined by an unknown number of polygenes with additive effects, though not necessarily equally additive. These traits

Figure 4.7
Distribution of acid phosphatase activities in red blood cells. The dashed curve shows a continuous distribution of the enzyme in the general population. The other curves show activity values for 5 separate genotypes (A, BA, B, CA, CB). The 5 genotypes cannot be distinguished on the basis of their enzyme activity alone, because of their overlapping distributions. (From Harris, *Proc. Roy. Soc.*, Ser. B, 164:298–310, 1966)

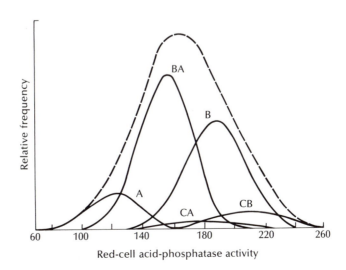

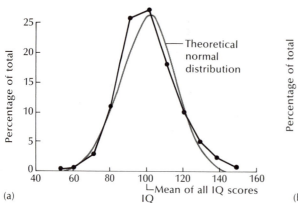

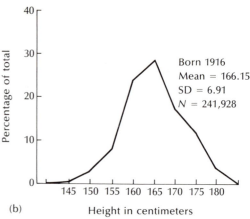

(a) IQ
Mean of all IQ scores

(b) Height in centimeters

Born 1916
Mean = 166.15
SD = 6.91
N = 241,928

Figure 4.8
The distribution of IQ and height among two different European populations. The IQ study was done on a Scottish population and the height study on a population of Italian males.

show a continuous variation and are not organizable into genotypic/phenotypic classes. We can look at human height or IQ as examples (Figure 4.8). In both instances, polygenes and environment interact to produce the final phenotype.

If these represent some of the complexities of quantitative traits, how can we best assess the genetic components of these traits? One way is to quantitatively describe the trait. To do this, one usually takes random samples from a population, analyzes them, and then extrapolates to the larger population. Values calculated from samples are called *statistics*. Values calculated from the entire population are called *parameters*. Thus quantitative traits are usually analyzed using statistics, and we shall now look at some of those statistical measurements.

Mean, median, and mode In the analysis of a sample of individuals expressing a quantitative trait, we need to know where a "typical" value is on the value scale of the trait. In other words, we need values that give us *locations* in our sample. Say, for example, we randomly pick 13 male sophomores in a college class and measure their height in centimeters. We obtain the following measurements:

161	183	177	157	181	176	180
162	163	174	179	169	187	

Arranging these values in sequence, we get

157	161	162	163	169	174	176
177	179	180	181	183	187	

Location points that help characterize this sample are the mean and the median. The *mean* ($\bar{x}$) is the sum of an array of quantities divided by the number of quantities in the sample. The *median* is the middle value in a group of numbers arranged in order of size. In the foregoing sample, the mean is:

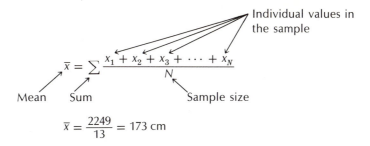

$$\bar{x} = \sum \frac{x_1 + x_2 + x_3 + \cdots + x_N}{N}$$

Individual values in the sample

Mean   Sum   Sample size

$$\bar{x} = \frac{2249}{13} = 173 \text{ cm}$$

The median is 176 cm.

Though this is a small sample, we can organize the data into *classes* and form a *frequency distribution*.

Class	Individuals in each class	Class mean	Frequency
156–160	157	—	1
161–165	161, 162, 163	162	3
166–170	169	—	1
171–175	174	—	1
176–180	176, 177, 179, 180	178	4
181–185	181, 183	182	2
186–190	187	—	1

The *modal class* is that class containing more individuals than any other in a frequency distribution. In this distribution, it is the 176–180 class.

If it is done with a large-enough sample, a frequency distribution can provide useful information about the sample, especially when converted to graphlike form. Figure 4.9 graphically depicts these data. (Figure 4.6 is a frequency distribution of PTC tasters.)

Variance and standard deviation   The location points we have just described tell us nothing about the variation between the individuals in the sample. In other words, we do not know how the values are dispersed in a sample. One way of characterizing dispersion in a sample is by calculat-

Figure 4.9
Histogram of the class frequencies for height, as discussed in the text

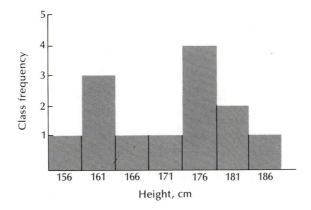

Table 4.1

Two sets of scores that have the same mean but differ in variability

Set A	Set B
32	46
32	42
31	38
31	34
30	30
30	30
29	26
29	22
28	18
28	14
$\Sigma X = 300$	$\Sigma X = 300$
$N = 10$	$N = 10$
$\bar{x} = \dfrac{300}{10} = 30$	$\bar{x} = \dfrac{300}{10} = 30$
Range $= 32 - 28 = 4$	Range $= 46 - 14 = 32$

ing the *variance*. When all values in a sample are expressed as plus and minus deviations from the mean, the variance is the mean of the squared deviations.

Each sample component

$$\sigma^2 = \frac{(x_i - \bar{x})^2}{N - 1}$$

Variance   Sample size less one

In the previous height example, the variance is 92.33 cm².

The variance can be used by itself as the variability measure, but because it is expressed in square units, it is usually more convenient to take the square root of the variance. Doing so returns us to the original scale of measurement. The square root of the variance is called the *standard deviation* (σ), and from our previous height example,

$$\sigma^2 = 92.33 \text{ cm}^2$$
$$\sigma = \sqrt{92.33} = 9.61 \text{ cm}$$

The larger the standard deviation, the larger the variability in our sample. The importance of knowing about the dispersion of values in a sample is apparent in Table 4.1, where the two samples shown have identical means but differ greatly in their dispersion properties (variance and standard deviation).

Before we move on, we should explore another aspect of the meaning of standard deviation and variance. If we were to examine most continuously varying traits in a population, we would see that they are distributed in a bell-shaped pattern called the *normal distribution* or *Gaussian distribution,* in which the distribution of values is fairly symmetrical about the mean (Figure 4.10). In such a normal distribution, about two-thirds of all individuals fall between 1 standard deviation above and 1 standard deviation below the mean. Ninety-five percent of all individuals fall be-

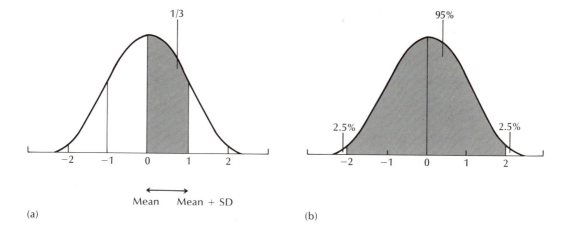

(a)

(b)

Mean   Mean + SD

Figure 4.10
Meaning of standard deviation and variance. The distributions are normal in shape. The scale of the horizontal axis in graphs (a) and (b) is such that the mean is zero and the SD = 1 (as is the variance). About one-third of all individuals in the distribution fall between the mean and one SD above the mean. Since the distribution is symmetrical, two-thirds of the individuals fall between one SD below the mean and one SD above the mean. 95% of all individuals in the distribution fall between two SD's on either side of the mean.

tween 2 standard deviations on both sides of the mean. In Figure 4.10, the mean is 0, so the variance and standard deviation are both 1.

Heritability   When we examine a population for a particular trait, the variations we observe between individuals may be the result of genetic differences, environmental differences, and/or the interaction between the genotype and the environment. Stated another way, the *total phenotypic variance* (V_p) observed in a population is the sum of three factors: the *environmental variance* (V_E), the *genetic variance* (V_G), and the *variance due to genetic and environmental interactions* (V_{GE}). We express this in the following formula:

$$V_p = V_G + V_E + V_{GE}$$

Let us analyze these components of phenotypic variability. The V_E component is an expression of all nongenetically based differences. The V_G component is an expression of all genetically based differences. The V_{GE} component is an expression of the way the genotypic expression varies as a function of the environment in which the genotype is placed. The V_{GE} component is difficult to analyze and quantify, so it is usually ignored. *Heritability* (H or sometimes h^2) is expressed as

$$H = \frac{V_G}{V_p}$$

In words, this equation says that heritability is a measure of the degree to which a phenotype is genetically determined and the degree to which it can be changed by selection. But V_G is so complex that it requires further analysis. It is composed of three subcomponents: Some genes are additive in their effects (V_A), some genes are dominant (V_D), and some genes are epistatic (V_I), where the subscript I stands for interaction. Thus

$$V_G = V_A + V_D + V_I$$

The most important of all these components (or the only one usually considered) is V_A. Heritability is thus often expressed as

$$H = \frac{V_A}{V_p}$$

A trait with a heritability of 1 is, for all intents and purposes, not influenced by the environment. For example, if you inherit the blood alleles A and B, your phenotype will be AB, irrespective of the environment. The heritability of the A-B-O phenotypes is essentially 1. Other traits, however, such as bristle number in *Drosophila*, have a heritability of about 0.5. A trait with a heritability of 0 has no genetic basis.

It is important to understand that heritability is usually only an approximation. We stated already, for example, that the V_I and V_D components are usually not included in the V_G value. Also we usually ignore V_{GE}, because we cannot quantify it. It may be important, however, because a given gene product, unable to function under one set of environmental conditions, is perfectly functional under another.

Failure to consider V_{GE} is very serious in estimates of IQ heritability (see Chapter 3, page 125). Though there is no evidence pointing to specific genes that determine IQ (at least in part), it is presumptuous to suggest that any genes that might be involved in intelligence are independent of the environment in their expressivity.

Extranuclear inheritance

The patterns of inheritance we have discussed so far depend almost entirely on nuclear genes and their behavior in meiosis. Except for genes located on the sex chromosomes, it made no difference to the genetic constitution of the F_1 whether the mutant parent was male or female. This, according to Mendelian principles, is precisely what is to be expected. The implication of this observation is that, since the nuclear contribution of both sexes is essentially the same and since the cytoplasmic contribution is unequal, the cytoplasm plays no role in inheritance. But this inference has not stood up. As we have already seen, genes and the cellular environment form a closely integrated interacting system.

In this section, we shall examine inheritance schemes that are caused by extranuclear inclusions. We can refer to these schemes as non-Mendelian inheritance patterns. In some cases, the phenotype may be determined by a gene product present in the cytoplasm of the egg. This is called a *maternal effect*. Sometimes a phenotype is the consequence of intracellular symbionts; and sometimes it is the result of intracellular organelles, such as mitochondria and chloroplasts.

The maternal effect   We have already mentioned one example of the maternal effect in connection with the inheritance of tumorous head (*tu-h*) in *Drosophila*. The expression of tumorous head is influenced by cytoplasm in the egg that has in turn been modified by chromosomal genes. This maternal effect in *tu-h* inheritance produces differing results in reciprocal crosses (see pages 138–139). The maternal effect is a consequence of the product of chromosomal genes. Unlike genes, this product is not self-regenerative. There are other types of extranuclear inheritance in which the cytoplasmic factors exhibit replication and independent transmission. These factors contain their own genetic material, but the traits encoded by these genes are not transmitted in Mendelian ratios.

Another interesting example of the maternal effect is the coiling of the shell of *Limnaea peregra*, a small snail, whose shell may coil to the right (*dextral*) or left (*sinistral*). This coiling direction is genetically determined (Figure 4.11) except that the female's genotype determines the phenotype of all her offspring. A heterozygous dextral female (*Dd*), on self-fertilization, produces offspring that are all dextral, even though one-fourth of the offspring are *dd* (sinistral) genotypically. A heterozygous sinistral female (*Dd*) also produces offspring that are all dextral. The basis of this maternally determined trait lies in the early cleavage pattern of the embryo. The egg cytoplasm determines the cleavage pattern.

Figure 4.11
Inheritance of coiling in
Limnaea peregra

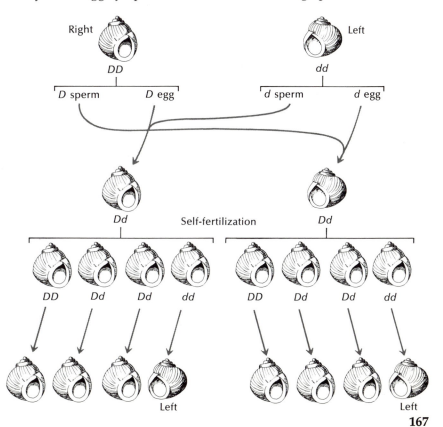

Two major groups of cytoplasmic factors have been studied as replicating genetic systems. The first group, *intracellular symbionts*, consists of organisms such as bacteria and viruses that live within cells in a symbiotic relationship with a host organism. They replicate along with the host and

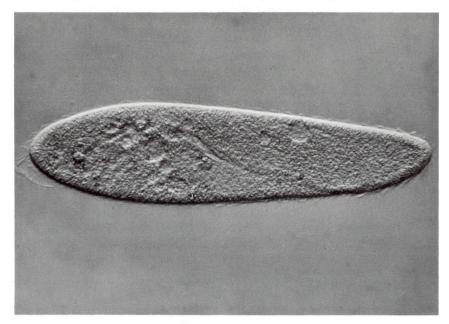

Figure 4.12
Paramecium (Courtesy of Carolina Biological Supply Company)

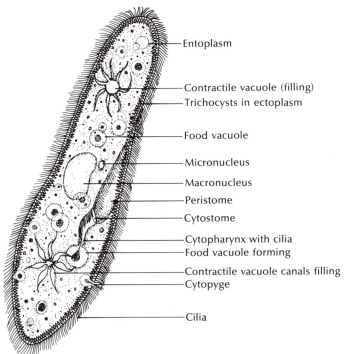

Entoplasm

Contractile vacuole (filling)
Trichocysts in ectoplasm

Food vacuole

Micronucleus

Macronucleus

Peristome

Cytostome

Cytopharynx with cilia
Food vacuole forming

Contractile vacuole canals filling
Cytopyge

Cilia

by their presence cause a special modification of the phenotype. The intracellular symbionts we will consider are the kappa particle in the protozoan *Paramecium* and the sigma factor in *Drosophila*. The second group of cytoplasmic factors involves *intracellular organelles,* such as mitochondria and chloroplasts, that have gene systems of their own.

In 1938, T. M. Sonneborn discovered killer strains of *Paramecium aurelia.* He later found that the killer trait depended on the presence of a cytoplasmic factor called *kappa,* which in turn depended on a dominant chromosomal gene (*K*) for its maintenance. A killer strain (that is, one possessing kappa) destroys sensitive strains, which have no kappa particles, by liberating toxic particles. Killer cells are immune to their own toxin—as are sensitive cells in the process of *conjugation,* a system of parasexual reproduction in *Paramecium* (Figure 4.12) characterized by the exchange and fusion of haploid nuclei (Figure 4.13).

Figure 4.13
Stages in sexual reproduction in *Paramecium aurelia*

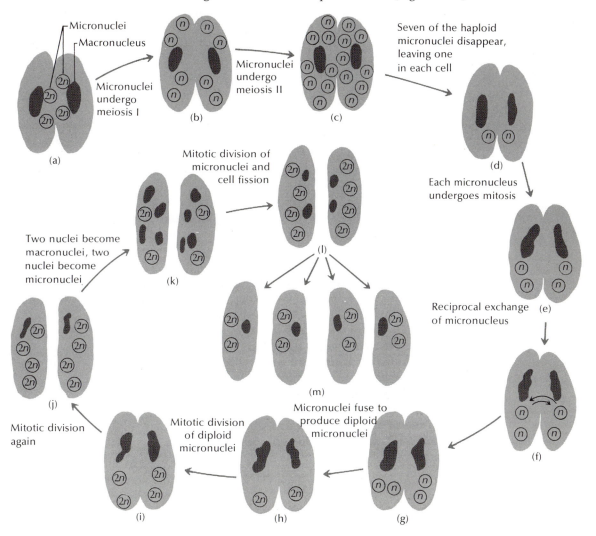

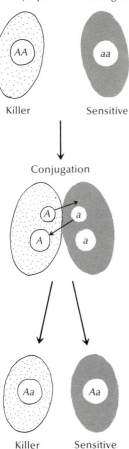

Normal conjugation
(no cytoplasmic exchange)

Killer   Sensitive

Conjugation

Killer   Sensitive

Figure 4.14
Mating of a ''killer'' strain
(*AA*) of *Paramecium* with a
sensitive strain (*aa*)

The inheritance of the killer trait does not follow a Mendelian pattern, because according to that pattern, heterozygous cells produced by the conjugation of homozygotes should be phenotypically identical, and they are not (Figure 4.14).* The *Aa* heterozygotes, which are equally divided between killers and sensitives, can undergo another form of sexual reproduction called *autogamy* (Figure 4.15). During autogamy, the diploid nuclei undergo meiosis, seven of the eight meiotic products are destroyed, the one remaining haploid nucleus undergoes mitosis, and the two identical haploid nuclei fuse to form a homozygous diploid nucleus. Autogamy of heterozygous *Paramecia* (*Aa*) results in segregation, with some organisms *AA* and some *aa*. If killer strains were determined by chromosomal genes following a Mendelian pattern, autogamy should produce two genotypes and two phenotypes. But all sensitive heterozygotes produce only sensitive offspring, and all killer heterozygotes produce only killer offspring (Figure 4.16). This pattern suggests nonchromosomal inheritance rather than a Mendelian mechanism.

Confirmation of the cytoplasmic basis of killer strains came about when conjugation between a killer and a sensitive strain was prolonged, allowing cytoplasm to be exchanged (Figure 4.17). All heterozygotes were then killer, confirming the suspicion that this trait is determined cytoplasmically.

A dominant chromosomal gene (*K*) is required to maintain the cytoplasmic killer particle (kappa). This means that the particle can be transmitted cytoplasmically but will disappear without a *K* gene in the host. Figure 4.18 summarizes the relationship between kappa and the *K* gene.

Kappa appears to be a bacterium that contains its own genetic information (DNA) and is autonomously replicating. It has been suggested that the toxic substance is a defective virus particle housed inside kappa, which makes this symbiotic association a complex triangle.

Another trait that follows an extranuclear pattern of inheritance and is caused by an intracellular symbiont is *CO_2-sensitivity* in *Drosophila*. Normally, *Drosophila* is immobilized rapidly by exposure to high concentrations of CO_2 but recovers fully within a few minutes of the removal of the CO_2. In 1937, L'Héritier and Tessier discovered a strain of *Drosophila* that was permanently paralyzed by CO_2. Crossing a CO_2-sensitive female with a normal male produced almost all sensitive offspring. But the reciprocal cross of a sensitive male with a normal female produced almost all normal offspring, which suggests extranuclear inheritance.

The cytoplasmic entity responsible for CO_2-sensitivity was found to be an intracellular symbiotic virus, called *sigma* (Figure 4.19). The trait is generally transmitted by the female because of the volume of cytoplasm in

* Figures 4.14 through 4.18 and 4.20 through 4.22 are adapted by permission of The Macmillan Co., Inc., from *Genetics* by M. W. Strickberger, copyright © 1976 by Monroe W. Strickberger. After G. H. Beale, 1954. *The Genetics of Paramecium aurelia*. London, Cambridge University Press.

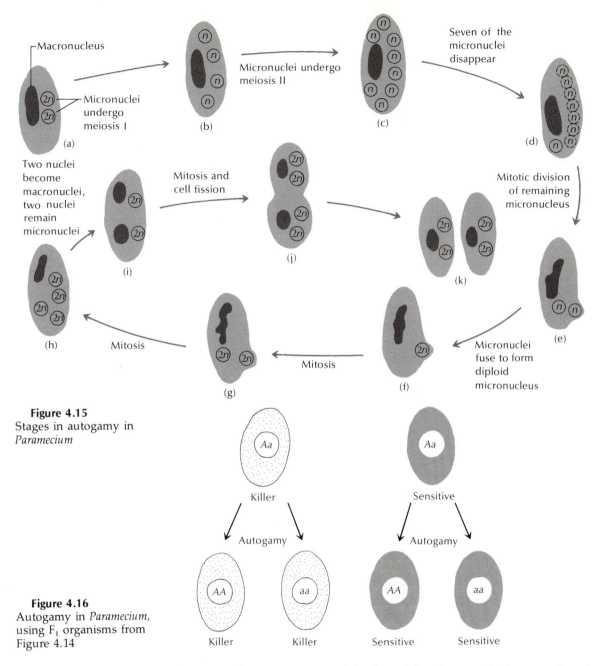

Figure 4.15
Stages in autogamy in *Paramecium*

Figure 4.16
Autogamy in *Paramecium*, using F₁ organisms from Figure 4.14

the egg. Males too can transmit it, though less frequently, because there is so little cytoplasm in sperm cells. Other intracellular symbionts are known, and although their effects vary, their inheritance patterns do not. They are generally passed only from mother to offspring.

Mitochondria are cytoplasmic organelles that enable cells to respire aerobically. They contain respiratory enzymes as well as their own genetic information (DNA), and although most of the mitochondrial protein and enzymes are produced by nuclear genes, mitochondrial genes account for nearly 20 percent of them. Mitochondria are not autonomous; they require both their own genes and nuclear genes in order to exist. But the fact that they contain their own DNA has led to the theory that they arose from symbiotic bacteria that had lost some of their own genes.

The first genetically oriented study of a mitochondrial trait was the analysis of the "petite" mutation in baker's yeast *Saccharomyces cerevisiae*. Because these mutants lack some respiratory enzymes, they are defective in their ability to use oxygen in metabolizing glucose. This defect causes the yeast colonies to be very small in size; hence the name "petite."

Petite mutants can result from mutant nuclear genes or from mitochondrial genes. A mutation in a nuclear gene that produces defec-

Figure 4.17
Conjugation with the passage of cytoplasm in *Paramecium,* demonstrating cytoplasmic inheritance of killer

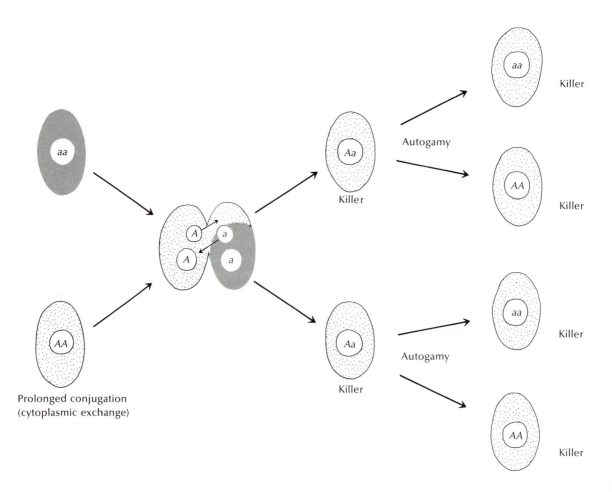

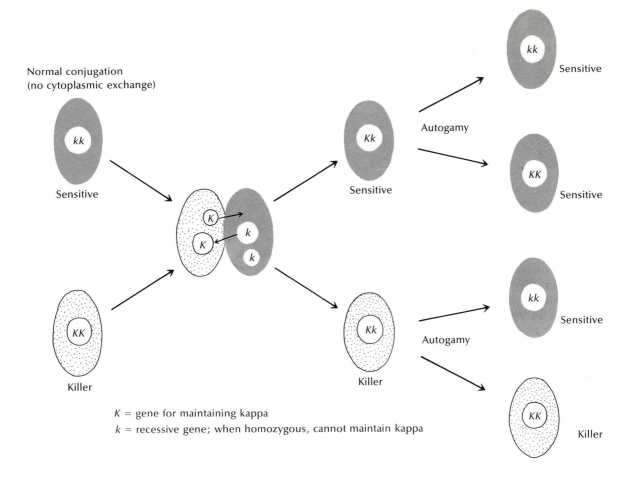

Normal conjugation
(no cytoplasmic exchange)

K = gene for maintaining kappa
k = recessive gene; when homozygous, cannot maintain kappa

Figure 4.18
Results of crosses
between killer and
sensitive strains of
Paramecium aurelia when
the sensitive strain is
homozygous for a gene
(*k*) unable to maintain
killer particles

tive mitochondria follows a Mendelian pattern of inheritance, with segregation occurring in the heterozygous zygote. This type of a petite mutant is called a *segregational petite* (Figure 4.20). A second type arises from mutations in mitochondrial genes leading to defective mitochondria. This category consists of two classes, *neutral* and *suppressive*.

When crossed with a wild-type strain, the *neutral petite* produces offspring that are all normal. No matter how many generations of mating are carried out, the trait never reappears. The neutral petite mutant is self-sterile, but if mutant haploid cells are mated to haploid vegetative cells of a normal strain, the diploid vegetative cell is normal and fertile (Figure 4.21). These diploid cells can multiply indefinitely by the asexual budding process, but if the proper environmental conditions allow, the diploid cells can undergo meiosis to form four haploid spores, which can germinate to form haploid colonies of vegetative cells. The petite mutant found in the original haploid cell does not reappear in any of the haploid

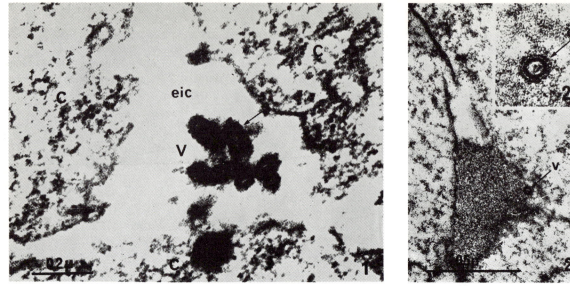

The viruses (V) are observed only in the intercellular spaces (eic) between the follicular cells (C). The virus indicated by the arrow demonstrates the characteristic "glove-finger" form. The enlargements show a cross section of a virus. Photograph 2a shows the virus attached to a cytoplasmic membrane; 2b is an enlargement of part of 2a.

Figure 4.19
Electron micrograph of sigma infection in *Drosophila*.
[A. Berkaloff et al., *Compt. Rend. Acad. Sci.* 260:5956–5959 (1965)]

Figure 4.20
(a) Yeast life cycle, showing aspects of its sexual and asexual phases
(b) inheritance pattern for segregational petites

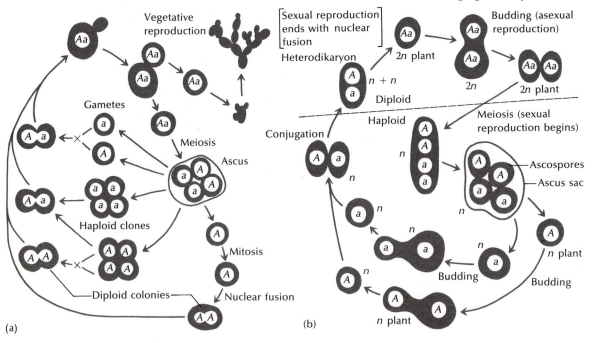

Figure 4.21

Inheritance pattern for neutral petite in yeast (the genetic defect is a lack of mitochondrial DNA) (Based on Ephrussi, Hottinguer, and Roman, 1955, *PNAS* 41)

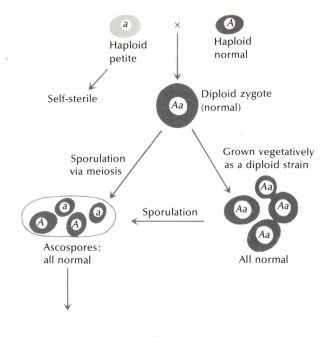

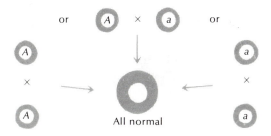

or diploid progeny cells. If the determinant of this trait were chromosomal, one would expect the trait to reappear in a 1:1 ratio in the haploid spore cells. The fact that it does not reappear suggests non-chromosomal inheritance.

The genetic basis of this trait is a cytoplasmic factor (p^+) present in wild-type strains of yeast but missing in petites (p^-). The p^- mutants usually contain no mitochondrial DNA, which also supports the non-chromosomal basis of the trait.

Having established a basis for the neutral petite mutation, we can better interpret its pattern of inheritance. If the mutant were crossed to a wild-type strain, the diploid cell would be normal because normal mitochondria would have been contributed by the wild-type haploid cell. These normal mitochondria would replicate and be passed on to the haploid spores formed from meiosis of the diploid cell, so all of the spores would be normal—as would all of their descendants.

The second class of nonnuclear petite mutations, *suppressive petite*, differs from the neutral petite in that the mutant phenotype reappears in the progeny diploid and haploid cells, though in non-Mendelian ratios. The diploid cells produced by the mating of haploid cells of a suppressive petite strain with haploid cells of a normal strain are of two types, normal and petite. The diploid normal cells give rise to all normal haploid spores. But the diploid petite cells, when successfully induced to sporulate, produce haploid spores that may be all normal, all petite, or some normal and some petite. The pattern of inheritance of suppressive petite does not follow a Mendelian scheme (Figure 4.22).

The genetic basis for the inheritance of suppressive petite has been traced to the mitochondria. Suppressive petite mutants have mitochondria that contain DNA, but the DNA is mutant; they therefore express a mutant phenotype. The mutant mitochondria can replicate, however, and therefore can be passed on to progeny cells that can in turn express the mutant phenotype. In the original mating of a suppressive petite with a

Figure 4.22
Inheritance pattern for suppressive petite in yeast (the genetic defect is a mutant gene carried by mitochondrial DNA) (Based on Ephrussi, Hottinguer, and Roman, 1955, *PNAS* 41)

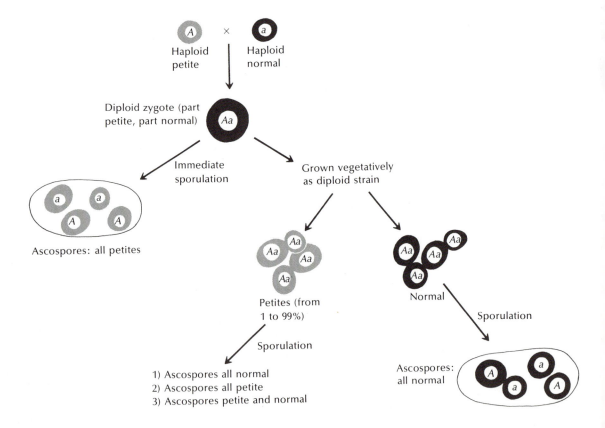

normal, the diploid cell would have received mitochondria from both strains. If mutant mitochondria predominated, the phenotype would be mutant. If normal mitochondria predominated, the phenotype would be normal. The phenotype of the haploid spores would also depend on the proportion of normal and mutant mitochondria they received from the parent diploid cell.

Since the discovery of petite mutations, mitochondrial mutants have been found in *Neurospora, Paramecium, Chlamydomonas,* and *Trypanosoma.* Similar mutants will no doubt be found in many organisms as detection techniques are refined. One important technique in studies of mitochondrial genetics is microinjection. J. Beisson and his colleagues isolated mitochondria from one *Paramecium* species and injected them into the cytoplasm of another species. Their object was to study the interaction between nuclear and cytoplasmic genes, and they found an interesting relationship. A type of incompatibility is frequently observed between a specific type of mitochondria (one carrying identifiable mutant genes) and the host cell into which they are injected. This incompatibility is manifested when the foreign mitochondrion has difficulty reproducing or when the host cell actually modifies the foreign mitochondrion.

What mechanisms might account for this observed incompatibility? To date there is no answer, but there are some intriguing ideas. One hypothesis is that genes in the host cell modify genes in the mitochondria so that the foreign mitochondria's gene products are more compatible with the host cell. The unmodified mitochondria would be selected against, while the modified mitochondria would be selected for. Another hypothesis suggests that the foreign mitochondrial membranes are altered by the interaction of membrane subunits coded for by two different sets of genes: host genes and foreign mitochondria genes. These altered surface properties may well account for the incompatibility. But so far, the experiments needed to differentiate between these two hypotheses have not been done.

The genetics of chloroplasts

Chloroplasts are plant organelles that carry out photosynthesis. Like mitochondria, they have their own DNA and perhaps originated as intracellular symbiotic bacteria. In 1909, Carl Correns was the first to discover in *Mirabilis* (the plant we call the four o'clock) a trait ascribable to chloroplast inheritance. The leaves of *Mirabilis* can be green, white, or green-and-white striped. When flowers from a green-leaved branch are pollinated by flowers from a white-leaved branch, all the offspring are green-leaved, and subsequent inbreeding produces all green-leaved plants. But a reciprocal cross ("green" pollen on "white" flowers) produces white offspring only. Since white plants cannot survive, further matings are not possible. If the female is green-and-white striped, its offspring will also be green-and-white striped, irrespective of the male parent. Thus the leaf-color phenotype of the *Mirabilis* offspring is solely a

Figure 4.23
Inheritance of plastids

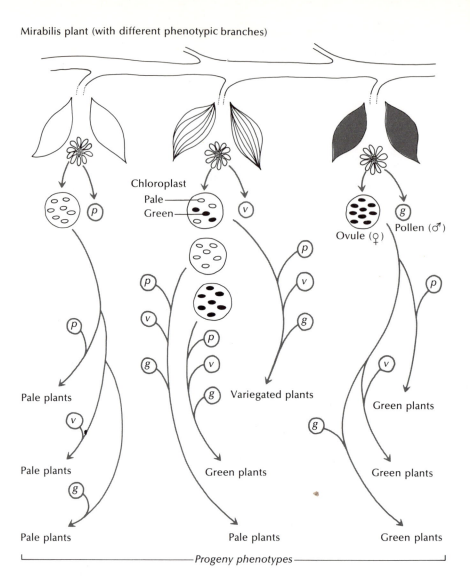

Mirabilis plant (with different phenotypic branches)

function of the maternal plant and more specifically of chloroplasts found in the egg cytoplasm (Figure 4.23). In some plant genera, such as *Oeno-thera* and *Pelargonium*, pollen does contribute chloroplasts to the offspring, although offspring still tend to resemble the female more than they do the male.

The genetics of chloroplasts and mitochondria is complex, because nuclear genes are often involved in the maintenance of both these organelles. Indeed, chloroplast genes may affect the characteristics of *both* chloroplasts and mitchondria. S. J. Surzycki and N. W. Gillham, who presented evidence to support this, also presented evidence that mitochondrial genes affect chloroplast characters. Their report implies that great care must be exercised in discussing the site of cytoplasmic mutations affecting chloroplasts or mitochondria, or both.

Thus, although most genetic traits are determined only by chromosomal genes interacting with environmental factors, some are not. Those that do not follow Mendelian patterns are usually due to cytoplasmic factors such as intracellular symbionts or cytoplasmic organelles. These cytoplasmic factors generally contain DNA, are self-perpetuating, and may have originated as symbionts.

Overview

Every important discovery that sets forth a new paradigm is followed by a period of intense research activity during which the paradigm's limits are tested. This was certainly true for Mendelism. The two basic principles formulated by Mendel were challenged in many ways, but they withstood these challenges and emerged stronger.

In this chapter, we focused on two types of inheritance that at first did not seem consistent with Mendelism: quantitative inheritance and extranuclear inheritance.

Quantitative inheritance does not yield Mendelian ratios. Rather it commonly yields a continuous range of variation for a trait. The work of Johannsen, Nilsson-Ehle, and East showed that more than one gene pair can determine a phenotype. In fact, if several gene pairs having an additive effect determine a phenotype, and if the environment affects genotypic expression, then a normal distribution of variation results. But it is important to realize that segregation and independent assortment are still operating, even though Mendelian ratios are not obtained.

In the analysis of quantitative inheritance patterns, we are often dealing with an unknown number of polygenes with additive effects, effects that can be modified by the environment. To analyze such traits, we describe them statistically. To pinpoint locations in our sample, we can calculate a mean, a median, and a mode. But location points tell us nothing about variation between individuals in the sample. To find this variation, we calculate variance and standard deviation. Finally, we examined the procedure commonly employed to determine how much of the variation we observe is the result of genetic differences, environmental differences, or genotype–environment interaction. Heritability expresses the degree to which a particular phenotype is genetically determined and changeable by selection.

Extranuclear inheritance patterns generally follow non-Mendelian rules, because they follow rules for cytoplasmic inheritance. We explored this type of inheritance pattern by looking first at the maternal effect: nuclear genes producing a cytoplasmic product that is transmitted only through the egg to the offspring and determines a specific phenotype. Intracellular symbionts, such as viruses or the kappa particle in *Paramecium,* also are cytoplasmically transmitted particles that often determine specific phenotypes.

Intracellular organelles such as mitochrondria and chloroplasts have their own DNA (genes) and so generate many of the products required for their existence. But nuclear genes are also required. Mutations in genes in the intracellular organelles are transmitted cytoplasmically in non-Mendelian patterns. Nuclear gene mutations that affect the structure and function of intracellular organelles follow Mendelian patterns.

Questions and problems

4.1 How would you demonstrate the existence of genes in mitochrondria?

4.2 What criteria should be used to detect extranuclear inheritance?

4.3 How would you distinguish between a phenotype controlled by polygenes and one controlled by a single gene pair with variable expressivity?

4.4 In a sample of *Drosophila mojavensis,* the following thorax lengths were determined (in millimeters).

2.1, 1.6, 1.5, 1.4, 1.9, 1.8, 2.3, 2.2, 2.0, 2.6, 1.1, 1.2, 2.2, 1.5, 1.6, 1.4, 1.2, 2.4, 2.0, 1.1

Calculate $\bar{x}$, σ^2, and σ.

4.5 In a genetics class, the following distribution of grades was obtained in two sections.

Section 1		Section 2	
Range	Frequency	Range	Frequency
80–89	4	90–99	2
70–79	10	80–89	6
60–69	18	70–79	12
50–59	9	60–69	20
40–49	5	50–59	10
30–39	2	40–49	2

(a) Determine the mean of Section 1.
(b) Determine the mean of Section 2.
(c) Determine the mean of the whole group.
(d) Determine the variance and the standard deviation of the whole group.

4.6 Emerson and East studied the inheritance of ear length in corn. In one of their experiments they obtained the results shown in the following table.

Crosses	5	6	7	8	9	10	11	12	13	14	15	16	17	18	19	20	21
(1) $P_1 \times P_1$:	4[a]	21	24	8													
(2) $P_2 \times P_2$:									3	11	12	15	26	15	10	7	2
(3) $P_1 \times P_2$:					1	12	12	14	17	9	4						
(4) $F_1 \times F_1$: from (3)					1	10	19	26	47	73	68	68	39	25	15	9	1

Ear length (cm)

[a]There were 4 plants producing ears of corn 5 cm in length. This tells you how to read the table.
Data from R. A. Emerson and E. M. East, *Nebr. Agricult. Exp. Res. Bull.* 2, (1913).

(a) Calculate $\bar{x}$, σ^2, and σ for each parent, the F_1, and the F_2.
(b) Discuss evidence for transgressive or regressive variation.
(c) Make an estimate of the minimal number of genes involved.

4.7 Nilsson-Ehle studied the inheritance of kernel color in wheat. In one cross, he crossed a true-breeding red variety to a true-breeding white variety. The F_1 was red, as was the F_2. But when he self-fertilized the 78 F_2 plants, he obtained the following results.

Number of F_2 plants	Progeny from self-fertilized plants
50	All red
15	15 red : 1 white
8	3 red : 1 white
5	63 red : 1 white
78	

(a) How would you interpret the data with respect to the number of gene pairs involved in this trait?
(b) What are the parental and F_1 genotypes?

4.8 The following table gives data on the variances of two phenotypic traits in sparrows (wing span and beak length).

Wing span		Beak length	
V_p	271.4	V_p	627.8
V_E	71.2	V_E	107.3
V_A	102.0	V_A	342.9
V_{GE}	98.9	V_{GE}	177.6

Calculate the heritability for each trait, and tell why one of these two traits is more susceptible than the other to selection pressure.

4.9 In corn, there is a gene we shall call i that maps on chromosomes VII. In the homozygous state, ii plants are either inviable white seedlings or green-and-white striped. The I allele specifies green plants. The following crosses were performed:

Pollen	Ovule	Offspring
1. Striped (*ii*)	Green (*II*)	All green
2. Green (*II*)	Striped (*ii*)	Green, white, striped
3. Green (*II*)	Striped (*Ii*) (from cross 2)	Green,* white,* striped*

* Some of these plants are II, some Ii.

(a) How would you interpret this inheritance pattern?

(b) Based on your interpretation, predict the genotypes and phenotypes of the offspring from the following crosses involving ovules from offspring in cross 3.

1. *II* white ♀ × *II* green ♂
2. *Ii* white ♀ × *II* green ♂
3. *II* green ♀ × *Ii* striped ♂
4. *II* striped ♀ × *ii* striped ♂

4.10 In *Neurospora,* there is a strain that grows very slowly. When such a "poky" strain is used as a female and crossed to a normal strain that is used as a male, all of the spores produce "poky" offspring. In the reciprocal cross, all of the offspring are normal. Interpret this cross.

4.11 The following cross is performed in *Drosophila:*

$$\text{(male) } Aa \times Aa \text{ (female)}$$
$$\downarrow$$
$$aa$$
$$\text{male and female}$$

where *a* is an autosomal gene. When *aa* females are crossed to *AA* males, only male offspring are produced, all *Aa*. If *aa* males are crossed to *AA* females, *Aa* males and females are produced. Interpret this cross.

4.12 In *Chlamydomonas,* there are strains resistant to streptomycin (*sr*) and strains sensitive to the drug (*ss*). In a cross involving *sr mt$^+$* and *ss mt$^-$* (where *mt* = mating type), all of the offspring are *sr*. But *ss mt$^+$* crossed to *sr mt$^-$* produces all *ss* progeny. Interpret this cross.

4.13 In Figure 4.19, we showed that a cross between a killer strain of *Paramecium* and a sensitive strain produced killer and sensitive offspring. Assume normal conjugation.

(a) What would you predict from a cross of two F_1 killers? two F_1 sensitives?

(b) What would you predict from a cross of a *KK* sensitive produced by autogamy and an F_1 killer?

4.14 Distinguish between a polygene and a multiple allele.

4.15 A certain trait is determined by five unlinked pairs of alleles, with each dominant allele having an equal additive effect on the phenotype. What phenotypic classes would you predict, and in what frequencies, from a cross between *AaBbCcDdEe* × *AaBbCcDdEe*?

References

Biesson, J., A. Sainsard, A. Adoutte, G. H. Beale, J. Knowles, A. Tait. 1974. Genetic control of mitochondria in paramecium. *Genetics* 78:403–413.

Borst, P. 1977. Structure and function of mitochondrial DNA. *Trends in Biochem. Sci.* 2:31–34.

Carlson, E. A. 1966. *The Gene: A Critical History.* Philadelphia, Saunders.

Correns, C. 1909. Vererbungsversuche mit blass (gelb) grünen und bunt blättrigen Sippen bei Mirabilis, Urtica, and Lunaria. *Zeit. Induk. Abst. v. Vererbung* 1:291–329.

Corwin, H. O., and J. B. Jenkins. 1976. *Conceptual Foundations of Genetics*. Boston, Houghton Mifflin.

East, E. M. 1910. A Mendelian interpretation of variation that is apparently continuous. *Amer. Nat.* 44:65–82.

Ephrussi, B. 1953. *Nucleo-Cytoplasmic Relations in Micro-Organisms*. London, Oxford University Press.

Falconer, D. S. 1960. *Introduction to Quantitative Genetics*. Edinburgh, England, Oliver and Boyd.

Hallick, R. B., C. Lipper, O. C. Richards, and W. J. Rutter. 1976. Isolation of a transcriptionally active chromosome from chloroplasts of *Euglena gracilis*. *Biochemistry* 15:3039–3045.

Johannsen, W. 1909. *Elemente der exakten Erblichkeitslehre*. Jena, East Germany, G. Fisher.

Kempthorne, O. 1957. *An Introduction to Genetic Statistics*. New York, Wiley.

L'Héritier, Ph. 1957. The hereditary virus of *Drosophila*. *Adv. Virus Res.* 5:195–245.

Linane, A. W., H. B. Lukins, P. L. Molloy, P. Nagley, J. Rytka, K. S. Sriprakash, and M. K. Trembath. 1976. Biogenesis of mitochondria: Molecular mapping of the mitochondrial genome of yeast. *Proc. Natl. Acad. Sci. U.S.* 73:2082–2085.

Mather, K., and J. L. Jinks. 1971. *Biometrical Genetics*. London, Chapman and Hall.

Mather, K., and J. L. Jinks. 1977. *Introduction to Biometrical Genetics*. Ithaca, Cornell University Press.

Mather, W. B. 1964. *Principles of Quantitative Genetics*. Minneapolis, Minn., Burgess.

Nilsson-Ehle, H. 1909. Kreuzungsuntersuchunger an hafter und weizen. *Lunds Univ. Aarskr. N.F. Afd.* 2, 5, 2:122.

Pearson, K. 1904. A Mendelian view of the law of ancestral heredity. *Biometrika* III: 109–112.

Preer, J. R. 1971. Extrachromosomal inheritance: Hereditary symbionts, mitochondria, chloroplasts. *Ann. Rev. Genetics* 5:361–406.

Sager, R. 1977. Genetic analysis of chloroplast DNA in Chlamydomonas. *Adv. in Genetics* 19:287–340.

Sager, R., and Z. Ramanis. 1976. Chloroplast genetics of Chlamydomonas. I. Allelic segregation ratios. *Genetics* 83:303–321.

Sager, R., and Z. Ramanis. 1976. Chloroplast genetics of Chlamydomonas. II. Mapping by cosegregation frequency analysis. *Genetics* 83:323–340.

Singer, B., R. Sager, and Z. Ramanis. 1976. Chloroplast genetics of Chlamydomonas. III. Closing the circle. *Genetics* 83:341–354.

Sonneborn, T. M. 1938. Mating types in *Paramecium aurelia*. *Proc. Amer. Phil. Soc.* 79:411–434.

Stern, C. 1970. Model estimates of the number of gene pairs involved in pigmentation variability of the Negro American. *Hum. Hered.* 20:165–168.

Stern, C. 1973. *Principles of Human Genetics*. San Francisco, Freeman.

Surzycki, S. J., and N. W. Gillham. 1971. Organelle mutations and their expression in *Chlamydomonas reinhardi*. *Proc. Natl. Acad. Sci. U.S.* 68:1301–1306.

Wilkie, D. 1964. *The Cytoplasm in Heredity*. London, Methuen.

Willets, N. 1972. The genetics of transmissible plasmids. *Ann. Rev. Genetics* 6:257–268.

Five

Variation in chromosome number and structure

Change in chromosome number

The origin and evolutionary significance of euploidy
Aneuploidy
Aneuploidy in humans
Sex-chromosome aneuploidy
Autosomal aneuploidy in humans
Aneuploid mosaics

Change in chromosome structure

Duplications
Deficiencies
Inversions
Translocations

Overview

Questions and problems

References

Genetic mutation in its broadest sense can include any change in the genetic material, but most geneticists use the term to refer to qualitative changes within a gene. All other changes within a genome involve structural and numerical alterations of chromosomes, and these changes are the subject of this chapter. Intragenic mutations will be dealt with in Chapter 10.

Perhaps the easiest chromosomal alterations to identify cytologically are of the numerical variety, occurring when single chromosomes are added or deleted (*aneuploidy*) or when there are additional sets of chromosomes (*euploidy*)—a set being one member of each homologous chromosome pair. Structural alterations—changes in gene sequence within chromosomes—are more difficult to identify cytologically. These include *inversions* (intrachromosomal gene-sequence alteration), *deletions* or *deficiencies* (loss of genetic information), *translocations* (transposition of genes from one chromosome to another nonhomologous chromosome), and *duplications* (addition of more than the normal amount of genetic information).

Change in chromosome number

The addition or deletion of entire chromosomes or chromosome sets has been instrumental in the evolution of many plant species and is the cause of many human afflictions.

The origin and evolutionary significance of euploidy

The origin of euploidy is rooted in life cycles, which generally involve changes in the chromosome constitution from haploid to diploid, and vice versa. A *haploid,* or *monoploid,* organism carries a single set of chromosomes through most of its life cycle. A *diploid* organism possesses two sets of chromosomes, one set usually contributed by each parent. It is especially common for plants to possess more than the diploid number of chromosome sets, and those that do are termed *polyploids.* Following is a list of the various types of euploidy.

Euploidy type	Chromosome sets
1. Haploid, or monoploid	One (n)
2. Diploid	Two ($2n$)
3. Polyploid	More than two
a. Triploid	Three ($3n$)
b. Tetraploid	Four ($4n$)
c. Pentaploid	Five ($5n$)
d. Hexaploid	Six ($6n$)
e. Septaploid	Seven ($7n$)
f. Octaploid	Eight ($8n$)

Monoploidy and diploidy are typically found throughout the animal and plant kingdoms, and most organisms experience both phases during their normal life cycles. Human beings, for example, are diploid organisms, having one set of paternal and one set of maternal chromosomes. Part of our life cycle, however, entails the production of haploid cells—sperm and eggs—used in the formation of the next generation. The higher plants (the angiosperms, or flowering plants) are similar in that they too are primarily diploid, but they also generate haploid nuclei in pollen and embryo sacs to initiate a new generation.

Polyploidy is found primarily in the plant kingdom. It was the basis for the origin of such important cultivated plants as cotton, wheat, oats, tobacco, potatoes, bananas, apples, and many ornamental flowers. Although some animal tissues are commonly polyploid (the liver, for example), polyploid animals are rare. Most animal polyploids reproduce either *hermaphroditically* (a single individual produces both male and female gametes, as do leeches, some turbellarian worms, and some earthworms) or *parthenogenetically* (the ovum develops without fertilization, as in brine shrimp, aphids, and rotifers). Polyploidy has been found to occur spontaneously in certain salamanders, but these individuals do not become established in the population.

Animals in general are less tolerant than plants of alterations in chromosome number. Perhaps this is because sex determination in animals generally depends on a delicate balance between the numbers of sex chromosomes and the numbers of autosomes. It may also be a reflection of different life styles. Plants are in the main sessile (not free to move); hence, they must be able to tolerate more extreme environmental changes. Animals, being generally more motile, can move away from a hostile or stressful environment. Mechanisms that can increase genetic—and hence phenotypic—variability are advantageous to organisms that are unable to change their environment, and polyploidy is one such mechanism.

Autopolyploidy is a polyploid condition in which all the chromosome sets originate from the same species. Because of this common source, the extra chromosome sets are largely homologous and therefore may pair during meiosis. Autopolyploidy may result from a number of sources. For example, it may result from fertilization of an egg (n) by more than one sperm, which could lead to zygotes that are triploid, tetraploid, and so on. Another means of generating autopolyploidy is through a mitotic aberration in which chromosomes duplicate but the nucleus fails to divide, thus creating a cell with a $4n$ nucleus. Regarding this latter mechanism, if the mitotic failure occurs at the first cleavage of the zygote, a tetraploid organism results that produces only $2n$ gametes. If the mitotic failure occurs later in the organism's development, only a portion of the cells are $4n$. If these cells happen to be germ cells, then $2n$ gametes are produced. But if the mitotic failure occurs only in somatic tissue, the tetraploidy would go no further. Figure 5.1 illustrates some of the mechanisms that induce autopolyploidy.

Fertilization involving a $2n$ gamete and a haploid (n) gamete produces a triploid organism ($3n$), which in most cases is sterile, as are almost all

Figure 5.1
Autopolyploidy, which can occur (a) when two or more sperms (all of the same species) fertilize an egg or (b) through a mitotic failure

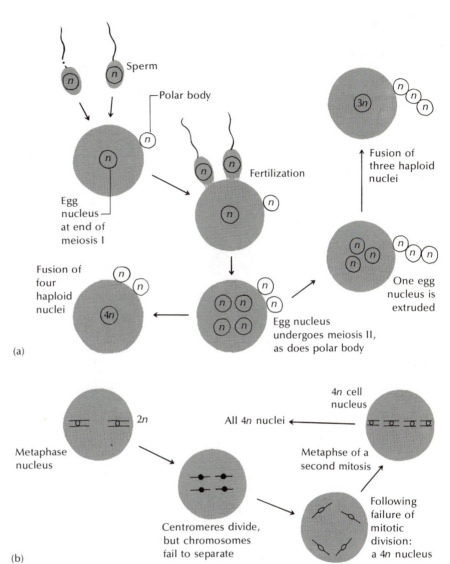

sexually reproducing organisms with odd-numbered sets of chromosomes (5n, 7n, and so on). Sterility in such polyploids is usually the result of gametes with abnormal chromosome numbers, because in the pairing of homologous chromosomes during meiosis, there can be only two homologs per chromosomal region. The third chromosome, whether or not it participates in pairing, is usually randomly distributed to the gametes, as noted in Figure 5.2. Of any three homologous chromosomes in a triploid species, two are often included in one gamete and one in the other. This distribution is random for each homologous set so that the number of chromosomes found in gametes generally varies from n to 2n with all integral values in between. Thus a triploid organism whose hap-

187

Figure 5.2
Types of meiotic
chromosome pairing
possible in a triploid
nucleus with three
homologous
chromosomes

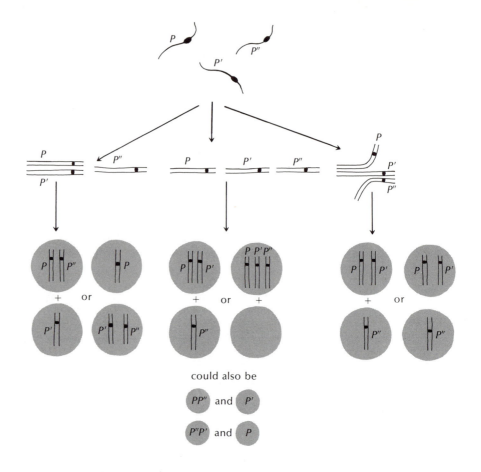

loid chromosome number (*n*) is 4 produces gametes containing 4, 5, 6, 7, and 8 chromosomes. It is this imbalance in chromosome number that makes triploids—indeed all individuals with odd-numbered chromosome sets—essentially sterile. Their sterility does not, however, mean they cannot be used. Bananas, winesap apples, and European pears are all triploid species that are propagated through asexual cuttings.

Autotetraploids are seldom sterile, because all chromosomes can pair up and segregate equally during meiosis to give 2*n* gametes. Potatoes, coffee, peanuts, and alfalfa are examples of useful autotetraploid species.

A more common type of polyploidy is *allopolyploidy*, in which chromosome sets are contributed by different species. A gamete from species A fertilizes a gamete from species B, producing an AB hybrid zygote. If the chromosomes of the species differ, chromosome pairing is for all practical purposes absent at meiosis. The result is a random distribution of chromosomes to the gametes and subsequent sterility. However, if the hybrid zygote experiences a mitotic failure leading to tetraploidy, meiosis may then occur normally, since each chromosome has a homolog with which to pair.

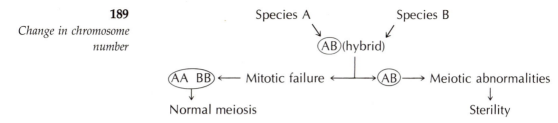

Allopolyploidy has played a role in the evolution of perhaps one-fourth of all our modern plant species, wild and domesticated. It is often difficult, though, to establish whether one is observing an autopolyploid or an allopolyploid produced by a union between morphologically similar chromosome sets. Among the noncultivated plants, the genus of ferns known as *Asplenium* (spleenwort) exemplifies speciation by allopolyploidy. *Asplenium*'s life cycle (Figure 5.3) includes a diploid sporophyte generation and a haploid gametophyte generation. In the genus

Figure 5.3
Fern life cycle, typified by
Asplenium (spleenwort)

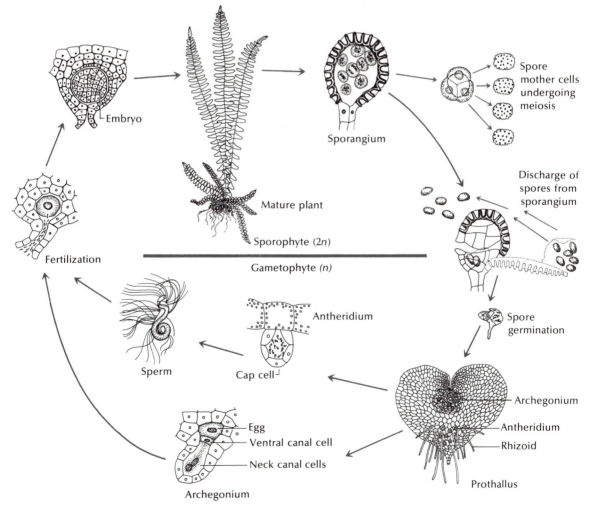

1. *A. rhizophyllum*   ×   *A. platyneuron*
     $n = 36_r$ $n = 36_p$

Zygote
$36_r + 36_p \; (2n = 72)$

Mitotic failure ⟷ No homologous chromosome
pairing, therefore sterility

$36_r + 36_r + 36_p + 36_p$
$(4n = 144)$

Normal meiosis ⟶ *A. ebenoides*
$4n = 144$
72 pairs of chromosomes observed

2. *A. rhizophyllum*   ×   *A. montanum*
     $n = 36_r$ $n = 36_m$

Zygote
$36_r + 36_m \; (2n = 72)$

Mitotic failure ⟷ No homologous chromosome
pairing, therefore sterility

$36_r + 36_r + 36_m + 36_m$
$(4n = 144)$

Normal meiosis ⟶ *A. pinnatifidum*
$4n = 144$
72 pairs of chromosomes observed

3. *A. platyneuron*   ×   *A. montanum*
     $n = 36_p$ $n = 36_m$

Zygote
$36_p + 36_m \; (2n = 72)$

Mitotic failure ⟷ No homologous chromosome
pairing, therefore sterility

$36_p + 36_p + 36_m + 36_m$
$(4n = 144)$

Normal meiosis ⟶ *A. bradleyi*
$4n = 144$
72 pairs of chromosomes observed

Asplenium, there are three primary species: *A. rhizophyllum, A. platyneuron,* and *A. montanum.* In each case, the haploid chromosome number (n) is 36. Three other *Asplenium* species (*A. ebenoides, A. pinnatifidum,* and *A. bradleyi*) are considered allotetraploids that perhaps arose as hybrids of the three primary *Asplenium* species.

The supposed origin of *A. ebenoides, A. pinnatifidum,* and *A. bradleyi* is supported by observations of chromosome pairing when the allotetraploid species are backcrossed to the primary species. Chromosome pairing is expected to occur only if chromosomes are homologous. Such pairing can be interpreted as an indication of species relatedness (noting that a pair of homologous chromosomes = bivalence and an unpaired chromosome = monovalence).

Two other *Asplenium* species, *A. gravesii* and *A. trudelli* are complicated ones. One possibility is that *A. gravesii* arose as a hybrid between *A. bradleyi* and *A. pinnatifidum.* This is deduced from the chromosome-pairing properties exhibited by this species when it is backcrossed to *A. bradleyi* and *A. pinnatifidum:* 36 paired chromosomes and 72 unpaired. The paired chromosomes in this case would probably be *A. montanum,* because *pinnatifidum* and *bradleyi* both have two sets of *montanum* chromo-

Cytological observations of triploid hybrid at meiosis:

1. *A. ebenoides* × *A. rhizophyllum*
   $n = 72$ $n = 36_r$
   $(36_r + 36_p)$

   36 pairs of chromosomes
   $(36_r + 36_r)$
   36 unpaired chromosomes (36_p)

2. *A. ebenoides* × *A. platyneuron*
   $n = 72$ $n = 36_p$
   $(36_r + 36_p)$

   36 pairs of chromosomes
   $(36_p + 36_p)$
   36 unpaired chromosomes (36_r)

3. *A. ebenoides* × *A. montanum*
   $n = 72$ $n = 36_m$
   $(36_r + 36_p)$

   108 unpaired chromosomes
   $(36_r + 36_p + 36_m)$

4. *A. pinnatifidum* × *A. rhizophyllum*
   $n = 72$ $n = 36_r$
   $(36_r + 36_m)$

   36 pairs of chromosomes
   $(36_r + 36_r)$
   36 unpaired chromosomes (36_m)

5. *A. pinnatifidum* × *A. montanum*
   $n = 72$ $n = 36_m$
   $(36_r + 36_m)$

   36 pairs of chromosomes
   $(36_m + 36_m)$
   36 unpaired chromosomes (36_r)

6. *A. pinnatifidum* × *A. platyneuron*
   $n = 72$ $n = 36_p$
   $(36_r + 36_m)$

   108 unpaired chromosomes
   $(36_r + 36_p + 36_m)$

7. *A. bradleyi* × *A. platyneuron*
   $n = 72$ $n = 36_p$
   $(36_p + 36_m)$

   36 pairs of chromosomes
   $(36_p + 36_p)$
   36 unpaired chromosomes (36_m)

8. *A. bradleyi* × *A. montanum*
   $n = 72$ $n = 36_m$
   $(36_p + 36_m)$

   36 pairs of chromosomes
   $(36_m + 36_m)$
   36 unpaired chromosomes (36_p)

9. *A. bradleyi* × *A. rhizophyllum*
   $n = 72$ $n = 36_r$
   $(36_p + 36_m)$

   108 unpaired chromosomes
   $(36_r + 36_p + 36_m)$

somes. The 72 nonpairing chromosomes could have come from *rhizophyllum* (36) and *platyneuron* (36), nonhomologous sets. *A. trudelli* appears to be a hybrid between *A. montanum* and *A. pinnatifidum*. In backcrosses with these two species, 36 bivalents and 36 univalents are observed. The bivalents are probably *A. montanum* and the univalents *A. rhizophyllum*.

Among the cultivated plants, none has been more essential to our cultural growth than wheat, another example of evolution through allopolyploidy. The 14 species of wheat fall into three general groups, one containing 14 chromosomes, the second 28, and the third 42. The most ancient is the 14-chromosome group, which includes two species, *Triticum monococcum*, known as einkorn (one seed), and *Triticum boeoticum*, or wild einkorn. Neither species is very useful for human consumption, because the wheat grain is tightly enclosed in glumes, or bracts, which makes hulling difficult. The two einkorn species are now used principally as animal feed, although they are still cultivated in the hilly regions of southern Europe and the Middle East for use in some dark breads. Einkorn probably evolved from wild einkorn, the more primitive species, which originated in the Middle East over 10,000 years ago.

The 28-chromosome group, the emmer wheats, consists of seven species. It was apparently derived from the hybridization of wild einkorn with a wild grass of a different genus called *Aegilops* (goat grass). Such a cross produces a vigorous but sterile hybrid. However, if chromosome doubling occurred, a vigorous *fertile* allotetraploid hybrid could have resulted. An allotetraploid species so formed could be the progenitor of the wild emmer wheat, *T. dicoccoides*. The "true emmer," *T. dicoccum*, could have evolved from the wild emmer without further changes in chromosome number. The emmer wheats, which originated in the Middle East over 8000 years ago, also have clinging glumes and are therefore principally used as livestock feed. But one of them, durum wheat (*T. durum*), has a high gluten content (a substance that makes flour sticky when wet), which makes it particularly suitable for the production of macaroni and spaghetti. Durum glumes are not as tenacious as those of other emmer wheats and einkorn wheat.

Bread wheat, a 42-chromosome group ($6n$) that includes five species, is by far the most useful of the wheats. The bread wheats probably originated through the hybridization of a tetraploid emmer wheat with yet another species of *Aegilops*, followed by the doubling of the chromosomal complement. Goat grass was and continues to be a common weed in fields of tetraploid wheat, so opportunities for hybridization would have been numerous. Hybridization was probably accidental, but once it occurred, humans selected the new, more desirable hybrid. The most common bread wheat, *T. aestivum*, may have originated through hybridization between the tetraploid Persian emmer wheat (*T. dicoccum*) and the goat grass of eastern Turkey and northwestern Iran. Figure 5.4 summarizes wheat evolution.

In conclusion, both autopolyploidy and allopolyploidy are means by which new species can evolve essentially in one generation.

Figure 5.4
Left to right, einkorn (*Triticum monococcum*), *Aegilops speltoides*, wild emmer (*T. dicocoides*), emmer (*T. dicoccum*), *A. squarrosa*, and common bread wheat (*T. aestivum*). The diagram shows the history of diploid, tetraploid, and hexaploid cultivated wheat. [Reprinted by permission of the publisher from *Plants and Civilization*, 2nd edition, by Herbert G. Baker, copyright © 1970 by Wadsworth Publishing Company, Belmont, Calif. 94002]

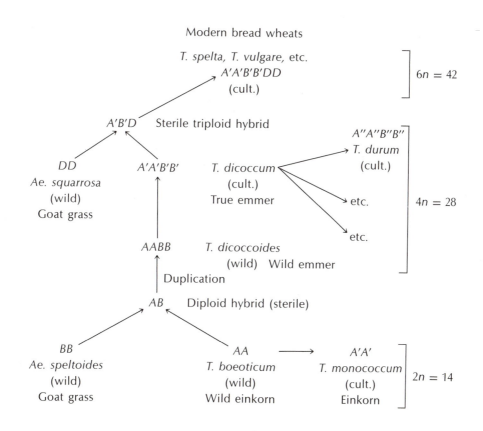

Modern bread wheats

T. spelta, T. vulgare, etc.
A′A′B′B′DD
(cult.)

$6n = 42$

A′B′D   Sterile triploid hybrid

A″A″B″B″
→ *T. durum*
(cult.)

DD
Ae. squarrosa
(wild)
Goat grass

A′A′B′B′

T. dicoccum
(cult.)
True emmer

etc.

$4n = 28$

etc.

AABB   *T. dicoccoides*
(wild) Wild emmer

Duplication

AB   Diploid hybrid (sterile)

BB
Ae. speltoides
(wild)
Goat grass

AA
T. boeoticum
(wild)
Wild einkorn

→   *A′A′*
T. monococcum
(cult.)
Einkorn

$2n = 14$

During meiosis in a diploid organism, homologous pairs of chromosomes usually synapse, later to separate and be distributed equally into gametes and spores. During mitosis, daughter cells are formed that contain identical genomes. Occasionally, however, these processes go awry, and cells are produced that are deficient or duplicated for a particular chromosome. Such cells are said to be *aneuploid* and generally result from chromosome nondisjunction during meiosis (see pp. 60–64). Bridges suggested that, in the meiotic process, aneuploids arise when homologous chromosomes fail to synapse properly. Such a failure could cause individual chromosomes to arrive at the metaphase plate separately and then randomly proceed to the poles during anaphase. The end result of this is often the formation of gametes or spores with one too many or one too few chromosomes. In mitosis, an aneuploid cell can be formed when a chromosome fails to form an attachment with the spindle fiber apparatus and is thus excluded from the telophase nuclei.

An $n + 1$ gamete (haploid with one extra chromosome) can fertilize a normal gamete (n) to give rise to a $2n + 1$ individual, *trisomic* for the chromosome involved. An $n - 1$ gamete (haploid with one chromosome missing) can fertilize a normal gamete to give a $2n - 1$ individual, a *monosomic* for the chromosome involved. The loss of a pair of homologous chromosomes ($2n - 2$) is termed *nullisomy*, and the normal condition ($2n$) is often termed *disomy*.

Monosomy is rare in nature, because a normally diploid organism can rarely survive the loss of all the genetic information contained in a chromosome. If the chromosome involved is small, however, and is not crucial to survival, its loss might not be lethal. In *Drosophila*, for example, the fourth chromosome is sufficiently minute that its loss can be tolerated. The phenotype of a *Drosophila* fly containing only a single fourth chromosome (a haplo-IV) is shown in Figure 5.5. The fly is pale, has a prominent trident pattern on the thorax, is usually sterile, and exhibits erratic viability. A sex chromosome, on the other hand, can be lost in organisms such as *Drosophila* and humans without being lethal. **XO** in *Drosophila*, for instance, is a morphologically normal but sterile male. As we shall soon discuss, **XO** in humans is a morphologically abnormal female. However, **YO** is lethal in both species.

Trisomy is generally not so detrimental. A triplo-IV *Drosophila* female is almost indistinguishable from a normal female. In humans, trisomics are not uncommon and are responsible for many maladies, some of which we will discuss shortly.

Aneuploidy is an important source of intraspecific variation, as evidenced in studies of *Datura stramonium*, or jimsonweed, and its various types. *Datura* has a haploid set of 12 chromosomes, so 12 different trisomics are possible. Indeed, precisely 12 trisomics are known in *Datura*, each trisomic involving a different chromosome pair and each phenotypically unique (Figure 5.6). Trisomy has also been found to be an important source of variation in *Clarkia*.

Figure 5.5
Haplo-IV *Drosophila* (only one chromosome IV)

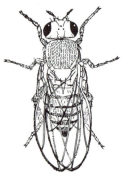

Normal adult

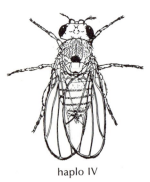

haplo IV

Figure 5.6
Variations in *Datura* (jimsonweed) due to changes in chromosome number [Reprinted from A. F. Blakeslee, "Variations in *Datura* Due to Changes in Chromosome Number," *American Naturalist* 56 (1922), by permission of the University of Chicago Press]

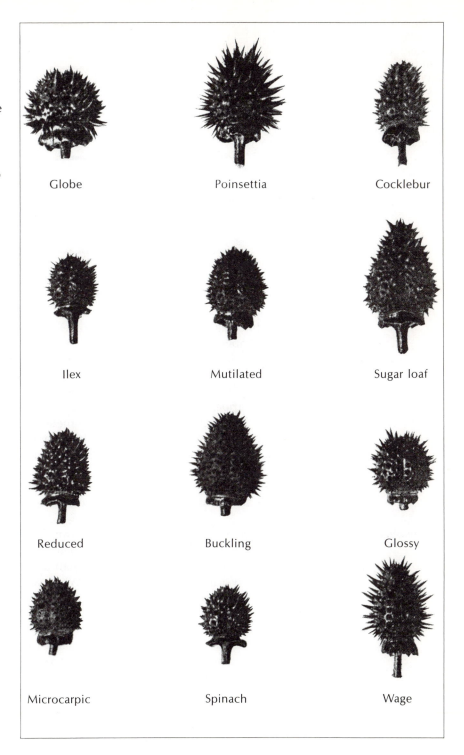

Globe   Poinsettia   Cocklebur

Ilex   Mutilated   Sugar loaf

Reduced   Buckling   Glossy

Microcarpic   Spinach   Wage

Figure 5.7
Formation of a
metacentric chromosome
from (a) two telocentric
chromosomes and (b) two
acrocentric chromosomes.
(c) Formation of two
telocentric chromosomes
from a metacentric
chromosome.

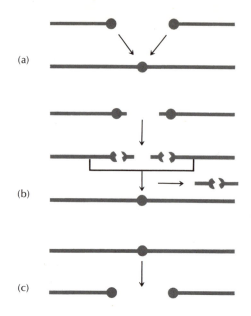

Abnormal numbers of chromosomes are expected in aneuploidy, but it can happen that chromosome numbers change even though the amount of genetic information changes little, if at all. For example, two chromosomes with centromeres located near their tips (*acrocentric*) or at their tips (*telocentric**) might join, giving rise to a larger chromosome whose centromere appears to be located in the middle (*metacentric*, as shown in Figures 5.7a and 5.7b). Conversely, a metacentric chromosome could become two telocentric chromosomes by breaking in the middle of the centromere (Figure 5.7c). Such changes do not modify the amount of genetic information, only its arrangement. The joining and dissociation of centromeres is common in such organisms as the common shrew and the marine snail, but the significance of such events has yet to be determined.

Aneuploidy in humans   Although aneuploidy may be a significant force in the speciation of some organisms, its effects are generally deleterious. Humans are certainly no exception to this generalization. Human aneuploids sometimes express severe physical and mental defects. While we must provide the afflicted with proper care, we must also try to learn more about the mechanisms that lead to such chromosomal diseases in order to prevent their future occurrence.

Techniques enabling us to trace a disease or a syndrome in humans to a chromosome anomaly were first announced in 1956 by J. H. Tjio and A. Levan. They treated human embryonic lung cells with colchicine (a chemical that prevents cells from dividing but permits chromosomes to replicate) followed by hypotonic salt solution (water enters the nuclei, causing them to swell and the chromosomes to move apart). The cells were then stained, thus prepared for cytological observation (Figure 5.8). Using this

* There is currently a great deal of debate about the existence of telocentric chromosomes.

Figure 5.8
Technique for preparing a slide of human chromosomes (Photographs courtesy of Dr. J. H. Tjio)

White blood cells cultured in a tissue culture medium

↓

White blood cells treated with colchicine, resulting in the accumulation of dividing cells in metaphase, assuring, therefore, an adequate number of cells at a stage of division satisfactory for studying chromosomes

↓

Treatment of cells with a hypotonic solution producing swelling of the nuclei and the spreading of the chromosomes, thus facilitating their study

↓

Chromosomes treated with a protein denaturing agent such as trypsin, then stained with Geimsa stain, producing the pattern of light and dark bands

↓

Chromosomes photographed, cut out, and arranged according to size:

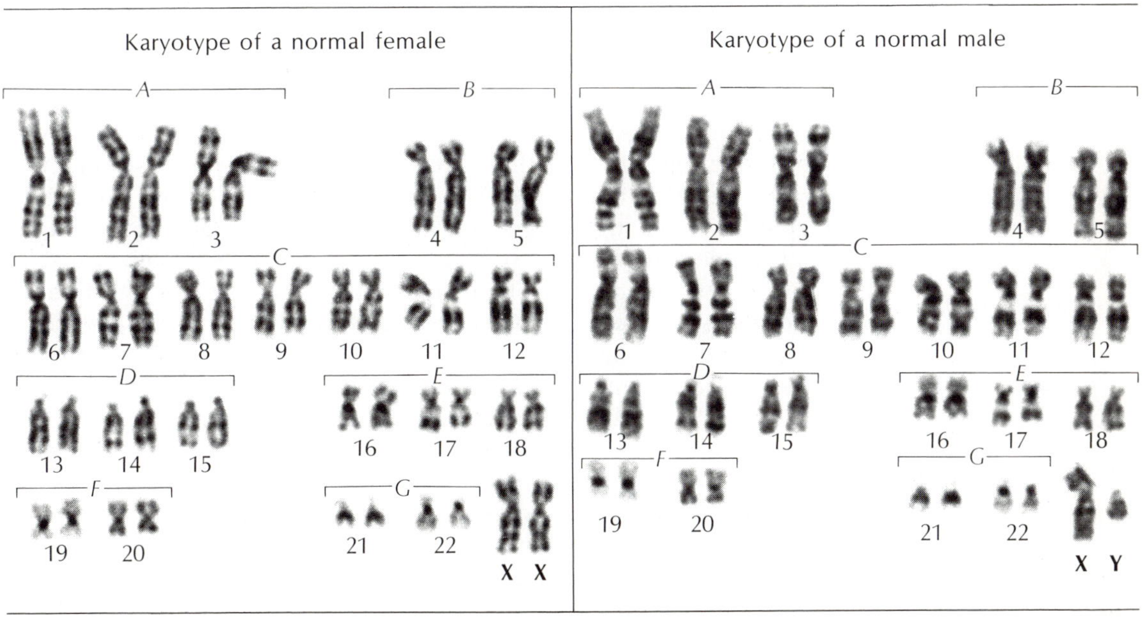

Karyotype of a normal female	Karyotype of a normal male

technique, three human chromosome anomalies were discovered in 1959: Klinefelter's syndrome, an **XXY** condition; Down's syndrome, a trisomy for chromosome 21; and Turner's syndrome, an **XO** condition. Since then, several other chromosome anomalies have been identified. Human diseases caused by aneuploidy can be divided into sex-chromosome aneuploids and autosomal aneuploids.

Sex-chromosome aneuploidy Individuals with *Klinefelter's syndrome* are phenotypically male but somewhat feminized physically. In the more extreme cases, they have high-pitched voices, long legs, female breast and hip development, and sparse body hair (Figure 5.9). Jacobs and Strong showed that this syndrome is due to the presence of an extra **X** chromosome in otherwise normal males. Such **XXY** individuals thus have 47 chromosomes. They are infertile and usually mentally subnormal, although a few with normal male phenotypes and normal intelligence have been described. The frequency of Klinefelter's syndrome is about 1 in 800 live male births. It is difficult to recognize this anomaly at birth, because the departure from normal phenotype is not pronounced. In Klinefelter's syndrome, though the phenotypic sex is male, the nuclei have a Barr body (see page 48).

The Lyon hypothesis and Barr bodies raise an interesting problem in human aneuploids. Why should an **XXY** be abnormal if in fact one of the **X** chromosomes is inactivated? And why should increasing numbers of **X** chromosomes produce increasingly severe defects if all but one are inactivated? The answer to this paradox may lie in *when* inactivation takes place. If it occurs later in development, then in the early crucial stages of development, the sex-chromosome imbalance exists and leads to the observed abnormalities associated with this imbalance. In other words, the damage is done *before* the extra **X** chromosome(s) is (or are) inactivated.

C. E. Ford and his colleagues demonstrated that another human affliction—*Turner's syndrome*—is due to the absence of a sex chromosome. Individuals with Turner's syndrome have 45 chromosomes and are monosomic for the **X** chromosome; that is, they are **XO**. These persons are phenotypic females, less than five feet tall, web-necked, skeletally abnormal, sexually infantile, sterile, and usually have below average IQs (Figure 5.10). Though phenotypically female, they do not have a Barr body. Turner's syndrome occurs about once per 2000 female births. The loss of genetic information should make this a more serious affliction than Klinefelter's syndrome, but its defects are relatively minor. Interestingly, though its defects are relatively minor, **XO** is the most frequent chromosome anomaly found in first-trimester spontaneous abortions. It is estimated that fewer than 2 percent of all **XO** zygotes go to full term.

Studies of the **X** chromosomes in **XO** and **XXY** syndromes reveal information about the origin of the aneuploidy event. For example, from studies of certain blood-group characters determined by **X**-linked genes, we know that about 75 percent of the **XO** females have their maternal **X**. This means that 75 percent of Turner's syndrome females are the result of

Figure 5.9
Klinefelter's syndrome
(**XXY**) [Photos courtesy of
Dr. D. S. Borgoankar,
Johns Hopkins School of
Medicine]

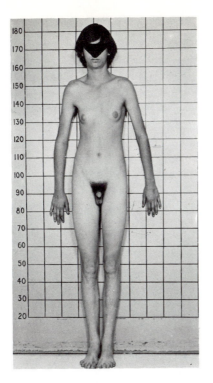

An individual with Klinefelter's
syndrome has normal male external
genitalia but small testes and sparse
body hair. There is usually some
female breast development and
femaleness in body conformation.
There are 47 chromosomes (an
extra **X**).

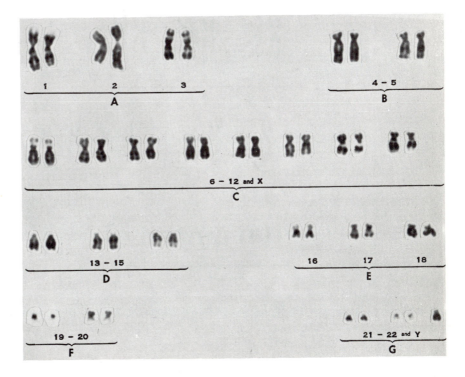

Figure 5.10
Turner's syndrome (**XO**)
[Photos courtesy of
Dr. D. S. Borgoankar,
Johns Hopkins School of
Medicine]

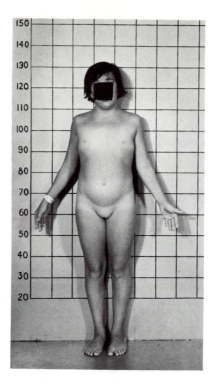

An individual with Turner's syndrome
has female external genitalia, is of
short stature, and has a webbed neck,
shield-like chest with widely spaced
nipples, undeveloped breasts, and
rudimentary ovaries. There are 45
chromosomes (a missing **X**).

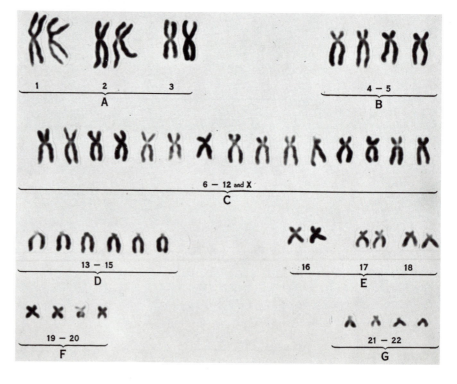

paternal nondisjunction. For Klinefelter's syndrome, the opposite is true: About 65 percent of all **XXY** males are the result of maternal nondisjunction. There is no explanation for this discrepancy.

Females with an **XXX** constitution are known too. Usually, they are physically normal and fertile, though somewhat below average in IQ. They occur with an estimated frequency of about once per 500 female births. Gametes produced by these females are **XX** and **X** and hence can theoretically produce **XXY**, **XXX**, **XY**, and **XX** offspring. However, some **XXX** females (termed *metafemales*) produce only normal offspring. The reason for the lack of **XXY** and **XXX** offspring is unknown.

The recent discovery of **XYY** males has generated a great deal of interest and controversy because of the possible link between the extra **Y** chromosome and aberrant behavior. Many such males, who were first discovered in penal institutions, have been found to be aggressive, taller than average, and lower than average in IQ. The presence of the extra **Y** chromosome may be in part responsible for these qualities, but supportive evidence for this contention may prove difficult to obtain, because many **XYY** males have been found who appear to be perfectly normal by all standards of evaluation. Much more research into the **XYY** phenotype is needed before it can be properly characterized. At this writing, it appears that the presence of an additional **Y** chromosome may predispose an individual to a certain degree of aggressive behavior, and to an extended growth period. But another possibility is that the extra **Y** causes extended growth, low IQ, acne, and thus leads to poor social adjustment and hence aggressiveness.

The occurrence of **XYY** males in the general population has created a number of difficult ethical issues. Should **XYY** males be identified at birth, or even prenatally? If identified prenatally, should they be aborted? Should the parents of an **XYY** boy be informed of his condition and told that he may be a behavior problem, even though there is no evidence to prove that he will? How will the knowledge that the child is **XYY** affect the parent–child relationship? Such difficult questions pervade this and other genetic problems.

Figure 5.11 summarizes sex-chromosome variants in human beings.

Autosomal aneuploidy in humans   The first chromosomal abnormality identified in humans was *Down's syndrome* (formerly referred to by the inappropriate term *Mongoloid idiocy*). J. Lejeune and his colleagues (1959) ascribed this syndrome to the presence of an extra chromosome 21 (trisomy 21). In addition to mental defectiveness, individuals with Down's syndrome have short stature, short, stubby fingers and toes, round faces, large protruding tongues, and characteristic fingerprints and palmprints (Figure 5.12). They are also very susceptible to infectious diseases and circulatory illnesses and consequently have a life expectancy of only about 30 years. The frequency of Down's syndrome is about 1 in every 600 births, but there is a marked variation in this frequency as a function of maternal age (Figure 5.13). The older the mother, the greater

Figure 5.11
Human sex-chromosome anomalies related to the quantity of chromosomal material [Adapted from D. D. Federman, M. D., *Abnormal Sex Development*, Philadelphia: W. B. Saunders Company, 1967]

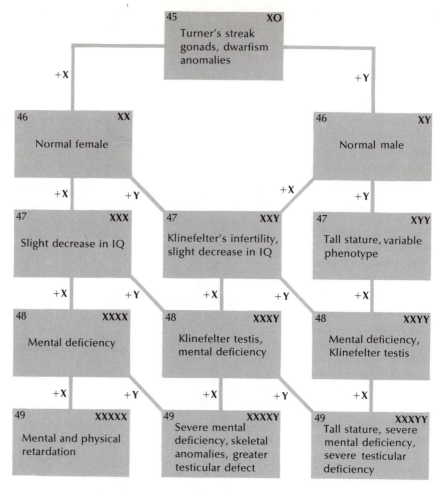

the chance that her child will be born with Down's syndrome. Recent evidence indicates that the nondisjunctional event that accounts for Down's syndrome probably occurs in the ovum during meiosis. Although trisomy 21 is the common cause of Down's syndrome, the condition is also caused by a translocation between chromosome 21 and either chromosome 14 or 15 (Figure 5.14a) in about 4 percent of the cases. Translocation 14-21 consists of most of chromosome 14 with a segment of chromosome 21 joined to it. The gametes formed from a normal cell that has undergone a 14-21 translocation are shown in Figure 5.14b, along with the zygotes formed after fertilization by normal gametes. Translocation Down's syndrome occurs when three doses of chromosome 21 are present, but a study of the figures shows that one of the zygotes will be normal but a carrier. In general, autosomal trisomics have more severe defects than sex-chromosome trisomics. Consequently, only a few survive to birth that involve chromosomes larger than chromosome 21, among which are *trisomy 18* and *trisomy 13*.

Figure 5.12
Down's syndrome
(trisomy 21). The
individual shown has a
partly shaved head.
[Photos courtesy of
Dr. D. S. Borgoankar,
Johns Hopkins School of
Medicine]

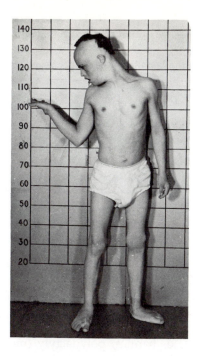

This anomaly is characterized by
mental retardation, an eyelid
abnormality, short stature, stubby
hands and feet, characteristic palm
prints, and internal abnormalities,
especially of the heart.

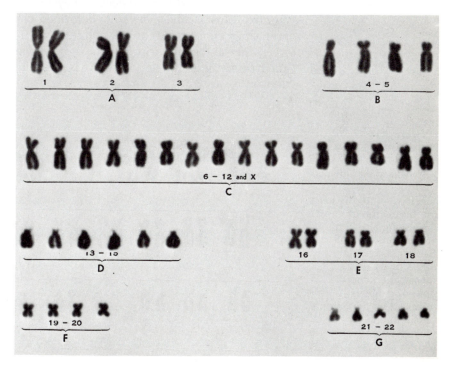

Trisomy 18, discovered in 1960, is a syndrome characterized by ear
deformities, spasticity (marked by involuntary, sometimes convulsive,

Figure 5.13
Age distribution of mothers of patients with Down's syndrome compared with that of all mothers [Data from L. S. Penrose, *Int. Copenhagen Cong. Sci. Study Ment. Retard.* I, 165 (1964)]

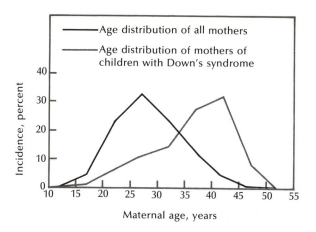

muscle contractions) caused by defects in the central nervous system, "rocker-bottom" feet, flexion deformity of the fingers, small lower jaw, umbilical hernia, and congenital heart defects. Afflicted individuals rarely survive more than a few months (Figure 5.15a). Trisomy 13 (Figure 5.15b) has been found in a number of newborn infants, who express severe eye defects, cleft palate, harelip, polydactyly (more than 20 digits), severe brain damage, and heart defects. They too rarely live more than a few months. Most of these severely deformed children are aborted early in pregnancy, so statistics on the condition are sketchy.

Other autosomal trisomies have been reported, such as trisomy 12 and trisomy 22, but their occurrence is rare. Until recently, they have been difficult to study.

Figure 5.14
Translocation Down's syndrome. (a) Karyotype of translocation Down's syndrome. Note translocation of the long arm of a chromosome 21 to the long arm of a chromosome 14. (b) (See opposite page.) Origin of the 14-21 translocation and the types of gametes produced by the original translocation heterozygote. (See Figure 5.36 for meiosis in a translocation heterozygote.)

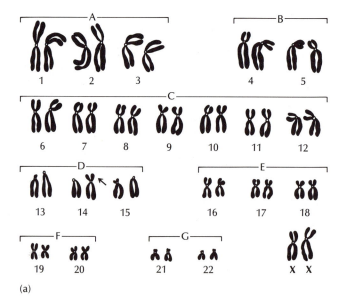

(a)

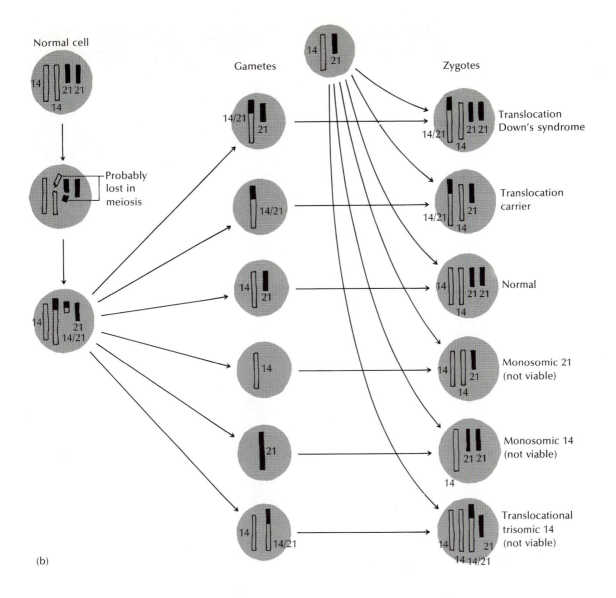

Normal cell

Gametes

Zygotes

Probably lost in meiosis

14/21 → Translocation Down's syndrome

14/21 → Translocation carrier

Normal

Monosomic 21 (not viable)

Monosomic 14 (not viable)

Translocational trisomic 14 (not viable)

(b)

Aneuploid mosaics A chromosomal mosaic is an organism that has cells of different chromosomal constitutions. They generally arise because the fusion of two normal haploid gametes is accompanied by a mitotic aberrancy that produces daughter cells with varying chromosomal constitutions. These individuals are called mosaic aneuploids. Many examples of chromosomal mosaics are known in which some cells have one genotype (**XO**, for example) and other cells have another genotype (**XX**). Mosaics can arise in two ways. Let us initially examine the events that could lead to a mosaic with **XO** and **XX** cells. First, the two **X** chromo-

205

Figure 5.15
(a) An infant with trisomy 18 (From J. S. and M. W. Thompson, *Genetics in Medicine,* 2nd ed., Philadelphia, W. B. Saunders, 1973)
(b) An individual with trisomy 13 (Courtesy of Dr. Irene Uchida)

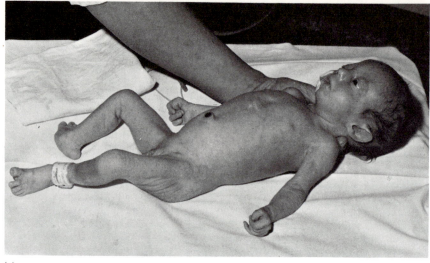

(a)

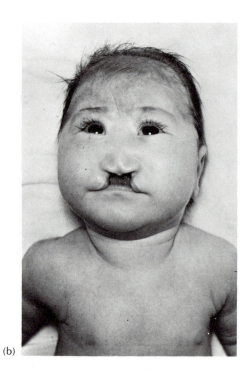

(b)

somes of a normal female zygote are aligned at the metaphase plate during mitosis. But only one of the two chromosomes replicates, which results in one cell having only one **X** chromosome (**XO**) and one having two (**XX**). Another explanation may be that a centromere fails to divide, producing an **XX–XO** mosaic:

One explanation of an
aneuploid mosaic: only one
centromere divides

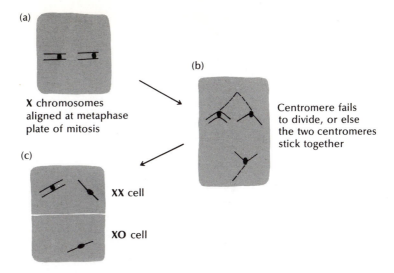

(a)

X chromosomes
aligned at metaphase
plate of mitosis

(b)

Centromere fails
to divide, or else
the two centromeres
stick together

(c)

XX cell

XO cell

In other cases, chromosome alignment and centromere division proceed normally, but the attachment of the spindle fiber fails, leaving one or
more chromosomes to float randomly. In that case, either a mosaic* or a
normal individual can result, as indicated in the diagram below.

* The mosaic would be **XXX/XO** or **XX/XO**.

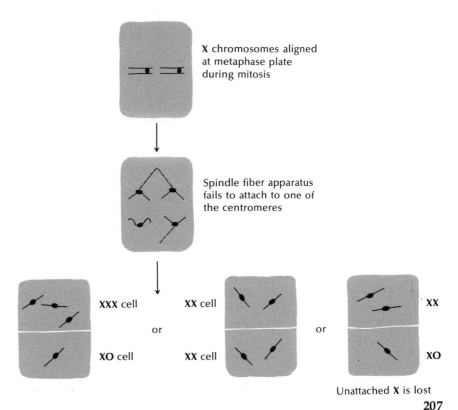

X chromosomes aligned
at metaphase plate
during mitosis

Spindle fiber apparatus
fails to attach to one of
the centromeres

XXX cell

XO cell

or

XX cell

XX cell

or

XX

XO

Unattached **X** is lost

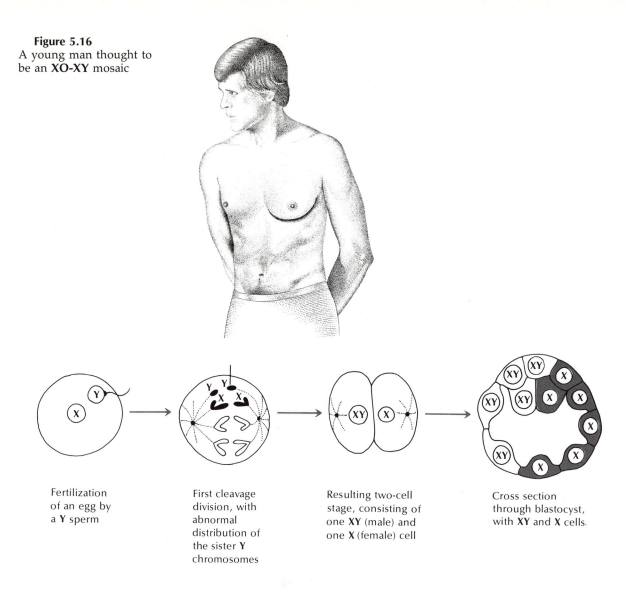

Figure 5.16
A young man thought to be an **XO-XY** mosaic

Fertilization
of an egg by
a **Y** sperm

First cleavage
division, with
abnormal
distribution of
the sister **Y**
chromosomes

Resulting two-cell
stage, consisting of
one **XY** (male) and
one **X** (female) cell

Cross section
through blastocyst,
with **XY** and **X** cells.

Figure 5.16 shows a young man thought to be an **XY–XO** mosaic. Such a condition is perhaps caused by the failure of a **Y** chromosome to move to a pole during mitosis. Thus it is lost.

In *Drosophila*, a striking sex-chromosome mosaic exists called a *gynandromorph*. Discovered by Morgan, this mosaic is phenotypically half male and half female (Figure 5.17). It can occur because *Drosophila* lacks sex hormones, so tissues develop autonomously. Since sex in *Drosophila* is determined by the number and ratio of **X** chromosomes to autosomes, the generation of gynandromorphs is, given nondisjunction, a reasonable expectation. The occurrence of the gynandromorph shown in Figure 5.17 might be explained by the events in the diagram that is below it.

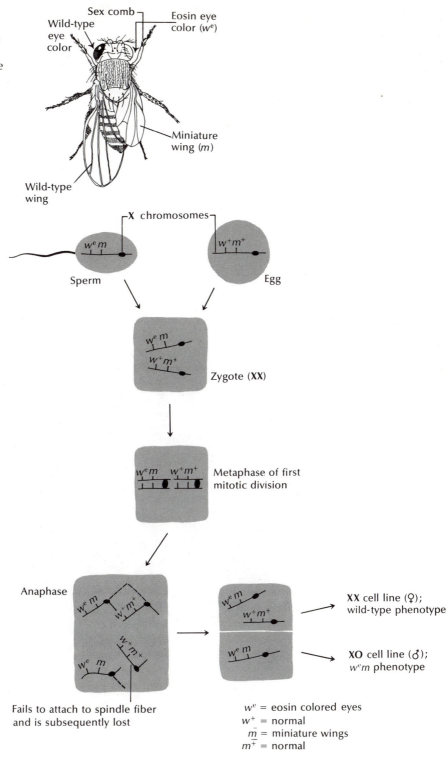

Figure 5.17
Drosophila gynandromorph: The left side of the body is female, and the right side is male

Sex comb

Wild-type eye color

Eosin eye color (w^e)

Miniature wing (m)

Wild-type wing

X chromosomes

$w^e\,m$ Sperm

$w^+\,m^+$ Egg

$w^e\,m$
$w^+\,m^+$ Zygote (**XX**)

$w^e\,m$ $w^+\,m^+$ Metaphase of first mitotic division

Anaphase

$w^e\,m$ $w^+\,m^+$

$w^+\,m^+$

$w^e\,m$

Fails to attach to spindle fiber and is subsequently lost

$w^e\,m$
$w^+\,m^+$

$w^e\,m$

XX cell line (♀); wild-type phenotype

XO cell line (♂); $w^e m$ phenotype

w^e = eosin colored eyes
w^+ = normal
$\underline{m}$ = miniature wings
m^+ = normal

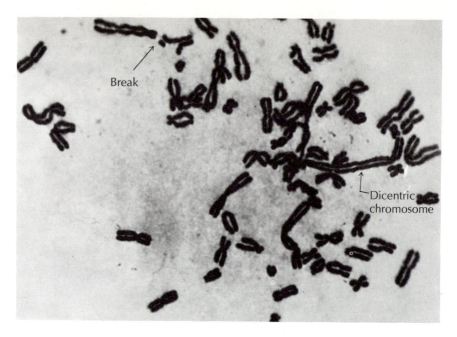

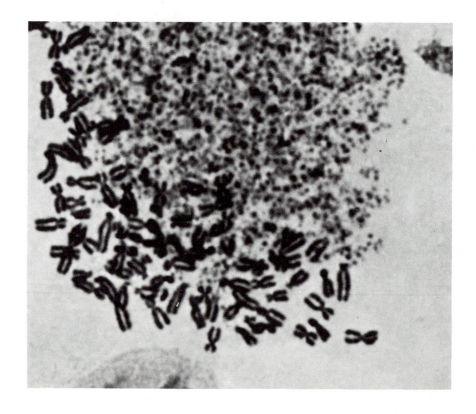

The factors that cause nondisjunction are not well understood. An interesting, though highly speculative, explanation of the increased incidence of the nondisjunction that causes Down's syndrome in the offspring of older mothers is the decreasing frequency of sexual intercourse as age increases. The eggs available for fertilization might have to wait longer for contact with spermatozoa, and this longer waiting period may entail some degenerative processes in the egg that could interfere with meiosis (meiosis II in an oocyte does not occur until *after* fertilization). This hypothesis is poorly substantiated by experimental data, however. Viruses have also been implicated in nondisjunction. The mumps and herpes viruses have been seen associated with the spindle fiber apparatus of cells and may increase the incidence of nondisjunction by disrupting the attachment of centromeres to spindle fibers. The parainfluenza virus seems to cause the actual disintegration of some chromosomes, and other viruses are known to induce a variety of chromosomal aberrations (Figure 5.18).

In summary, both euploidy and aneuploidy have important biological consequences. Euploidy, the addition or deletion of entire chromosome sets, is much more common in plants than in animals and has been instrumental in the evolution of many important agricultural crop plants. Aneuploidy, the addition or deletion of individual chromosomes, may contribute to the process of speciation of some plants, but its impact on animals appears to be largely negative. Most animal aneuploids have severe physical and mental defects. The addition and deletion of chromosomes often results from meiotic abnormalities.

Change in chromosome structure

Structural changes in chromosomes entail two basic types of events. The first type, the addition or deletion of chromosome segments, is analogous to aneuploidy but occurs at the level of the gene or chromosome segments rather than involving the entire chromosome. The second type is characterized by the rearrangement of the linear sequence of genes.

Duplications   The presence of an extra, repeated piece of chromosomal material beyond the normal complement is known as a *duplication*. The extra piece may be integrated into a chromosome, or it may remain free with its own centromere. If it is integrated, the duplication could occur in one of several configurations, depending on the position and sequence of the duplicated genes, as shown at the top of the next page.

ABCDEFGHIJKLMNOPQRS   Normal chromosome
→

ABCDEDEFGHIJKLMNOPQRS   Direct tandem duplication
→→

ABCDEEDFGHIJKLMNOPQRS   Reversed tandem duplication
→←

ABCDEFGHIJKLDEMNOPQRS   Transposed direct duplication
→   →

ABCDEFGHIJKLEDMNOPQRS   Transposed reversed duplication
→   ←

There are several methods by which duplications arise. One of the most common is unequal crossing-over, a meiotic crossover that produces one chromosome with a duplicated segment and one with a deficiency for the same segment. An example of this, the bar-eye mutation (*B*) in *Drosophila*, has had a significant impact on genetic theory. Bar eye is a sex-linked, incompletely dominant mutation that causes a reduced number of facets in the eye, which thereby takes on a bar appearance. Homozygous stocks of bar-eye mutants have occasionally produced flies with normal eyes and flies even more extremely bar (double bar) in approximately equal frequencies (about 1 in every 600 flies). The frequency suggests that the bar locus is very unstable, but the appearance in equal numbers of normal *and* double-bar flies is difficult to explain. Furthermore, normal and double-bar flies appear only in the offspring of bar females crossed to normal males, never in the offspring of bar males crossed to normal females. And *Drosophila* is unusual in that meiotic crossing-over occurs only in females, which implies that bar might somehow be related to meiotic crossing-over. To detect the crossing-over, however, genes to the left and right of the bar region would have to be followed.

In 1925, Sturtevant presented evidence that these normal and double-bar flies are indeed a result of crossing-over in the homozygous bar stock. Females homozygous for bar but heterozygous for forked bristles (*f*, located 0.2 map units to the left of bar) and for fused wing veins (*fu*, located 2.5 map units to the right of bar) were crossed with *f B fu* males. In the subsequent generation, Sturtevant obtained some normal and some double-bar progeny, both associated with recombinant chromosomes. He proposed that the bar-eye phenotype arose from a duplicated segment of the chromosome that had resulted from unequal crossing-over. The mutation of bar both to normal and to double bar was also due to unequal crossing-over, as we see at the top of the next page.

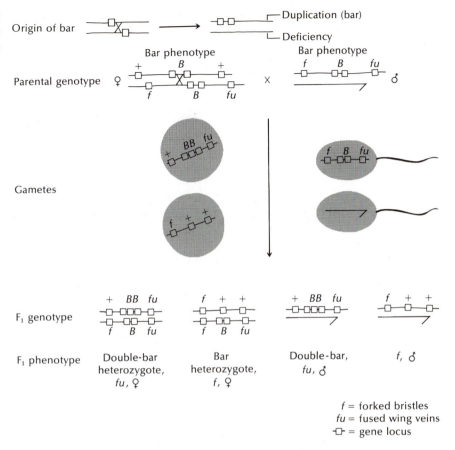

Origin of bar — Duplication (bar) — Deficiency

Bar phenotype

Bar phenotype

Parental genotype ♀ × ♂

Gametes

F₁ genotype

F₁ phenotype | Double-bar heterozygote, fu, ♀ | Bar heterozygote, f, ♀ | Double-bar, fu, ♂ | f, ♂

f = forked bristles
fu = fused wing veins
-⊡- = gene locus

Sturtevant adduced his hypothesis of unequal crossing-over from his recombinational studies. It was strengthened in 1936, when Bridges made a *cytological* investigation of the chromosomes of bar, double-bar, and normal *Drosophila.* The segment of the **X** chromosome known as 16A was present once in normal, twice in bar, and three times in double-bar flies (Figure 5.19). This was cytological verification of genetic recombination.

The bar duplication illustrates yet another important genetic phenomenon—*position effect,* a phenotypic alteration based solely on a

Figure 5.19
Diagram illustrating the number of 16A segments of the **X** chromosome in normal, bar, and double-bar *Drosophila*

Normal

Bar

Double-bar

Figure 5.20
Phenotype of the eyes of a female *Drosophila* carrying different numbers and arrangements of segment 16A of the **X** chromosome

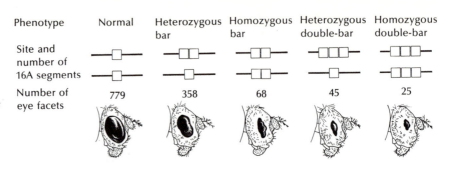

Phenotype	Normal	Heterozygous bar	Homozygous bar	Heterozygous double-bar	Homozygous double-bar
Site and number of 16A segments					
Number of eye facets	779	358	68	45	25

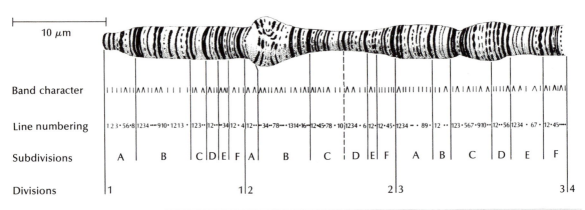

10 μm

Band character

Line numbering

Subdivisions

Divisions

Figure 5.21
Diagram of the **X** chromosome of *Drosophila* and a photograph of the salivary-gland chromosomes showing bands (chromomeres) [By permission of the American Genetic Association, courtesy of Dr. Helen Gay]

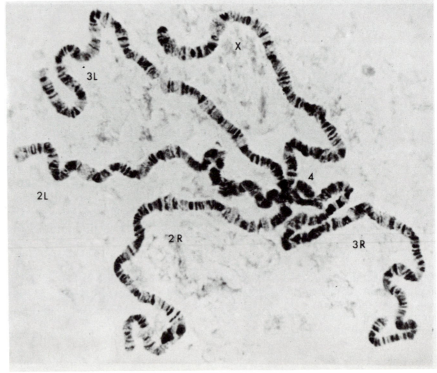

change in gene location (Figure 5.20). For example, four doses of segment 16A give varying phenotypes depending on how the four segments are distributed to the two homologous chromosomes. Three 16A segments on one homolog and one segment on the other give a bar phenotype averaging only 45 facets per eye, whereas two 16A segments on each homolog produce a bar phenotype averaging 68 facets per eye. This difference is due to the position and location of the segments, not their number.

Duplications are significant in part because of their impact on speciation and on the process of genetic divergence. Organisms belonging to different species often have different quantities of genetic information. For example, the relatively simple bacterium *Escherichia coli* has about 3000 genes, whereas *Drosophila melanogaster* is estimated to have from 5000 to 10,000. Evolution from the earliest of living organisms to more advanced and complex organisms is undoubtedly characterized by an increase in the amount of genetic information, but since genetic information does not appear out of nowhere in a living system, an alternative mechanism must be sought. In addition to polyploidy and aneuploidy, which add large amounts of genetic materials, duplication is a plausible vehicle for adding small increments of genetic information. For example, the duplication of genes through unequal crossing-over would ultimately lead to duplicated genes acquiring unique gene functions. The duplicated segments could subsequently undergo functional divergence through the accumulation of unique mutations, which would indeed lead to the acquisition of new functions.

Evidence supporting such a scheme is found in *Drosophila.* The chromosomes in the *Drosophila* salivary gland are large and morphologically distinguishable. The chromosomes are large because they replicate several times but do not separate, so that each chromosome is actually a "rope" of about 1000 chromatids tightly coiled around each other—a *polytene chromosome.* When properly stained, polytene salivary gland chromosomes are fairly easy to examine. With careful analysis, genes may be cytologically pinpointed to specific bands or chromomeres (Figure 5.21). The dumpy (*dp*) pseudoalleles are located in the "shoebuckle" region of the left arm of the second chromosome in bands 25E2 to 25A2 (Figure 5.22). These pseudoalleles are separable by crossing-over, they are functionally related, and they produce unique phenotypes. A study of the banding pattern in this region suggests the possibility of duplication and subsequent functional divergence.

Another indication that unequal crossing-over can occur and lead to the acquisition of new genetic traits is seen in the discovery of a new class of hemoglobin, called hemoglobin lepore (Hb$_{lepore}$). Hemoglobin A (HbA) is a blood protein, the kind most commonly found in humans. It is composed of four polypeptide chains: two alpha (α_2) chains and two beta (β_2) chains. The protein HbA$_2$ is a variant that has two alpha and two delta (δ_2) chains. In only three people has Hb$_{lepore}$ been found, and in all instances, the hemoglobin consisted of two normal α-chains and two chains that were partly β and partly δ. It has been suggested that Hb$_{lepore}$ originated

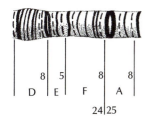

Figure 5.22
"Shoebuckle" in the left arm of chromosome II (bands 24E2–25A2); the dumpy region is in this series of bands

Figure 5.23
Postulated unequal
crossing-over between
the genes for the β
polypeptide chain and
the δ polypeptide chain,
leading to the hemoglobin
lepore disease (an anemia)
[Adapted from V. M.
Ingram, *The Biosynthesis of
Macromolecules,* copyright
© 1965 by W. A.
Benjamin, Inc.]

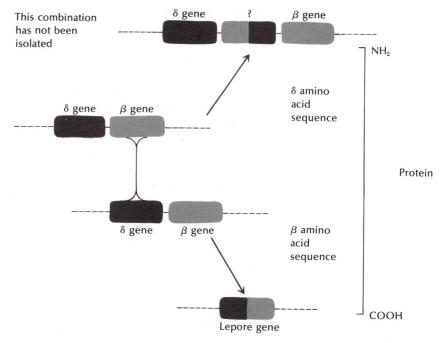

from unequal crossing-over between the β and δ genes (Figure 5.23).
Such crossing-over would produce one chromosome containing a duplication (β, β/δ, δ) and one containing a deficiency (β/δ), the new "lepore"
gene.

Deficiencies   The loss of a block of genetic information is termed a *deficiency,* or *deletion.*
Like duplications, deficiencies can occur in a number of ways, but one of
the most common is the *reciprocal production of a deficiency with a duplication*
resulting from unequal crossing-over:

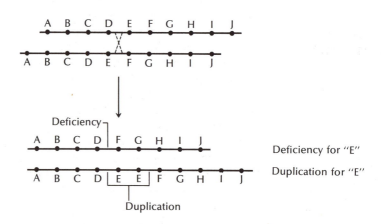

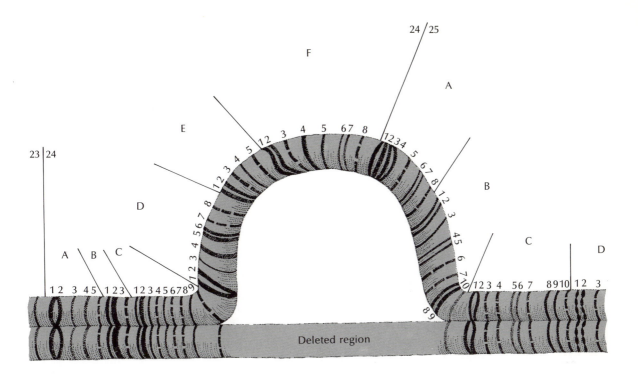

Deleted region

Figure 5.24
Diagram of a loop formed in a region of a deletion in chromosome II of *Drosophila*. Lower chromosome has a deletion in left arm running from band 24D1 through 25B9. Upper chromosome is a normal homolog with all bands present. Deleted region includes the genes for echinoid eyes (*ed*) and dumpy wings and thorax (*dp*).

We can often detect deficiencies and duplication cytologically in *Drosophila* by examining the polytene chromosomes of deficiency and duplication heterozygotes. A normal, nondeficiency chromosome that is synapsed with a deficient one will loop, or buckle, at the site of a deficiency, and the size of the loop indicates the size of the deficiency (Figure 5.24). The loop in a duplication heterozygote forms on the duplicated chromosome. The looping occurs because of another unusual property of salivary gland chromosomes. In addition to being polytene, the homologous salivary-gland chromosomes undergo pairing, something usually found only in meiotic cells.

Deficiencies have been useful in assigning genes to specific regions of the chromosome. In a chromosome containing a number of dominant alleles, it is possible to induce a deletion by using ionizing radiation. *Pseudodominance* occurs when the chromosome that carries the deletion of the dominant allele is paired with a normal chromosome that carries all recessive alleles. The phenotype caused by the expression of the recessive alleles would normally not appear, but in the absence of the dominant allele, the recessive allele is expressed, giving rise to what we call the pseudodominant effect. Knowing where the deletion occurs on the chromosome allows us to map the recessive allele cytologically, as we see at the top of the next page.

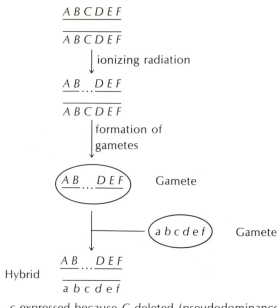

c expressed because C deleted (pseudodominance)

Phenotype: ABcDEF

The white-eye (*w*) and notch-wing (*n*) mutations are among the many genes that have been cytologically mapped (Figure 5.25). A series of deletions were characterized cytologically by the missing chromosome bands, then numbered. The white-eye mutant allele was pseudodominant in deletions numbered 258-11, 258-14, N-8 Mohr, and 264-31. The only bands common to all the deletions appear to be 3C1 and 3C2; hence the *w* locus should be in 3C1 or 3C2. And recent studies place the *w* locus in band 3C2. Likewise, the deletions numbered 264-33, 264-37, 264-2, and 264-19 all involve losses of band 3C7 only, and all express a notch-wing (*N*) phenotype. Therefore the notch locus must be in or very near 3C7.

Deficiencies are more than simply genetic tools in the cytological mapping of genes. They are important as causative agents in some serious human afflictions. For example, partial deletion of the short arm of chromosome 5 results in severe facial malformations and brain damage in infants; the afflicted individual utters a sound reminiscent of a cat meowing—hence its name, the *cri-du-chat* syndrome (Figure 5.26). The severity of the condition varies according to the amount of material deleted. A similar deletion in chromosome 4 produces an almost identical syndrome, suggesting that both chromosome segments are involved in head and brain development. The victims of most deficiencies, however, are probably not viable.

The etiology of chronic granulocytic leukemia is traceable to a deletion of the long arm of chromosome 22. (Such a deleted chromosome 22 has been called the Philadelphia chromosome, because it was first discovered in a patient hospitalized in Philadelphia.) Chromosome 22 is one of the smallest chromosomes, so it is often difficult to ascertain whether it has undergone an actual deletion or simply a translocation of the long arm to one of the larger chromosomes. Some recent studies suggest that the segment lost from chromosome 22 is in most cases translocated to chromosome 9.

The fact that children with Down's syndrome (trisomy 21) have an increased probability of being afflicted with chronic myelocytic leukemia suggests that chromosomes 21 and 22 share a function the disruption of which leads to this rare disease.

Figure 5.25
Fourteen deficiencies responsible for the white-notch phenotype in *Drosophila*, with a genetic map and the corresponding segment of the salivary-gland **X** chromosome. Black bars in chart indicate extent of each deficiency. Hatched segments represent possible missing bands (Slizynska, 1938).

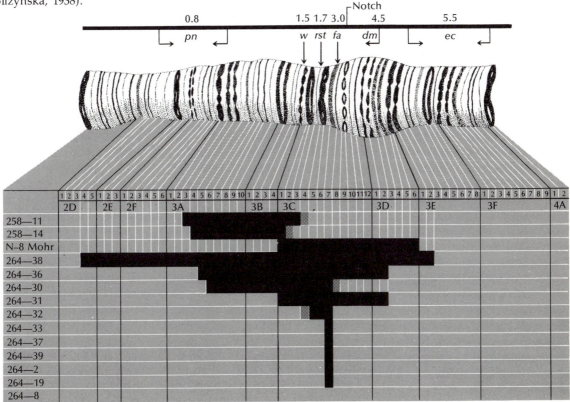

Figure 5.26
(a) Patient with *cri-du-chat* syndrome, associated with a deletion of part of the short arm of chromosome 5. Patients carrying this deletion express microcephaly and utter catlike cries.
(b) Chromosomes showing partial deletion of the short arm of number 5. [From J. S. and M. W. Thompson, *Genetics in Medicine*, 2nd edition, Philadelphia: W. B. Saunders Company, 1973]

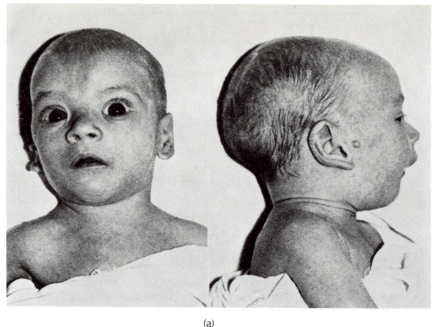

(a)

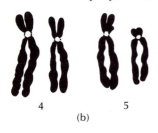

4   5

(b)

Inversions   Inversions involve alterations in the order of genes but not in their number. They usually form when two breaks occur in a chromosome. The segment between break points is reversed and reinserted at the same location.

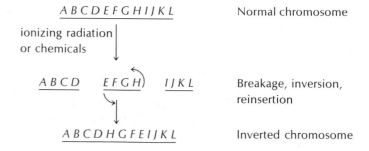

A B C D E F G H I J K L   Normal chromosome

ionizing radiation
or chemicals

A B C D   E F G H)   I J K L   Breakage, inversion,
reinsertion

A B C D H G F E I J K L   Inverted chromosome

Inversions are classified according to whether the inverted segment carries a centromere. An inversion that does not involve the centromere is termed a *paracentric inversion,* and one that does is a *pericentric inversion.* Pericentric inversions can change the position of the centromere and thus the morphology of the chromosome with respect to the location of its centromere.

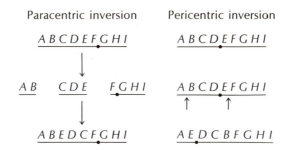

Paracentric inversion

A B C D E F G H I

↓

A B   C D E   F G H I

↓

A B E D C F G H I

Pericentric inversion

A B C D E F G H I

A B C D E F G H I
↑   ↑

A E D C B F G H I

When an inversion chromosome pairs with a standard, or noninverted, chromosome, a characteristic *inversion loop* forms. The size of the loop is a function of the size of the inversion—the larger the inversion, the larger the loop (Figure 5.27).

Crossing-over within a pericentric inversion (Figure 5.28) produces two normal meiotic products and two abnormal products. The latter con-

Figure 5.27
Diagrammatic view of chromosome pairing in an inversion heterozygote

Figure 5.28
Results of crossing-over within a pericentric inversion: two normal chromatids, one with a duplication for the *a* and *b* regions and a deletion for the *e* and *f* regions, and one with a duplication for the *e* and *f* regions and a deficiency for the *a* and *b* regions

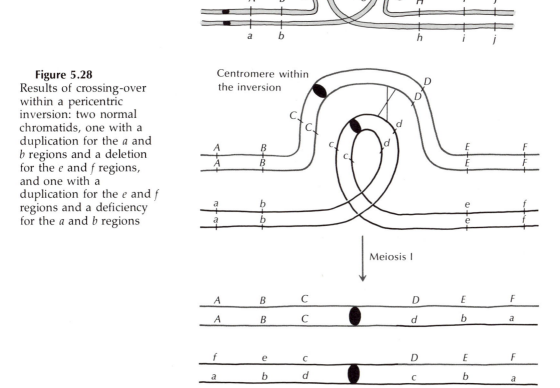

Figure 5.29
Results of crossing-over within a paracentric inversion: a dicentric chromatid (two centromeres) and an acentric fragment (no centromere)

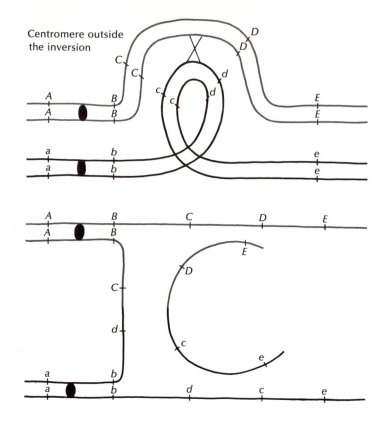

tain chromosomes that are either duplicated or deficient for some segment of the chromosome. In plants, gametes containing the duplicated and deficient chromosomes are generally not viable, so pericentric inversion heterozygotes are semisterile, although more than 50 percent viable. In animals, gametes with the duplicated and deficient chromosomes usually function normally, but the zygote usually does not.

Crossing-over within a paracentric inversion has more complex consequences (Figure 5.29). A crossover within a paracentric inversion produces one chromosome with two centromeres (*dicentric chromosome*) and one with no centromere (*acentric fragment*). The dicentric chromosome is pulled in two directions during meiosis. The acentric fragment, because it has no spindle fiber attachment, floats randomly and is eventually lost. In general, females heterozygous for a paracentric inversion manifest no serious sterility, because in most cases the chromosomes are all oriented in a specific direction during gametogenesis, which facilitates the exclusion of the dicentric and acentric chromosomes from the functional gamete. (Recall that in both higher plants and higher animals, female gametogenesis terminates with only one of the four meiotic products being functional. See Figure 5.30.) At the anaphase of meiosis I, standard and inverted chromatids are so oriented that they will be included in the outer megaspores (sex cells of the female plant), and the abnormal chromatids will be included in the inner megaspores. The functional megaspore is almost always one of the outer ones, and since these always receive either a standard or an inverted chromosome, the megaspore is viable. In animal oogenesis, the chromosomal orientation is such that the acentric and dicentric chromosomes will be included in the polar bodies, and the functional egg will receive either a standard or an inverted chromosome.

An inversion heterozygote acts as a suppressor of measured crossing-over for genes mapping within the inverted segment. Crossing-over does occur within the inversion, but the crossover prod-

Figure 5.30
Explanation of the fertility of females heterozygous for a paracentric inversion

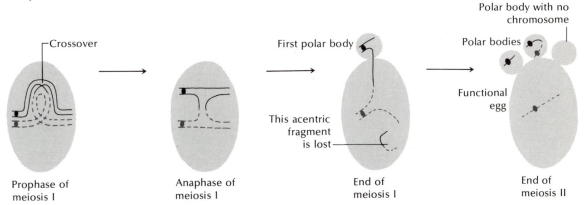

Higher Animals Since meiosis is initiated when the nucleus is eccentric, the first polar body receives either a standard or an inverted chromatid at the end of meiosis I. In meiosis II the dicentric chromatid is assumed to be included in the second polar body, the egg nucleus contains either a standard or an inverted chromosome, and the acentric fragment disintegrates in the egg cytoplasm.

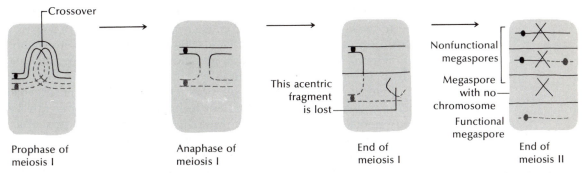

Higher Plants At the end of meiosis I the dicentric chromatid is oriented so that the terminal cells of the linear quartet of megaspores, one of which will develop into the gametophyte, receive a standard or inverted chromosome, but not the dicentric or acentric fragments.

ucts do not usually contribute to the next generation, either because the gametes or zygotes are nonviable or because the crossover chromosomes are eliminated into nonfunctional megaspores or polar bodies. Muller (1928) was the first to take advantage of these crossover-suppressing qualities of inversion heterozygotes. He used a chromosome with an inverted segment flanked on either side by two marker genes—bar eye (*B*) and a recessive lethal (*l*). This *ClB* chromosome (*C* = inversion) was used to detect sex-linked, recessive, lethal mutations in *Drosophila* induced by x rays.

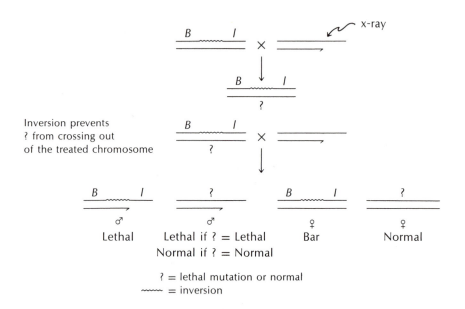

The inversion between *B* and *l* ensured that the chromosomes would remain intact in this region, as diagrammed, since the crossover chromosomes would give rise to nonviable gametes or zygotes. Any lethal mutation induced in the male **X** chromosome in the inversion region would remain on the male **X** and not be lost by transfer to the homologous **X** through crossing-over. If a lethal mutation has been induced, no males will appear in the third generation.

Inversions, though not commonly observed in other organisms (perhaps owing to limitations in our technology), have apparently been instrumental in the evolution of three *Drosophila* species, *D. persimilis*, *D. pseudoobscura*, and *D. miranda*. Using the banding patterns in the salivary-gland chromosomes, investigators have precisely defined the inversions found in various populations of *D. pseudoobscura*, *D. persimilis*, and *D. miranda* in terms of the length of the inversion and the chromosome segment involved. A "standard" chromosome was arbitrarily

Figure 5.31
The relationship of gene sequences found in chromosome 3 of various populations of *Drosophila pseudoobscura, D. miranda,* and *D. persimilis* by overlapping inversions. (Adapted from T. Dobzhansky, *Genetics of the Evolutionary Process,* copyright © 1970 by Columbia University Press.)

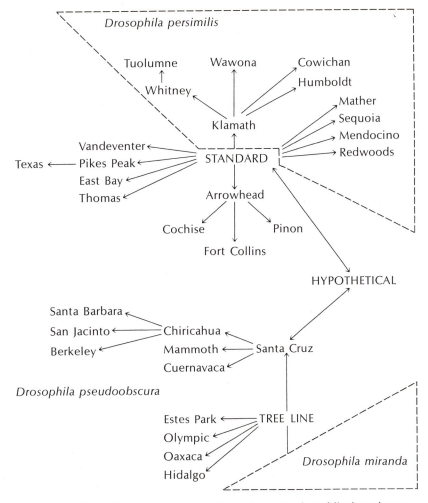

The sequences joined by a single arrow in this phylogenetic chart differ by only one paracentric inversion. The "standard" sequence is found in both *D. pseudoobscura* and *D. persimilis;* sequences located above the standard one occur only in *D. pseudoobscura,* whereas those below occur only in *D. persimilis.*

selected to represent the hypothetical normal, uninverted arrangement from which all the other chromosome arrangements could be derived by inversion (Figure 5.31). For example, both the "Arrowhead" and "Pikes Peak" populations of *D. pseudoobscura* carry different single inversions on their third chromosome. The "Cochise" and "Texas" populations contain double overlapping inversions compared with the standard. In each case, the double overlapping inversion can be traced back to a secondary inversion occurring in the Pikes Peak and Arrowhead single inversion. Using

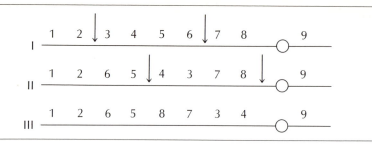

I   1 2 ↓ 3 4 5 6 ↓ 7 8   ◯   9

II   1 2 6 5 ↓ 4 3 7 8 ↓   ◯   9

III 1 2 6 5 8 7 3 4   ◯   9

The order of the sequences is III → II → I or I → II → III or I ← II → III. The arrows indicate the sites of breaks that caused the inversions.

the technique of relating inversions to each other, Dobzhansky constructed a phylogenetic tree of the various populations of *D. pseudoobscura*, *D. persimilis*, and *D. miranda*. He showed that the two species are closely related and that populations with two or more inversions can be traced back to the standard via single inversion. Figure 5.32 illustrates three chromosomes, two of which are assumed to have been sequentially derived from the first. Chromosome III could not have been derived from chromosome I in one step by a single inversion; the only sequences possible are III → II → I, I → II → III, or I ← II → III.

Inversions would seem to be an important evolutionary force in the divergence of the two *Drosophila* species we have just discussed. For example, subpopulations of *D. pseudoobscura* differ dramatically in their frequencies of three inversions (Standard, Arrowhead, and Pike's peak). This suggests that the inversions offer differentially adaptive features. The inversions in *D. pseudoobscura* have different frequencies in different geographic areas, presumably because some inversion combinations are better adapted than others to a specific set of environmental parameters (Figure 5.33). The inversions themselves may not be the direct cause of this distribution, but they may well contribute to it. As mutant genes accumulate in the populations, the inversions keep these new mutations from being exchanged by crossing-over. Hence mutant genes accumulate in groups in the inversions and form blocks of new genetic information ("supergenes").

Translocations   In contrast to duplications, deficiencies, and inversions—all *intra*chromosomal structural changes—translocations are *inter*chromosomal structural changes characterized by the detachment of chromosome segments and their reunion with nonhomologous chromosomes. There are three major types of translocations. In *simple translocation*, a piece of one chromosome is transferred to the end of a nonhomologous, unbroken chromosome (Figure 5.34a). This is not commonly observed, because the ends (telomeres) of unbroken chromosomes are not "sticky," whereas

Figure 5.33
Percentages of different
kinds of chromosomes in
populations of *Drosophila
pseudoobscura* [Adapted
from T. Dobzhansky,
*Genetics and the Origin of
Species,* 3rd edition,
copyright © 1951 by
Columbia University
Press]

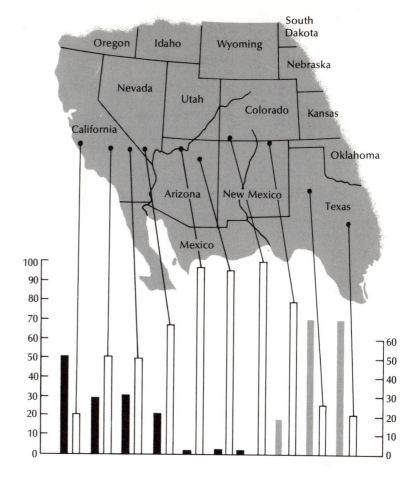

 = Standard
☐ = Arrowhead
▨ = Pike's peak

broken chromosome ends often are. In *translocation shift,* an interior segment of a chromosome that was induced by two breaks is transferred to
the interior of a nonhomologous, broken chromosome that was induced
by one break (Figure 5.34b). This is more common than the simple translocation, perhaps because only sticky ends are involved. But the most
common and most studied type is the *reciprocal translocation.* This type
requires a single break in each of two nonhomologous chromosomes and
the reciprocal exchange of the pieces (Figure 5.34c).

The genetic consequences of translocations are profound. To begin
with, translocations alter linkage groups so that a gene or series of genes
forming part of one chromosomal linkage group can be translocated to a
nonhomologous chromosome, thereby forming a new linkage group.
Second, translocation heterozygotes are generally semisterile, because
they produce gametes containing duplicated and deficient chromosomes.

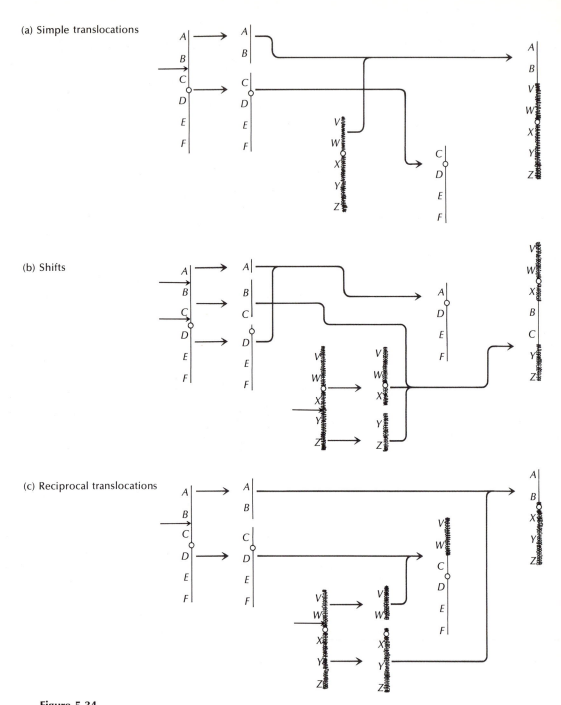

(a) Simple translocations

(b) Shifts

(c) Reciprocal translocations

Figure 5.34
Three possible types of translocation between two nonhomologous chromosomes, *ABCDEF* and *VWXYZ*

Figure 5.35
Chromosome pairing in a translocation heterozygote in maize [M. M. Rhoades, in *Corn and Corn Improvement*, New York: Academic Press, 1955]

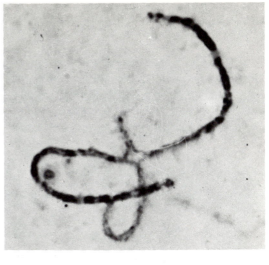

——————o—————————— Normal 10

– – – – –o– – – – – – – – – –Normal 8

——————o———— – – – –Interchanged 10—8

– – – – –o– – – – ——————Interchanged 8—10

This in turn is the result of the crossover and segregation caused by the characteristic crosslike conformation of paired chromosomes in translocation heterozygotes (Figure 5.35).

When a reciprocal translocation heterozygote undergoes meiosis, a cross configuration characterizes the prophase and metaphase of the first meiotic division. The segregation of the chromosomes in meiosis I occurs in three different patterns (Figure 5.36). *Alternate segregation* refers to the movement to the same pole of the nonhomologous translocated chromosomes in one case and the nonhomologous nontranslocated chromosomes in the other. *Adjacent-1 segregation* is the movement to the same pole of nonhomologous chromosomes, one translocated and one nontranslocated chromosome going to each pole. *Adjacent-2 segregation* is the movement of homologous chromosomes to the same pole. Only in the case of alternate segregation do the gametes get a full complement of genes. In the two types of adjacent segregation, the gametes contain duplications and deficiencies. Usually, they are either nonviable or they give rise to nonviable zygotes, the basis for the semisterility of translocation heterozygotes.

Crossing-over in a translocation heterozygote can occur in any of the four arms that comprise the crosslike configuration, but the consequences vary according to the crossover site relative to the centromeres and break-

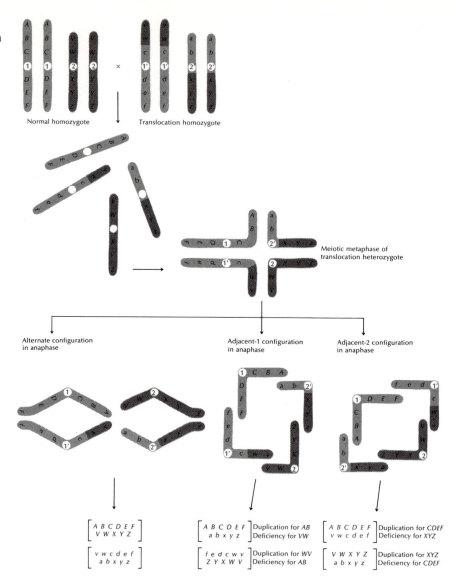

Figure 5.36
Meiosis in a translocation heterozygote

Normal homozygote   Translocation homozygote

Meiotic metaphase of translocation heterozygote

Alternate configuration in anaphase

Adjacent-1 configuration in anaphase

Adjacent-2 configuration in anaphase

$$\begin{bmatrix} A\ B\ C\ D\ E\ F \\ V\ W\ X\ Y\ Z \end{bmatrix}$$

$$\begin{bmatrix} v\ w\ c\ d\ e\ f \\ a\ b\ x\ y\ z \end{bmatrix}$$

$$\begin{bmatrix} A\ B\ C\ D\ E\ F \\ a\ b\ x\ y\ z \end{bmatrix}$$ Duplication for *AB* / Deficiency for *VW*

$$\begin{bmatrix} f\ e\ d\ c\ w\ v \\ Z\ Y\ X\ W\ V \end{bmatrix}$$ Duplication for *WV* / Deficiency for *AB*

$$\begin{bmatrix} A\ B\ C\ D\ E\ F \\ v\ w\ c\ d\ e\ f \end{bmatrix}$$ Duplication for *CDEF* / Deficiency for *XYZ*

$$\begin{bmatrix} V\ W\ X\ Y\ Z \\ a\ b\ x\ y\ z \end{bmatrix}$$ Duplication for *XYZ* / Deficiency for *CDEF*

points of the translocation (Figure 5.37). Crossing-over in the regions between the centromere and the breakpoint (*interstitial region*) will result in duplicated and deficient chromosomes irrespective of alternate or adjacent segregation patterns. This may be a major cause of sterility in those translocation heterozygotes that have primarily alternate segregation patterns, but crossing-over in these interstitial regions appears to be restricted, perhaps because of inefficient chromosome pairing. This region where crossing-over is restricted can lead to "supergenes," just as in inversions. Crossing-over outside interstitial regions has no effect on

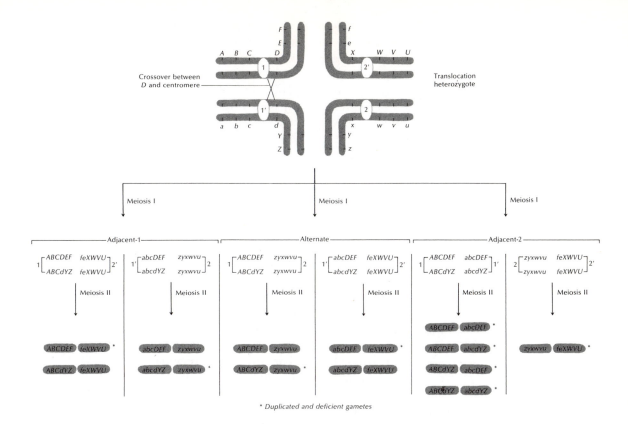

Figure 5.37
Results of crossing-over in an interstitial region in a translocation heterozygote. The interstitial region is between centromere 1 and gene *D*. Half the gametes containing crossover products from alternate segregation are balanced, and half the crossover products from adjacent-1 segregation are balanced. (Adapted with the permission of The Macmillan Co., Inc., from *Genetics*, by M. W. Strickberger, copyright © 1968 by Monroe W. Strickberger)

segregation patterns, because one homologous section is exchanged for another.

Like inversions, translocations appear to be important in the process of speciation. In some plant genera, such as *Campanula* and *Paeonia*, translocations are common. The same is true in such animal groups as scorpions and cockroaches. There are even some *Drosophila* groups in which translocations appear to be important factors in the differentiation of populations into new subspecies. In many other plant and animal species, translocations occur commonly and are undoubtedly important, but we do not often understand how the translocations confer selective advantage on these species.

Overview

In this chapter, we examined numerical and structural changes in chromosomes. We discussed these changes in terms of their origin and their consequences, placing special emphasis on their evolutionary significance.

The first category of numerical alterations we discussed was euploidy, the addition and deletion of entire chromosome sets. Animal and plant groups commonly experience a haploid and diploid phase in their life cycles, but it is of special interest that we note the occurrence of polyploidy, especially in plants. Polyploidy has been instrumental in the evolution of many naturally occurring and cultivated plant species. This polyploidy has been of two types. Autopolyploidy is polyploidy that originates between members of the same species, and allopolyploidy originates between members of different species. We suggested that polyploidy may originate when two or more sperm (pollen) fertilize an egg (ovule) or that it may occur through a mitotic failure.

The second category of numerical alterations is aneuploidy, the addition or deletion of single chromosomes. Aneuploidy usually develops from a meiotic defect. Aneuploidy may be a significant evolutionary force in some species, but it is definitely a problem in humans, in whom the addition or deletion of chromosomes creates a spectrum of defects.

We discussed four main classes of structural alterations: duplications, deficiencies, inversions, and translocations. Duplications and deficiencies can occur through unequal crossing-over. While deficiencies are generally not viable, duplications are, and more importantly, they are primary sources of new genetic information. We discussed this aspect of duplications in the context of the bar and dumpy regions in *Drosophila*. Deficiencies have been valuable tools, helping us to map genes cytologically. They are, however, debilitating in humans, as the *cri-du-chat* syndrome and Philadelphia chromosome attest.

Inversions, intrachromosomal inversions of gene sequences, provide an important source of genetic variation in a population and thus contribute to the overall process of speciation. We discussed how inversions create "supergenes" by preventing crossing-over within blocks of genes, thus keeping them intact. We distinguished between inversions that contain the centromere (pericentric), and inversions that exclude the centromere (peracentric).

Translocations are interchromosomal structural changes characterized by the detachment of chromosomal segments and their reunion with nonhomologous chromosomes. Because of their pairing characteristics, translocated chromosomes in the heterozygous state have serious sterility problems. Nevertheless, they occur commonly in many plants and animals and, like inversions, appear to be evolutionarily significant.

Questions and problems

5.1 It was suggested on page 191 that *Asplenium gravesii* could have arisen as a hybrid of *A. bradleyi* and *A. pinnatifidum*. Suggest two other possibilities and defend them.

5.2 Show how the addition and deletion of chromosomes leads to exceptions in Mendelian ratios.

5.3 How would you determine whether a particular tetraploid was the result of autopolyploidy or allopolyploidy?

5.4 Chromosomal aberrations are generally considered less significant in evolution than gene mutations. How would you explain this?

5.5 In what ways does variation in chromosome number and structure contribute to the development of the chromosome theory of inheritance?

5.6 In a particular cross of $AAbb \times aaBB$ bean plants, a triploid plant was obtained. When this triploid was crossed to aa bean plants, it produced a phenotypic ratio of five A plants to one a plant. When crossed to bb, it produced a 1:1 phenotypic ratio of B to b plants. What was the triploid plant's genotype? What were the genotypes of the gametes that formed it?

5.7 When a triploid plant that is BBb undergoes self-fertilization, what phenotypic ratio of B to b do you expect among the offspring?

5.8 An autotetraploid plant is $AAaa$. If it undergoes self-fertilization, what phenotypic ratio do you expect among the progeny?

5.9 In a strain of *Drosophila*, the flies are trisomic for the fourth chromosome, and one of the chromosomes carries the eyeless mutation (ey). The other two chromosomes carry normal alleles. If this trisomic fly is mated to an ey/ey fly, what ratio would you predict among the offspring?

5.10 If an **XYY** human male marries a normal female, what are the possible genotypes and phenotypes of the offspring produced by this couple?

5.11 A child with Klinefelter's syndrome is born to phenotypically normal parents. The child expresses a sex-linked recessive trait, hemophilia. On the basis of this information, can you tell which of the two parents produced the nondisjunctional gamete?

5.12 You have a plant that is monosomic for a certain chromosome. This plant is crossed to a normal diploid plant carrying the recessive alleles aa. What kinds of offspring would you expect from this cross (genotypic and phenotypic ratios) if the monosomic plant carried the dominant A allele on the monosomic chromosome?

5.13 Bridges studied duplications in *Drosophila*. In one of his studies, he developed a strain of flies that had an **X** chromosome carry the vermilion eye-color mutation (v) and a normal v^+ allele: $\underset{\rule{0pt}{0.6em}}{\overset{\displaystyle v^+ \quad v}{\rule{1.4em}{0pt}\!\!+\!\!+\!\!\rule{1em}{0pt}}}$. This stock was wild type for eye color. When females of this strain were crossed to v males, the offspring consisted of phenotypically v females and wild-type males. Interpret the results of this experiment.

5.14 Dobzhansky also studied duplications. He employed attached-**X** chromosomes homozygous for various recessive traits. He crossed these flies to males carrying an **X**-chromosome fragment with some of the dominant wild-type alleles. In general, the offspring from this cross expressed the wild-type phenotype as defined by the alleles on the **X** fragment contributed by the father. Interpret the results of this experiment.

5.15 When a paracentric inversion is long, double crossovers may appear with such high frequency that the crossover-suppressing qualities of the inversion for single crossover events are considerably lessened. How do you account for this?

5.16 You are given a stock of *Drosophila* homozygous for the second chromosome gene, *dp* (dumpy), and homozygous for the third chromosome gene, *e* (ebony). You also have a wild-type strain ($dp^+/dp^+\ e^+/e^+$). The wild-type strain is exposed to x rays, and you want to know if a reciprocal translocation occurred between the II and III chromosomes. Using only the x-rayed wild-type stock and the *dp/dp e/e* stock, show how you would detect a translocation.

5.17 In a cytological examination of a meiotic cell, you observe the following chromosome-pairing configuration:

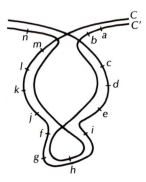

Assume that C′ is the original sequence and that C is derived from it. Show how that derivation occurred.

Bridges, C. B. 1922. The origin of variations in sexual and sex-limited characters. *Amer. Nat.* 56:51–63.

Brown, W. V. 1972. *Textbook of Cytogenetics*. St. Louis, Mo., Mosby.

Carr, D. H. 1971. The genetic basis of abortion. *Ann. Rev. Genetics* 5:65–80.

Demerec, M., and B. P. Kaufmann. 1967. *Drosophila Guide*. Washington, D.C., Carnegie Institution of Washington.

Dobzhansky, T., F. J. Ayala, G. L. Stebbins, J. W. Valentine. 1977. *Evolution*. San Francisco, Freeman.

Ehrlich, P. R., R. W. Holm, and D. R. Parnell. 1974. *The Process of Evolution*. New York, McGraw-Hill.

Ford, C. E., K. W. Jones, P. E. Polani, J. C. de Almeida, and J. H. Briggs. 1959. A sex chromosome anomaly in a case of gonadal dysgenesis (Turner's syndrome). *Lancet* 1:711–713.

Galinat, W. C. 1971. The origin of maize. *Ann. Rev. Genetics* 5:447–478.

Garber, E. D. 1972. *Cytogenetics: An Introduction*. New York, McGraw-Hill.

Grant, V. 1977. *Organismic Evolution*. San Francisco, Freeman.

Gustafson, J. P. 1976. Evolutionary development of Triticale: The wheat–rye hybrid. In *Evolutionary Biology* (M. K. Hecht, W. C. Steere, and B. Wallace, eds.) 7:107–136.

Hook, E. B. 1973. Behavioral implications of the human *XYY* genotype. *Science* 179:139–150.

Jacobs, P. A., and J. A. Strong. 1959. A case of human intersexuality having a possible *XXY* sex-determining mechanism. *Nature* 183:302–303.

John, B., and K. R. Lewis. 1975. *Chromosome Hierarchy*. London, Oxford University Press.

Khush, G. H. 1973. *Cytogenetics of Aneuploids*. New York, Academic.

Lejeune, J., R. Turpin, and M. Gautier. 1959. Le mongolisme, premier exemple d'aberration autosomique humaine. *Ann. Genet. Hum.* 1:41–49.

Levitan, M., and A. Montagu. 1977. *Textbook of Human Genetics*. New York, Oxford University Press.

Moody, P. A. 1975. *Genetics of Man*. New York, Norton.

Muller, H. J., and I. I. Oster. 1963. Some mutational techniques in Drosophila. In W. J. Burdette (ed.), *Methodology in Basic Genetics*. San Francisco, Holden-Day.

Nayar, N. M. 1973. Origin and cytogenetics of rice. *Adv. Genetics* 17:153–192.

Slizynska, H. 1938. Salivary gland analysis of the *white-facet* region of *Drosophila melanogaster*. *Genetics* 23:291–299.

Stebbins, G. L. 1971. *Processes of Organic Evolution*. Englewood Cliffs, N.J., Prentice-Hall.

Stebbins, G. L. 1976. Chromosome, DNA, plant evolution. In *Evolutionary Biology* (M. K. Hecht, W. C. Steere, and B. Wallace, eds.) 7:1–34.

Stern, C. 1973. *Principles of Human Genetics*. San Francisco, Freeman.

236
*Variation in chromosome
number and structure*

Sturtevant, A. H. 1925. The effects of unequal crossing over at the bar locus in *Drosophila. Genetics* 10:117–147.

Sutton, H. E. 1975. *Introduction to Human Genetics*. New York, Holt.

Swanson, C. P. 1957. *Cytology and Cytogenetics*. New York, Prentice-Hall.

Tjio, J. H., and A. Levan. 1956. The chromosome number in man. *Hereditas* 42:1–6.

Turpin, R., and J. Lejeune. 1969. *Human Afflictions and Chromosomal Aberrations*. Oxford, England, Pergamon.

Wilson, C. L., W. E. Loomis, and T. A. Steeves. 1971. *Botany*. New York, Holt.

Six

The identification and structure of the genetic material

The chemical basis of inheritance

Miescher isolates nuclein

Griffith points the way toward DNA

Griffith's experiments refined

Avery, MacLeod, and McCarty identify the chemical
 nature of the transforming principle

Hershey and Chase demonstrate DNA as the genetic
 material in viruses

Ribose nucleic acid (RNA) as the carrier of the
 genetic information

The discovery of the structure of DNA

Physical and chemical evidence on DNA structure

The chemical composition of DNA

The Watson–Crick model of DNA

Overview

Questions and problems

References

The preceding chapters have dealt primarily with patterns of inheritance—Mendelian and non-Mendelian—between individuals. The common thread uniting the material was the gene, but we have so far concentrated on the transmission and localization of these units positioned on chromosomes. This chapter will explore the experiments and observations that led to identification of the genetic material and understanding of its structure.

The chemical basis of inheritance

The fragile link between one generation and the next is the chromosome, which all organisms possess in one form or another. Our first concern in this chapter will be to demonstrate that a specific chemical component of the chromosome—nucleic acid—is responsible for carrying the genetic information.

The chemical composition of nucleic acids (DNA and RNA) was known long before their structure was determined. The nucleic acid DNA (deoxyribose nucleic acid) is composed of inorganic phosphate, a pentose (5-carbon) sugar called deoxyribose, and four nitrogenous bases: adenine (A), thymine (T), guanine (G), and cytosine (C). RNA (ribose nucleic acid) contains the sugar ribose instead of deoxyribose and the base uracil (U) instead of thymine (T). Figure 6.1 depicts the molecular components of DNA and RNA, and Table 6.1 presents the terminology commonly used in describing the nucleic acid subunits. For example, when the base adenine is covalently bound to the deoxyribose moiety, the combination is called deoxyadenosine; and when inorganic phosphate is covalently added to deoxyadenosine, deoxyadenylic acid is formed.

Miescher isolates nuclein   The substance that was eventually established as the genetic material of most organisms, DNA, was discovered in 1869, only 4 years after Mendel announced his findings and 80 years before DNA became the focal point of modern molecular biology. The discoverer, a Swiss physician named Friedrich Miescher, did not link DNA with inheritance, although some of his contemporaries did make tenuous associations.

While a medical student, Miescher became absorbed in the chemical composition of living tissues. Undertaking a study of the types of biological macromolecules found in living systems, he selected pus as his experimental material. It was readily available from bandages in a nearby hospital, and it contained white blood cells, which he considered to be among the simplest of cells. He treated the cells with salt solutions, alcohol, acidic solutions, and alkaline solutions and made two observations: that cells treated with a salt solution give a gelatinous precipitate and that cells placed in an alkaline solution give a precipitate when the solution is acidified. Miescher speculated that the precipitates he observed might be associated with cell nuclei. To test this possibility, he set about isolating nuclei—a feat he was the first to accomplish. When he treated isolated nuclei with an alkaline solution and then acidified them, he observed a precipitate. Analysis of this precipitate established that it was a complex material containing, among other things, nitrogen and phosphorus. The proportions of nitrogen and phosphorus were unlike any biological material yet studied, so Miescher concluded that he had isolated a new, previously undescribed biological compound associated almost exclusively with the nucleus. He called this substance *nuclein.* In retrospect, we know that Miescher had actually isolated a DNA–protein complex. The initial treatment in alkaline solution dissolved the DNA and some of the small, low-molecular-weight proteins closely linked to the DNA of higher organisms. The acidic character of the solution made the DNA–protein complex insoluble so that it precipitated out of solution.

Nuclein became Miescher's one absorbing interest, and he devoted his life to studying it. He changed his experimental material from white blood cells to salmon sperm, because the sperm were mainly nuclei and had very little cytoplasm. He chose salmon because his laboratory was on the Rhine River, a main avenue of salmon migration.

Miescher was reluctant to ascribe a functional role to nuclein, although he intimated at one point that it was involved in fertilization. Some of his contemporaries were not so reticent, however. In 1881, a young botanist named Zacharias extended Miescher's study to a wide range of cell types and stated that not only was nuclein found in the nucleus of all cells studied but that it was also specifically associated with chromosomes. In 1884, Hertwig, tying together his own observations on nuclei and chromosomes, made the perceptive statement:

> I believe that I have at least made it highly probable that nuclein is the substance responsible not only for fertilization, but also for the transmission of the hereditary characteristics.

The work of Zacharias and Hertwig on nuclein might have been the immediate forerunner of great developments, for in it lay the key to modern molecular genetics. However, neither of these investigators pursued the issue, and it was subsequently ignored.

By 1885, at least two reputable biologists were loosely ascribing a hereditary role to nuclein. As discussed in Chapter 1, the phenomenon of inheritance had by this time been traced to the nucleus of a cell and, more specifically, to the chromosomes housed within it. And nuclein had been shown to be a component of chromosomes—all of this almost 70 years before the biology of DNA became the consuming interest of so many geneticists. In the first edition of E. B. Wilson's classic treatise on the cell, published in 1895, he suggested that the chemical nature of the genetic

Figure 6.1
(a) Molecular components
of DNA

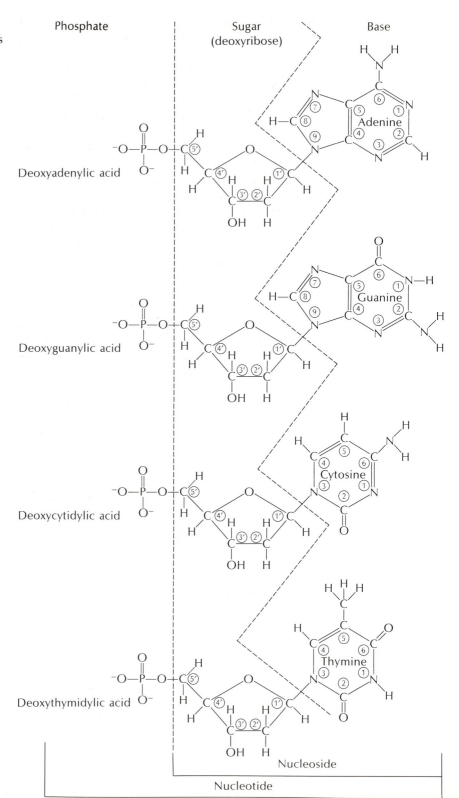

Figure 6.1 (cont.)
(b) Molecular components of RNA (the same as those of DNA except that the sugar is ribose and uracil replaces thymine)

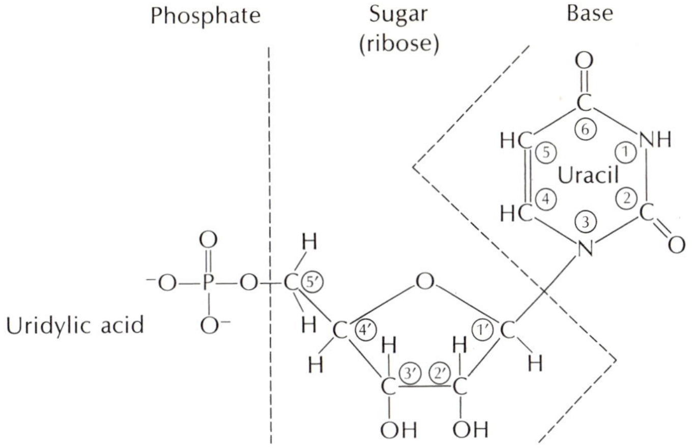

material might be nucleic acid. This idea persisted into the second edition of the book in 1910 but was dropped in the third edition (1926) in favor of the argument that protein was the genetic material. Perhaps if Hertwig, Miescher, or Weismann had expressed more enthusiasm for nuclein and nucleic acid, it would have ensured an earlier birth of molecular genetics. But they did not, and consequently nucleic acids were not central issues in the study of inheritance until 1944, when O. T. Avery, C. M. MacLeod, and M. McCarty identified DNA as the genetic material. However, between Miescher's discovery of nucleic acid (DNA) in 1869 and 1944, several observations and experiments pointed the way to the biochemical basis of inheritance.

Between the publication of Mendel's work in 1865 and its discovery in 1900, a number of parallel developments in cell biology, biochemistry, and genetics were taking place, although their fusion into a unified gene concept would be incomplete until 1952, when it was finally accepted by

Table 6.1
DNA and RNA nomenclature

	Base	Nucleoside (base + deoxyribose sugar in DNA, ribose sugar in RNA)	Nucleotide (base + sugar + phosphate)
DNA	Adenine (A)	Deoxyadenosine	Deoxyadenylic acid
	Guanine (G)	Deoxyguanosine	Deoxyguanylic acid
	Cytosine (C)	Deoxycytidine	Deoxycytidylic acid
	Thymine (T)	Deoxythymidine	Deoxythymidylic acid
RNA	Uracil (U)	Uridine	Uridylic acid
	Adenine (A)	Adenosine	Adenylic acid
	Cytosine (C)	Cytidine	Cytidylic acid
	Guanine (G)	Guanosine	Guanylic acid

Figure 6.2
Thematic view of the threads of research carried out between 1831, when the cell nucleus was discovered, and 1944–1952, when the biochemical nature of the genetic information contained in the nucleus was described

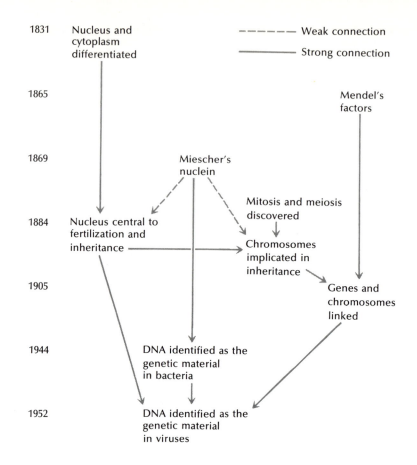

most geneticists that DNA is the genetic material. Figure 6.2 is a diagrammatic view of the various developments and their relationships to each other.

Griffith points the way toward DNA

Miescher was the first to isolate DNA from the nucleus of a cell, yet there were no sustained follow-up efforts to associate this material with the phenomenon of inheritance. Its role was viewed by most as primarily structural. It was F. Griffith in 1928, probably unaware of Miescher's discoveries, who laid the foundation for the ultimate identification of DNA as the genetic material.

Between 1920 and 1950, biologists were attempting to induce predictable and specific changes in genes, using chemicals and x rays. These mutations had to be stable and passed on from generation to generation. Among microorganisms, one of the most striking mutations that could be chemically induced and give reproducible results was the transformation

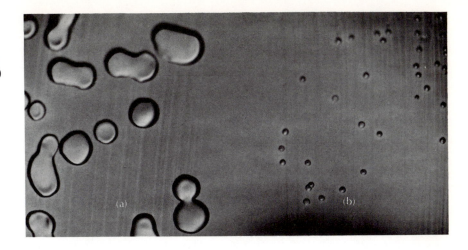

of an *avirulent* strain (having no disease-inducing capabilities) of *Diplococcus pneumoniae* (also called *Pneumococcus*, a bacterium that causes pneumonia) into a stable *virulent* strain (having disease-inducing capabilities). This transformation was first accomplished by Griffith in 1928, and the chemical responsible for the transformation phenomenon was later identified as DNA.

Virulence in *Pneumococcus* depends on the presence or absence of a carbohydrate capsule around the bacterium. If no capsule is present, the bacterium is avirulent; if the capsule is present, the bacterium is virulent. The presence or absence of this capsule is an inherited characteristic, as is its molecular organization. The capsules are classified as type II or type III, reflecting different compositions, and each is immunologically distinct.

A virulent bacterium with either type of carbohydrate capsule produces a colony of bacteria that has a smooth, pearllike appearance (Figure 6.3a). The bacterium may be classified as IIS or IIIS, where S denotes the smooth appearance of the colony. A IIS bacterium can mutate to an avirulent, nonencapsulated strain that is designated as IIR—II because it was derived from a type II cell and R because colonies of IIR cells are rough in appearance (Figure 6.3b). Mutation to IIR occurs about once in every million cells, is stable, and is passed on to all progeny cells (the same would also hold true for IIIS mutating to stable IIIR).

A IIR cell can mutate back to IIS with a very low frequency, but it is important to bear in mind that IIR can mutate back to IIS *only, never* to IIIS. Likewise, IIIR cells can mutate back to IIIS only.

Knowing that heating *Pneumococcus* cells to 65°C killed them and made them avirulent, Griffith mixed *Pneumococcus* type IIR (the nonpathogenic derivative of IIS) with heat-killed IIIS cells (nonpathogenic)

Figure 6.4
Summary of Griffith's
transformation
experiment

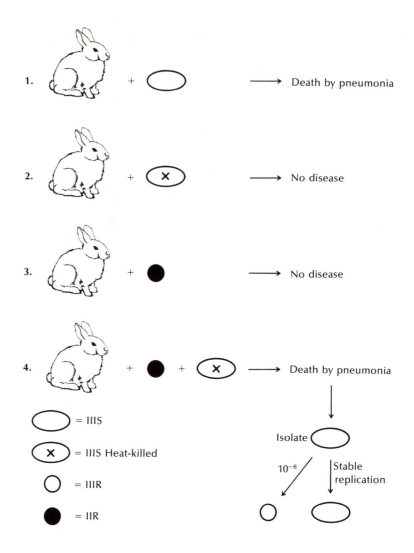

1. + ⬭ ⟶ Death by pneumonia

2. + ⊗ ⟶ No disease

3. + ⬤ ⟶ No disease

4. + ⬤ + ⊗ ⟶ Death by pneumonia

⬭ = IIIS

⊗ = IIIS Heat-killed

○ = IIIR

⬤ = IIR

Isolate ⬭

10^{-6} / | Stable replication

○   ⬭

and injected them into a host animal (Figure 6.4). Since neither component of the inoculum was pathogenic, no disease was expected. But many
animals did contract pneumonia, and when their blood was sampled,
living IIIS cells were found. These cells were stable and maintained the
IIIS characteristic generation after generation, rarely mutating to IIIR. The
IIIS cells so obtained could not have been derived from the back mutation
of IIR cells. Ascribing this phenomenon to some sort of "vital force"
emanating from the living IIR cells or host animal body and revitalizing
the heat-killed IIIS or converting IIR to IIIS was not to be taken seriously,
but such a "theory" was certainly difficult to disprove experimentally.

Griffith hypothesized that some entity—a "transforming principle"—is released from the heat-killed IIIS cells and taken up by the avirulent IIR cells, thus transforming them genetically.

Griffith's experiments were confirmed between 1928 and 1930 and refined during the early 1930s. Some of the experiments, specifically those of M. H. Dawson and R. H. P. Sia in 1931 and J. L. Alloway in 1933, sought to refine Griffith's experimental procedure by investigating the possibility of obtaining successful transformation outside the living body and by determining whether intact heat-killed IIIS cells (as opposed to simply a IIIS extract) were required in transformation.

To show that transformation could occur outside the living body, Dawson and Sia grew IIR cells together with heat-killed IIIS cells *in vitro* in a medium containing anti-R serum. This serum has R-cell-specific antibodies that will complex with and precipitate out only the R cells, leaving S cells unaffected. The R cells clump at the bottom, thus leaving the medium clear. If any transformation of IIR to IIIS occurs, the transformed cells—having no affinity for the anti-R antibody—float free in the medium and cloud it as they multiply. Dawson and Sia observed this clouding and subsequently isolated IIIS cells. Thus they showed that transformation could take place in a test tube as well as in the body of the host animal.

Alloway performed a refined version of the Dawson–Sia experiment. He ground up heat-killed IIIS cells and filtered out all the cellular debris. Using this filtrate and IIR cells in a medium containing the anti-R serum, he was able to show that *in vitro* transformation could take place with the IIIS extract alone.

The major conclusion drawn from these three experiments was that a substance called the "transforming principle" is transferred from one cell to another and that it permanently alters the recipient cell's "inherited characteristics." Between 1933 and 1944, experimenters confirmed the work of Griffith, Dawson and Sia, and Alloway and sought to demonstrate the same phenomenon in other organisms.

Figure 6.5
O. T. Avery (From the collections of the archives of the National Academy of Sciences)

**Avery, MacLeod, and
McCarty identify the
chemical nature of the
transforming principle**

In 1944, three medical researchers, Avery (Figure 6.5), MacLeod, and McCarty, attempted to identify the chemical nature of the transforming principle described by Griffith. These three investigators may be considered the direct intellectual descendants of Griffith, because Dawson, who extended Griffith's experiments to an *in vitro* system, was part of their laboratory at the Rockefeller Institute in New York City. Dawson was undoubtedly a major influence on the research direction taken by Avery, MacLeod, and McCarty.

Many properties of the transforming principle had been known before experimentation was begun. Griffith had shown that heating the transforming principle to 65°C did not inactivate it, though the heat did destroy enzymes that could in turn inactivate it. This suggested that the transforming principle was not protein, because enzymes are protein. Preliminary experimentation by Avery and his colleagues also established that alcohol sterilization had no effect on the principle, making contamination an unlikely source of the transformation. To isolate and identify the active principle from crude bacterial extracts, they chose Griffith's basic system of the transformation of *Pneumococcus* from IIR to IIIS, with some modifications. The components of their system were as follows: nutrient broth to grow the cells, anti-R serum to screen out all but transformed cells, IIR cells, and a DNA extract from IIIS cells. A DNA extract seemed the logical choice, because the nucleus contained primarily protein and DNA, and, as we just noted, protein did not appear as strong a candidate as DNA.

The DNA extract was the key component of the system; its method of preparation is diagrammed in Figure 6.6. The extracted DNA was free of any other biological molecules, and when used in the transformation detection system described above, it transformed IIR cells to IIIS. The capacity of highly purified DNA to transform cells was the same as that of

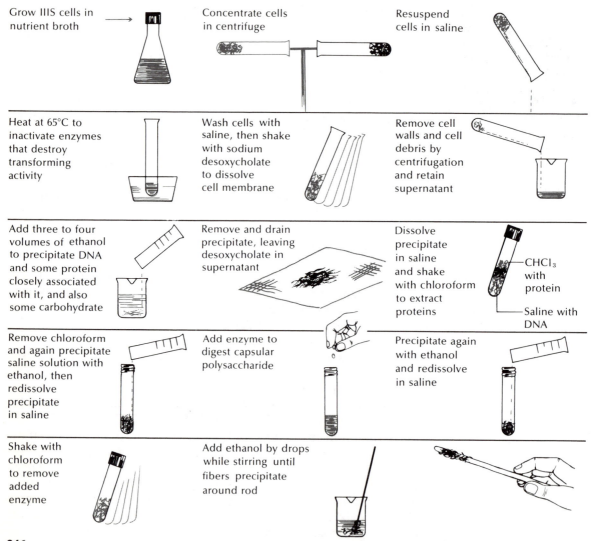

Grow IIIS cells in nutrient broth

Concentrate cells in centrifuge

Resuspend cells in saline

Heat at 65°C to inactivate enzymes that destroy transforming activity

Wash cells with saline, then shake with sodium desoxycholate to dissolve cell membrane

Remove cell walls and cell debris by centrifugation and retain supernatant

Add three to four volumes of ethanol to precipitate DNA and some protein closely associated with it, and also some carbohydrate

Remove and drain precipitate, leaving desoxycholate in supernatant

Dissolve precipitate in saline and shake with chloroform to extract proteins

CHCl$_3$ with protein

Saline with DNA

Remove chloroform and again precipitate saline solution with ethanol, then redissolve precipitate in saline

Add enzyme to digest capsular polysaccharide

Precipitate again with ethanol and redissolve in saline

Shake with chloroform to remove added enzyme

Add ethanol by drops while stirring until fibers precipitate around rod

the crude extract. Moreover, protein, carbohydrates, or lipids did not transform the pneumococcal cells. The investigators deduced, therefore, that DNA was the transforming principle. In their conclusion, Avery, MacLeod, and McCarty remark that perhaps something undetected was isolated along with the DNA. But, they add,

> If . . . the biologically active substance isolated in highly purified form as . . . DNA actually proves to be the transforming principle, as the available evidence strongly suggests, then nucleic acids of this type must be regarded not merely as structurally important* but as functionally active in determining the biochemical activities and specific characteristics of pneumococcal cells.

The importance of the work of Avery, MacLeod, and McCarty was grasped less quickly than one might have thought. Many argued that this was only a special case involving *Pneumococcus* and had little or no relevance to inheritance in other organisms. Some argued that their technique left room for doubt because of the possibility that some undetected protein could have contaminated the preparation. Indeed, many scientists interested in identifying the chemical nature of the genetic material seemed unaware of this work and its significance. In 1946, for example, two years after the Avery–MacLeod–McCarty work was published, A. W. Pollister and A. E. Mirsky published an extensive paper on the nucleic acid–protein complex found in sperm nuclei. They extended the observations made by Miescher and refer to him extensively, but they do not refer to Avery, MacLeod, and McCarty. A. Boivin, R. Vendrely, and C. Vendrely published an important paper in 1948 suggesting that DNA was an essential component of the gene, but they too failed to cite the *Pneumococcus* work. In 1950, Hewson Swift published a key paper showing that in cells undergoing meiosis the DNA content is reduced by half, whereas in cells undergoing mitosis the amount of DNA in the nucleus first doubles and then, as the nuclei divide, is halved, thus returning to the original state. Cellular protein levels showed no such fluctuation. Swift's paper is perfectly consistent with the idea that DNA is the genetic material—and he suggests this—but again fails to cite the *Pneumococcus* paper.

Gunther Stent has suggested that the work of Avery and his colleagues, like the work of Mendel, was premature or ahead of its time. Stent suggests that this is the reason why the transformation work failed to ignite the scientific world's interest in DNA as the genetic material. This notion of prematurity has been attacked on two main grounds. First, at a meeting of molecular biologists in Cold Spring Harbor in 1946, two years after the publication of Avery, MacLeod, and McCarty's paper, DNA was the focal point of discussion and their transformation work occupied a key position in that discussion. So the work was not completed and then ignored, as was the case with Mendel's experiments. A second attack against Stent's position is more profound in its implica-

---

* Remember that DNA was then thought by most to play only a structural role in chromosomes.

tions, because it asserts that the discussion of prematurity is useless. If a discovery is premature, it is not appreciated. Therefore a discovery is premature because it is not appreciated, and it is not appreciated because it is premature—a circular tautology. Darwin's theory of evolution by natural selection is also a circular tautology. Survival of the fittest means that those that survive are the fittest and the fittest are those that survive. These arguments against Stent's suggestion of prematurity are important and need to be answered.

Stent's position on the prematurity of the transformation work can be defended. In response to the first attack (that the work was not ignored after all), we can argue that, though the work was discussed, only a relatively small group of scientists interested in transformation and related phenomena were involved. The work was certainly appreciated, but most people did not grasp its real significance. It was still widely believed that protein was the genetic material, because nucleic acid was far too simple a molecule. Nucleic acid researchers were burdened by a theory called the *tetranucleotide theory*, wherein DNA was seen simply as a chain of repeating bases: ATGCATGCATGC, and so on. This monotonous structure could not possibly carry the complex information required of the genetic material. Proteins, on the other hand, were composed of 20 amino acids arranged in numerous sequences; they could easily accommodate biological diversity. The nucleic acids were viewed as serving a structural role and nothing more. In short, the work was appreciated but not understood.

The tautology argument is more difficult to deal with. Even though an argument is tautological, it does not necessarily follow that it is useless. One would hardly consider Darwin's theory of evolution as anything less than a great insight. Freud's insight into the subconscious and Marx's theory of economic determinism can also be considered tautological, but certainly they are extremely important. The notion of prematurity, as Stent argues, has more to it than lack of appreciation. Can an insightful discovery be tied in by a series of logical steps to contemporary ideas? If it cannot, it is premature whether or not it is appreciated. In Mendel's case, his work was both unappreciated and unconnected to any existing scientific canons. The Avery, MacLeod, and McCarty work was appreciated but unconnected to existing scientific canons.

The concept that would probably have led to understanding *and* appreciation was the idea that DNA was not a simple tetranucleotide. Instead it is composed of an almost infinite variety of sequences derived from its four component bases. But this fact was not fully illuminated until 1950.

The idea that DNA was the genetic material became firmly entrenched in 1952, when Hershey and Chase published the results of their studies on the life cycle of the bacterial viruses. Their studies could be easily related to current knowledge.

Hershey and Chase demonstrate DNA as the genetic material in viruses

The work that convinced most biologists that DNA is, in fact, the genetic material emerged from the studies of A. Hershey and M. Chase, who focused on the ways in which bacterial viruses (called bacteriophages) reproduce. These viruses were known to be composed exclusively of protein and DNA (Figure 6.7). In the course of their life cycle, they first

Figure 6.7
Structure of a T-even bacteriophage. [Courtesy of E. Kellenberger; photo by F. Eiserling]

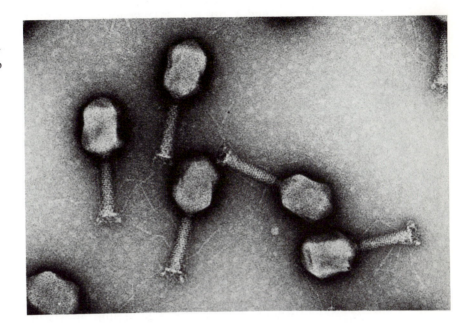

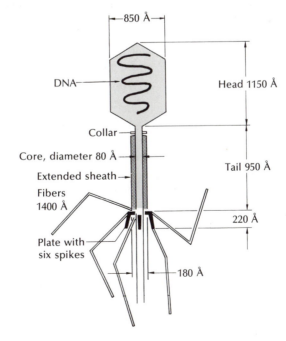

250

*The identification and
structure of the genetic
material*

attach to a bacterium, penetrate the cell, multiply inside the cell in a noninfective form, and subsequently release infective progeny from the host cell. Hershey and Chase were chiefly concerned with the events that transpired from the time the virus attached to the cell to the time the progeny viruses were released. Watson and others had reported that when the DNA of the parent viruses was labeled with ^{32}P (the radioactive isotope of ^{31}P), the ^{32}P label appeared in the host bacterial cells and was distributed to the next virus generation, after completion of the life cycle, in the following manner:

1. Fifty percent of the label was found in the DNA of infective progeny viruses, strongly suggesting that DNA is passed on from generation to generation.

2. Forty percent of the label was found free in the medium after the bacteria released the progeny viruses, indicating that this was probably labeled DNA that was not incorporated into mature, infective virus particles.

3. Ten percent of the label stayed with the parent virus, suggesting that some of the parental viruses failed to part with their DNA.

Armed with this information, Hershey and Chase set out to study the role of viral protein and viral DNA in the intracellular phase of the virus life cycle. They performed a series of experiments in which the protein component of the T2 virus was selctively labeled with ^{35}S (a radioactive sulfur isotope of ^{32}S), because sulfur is found in protein but not in DNA; the DNA component was labeled with ^{32}P, because phosphorus is a part of DNA but not of protein. Their experiments (Figure 6.8) showed that when a virus infects a bacterium, it attaches to a receptive cell by its protein coat and injects its DNA inside the cell. The protein component remains outside and apparently plays no further role in the subsequent life cycle of the virus. Figure 6.8 shows viruses labeled with either ^{32}P or ^{35}S that infected *E. coli* cells.

The Hershey and Chase paper is a collection of experiments rather than just a single experiment. These experiments can be summarized as follows:

1. By osmotically shocking the phage, the DNA separates from the protein, leaving what we call "phage ghosts" (phage coats without DNA).

2. After attachment to a host cell, phage DNA becomes sensitive to DNase.

3. Phage can attach to bacterial fragments, and this triggers the phage to release their DNA into the medium.

4. Phage coats can be removed from infected bacteria by using a blender.

5. About 1 percent of the labeled protein in a phage ends up in the progeny phage.

6. About 30 percent of the labeled DNA in a phage is transferred to the progeny phage.

7. Phage inactivated with formalin can attach to bacteria, but they cannot inject their DNA.

Figure 6.8
Summary of the 1952
Hershey–Chase
experiment

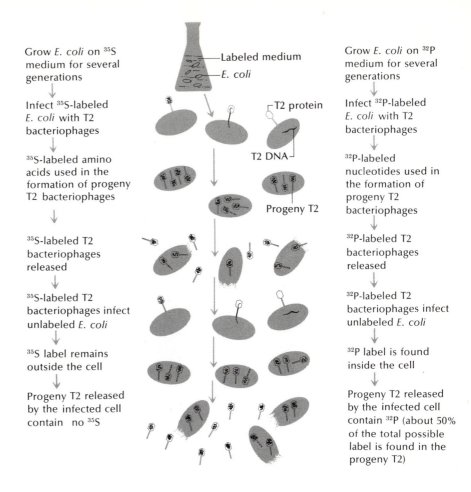

Grow *E. coli* on ³⁵S medium for several generations

↓

Infect ³⁵S-labeled *E. coli* with T2 bacteriophages

↓

³⁵S-labeled amino acids used in the formation of progeny T2 bacteriophages

↓

³⁵S-labeled T2 bacteriophages released

↓

³⁵S-labeled T2 bacteriophages infect unlabeled *E. coli*

↓

³⁵S label remains outside the cell

↓

Progeny T2 released by the infected cell contain no ³⁵S

Labeled medium
E. coli

T2 protein

T2 DNA

Progeny T2

Grow *E. coli* on ³²P medium for several generations

↓

Infect ³²P-labeled *E. coli* with T2 bacteriophages

↓

³²P-labeled nucleotides used in the formation of progeny T2 bacteriophages

↓

³²P-labeled T2 bacteriophages released

↓

³²P-labeled T2 bacteriophages infect unlabeled *E. coli*

↓

³²P label is found inside the cell

↓

Progeny T2 released by the infected cell contain ³²P (about 50% of the total possible label is found in the progeny T2)

The first three experiments show that the phage attaches to the bacterial cell and injects its DNA. Experiments 4, 5, and 6, the most important of all, show that DNA and not protein is transferred from parent to progeny. Experiment 7 shows that no progeny are formed without the injection of DNA.

The blender experiments are the most important, so we will discuss them in more detail. The phage-infected cells were placed in a blender long enough to ensure that the virus particles were knocked off the surface of the infected cells (Figure 6.9a). The data from this experiment, as seen in Figure 6.9b, argue two points. First, the treatment in the blender does not break open more than 7 percent of the cells. Second, it is evident from this graph that the blender removes 80 percent of the ³⁵S but only 18 percent of the ³²P, which in turn argues that most of the protein remains

Figure 6.9
(a) Blender technique for removing phage "ghosts" from the infected cell (b) Graph showing percentages of ^{35}S and ^{32}P removed from bacteria infected with radioactive phages by agitation in a blender [From A. D. Hershey and M. Chase, *J. Gen. Physiol.* 36, 1 (1952)]

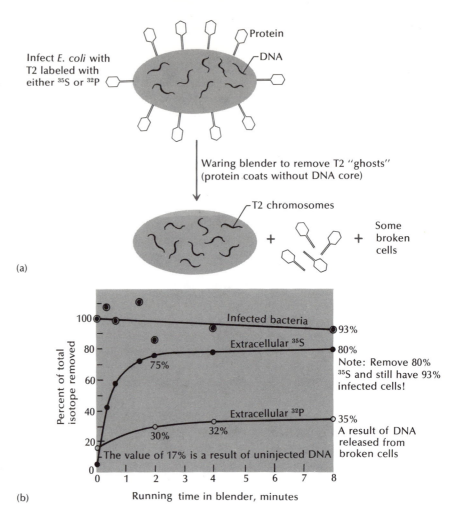

(a)

(b)

outside the infected cell and that most of the DNA goes inside. Table 6.2 confirms this inference by showing that using the blender technique to remove 82 percent of the viral protein from the infected cells still leaves 85 percent of the cells intact, suggesting that it is DNA that a bacteriophage injects into a cell. Hershey and Chase deduced that DNA is the genetic material that directs the process of assembling progeny bacteriophages and that protein does not appear to contain genetic information.* It should be pointed out, however, that the work of Avery, MacLeod, and McCarty is cleaner and less ambiguous than this study.

With the publication of the Hershey–Chase experiment, DNA became generally recognized as the genetic material. There was, and still is, some question about the small amount of protein that seems to enter the bacterial cell when it is infected by a virus. Most interested parties see this

---

* A. D. Hershey, S. Luria, and M. Delbrück were awarded the Nobel Prize in 1969 for their contribution to genetics.

Table 6.2
Effect of multiplicity of infection on elution of phage membranes from infected bacteria [Adapted from A. D. Hershey and M. Chase, "Independent functions of viral protein and nucleic acid in growth of bacteriophage," *J. Gen. Physiol.* 36:39–56 (1952).]

Running time in blender, minutes	Multiplicity of infection (phage/bacterium)	^{32}P-labeled phage		^{35}S-labeled phage	
		Percent of viral protein removed	Infected bacteria surviving, percent	Percent of viral protein removed	Infected bacteria surviving, percent
0	0.6	10	120	16	101
2.5	0.6	21	82	81	78
0	6.0	13	89	46	90
2.5	6.0	24	86	82	85

as either a procedural problem of contamination, such as ^{35}S-protein binding to cells throughout the blender process and thus being classified as intracellular ^{35}S-protein, or as a viral protein species that functions other than as a carrier of genetic information. For example, it may be a protein assisting with the injection of the virus DNA into the host cell, a possibility to be explored in Chapter 8.

Ribose nucleic acid (RNA) as the carrier of the genetic information

The experiments of Avery, Hershey and Chase, and their colleagues are certainly consistent with the notion that DNA is the genetic material, but they did not prove that DNA is the only genetic material. In the early 1950s, viruses that contained no DNA were being investigated. These viruses (Figure 6.10), of which the tobacco mosaic virus (TMV) is among

Figure 6.10
Tobacco mosaic virus (a) Electron micrograph and model of TMV (tobacco mosaic virus) (b) Diagram of TMV (From H. Fraenkel-Conrat, "The Genetic Code of a Virus," *Scientific American*, October 1964. Copyright © 1964 by Scientific American, Inc. All rights reserved)

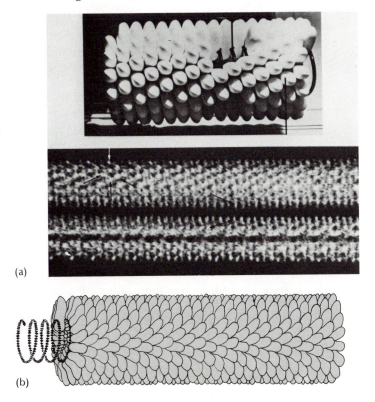

(a)

(b)

Figure 6.11
Summary of Fraenkel-Conrat's experimentation with TMV

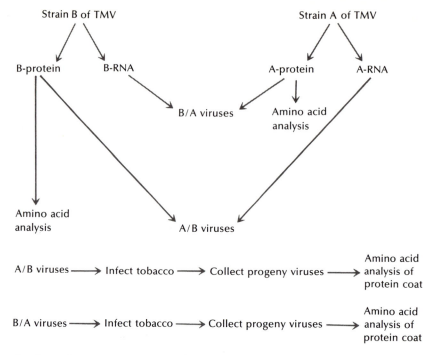

A/B viruses ⟶ Infect tobacco ⟶ Collect progeny viruses ⟶ Amino acid analysis of protein coat

B/A viruses ⟶ Infect tobacco ⟶ Collect progeny viruses ⟶ Amino acid analysis of protein coat

the best known, contain only protein and RNA, an analog of DNA. Fraenkel-Conrat and his colleagues at the University of California disaggregated the TMV particles and separated the TMV-RNA from the TMV-protein. Mixing the two components, they were able to reconstitute fully infectious TMV (Figure 6.11). In doing so, they demonstrated that it is indeed the RNA that carries the genetic information in TMV and not the protein. The genetic nature of RNA was explicated when proteins from three different TMV strains and RNA's from four different strains were isolated. Protein and RNA were then mixed, and reconstituted viruses were isolated. The reconstituted TMV, in various combinations containing RNA from one strain and protein from another, were then used to infect tobacco leaf cells. The progeny TMV were collected and the protein coats analyzed. In each case, the protein coat of the progeny TMV reflected the composition of the protein of the parental strain that had contributed the RNA and not the one that had contributed the protein (Table 6.3). Thus it was shown that the RNA determined the amino acid composition of the protein coat, confirming that where DNA is absent it is RNA, not DNA, that serves as the conveyer of the genetic information.

In summary, identification of nucleic acids as the carriers of the genetic information came only after decades of careful observation. Miescher, who first identified DNA as a unique compound in cell nuclei, was not a major link in the chain of events leading to the association between DNA and genetic traits. His research was more significant for its contributions to DNA biochemistry. The first important step toward the identification of the genetic material came from the transformation exper-

Table 6.3
Amino acid composition of two strains of normal and reconstituted TMV (numbers represent grams of amino acid per 100 grams of virus)

Amino acid	B strain of TMV: experimental values for protein coat	Progeny TMV: from B RNA/A protein parents	A strain of TMV: experimental values for protein coat	Progeny TMV: from A RNA/B protein parents
Glycine	1.6	1.8	2.3	2.3
Alanine	8.5	8.5	6.5	6.9
Valine	5.9	6.3	9.6	9.0
Leucine-isoleucine	12.2	12.2	14.2	14.3
Proline	5.0	5.1	5.0	5.1
Serine	8.1	8.1	9.0	8.8
Threonine	7.2	7.5	8.9	8.9
Lysine	2.4	2.3	1.9	1.8
Arginine	8.9	8.5	9.5	9.7
Histidine	0.7	0.7	0.0	0.0
Phenylalanine	5.3	5.4	7.2	7.1
Tyrosine	6.3	6.2	4.1	4.3
Tryptophan	2.2	2.2	2.8	2.6
Methionine	2.0	2.2	0.0	0.0
Glutamic acid	16.4	17.3	12.4	12.1
Aspartic acid	15.0	14.8	13.8	14.2

iments of Griffith in 1928, which led to the more definitive study by Avery and his colleagues, who in 1944 showed that DNA is the carrier of the genetic information and the key to bacterial transformation. In 1952, Hershey and Chase established beyond reasonable doubt that DNA is the genetic material, except in certain RNA viruses. This series of revelations steadily eroded the belief that DNA served a purely structural purpose in chromosomes—a misconception embraced by the "tetranucleotide theory." This theory was abandoned when it was demonstrated that the A, T, G, and C content of DNA varied among species, suggesting that a regularly repeating sequence of the four DNA subunits was too simple a model. This point will be dealt with more extensively in our discussion of the chemical composition of DNA.

The discovery of the structure of DNA

Once the nucleic acids, in the form of DNA or RNA, were generally recognized to be the genetic material, structural and functional description of these critically important biological macromolecules became imperative tasks. It had been known that the genetic material was replicated and passed on from generation to generation virtually error-free—though mistakes, or mutations, did on occasion occur. It was known that the genetic material must contain the information required to generate a complex organism from a single cell. Among the questions being asked about DNA were the following:

Figure 6.12
Maurice Wilkins (Wide
World Photos)

1. How is DNA replicated in such a precise manner?
2. How do mutations occur?
3. How is the biological information stored in the DNA structure, and how is this stored information ultimately used?

These three questions were directives for frenzied activity in laboratories all over the world. From 1952 (when Hershey and Chase focused attention on DNA as the genetic material) to the present, answers to these and related questions have been actively sought. The first major step toward answering them came in 1953, when Watson and Crick announced a model for the structure of DNA, a key achievement the background of which will be developed in this section. The model of the structure of DNA suggested a means by which DNA is replicated. It also suggested a mechanism for mutation, and it held further promise as a vehicle for storage of biological information. The remainder of this chapter will discuss the research leading to the successful discovery of the structure of DNA.

The attempts to discover the structure of DNA make a fascinating story, in part because they involve such disparate approaches and in part because of the people involved in the research. To understand the structure of DNA, Erwin Chargaff at Columbia studied its chemical composition. Maurice Wilkins and Rosalind Franklin at Kings College analyzed the diffraction patterns of DNA crystals exposed to x rays. Linus Pauling at Cal Tech developed models of DNA based on his research with hydrogen bonds and the helical structure of proteins. At Cambridge University, James Watson and Francis Crick built DNA models based on the research of Chargaff, Wilkins, Franklin, Pauling, and others. The Watson–Crick approach proved successful, perhaps because they brought a wider array of facts to bear on the problem. But whatever the reasons, a fascinating and controversial account of the personalities and events leading to that success—as viewed by one participant—is given in Watson's book, *The Double Helix*. An alternative viewpoint, one more sympathetic to Franklin's contributions, is presented in Sayre's book, *Rosalind Franklin and DNA*.

Figure 6.13
Rosalind Franklin
(Drawing by William
Carroll)

Physical and chemical evidence on DNA structure

Wilkins (Figure 6.12) and Franklin (Figure 6.13) believed that the structure of DNA could be revealed through the study of DNA's x-ray diffraction pattern, and they were in large part correct. Their studies were the most crucial to the thinking of Watson and Crick. X-ray diffraction studies are used to elucidate the geometry of molecules—that is, their arrangement in space—and to determine the relative positions of the atoms that make up the molecule. X rays are a form of electromagnetic radiation that can be produced by the bombardment of a metallic target by a stream of high-speed electrons; they are especially useful for studying systems that have a regular spatial geometry. A system that has a completely ordered three-dimensional organization is called a *crystal*, which is generated by

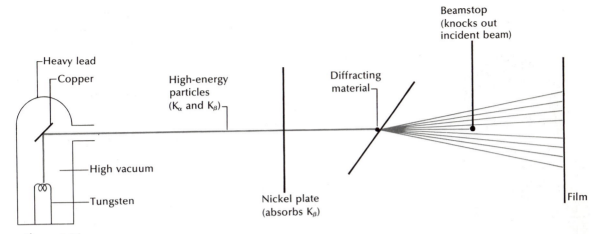

Figure 6.14
Schematic view of x-ray
diffraction setup

the indefinite repetition of a small three-dimensional unit termed the *unit cell*.

If a crystal is placed in the path of a beam of x rays and rotated on an axis perpendicular to the direction of the beam, a photographic plate positioned behind the crystal will register both the beams diffracted by the crystal and also the incident, or nondiffracted, beam (Figure 6.14). The x-ray diffraction pattern for a crystal of DNA is depicted in Figure 6.15.

Figure 6.15
X-ray diffraction
photograph of DNA fibers
in crystalline form
(Courtesy of M. H. F.
Wilkins)

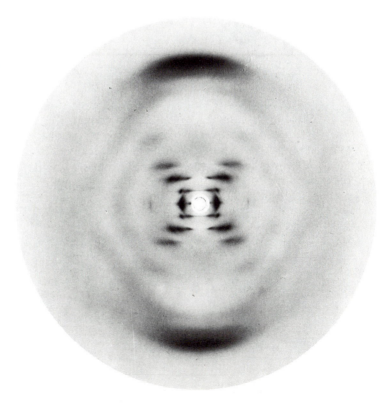

257

The pattern consists of a series of discrete exposures of varying spacing and intensity. The position of the exposures and their intensities give crucial information. Watson and Crick saw the diffraction pattern obtained by Wilkins and Franklin as suggesting that the DNA molecule was a *double-stranded* structure, in marked contrast to the three-stranded structure that Pauling (Figure 6.16) proposed earlier. The diffraction patterns also indicated that the DNA subunits (A, T, G, and C) were stacked in a *precise configuration* and at *regular intervals*.

The α helix (Figure 6.17) is Pauling's major contribution to knowledge of the three-dimensional configurations of biological macromolecules. Of the large number of configurations that have been proposed for proteins, the most important is the α helix. Pauling's α-helix model was stimulated by the desire to present a plausible model for a stable structure of proteins—that is, *the conformation of minimal potential energy*. Thus the model had to permit the formation of the *maximum number of hydrogen bonds* while requiring the *least distortion of covalent bonds and bond angles*.

The spatial geometry of DNA was determined to be in the form of an alpha helix, a spiral configuration maintained by *intra*molecular hydrogen bonds, a *hydrogen bond* being a link between a covalently bound hydrogen

Figure 6.17
The α helix. (a) Diagram of the α helix with 3.7 amino acid residues per turn (the NH and CO of each peptide bond are linked by hydrogen bonds). (b) Cross section of the α helix.

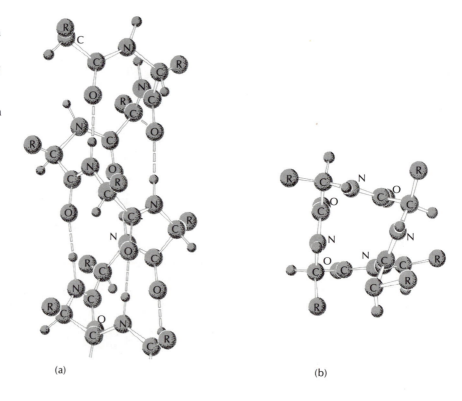

(a)

(b)

Figure 6.18
The hydrogen bond

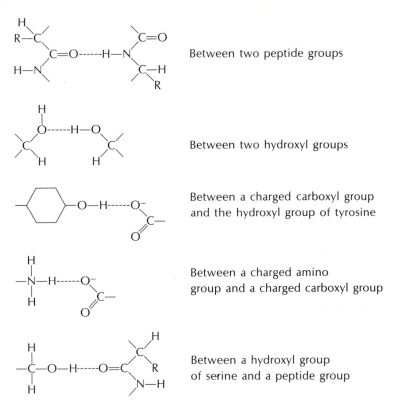

Between two peptide groups

Between two hydroxyl groups

Between a charged carboxyl group
and the hydroxyl group of tyrosine

Between a charged amino
group and a charged carboxyl group

Between a hydroxyl group
of serine and a peptide group

atom with a partially positive character and a partially negative, covalently bound acceptor atom (Figure 6.18). [An alternative to the α helix is the β configuration found in some proteins (Figure 6.19). This is an extended configuration stabilized by *inter*molecular bonds involving different protein chains.]

Figure 6.19
The β configuration of a
polypeptide chain

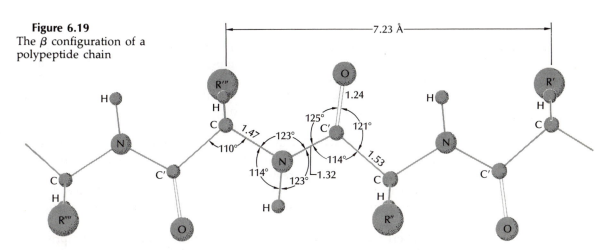

The chemical composition of DNA

Chargaff (Figure 6.20) approached the problem of DNA structure by analyzing its components. He revealed some profoundly significant relationships among the DNA components, but at first they were not fully appreciated. He demonstrated that the molar concentration of *purines*, A + G, was equal to the molar concentration of *pyrimidines*, C + T, or

$$A + G = C + T$$

Furthermore, the molar concentration of A was equal to that of T, and G was equal to C. That is,

$$A = T, \qquad G = C$$

Figure 6.20
E. Chargaff

Some of Chargaff's data supporting these relationships are given in Table 6.4. Finally, he observed that the A-T and G-C content varied from species to species but that for any one species, these (A + T)/(G + C) ratios were constant. This required that the concept of DNA as a simple repeating sequence of the four bases had to be rejected. The tetranucleotide theory was unacceptable in the face of this information. Chargaff's data supplemented the x-ray diffraction data of Wilkins and Franklin and permitted Watson and Crick to interpret the internal structure of DNA.

Table 6.4
Chargaff's data on composition of two organisms' DNA

Constituent	Yeast preparation 1	Yeast preparation 2	Avian tubercle bacilli
Adenine	0.24	0.30	0.12
Guanine	0.14	0.18	0.28
Cytosine	0.13	0.15	0.26
Thymine	0.25	0.29	0.11
Recovery	0.76	0.92	0.77

The Watson–Crick model of DNA

Watson and Crick (Figure 6.21) assembled all the information provided by Chargaff, Wilkins, Franklin, and Pauling and began to construct physical, scale models of DNA that would satisfy the data. Assuming that DNA had a helical configuration, one of the first decisions they had to make was whether to locate the sugar–phosphate molecules in the center of the helix or on the outside. Pauling had earlier presented a model with the sugar–phosphate units in the center, but this seemed unstable because the high-density negative charges on the phosphate might repel each other. Watson and Crick reasoned, therefore, that the sugar–phosphate units were on the outside of the DNA molecule, where negative charges on phosphate could be balanced by positively charged cations, such as Mg^{++}.

There were no conclusive data on the number of strands comprising a DNA molecule. And apparently unaware of the Wilkins–Franklin x-ray

Figure 6.21
J. D. Watson and F. H. C. Crick

diffraction evidence, Pauling suggested a three-stranded molecule. Watson and Crick toyed with such a model but ultimately rejected it, because a two-stranded structure seemed more consistent with the x-ray data and also because a three-stranded helix with the sugar–phosphate units interior and the bases exterior required too much stretching of covalent bond angles to be a stable configuration.

The last problem to be worked out was the arrangement of the bases in the molecule in a way that would satisfy the relationships established by Chargaff. Donahue, a postdoctoral fellow in the laboratory, suggested to Watson and Crick that DNA bases underwent changes in their pattern of electron and H-atom distribution (*tautomeric shifts*) and thus altered their hydrogen bonding capacities (Figure 6.22). Donahue also indicated

Figure 6.22
Tautomeric shifts of nitrogen bases of DNA

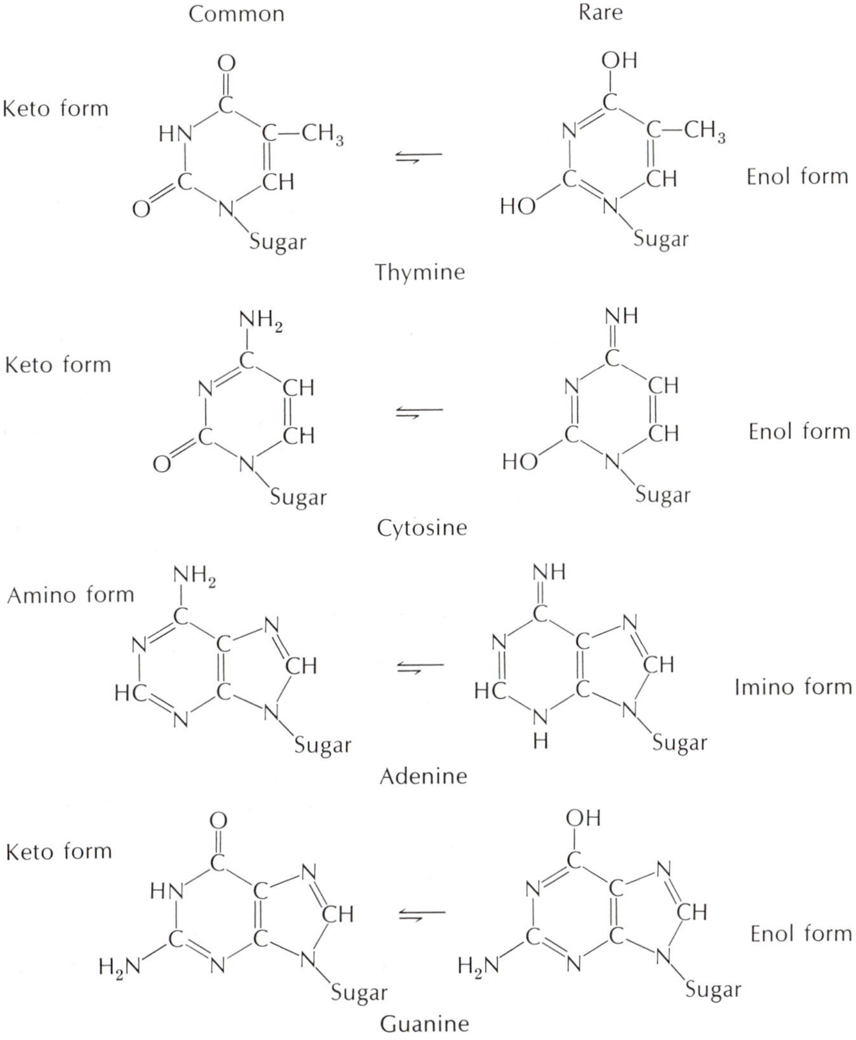

Common   Rare

Keto form   Enol form

Thymine

Keto form   Enol form

Cytosine

Amino form   Imino form

Adenine

Keto form   Enol form

Guanine

Figure 6.23
DNA base pairs

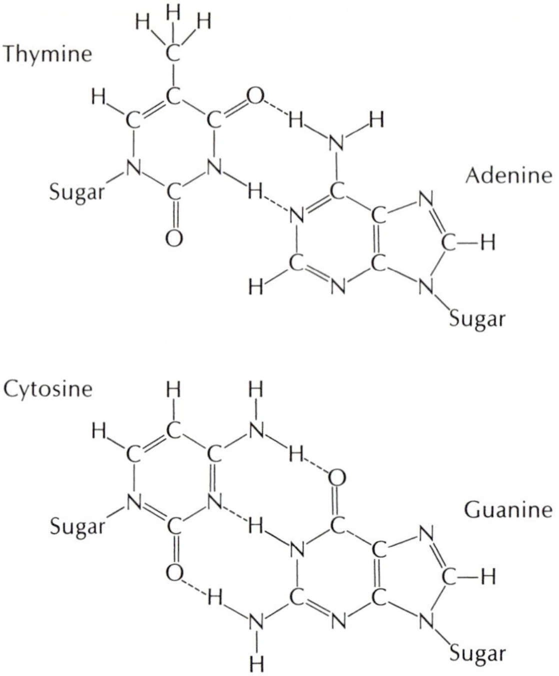

that one form—the *keto* (and amino) form—was more stable and thus predominated. Armed with these insights into hydrogen-bonding capabilities, Watson shuffled the models of various base-pair combinations around on his desk and observed that an A hydrogen bonded perfectly with a T and a G with a C (Figure 6.23). He further noted that the hydrogen-bonded ring system of the A-T pair had the same·dimensions as the hydrogen-bonded ring system of the G-C pair, and that these dimensions were perfectly compatible with those derived from the x-ray data. This highly specific base pairing could explain Chargaff's data, and it allowed Watson and Crick to confidently construct their double helix. Their model (Figure 6.24) for the structure of DNA was hurriedly published in April, 1953—hurriedly because of their apprehension that Pauling would recognize the mistakes in his own model and quickly arrive at the correct one.

The Watson–Crick model for the structure of DNA has withstood the test of time and has required but little alteration. Its effect on biological thinking was profound, for it immediately suggested mechanisms by which genes replicate, mutate, and store information. DNA replication involves the separation of the two strands of the DNA molecule, with each strand serving as a template for the synthesis of a new strand; the specificity of the base pairing gives this model a high degree of precision. Mutations may occur when bases occasionally mispair because of

Figure 6.24
Diagram of the structure of DNA proposed by Watson and Crick. (a) Chemical formula for a single DNA chain and diagram of a double-stranded DNA molecule. (b) Chemical formula for a double-stranded DNA molecule.

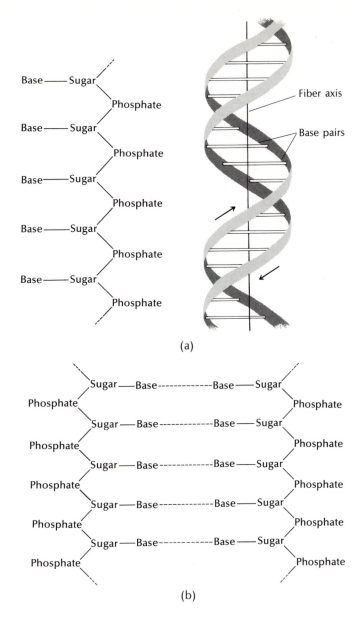

(a)

(b)

tautomeric shifts, such as the pairing of an A with a C instead of an A with the normal T. The information contained in genes is coded as the almost infinitely variable sequence of base pairs in the DNA molecule.

For their work, Watson, Crick, and Wilkins were awarded the Nobel Prize in 1962. Rosalind Franklin died tragically in 1958 at the age of 37. Had she not died, she would probably have shared the prize.

Overview

The experiments and observations culminating in the identification of nucleic acids as the genetic material developed along two main lines. One line of development began with the work of Miescher, the first to isolate DNA (complexed with protein) from cell nuclei. His important observations led to experiments on the chemical composition of this "nuclein." These in turn led to quantitative studies of nucleic acids in nuclei, which showed that the quantitative changes in nuclear DNA followed a pattern that would be expected of the genetic material. Unfortunately, the lines of investigation emanating from Miescher did not culminate in the unequivocal identification of nucleic acids as the genetic material.

The experiments of Griffith, studying what he called a transformation phenomenon in bacteria, mark the beginning of a series of experiments that culminated in the identification of DNA as the genetic material. Griffith's work led directly to the experiments of Avery, MacLeod, and McCarty. They showed that DNA was responsible for the transformation phenomenon and hence was the genetic material. We discussed the premature aspects of this work: The real significance of the findings eluded investigators. Hershey and Chase, in a crucial series of experiments employing bacteriophage, showed that DNA, not protein, dictates the structure of progeny. Their work can be said to have marked the beginning of the age of molecular biology. Interest focused rapidly on DNA, specifically its structure.

In 1953, one year after the Hershey and Chase study, Watson and Crick suggested a structure for DNA. They based their model on x-ray diffraction analysis of DNA crystals, the A, T, G, and C composition, and hydrogen bonding properties of the bases. The model they proposed, one of the most important models in biology, suggested a mode of replication, a mutational mechanism, and a scheme for storing coded information.

Questions and problems

6.1 If the base ratios of the DNA from two distinct species are the same, can you infer that single DNA strands from each species are identical in sequence?

6.2 An extraterrestrial organism is discovered during the course of an interplanetary exploration. From preliminary observations, it is clear that this organism reproduces and is composed of the same elements as Earth creatures. How would you go about determining the genetic material of the organism?

6.3 What lines of evidence would you cite to show that RNA is not the genetic material in most organisms?

6.4 Suppose you extract the DNA from a particular species and note that $(A + T)/(G + C) = 1$. Can you conclude that this DNA is double-stranded?

6.5 You have succeeded in extracting from a cell a material that you suspect is DNA. How would you prove that it is indeed DNA?

6.6 What lines of evidence would you cite to show that DNA is the genetic material?

6.7 What are the structural differences between DNA and RNA?

6.8 Why are the blender experiments of Hershey and Chase so important?

6.9 What is the significance of the tetranucleotide theory, and how did it relate to Chargaff's investigations?

6.10 Table 6.2 shows that Hershey and Chase found 120 percent infected bacteria surviving under certain circumstances. How do you explain this?

References

Alloway, J. L. 1933. Further observations on the use of pneumococcus extracts in effecting transformation of type *in vitro*. *J. Exp. Med.* 57:265.

Avery, O. T., C. M. MacLeod, and M. McCarty. 1944. Studies on the chemical nature of the substance inducing transformation of pneumococcal types. *J. Exp. Med.* 79:137–158.

Boivin, A., R. Vendrely, and C. Vendrely. 1948. L'acide desoxyribonuléique du noyau cellulaire, dépositaire des caractères héréditaires; arguments d'ordre analytique. *Compt. Rend. Acad. Sci.* 226:1061–1063.

Cairns, J., G. S. Stent, and J. D. Watson (eds.). 1966. *Phage and the Origins of Molecular Biology.* Cold Spring Harbor, N.Y., Cold Spring Harbor Laboratory of Quantitative Biology.

Chargaff, E. 1950. Chemical specificity of the nucleic acids and mechanism of their enzymatic degradation. *Experientia* 6:201–209.

Chargaff, E. 1971. Preface to a grammar of biology: A hundred years of nucleic acid research. *Science* 172:637–642.

Chargaff, E. 1975. A fever of reason: The early way. *Ann. Rev. Biochemistry* 44:1–18.

Coleman, W. 1965. Cell, nucleus, and inheritance: An historical study. *Proc. Amer. Phil. Soc.* 109:142–158.

Corwin, H. O., and J. B. Jenkins. 1976. *Conceptual Foundations of Genetics: Selected Readings.* Boston, Houghton Mifflin.

Dawson, M. H., and R. H. P. Sia. 1931. *In vitro* transformation of pneumococcal types: I. A technique for inducing transformation of pneumococcal types *in vitro*. *J. Exp. Med.* 54:681–699.

Fraenkel-Conrat, H., and B. Singer. 1957. Virus reconstitution: II. Combination of protein and nucleic acid from different strains. *Biochim. Biophys. Acta* 24:540–548.

Franklin, R. E., and R. Gosling. 1953. Molecular configuration of sodium thymonucleate. *Nature* 171:740–741.

Glass, B. 1965. A century of biochemical genetics. *Proc. Amer. Phil. Soc.* 109:227–236.

Griffith, F. 1928. The significance of pneumococcal types. *J. Hygiene* 27:113–159.

Hershey, A., and M. Chase. 1952. Independent functions of viral protein and nucleic acid in growth of bacteriophage. *J. Gen. Physiol.* 36:39–56.

Hess, E. L. 1970. Origins of molecular biology. *Science* 168:664–669.

Hotchkiss, R. D. 1966. Gene, transforming principle, and DNA. In J. Cairns, G. Stent, and J. D. Watson (eds.), *Phage and the Origins of Molecular Biology.* Cold Spring Harbor, N.Y., Cold Spring Harbor Laboratory of Quantitative Biology.

Mirsky, A. E. 1947. Chemical properties of isolated chromosomes. *Cold Spring Harbor Symp. Quant. Biol.* 12:143–146.

Mirsky, A. E. 1968. The discovery of DNA. *Sci. Amer.* 218:78–88.

Olby, R. 1974. *The Path to the Double Helix,* London, Macmillan.

Pollister, A. W., and H. Ris. 1947. Nucleoprotein determinations in cytological preparations. *Cold Spring Harbor Symp. Quant. Biol.* 12:147–157.

Sayre, A. 1976. *Rosalind Franklin and DNA.* New York, Norton.

Stent, G. S. 1968. That was the molecular biology that was. *Science* 160:390–393.

Stent, G. S. 1970. The premature and unique in scientific discovery. In *Proc. of the Conf. on the History of Biochem. and Molec. Biol. Amer. Acad. Arts and Sci.*

Swift, H. 1950. The constancy of DNA in plant nuclei. *Proc. Natl. Acad. Sci. U.S.* 36:643–654.

Twenty-one years of the double helix. 1974. *Nature* 248:721–788 (Various authors).

Watson, J. D. 1968. *The Double Helix.* New York, Atheneum.

Watson, J. D., and F. H. C. Crick. 1953. Molecular structure of nucleic acids: A structure for deoxyribose nucleic acid. *Nature* 171:737–738.

Watson, J. D., and F. H. C. Crick. 1953. Genetical implications of the structure of deoxyribose nucleic acid. *Nature* 171:964–967.

Wilkins, M. H. F. 1963. The molecular configuration of nucleic acids. *Science* 140:941–950.

Wilson, E. B. 1928. *The Cell in Heredity and Development.* New York, Macmillan.

Wyatt, H. B., 1972. When does information become knowledge? *Nature* 235:85–88.

Wyatt, H. B. 1974. How history has blended. *Nature* 249:803–805.

Seven

The replication of the genetic material

If one characteristic of life had to be singled out as the most important, it would be its ability to perpetuate itself, to replicate. When DNA was identified as the repository for all the genetic information required to build a functioning organism, understanding replication or reproduction was in many ways reduced to understanding DNA replication. But when DNA was identified as the genetic material in 1944 (and again in 1952), it was not at all clear how this process of replication took place. How could the genetic material be accurately replicated and passed on to the next generation?

When Watson and Crick proposed their model for the structure of DNA, a model of how DNA molecules are copied during chromosomal duplication was apparent. The specificity of the base pairs and the double-stranded nature of DNA suggested that one strand of the DNA molecule would serve as the template for the synthesis of a new strand. The sequence of bases in one strand would automatically determine the sequence in the complementary strand. For example, an A-G-C-C-T sequence in one strand would dictate a T-C-G-G-A sequence in the complementary strand. This occurs because A must pair with T and G with C.

A-G-C-C-T
↓ ↓ ↓ ↓ ↓
T-C-G-G-A

The model for DNA replication (Figure 7.1) suggested by Watson and Crick depicted simultaneous unwinding of the two DNA strands and new DNA synthesis. The model was termed *semiconservative replication:* half (that is, one strand) of a DNA parent molecule is conserved in each daughter molecule. Though attractive, the semiconservative model had

Figure 7.1
Scheme of DNA replication proposed by Watson and Crick

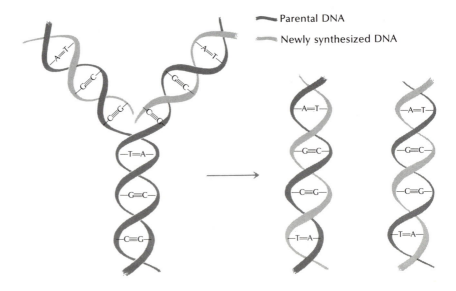

Parental DNA

Newly synthesized DNA

no experimental support when first presented (1953). Thus other possibilities were considered.

Foremost among these possibilities were conservative replication and dispersive replication. According to *conservative replication,* the original double helix remained intact and somehow produced a completely new, two-stranded molecule (Figure 7.2). In this model, the DNA is shown bound to proteins that could hold the DNA molecule in place for replication. The hydrogen bonds between base pairs are broken, the bases rotate to form hydrogen bonds with new complementary bases, and new DNA molecules are generated that contain all new DNA or all parental DNA. A problem with this scheme lies in the mechanics of the change from configuration (c) to configuration (d). A mechanism that would separate the pairs containing one parental base and one new base into new pairs composed of only parental or only new DNA is not easy to visualize.

Dispersive replication described a complicated series of breaks, transfers, and reunions within strands (Figure 7.3a). The parental strands break and reattach to the newly synthesized ones. New DNA is added to the ends of the broken parental strands, using the translocated parental strands as templates. In time, another pair of breaks occurs in the parental strands, and they are again translocated to the ends of the strands containing newly synthesized DNA. This switching of strands continues until a new DNA molecule is synthesized. The result of this type of replication would be hybrid DNA molecules with some parental DNA present in all four of the strands (Figure 7.3b).

One of the features of the Watson–Crick model of DNA was its predictive value: The semiconservative scheme portrayed daughter DNA molecules with one parental strand and one newly synthesized strand. The conservative scheme depicted daughter DNA molecules as containing all parental or all newly synthesized DNA. The dispersive scheme showed each strand of the daughter DNA molecules with a mixture of newly synthesized and parental DNA. The three schemes awaited a test that could distinguish parental DNA from newly synthesized DNA.

Figure 7.2
Model for conservative DNA replication. (Figure continued on page 270.)

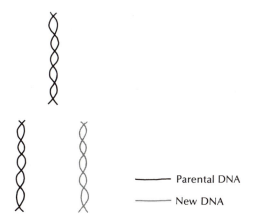

———— Parental DNA

———— New DNA

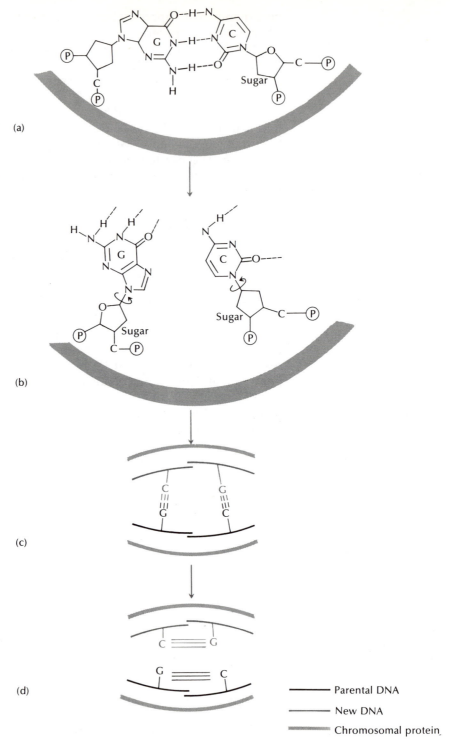

Figure 7.2 (cont.)
Model for conservative DNA replication. (a) Cross section through pair of DNA strands in standard configuration. The chromosomal protein holds the DNA molecule in place. (b) The base pairs break apart and the bases rotate 180° to allow hydrogen bonding with other bases. (c) Formation of new DNA. (d) The base pairs split apart and re-form.

(a)

(b)

(c)

(d)

Parental DNA

New DNA

Chromosomal protein

Figure 7.3
Model for dispersive
replication

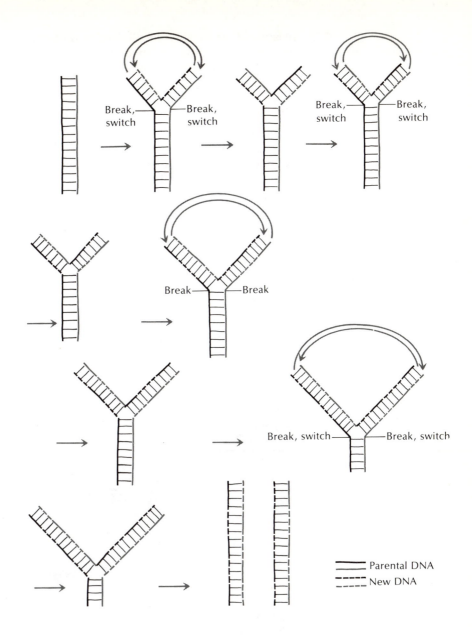

Break,
switch

Break,
switch

Break,
switch

Break,
switch

Break — Break

Break, switch — Break, switch

Parental DNA
New DNA

Evidence supporting semiconservative replication

The first clear indication of a semiconservative replication scheme came in 1957. J. H. Taylor had prepared thymidine that contained a radioactive hydrogen called *tritium* (^{3}H). Since thymidine is incorporated into DNA, and not into protein or RNA, ^{3}H-thymidine is a selective label. Taylor reasoned that if cells were allowed to undergo one round of DNA replication in ³H-thymidine followed by several replications in the presence of normal, nonradioactive thymidine, then conservative, semiconservative, and dispersive replication could be distinguished from each other. The technique used to monitor the distribution of label in the chromosomes is called *autoradiography*. Cells containing the labeled DNA are covered with a photographic film. As the radioactive atoms decay, they create black, exposed spots over the point of their incorporation. (Figure 7.4 is an autoradiogram of root cell chromosomes at division after one round of replication in ³H-thymidine.)

If the cells are allowed to undergo one round of DNA replication in ³H-thymidine—and assuming that a eukaryotic chromosome has either one long molecule of double-stranded DNA or a linear sequence of linked DNA molecules—a conservative scheme of DNA replication would predict one chromosome with label and one without (Figure 7.5a). The presence of label in both subunits of the chromosome would indicate either a dispersive or a semiconservative scheme. If the cells were washed free of excess ³H-thymidine after one round of DNA replication and allowed to replicate (this time in the presence of nonlabeled thymidine), a dispersive scheme of DNA replication would predict the presence of ³H label in all

Figure 7.4
Autoradiogram of *Vicia* chromosomes at division after one replication in which ³H-thymidine was incorporated into the DNA (Courtesy of J. Herbert Taylor)

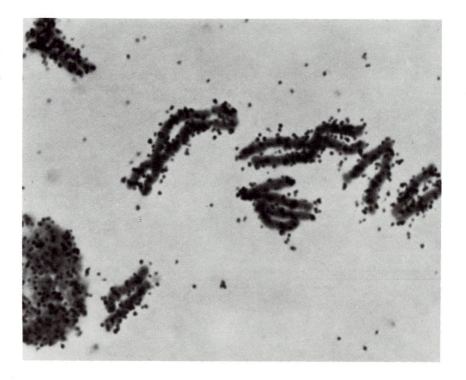

Figure 7.5
Predictions of
chromosome labeling
from conservative,
dispersive, and
semiconservative
replication; the results
appear most consistent
with a semiconservative
scheme

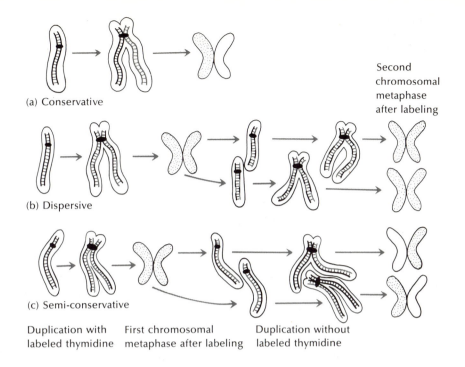

(a) Conservative

(b) Dispersive

(c) Semi-conservative

Second
chromosomal
metaphase
after labeling

Duplication with   First chromosomal   Duplication without
labeled thymidine   metaphase after labeling   labeled thymidine

DNA molecules, hence in all chromatids. A semiconservative scheme would predict ³H label in only half the molecules, hence in half the chromatids (Figures 7.5b and 7.5c). Results have clearly shown the presence of chromosomes without label, which supports the semiconservative scheme. But some uncertainties about the structure and organization of DNA in the chromosomes in higher organisms make it difficult to say that DNA replication at this level is definitely semiconservative.

The first clear evidence at a molecular level that DNA replication might be semiconservative was provided by M. Meselson and F. W. Stahl in 1958. The key to their experiment was a technique developed in 1957 by Meselson, Vinograd, and their colleagues to allow the separation of DNA molecules of differing densities. *Escherichia coli* grown in a medium supplemented with ^{15}N (the heavy isotope of normal nitrogen, ^{14}N) contains denser DNA than *E. coli* grown in ^{14}N medium. The ^{15}N-DNA can be separated from ^{14}N-DNA by centrifugation in a concentrated solution of cesium chloride (CsCl). When a solution of CsCl is subjected to intense centrifugal force, it becomes more dense at the bottom of the centrifuge tube and less dense at the top, and the CsCl density throughout the tube increases linearly from top to bottom (Figure 7.6). In such a density gradient, DNA molecules will move to positions in the tube where their density equals that in the CsCl gradient. Molecules containing ^{15}N will locate at a denser region of the tube than will molecules with ^{14}N.

Figure 7.6
Cesium chloride
centrifugation technique

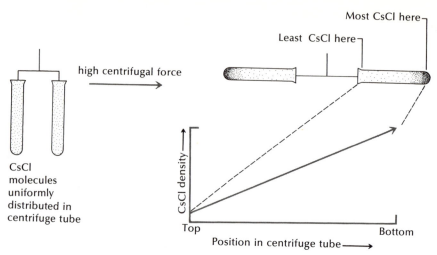

high centrifugal force

Most CsCl here

Least CsCl here

CsCl
molecules
uniformly
distributed in
centrifuge tube

CsCl density

Top

Bottom

Position in centrifuge tube

Figure 7.7
Meselson–Stahl
experiment: technique

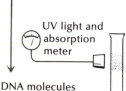

E. coli cells
growing at least
14 generations
in an ^{15}N medium

^{15}N medium
exchanged for
^{14}N medium

Cells sampled at
regular intervals
and DNA extracted

DNA molecules
spun in a CsCl
density gradient

UV light and
absorption
meter

DNA molecules
separate according
to their density; their
location in the gradient
determined by UV
absorption

Meselson and Stahl grew *E. coli* for several generations in an ^{15}N medium. The result was that the ^{15}N replaced the ^{14}N in the cells. When the ^{14}N had been so replaced, an excess of ^{14}N was added to the medium to ensure that from the time of addition, the synthesis of DNA would involve ^{14}N components only. Just prior to the addition and at regular intervals thereafter, an aliquot of cells was taken and their DNA extracted. The extracted DNA was centrifuged in a CsCl gradient and its density determined (Figure 7.7).

Before the addition of ^{14}N, all the DNA should band at the ^{15}N-DNA density. After one generation of growth in ^{14}N medium, a semiconservative scheme of DNA replication predicts that the DNA molecules will contain one light DNA strand and one heavy and that they will band at a density intermediate between the light ^{14}N-DNA and the heavy ^{15}N-DNA. After two rounds of replication, half the DNA molecules will be all light ^{14}N-DNA and half will be hybrid (Figure 7.8a). The conservative replication scheme predicts, after one round of DNA replication, all light ^{14}N-DNA and all heavy ^{15}N-DNA in a 1:1 ratio; two rounds of conservative DNA replication will produce a 3:1 ratio of ^{14}N-DNA to ^{15}N-DNA (Figure 7.8b). Dispersive replication predicts DNA molecules of intermediate density after one round of DNA replication. After two rounds of replication, it predicts DNA molecules that band in the gradient at a point less dense than a half-and-half hybrid but more dense than ^{14}N-DNA (Figure 7.8c).

The results obtained by Meselson and Stahl (Figure 7.9) support a semiconservative scheme only. Their paper is a landmark in genetic research. Their findings are unequivocal and represent the first clear verification of a prediction made by the Watson–Crick model of DNA.

Since the publication of the work of Taylor and his colleagues and that of Meselson and Stahl, the semiconservative replication scheme has been verified in organisms throughout the plant and animal kingdoms, and in many prokaryotes.

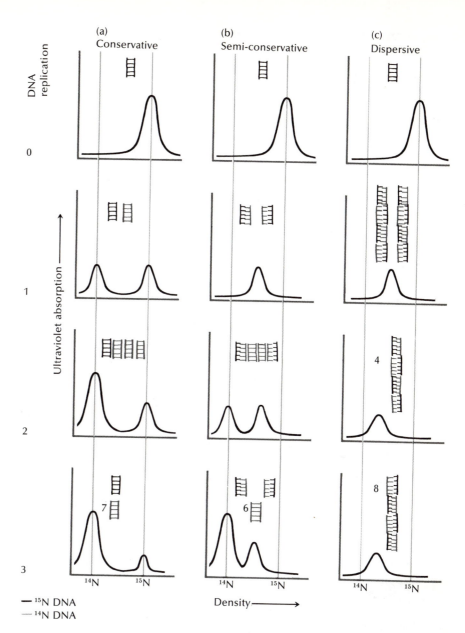

Figure 7.8
Meselson–Stahl experiment: predictions

DNA replication

Ultraviolet absorption →

(a) Conservative

(b) Semi-conservative

(c) Dispersive

0

1

2

3

^{14}N ^{15}N ^{14}N ^{15}N ^{14}N ^{15}N

——— ^{15}N DNA
- - - - ^{14}N DNA

Density ——→

DNA polymerase

DNA polymerase is discovered

The first major breakthrough in our understanding of how DNA is repli-cated occurred in 1956 when A. Kornberg (Figure 7.10) discovered an enzyme called DNA polymerase. This enzyme is able to catalyze the step-by-step addition of deoxyribonucleotides into a DNA chain. Other

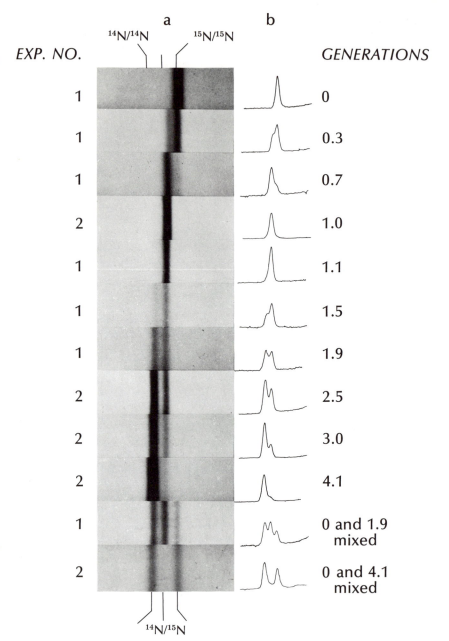

Figure 7.9
Meselson–Stahl
experiment: results
(Courtesy of M. Meselson)

EXP. NO.

a   b

¹⁴N/¹⁴N   ¹⁵N/¹⁵N

GENERATIONS

EXP. NO.	GENERATIONS
1	0
1	0.3
1	0.7
2	1.0
1	1.1
1	1.5
1	1.9
2	2.5
2	3.0
2	4.1
1	0 and 1.9 mixed
2	0 and 4.1 mixed

¹⁴N/¹⁵N

Figure 7.10
Arthur Kornberg
(Courtesy of A. Kornberg)

enzymes that can do the same thing have since been discovered, so we now refer to Kornberg's enzyme as *DNA polymerase I.*

The *in vitro* synthesis of DNA using DNA polymerase I requires the following components.

276

1. All four deoxyribonucleotide 5′-triphosphates (dATP, dTTP, dGTP, dCTP)

2. Mg^{++}

3. A *primer strand* of DNA (or RNA) with a free 3′ OH (the nucleotides are added to the 3′ end of the strand)

4. A DNA *template* to specify the base sequence

When all of these components are present in an *in vitro* (outside the living cell) system, nucleotides are polymerized into DNA chains:

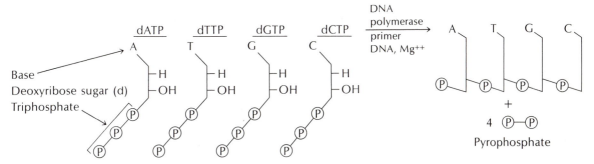

The primer-template DNA used by Kornberg is partially denatured so that there would be a free 3′ end for DNA growth (primer activity) and a strand to specify the sequence of bases added (template activity):

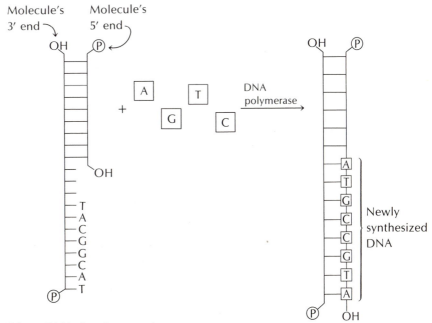

Primer DNA showing a region
of single-strandedness
(partial denaturation)

Note that the functions we have described so far do not suggest a replication scheme so much as they suggest gap filling. Indeed, DNA polymerase I seems to be responsible mainly for repairing gaps in DNA molecules. But it is important to know that the properties of the enzyme enabled Kornberg and others to begin to unravel the mechanisms of DNA replication. DNA polymerase I was crucial to efforts aimed at defining the process of DNA replication.

DNA polymerase I can accurately replicate DNA

Further studies with DNA polymerase I showed that the newly synthesized DNA was similar in base composition to the primer-template DNA, but it was structurally abnormal. The newly synthesized DNA contained numerous branch points, unlike normal DNA, which is either linear or circular and has no such branch points (Figure 7.11). The branches are the result of the polymerase switching to the displaced DNA strand and then folding back upon itself (Figure 7.12).

How do we know that the newly synthesized DNA is made using the primer-template DNA? In a very simple test, we can look at the A, T, G, and C content of the newly synthesized DNA and compare it to the A, T, G, and C content of the primer-template DNA. This has been done, and the values agree. But agreement does not establish that the new DNA is

Figure 7.11
(a) Branched bacteriophage T7 DNA synthesized by Kornberg; normal T7 DNA is shown in part (b) on page 279 (From R. B. Inman, C. K. Schildkraut, and A. Kornberg, in *J. Gen. Microbiol.* (1965); courtesy of R. B. Inman)

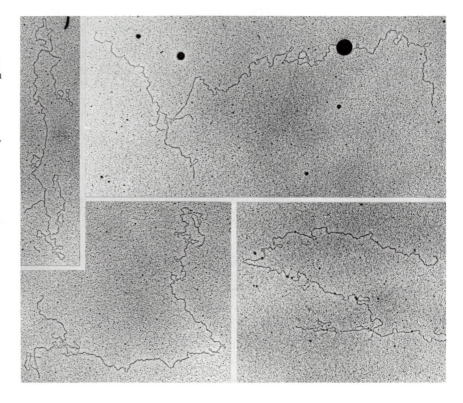

Figure 7.11 (cont.)
(b) Normal T7 DNA
(Courtesy of L. A.
MacHattie)

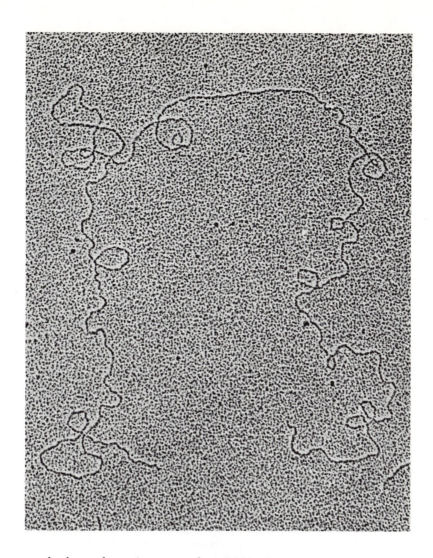

made from the primer-template DNA. For example, consider the two DNA molecules

```
AAATTTGGGCCC              ATGCATGCATGC
||||||||||||    and       ||||||||||||
TTTAAACCCGGG              TACGTACGTACG
```

Both have the same ATGC content, yet they are quite different from each other.

A test that more accurately assesses the fidelity of replication mediated by DNA polymerase I is called the *nearest-neighbor analysis*. Developed by Kaiser, Kornberg, Josse, and colleagues, this test determines

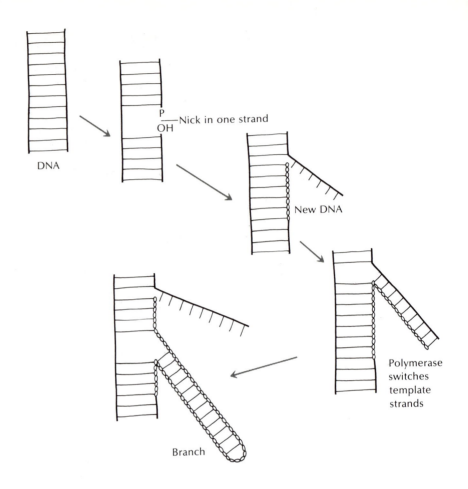

Figure 7.12
Formation of branch points in a replicating DNA molecule

P
—Nick in one strand
OH

DNA

New DNA

Polymerase switches template strands

Branch

the frequency with which any two DNA bases occur as adjacent or nearest neighbors in a particular DNA molecule. Each of the four DNA bases can have four possible nearest neighbors, which means that there are 16 total nearest-neighbor pairs.

AA	TA	GA	CA
AT	TT	GT	CT
AG	TG	GG	CG
AC	TC	GC	CC

It is reasonable to expect that the frequencies of the nearest neighbors will vary among different species and that, for a particular species, nearest-neighbor frequencies will be characteristic and constant for all members of the species. Furthermore, by comparing the nearest-neighbor frequencies of the primer-template DNA with those of the newly synthesized DNA, we can get an accurate indication of the fidelity of DNA replication as it is mediated by DNA polymerase I.

The nearest-neighbor analysis is an elegant experiment. It is designed to determine the frequencies of the four possible bases that can come to be next to any given base introduced into a DNA molecule by the action of DNA polymerase. Figure 7.13 summarizes the technique. DNA

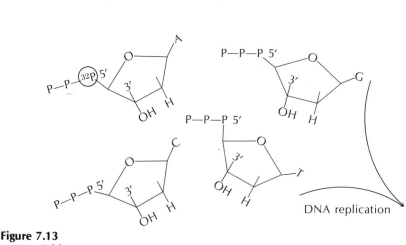

Figure 7.13
Nearest-neighbor
analysis. In this analysis,
cells are grown in a
medium containing the
four DNA bases, one of
them labeled with ^{32}P at
the 5' position. The
synthesized DNA,
containing the ^{32}P label, is
extracted from the cells
and degraded to 3'
monophosphate
nucleotides. The ^{32}P is
now attached to the 3'
carbon of the adjacent
base.

DNA replication

Degradation to 3' nucleotides

is synthesized using one of the four DNA bases with a radioactive phosphate (^{32}P). This radioactive phosphate group serves as a label. In one experiment, for example, all of the A bases are labeled. The labeled A is inserted into a DNA molecule opposite a T and covalently linked to the 3' end of the primer strand. The ^{32}P label is thus between the 3' carbon of the nearest neighbor and the 5' carbon of the A that brought it in. Other bases are added to the chain according to the specifications of the template strand.

When synthesis is completed, a specific nuclease is added that cleaves the DNA molecule into nucleotides with phosphates attached to the 3' carbon instead of the 5' carbon. The ^{32}P label is thus transferred from A to one of the four possible nearest neighbors. The experiment is repeated using another labeled base. For each experiment, four nearest-neighbor frequencies are determined, and when the experiment is done for each labeled base, there are 16 nearest-neighbor frequencies. Table 7.1 gives the nearest-neighbor frequencies for DNA synthesized by polymerase using *Mycobacterium phlei* DNA as the primer-template DNA.

The next step in the analysis was comparison of the nearest-neighbor frequencies of the primer-template DNA with those of the newly synthesized DNA. The results showed that the nearest-neighbor frequencies of the primer-template DNA and the newly synthesized DNA were essentially identical.

Certain important conclusions can be drawn from these experiments. First, it is reasonable to suggest that DNA polymerase I accurately repli-

Table 7.1
Nearest-neighbor frequencies of *Mycobacterium phlei* DNA

Reaction no.	Labeled triphosphate	Isolated 3'-deoxyribonucleotide			
		T	A	C	G
1	A	A⟶T 0.012	A⟶A 0.024	A⟶C 0.063	A⟶G 0.065
2	T	T⟶T 0.026	T⟶A 0.031	T⟶C 0.045	T⟶G 0.060
3	G	G⟶T 0.063	G⟶A 0.045	G⟶C 0.139	G⟶G 0.090
4	C	C⟶T 0.061	C⟶A 0.064	C⟶C 0.090	C⟶G 0.122
	Sums	0.162	0.164	0.337	0.337

Adapted from J. Josse, A. D. Kaiser, and J. Kornberg, "Enzymatic synthesis of DNA. VIII: Frequencies of nearest neighbor base sequences in DNA," *J. Biol. Chem.* 236:864–875 (1961).

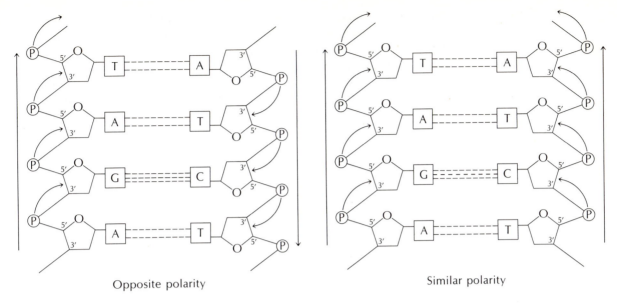

Opposite polarity

Similar polarity

⟶ Direction of ³²P transfer

Figure 7.14
DNA molecules having strands with opposite polarity and strands with similar polarity; the nearest-neighbor analysis is consistent only with DNA strands of opposite polarity. The frequency of ³²P transfer from A to G (A → G) would be the same as ³²P transfer from C to T (C → T) if the strands were of opposite polarity, and T to C (T → C) if the strands were similar in their polarity.

cates the primer-template DNA. Second, the data show that the two strands of a DNA molecule are arranged in an antiparallel fashion. Such an arrangement was clear from the model of DNA postulated by Watson and Crick, but the nearest-neighbor analysis provided experimental verification of the idea. Figure 7.14 and Table 7.2 demonstrate how the experiments verified the antiparallel structure of DNA. If the strands ran parallel, certain nearest-neighbor frequencies of the template strand and new strand would not be equivalent; they are equivalent only when the strands are arranged in an antiparallel way.

We can conclude, then, that DNA polymerase I at least has the potential for replicating DNA in an accurate fashion. But is this the enzyme that replicates DNA inside the living cell?

Table 7.2
Nearest-neighbor analysis: frequency predictions based on arrangement of strands of the DNA molecule (see also Figure 7.14)

Opposite polarity	Similar polarity
(0.065) A ⟶ G = C ⟶ T (0.061)	(0.065) A ⟶ G = T ⟶ C (0.045)
(0.045) G ⟶ A = T ⟶ C (0.045)	(0.045) G ⟶ A = C ⟶ T (0.061)
(0.012) A ⟶ T = A ⟶ T (0.012)	(0.012) A ⟶ T = T ⟶ A (0.031)

The values in parentheses were gathered from Table 7.1.

DNA ligase The question of whether DNA polymerase I could do anything other than fill gaps in a DNA molecule was important. It was necessary to know if the enzyme could faithfully replicate an entire DNA molecule, and a key element in the investigations that provided an answer to this question was the discovery of *DNA ligase*. This enzyme catalyzes the formation of a covalent bond (phosphodiester bond, to be precise) between two DNA strands.

The enzyme has been found to exist in many cell types, prokaryotes as well as eukaryotes. It is crucial to the replication scheme, because DNA polymerase I cannot perform that final function of tying together two DNA strands once all of the complementary bases have been inserted. It is also indispensable in recombination and in the creation of circular DNA molecules from linear molecules.

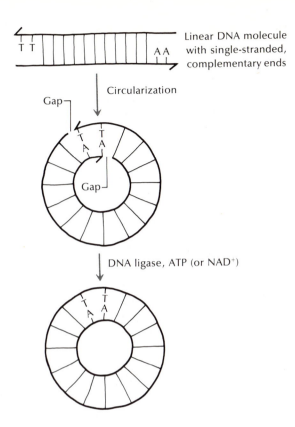

Linear DNA molecule with single-stranded, complementary ends

Circularization

Gap

Gap

DNA ligase, ATP (or NAD⁺)

DNA synthesis using DNA polymerase I and DNA ligase

In 1967, M. Goulian, Kornberg, and R. Sinsheimer established that Kornberg's DNA polymerase could indeed synthesize normal, biologically active DNA that was identical in all ways to a native DNA molecule used as a template. They announced the *in vitro* synthesis of a biologically active φX174 bacterial virus chromosome (Figure 7.15a). This synthetic chromosome could enter an *E. coli* cell, replicate, and give rise to normal φX174 progeny bacteriophage. The *in vitro* system employed by these investigators was composed of the following.

1. dATP, dCTP, dGTP, and dBUTP (bromouracil, denoted by BU). 5-bromouracil is an analog of T, containing Br instead of CH_3 at the 5 position of the nucleotide (Figure 7.15b). And since dBUTP is heavier than dTTP, it serves as a density label.

2. φX174 DNA to serve as a template. This virus is rare in nature, because its chromosome is a single strand of DNA arranged in a circle. For Kornberg's DNA polymerase, this single-stranded quality was important, because the enzyme works most efficiently on single-stranded regions of DNA.

3. Highly purified DNA polymerase I.

4. DNA ligase.

5. Boiled extract of *E. coli*, which contains small, single-stranded pieces of DNA used to start the synthesizing process.

6. Mg^{++} ions, perhaps to neutralize the heavy preponderance of negative charges found on the DNA's phosphate moiety or perhaps to interact with DNA polymerase as a cofactor. The precise function of this cation is not known.

When these six components are present in an *in vitro* system, biologically active DNA is synthesized (Figure 7.15a). An intact ϕX174 chromosome, arbitrarily designated as a "+" strand, is used as the starting template. A small, single-stranded piece of DNA base-pairs with a region of the ϕX174 chromosome, and new DNA is added to the 3' end of that piece. Synthesis proceeds in the presence of 5-bromouracil (BU), which acts as a density label. And DNA ligase completes the circle to form intact double-stranded, circular DNA molecules—one strand ("−") BU-labeled and one parental strand. These double-stranded DNA molecules are treated with DNase (which degrades DNA) just long enough to generate single-stranded breaks in half of them. The nicked DNA molecules are heated to break the hydrogen bonds holding the two strands together, then centrifuged to separate BU-labeled DNA strands from native DNA and circular DNA from linear DNA (circular DNA is less dense than linear DNA). Circular, single-stranded BU-DNA ("−") is isolated and used as a template to synthesize a normal ("+") strand containing T instead of BU. The fully synthetic double-stranded replicative form (RF) is as infective as the native double-stranded replicative form. That is to say, the synthetic RF is taken up by *E. coli* cells, and these infected cells ultimately produce ϕX174 progeny. This experiment unequivocally demonstrates that Kornberg's DNA polymerase can indeed synthesize normal, biologically active DNA, at least *in vitro*.

DNA polymerase I has nuclease activity

Even though DNA polymerase I can synthesize biologically active DNA, its primary function appears to be repair. It has long been known that the enzyme, in addition to adding bases to a growing chain, can also remove them. This is called nuclease activity, and DNA polymerase I has two distinct kinds of nuclease activities. It can start at the 3' end of a DNA strand and remove bases in a 3' to 5' direction. And it can start at the 5' end of a DNA strand and remove bases in a 5' to 3' direction. We can inquire into the significance of this nuclease activity.

The 3' to 5' nuclease activity appears to have an "editorial" function as the enzyme adds bases complementary to the template. In other words, if the enzyme incorporates an incorrect base, it backs up and clips

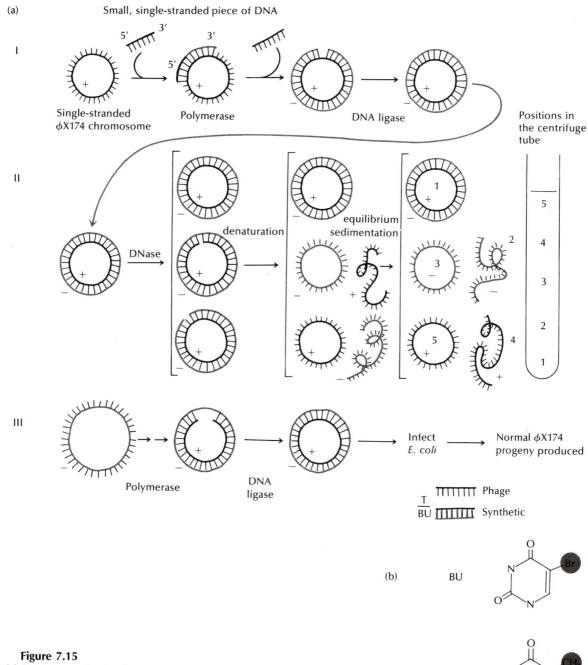

Figure 7.15
(a) *In vitro* synthesis of φX174 (b) Formulas for bromouracil (BU) and thymine (T)

287

that base out. Then it reinserts the correct base and continues on its way, adding bases in a 5' to 3' direction.

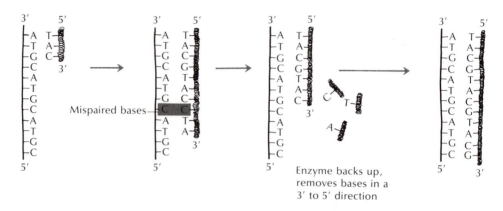

Enzyme backs up, removes bases in a 3' to 5' direction

The 5' to 3' nuclease activity operates under a different set of circumstances. When DNA is damaged (for example, by ultraviolet light or certain chemicals), that damage is often removed by the 5' to 3' nuclease activity of DNA polymerase I.

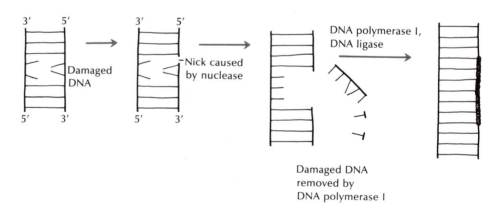

Damaged DNA removed by DNA polymerase I

DNA polymerase I is a complex enzyme performing several functions. Figure 7.16 is a model of DNA polymerase I showing its various active sites.

Other DNA polymerases   A major breakthrough in our understanding of DNA replication came in 1969 when Paula DeLucia and John Cairns isolated a mutant of *E. coli* that was essentially lacking DNA polymerase I activity. This mutant strain was called po1A1. Though lacking DNA polymerase I activity, the cells replicated their DNA normally but were unable to repair damage to their DNA

Figure 7.16
Functional sites in the active center of DNA; the small arrow in the triphosphate site shows the phosphodiester linkage that will form

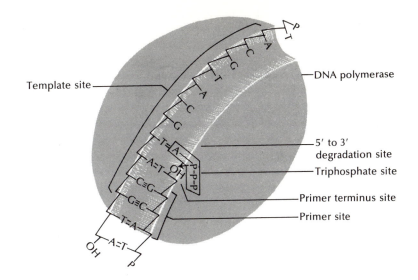

caused by ultraviolet radiation. The clear implication is that DNA polymerase I is a repair enzyme and that other polymerases are responsible for replication.

Soon after the discovery of the mutants of *E. coli* defective in DNA polymerase I, two new DNA polymerases were discovered, DNA polymerase II and DNA polymerase III. DNA polymerase II remains something of an enigma, and we do not yet fully understand its function in a living cell. It can polymerize DNA, but it does so more slowly than DNA polymerase I. It has 3' to 5' nuclease activity, but no 5' to 3' activity.

DNA polymerase III was discovered by T. Kornberg (A. Kornberg's son) and M. L. Gefter in 1972. It is by far the most active of the three polymerizing enzymes. From various studies of DNA polymerase III, it is clear that the enzyme is essential for DNA replication. In addition to its polymerizing activity, DNA polymerase III has 3' to 5' *and* 5' to 3' nuclease activity.

When it is involved in the replication process, DNA polymerase III is in a complex form called DNA polymerase III*. This form consists of two molecules of DNA polymerase III complexed with two molecules of an auxiliary protein called copolymerase III*. It is in this form that DNA polymerase III apparently replicates DNA.

We do not want to leave the impression that DNA polymerase I functions only in repair and gap-filling. Though that may turn out to be true, such a conclusion is premature for the present. It has not yet been possible to isolate an *E. coli* mutant that is viable and that has absolutely no DNA polymerase I activity. So we can suggest that the enzyme has an essential role in the replication of the cell's genetic material.

Figure 7.17
(a) Autoradiograph of the chromosome of *E. coli* K12 Hfr, labeled with tritiated thymidine for two generations (the scale shows 100 μm); the inset shows the same structure diagrammatically.

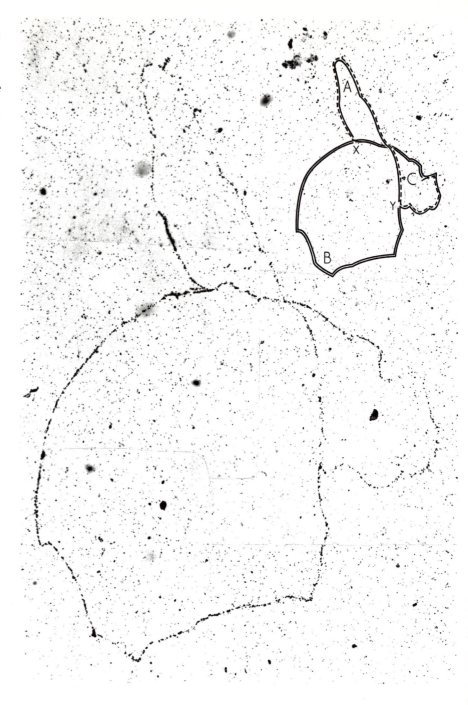

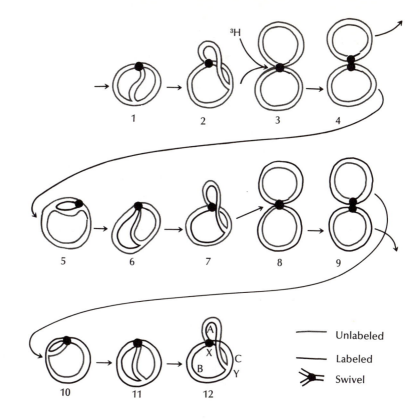

Figure 7.17 (cont.)
(b) Diagrammatic representation of replication of a circle, based on the assumption that each round of replication begins at the same place and proceeds in the same direction (From Cold Spring Harbor Symposium, 1963, courtesy of J. Cairns)

³H

1   2   3   4

5   6   7   8   9

10   11   12

A
X   C
B   Y

——— Unlabeled

——— Labeled

✳ Swivel

The replication of circular DNA

So far, we have discussed DNA replication as primarily involving linear molecules, but an understanding of how double-stranded circular DNA molecules replicate is also important. Cairns's studies with *in vivo* replication in *E. coli* provided a key insight into this problem.

Cairns's studies with the *E. coli* chromosome showed that replication is semiconservative and involves a structure called a *replication fork*. He also showed that the DNA molecule retains its circular form while it is being replicated. His technique employed autoradiography (see pages 271–273). He grew the *E. coli* cells in ³H-thymidine medium and analyzed chromosomes at various stages of replication. His results are shown in Figure 7.17, along with an interpretation of *E. coli* chromosome replication.

Cairns's experiments raised a number of questions. For example, does replication begin at a unique site on the *E. coli* chromosome, or can it begin anywhere? And is replication bidirectional or unidirectional? A number of experiments address themselves to this problem. In one experiment, Inman partially denatured DNA by exposing it to high pH. The regions that were rich in AT pairs denatured first, because AT pairs have only two H bonds holding them together, while GC pairs have three. Once they were denatured, Inman prevented the two strands from coming back together by blocking the H-bond formation through the use of formaldehyde. This procedure resulted in denatured or single-stranded regions at AT-rich sections of the molecule. When viewed with the electron microscope, these denatured regions appeared as "bubbles" in the molecule.

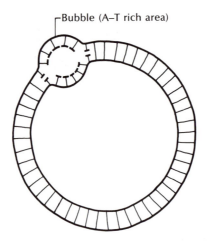

Bubble (A–T rich area)

The location of these bubbles relative to each other was constant for any one species' DNA.

The bubbles were now used as reference points in the examination of replicating DNA molecules. Inman isolated bacteriophage λ DNA at various points in its replication and then denatured it to form these bubbles. If replication began at a specific point and proceeded unidirectionally, then the position of the replication fork relative to the reference bubble should change as replication proceeds.

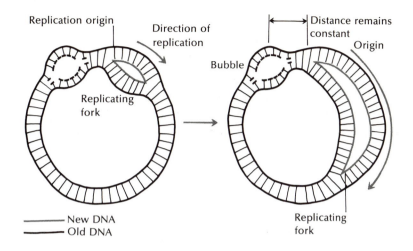

New DNA
Old DNA

But if replication proceeds in a bidirectional way from a unique origin, then the relative position of the bubble and the *two* replication forks should change.

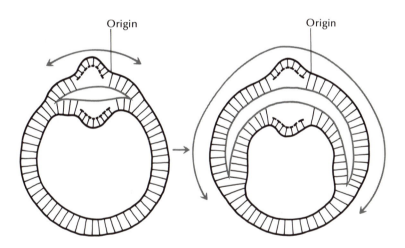

The evidence supported the latter model, suggesting that there is a specific starting point for replication with the fork moving bidirectionally from that point.

An interesting experiment by R. E. Bird and his colleagues confirmed that replication in *E. coli* begins at a specific site. They studied the relative amounts of different genes in *E. coli* under conditions in which the cells were synthesizing DNA at a maximal rate. Consider two genes *x* and *y*. Now suppose that replication begins at a unique site and is bidirectional.

If x is near the origin and y near the end, then at any time, there should be twice as many copies of the x gene as the y gene. If replication begins at random points, then the copies of x and y genes in the population will be about equal. Figure 7.18a illustrates these contrasting predictions. The data from these experiments (Figure 7.18b) show that DNA replication begins at a specific point at or near the *ilv* locus and proceeds in a bidirectional way until the two replicating forks meet about half-way around at the *trp* locus.

Figure 7.18
Two views (a) and (b) of the origin of DNA replication

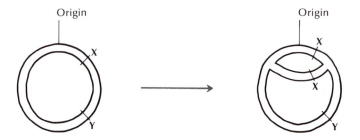

Replication begins at a specific origin point and moves bidirectionally. Most molecules have 2**X** genes for each **Y** gene.

(a)

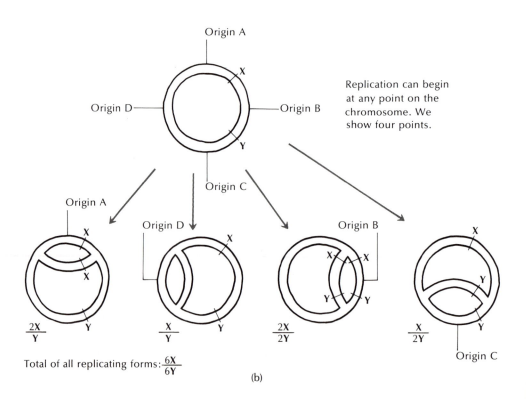

Replication can begin at any point on the chromosome. We show four points.

Total of all replicating forms: $\frac{6X}{6Y}$

(b)

Figure 7.18 (cont.)
(c) An experiment that shows that replication in *E. coli* begins at a specific origin point (see text)

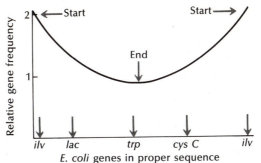

E. coli genes in proper sequence

From these data, replication begins at or near the *ilv* locus, proceeds bidirectionally, and ends halfway around the circle at the *trp* locus. (From Bird, *et al.*, *J. Mol. Biol.* 70:549, 1972).

(c)

We can conclude then, that DNA replication in *E. coli* has a unique starting point and proceeds bidirectionally. In higher organisms, we shall see that there may be several sites where DNA synthesis may begin. However, they are all unique and certainly not random.

DNA replication is discontinuous Cairns's studies and those of others focused attention on the replication fork. It appeared that replication proceeded along the antiparallel strands in the same direction. That is to say that one DNA strand was growing in a 5′ to 3′ direction and the other strand was growing in a 3′ to 5′ direction.

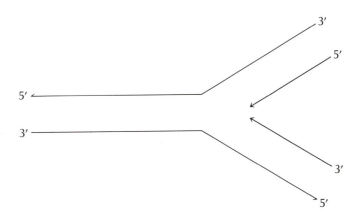

The one problem with this type of replication was that it contradicted the properties of the three DNA polymerases. Each of these polymerizing enzymes can synthesize DNA in a 5′ to 3′ direction only. They cannot synthesize DNA in a 3′ to 5′ direction. How can we resolve the observation with the enzyme properties?

A resolution was suggested by R. Okazaki, who observed that newly synthesized DNA in the region of the replication fork was present as small fragments (called *Okazaki fragments*). As synthesis continues, these fragments become covalently linked to each other forming longer DNA strands. Our model, then, is one in which DNA synthesis proceeds in a 5′ to 3′ direction along one template strand, then switches strands and proceeds in a 5′ to 3′ direction along the other template strand. DNA ligase ties the fragments into longer strands, and the result is the *appearance* of 5′ to 3′ and 3′ to 5′ growth at the replication fork.

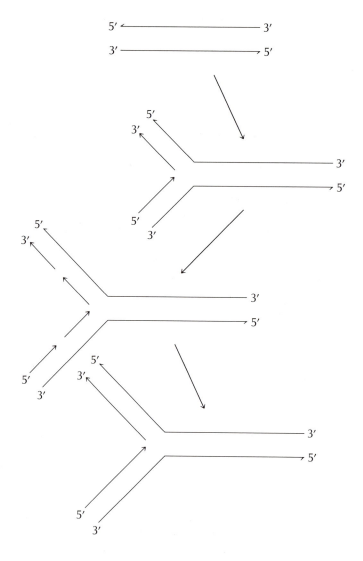

We are now in a position to try to tie together, in a general model, some of the many things we have discussed about DNA replication. The model that we develop may not apply to all bacteria, viruses, and eukaryotic organisms, but it does serve as a unifying scheme, especially for bacteria. Figure 7.19 presents a model of DNA replication.

Replication begins at a specific initiation point. An initiator protein recognizes this point and perhaps complexes with an RNA polymerase. The initiation point may be the point at which the DNA molecule is attached to the cell membrane, as it so often is in prokaryotic cells.

The next step is the attachment of unwinding proteins to the area in front of the initiation point. These proteins uncoil the DNA in small

Figure 7.19
A model for DNA synthesis

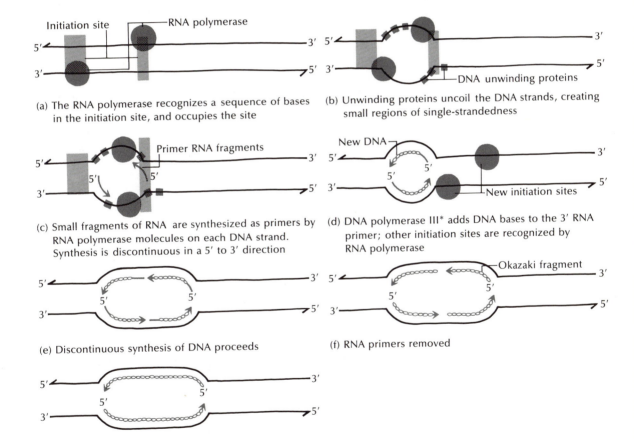

(a) The RNA polymerase recognizes a sequence of bases in the initiation site, and occupies the site

(b) Unwinding proteins uncoil the DNA strands, creating small regions of single-strandedness

(c) Small fragments of RNA are synthesized as primers by RNA polymerase molecules on each DNA strand. Synthesis is discontinuous in a 5' to 3' direction

(d) DNA polymerase III* adds DNA bases to the 3' RNA primer; other initiation sites are recognized by RNA polymerase

(e) Discontinuous synthesis of DNA proceeds

(f) RNA primers removed

(g) DNA polymerase I and DNA ligase fill in the gaps and covalently link the fragments. Synthesis proceeds in this manner until entire molecule is replicated

regions in order to facilitate the bonding of bases to the template.

Recall that DNA polymerase requires a free 3' OH group to begin adding bases to. This was one of the reasons for using denatured DNA in the *in vitro* experiments. RNA polymerase requires no such 3' OH group as a primer. The RNA polymerase that is complexed at the initiation site adds a small chain of RNA bases (50 to 100) complementary to the template strands. These small RNA fragments provide the 3' OH end for the DNA polymerase to add bases to.

DNA polymerase III* now extends the DNA chains in a discontinuous manner. When an Okazaki fragment has been synthesized, the RNA primer is removed, perhaps by DNA polymerase I. The gaps left by the removal of the RNA primer are filled in by DNA polymerase I, and DNA ligase ties the fragments together.

An interesting model for the replication of circular DNA molecules is called the *rolling-circle model* (Figure 7.20). This model eliminates the need for an RNA primer end, and it accounts nicely for the replication of circular molecules and bacterial mating (see the discussion of conjugation in Chapter 12). Figure 7.20 shows how the rolling-circle model accounts for the replication of the φX174 chromosome and the T4 chromosome. Let us examine the way this model works.

For a chromosome such as the T4 chromosome, the linear molecule is circularized, then a nick is induced in one of the two strands. We suggest that the 5' end is attached to the cell membrane and new bases are added to the 3' end. As replication proceeds, the circular strand rolls out with bases added accordingly. T4 proteins condense around the newly synthesized DNA, and when the T4 head is full of chromosomal material, the DNA is clipped off.

For a single-stranded chromosome such as the φX174 chromosome, replication can proceed from a double-stranded replicative form of the chromosome. The only difference we show here is that the formation of a double-stranded molecule is prevented by the attachment of phage proteins that maintain single-strandedness.

There are many unanswered questions about the rolling-circle model, but there is also considerable evidence supporting it. Additional work is certainly needed.

DNA replication in eukaryotes

Replication in eukaryotes is similar in many ways to replication as we have described it in prokaryote systems. There are also important differences. So far, we know of four or five distinct species of eukaryotic DNA polymerases (they are found in nuclei, cytoplasm, mitochondria, and chloroplasts). We do not know how they compare functionally with DNA polymerases I, II, III, and III*. There is also a eukaryotic DNA ligase.

Studies of DNA replication in *Drosophila* eggs have shown the presence of several points where replication originates. Replication at these

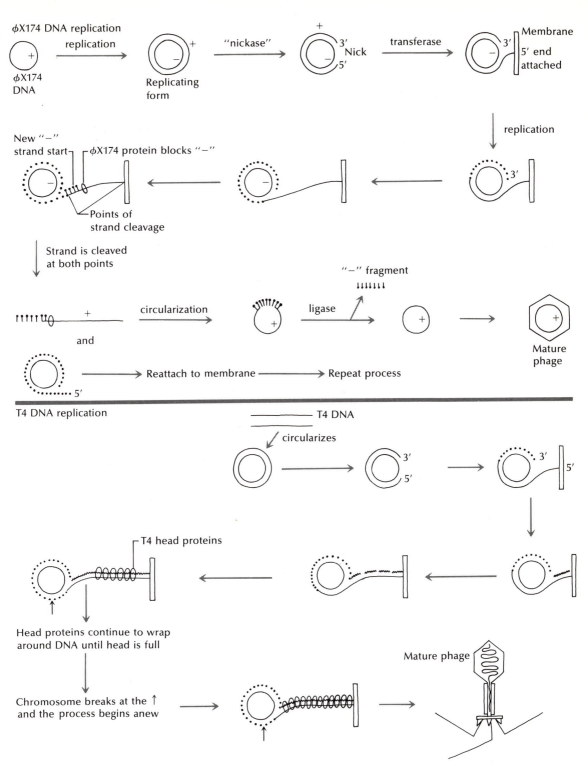

Figure 7.20 Rolling-circle model of DNA replication as it might apply to the bacteriophages φX174 and T4

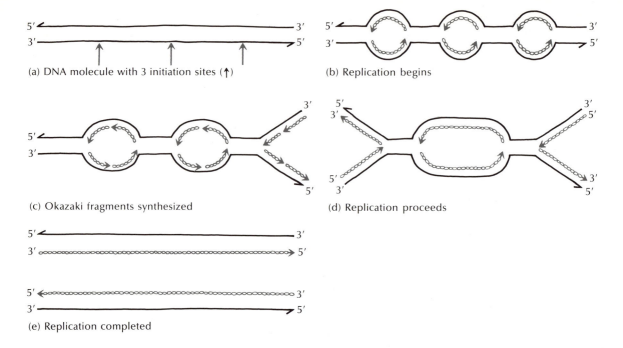

(a) DNA molecule with 3 initiation sites (↑)

(b) Replication begins

(c) Okazaki fragments synthesized

(d) Replication proceeds

(e) Replication completed

Figure 7.21
A model for DNA replication in eukaryotic organisms, showing multiple initiation points

points is bidirectional, creating a series of expanding bubbles. As the replication proceeds, the bubbles fuse, and eventually the daughter molecules separate (Figure 7.21). As in some prokaryotes, RNA primes replication in eukaryotes, and synthesis proceeds via the formation of Okazaki fragments.

Eukaryotic replication is much slower than, say, *E. coli* replication. It proceeds at about one-sixth the rate. This slow rate may account for the existence of the several replication origins, for with replication beginning at many points, the slow rate can be compensated for. In general outline, however, there are similarities in the replication process whatever the organism being discussed. But time and evolutionary divergence have also produced variations in the replication schemes employed by different species.

DNA repair

Ultraviolet light, a nonionizing form of radiation, damages DNA and RNA. DNA and RNA have a specific absorption capacity for UV light, and this led to a demonstration in 1960 that UV irradiation of nucleic acids generates *pyrimidine dimers*. For example, two contiguous pyrimidine bases, such as two thymines, become covalently linked to each other (Figure 7.22). These dimers are very stable and can be easily removed and

Figure 7.22
Photoproducts of
ultraviolet irradiation of
DNA

(a) Dimerization

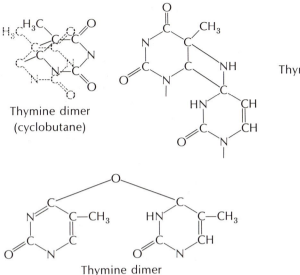

Thymine dimer
(cyclobutane)

Thymine-cytosine dimer
(TNC)

Thymine dimer
(TOT)

Figure 7.23
Cross-linking of
pyrimidines on different
DNA strands

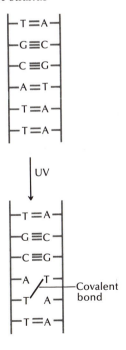

isolated from the intact DNA molecule. Other, less stable dimers also result from irradiation of DNA, such as cytosine–thymine, cytosine–cytosine, thymine–uracil, and uracil–uracil (the uracil being formed by the removal of NH_2 from dimerized cytosine).

Dimers affect the secondary and tertiary structure of DNA and therefore interfere with DNA replication and RNA transcription. Dimerization can also occur between pyrimidines on *different strands* of the DNA molecule (*cross-linking*). This seriously interferes with DNA replication, because it inhibits separation of DNA strands (Figure 7.23). Indeed, perhaps the major effect of UV irradiation is the inhibition of DNA synthesis. Whether cross-linking requires the distortion caused by adjacent dimers before it can occur or is induced by a separate and independent mechanism remains to be seen. The latter account is suggested by the observation that cross-linking of pyrimidines on different DNA strands is induced under conditions that do not favor dimerization.

In summary, the single most important event in the irradiation of DNA by UV light is the generation of thymine dimers. These dimers distort and weaken hydrogen bonding and also modify the secondary and tertiary structure of the DNA molecule, which interferes with its replication and template activity in RNA formation. Another consequence of UV irradiation, cross-linking, can result in the disruption of DNA replication. In Chapter 10, we shall see how UV irradiation induces gene mutations.

Repair of UV-induced genetic damage

Photoreactivation and dark repair are two classes of repair systems known to be associated with UV light irradiation, each involving a different enzyme system. Both systems repair the genetic damage induced by UV irradiation.

Light repair was discovered in 1949, when it was demonstrated that exposing a previously irradiated organism, such as *Escherichia coli* or *Bacillus subtilis,* to visible light lowered the mutation frequency and the killing rate. The basis for this decrease was shown to be an enzyme that, in the presence of intense visible light (blue range) and irradiated DNA, can repair the damage caused by the previous exposure to UV light. The enzyme is not species-specific; for example, the enzyme isolated from *E. coli* will repair damaged DNA from the thymus gland of a calf. The repair process is called *photoreactivation* and is summarized as follows.

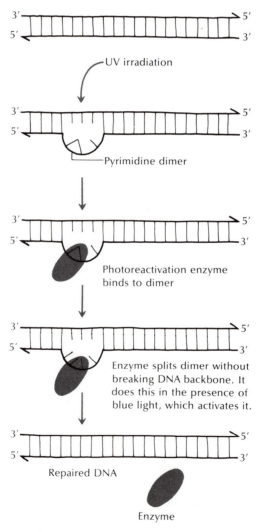

Unirradiated DNA does not complex with the photoreactivating enzyme, but in the presence of irradiated DNA and visible light, it has been demonstrated that the enzyme splits dimers *without* releasing bases or cleaving the DNA backbone.

Dark repair was discovered in *E. coli* when cells highly sensitive to UV light were isolated. The cells were unable to repair damage to their own DNA or to the DNA of irradiated infecting phages. In contrast, wild-type *E. coli* can dark-repair most of the UV-induced damage—but in a way different from photoreactivation. The dark-repair process was diagrammed on page 288. An alteration such as a thymine dimer is induced in a DNA molecule by UV light. Repair enzymes—including DNA polymerase I—recognize the damage, excise the aberrancy, and fill in the gap by synthesizing new DNA. The excised dimer can be isolated and identified, which shows that unlike photoreactivation, dark repair involves excision and synthesis. The repaired molecule can also be isolated if 5-bromouracil (a heavier analog of thymine that can replace thymine in newly synthesized DNA) is used in the synthesis medium. The newly synthesized DNA is heavier than normal DNA and hence separable in a cesium chloride gradient.

Mutants of *E. coli* that are unable to carry out a dark-repair process are called *uvr* (ultraviolet-repair-deficient) mutants. There appear to be three different *uvr* gene loci: *uvr A*, *uvr B*, and *uvr C*. All three must be present in a nonmutant condition in order for repair to proceed. If each gene codes for the synthesis of an enzyme, there may be three enzymes participating in dark repair. If DNA polymerase I is assumed to be one of the repair enzymes, the three gene products may be enzymes that somehow recognize the photoproduct distortion and get the area "ready" for DNA polymerase I repair. An alternative idea would have each gene specifying one polypeptide chain of a complex, multicomponent enzyme.

The importance of DNA repair is dramatically evident in a rare and tragic human skin disease. Xeroderma pigmentosum, inherited as an autosomal recessive trait, is characterized by the extreme sensitivity of the skin to sunlight and UV irradiation. The disease is apparent in early childhood and progressively worsens with age, the skin becoming dry and the eyes ulcerated. Most people afflicted with this disease die before they reach the age of 30 from the several cancerous skin tumors that develop on their bodies.

The cause of the disease is a faulty DNA-repair system in the skin tissue. Normally the pyrimidine dimers induced by UV irradiation are excised from human DNA within 24 hours. But in xeroderma pigmentosum, there is no such removal. The specific component that is defective in xeroderma pigmentosum is the nuclease that hydrolyzes the DNA backbone in the vicinity of a pyrimidine dimer.

Gene conversion

Our knowledge of DNA structure and replication now enables us to examine the crossing-over phenomenon more carefully. We noted in Chapter 2 that recombination in *Neurospora* is evaluated by the sequence of ascospores in the ascus sac. By this method, we can assess first-division and second-division segregation and the strands involved in crossing-over. But regardless of the sequence, there should be four representatives of each member of each allelic pair.

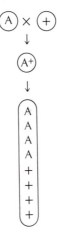

Occasionally, however, departures from this 4:4 ratio are uncovered, suggesting that some alleles have been "converted" to the opposite allele. This process has been termed "gene conversion" or "nonreciprocal recombination." Some of the non-Mendelian ratios are

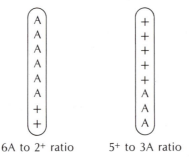

6A to 2+ ratio   5+ to 3A ratio

The remarkable feature of these ratios is that they seem to contradict everything we have said about Mendelian principles—or do they? Let us look more carefully at them to see whether our knowledge of DNA structure and replication can help us.

Studies of gene conversion have provided some important clues to the mechanism. (1) Mutation does not account for gene conversion. (2) The phenomenon is usually restricted to a limited locus, because other heterozygous loci show the usual 4:4 segregation pattern. (3) Gene conversion is usually associated with recombination between alleles that flank the locus at which conversion has occurred. Conversion thus appears to be connected with recombination in very small regions of the chromosome.

Holliday has suggested a model for gene conversion in *Neurospora* (Figure 7.24). The model is based on the repair and replication of the

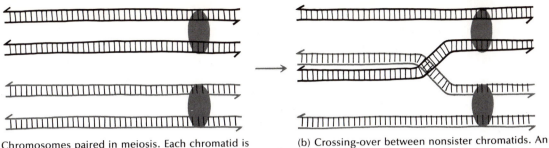

(a) Chromosomes paired in meiosis. Each chromatid is diagrammed as a double-stranded DNA molecule.

(b) Crossing-over between nonsister chromatids. An exchange of single DNA strands is postulated to occur here.

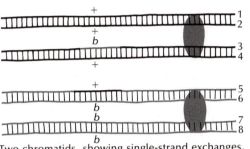

(c) Two chromatids, showing single-strand exchanges. There is some base mismatching because each DNA strand carries a different allele. Each DNA strand is numbered.

(d) A repair process corrects the mismatched bases by clipping bases from one strand, then filling in the gap.

Clip bases from strands numbered	As templates, use strands numbered	Ascospore sequence after meiosis and one mitosis									Comments
		1	2	3	4	5	6	7	8		
4 and 6	3 and 5	+	+	b	b	+	+	b	b	4b:4+;	Appears as a reciprocal recombination
3 and 5	4 and 6	+	+	+	+	b	b	b	b	4b:4+;	Appears as a nonrecombinant
3 and 6	4 and 5	+	+	+	+	+	+	b	b	6+:2b;	Gene conversion
4 and 5	3 and 6	+	+	b	b	b	b	b	b	2+:6b;	Gene conversion
4*	3	+	+	b	b	+	b	b	b	3+:5b;	Gene conversion
3*	4	+	+	+	+	+	b	b	b	5+:3b;	Gene conversion
5*	6	+	+	b	+	b	b	b	b	3+:5b;	Gene conversion
6*	5	+	+	b	+	+	+	b	b	5+:3b;	Gene conversion

*Only one DNA molecule is repaired. The mismatched bases in the other molecule are fixed during mitosis:

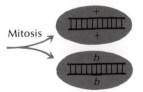

Spore after meiosis

Mitosis

Two ascospores: 1+ and 1b

Figure 7.24
A model for gene conversion in *Neurospora;* only DNA molecules are shown along with centromeres

305

chromosomal DNA, and it accounts for the various allelic ratios that have been observed in *Neurospora*.

Gene conversion occurs in most eukaryotes. It does not contradict Mendelian principles of segregation and independent assortment, because, though alleles have been converted, segregation and independent assortment still go on. When we discuss virus genetics, we shall see other examples of nonreciprocal recombination.

Overview

The faithful replication of the genetic material is a key feature—if not *the* key feature—of a living system. Speculation on how this replication occurred was without any foundation until Watson and Crick's model of the structure of DNA emerged. With this model came the suggestion that replication was semiconservative, with a parental strand serving as a template for the generation of a daughter strand. As attractive as this replication scheme was, however, it did not garner any experimental support until four years after it first appeared, when the replication of eukaryotic chromosomes was studied using autoradiography. Shortly after this, Meselson and Stahl published their classic paper on the replication of DNA, showing that it indeed followed a semiconservative scheme.

With the general scheme of replication identified, the next task was to describe the molecular details of the process. This was no easy task. A crucial milestone was Kornberg's discovery of DNA polymerase I, an enzyme that can polymerize nucleotide triphosphates into DNA strands. But Kornberg's enzyme had some rather strange properties that raised doubts about whether it was the true *in vivo* replicating enzyme. The enzyme, for example, required partially denatured DNA as the primer template; it also degraded DNA, in addition to synthesizing it. These properties suggested that the enzyme was involved in the repair of damaged DNA.

One thing was clear from the early studies of DNA polymerase I: It could faithfully replicate a DNA molecule. This fact emerged when a clever procedure called nearest-neighbor analysis was developed, and showed that the newly synthesized DNA is identical in nearest-neighbor base pairs to the primer-template DNA. Nearest-neighbor analysis also confirmed the antiparallel arrangement of the two DNA strands.

But could DNA polymerase replicate an intact DNA molecule such that the replica was biologically active? This was a problem of no small magnitude, for many studies with the enzyme showed that the DNA it synthesized was not normal in appearance: It contained many forks or branch points. But using DNA polymerase I and a newly discovered enzyme called DNA ligase, which catalyzes the formation of a covalent bond between two DNA strands, investigators showed that DNA polymerase I could faithfully replicate a complete ϕX174 chromosome.

This unusual chromosome is a single-stranded circular DNA molecule housed within the confines of the virus protein coat. So Kornberg's enzyme clearly had the capability of faithfully replicating a complete DNA molecule.

But the nuclease activity, as well as some other properties, made it likely that Kornberg's enzyme was primarily a repair enzyme. The nuclease activities expressed by the enzyme are involved in two quite different functions. One of the functions is to edit the new DNA: If the enzyme inserts an incorrect base, it backs up and removes it. The other function is the repair of DNA damaged by UV irradiation or certain chemicals.

About 13 years after the discovery of DNA polymerase I, two other DNA polymerases were discovered: II and III. It now appears that DNA polymerase III, complexed with auxiliary proteins called copolymerase III*, is the main replicating enzyme. DNA polymerase II's function is still not well understood. DNA polymerase I appears to be important to the replicating process, but it is still not clear how.

The study of replicating circular double-stranded DNA molecules provided additional insights into the mechanisms of DNA replication. Cairns showed that the *E. coli* chromosome replicates in a semiconservative fashion and has a structure called the *replication fork*. This replication fork has attracted a great deal of attention. Though our understanding of events at the fork is far from complete, it is not as obscure as it once was.

Studies on the origin of replication showed that replication has a specific point of origin and then proceeds in a bidirectional manner. The studies also showed that replication at the fork is discontinuous, proceeding in a 5' to 3' direction along both template strands.

We discussed models of DNA replication that serve to unify many of the details of the replication process. A specific initiation site binds RNA polymerase, which synthesizes a small fragment of RNA to serve as a primer for DNA synthesis. Unwinding proteins uncoil the DNA molecule, and DNA polymerase III* extends the DNA chain in a discontinuous manner, 5' to 3'. The RNA is removed, the gap filled in (perhaps by DNA polymerase I), and the fragments covalently linked by DNA ligase. An interesting variation of this replication scheme is the rolling-circle model. It does not require the RNA as a primer.

Studies of replication in eukaryotes lag behind those in prokaryotes. We know of several distinct DNA polymerase species. We know that there are several points of origin for replication, and that this replication proceeds in a bidirectional fashion. The rate of replication is slower, however.

Replication and repair go hand in hand. DNA that is damaged by agents such as UV irradiation is repaired in *E. coli* by a light-activated system and by a dark-repair system. The light-activated system repairs bases without breaking the DNA backbone. The dark-repair system clips out the damaged DNA, then fills in the gap with new synthesis. DNA polymerase I is involved in this latter system.

Finally, we discussed a phenomenon called gene conversion that has been observed in eukaryotes. It involves the conversion of one allele to another. The mechanism proposed to account for gene conversion requires the repair and replication of very limited regions of the chromosome following synapsis. This mechanism accounts for the aberrant Mendelian ratios that have been observed.

Questions and problems

7.1 In the Kornberg *in vitro* DNA-synthesizing system without DNA primer and without the bases A and T, the synthesis of a GC polymer occurs. How would you determine that the base sequence in this polymer is

$$
\begin{array}{ccc}
\text{G}\equiv\text{C} & & \text{G}\equiv\text{C} \\
\text{G}\equiv\text{C} & & \text{C}\equiv\text{G} \\
\text{G}\equiv\text{C} & \text{and not} & \text{G}\equiv\text{C} \quad ? \\
\text{G}\equiv\text{C} & & \text{C}\equiv\text{G} \\
\text{G}\equiv\text{C} & & \text{G}\equiv\text{C}
\end{array}
$$

7.2 Compare DNA replication in prokaryotes and in eukaryotes.

7.3 Why must the DNA of eukaryotes be double-stranded?

7.4 Design an experiment that would show that the sequence of bases in DNA is determined by template DNA, not by the DNA polymerase.

7.5 A ϕX174 bacteriophage is a single-stranded DNA organism. When it infects a host cell, it first generates a double-stranded DNA molecule called the replicative form (RF). If the base composition of the DNA before RF formation is 0.27 A, 0.31 G, 0.22 T, and 0.20 C, what is the base composition of the RF and the DNA strand complementary to the parental strand?

7.6 How would you determine whether a DNA molecule had parallel or antiparallel strands, if one strand was 5'-ATGCCTGATGGC-3'?

7.7 RNA is transcribed from DNA templates in most cases. Since the RNA is single-stranded, does it make any difference to its informational content which DNA strand it is transcribed from?

7.8 If the G content of a double-stranded DNA molecule is 0.24, what is the T content?

7.9 If x represents the amount of DNA in a cell before it enters the division cycle, how much DNA is there in the cell during
   (a) mitotic prophase?
   (b) mitotic telophase?
   (c) meiotic prophase I?
   (d) meiotic telophase I?
   (e) meiotic metaphase II?
   (f) meiotic telophase II?

7.10 Using the same experimental setup as Taylor when he did his autoradiography experiments with Vicia, and with chromosomes genetically labeled as follows,

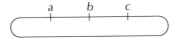

(a) How would the labeling pattern on the chromatids appear at the second metaphase after labeling if a reciprocal exchange occurred between nonsister chromatids between *b* and *c*?

(b) How would you explain the following chromosome?

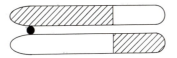

7.11 What evidence made it seem unlikely that the Kornberg enzyme was the replicating enzyme?

7.12 It has been argued that, although DNA replication is semiconservative, the conserved unit is the double helix, so that a DNA molecule would have four strands, not two. How would you test for this possibility?

7.13 How would Figure 7.18(b) change if DNA replication started just to the right of the *ilv* locus and proceeded unidirectionally in a clockwise fashion? in a counterclockwise fashion?

7.14 Why is it important for the DNA-replicating enzyme to have 3' to 5' nuclease activity?

7.15 Why is RNA polymerase so important to the DNA replication process?

References

Alberts, B. M., and L. Frey. 1970. T4 bacteriophage gene 32: A structural protein in the replication and recombination of DNA. *Nature* 227:1313–1318.

Bird, R. E., J. Lovarn, J. Martuscelli, and L. Caro. 1972. Origin and sequence of chromosome replication in *E. coli*. *J. Mol. Biol.* 70:549–566.

Bollum, F. J. 1975. Mammalian DNA polymerases. *Prog. Nuc. Acid Res. Molec. Biol.* 15:109–144.

Brutlag, D., and A. Kornberg. 1972. Enzymatic synthesis of deoxyribonucleic acid: XXXVI. A proofreading function for the 3' $\longrightarrow$ 5' exonuclease activity in deoxyribonucleic acid polymerases. *J. Biol. Chem.* 247:241–248.

Brutlag, D., R. Schekman, and A. Kornberg. 1971. A possible role for RNA polymerase in the initiation of M13 DNA synthesis. *Proc. Natl. Acad. Sci. U.S.* 68:2826–2829.

Cairns, J. 1963. The bacterial chromosome and its manner of replication as seen by autoradiography. *J. Molec. Biol.* 6:208–213.

Callan, H. G. 1972. Replication of DNA in the chromosome of eukaryotes. *Proc. Roy. Soc. Lond.* 181:19–41.

Chargaff, E. 1976. Initiation of enzymatic synthesis of DNA by RNA primers. *Prog. Nuc. Acid Res. Molec. Biol.* 16:1–24.

Cleaver, J. 1969. Xeroderma pigmentosum: A human disease in which an initial stage of DNA repair is defective. *Proc. Natl. Acad. Sci. U.S.* 63:428–435.

Corwin, H. O. and J. B. Jenkins. 1976. *Conceptual Foundations of Genetics: Selected Readings.* Boston, Houghton Mifflin.

Davidson, J. N. 1972. *The Biochemistry of the Nucleic Acids*, 7th ed. New York, Academic.

DeLucia, P., and J. Cairns. 1969. Isolation of an *E. coli* strain with a mutation affecting DNA polymerase. *Nature* 224:1164–1166.

Doerman, A. H. 1973. T4 and the rolling circle model of replication. *Ann. Rev. Genetics* 7:325–342.

Dressler, D. 1970. The rolling circle for ϕXDNA replication: II. Synthesis of single-stranded circles. *Proc. Natl. Acad. Sci. U.S.* 67:1934–1942.

Gefter, M. L. (ed.). 1972. DNA synthesis in vitro. *PAABS Rev.* 1:495–498.

Gefter, M. L. 1974. DNA polymerases II and III of *Escherichia coli. Prog. Nuc. Acid Res. Molec. Biol.* 14:101–116.

Gefter, M. L. 1975. DNA replication. *Ann. Rev. Biochem.* 44:45–78.

Gefter, M. L., Y. Hirota, T. Kornberg, J. A. Wechsler, and C. Barnoux. 1971. Analysis of DNA polymerases II and III in mutants of *Escherichia coli* thermosensitive for DNA synthesis. *Proc. Natl. Acad. Sci. U.S.* 68:3150–3153.

Gellert, M., J. W. Little, C. K. Oshinsky, and S. B. Zimmerman. 1968. Joining of DNA strains by DNA ligase of *E. coli. Cold Spring Harbor Symp. Quant. Biol.* 33: 21–26.

Gilbert, W., and D. Dressler. 1968. DNA replication: The rolling circle model. *Cold Spring Harbor Symp. Quant. Biol.* 33:473–484.

Goulian, M. 1971. Biosynthesis of DNA. *Ann. Rev. Biochem.* 40:855–898.

Goulian, M. 1972. Some recent developments in DNA enzymology. *Prog. Nuc. Acid Res. Molec. Biol.* 12:29–49.

Goulian, M., A. Kornberg, and R. Sinsheimer. 1967. Enzymatic synthesis of DNA: XXIV. Synthesis of infectious phage ϕX174 DNA. *Proc. Natl. Acad. Sci. U.S.* 58:2321–2328.

Grossman, L., A. Braun, R. Feldberg, and I. Mahler. 1975. Enzymatic repair of DNA. *Ann. Rev. Biochem.* 44:19–43.

Gross, J., and M. Gross. 1969. Genetic analysis of an *E. coli* strain with a mutation affecting DNA polymerase. *Nature* 224:1166.

Huberman, J. A., and A. D. Riggs. 1968. On the mechanism of DNA replication in mammalian chromosomes. *J. Molec. Biol.* 32:327–341.

Inman, R. B., and M. Schnös. 1972. Structure of branch points in replicating DNA: Presence of single-stranded circles. *Proc. Natl. Acad. Sci. U.S.* 67:1934–1942.

Josse, J., A. D. Kaiser, and J. Kornberg. 1961. Enzymatic synthesis of DNA: VIII. Frequencies of nearest neighbor base sequences in DNA. *J. Biol. Chem.* 236:864–875.

Keller, W. 1972. RNA-primed synthesis *in vitro. Proc. Natl. Acad. Sci. U.S.* 69:1560–1564.

Klien, A., and F. Bonhoeffer. 1972. DNA replication. *Ann. Rev. Biochem.* 41:301–332.

Knippers, R. 1970. DNA polymerase II. *Nature* 228:1050–1053.

Kornberg, A. 1960. The biologic synthesis of DNA. *Science* 131:1503–1508.

Kornberg, A. 1969. The active center of DNA polymerase. *Science* 163:1410–1418.

Kornberg, A. 1974. *DNA Synthesis.* San Francisco, Freeman.

Kornberg, A. 1977. Why work on the enzymology of DNA replication? *Trends in Biochem. Sci.* 2:N56–N58.

Kornberg, T., A. Lockwood, and A. Worcel. 1974. Replication of the *E. coli* chromosome with a soluble enzyme system. *Proc. Natl. Acad. Sci. U.S.* 71:3189–3193.

Kriegstein, H. J., and D. S. Hogness. 1974. Mechanism of DNA replication in *Drosophila* chromosomes: Structure of replication forks and evidence for bidirectionality. *Proc. Natl. Acad. Sci. U.S.* 71:135–139.

Lehman, I. R. 1974. DNA ligase: Structure, mechanism, and function. *Science* 186:790–797.

Lehman, I. R. and D. G. Uyemura. 1976. DNA polymerase I: Essential replication enzyme. *Science* 193:963–969.

Lehninger, A. L. 1975. *Biochemistry,* 2d ed. New York, Worth.

McKenna, W. G., and M. Masters. 1972. Biochemical evidence for the bidirectional replication of DNA in *E. coli. Nature* 240:536–539.

Meselson, M., and F. W. Stahl. 1958. The replication of DNA in *E. coli. Proc. Natl. Acad. Sci. U.S.* 44:671–682.

Meselson, M., and R. Yuan. 1968. DNA restriction enzyme from *E. coli. Nature* 217:1110–1114.

Mizutani, S., and H. M. Temin. 1970. An RNA-dependent DNA polymerase in virions of Rous Sarcoma Virus. *Cold Spring Harbor Symp. Quant. Biol.* 35:847–849.

Moses, R. E., and C. C. Richardson. 1970. Replication and repair of DNA in cells of *Escherichia coli* treated with toluene. *Proc. Natl. Acad. Sci. U.S.* 67:674–681.

Okazaki, R., T. Okazaki, K. Sakabe, K. Sugimoto, and A. Sugino. 1968. Mechanism of DNA chain growth: I. Possible discontinuity and unusual secondary structure of newly synthesized chains. *Proc. Natl. Acad. Sci. U.S.* 59:598–905.

Prescott, D. M. 1970. Structure and replication of eucaryotic chromosomes. *Adv. Cell Biol.* 1:57–118.

Prescott, D. M., and P. L. Kuempel. 1972. Bidirectional replication of the chromosome in *E. coli. Proc. Natl. Acad. Sci. U.S.* 69:2842–2845.

Quinn, W. G., and N. Sueoka. 1970. Symmetric replication of the *Bacillus subtilis* chromosome. *Proc. Natl. Acad. Sci. U.S.* 67:717–723.

Schaller, H., B. Otto, V. Nüsslein, J. Huf, R. Herrman, and F. Bonhoeffer. 1972. Deoxyribonucleic acid replication *in vitro. J. Molec. Biol.* 63:183–200.

Schekman, R., A. Weiner, and A. Kornberg. 1974. Multienzyme systems of DNA replication. *Science* 186:987–993.

Setlow, R. B., and J. K. Setlow. 1972. Effects of radiation on polynucleotides. *Ann. Rev. Biophys. Bioeng.* 1:293–346.

Shenin, R., J. Humbert, and R. E. Pearlman. 1978. Some aspects of eukaryotic DNA replication. *Ann. Rev. Biochem.* 47:277–316.

Stent, G. S. 1971. *Molecular Genetics: An Introductory Narrative.* San Francisco, Freeman.

Stryer, L. 1975. *Biochemistry.* San Francisco, Freeman.

Sugino, A., S. Hirose, and R. Okazaki. 1972. RNA-linked nascent DNA fragments in *E. coli. Proc. Natl. Acad. Sci. U.S.* 69:1863–1897.

Tait, R. C., and D. W. Smith. 1974. Roles for *E. coli* DNA polymerases I, II, and III in DNA replication. *Nature* 249:116–119.

Taylor, J. H. 1959. The organization and duplication of the genetic material. *Proc. 10th Internat. Cong. Genetics* 1:63–78.

Taylor, J. H. 1965. *Selected Papers on Molecular Genetics.* New York, Academic.

Watson, J. D. 1976. *Molecular Biology of the Gene,* 3d ed. New York, Benjamin.

Werner, R. 1971. Nature of DNA precursors. *Nature* 233:99–103.

Wickner, S. H. 1978. DNA replication proteins of *E. coli. Ann. Rev. Biochem.* 7:1163–1191.

Wintersberger, E. 1977. DNA-dependent DNA polymerases from eukaryotes. *Trends in Biochem. Sci.* 2:58–61.

Zubay, G., and J. Marmur. 1973. *Papers in Biochemical Genetics.* New York, Holt.

Eight

Gene function

Genes are shown to direct the synthesis of proteins
Inborn errors of metabolism
The genetic control of insect pigments
The one gene–one enzyme theory
Genetic control of protein structure
Colinearity

Protein structure and function
Amino acids
Peptide bonds
General properties of proteins

Transfer of information from gene to protein

The transcription of RNA
RNA structure
Transcription of RNA by RNA polymerase
RNA polymerase: Its ability to initiate, elongate,
 and terminate an RNA chain

RNA and protein synthesis
Components of the protein-synthesizing system
tRNA
mRNA
rRNA and ribosomes
Protein synthesis
Polypeptide initiation
Polypeptide elongation
Polypeptide termination
The polyribosome

Overview

Questions and problems

References

Research into the replication of the genetic material, discussed in Chapter 7, proceeded concurrently with research into functions of genes. A central issue is: How do genes produce gene products that in turn determine specific phenotypes? The first link between a gene and a gene product was established in 1909 when A. E. Garrod suggested that gene products were proteins. His hypothesis went largely ignored until around 1940, when it was conclusively demonstrated by Beadle and Tatum that genes direct the synthesis of proteins.

Encouraged by this association between genes and proteins, investigators sought the mechanisms by which information housed in the DNA base sequence is translated into the structure of proteins. An early clue came when it was shown that DNA is first transcribed into RNA, which in many cases is subsequently translated into protein. Attention then focused on how RNA is transcribed from DNA templates and how RNA is translated into the sequences of amino acids that make up proteins.

This chapter will first review the studies that linked genes with proteins, then explore the role of RNA in protein synthesis, and finally analyze how proteins are actually constructed in a cell.

Genes are shown to direct the synthesis of proteins

One of the first steps toward understanding how genes affect phenotypes occurred with the identification of the products for which genes are responsible. Garrod, as well as some later investigators, surmised that genes produce proteins that act as enzymes.

Inborn errors of metabolism Garrod, analyzing a disease in humans in which the cartilaginous tissues are dark and the urine turns black on exposure to air, suggested that it might be caused by an abnormality in the metabolism of the amino acid phenylalanine. He thought this because feeding phenylalanine to patients with the syndrome increased the excretion of homogentisic acid (alcapton)—a breakdown product of phenylalanine and a substance that, when oxidized, turns the urine black and darkens the cartilage. This disease, called *alcaptonuria,* is inherited as a simple Mendelian recessive gene. Since it results in a metabolic disturbance, Garrod supposed that the mutant gene produced a defective form of an enzyme involved in the utilization of phenylalanine.

In 1914, Garrod's observations were confirmed by Gross, though it was not until 1958 that the precise nature of the defect was elucidated. At that time, LeDu and his colleagues showed that alcaptonurics are deficient in homogentisic acid oxidase, the enzyme responsible for converting homogentisic acid to maleylacetoacetic acid. They are normal with respect to the other enzymes involved in phenylalanine metabolism (Figure 8.1).

Since Garrod's pioneering work on the relationship between genes and enzymes, the metabolism of phenylalanine has been extensively in-

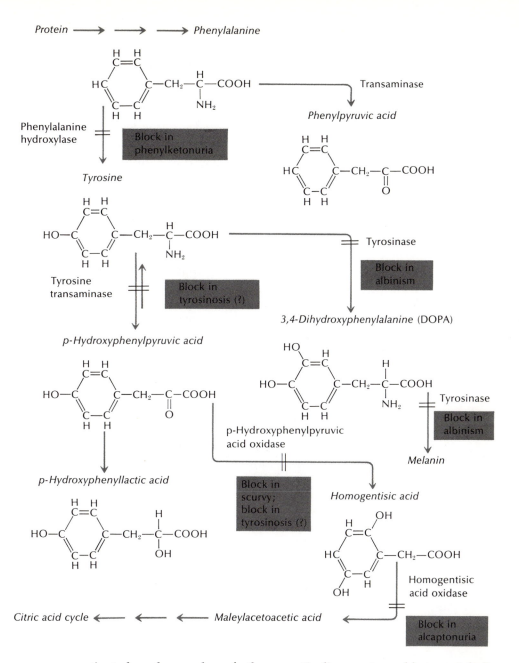

Protein ⟶ ⟶ ⟶ Phenylalanine

Transaminase

Phenylpyruvic acid

Phenylalanine hydroxylase

Block in phenylketonuria

Tyrosine

Tyrosine transaminase

Block in tyrosinosis (?)

Tyrosinase

Block in albinism

3,4-Dihydroxyphenylalanine (DOPA)

p-Hydroxyphenylpyruvic acid

p-Hydroxyphenylpyruvic acid oxidase

Tyrosinase

Block in albinism

Melanin

p-Hydroxyphenyllactic acid

Block in scurvy; block in tyrosinosis (?)

Homogentisic acid

Homogentisic acid oxidase

Citric acid cycle ⟵ ⟵ ⟵ Maleylacetoacetic acid ⟵

Block in alcaptonuria

Figure 8.1
Phenylalanine metabolism and some genetically caused disruptions in the process

vestigated, and a number of other genetic diseases traceable to metabolic disruptions in its metabolism have been recognized (Figure 8.1). Phenylalanine is converted to tyrosine by phenylalanine hydroxylase. If phenylalanine hydroxylase is missing or defective, *phenylketonuria* results, a major disease that, if not treated, can cause severe mental retardation and early death. Phenylketonurics have an inordinately high level of phenylalanine in their blood and phenylpyruvic acid in their urine and cerebrospinal fluid. This metabolic defect (phenylketonuria) is inherited as a simple Mendelian recessive. Heterozygotes often express symptoms of the disease, though to a much lesser degree, which suggests incomplete dominance.

Another disease caused by the disruption of phenylalanine metabolism and inherited as a Mendelian recessive is *albinism,* a condition characterized by an organism's lack of a dark pigment called *melanin.* Normally, tyrosine is converted through a series of reactions to melanin, which causes brown coloring in skin, hair, and eyes. Albinism is caused by an enzyme deficiency for the reaction that converts dopa to dopaquinone; the condition therefore blocks the production of melanin pigments. The inability to convert the amino acid tyrosine into products that feed into the energy-producing citric acid cycle results in a disease called *tyrosinosis,* which may be due to a deficiency in either tyrosine transaminase or hydroxyphenylpyruvic acid oxidase. It is not yet known which applies, but either way, the urine in patients with this disease has abnormal reducing properties.

Numerous other human metabolic diseases have been well documented, but Garrod's work pioneered the association of Mendelian genetics with proteins. Nevertheless, he was not accorded wide recognition for his contributions until Beadle and Tatum hypothesized 26 years later that one gene produced one enzyme.

The genetic control of insect pigments

In 1935, Beadle and B. Ephrussi suggested that mutants that affected eye color in *Drosophila* were deficient in pigment precursors. Both vermilion (*v*) and cinnabar (*cn*) mutants have bright, red-orange eyes owing to their inability to form the brown pigments necessary for wild-type eye color. Beadle and Ephrussi developed a technique whereby embryonic eye tissue from one larva was transplanted into the body cavity of another and allowed to develop into mature eye tissue. The pattern of eye tissue development as a function of genotype is shown in Figure 8.2.

They observed that when embryonic *v* or *cn* eye tissue was transplanted into wild-type larvae, eyes developed with normal eye color in the adult abdomen. From this, they concluded that the host larvae supplied to the transplanted eye tissue a substance or substances that allowed the eye tissue to bypass the metabolic block and to synthesize brown pigments.

However, reciprocal transplantation experiments using *v* and *cn* larvae showed an interesting phenomenon: *v* eye tissue transplanted into *cn* larvae developed into wild-type eyes, but *cn* eye tissue transplanted into *v* larvae developed into *cn*-colored eyes only.

The conclusion Beadle and Ephrussi drew was that the *cn* host tissue supplied the *v* eye tissue with a diffusable substance that enabled the *v* tissue to bypass the metabolic block in the brown pigment pathway and thus synthesize a normal eye color. On the other hand, the *v* host was

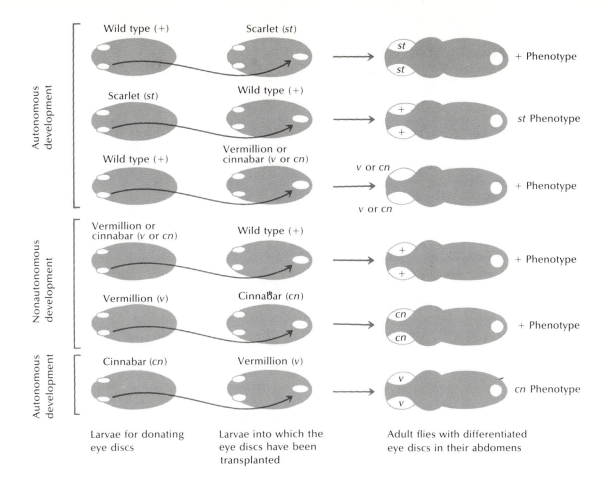

Larvae for donating
eye discs

Larvae into which the
eye discs have been
transplanted

Adult flies with differentiated
eye discs in their abdomens

Figure 8.2
Eye disc transplants in
Drosophila, showing
autonomous and
nonautonomous
development (From
Ephrussi, *Quart. Rev. Biol.*
17:327, 1942)

unable to supply the *cn* eye tissue with such a diffusable substance. From
this evidence, it was possible to arrange the *v* and *cn* mutants in a sequence in terms of the biosynthesis of brown pigment.

$$\text{Pigment precursor} \xrightarrow{\quad} \underset{\substack{A \\ \text{enzyme} \\ \uparrow \\ v^+ \\ \text{gene}}}{} \text{Substance X} \xrightarrow{\quad} \underset{\substack{B \\ \text{enzyme} \\ \uparrow \\ cn^+ \\ \text{gene}}}{} \text{Substance Y} \xrightarrow{\quad} \text{Brown pigment}$$

Enzyme A would be coded by the v^+ gene, enzyme B by the cn^+ gene.
Applying this to the reciprocal transplantation experiments, *v* eye tissue
transplanted into *cn* host larvae does not produce X but has enzyme B.
The *cn* host produces X but not Y, because it lacks enzyme B. The substance X produced by the *cn* mutant host diffuses into the *v* eye tissue and

Figure 8.3
Formation of brown
pigment in *Drosophila*

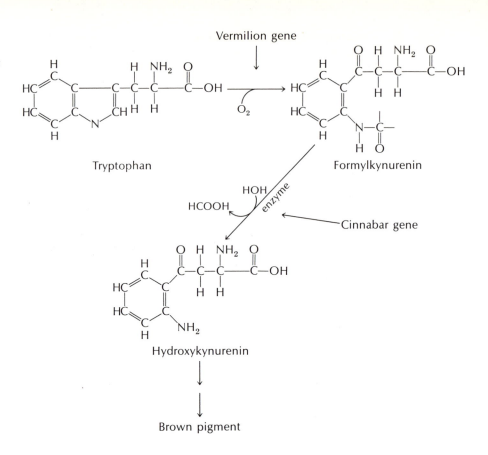

Vermilion gene

Tryptophan → Formylkynurenin

Cinnabar gene

HCOOH, HOH, enzyme

Hydroxykynurenin

↓

Brown pigment

results in normal eye color development because the X can be converted to Y and then to brown pigment. The *cn* eye tissue in the *v* host presents a different problem in that the *cn* eye tissue has X but no enzyme for converting X to Y; and the host larva *v* has no Y to bypass the block, so the eye tissue retains its *cn* phenotype.

A few years after this proposal was made, the chemical substances were identified (Figure 8.3). The pigment precursor was tryptophan; substance X was formylkynurenin, and substance Y was hydroxykynurenin—all intermediates in the formation of brown pigment.

This work of Beadle and Ephrussi was important, because it indicated a firm link between a phenotype (eye color), a genotype (the *v* and *cn* mutants), and gene products. Eye pigment studies led directly to investigations of biochemical mutants in *Neurospora* and microorganisms that resulted in the hypothesis that each gene produced one enzyme.

The one gene–one enzyme theory   The work with *Drosophila* pointed toward a way of studying gene function, but fruit flies are biochemically cumbersome to work with, partly because of technical problems and partly because of the complicating factor of dominance. Beadle and Tatum, seeking a simpler organism to work with, selected the common bread mold *Neurospora*, which is simpler to handle, grows on a chemically well-defined medium, is not complicated by dominance relationships because it is haploid, and easily yields biochemical mutants when it is exposed to mutagenic agents, such as ultraviolet (UV) light. Their work was the first clear, well-documented study to link genes and enzymes.

An analysis of the UV-induced nutritional mutants showed them to be single gene mutations inherited in accordance with Mendelian principles of segregation. From the knowledge that each mutant was usually shown to be defective in one enzyme and that each mutant followed a Mendelian scheme of inheritance, Beadle and Tatum set forth the *one gene–one enzyme theory,* the theory that one gene produces one enzyme.

Though it was tremendously important as a stimulus to a biochemical approach to gene function, the value of the one gene–one enzyme hypothesis is limited today. The idea that each gene codes for one enzyme is too simple. Genes are also known to code for single polypeptide chains, enzymes, antibodies, hormones, and various types of nontranslated RNA, such as rRNA and tRNA.

Genetic control of protein structure

Perhaps the most exciting and definitive work on the relationship between genes and proteins has been that of C. Yanofsky and his colleagues on the tryptophan synthetase system of *Escherichia coli.* They showed that the enzyme was composed of two different polypeptide chains, each produced by a *different* gene. In this instance, *two* genes produce one enzyme.

Wild-type *E. coli* synthesizes the amino acid tryptophan from the ingredients found in a minimal medium. The tryptophan mutant (*trp⁻*), on the other hand, is unable to accomplish the synthesis, so the amino acid must be supplied in the medium if the organism is to survive. Such a mutant could be defective at any of a number of places (Figure 8.4). The problem is to determine which step in the process is interrupted. Yanofsky studied various *trp⁻* mutants in *E. coli* and found that he was able to bypass the metabolic block by adding compounds other than tryptophan (Table 8.1). Certain of these compounds negated the effects of the

Table 8.1
The ability of various *trp⁻* mutants to grow on supplemented media

| Group | Growth on minimal medium plus | | | | Substance accumulated |
	No supplement	Anthranilic acid	Indole	Tryptophan	
trp⁺	+	+	+	+	none
trpE⁻	−	+	+	+	none
trpD⁻	−	−	+	+	anthranilic acid
trpC⁻	−	−	+	+	CDRP[a]
trpA⁻	−	−	+	+	IGP[b]
trpB⁻	−	−	−	+	indole

[a] CDRP, carboxyphenylamino deoxyribulose phosphate.
[b] IGP, indole glycerol phosphate.

From *Molecular Genetics* by Gunther S. Stent. Copyright © 1971 by W. H. Freeman and Company.

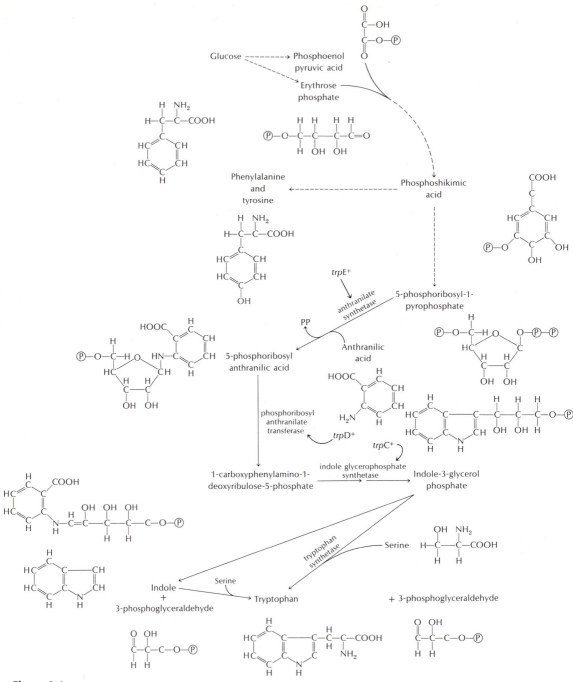

Figure 8.4
Biosynthesis of tryptophan

metabolic block, which implied that they were precursors in the biosynthesis of tryptophan. For example, the *trpE* mutant was probably a mutation in the anthranilate synthetase enzyme, *trpD* was probably defective in phosphoribosyl anthranilate transferase, and *trpC* probably involved indole glycerol phosphate synthetase. These conclusions are based not only on the compounds that will by-pass the metabolic block, but also on the compounds accumulated in the cell. That is, when a reaction in a sequence is stopped, the reactants preceding the block will either accumulate or be shunted elsewhere.

The *trpA* and *trpB* mutants presented a problem in that *trpA* was reversed only by tryptophan itself or indole, a compound that is not normally a tryptophan precursor. In *E. coli*, tryptophan is synthesized by condensing indole glycerol phosphate with serine (although the use of indole is an alternative route not usually followed); indeed, all the mutants except *trpB* were capable of using indole. The role of indole in tryptophan synthetase was clarified when it was demonstrated that the enzyme is composed of two separate polypeptide chains, α (Greek alpha) and β (Greek beta), and that each has partial and unique enzymatic capabilities (Figure 8.5). The complete tryptophan synthetase enzyme consists of four polypeptide chains, two α and two β, and the complete enzyme synthesizes tryptophan from serine and indole glycerol phosphate. The α and β chains, each produced by a different gene (contrast

Figure 8.5
Conversion of indole glycerol phosphate to tryptophan

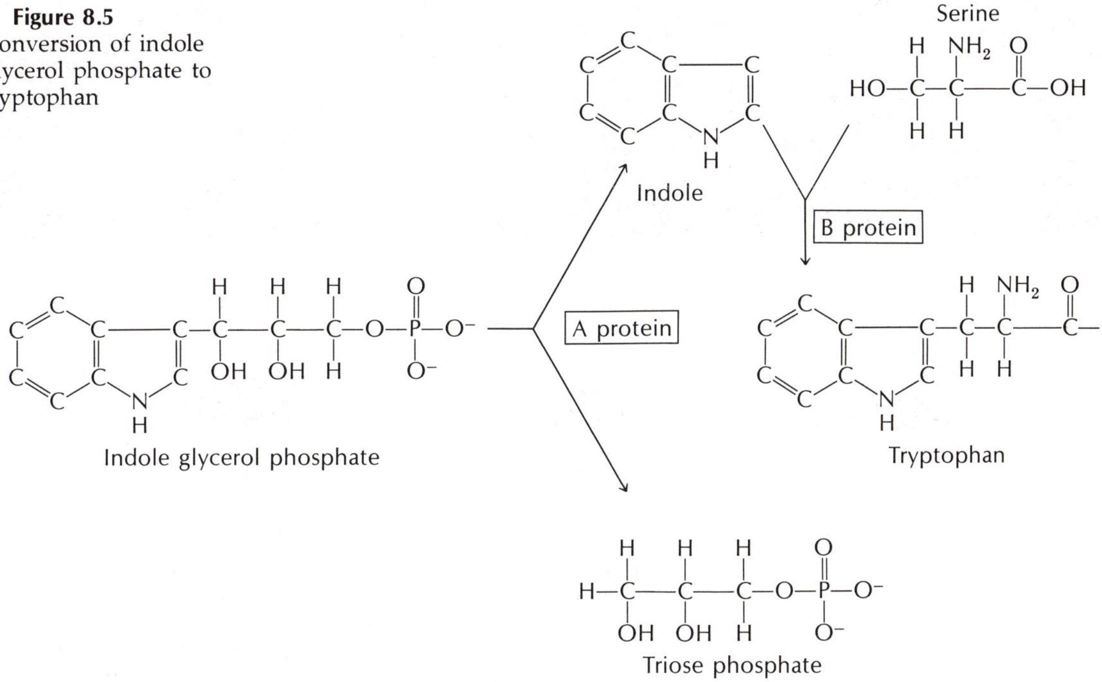

this with the one gene–one enzyme hypothesis), function uniquely. That is, the α chain can cleave indole glycerol phosphate to indole plus phosphoglyceraldehyde, and the β chain can condense indole with serine to form tryptophan. Maximum biological efficiency in *E. coli* is achieved only when the chains exist together as the complete enzyme.

Work on the tryptophan synthetase enzyme proved that more than one gene can code for a single enzyme; genes usually code specifically for polypeptide chains, which may or may not be active enzymes. This point was demonstrated in studies of the tobacco mosaic virus (TMV), an RNA virus with a protein coat that contains multiple copies of a single polypeptide. Mutations in the coat protein gene are manifest in amino acid alterations in the coat protein polypeptides.

Sickle cell anemia is an inherited human disease caused by a recessive gene that results in severe anemia when the gene is homozygous. The disease is named for the sickle shape assumed by the red blood cells as a result of low oxygen tension (Figure 8.6). Analysis of the hemoglobin

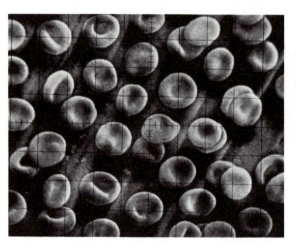
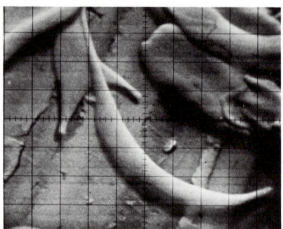

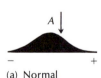
(a) Normal

(b) Homozygous for
sickle cell anemia

(c) Heterozygous for
sickle cell trait

Figure 8.6
(Photos courtesy of Irene Piscopo Rodgers of Philips Electronic Instruments, Inc.)

Normal red blood cells are shown in the photograph on the left; the photograph on the right shows the red blood cells of a patient with sickle cell anemia. The three curves below the photographs indicate a test for sickle cell anemia. Solutions of hemoglobin from various genotypes are placed at a certain point in an electrophoretic field (arrow). Normal hemoglobin migrates toward the cathode (the negative terminal); hemoglobin from individuals homozygous for sickle cell anemia migrates toward the anode (positive). Individuals heterozygous for the sickle cell trait have both types of hemoglobin. (A = hemoglobin A, normal; S = hemoglobin S, sickle cell)

of sickle cell anemic individuals has shown that valine has replaced glutamic acid at site number 6. The replacement occurs in the β polypeptide chain of the hemoglobin molecule, a molecule that in the adult consists of two α chains and two β chains. This demonstration that a single polypeptide is altered by a single mutation is another demonstration that a gene codes for a polypeptide chain.

Colinearity

The work described so far can be summarized by stating that genes code for polypeptides. If a gene produces a polypeptide, the sequence of mutant sites in a gene should correspond to the sequence of amino acid alterations in the respective polypeptide. In other words, the gene and the protein should be *colinear*. If through fine-structure analysis, a series of mutant sites maps from left to right in the sequence *a-d-c-b* and if the mutations are in separate codons (a *codon* is a sequence of three nucleotides that code for an amino acid), the isolated mutant polypeptide chains should also show a sequence of amino acid alteration sites in the order *a-d-c-b* (Figure 8.7).

Figure 8.7
Colinear sequence of mutant sites in the gene and amino acids in the polypeptide

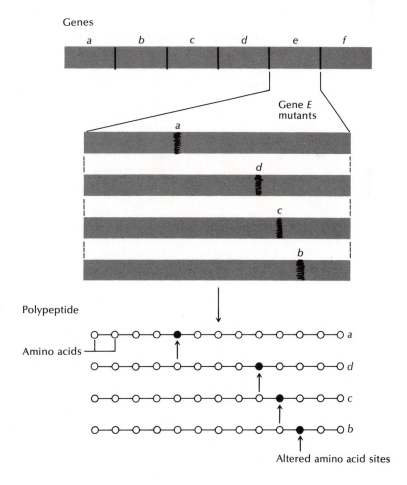

Figure 8.8
Amino acid sequence of
the polypeptide α chain
of tryptophan synthetase

```
        1                                       10
Met-Gln-Arg-Tyr-Glu-Ser-Leu-Phe-Ala-Gln-Leu-Lys-Glu-Arg-Lys-Glu-Gly-

        20                                      30
Ala-Phe-Val-Pro-Phe-Val-Thr-Leu-Gly-Asp-Pro-Gly-Ile-Glu-Gln-Ser-Leu-Lys-

        40                                      50
Ile-Asp-Thr-Leu-Ile-Glu-Ala-Gly-Ala-Asp-Ala-Leu-Glu-Leu-Gly-Ile-Pro-Phe-

        60                                      70
Ser-Asp-Pro-Leu-Ala-Asp-Gly-Pro-Thr-Ile-Gln-Asn-Ala-Thr-Leu-Arg-Ala-Phe-

                        80
Ala-Ala-Gly-Val-Thr-Pro-Ala-Gln-Cys-Phe-Glu-Met-Leu-Ala-Leu-Ile-Arg-Gln-

90                              100
Lys-His-Pro-Thr-Ile-Pro-Ile-Gly-Leu-Leu-Met-Tyr-Ala-Asn-Leu-Val-Phe-Asn-

        110                             120
Lys-Gly-Ile-Asp-Glu-Phe-Tyr-Ala-Gln-Cys-Glu-Lys-Val-Gly-Val-Asp-Ser-Val-

        130                             140
Leu-Val-Ala-Asp-Val-Pro-Val-Gln-Glu-Ser-Ala-Pro-Phe-Arg-Gln-Ala-Ala-Leu-

                150                             160
Arg-His-Asn-Val-Ala-Pro-Ile-Phe-Ile-Cys-Pro-Pro-Asn-Ala-Asp-Asp-Asp-Leu-

                170
Leu-Arg-Gln-Ile-Ala-Ser-Tyr-Gly-Arg-Gly-Tyr-Thr-Tyr-Leu-Leu-Ser-Arg-Ala-

180                             190
Gly-Val-Thr-Gly-Ala-Glu-Asn-Arg-Ala-Ala-Leu-Pro-Leu-Asn-His-Leu-Val-Ala-

        200                             210
Lys-Leu-Lys-Glu-Tyr-Asn-Ala-Ala-Pro-Pro-Leu-Gln-Gly-Phe-Gly-Ile-Ser-Ala-

                220                             230
Pro-Asp-Gln-Val-Lys-Ala-Ala-Ile-Asp-Ala-Gly-Ala-Ala-Gly-Ala-Ile-Ser-Gly-

                240                                     250
Ser-Ala-Ile-Val-Lys-Ile-Ile-Glu-Gln-His-Asn-Ile-Glu-Pro-Glu-Lys-Met-Leu-

                        260                             267
Ala-Ala-Leu-Lys-Val-Phe-Val-Gln-Pro-Met-Lys-Ala-Ala-Thr-Arg-Ser
```

Although this principle may appear obvious, it was difficult to prove, and it was not until 1966 that Yanofsky and his colleagues verified it. To do so, they concentrated on gene A (α chain) of tryptophan synthetase. Comparing the sequence of amino acids in the polypeptide α chain (Figure 8.8) with the sequence of mutant sites in the A gene, it can be seen that each mutational site corresponds in sequence to an amino acid site (Figure 8.9). This is strong evidence for the principle of colinearity.

Colinearity can also be demonstrated in the T4 phage by analyzing the synthesis of the head protein. A number of T4 head-protein mutants are known that result in the premature termination of the growing head polypeptide chain. The mutants, referred to as *termination mutants,* can be cultured and maintained, because they will grow on certain strains of E. *coli* that can suppress the termination mutation by inserting an amino acid at the premature termination mutant site. This allows elongation of the polypeptide to continue past the mutant site until a complete head protein is formed. If a head-protein termination mutant is grown in a non-suppressing host cell, growth of the polypeptide chain ends at the site of

Figure 8.9
Colinearity in the
tryptophan synthetase α
cistron and α polypeptide

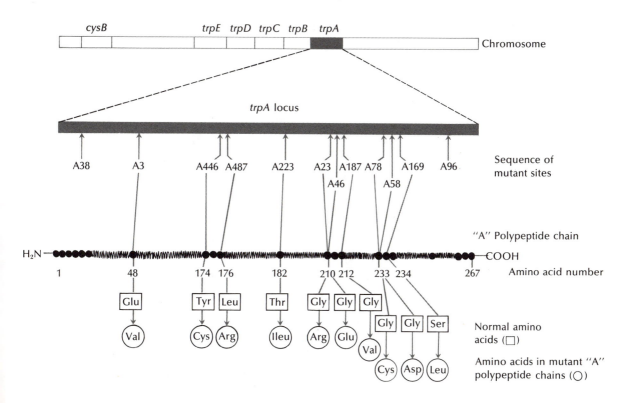

the mutation, and an incomplete head protein results. The protein fragment can be isolated and its length determined. The results show that incomplete head proteins of varying length are produced by different termination mutants, but that each type of mutant produces polypeptide molecules of only one length. When these mutants are mapped, a perfect

Figure 8.10
Colinearity in head-protein mutants of T4 [From A. S. Sarabhai et al., *Nature* 201:13 (1964)]

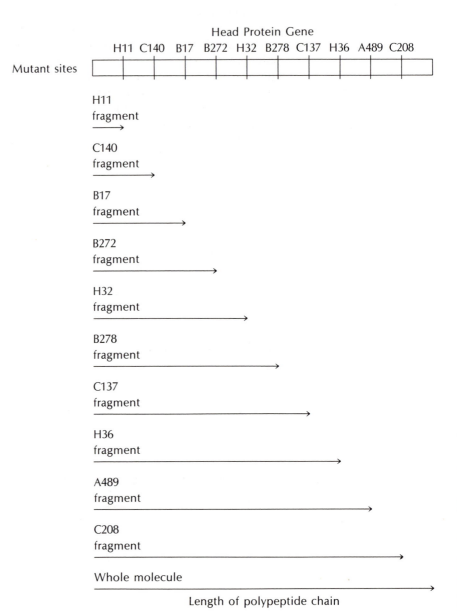

correlation is found between the position of the mutant site and the size of the protein (Figure 8.10). The closer the mutant site is to the left end of the head-protein gene, the shorter the protein fragment. Conversely, when the mutant site lies farther toward the right end of the gene, a larger head-protein fragment is produced. This is further evidence for colinearity between gene and polypeptide.

Protein structure and function

Genes usually code for proteins, and proteins, especially enzymes, largely determine phenotypes, so it will be profitable to review protein structure and function.

Amino acids The basic building unit of a protein is an amino acid, a molecule that has an acidic carboxyl group (—COOH) and a basic amino group (—NH$_2$) attached to a common structure, a carbon atom called the α *carbon:*

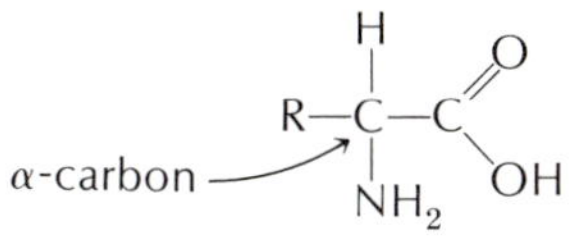

Each of the twenty common amino acids has a unique R group—a substitute for one of the hydrogens of glycine, the simplest amino acid. Aside from H, the simplest R group is a methyl (—CH$_3$) group as found in the amino acid alanine (Figure 8.11). The various R groups contribute to the three-dimensional configuration, and hence to the function of the protein, by forming specific types of bonds as a consequence of R-group interaction (Figure 8.12). For example, two nonadjacent cysteine molecules can link to form a covalent disulfide bridge. Hydrogen bonds can form between such amino acids as tyrosine and aspartic acid. Nonpolar side groups—those with neither a positive nor a negative charge—can interact with each other. Salt linkages can form between basic and acidic R groups. Van der Waals forces—those that involve attraction between small nonpolar aliphatic R groups—are generally much weaker than any of the other forces. The net result of the various types of R-group interactions is the conversion of a linear sequence of amino acids into a characteristic three-dimensional configuration, and the nature and sequence of amino acids determine that specific configuration.

Peptide bonds Adjacent amino acids in the primary structure are linked together by *peptide bonds*. A peptide bond forms when the —NH$_2$ group attached to

R-group characteristic	Chemical structure	Amino acid
Aliphatic, nonpolar	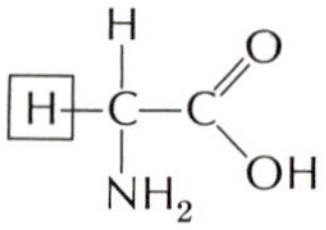	Glycine
	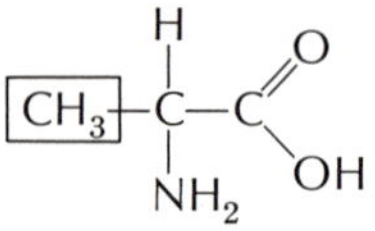	Alanine
	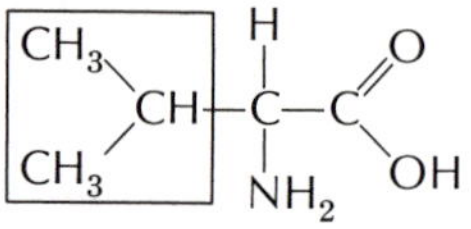	Valine
	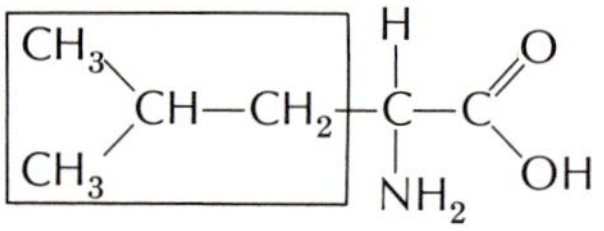	Leucine
	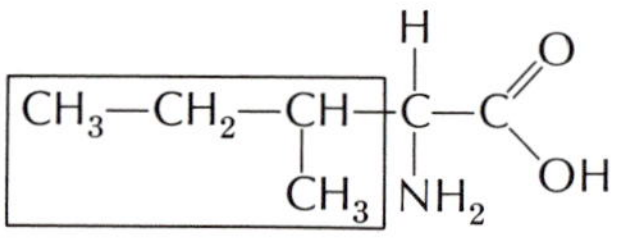	Isoleucine
Alcoholic	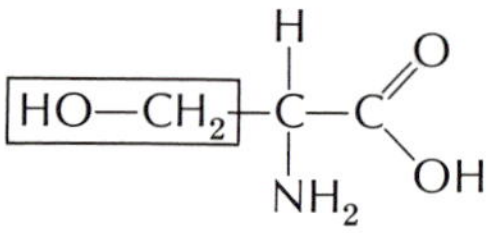	Serine
	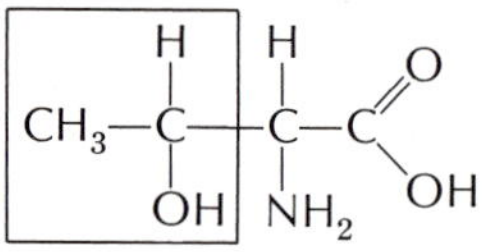	Threonine
Aromatic	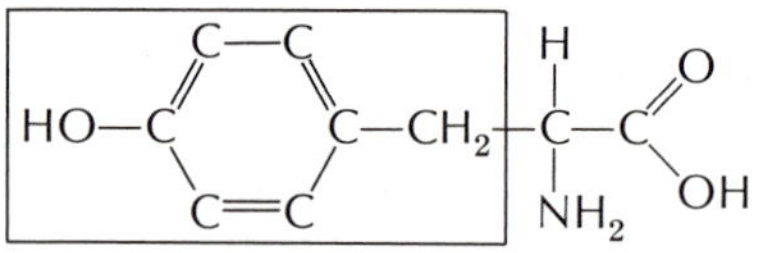	Tyrosine
	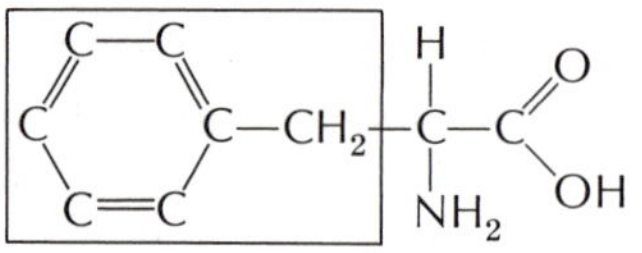	Phenylalanine
	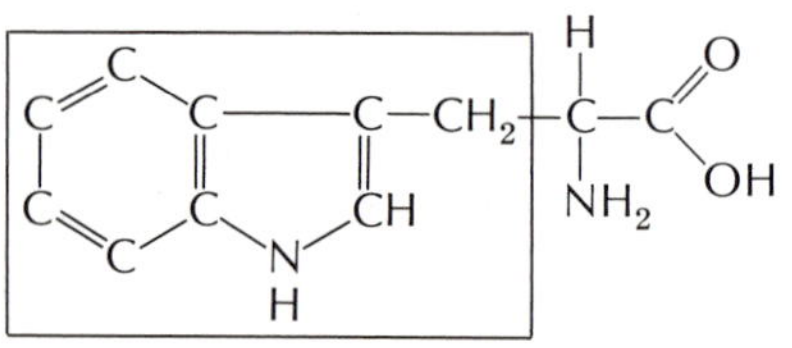	Tryptophan

Figure 8.11
The twenty common amino acids found in proteins

R-group characteristic	Chemical structure	Amino acid
Carboxylic (acidic)	HO—C(=O)—CH₂—C(H)(NH₂)—C(=O)OH	Aspartic
	HO—C(=O)—CH₂—CH₂—C(H)(NH₂)—C(=O)OH	Glutamic
Amine bases (basic)	NH₂—CH₂—CH₂—CH₂—CH₂—C(H)(NH₂)—C(=O)OH	Lysine
	NH₂—C(=NH)—NH—CH₂—CH₂—CH₂—C(H)(NH₂)—C(=O)OH	Arginine
	Histidine imidazole ring—CH₂—C(H)(NH₂)—C(=O)OH	Histidine
Sulfur-containing	HS—CH₂—C(H)(NH₂)—C(=O)OH	Cysteine
	CH₃—S—CH₂—CH₂—C(H)(NH₂)—C(=O)OH	Methionine
Amides	NH₂—C(=O)—CH₂—C(H)(NH₂)—C(=O)OH	Asparagine (AspNH₂)
	NH₂—C(=O)—CH₂—CH₂—C(H)(NH₂)—C(=O)OH	Glutamine (GluNH₂)
Imino	Proline ring structure	Proline

Figure 8.12
R-group interaction in the
establishment of a
three-dimensional
configuration

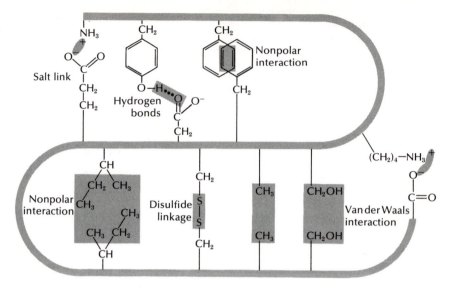

the α carbon of one amino acid is joined covalently to the —COOH group
attached to the α carbon of a second amino acid. This is accompanied by
the elimination of water.

Two amino acids joined by a peptide linkage comprise a *dipeptide,* three
make a *tripeptide,* and more than three make a *polypeptide* (Figure 8.13).

**General properties
of proteins** All proteins are composed of amino acids and exist on four levels of structural organization.

1. The sequence of amino acid molecules that compose the polypeptide
   chains is called the *primary structure.*
2. The structural configuration of the polypeptide, which usually assumes
   the form of a helix, is called the *secondary structure.* Hydrogen bonding,
   which results from the interaction between nonadjacent —CO groups and
   —NH groups, is the chief force in maintaining secondary structure.
3. The helical polypeptide folds and bends by means of R-group interaction
   (nonpolar interactions, salt linkages, and van der Waals forces) and assumes a third level of structural organization called the *tertiary structure.*
4. *Quaternary structure* results from the aggregation of polypeptide chains
   into superprotein structures, such as those examplified by the tryptophan
   synthetase enzyme and the hemoglobin molecule.

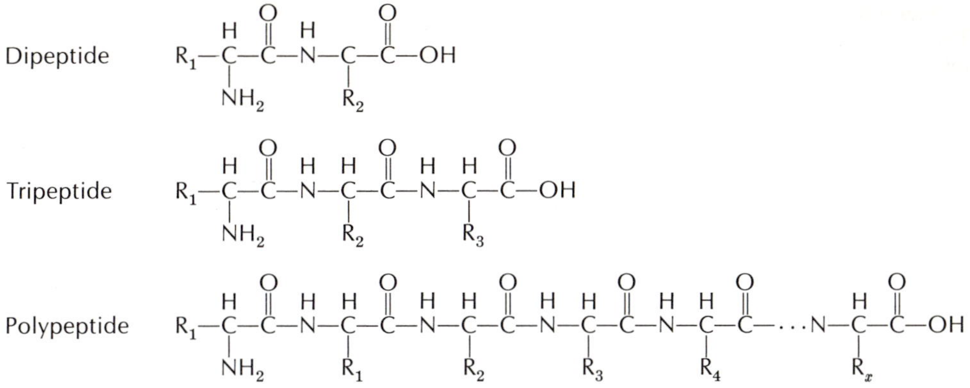

Dipeptide

$$R_1 - \overset{\underset{\displaystyle NH_2}{|}}{\underset{\displaystyle H}{C}} - \overset{\displaystyle O}{\overset{\|}{C}} - \overset{\displaystyle H}{N} - \overset{\underset{\displaystyle R_2}{|}}{\underset{\displaystyle H}{C}} - \overset{\displaystyle O}{\overset{\|}{C}} - OH$$

Tripeptide

$$R_1 - \overset{\underset{\displaystyle NH_2}{|}}{\underset{\displaystyle H}{C}} - \overset{\displaystyle O}{\overset{\|}{C}} - \overset{\displaystyle H}{N} - \overset{\underset{\displaystyle R_2}{|}}{\underset{\displaystyle H}{C}} - \overset{\displaystyle O}{\overset{\|}{C}} - \overset{\displaystyle H}{N} - \overset{\underset{\displaystyle R_3}{|}}{\underset{\displaystyle H}{C}} - \overset{\displaystyle O}{\overset{\|}{C}} - OH$$

Polypeptide

$$R_1 - \overset{\underset{\displaystyle NH_2}{|}}{\underset{\displaystyle H}{C}} - \overset{\displaystyle O}{\overset{\|}{C}} - \overset{\displaystyle H}{N} - \overset{\underset{\displaystyle R_1}{|}}{\underset{\displaystyle H}{C}} - \overset{\displaystyle O}{\overset{\|}{C}} - \overset{\displaystyle H}{N} - \overset{\underset{\displaystyle R_2}{|}}{\underset{\displaystyle H}{C}} - \overset{\displaystyle O}{\overset{\|}{C}} - \overset{\displaystyle H}{N} - \overset{\underset{\displaystyle R_3}{|}}{\underset{\displaystyle H}{C}} - \overset{\displaystyle O}{\overset{\|}{C}} - \overset{\displaystyle H}{N} - \overset{\underset{\displaystyle R_4}{|}}{\underset{\displaystyle H}{C}} - \overset{\displaystyle O}{\overset{\|}{C}} - \cdots - \overset{\displaystyle H}{N} - \overset{\underset{\displaystyle R_x}{|}}{C} - \overset{\displaystyle O}{\overset{\|}{C}} - OH$$

Figure 8.13
Formulas for a dipeptide,
a tripeptide, and a
polypeptide

The levels of structural organization are illustrated in Figure 8.14.

Once a protein assumes the final three-dimensional configuration determined by its primary structure, it functions as a structural protein (cell wall, viral coat, and so on), an enzyme, an antibody, or a hormone. It is beyond the scope of this book to discuss enzyme functions in detail. For further information, the reader is advised to consult the biochemistry texts referred to at the end of this chapter.

Transfer of information from gene to protein

With the knowledge that genes code for proteins, we can now approach the question of how the sequence of nucleotides in DNA dictates the sequence of amino acids in polypeptides. As might be anticipated, trans-

Figure 8.14
Levels of protein
structural organization

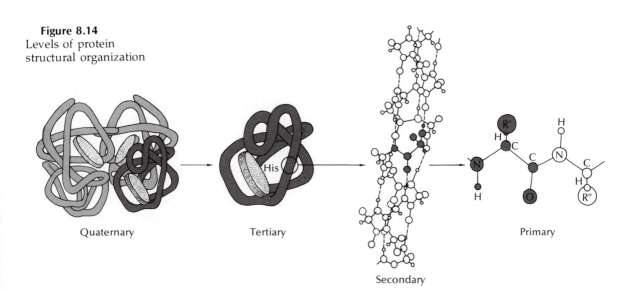

Quaternary   Tertiary   Primary

Secondary

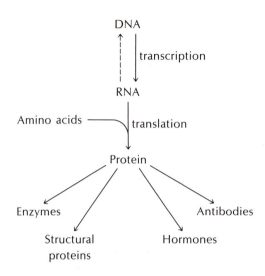

Figure 8.15
Transfer of genetic information

lating the alphabet of the nucleotides into the alphabet of the amino acids is indeed complex, and unraveling the mechanisms behind this flow of information required highly sophisticated techniques. A scheme of the entire process of gene function is given in Figure 8.15 and will be developed throughout the rest of the chapter.

One of the first discoveries made about the information flow from gene to protein was that two distinct classes of nucleic acids were involved: DNA and RNA. By the late 1950s, it was apparent that RNA—a single-stranded polynucleotide chain composed of adenine, uracil, guanine, and cytosine, and containing the sugar ribose instead of the deoxyribose found in DNA—served as an intermediate between DNA and protein. Several observations supported such a conclusion. First, proteins can be synthesized in the complete absence of DNA. Cells whose nuclei have been removed carry on active protein synthesis, whereas the nucleus, which contains most of the cellular DNA carries out no appreciable protein synthesis. (The DNA present in mitochondria and chloroplasts has a negligible influence on total protein synthesis.) Autoradiography experiments using labeled RNA show that RNA is synthesized in the nucleus and then transported to the cytoplasm. This supports the idea that DNA makes RNA, which, when transported to the cytoplasm, engages in the synthesis of proteins (Figure 8.16). Second, the amount of protein synthesis is directly proportional to the amount of RNA present and is *not* correlated with the amount of DNA. In cells with low levels of RNA, protein synthesis is low. In cells with high levels of RNA, protein synthesis is high. Variations in the levels of DNA are not well correlated with levels of protein synthesis.

In 1958, Crick proposed the generalization that there was a sequential flow of information from DNA to RNA to protein. This generalization came to be called the *central dogma*. At that time, it was based on much

Figure 8.16
Nuclear origin of RNA.
(From Zalokar, *Nature*
183:1330, 1959)

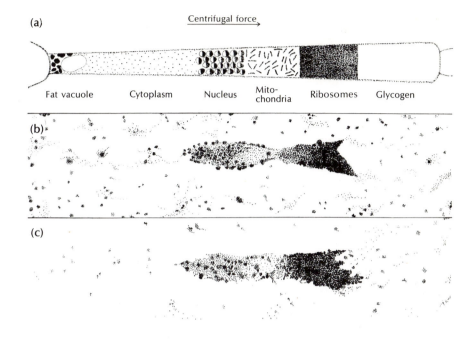

(a)

Centrifugal force

Fat vacuole   Cytoplasm   Nucleus   Mito-chondria   Ribosomes   Glycogen

(b)

(c)

(a) Schematic drawing of a centrifuged hypha of *Neurospora*. (b) Autoradiograph of a centrifuged hypha that has been fed tritiated uridine for one minute prior to centrifugation. Uridine is incorporated into RNA, not other cellular macro-molecules. Note that the label is located primarily in the nucleus, where the RNA is synthesized. (c) Autoradiograph of a centrifuged hypha that has been fed tritiated uridine for one minute and unlabeled uridine for one hour prior to centrifugation. Note that the label has become more dense in the cytoplasm, suggesting that RNA is transferred to the cytoplasm from the nucleus.

speculation and little fact. Two central concepts formed the core of the central dogma: the sequential flow of genetic information and the translation of that information. It has occasionally been schematized as follows.

$$DNA \xrightarrow{\text{transcription}} RNA \xrightarrow{\text{translation}} Protein$$

replication

Actually, the scheme shown here is incomplete, as will be seen shortly. Crick, with his central dogma, had set out to establish some general rules for the transfer of genetic information from one polymer (nucleic acid) to another polymer (protein). In so doing, he made a number of assumptions that by and large have proved true. For example, he assumed that the secondary, tertiary, and quaternary structures of proteins were determined by the protein's primary structure. This

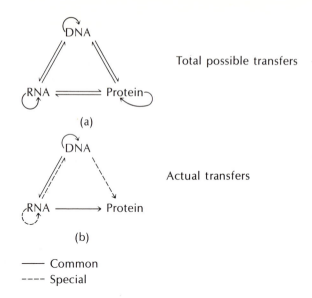

Figure 8.17
Informational transfers

Total possible transfers

(a)

Actual transfers

(b)

—— Common
---- Special

simplified matters considerably and made the informational content of nucleic acids a one-dimensional problem—nucleic acids to proteins. Crick further assumed that the rare amino acids occasionally found in nature were modifications of the 20 common ones and that the modifications occurred after or during protein biosynthesis. Finally, he assumed that as far as translation was concerned, thymine in DNA and uracil in RNA were equivalent. This reduced the problem to translating a nucleic acid alphabet of 4 letters (A, U, G, C) into one with 20 letters (amino acids).

The central dogma, as Crick hypothesized it, can be viewed in two ways: possible transfers and actual transfers (Figure 8.17). So far, only a one-directional flow of information is known, that from nucleic acids to protein. No informational transfer from protein to nucleic acids has ever been shown. Temin, Baltimore, and their colleagues proved that RNA can serve as a template for the synthesis of DNA. Certain RNA tumor viruses in chickens and other organisms produce progeny viruses through a DNA intermediate. This revelation modifies the central dogma but in no way invalidates it. We shall explore this RNA-to-DNA transcription in Chapter 11.

In summary, the central dogma retains an important function as a basis for current and future research. It gives us a framework within which we can visualize the flow of genetic information from nucleic acids to proteins.

The transcription of RNA

A general model for the process of gene expression as presented by Crick's central dogma specifies that the genetic information encoded in the DNA nucleotide sequences is first *transcribed* into complementary sequences of RNA (thymine and uracil being equivalent) that are in turn *translated* into polypeptides. The remainder of this chapter and the next chapter will explore how RNA is transcribed from DNA, what role RNA plays in protein synthesis, and how the information encoded in DNA is translated into protein structure.

RNA structure

Earlier, we examined the similarities and differences between DNA and RNA. To review briefly, RNA contains the sugar ribose rather than deoxyribose, has the base uracil rather than thymine, and is single-stranded rather than double-stranded. Because of this single-stranded quality, the purine–pyrimidine ratio is variable. This contrasts with the double-stranded DNA, in which the ratio is 1:1. X-ray analysis of RNA molecules shows that although they are single-stranded polymers of ribonucleotides, they do have regions of double-strandedness where the polymer folds back on itself, stabilized by hydrogen bonds between certain base pairs. This type of secondary and tertiary structure gives RNA molecules a variety of structural configurations. It should be noted that RNA can be much modified (with respect to its nitrogenous bases) after translation is effected.

Transcription of RNA by RNA polymerase

Protein biosynthesis is preceded by the transcription of single-stranded RNA from a double-stranded DNA template. Just as DNA synthesis requires specific enzymes (DNA polymerases I, II, and III*, for example), the synthesis of RNA from a DNA template also requires a specific enzyme—DNA-dependent *RNA polymerase.* One of the first things discovered about RNA polymerase is that it synthesizes RNA in a 5' to 3' direction; that is, ribonucleotides are added to a growing RNA strand at the 3' end (Figure 8.18).

To prove that RNA synthesis, like DNA synthesis, occurs in a 5' to 3' direction, an *in vitro* RNA-synthesizing system containing DNA, RNA polymerase, and unlabeled ribonucleotides was used. Synthesis of RNA was allowed to proceed long enough to ensure that the assembly of all RNA molecules had at least started. Then ^{3}H-labeled ribonucleotides were added to the system for a brief period, and shortly thereafter the newly synthesized RNA was isolated and analyzed. The results showed that only the 3' end of the RNA strands contained any label and that the 5' ends were nonradioactive, thus establishing the 5' to 3' direction of RNA synthesis.

Investigators have used a variety of approaches to analyze the RNA product as it relates to the DNA template. Three will be briefly discussed. All point to one main conclusion: The base composition of the newly synthesized RNA reflects the composition of the template DNA, suggesting that DNA is the template for RNA synthesis. Further, for any one cistron (a DNA segment that specifies a polypeptide sequence), or gene, only one strand of the DNA serves as a template.

E. Volkin and L. Astrachan (1956) were the first to demonstrate that newly synthesized RNA was essentially identical to the DNA templates in

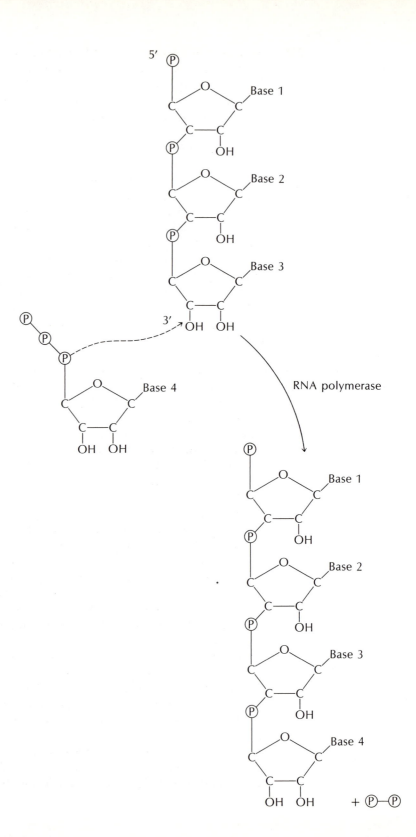

Figure 8.18
RNA polymerase adds ribonucleoside triphosphates to the 3′ end of a growing RNA chain

Table 8.2
RNA and DNA compositions from various sources

Source of DNA template	Composition of the RNA bases				$\dfrac{A+U}{G+C}$ observed	$\dfrac{A+T}{G+C}$ in DNA
	Adenine	Uracil	Guanine	Cytosine		
T2	0.31	0.34	0.18	0.17	1.86	1.84
Calf thymus	0.31	0.29	0.19	0.21	1.50	1.35
E. coli	0.24	0.24	0.26	0.26	0.92	0.97
Micrococcus lysodeikticus (a bacterium)	0.17	0.16	0.33	0.34	0.49	0.39

From J. D. Watson, *Molecular Biology of the Gene,* 3d edition, Menlo Park, Calif.: W. A. Benjamin, Inc., 1976. Copyright © 1976 by J. D. Watson. The data included in this table were obtained from various sources.

base composition (Table 8.2). They examined T2-infected *E. coli* cells and showed that T2 RNA was nearly identical in base composition to T2 DNA but not to *E. coli* DNA. Others have since confirmed this relationship in other cell types. The basis for the identical ratios is seen in Figure 8.19. Although the individual base composition may vary, the $(A + T)/(G + C)$ ratio of the template DNA is identical to the $(A + U)/(G + C)$ ratio of RNA.

The nearest-neighbor analysis has also been used to compare RNA and DNA composition. The results support the premise that RNA is synthesized on DNA templates (Table 8.3). In T4 RNA, for example, the frequency of the sequence ApU is 1.03, whereas in T4 DNA, the frequency of the complementary sequence ApT is 1.04.

Some of the early concepts of RNA transcription from DNA templates suggested that perhaps *both* DNA strands in a cistron were transcribed into RNA. This would mean that for any one cistron, there would be two complementary populations of RNA, synthesized from different but complementary DNA templates. If both RNA species were used, this

Figure 8.19
The basis for the identical base ratios in DNA and RNA

DNA:
A T T G G C T A G C C
T A A C C G A T C G G

RNA: U A A C C G A U C G G

DNA	RNA
A = 5	A = 3
T = 5	U = 2
G = 6	G = 3
C = 6	C = 3
$A + T = {}^{10}\!/_{22} = 0.455$	$A + U = {}^{5}\!/_{11} = 0.455$
$G + C = {}^{12}\!/_{22} = 0.545$	$G + C = {}^{6}\!/_{11} = 0.545$
$A + T/G + C = 0.835$	$A + U/G + C = 0.835$
$A = {}^{5}\!/_{22} = 0.23$	$A = {}^{3}\!/_{11} = 0.27$
$T = {}^{5}\!/_{22} = 0.23$	$U = {}^{2}\!/_{11} = 0.19$
$G = {}^{6}\!/_{22} = 0.27$	$G = {}^{3}\!/_{11} = 0.27$
$C = {}^{6}\!/_{22} = 0.27$	$C = {}^{3}\!/_{11} = 0.27$

Table 8.3
Comparison of nearest-neighbor frequencies between T4 mRNA and T4 DNA

Nearest RNA neighbors	Frequency of nearest neighbors in T4 mRNA[a]	Frequency of nearest neighbors in T4 DNA[b]
HO–(P)–(P) A C	0.87	0.90
HO–(P)–(P) A U	1.03	1.04
HO–(P)–(P) G C	1.11	1.13
HO–(P)–(P) G U	0.87	0.84

[a] Data from Bautz and Heding (1964).
[b] Data from Josse, Kaiser, and Kornberg (1961).

could lead to two different gene products for each gene. We now know this is incorrect. For any one cistron, only one strand of the DNA duplex is transcribed into RNA. Let us now examine evidence that has led to this conclusion.

One of the most convincing studies showing RNA transcription from only one DNA strand is that of S. Spiegelman and his colleagues. The technique that enabled them to reach this conclusion is called *hybridization* and is described as follows. If native double-stranded DNA is heated to 95°C, the hydrogen bonds holding the two strands together break, and the molecule dissociates or denatures into single strands. If the temperature of this solution of single-stranded DNA is slowly lowered (over several hours) to 25°C, the single strands renature to form double-stranded molecules. In 1962, Spiegelman and his colleagues applied the technique to DNA and RNA, hoping to show that RNA complementary to DNA hybridizes with it. They isolated RNA that was synthesized after *E. coli* cells had been infected by T2 bacteriophages. They hybridized the RNA, presumably T2 RNA, with heat-denatured T2 DNA. This technique, summarized in Figure 8.20, shows that the RNA is complementary to T2 DNA but not to *E. coli* DNA, confirming that only complementary nucleotide sequences hybridize. The sensitivity of the hybridization technique is seen when the bacteriophages T7 and lambda (λ) are examined. The nucleotide composition of T7 DNA and λ DNA are nearly identical. But RNA synthesized from λ DNA will not hybridize to any appreciable extent with T7 DNA, and RNA synthesized from T7 DNA will likewise not hybridize with λ DNA (Table 8.4). This suggests that although the DNA compositions may be nearly identical, the sequences definitely are not.

The DNA–RNA hybridization studies have enabled us to demonstrate that only one DNA strand is transcribed into RNA. For example, it

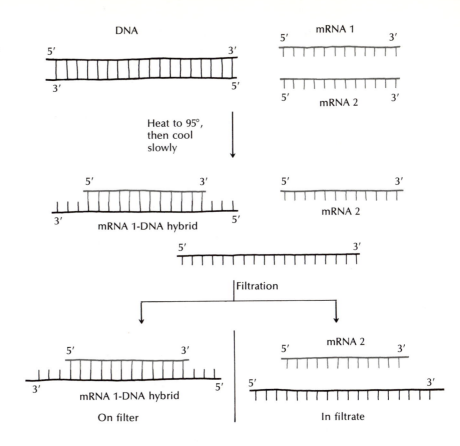

Figure 8.20
Technique of DNA–RNA
hybridization

DNA

mRNA 1

mRNA 2

Heat to 95°,
then cool
slowly

mRNA 1-DNA hybrid

mRNA 2

Filtration

mRNA 1-DNA hybrid

On filter

mRNA 2

In filtrate

has been shown that T4 RNA will not renature with itself, which suggests that the RNA population consists, not of complementary sequences, but of noncomplementary sequences. Furthermore, when two T7 bacteriophage DNA strands are physically separated and set up in a hybridization system with T7 RNA, the DNA–RNA hybrids form with only one of the two DNA strands.

Another observation consistent with the notion that only one DNA strand serves as a template for RNA comes from studies of RNA synthe-

Table 8.4
Hybridization of RNA
polymerase product RNA
with template and
nontemplate DNA

DNA template for RNA polymerase reaction	Percent of RNA hybridized with		
	T7 DNA	λ DNA	Blank filter
T7	50	0.2	0.3
λ	0.3	49	0.4

From. J. P. Richardson, "RNA polymerase and the control of RNA synthesis," *Prog. Nuc. Acid Res. and Molec. Biol.* 9:75–116 (1969).

sized from phage φX174 DNA. (Remember that mature φX174 DNA is single-stranded.) The RNA synthesized in φX174-infected *E. coli* cells hybridizes well with the denatured replicative form of DNA but fails to hybridize with the original single strand of parental DNA. This interesting observation suggests that the DNA strand that is complementary to the original parental strand—but not the infecting chromosome—is transcribed into RNA.

Perhaps one of the most elegant experiments showing that, for any one gene, only one of the DNA strands is transcribed was performed by Jayaraman and Goldberg. They used the T4 bacteriophage, looked at RNA produced from two different genes, and asked what DNA strands were used for RNA transcription at each gene locus.

In order to answer their question, Jayaraman and Goldberg selected a gene in T4 that was known to be transcribed early in the life cycle and another gene that was transcribed late. The "early" gene is *rIIB* and the "late" gene is *21*. After they collected "early" RNA and "late" RNA from cells infected with T4 (*rIIB*⁺ and *21*⁺), they took T4 DNA (*rIIB*⁺, *21*⁺), denatured it, and separated it into heavy strands and light strands. This separation is possible because the densities of the two DNA strands differ according to their purine and pyrimidine content.

The next steps in their experiment (Figure 8.21) were designed to see which of the two DNA strands transcribed the early and the late RNA. They mixed some light DNA strands with early RNA, and some light DNA strands with late RNA. The same thing was done with the heavy DNA strands. The conditions were such that RNA complementary to the DNA would hybridize with it. The DNA–RNA hybrid, because of its double-stranded nature, is not susceptible to the digestive attacks of an endonuclease that specifically digests single-stranded DNA. So, after the hybridized DNA–RNA molecules have been treated with the endonuclease, the only parts of the DNA that survive are those parts hybridized with RNA. If that RNA is early, we have essentially isolated the *rIIB* gene; if it is late, we have essentially isolated the *21* gene.

Next, they needed to identify the genotype of the RNA-protected DNA in both the heavy and light DNA fractions. To do this, they separated or denatured the strands in the DNA–RNA hybrid, digested the RNA using RNase, and used the remaining DNA fragments in a special kind of transformation system. This system involves a special bacterium (*Aerobacter aerogenes*) that has been made to carry T4 chromosomes with the mutant alleles *rIIB*⁻ and *21*⁻. The remaining DNA fragments from the light DNA were used to transform T4 (*rIIB*⁻, *21*⁻); the same procedure was employed for the DNA fragments from the heavy DNA.

The predictions of this experiment are as follows: If *rIIB* RNA is transcribed from the light DNA strand, then light DNA fragments will transform T4 (*rIIB*⁻, *21*⁻) into T4 (*rIIB*⁺, *21*⁻). If the *rIIB* RNA is transcribed from the heavy DNA strand, then only the heavy DNA fragments will transform T4 (*rIIB*⁻, *21*⁻) into T4 (*rIIB*⁺, *21*⁻). The same holds true for the

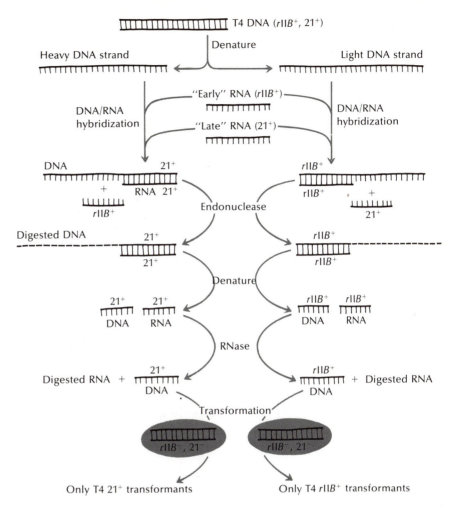

Figure 8.21
A diagrammatic summary of the Jayaraman and Goldberg experiment

21 RNA. If the heavy or light DNA transcribes both *rIIB* and *21* RNA, then double transformation (*rIIB⁻*, *21⁻* → *rIIB⁺*, *21⁺*) will occur from heavy or light DNA. If both DNA strands in a gene are transcribed into RNA, then transformation of a particular gene will occur from *both* the heavy and light DNA fragments.

The results clearly showed that *rIIB* and *21* were transcribed from different DNA strands, and that only one DNA strand was transcribed per gene. Light DNA fragments transformed T4 (*rIIB⁻*, *21⁻*) only into T4 (*rIIB⁺*, *21⁻*). Heavy DNA fragments transformed T4 (*rIIB⁻*, *21⁻*) only into T4 (*rIIB⁻*, *21⁺*). Thus the light DNA strand of the *rIIB* gene and the heavy strand of the *21* gene were transcribed into RNA.

To summarize, the DNA message is transcribed to RNA by DNA-dependent RNA polymerase, and for any one gene, only one DNA strand is transcribed—the direction of RNA synthesis being 5' to 3'. A picture of

Figure 8.22
Picture of transcription
(Courtesy of O. L. Miller)

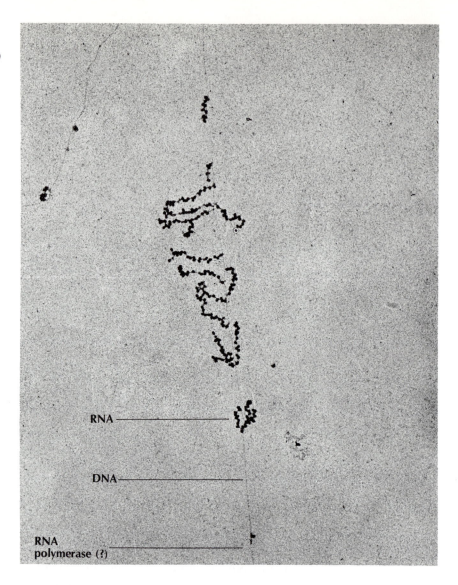

RNA ——————————————

DNA ——————————————

RNA
polymerase (?) ——————————————

a possible RNA transcription process (Figure 8.22) was published by Miller in 1970. It shows long strands, presumably of RNA, being transcribed from DNA. It may even be that the RNA polymerase can be seen.

RNA polymerase: Its ability to initiate, elongate, and terminate an RNA chain

A complex enzyme, RNA polymerase consists of five different polypeptide chains: β, β', α, σ, and ω. The respective molecular weights of these chains are 150,000, 160,000, 40,000, 90,000, and 10,000. The active enzyme has one copy of each chain except for α, which is present twice. The molecular weight of the complete enzyme is about 500,000.

Figure 8.23
RNA polymerase

Core enzyme
subunits

|——Core enzyme——|

Sigma factor

The polypeptide chains are held together by noncovalent bonds, but the attachment of σ is very weak indeed (Figure 8.23). This polypeptide is released from the core enzyme (β, β', α_2, and ω) after initiation of RNA synthesis and reassociates again at the beginning of a new cycle. R. R. Burgess and his colleagues isolated σ in 1969. Its function has since been extensively studied.

The role of the σ factor is apparently to recognize code words, or sequences of nucleotides, on the DNA molecule that determine where the synthesis of RNA will begin. It is not actually involved in the catalytic function of the enzyme, because the enzyme can synthesize RNA with or without σ. The RNA synthesized without σ is random, because synthesis can start anywhere along the DNA molecule and on either strand. The σ factor restricts the synthesis to specific starting points and, consequently, to one strand.

The σ factor stimulates the firm binding of RNA polymerase to DNA regions called *promoters,* regions in which RNA synthesis is initiated. These promoter regions are composed of three subregions. There is an *initial recognition site,* which is the area in which RNA polymerase first binds. There is the *binding site* to which the RNA polymerase moves and binds tightly. And there is the *initiation site,* the point at which RNA transcription begins. The recognition and binding sites are rich in AT base pairs and, interestingly, are not transcribed into RNA. The initiation site is located about seven base pairs away from the binding site.

Following initiation of RNA synthesis by σ plus the core RNA polymerase, the RNA chain elongates in a 5' to 3' direction under the influence of the core enzyme. The σ factor dissociates from the DNA–RNA polymerase–RNA complex. The final step in RNA transcription is termination, or completion, of the RNA molecule and its removal from the DNA template. RNA synthesis may be terminated in two ways: by a DNA-sequence-induced termination or by a protein-induced termination.

There appear to be specific DNA sequences, usually AAAA . . . A, that result in the termination of RNA synthesis when encountered by RNA polymerase. When such a DNA termination site is encountered, both the RNA polymerase and the RNA chain are released from the DNA template. A second method of RNA chain termination is mediated by a protein factor called *rho* (ρ), which has been isolated from *E. coli.* Rho is not part of the RNA polymerase molecule. It induces chain termination and release at sites *in addition to* the DNA termination sites we have mentioned, but it does not cause the release of the RNA polymerase from the DNA template unless its activity coincides with a DNA termination site. Rho-induced termination permits RNA synthesis to begin again only if no termination site is encountered and only if an initiation signal is present.

Although little is known about rho, it appears to bind to RNA alone,

Figure 8.24
A scheme summarizing
RNA transcription

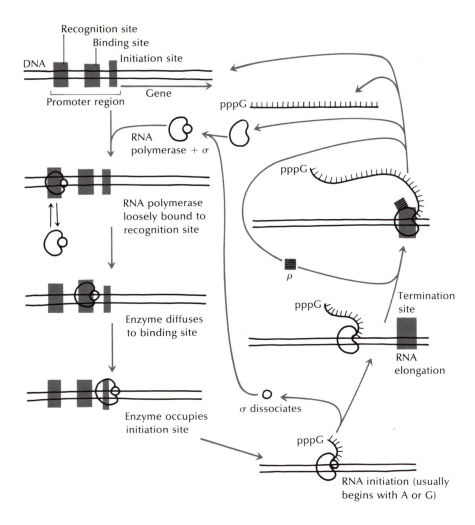

not to DNA or RNA polymerase. Figure 8.24 summarizes RNA transcription.

RNA and protein synthesis

RNA transcribed from a DNA template is either translated into a polypeptide or functions to support the translation process. We have seen that the synthesis of proteins is a two-step process. The DNA is transcribed into RNA, which in turn is translated into protein. Further, the amino acid sequence in a polypeptide chain is specified by the nucleotide sequence in a gene.

Three functional classes of RNA are involved in protein synthesis: *messenger RNA* (mRNA) contains the information encoded in the gene for the biosynthesis of a polypeptide. *Transfer RNA* (tRNA) binds to amino

Component	Function
tRNA	Transfers activated amino acids, recognizes mRNA
Messenger RNA (mRNA)	Contains genetic information that determines amino acid content and sequence in protein
Ribosomes	Particle on which the polypeptide is synthesized—2 subunits
Protein factors	Catalyze partial reactions in the initiation, elongation, and termination of polypeptides
Miscellaneous factors, such as ATP, GTP, Mg^{++}, K^+, NH_4^+	Required for activity of protein-soluble factors

Table 8.5
Components required for incorporation of amino acids into proteins

acids and to the mRNA, where the amino acids are linked to form a polypeptide. *Ribosomal RNA* (rRNA) and ribosomal protein function as a unit called a ribosome, on which the mRNA and tRNA—in conjunction with amino acids—aggregate during protein synthesis and function as the sites of polypeptide synthesis.

The translation of the mRNA into protein necessitates translating the alphabet of nucleotides into the alphabet of amino acids. The amino acids attached to tRNA molecules move to the ribosome–mRNA complex. A sequence of three nucleotides on mRNA specifies one amino acid. This nucleotide sequence is called a *codon* and will be discussed in Chapter 9. As noted, once the primary structure of a polypeptide has been determined, its secondary, tertiary, and quaternary structure follow automatically. A summary of the components required for amino acid incorporation into protein is given in Table 8.5.

Components of the protein-synthesizing system

Before considering the actual process of protein synthesis, we will first examine in more detail the roles of tRNA, mRNA, and rRNA.

tRNA   The affinity of a sequence of nucleotides for amino acids is essentially nonexistent. An mRNA molecule cannot determine a sequence of amino acids unless another RNA molecule intervenes as an adaptor. (An mRNA molecule may be viewed as a series of potential hydrogen bonds, and amino acids by themselves have no bonding potential with mRNA.) When tRNA was discovered in 1958, it was quickly recognized as just such an adaptor molecule. A specific tRNA first attaches to a specific amino acid and then unites that amino acid with a specific codon on the mRNA. The tRNA has an *anticodon,* which is a sequence of three bases complementary to an mRNA codon and which permits the tRNA to hydrogen-bond with the correct mRNA codon. There is at least one tRNA for each of the 20 amino acids, and since most amino acids are specified by more than one codon, there are more than 20 different species of tRNA.

Figure 8.25
Structure of alanine tRNA

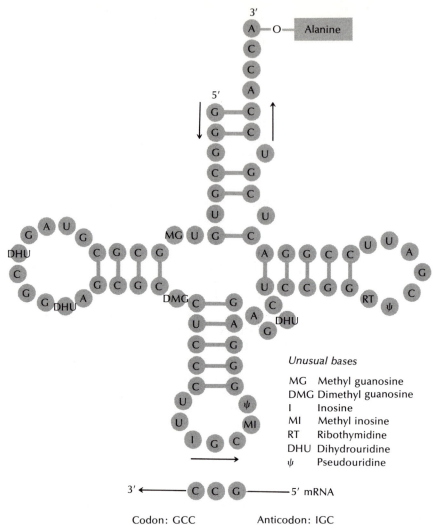

Codon: GCC   Anticodon: IGC

Compared with mRNA and rRNA, the tRNA molecule is small, being composed of only 75–85 ribonucleotides. The first nucleotide sequence of a tRNA molecule was determined by R. W. Holley, who in 1965 sequenced alanine tRNA (tRNA$_{ala}$, Figure 8.25). Since that time, the sequences of most of the other tRNA species have been determined. The tRNA molecules studied to date show regions of double-strandedness and a moderately high proportion (10–20 percent) of unusual bases (Figure 8.26). These unusual bases are formed by an alteration of the four common bases *after* the tRNA is transcribed from DNA. Because the unusual bases cannot easily form hydrogen bonds, their apparent function is to maintain regions of single-strandedness in the tRNA molecule, in the region of the anticodon, for example.

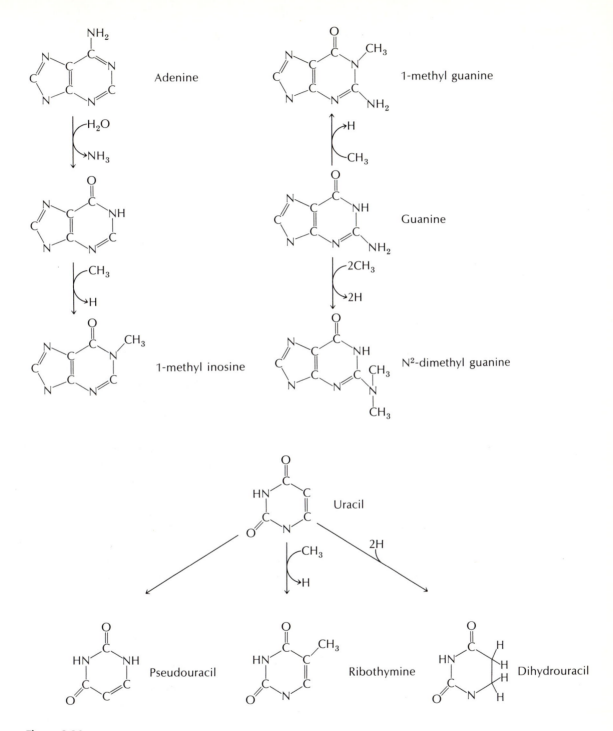

Figure 8.26
Unusual bases found in RNA

Figure 8.27
General diagram of a tRNA

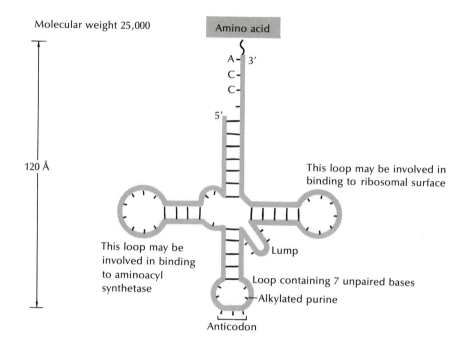

Molecular weight 25,000

Amino acid

A — 3'
C —
C —
5'

120 Å

This loop may be involved in binding to ribosomal surface

This loop may be involved in binding to aminoacyl synthetase

Lump

Loop containing 7 unpaired bases

Alkylated purine

Anticodon

Structural studies of various species of tRNA reveal a common clover-leaf pattern, with many configurations and regions in common among the various species (Figure 8.27). However, the only well-understood areas of these molecules are the anticodon site and the amino acid attachment site. What we call the *lump* and the two remaining loops may function in binding the tRNA to the ribosome and in "recognizing" the enzyme that attaches the amino acid to the 3' end. The C-C-A at the 3' end of each tRNA is added enzymatically *after* the tRNA has been transcribed from the DNA.

Great progress is being made in understanding the actual three-dimensional structure of tRNA. New developments in generating tRNA crystals and in x-ray crystallography are rapidly helping to solve this problem. Figure 8.28 depicts some recent work on the formation of tRNA crystals and the postulated three-dimensional configuration of the molecule as revealed by x-ray crystallography.

In *E. coli* and yeast, the tRNA molecules are coded for by genes scattered throughout the genome. Mutations can occur in these tRNA genes. One frequently encountered type of tRNA mutation is the *suppressor mutation*. This mutant tRNA species can recognize what is sometimes called a *nonsense codon*—a termination signal that specifies no amino acid but rather signals the termination of a polypeptide. The suppressor tRNA inserts an amino acid where none was specified. This can result in an abnormally long polypeptide (because the normal termination signal is ignored), or it can restore function to a nonsense mutation that prematurely terminates a polypeptide.

Figure 8.28
Photographs of tRNA crystals: tRNA$_{F-met}$ (upper photograph) and tRNA$_{leu}$ (lower photograph). The drawing shows a schematic model of yeast phenylalanine tRNA (the anticodon arm is the shaded lower part of the model). (Photos courtesy of R. M. Bock) [From S. H. Kim et al., *Science* 185:436 (2 August 1974)]

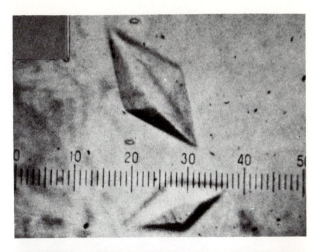

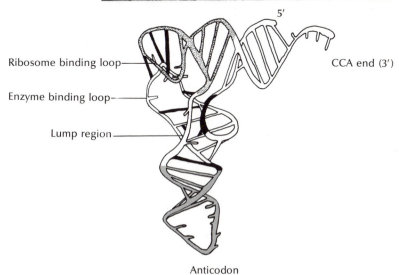

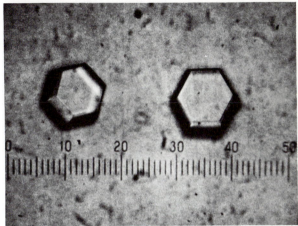

Ribosome binding loop

Enzyme binding loop

Lump region

5′

CCA end (3′)

Anticodon

Table 8.6
Purified aminoacyl-tRNA synthetases

Enzyme	Source	Molecular weight	Subunits	Binding sites Amino acid	ATP	Aminoacyl adenylate	tRNA
Arg	E. coli	75,000	1	—	—	—	—
Arg	B. stearo-thermophilus	78,000	1	1	1	0	—
Cys	Yeast	160,000	—	—	—	—	—
Gln	E. coli	69,500	1	—	—	—	—
Gly	E. coli	227,000	4	—	—	—	—
Glu	E. coli	102,000	2	—	—	—	1
His	E. coli	85,000	2	2	2	—	—
His	Salmonella typhimurium	100,000	—	—	—	—	—
Ile	E. coli	114,000	1	1	—	1	1
Ile	B. stearo-thermophilus	110,000	—	—	—	1	1
Leu	E. coli	105,000	1	—	1	1	1
Leu	Yeast	120,000	2	—	—	—	—
Leu	B. stearo-thermophilus	95,000	—	—	—	—	—
Lys	E. coli	104,000	2	2	2	—	—
Lys	Yeast	138,000	2	2	2	1	1
Met	E. coli native enzyme	173,000	2	2	4	2	2
Met	E. coli trypsin modified enzyme	64,000	1	1	1	1	1
Met	E. coli enzyme from strain CA 244 after autolysis	96,000	2	2	2	—	2
Phe	E. coli	180,000	4	1	—	1	1
Phe	Yeast	270,000	4	1	—	—	1
Phe	D. melanogaster	180,000	—	—	—	—	—
Phe	Rat liver	200,000	—	—	—	—	—
Phe	Euglena gracilis	65,000	—	—	—	—	—
Pro	E. coli	94,000	2	—	—	—	—
Ser	E. coli	95,000	2	2	2	—	1 or 2
Ser	Yeast	120,000 or 95,000	2	—	—	—	1 or 2
Ser	Rat liver	90,000	—	—	—	—	—
Ser	Hen liver	120,000	2	—	—	—	—
Thr	E. coli	117,000	—	—	—	—	—
Trp	E. coli	74,000	2	—	—	—	1
Trp	Beef pancreas	108,000	2	2	—	2	—
Tyr	Yeast	46,500	1	—	—	—	—
Tyr	E. coli	95,000	2	—	—	—	1
Tyr	B. subtilis	88,000	—	—	—	—	—
Val	E. coli	110,000	1	1	—	1	1
Val	Yeast	122,000	1	1	1	—	1

From D. Söll and P. R. Schimmel, "Aminoacyl-tRNA synthetases," in *The Enzymes*, 3d edition, Vol. 10, New York: Academic Press.

The attachment of each amino acid to its specific tRNA depends on a very specific enzyme called aminoacyl-tRNA synthetase. At least 20 such synthetase enzymes probably exist—one for each amino acid and a specific tRNA. Table 8.6 lists most of them. Synthetases vary widely in physical and chemical properties, but all perform the same two basic

functions. They activate a specific amino acid, and they couple the activated amino acid to a specific tRNA. Each activity seems to involve separate enzyme sites, as suggested by the following example.

P-hydroxy mercury benzoate
(interferes with enzyme
function)

Methionyl-tRNA synthetase ⟶ Could *activate*
methione but not
attach it to a
tRNA

The two activities are represented in Figure 8.29.

In the *activation reaction*, the amino acid, ATP, and the aminoacyl-
tRNA synthetase interact to form an amino acid–AMP–enzyme complex,
releasing inorganic pyrophosphate (P—P). The amino acid–AMP–enzyme complex has been isolated by chromatography, and the inorganic
pyrophosphate generated by the reaction has been determined through
the use of ^{32}P-labeling. The *transfer reaction* involves the coupling of the
activated amino acid–enzyme complex to a specific tRNA, forming a
tRNA–amino acid complex and releasing AMP and synthetase. The
amino acid–AMP–synthetase–tRNA complex has also been isolated.

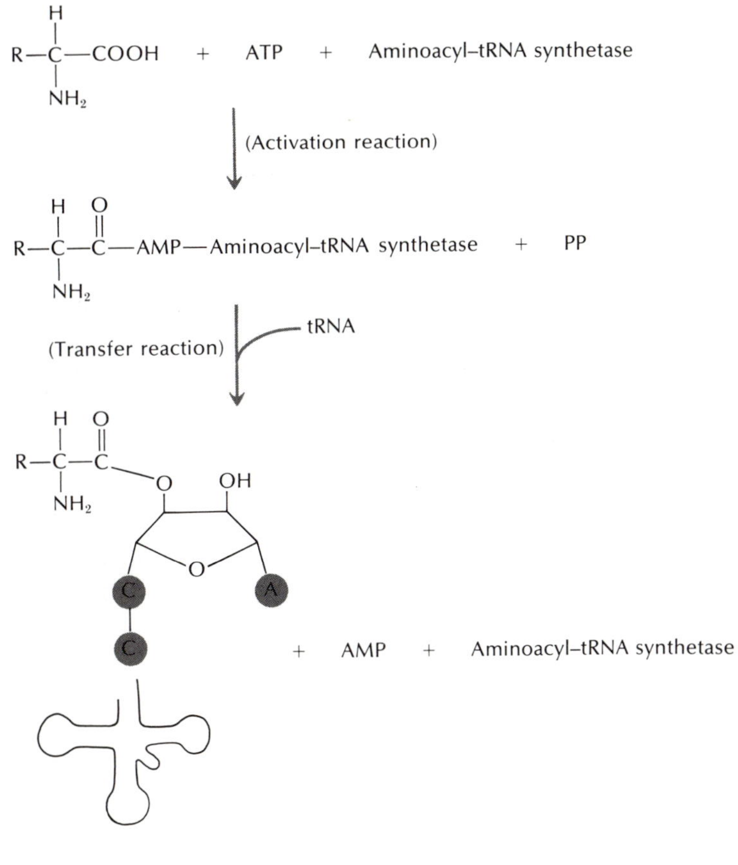

Figure 8.29
The activation and
transfer reactions of
aminoacyl-tRNA
synthetase

Table 8.7
Amino acid acceptor
capacity of E. coli tRNA

Amino acid	Nanomoles accepted per mg tRNA	Amino acid	Nanomoles accepted per mg tRNA
Alanine	2.97	Methionine	0.95
Arginine	2.97	Phenylalanine	0.88
Aspartate	0.76	Proline	1.50
Glutamate	2.33	Serine	1.49
Glycine	3.28	Threonine	1.86
Histidine	1.06	Tyrosine	0.94
Leucine	3.58	Valine	2.79
Lysine	2.59		

From C. D. Yegian, G. S. Stent, and E. M. Martin, "Intracellular condition of E. coli tRNA," Proc. Natl. Acad. Sci. U.S. 55:839–846 (1966).

As a group of enzymes, the synthetases are highly specific, though they do differ from each other in their ability to form the amino acid–AMP–synthetase complex (Table 8.7). However, amino acid activation is evidently less specific than the transfer reaction. For example, isoleucine synthetase will activate valine, and valine synthetase will activate threonine; but neither of these complexes can be combined with tRNA molecules. The specificity for the activation and transfer reactions lies in the ability of the synthetase to recognize a specific tRNA, but the recognition specificity is not wholly understood at the molecular level.

The *charged tRNA* (that is, the tRNA–amino acid) associates with the correct codon on the mRNA–ribosome complex (Figure 8.30), where the amino acid is linked by a peptide bond to a growing chain of amino acids and the tRNA is set free to pick up another amino acid. That the specificity for the correct location of the amino acid along the mRNA lies with the tRNA and not the amino acid attached to it was shown by an elegant experiment performed by G. Von Ehrenstein, B. Weisblum, and Benzer. They attached ^{14}C-cysteine to cysteine-tRNA (tRNA$_{cys}$). The —SH group of cysteine was then removed, thus converting ^{14}C-cysteine to ^{14}C-alanine. Using mRNA that produces a protein of known amino acid sequence, it was found that tRNA$_{cys}$-alanine was inserted into cysteine positions (Figure 8.31). This showed that the specificity for amino acid location in a polypeptide chain is a function of mRNA–tRNA interaction and that the amino acid itself is not involved.

mRNA The second major component of the protein-synthesizing system is messenger RNA (mRNA). It is transcribed from genes, associates with ribosomes, and is translated into polypeptide chains. Be-

Figure 8.30
Recognition between
codon and anticodon

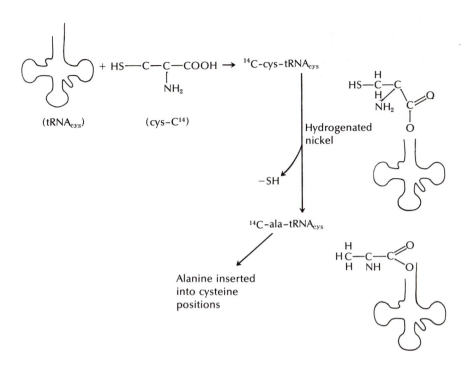

Figure 8.31
Specificity for mRNA location resides in tRNA, not amino acid

(tRNA$_{cys}$)   (cys–C^{14})

$+ \ HS—C—C—COOH \ \rightarrow \ ^{14}C\text{-cys-tRNA}_{cys}$

NH$_2$

$^{14}C\text{-ala-tRNA}_{cys}$

Hydrogenated nickel

–SH

Alanine inserted into cysteine positions

cause genes and polypeptides vary in size, so do the mRNA molecules for which they code.

Originally, ribosomes were thought to encode the genetic information for polypeptides. As a matter of fact, in the mid-1950s, prior to the discovery of mRNA, a "one gene–one ribosome–one protein" hypothesis was popular as a way to account for the intermediary role of RNA in protein synthesis. Even when it was postulated, however, evidence existed that appeared to contradict the hypothesis.

E. coli infected with T2 bacteriophages produces an entirely new series of proteins, so in keeping with this hypothesis, a new series of ribosomes should be synthesized. Contrary to expectations, there is no new synthesis of ribosomes in the T2-infected cells. The synthesized RNA is an unstable variety of RNA—rapidly synthesized and rapidly degraded, unlike rRNA and tRNA. This RNA is synthesized from the T2 DNA template, as evidenced by the fact that it hybridizes with T2 DNA but not with *E. coli* DNA.

In 1961, F. Jacob and J. Monod hypothesized the existence of mRNA and rejected the idea that rRNA or tRNA was translated into protein. They proposed this in conjunction with a theory that short-lived species of RNA (mRNA) are synthesized in the course of development and are crucial to the adaptation of an organism. These code for specific proteins as needed and disappear when the proteins are no longer needed. Both tRNA and rRNA are far too stable for this role.

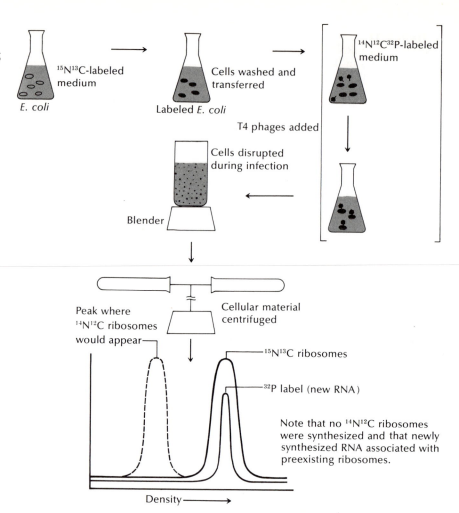

Figure 8.32
Brenner–Jacob–Meselson experiment demonstrating the existence of mRNA

$^{15}N^{13}C$-labeled medium

E. coli

Cells washed and transferred

Labeled *E. coli*

$^{14}N^{12}C^{32}P$-labeled medium

T4 phages added

Cells disrupted during infection

Blender

Cellular material centrifuged

Peak where $^{14}N^{12}C$ ribosomes would appear

$^{15}N^{13}C$ ribosomes

^{32}P label (new RNA)

Note that no $^{14}N^{12}C$ ribosomes were synthesized and that newly synthesized RNA associated with preexisting ribosomes.

Density

At about the same time, S. Brenner, Jacob, and Meselson established the existence of mRNA in an experiment summarized in Figure 8.32: They grew *E. coli* cells in a ^{15}N-^{13}C-labeled medium (heavy, nonradioactive) until all subcellular components were ^{15}N-^{13}C-labeled. Then the cells were washed and transferred to a ^{14}N-^{12}C-^{32}P medium (light, radioactive), in which they were immediately infected with T4 phages. The RNA synthesized after T4 infection was analyzed and compared to the preinfection ribosomes. The results of this experiment showed that new ^{32}P-labeled RNA was synthesized in the T4-infected *E. coli,* and this new RNA associated with *preexisting* ^{15}N-^{13}C ribosomes. That is, no ^{14}N-^{12}C-^{32}P ribosomes were synthesized. The conclusion drawn from this experiment was that the *E. coli* ribosomes serve as the sites of phage protein synthesis but that the information for new T4 proteins comes from a new, postinfection,

T4-coded RNA—messenger RNA (mRNA). The "one gene–one ribosome–one protein" hypothesis was invalidated.

Hence mRNA is the blueprint of the genetic information encoded in the gene. The sequence of bases in the mRNA is translated into a sequence of amino acids. There is an initiating codon composed of three bases (AUG) that signals the start of a polypeptide chain. After the initiating triplet, the bases are read in groups of three, in sequence, until a chain-terminating codon is reached (UAA, UGA, or UAG), at which time the polypeptide is released from the mRNA and ribosome.

rRNA and Ribosomes Before the discovery of mRNA there was, as noted earlier, a great deal of interest in ribosomes and rRNA, because so many investigators felt that all the genetic information was transcribed into rRNA. This was a reasonable suggestion; 80 percent of the cellular RNA is rRNA (tRNA accounts for about 15 percent and mRNA for about 5 percent). However, the discovery of mRNA in 1961 eclipsed the study of ribosomes until quite recently.

In 1959, K. McQuillen, R. B. Roberts, and R. J. Britten demonstrated that ribosomes are subcellular particles that serve as the sites of protein synthesis. These investigators grew *E. coli* cells for 15 seconds in a medium containing the radioactive label ^{35}S. This short pulse of ^{35}S labeled proteins in the process of synthesis. The ^{35}S was washed out using normal ^{32}S, and the cells were broken open after either 5 or 120 seconds. The broken cells were then centrifuged in a density gradient to separate cell walls, membranes, ribosomes, and molecules of various sizes. The results showed that at 5 seconds, some ^{35}S associates with ribosomes; but at 120 seconds, all the ^{35}S label is found in free protein molecules (Figure 8.33). One could conclude from this that protein is made on ribosomes and that it moves away from the ribosomes after synthesis.

Figure 8.33
McQuillen–Roberts–Britten experiment showing that new protein is synthesized on ribosomes and then moves away

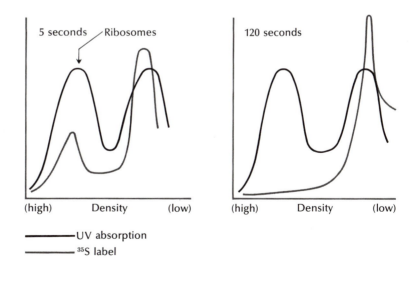

Structurally, *E. coli* ribosomes, the most extensively investigated ribosomes, are composed of two subunits—a 30*S* subunit and a 50*S* subunit (the *S* being the Svedberg coefficient, a measure of sedimentation velocity under standard conditions). Eukaryote ribosomes are composed of slightly larger subunits (40*S* and 60*S*), but since most of the research has been carried out on *E. coli* ribosomes, we shall base our discussion primarily on them. The complete *E. coli* ribosome has an *S* value of 70 (*S* values are not additive). Each subunit has its own unique composition

Figure 8.34
Structure of the *E. coli* and eukaryote ribosomes

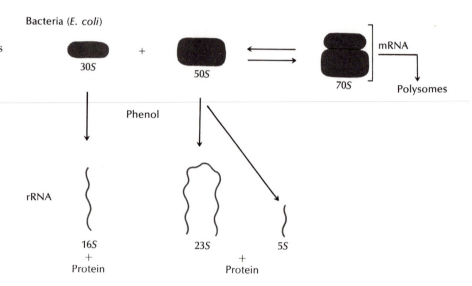

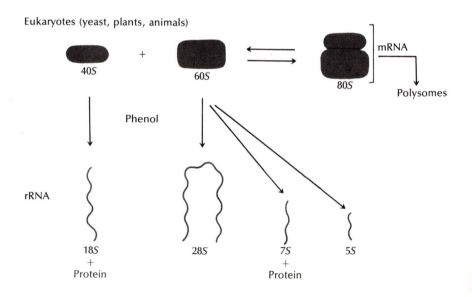

(Figure 8.34). The 30S subunit contains one molecule of 16S rRNA (~1500 nucleotides) and 21 different protein molecules. The 50S subunit contains one molecule of 23S rRNA (~3000 nucleotides), one molecule of 5S rRNA, and about 35 different protein molecules. Figure 8.34 also shows the composition of the eukaryote ribosome.

The function of rRNA is not really known. It is fairly clear now that it is not translated into protein. Our most reasonable suggestion is based on rRNA structure. Like tRNA, rRNA has many unusual bases and is folded in such a way as to produce regions of single-strandedness and regions of double-strandedness. The characteristic folding pattern with the specific ribosomal proteins result in an interaction, the outcome of which is the ribosome structure. The rRNA that is part of that structure almost certainly contributes to the binding of tRNA and mRNA to the ribosome during protein synthesis.

All three rRNA species (16s, 23s, and 5s) appear to be generated from a single larger 30s rRNA species. This 30s rRNA molecule is cleaved after transcription into the smaller rRNA molecules. The tRNA molecules are also generated from larger pre-tRNA molecules that go through a process of cleavage and base modification after transcription. The mRNA molecules of eukaryotes also seem to undergo a precise pattern of cleavages after transcription. These post-transcriptional cleavages are tightly regulated, and we shall explain them further in Chapter 13, when we discuss genetic regulation.

Note that we have rRNA and tRNA genes that are transcribed into RNA but not translated into protein. This emphasizes the inadequacy of the one gene–one protein idea.

The existence of the two subunits of the ribosome was initially difficult to explain. In 1960, A. Tissières suggested that ribosomes normally split into subunits after releasing the polypeptide chain but that the mechanism for such disjunction was not clear. In 1964, Watson concluded that ribosomes necessarily undergo a cycle of dissociation and reassociation during protein synthesis and that determining the ratio of 70S ribosomes to 30S + 50S subunits would give an important indication of the amount of protein synthesis. This idea was later confirmed when a small sample of density-labeled "heavy" ribosomes was mixed with an excess of non-labeled "light" ribosomes in an *in vitro* protein-synthesizing system. After a short period of time during which protein synthesis occurred, the ribosomes were isolated and their density determined. All the ribosomes were either light or hybrid (a mixture of heavy and light subunits). This argues convincingly that ribosomal subunits separate and combine during the process of protein synthesis. But if protein synthesis is inhibited, the subunits do not separate and recombine, so protein synthesis is critical to this subunit recycling.

Today, much research is devoted to the problem of ribosome cycling, or the ribosome life cycle. This will be explored later in more detail, but for

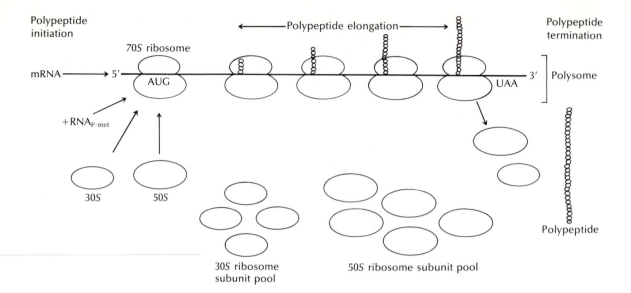

Polypeptide initiation

Polypeptide elongation

Polypeptide termination

70S ribosome

mRNA 5' AUG UAA 3' Polysome

+RNA_{F-met}

30S 50S

30S ribosome subunit pool

50S ribosome subunit pool

Polypeptide

Figure 8.35
Protein synthesis: a general view

now the generalized cycle scheme shown in Figure 8.35 will suffice. Ribosomes, far from being "simple" sites of protein synthesis, appear to have many functions.

1. They recognize a polypeptide initiation site on an mRNA.
2. They influence the accuracy of the codon–anticodon recognition and interaction.
3. They house within their structure sites for the binding of aminoacyl-tRNA.
4. They provide various protein factors involved in the initiation, transfer, and release of polypeptides.
5. They catalyze the reaction that forms the peptide bond between adjacent amino acids.
6. They move along the mRNA during translation.

Finally, just as mutations can occur in the genes coding for tRNA with a concomitant alteration of protein-synthesizing potential, similar consequences arise from mutations occurring in the genes that determine ribosome structure; that is, they can alter protein-synthesizing capacity. Resistance to drugs such as streptomycin is the result of a mutation in a gene that codes for a ribosomal protein.

We have thus far discussed the major components of the protein-synthesizing system: tRNA, rRNA and ribosomes, and mRNA. But many other factors are necessary to protein synthesis (see Table 8.5). These factors, in conjunction with the three types of RNA, participate in the synthesis of proteins, a matter we shall now investigate.

Protein synthesis For the sake of clarity and organization, the process of protein synthesis will be developed around three areas, arbitrarily partitioned from each

Figure 8.36
Structure of
N-formyl-methionine and
methionine

N-formyl-methionine   Methionine

other but in fact forming a continuum: polypeptide initiation, polypeptide elongation, and polypeptide termination. This area of genetic research has seen great progress in recent years, but the information presented here, although probably correct in its general outline, is not meant to foster the illusion that the mechanisms of protein synthesis are completely understood. Indeed, the details of many of the steps are poorly understood.

Most of the work discussed here will once again be concerned with *E. coli* since protein synthesis studies have focused on that organism.

Polypeptide initiation   The initiation of polypeptide biosynthesis involves interaction among mRNA, the 30S subunit of the ribosome, a special initiating species of aminoacyl-tRNA that apparently starts all polypeptide chains, and three initiating protein factors (IF$_1$, IF$_2$, and IF$_3$) that promote the binding of the three basic components. Let us examine this initiation complex and, more specifically, the role of the initiating tRNA.

About 40 percent of all completed *E. coli* proteins begin with the amino acid methionine (met), but methionine accounts for only about 2 percent of the total amino acids found in *E. coli* proteins. This suggests that methionine plays a key role in initiation of the protein synthesis. Further examination of *E. coli* proteins has shown that all of them are usually initiated by an analog of methionine called *N-formyl-methionine* (F-met), which is methionine with a formyl group attached (Figure 8.36). Since all proteins begin with F-met and since only 40% have methionine as the initial amino acid in their final structure, the formyl group must be removed from methionine in some cases, and in other cases F-met must be removed altogether. Separate enzymes that catalyze both reactions are known (Figure 8.37), but what determines whether F-met is merely de-

Figure 8.37
Loss of formyl group and
N-terminal methionine
during polypeptide
elongation

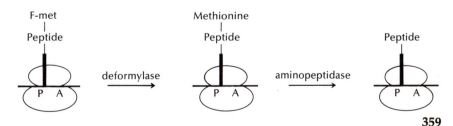

359

formylated or deformylated and then removed is not known. In eukaryotes, protein synthesis is initiated by methionine, not formyl methionine.

The existence of F-met raises the following question: Is there only one

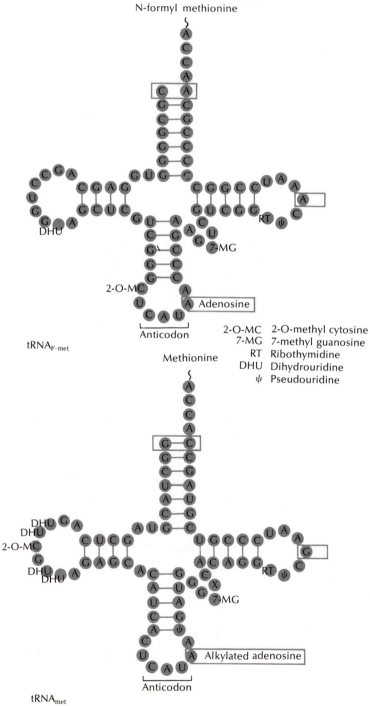

2-O-MC 2-O-methyl cytosine
7-MG   7-methyl guanosine
RT   Ribothymidine
DHU   Dihydrouridine
ψ   Pseudouridine

Figure 8.38
The main points of difference between F-met–tRNA and met-tRNA

species of tRNA$_{met}$, in which case it would be specific for the two kinds of amino acids, or are there two tRNA species, one for each amino acid? Evidence suggests the existence of two species of tRNA, tRNA$_{met}$ and tRNA$_{F-met}$, that have identical anticodons but differ in some of the other bases (Figure 8.38). A methionine attached to tRNA$_{met}$ cannot be formylated, whereas one attached to tRNA$_{F-met}$ can, an event that occurs *after* attachment of methionine to the tRNA$_{F-met}$.

The formation of F-met–tRNA$_{F-met}$ is summarized as follows.

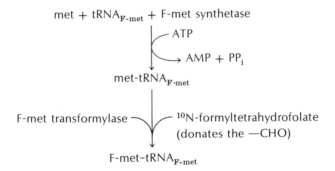

(See Figure 8.39.) With the formation of F-met–tRNA$_{F-met}$, a *30S initiation complex* can be formed.

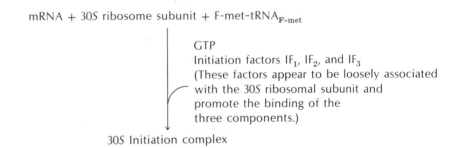

(See Figure 8.40.) The F-met–tRNA$_{F-met}$ associates with a specific codon, AUG, on the mRNA. (Chapter 9 will paint a more incisive portrait of genetic codons, but it seems worth noting here that AUG is an initiating codon for F-met, although this same codon—if found in an interior position in the coding sequence—codes for met-tRNA$_{met}$.) It now appears that there are specific base sequences on the mRNA that are *ribosome binding sequences* to which the 30S ribosome subunit binds. These sequences occur before the initiating codon, and there is one for each gene coded for on the mRNA.

Following formation of the initiation complex, the 50S ribosomal subunit attaches to the 30S subunit to form the 70S initiation complex. Two sites exist on the 50S ribosome subunit—one for the attachment of the amino acid–tRNA (called the aminoacyl, or *A site*) and one for the

Figure 8.39
Enzymatic formation of
F-met–tRNA$_F$ (E$_{met}$ is
methionyl-tRNA
synthetase)

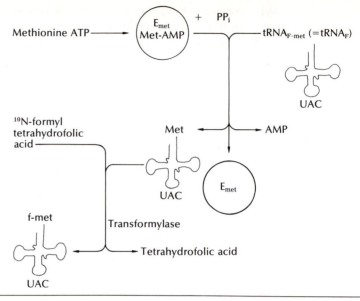

Figure 8.40
The polypeptide initiation
process

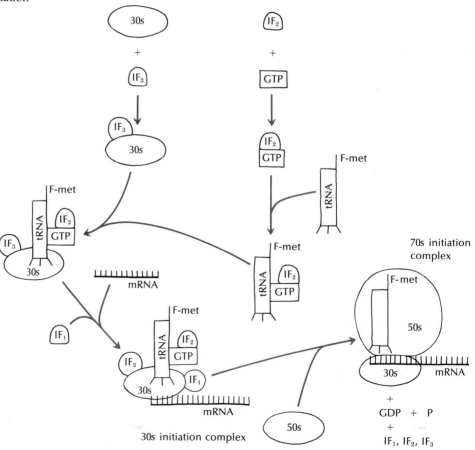

growing peptide chain attached to tRNA. The latter site is called the peptidyl, or *P site*. It currently appears that F-met–tRNA$_{F\text{-met}}$ sits in the P site. The initiation phase of protein synthesis ends with the formation of the initiation complex.

Polypeptide elongation When the A site is occupied by an incoming amino acid–tRNA and peptide bond formation is initiated between F-met and the incoming amino acid, the process of polypeptide elongation begins (Figure 8.41). The amino acid–tRNA that binds to the A site is specified by the triplet codon of the mRNA that occupies the A site on the ribosome.

The significance of F-met is seen in the formation of the peptide bond: The formylated amino group of F-met cannot participate in peptide bond formation, but the COOH group can. Thus, of necessity, F-met can serve as the amino acid only at the *beginning,* or amino terminal end of a polypeptide chain. To be an interior amino acid, it would have to be able to participate in two peptide linkages.

Several factors not associated with the ribosome are required for binding an incoming amino acid–tRNA to the A site and subsequent polypeptide elongation. Some of these factors are T_u, T_s, G, and GTP. The T_u complexes with the amino acid–tRNA and GTP and promotes the binding of the amino acid–tRNA to the A site. The GTP is cleaved to GDP and P in the process. The T_s catalyzes an exchange reaction.

$$T_u\text{-GDP} + \text{GTP} \xrightarrow{T_s} T_u\text{-GTP} + \text{GDP}$$

thus making more T_u-GTP available for additional amino acid–tRNA binding to the A site. Similar reactions are known to occur in eukaryotes, the difference being that only one factor—EF$_1$, which is equivalent in activity to T_u and T_s—is known at present.

Figure 8.41
Elongation of the peptide chain

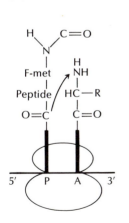

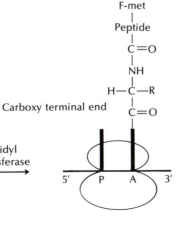

Amino terminal end

Carboxy terminal end

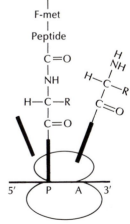

The next step in the elongation process is the formation of a peptide bond between F-met and the amino acid occupying the A site. This is catalyzed by an enzyme called *peptidyl transferase,* an integral part of the 50S ribosomal subunit. On completion of peptide bond formation, the A site is occupied by F-met–amino acid 2–tRNA$_{amino\ acid\ 2}$, and the P site is occupied by tRNA$_{F\text{-}met}$.

The *translocation* step follows. The tRNA$_{F\text{-}met}$ in the P site is discharged from the ribosome, the F-met-amino acid 2–tRNA$_{amino\ acid\ 2}$ is shifted from the A to the P site as the ribosome moves the length of one codon along the mRNA in a 5′ to 3′ direction, thus bringing a new codon to the A site. The *transfer factor,* G (EF$_2$ in eukaryotes), is essential to this translocation.

After translocation, amino acid 3–tRNA$_{amino\ acid\ 3}$ moves in to occupy the vacated A site; this is followed by peptide bond formation and translocation. Thus elongation of the polypeptide chain continues, one amino acid at a time, until a termination signal (a codon that specifies no amino acid) is reached and the completed polypeptide is released from both the ribosome and the tRNA. Hence we see that the mRNA is read as it is synthesized, in a 5′ to 3′ direction. The polypeptide chain is assembled beginning at the amino end (—NH$_2$) and ending at the carboxyl end (—COOH).

The process of chain elongation can be terminated prematurely by puromycin, a potent antibiotic. The drug resembles the 3′ end of a tRNA with an amino acid bound to it. It can occupy the A site and accept a growing polypeptide chain, but when the time comes for translocation to the P site, it detaches because it cannot be transferred. This results in the premature termination of a polypeptide chain—and, ultimately, cell death—making puromycin an effective agent for combating bacterial infection.

Polypeptide termination   As a polypeptide chain grows, it remains linked covalently to tRNA and bound to the ribosome. When complete, the polypeptide is released from both components. It appears that termination is triggered when, in the course of movement of the ribosome along the mRNA, a chain termination signal is reached at the A site of the 50S subunit. The termination signals are the nonsense codons UAA, UGA, and UAG.

Chain termination requires the participation of three *release-factor* proteins—RF$_1$, RF$_2$, and RF$_3$—in addition to the nonsense codons and

Figure 8.42
Schematic representation of the polypeptide release process

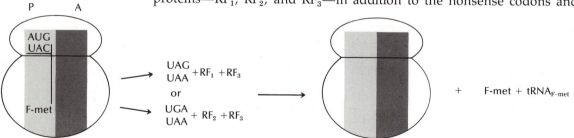

Figure 8.43
The polypeptide
termination process

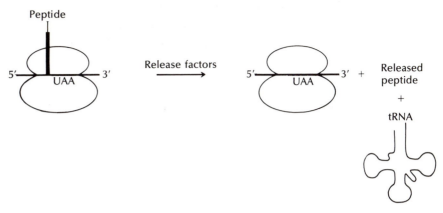

GTP. The RF_1 is required for the nonsense codon UAG, whereas RF_2 is required for UGA. The UAA can accept either RF_1 or RF_2. The RF_3 stimulates, or activates, the activity of RF_1 and RF_2 (Figure 8.42). For successful termination, there must be a peptidyl-tRNA on the P site and a termination codon on the A site (Figure 8.43).

The state of the ribosome at the completion of protein synthesis is a subject of current research. Figure 8.44 summarizes two concepts of the state of the ribosomes at the termination of protein synthesis. In Figure 8.44(a), chain termination is simultaneously accompanied by dissociation of the 70S ribosome into 30S and 50S subunits. To account for the presence of the 70S subunits in *E. coli* extracts, the 30S–50S subunit pool is said to be in equilibrium with 70S ribosomes. In Figure 8.44(b), at termination the 70S ribosome remains intact for a period of time before it dissociates into 30S and 50S subunits; 70S ribosomes here exist only during

Figure 8.44
Alternative models for the role of "free" 70S ribosomes in the ribosome cycle *in vivo*

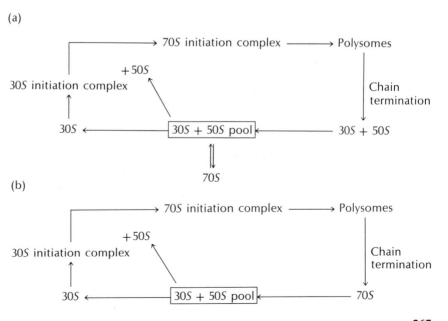

protein synthesis and shortly thereafter. Which of these models is correct has yet to be determined. (It may even be that both are correct.) The F_3 factor, important in the initiation reaction, is also required for the dissociation of the ribosome into 30S and 50S subunits.

The polyribosome Protein synthesis is an extremely efficient process. If only one ribosome attached to a single mRNA, about 90 percent of the cell's ribosomes would be idle—a most inefficient process. Instead a single mRNA usually has several ribosomes attached to it, as many as 1 for every 90 nucleotides. This mRNA with its attached ribosomes is called a *polyribosome,* or *polysome*. Ribosomes move in a 5' to 3' direction, synthesizing polypeptides in an H_2N to COOH direction. RNA transcription and translation can occur simultaneously (Figure 8.45).

Overview

Once it was established that genes produce proteins, it became necessary to describe the process by which the information encoded in DNA is translated into protein. The discovery of RNA and its identification as an intermediary in the process of protein synthesis led to the establishment of a central dogma, which suggested that there was a one-way flow of information from nucleic acids to protein. The study of RNA and its synthesis led to the discovery that three types of RNA participate in protein synthesis: messenger RNA (mRNA), transfer RNA (tRNA), and ribosomal RNA (rRNA). These three types interact with other factors for protein synthesis in a three-step process: initiation of protein synthesis,

Figure 8.45

The polyribosome in protein synthesis

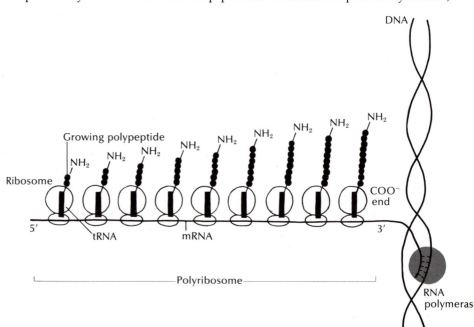

Figure 8.46
Protein synthesis: detailed outline

elongation of the protein molecule, and termination of protein synthesis. The process of protein synthesis, schematized superficially in Figure 8.35, is summarized in more detail in Figure 8.46.

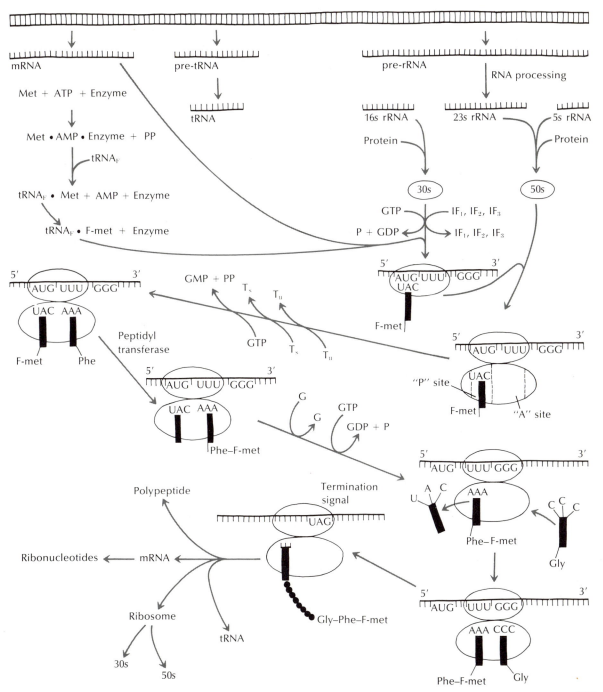

8.1 Describe experiments that demonstrate that proteins are not made directly from a DNA template.

8.2 If a cell is deprived of a source of thymine, it soon stops synthesizing DNA but continues synthesizing RNA and protein. How would you interpret this?

8.3 It has been shown that in the absence of protein synthesis, RNA synthesis continues temporarily and then ceases. How would you interpret this?

8.4 What is the significance of the ribosome cycle ($70S$ to $30S + 50S$ to $70S$)?

8.5 What is the function of the nonsense codons in protein synthesis?

8.6 Design an experiment that would show RNA to be single-stranded.

8.7 Design an experiment that would establish the existence of tRNA genes.

8.8 It is an established fact that proteins make nucleotides, fats, carbohydrates, vitamins, and amino acids. But proteins do not make themselves. What is the experimental proof that proteins do not make themselves?

8.9 Concerning the gene, H. J. Muller in 1937 stated that "the copying property depends upon some more fundamental feature of gene structure than does the specific pattern which the gene has and that it is the effect of the former to cause a copying not only of itself, but also of the latter, more variable features." What would you interpret today as
  (a) "Copying property"
  (b) "Fundamental feature of gene structure"
  (c) "Specific pattern"
  (d) "Variable features"

8.10 In the bacteriophage T4 there is a gene that codes for the synthesis of a head protein. There are 28 mutations known to occur in that gene. Four of these mutations are nonsense mutations causing the premature termination of the polypeptide chain. Genetic crosses between these four mutants gave the following recombination frequencies.

$A \times B = 0.002$
$C \times B = 0.003$
$D \times C = 0.002$
$A \times C = 0.005$

Assume that mutant D is closest to the origin point of transcription at the 3' end of the transcribed DNA strand.
  (a) What is the sequence of these four mutants relative to the direction of transcription?
  (b) Assuming colinearity, arrange the gene products from A, B, C, and D mutants in order of increasing size.

8.11 What is the sequence of bases in an mRNA molecule synthesized from the following DNA template strand.

5' A T T G C C A T G C T A 3'

8.12 Ribosomes were extracted from *E. coli* cells grown on ^{13}C-^{15}N medium ("heavy"), and from cells grown on ^{12}C-^{14}N medium ("light"). A sample of the "heavy" and "light" ribosomes was mixed and used in an *in vitro* protein-synthesizing system. After protein synthesis had progressed for a while, the ribosomes were again isolated and centrifuged in a density gradient. Describe the classes of ribosomes you would expect.

8.13 Why is the tRNA molecule the most functionally complex of all the translational elements?

8.14 What is colinearity and why is it so important?

8.15 Distinguish between transcription and translation.

References

Attardi, G., and F. Amaldi. 1970. Structure and synthesis of rRNA. *Ann. Rev. Biochem.* 39:183–226.

Adhya, S. and M. Gottesman. 1978. Control of transcription termination. *Ann. Rev. Biochem.* 47:967–996.

Baltimore, D. 1971. Expression of animal virus genomes. *Bact. Rev.* 35:235–241.

Bautz, E. K. F., and L. Heding. 1964. *Biochem.* 3:1010.

Beadle, G. W., and B. Ephrussi. 1937. Development of eye colors in *Drosophila:* Diffusable substances and their interrelations. *Genetics* 22:76–86.

Beadle, G. W., and E. L. Tatum. 1941. Genetic control of biochemical reactions in *Neurospora. Proc. Natl. Acad. Sci. U.S.* 27:499–506.

Bermek, E. 1978. Mechanisms in polypeptide chain elongation on ribosomes. *Prog. Nuc. Acid Res. Molec. Biol.* 21:64–100.

Brawerman, G. 1974. Eukaryotic mRNA. *Ann. Rev. Biochem.* 43:621–642.

Brenner, S., F. Jacob, and M. Meselson. 1961. An unstable intermediate carrying information from genes to ribosomes for protein synthesis. *Nature* 190:576.

Brimacombe, R., K. H. Nierhaus, R. A. Garrett, and H. G. Wittmann. 1976. The ribosome of *E. coli. Prog. Nuc. Acid Res. Molec. Biol.* 18:1–44.

Burgess, R. R. 1971. RNA polymerase. *Ann Rev. Biochem.* 40:711–740.

Burgess, R. R., A. A. Travers, J. J. Dunn, and E. K. F. Bautz. 1969. Factor stimulating transcription by RNA polymerase. *Nature* 221:43–46.

Capecchi, M. R. 1966. Initiation of *E. coli* proteins. *Proc. Natl. Acad. Sci. U.S.* 55:1517–1524.

Chakrabarti, S. L., and L. Gorini. 1977. Interaction between mutations of ribosomes and RNA polymerase: A pair of *str* A and *rif* mutants individually temperature-insensitive, but temperature-sensitive in combination. *Proc. Natl. Acad. Sci. U.S.* 74:1157–1161.

Chamberlin, M. J. 1974. The selectivity of transcription. *Ann. Rev. Biochem.* 43:721–775.

Chamberlain, M., J. McGrath, and L. Waskell. 1970. New RNA polymerase from *E. coli* infected with bacteriophage T7. *Nature* 228:227–231.

Chambers, R. E. 1971. On the recognition of tRNA by its aminoacyl-tRNA ligase. *Prog. Nuc. Acid Res. Molec. Biol.* 11:489–525.

Chambon, P. 1975. Eukaryotic nuclear RNA polymerases. *Ann. Rev. Biochem.* 44:613–638.

Clark, B. F. C. 1977. Correlation of biological activities with structural features of tRNA. *Prog. Nuc. Acid Res. Molec. Biol.* 20:1–19.

Corwin, H. O. and J. B. Jenkins. 1976. *Conceptual Foundations of Genetics.* Boston, Houghton Mifflin.

Crick, F. H. C. 1958. On protein synthesis. *Symp. Soc. Exp. Biol.* 12:138.

Davies, J., and M. Nomura. 1972. The genetics of bacterial ribosomes. *Ann. Rev. Genetics* 6:203–234.

Garrod, A. E. 1909. *Inborn Errors of Metabolism.* Oxford, England, Oxford University Press.

Geiduschek, E. P., and R. Haselkorn. 1969. Messenger RNA: *Ann. Rev. Biochem.* 38:647–676.

Grunberg-Manago, M. and F. Gros. 1977. Initiation mechanisms in protein synthesis. *Prog. Nuc. Acid Res. Molec. Biol.* 20:209–284.

Haselkorn, R., and L. B. Rothman-Denes. 1973. Protein synthesis. *Ann. Rev. Biochem.* 42:397–438.

Imamoto, F. 1973. Translation and transcription of the tryptophan operon. *Prog. Nuc. Acid Res. Molec. Biol.* 13:339–408.

Ingram, V. 1972. *Biosynthesis of Macromolecules.* New York, Benjamin.

Jacob, F., and J. Monod. 1961. Genetic regulatory mechanisms in the synthesis of proteins. *J. Molec. Biol.* 3:318.

Jacob, F., and J. Monod. 1961. On the regulation of gene activity. *Cold Spring Harbor Symp. Quant. Biol.* 26:193.

Jacob, S. T. 1973. Mammalian RNA polymerases. *Prog. Nuc. Acid Res. Molec. Biol.* 13:93–126.

Jayaraman, R., and E. B. Goldberg. 1969. A genetic assay for mRNA's of phage T4. *Proc. Natl. Acad. Sci. U.S.* 64:198–204.

Josse, J., A. D. Kaiser, and A. Kornberg. 1961. *J. Biol. Chem.* 236:864.

Leninger, A. L. 1970. *Biochemistry.* New York, Worth.

Lodish, H. F. 1976. Translational control of protein synthesis. *Ann. Rev. Biochem.* 45:39–72.

McQuillen, K., R. B. Roberts, and R. J. Britten. 1959. Synthesis of nascent protein by ribosomes of *E. coli. Proc. Natl. Acad. Sci. U.S.* 45:1437.

The Mechanism of Protein Synthesis. 1970. Cold Spring Harbor Symp. Quant. Biol. 34. Cold Spring Harbor, N.Y., Cold Spring Harbor Laboratory of Quantitative Biology.

mRNA: The Relation of Structure to Function. 1976. *Progress in Nucleic Acid Research and Molecular Biology,* Vol. 19 (proceedings of a symposium).

Noll, M., and H. Noll. 1972. Mechanism and control of initiation in the translation of R17 RNA. *Nature (New Biology)* 238:225–228.

Palade, G. 1975. Intracellular aspects of the process of protein synthesis. *Science* 189:347–358.

Pestka, S. 1976. Insights into protein synthesis and ribosome function through inhibitors. *Prog. Nuc. Acid Res. Molec. Biol.* 17:217–246.

Pnina, S-E. and D. Elson. 1976. Studies on the ribosome and its components. *Prog. Nuc. Acid Res. Molec. Biol.* 17:77–98.

Quigley, G. J., and A. Rich. 1976. Structural domains of transfer RNA molecules. *Science* 194:796–806.

Raacke, I. D. 1971. *Molecular Biology of DNA and RNA: An Analysis of Papers.* St. Louis, Mo., Mosby.

Rich, A., and U. L. RajBhandary. 1976. Transfer RNA: Molecular structure, sequence, and properties. *Ann. Rev. Biochem.* 45:805–860.

Roberts, J. W. 1969. Termination factor for RNA synthesis. *Nature* 224:1168–1174.

Schlessing, D. 1969. Ribosomes: Development of some current ideas. *Bact. Rev.* 33:445–453.

Smith, J. D. 1972. The genetics of tRNA. *Ann. Rev. Genetics* 6:235–256.

Smith, J. D. 1976. Transcription and processing of tRNA precursors. *Prog. Nuc. Acid Res. Molec. Biol.* 16:25–74.

Söll, D. 1971. Enzymatic modification of tRNA. *Science* 173:293–299.

Spiegelman, S. 1964. Hybrid nucleic acids. *Sci. Amer.* (May).

Stent, G. S. 1971. *Molecular Genetics: An Introductory Narrative.* San Francisco, Freeman.

Sussman, J. L., and S. H. Kim. 1976. Three-dimensional structure of a transfer RNA in two crystal forms. *Science* 192:853–858.

Thompson, R. C., and P. J. Stone. 1977. Proofreading of the codon–anticodon interaction on ribosomes. *Proc. Natl. Acad. Sci. U.S.* 74:198–202.

Tissières, A., D. Schlessing, and F. Gross. 1960. Amino acid incorporation into proteins by *E. coli* ribosomes. *Proc. Natl. Acad. Sci. U.S.* 46:1450.

Transcription of Genetic Material. 1971. Cold Spring Harbor Symp. Quant. Biol. 35. Cold Spring Harbor, N.Y., Cold Spring Harbor Laboratory of Quantitative Biology.

Volkin, E., and L. Astrachan. 1957. RNA metabolism in T2-infected *Escherichia coli.* In W. D. McElroy and B. Glass (eds.), *The Chemical Basis of Heredity.* Baltimore, Md., Johns Hopkins University Press.

Von Ehrenstein, G., B. Weisblum, and S. Benzer. 1963. The function of sRNA as amino acid adapter in the synthesis of hemoglobin. *Proc. Natl. Acad. Sci. U.S.* 49:669–675.

Wagner, R. P., and H. K. Mitchell. 1964. *Genetics and Metabolism.* New York, Wiley.

Watson, J. D. 1964. Involvement of RNA in the synthesis of proteins. *Science* 140:17–26.

Watson, J. D. 1976. *Molecular Biology of the Gene.* New York, Benjamin.

Weinberg, R. A. 1973. Nuclear RNA metabolism. *Ann. Rev. Biochem.* 52:329–354.

Yanofsky, C., B. C. Carlton, J. R. Guest, D. R. Helinski, and U. Henning. 1964. On the colinearity of gene structure and protein structure. *Proc. Natl. Acad. Sci. U.S.* 51:266–272.

Yanofsky, C., and J. Ito. 1966. Nonsense codons and polarity in the tryptophan operon. *J. Molec. Biol.* 21:313.

Zubay, G. 1973. *In vitro* synthesis of protein in microbial systems. *Ann. Rev. Genetics* 7:267–287.

Zubay, G. 1973. *Papers in Biochemical Genetics.* New York, Holt.

Nine

The genetic code

The genetic code is concerned with the processes involved in translating, or decoding, the information contained in the primary structure of DNA. In Chapter 8 on gene function, we discussed the flow of genetic information from gene (DNA) to RNA to protein, which is essential to elucidating the *mechanisms* of protein synthesis but goes only part of the way toward defining the *problems* of the genetic code.

The processes of the genetic code are the basis of life itself. That they have a mystical aura about them probably reflects the feeling that we are dealing with fundamental secrets of nature. These processes also raise fascinating questions. Which DNA code words specify which specific amino acids? And why?

The concept of the genetic code emerged in stages. First came the realization that a phenotype is a manifestation or expression of the genotype. This was followed by the discovery that genes usually direct the synthesis of proteins and that the gene itself is DNA. The relationship between DNA and protein was in a sense simplified when it was demonstrated that the secondary, tertiary, and quaternary structure of proteins are in essence defined by their primary structure. Thus one task stemming from the genetic code was to describe how the primary structure of DNA was translated into the primary structure of a polypeptide.

The approaches to describing the concept of the genetic code can be divided into three periods. The first is the pre-1961 era, when the approaches were mainly theoretical. Here it was hoped that clues to the nature of the code might emerge from the knowledge that there are 20 amino acids and only 4 DNA (or RNA) bases. The second period was that from 1961 to 1966, when the more theoretical and mathematical approaches gave way to experimental efforts to define the nucleic acid code words (*codons*) that determined specific amino acids. The third period is not yet over, nor is it likely to end in the near future. In it, we hope to witness the true "solving" of the genetic code—full elucidation of the basis and origin of the entire DNA–RNA–protein scheme. This chapter will concern itself with these three periods.

Theoretical approaches to the code

The first period we shall discuss is the theoretically oriented pre-1961 period, which was dominated by the ideas of the well-known physicist and cosmologist, George Gamow. This period led to the establishment of a conceptual framework that influenced and stimulated subsequent experimental efforts. The ideas put forth by Gamow in 1954 and 1955, soon after publication of the work by Watson and Crick, were later shown to be largely incorrect, but this does not detract from their value. As we have seen, the advancement of scientific principles often proceeds from incorrect but testable hypotheses.

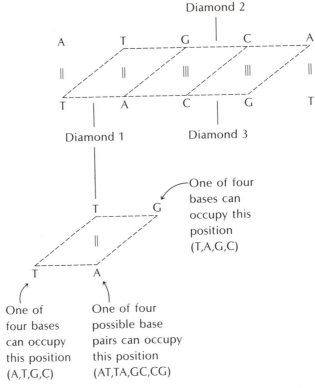

Figure 9.1
Gamow's diamond code

The Gamow era   Gamow formulated the first comprehensive scheme for a genetic code. Stimulated by research into the structure of DNA as set out by Watson and Crick, he devised what is referred to as the *diamond code*. Gamow visualized diamond-shaped pockets in the DNA duplex formed by a base on one DNA chain, an adjacent base pair spanning both chains, and the next adjacent base beyond the base pair on the opposite chain (Figure 9.1). He hypothesized that the amino acids, or more specifically their R groups, were fitted into the various diamonds and, subsequently, linked enzymatically to form polypeptides.

With this type of model, 64 different diamonds are possible. However, assuming that the rotation of the base pairs in a DNA double helix through 180° would not change the character of the diamond, the 64 possible diamonds are reduced to 20 categories—exactly the number of amino acids that must be accounted for! Figure 9.1 sets forth the rationale for Gamow's model. First, if the bases in diamond (1) are rotated 180°, we get

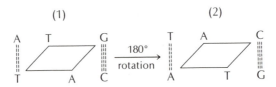

This type of rotation, then, reduces our 64 diamonds to 32 pairs of *equiva-lent* diamonds. Assuming further that

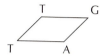

is equivalent to

the 64 diamonds reduce to 20 categories: 8 containing 2 equivalent diamonds each (these would be the diamonds with identical bases at the left and right corners of the diamond) and 12 containing 4 equivalent diamonds each. An example of a category containing 4 equivalent diamonds follows.

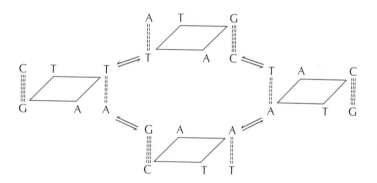

A key feature of Gamow's diamond code is its overlapping nature: Each diamond shares 2 bases with the diamonds immediately adjacent to it. In 1957, Brenner showed that a completely overlapping code was un-likely. He reasoned first that a sequence of 3 bases coded for 1 amino acid. This would give 64 possible sequences of 3 ($4 \times 4 \times 4$), whereas a se-quence of 2 bases would give only 16—not enough to account for all the amino acids. Assuming that 3 bases specified 1 amino acid, then in the nucleotide sequence ATGCA, an overlapping code would specify 3 amino acids (aa): aa_1 from ATG, aa_2 from TGC, and aa_3 from GCA. In this sequence of aa_1-aa_2-aa_3, certain predictions can be made from the overlapping code. If TGC codes for aa$_2$, there are only 4 possible amino acids that can form its left neighbor and only 4 on the right:

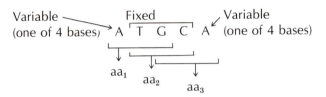

Possibilities for aa$_1$ with TG fixed = 4
Possibilities for aa$_3$ with GC fixed = 4

In other words, with any 1 amino acid forming the middle of a tripeptide, only 16 possible tripeptides could exist with that one in the middle. When all available data on amino acid sequences were analyzed, it was found that the tripeptide sequences with a fixed central amino acid far exceeded the theoretical number of 16. The finding that amino acid sequences are almost statistically random and that all possible nearest-neighbor relations occur forced Brenner to reject either the idea of an overlapping code (and hence Gamow's model) or the triplet nature of the code. He chose to reject Gamow's model.

Although Gamow was essentially wrong—in that the code is not overlapping and in that amino acids do not form polypeptides directly from a DNA template—he did describe a number of general features of the genetic code that remain valid. He established coding units (later to be called *codons*) in which each codon was constant in size and specified an amino acid. He also hypothesized colinearity between codons and amino acids. And he suggested that the code was degenerate (having more than one codon per amino acid) and was probably common to all organisms.

When the intermediary role of RNA in protein synthesis was discovered, Gamow suggested another model, the *triangle code*. This is based on single-stranded RNA that assumes triangular configurations with 3 bases in each triangle. Gamow grouped the total number of possible trinucleotide RNA sequences (64) into 3 classes and 20 categories without regard to the sequence of bases within any particular codon (Table 9.1). In the first class, each codon specifies 1 amino acid; in the second, each amino acid has 3 codons; and in the third, each amino acid has 6 codons. The degeneracy in this model is readily apparent. Although closer to the truth, the second model is also wrong, primarily because the *sequence* of nucleotides in a codon cannot be disregarded.

The adaptor hypothesis and the comma-free code Crick (1957) disagreed with Gamow's proposals and presented his own ideas, which supplanted Gamow's. Crick rejected the idea that the DNA molecule could act as a direct template for amino acids, because the required hydrogen-bonding potentialities were missing. He suggested that

Class I codons: nondegenerate, coding 4 amino acids					
	1. AAA				
	2. CCC				
	3. GGG				
	4. UUU				

Class II codons: triply degenerate, coding 12 amino acids								
1. AAC	ACA	CAA			**7.** GGA	GAG	AGG
2. AAG	AGA	GAA			**8.** GGC	GCG	CGG
3. AAU	AUA	UAA			**9.** GGU	GUG	UGG
4. CCA	CAC	ACC			**10.** UUA	UAU	AUU
5. CCG	CGC	GCC			**11.** UUC	UCU	CUU
6. CCU	CUC	UCC			**12.** UUG	UGU	GUU

Class III codons: sextuply degenerate, coding 4 amino acids					
1. ACG	AGC	CAG	CGA	GAC	GCA
2. ACU	AUC	CAU	CUA	UAC	UCA
3. AGU	AUG	GAU	GUA	UAG	UGA
4. CGU	CUG	GCU	GUC	UCG	UGC

Table 9.1
Gamow's triangle code

From C. R. Woese, *The Genetic Code: The Molecular Basis for Genetic Expression*, New York: Harper & Row, 1967.

each amino acid was enzymatically joined to a small *adaptor molecule* that would have the necessary hydrogen-bonding capabilities to combine with the nucleic acid template. In other words, each amino acid was fitted with an adaptor to go onto the nucleic acid template (*adaptor hypothesis*). Dounce actually preceded Gamow and Crick in that he proposed the idea of an adaptor in 1952; and he also recognized some of the coding problems. Strangely, though, his work was largely overshadowed by that of Gamow and Crick.

The discovery of tRNA and its role in protein synthesis verified Crick's prophetic speculations about adaptors. However, Crick's rejection of the idea that nucleic acids have no affinity for amino acids has perhaps been harmful in the long run. Nucleic acids, which serve directly as templates for amino acids, may at one time—perhaps billions of years ago—have been important in determining the actual amino acid code words and in the evolution of the protein-synthesizing machinery. Crick's line of thinking may have inhibited thought and experimentation in this direction.

With the verification of tRNA as the adaptor molecule, Crick and his colleagues turned toward interpreting the code words, or codons. Crick dismissed Gamow's overlapping code on the ground that if there was an adaptor sitting on one triplet codon, the bases comprising that codon would not be available to serve as members of other codons. However,

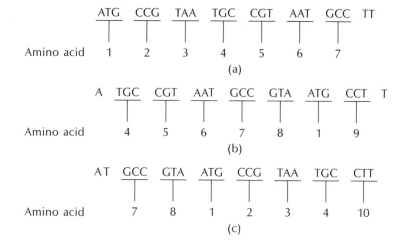

ATG CCG TAA TGC CGT AAT GCC TT

Amino acid   1   2   3   4   5   6   7

(a)

A TGC CGT AAT GCC GTA ATG CCT T

Amino acid   4   5   6   7   8   1   9

(b)

AT GCC GTA ATG CCG TAA TGC CTT

Amino acid   7   8   1   2   3   4   10

(c)

Figure 9.2
A nonoverlapping code can produce three different polypeptide sequences, depending on where translation begins

such a nonoverlapping code presented a serious problem. Translating a sequence of nucleotides comprising a cistron could begin anywhere in that cistron, and translating it in a nonoverlapping way could produce three very different polypeptide sequences (Figure 9.2). There are two possible resolutions of this difficulty. There may be a specific starting codon with synthesis starting from there, or there may be two classes of codons—those that specify amino acids and those that do not.

From Chapter 8, we know that there is a codon (the F-met codon, AUG) that begins each polypeptide chain and hence each cistron, but Crick and his colleagues argued that there existed two categories of codons. At this point, Crick developed his comma-free code.

The construction of the *comma-free code* centered on the concept that no amino-acid-specifying codon can occur as an overlap codon. The 64 possible nucleotide triplets were divided into two groups: codons specifying amino acids and codons specifying no amino acids. Two contiguous amino-acid-specifying codons, when read in the correct sequence, would specify a dipeptide. But if those same two codons were read in a different sequence, nothing would be specified:

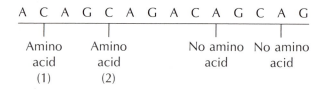

A C A G C A G A C A G C A G

Amino   Amino   No amino   No amino
acid   acid   acid   acid
(1)   (2)

Using concepts developed by information theorists and communications engineers, Crick and his colleagues showed that for a code with these restrictions, the maximum number of triplet codons that could be

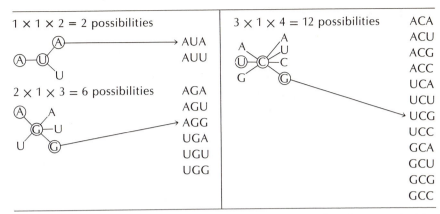

$1 \times 1 \times 2 = 2$ possibilities

$\rightarrow$ AUA
AUU

$2 \times 1 \times 3 = 6$ possibilities

AGA
AGU
$\rightarrow$ AGG
UGA
UGU
UGG

$3 \times 1 \times 4 = 12$ possibilities

ACA
ACU
ACG
ACC
UCA
UCU
$\rightarrow$ UCG
UCC
GCA
GCU
GCG
GCC

Table 9.2
Crick-Griffith-Orgel
model of the genetic code

Circled bases show how some of the codons are derived. Others follow the same principle.

assigned to amino acids was exactly 20 (Table 9.2). Thus the magic number 20 appeared as a consequence of crytographic game-playing, wholly lacking in data from biological systems. This code, because of its intellectual appeal and its magic number 20, was assumed by many to be the correct solution.

The comma-free code was distinguished from Gamow's ideas in three important ways. It was not degenerate, it was not overlapping, and it relied on adaptor molecules. The strong influence of the comma-free code was dissipated in 1961, when M. W. Nirenberg and J. H. Matthaei made the first experimental discovery of an amino-acid-specifying codon not on Crick's list. Between 1961 and 1966, efforts concentrated on finding out what the 64 possible codons specified.

Experimental determination of the genetic code words

After the discovery by Nirenberg and Matthaei, Crick, L. Barnett, Brenner, and R. J. Watts-Tobin presented some new thoughts on the general nature of the genetic code. Their ideas are significant, because they link the theoretical pre-1961 period to the experimental post-1961 period. They also establish a framework within which some aspects of the code can be interpreted.

Crick and his colleagues based their reassessment of the genetic code on their studies of a class of mutations called *frameshift mutations*. These are induced by certain chemicals, such as acridine dyes, that cause DNA bases to be deleted from or added to a DNA strand. If a DNA strand has an addition or a deletion, the mRNA transcribed from it will have a *shifted reading frame*. Translation occurs normally up to the point of addition or

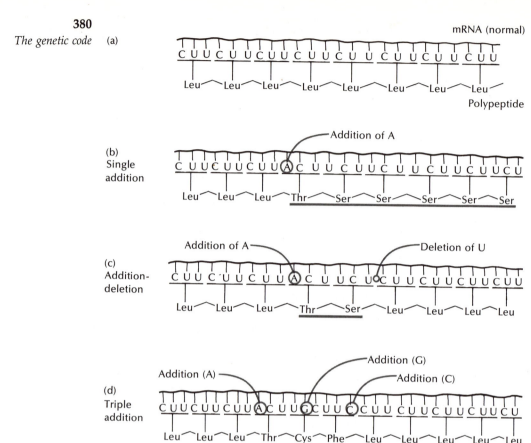

(a)

(b)
Single
addition

(c)
Addition-
deletion

(d)
Triple
addition

Figure 9.3
Frameshift mutations and
the genetic code

Heavy underlining indicates a mutant amino acid sequence

deletion, but all subsequent triplet codons code for incorrect amino acids, perhaps even for chain termination (see Figure 9.3a and 9.3b).

Crick and his colleagues characterized the genetic code as having four basic qualities: It is triplet in nature, or perhaps a multiple of three bases (though they considered this latter idea unlikely); it is not overlapping; there is a fixed starting point or triplet to determine the correct sequence of triplets (this eliminates the comma-free code proposed earlier by Crick); it is degenerate, because one amino acid can be coded for by more than one triplet. Let us see how they reached some of these conclusions.

Studying frameshift mutations, Crick and his colleagues found that if a single-deletion mutation were coupled to a single-addition (but distal to the deletion) mutation, the double mutant could be normal in phenotype (Figure 9.3). The reason for this is as follows. The deletion mutation shifts the reading frame one nucleotide to the right, but the distal-addition mutation (a pseudosuppressor) shifts the frame back one nucleotide to the left and thus restores the correct frame for reading codons. Further, two addition mutations or two deletion mutations retain their mutant character, but three addition mutations or three deletion mutations (indeed, any multiple of three—all of the same type, addition or deletion) can be non-mutant. Thus it appears that the codons are each composed of three bases. That is, they are triplet.

The analysis of the frameshift mutants also suggested that the code was degenerate, because all new codons would be produced between compensating addition–deletion mutations—and if these new codons were not translatable into amino acids, frameshift mutations would not generally be revertible. The revertibility of frameshift mutations also suggests that codons are all the same size. If they were not, restoration of the reading frame by compensating addition–deletion mutations would not occur.

With this evaluation of the nature of the genetic code, the conceptual framework necessary for interpreting the genetic code emerged in tandem with the biochemical means for defining it.

The first code word is discovered

It was in 1961 that Nirenberg and Matthaei synthesized a synthetic mRNA that contained only U (polyuracil, or simply poly-U); they ascertained that it directed the synthesis of a polypeptide chain containing only the amino acid phenylalanine. Thus it appeared that UUU was the code word for phenylalanine (note that UUU was not one of Crick's amino-acid-specifying codons). This exciting breakthrough was made possible by *in vitro* protein synthesis (a system composed basically of ribosomes, tRNA, aminoacyl-tRNA synthetases, ATP, and amino acids) and the discovery of the enzyme *polynucleotide phosphorylase*. This enzyme catalyzes the synthesis of RNA chains using ribonucleotide diphosphate molecules and *neither a DNA nor an RNA template*. The RNA synthesized by phosphorylase is made up of random sequences of whatever ribonucleotides are present.

The precise *in vivo* function of polynucleotide phosphorylase is not well understood. Under normal physiological conditions, the enzyme may degrade RNA rather than participate in its synthesis.

Intrigued by the possibilities of *in vitro* protein synthesis, Nirenberg and Matthaei developed a system whereby they could destroy the endogenous mRNA in a cell-free extract and control protein synthesis by adding their own synthetic mRNA or, indeed, RNA from any source (Figure 9.4). In 1961, they used polynucleotide phosphorylase to synthesize an mRNA that contained only poly-U. When poly-U was used as a messenger in the *in vitro* protein-synthesizing system, a polypeptide was synthesized that contained only one species of amino acid—phenylalanine. The main conclusion drawn from this result was that given the triplet nature of the code, a sequence of UUU must be a codon for phenylalanine. Thus the first codon was determined. The specificity

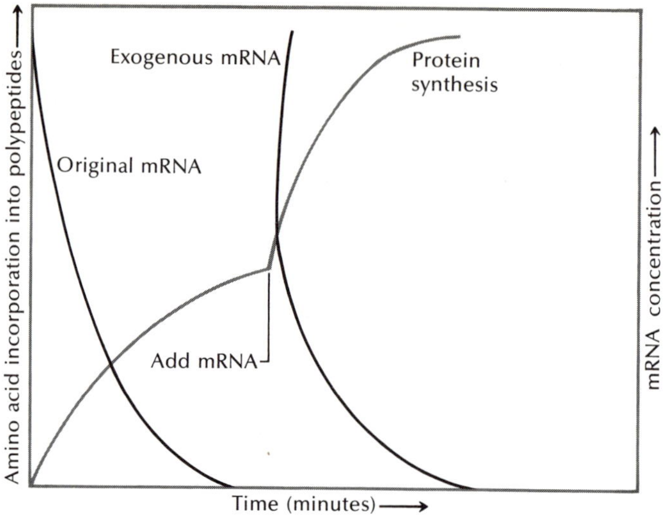

Figure 9.4
Effect of exogenous
mRNA on protein
synthesis (From
Nirenberg and Matthaei,
Proc. Nat. Acad. Sci., U.S.
47:1588, 1961)

that poly-U has for phenylalanine is shown in Table 9.3. Only poly-U
stimulates phenylalanine incorporation. The other polymers of A, C, inosine, and so on result in no phenylalanine incorporation.

Poly-C and poly-A were synthesized using the same technique, and
they produced polypeptide chains composed of one type of amino
acid—proline from poly-C and lysine from poly-A. Poly-G was unsuccessful as a messenger, perhaps because of secondary structural
anomalies. In a very short time, then, three codons were known, but
there remained 61 others.

**Codon assignments
using random copolymers
of mRNA**

To approach the problem of determining the other codons, Nirenberg
again used polynucleotide phosphorylase. But instead of synthesizing
homopolymers of all one nucleotide species, he synthesized copolymers
composed of two nucleotide species.

Table 9.3
Polynucleotide specificity
for phenylalanine
incorporation

	Additions	Counts/minute per mg protein
Labeled (^{14}C) phenylalanine +	None	44
	10 μg poly-U	39,800
	10 μg poly-A	50
	10 μg poly-C	38
	10 μg polyinosinic acid	57
	10 μg poly-AU (2:1 ratio)	53
	10 μg poly-U + 20 μg poly-A[a]	60

[a]All of the poly-U would be in a double helix with poly-A.

From M. W. Nirenberg and J. H. Matthaei, "The dependence of cell-free protein synthesis in *E. coli* upon
naturally occurring or synthetic polyribosomes," *Proc. Natl. Acad. Sci. U.S.* 47:1588–1602 (1961).

		Calculated codon frequency		Observed amino acid frequency	Inferred codon composition
Poly-AC	AAA	0.01		Asp = 0.03	2A + C
	2A + C	0.02	× 3 = 0.06	Glu = 0.03	2A + C
	A + 2C	0.12	× 3 = 0.36	His = 0.15	A + 2C (2A + C?)
	CCC	0.57		Lys = 0.01	3A
				Pro = 0.65	3C, A + 2C
				Thr = 0.14	A + 2C (2A + C?)
Poly-UC	UUU	0.06		Phe = 0.18	3U, 2U + C
	2U + C	0.10	× 3 = 0.30	Ser = 0.21	2U + C, U + 2C
	U + 2C	0.144	× 3 = 0.43	Pro = 0.36	3C, U + 2C
	CCC	0.21		Leu = 0.25	2U + C, U + 2C

Table 9.4
Amino acid incorporation
into polypeptides as
stimulated by random
copolymers (A/C = 1/5;
U/C = 2/3)

Knowing the proportion of each nucleotide and assuming random incorporation of the bases into mRNA, it is possible to make certain predictions about codon frequencies in the synthetic mRNA (Table 9.4). For example, if an AC copolymer is synthesized and if the A to C ratio is 1 to 5, then assuming random incorporation, a C would be incorporated into the mRNA five times more frequently than an A. Accordingly, we could predict the composition, though not the sequence, of triplet codons.

$$AAA = \frac{1}{6} \times \frac{1}{6} \times \frac{1}{6} = \frac{1}{216} \cong 0.01$$

$$2A, C = \frac{1}{6} \times \frac{1}{6} \times \frac{5}{6} \times 3 = \frac{15}{216} \cong 0.06$$

There are 3 possible sequences in a codon containing 2A and 1C—AAC, ACA, and CAA—and each has an individual probability of $\frac{5}{216}$, or about 0.02.

$$1A, 2C = \frac{1}{6} \times \frac{5}{6} \times \frac{5}{6} \times 3 = \frac{75}{216} \cong 0.36$$

Again, there are 3 possible sequences in a codon containing 1A and 2C—ACC, CAC, and CCA—and each has an individual probability of $\frac{25}{216}$, or about 0.12.

$$CCC = \frac{5}{6} \times \frac{5}{6} \times \frac{5}{6} = \frac{125}{216} \cong 0.57$$

The amino acid composition of the polypeptide synthesized from this copolymer was

Asparagine	0.03
Glutamine	0.03
Histidine	0.15
Lysine	0.01
Proline	0.65
Threonine	0.14

Comparing the codon frequencies with the amino acid frequencies, we can suggest the composition of the codons specifying each amino acid. Since lysine accounted for only 0.01 of the amino acids and since the probability of AAA is 0.01, we can assign the AAA codon to lysine. This is already evident, however, since we know that poly-A codes for a polypeptide containing only lysine.

Proline accounted for 0.65 of the amino acids, suggesting that proline is coded for by CCC (0.57) and perhaps by one of the codons containing 2C and 1A (0.12), which would give an expected frequency of 0.69.

Histidine and threonine account for 0.15 and 0.14, respectively, of the total amino acids. This suggests that both are coded for by a codon containing 2C and 1A (0.12) and perhaps that one is also coded for by a codon composed of 1C and 2A (0.02), which would give probabilities of 0.12 and 0.14 for these amino acids.

Asparagine and glutamine each account for 0.03 of the total amino acids, suggesting that they are both specified by codons composed of 1C and 2A (0.02).

We have accounted for each codon of the eight codon possibilities and can assign them to specific amino acids, but it is important to realize that sequencing nucleotides within the codons is generally not possible using random copolymers. Nirenberg and S. Ochoa, the discoverer of polynucleotide phosphorylase, continued to investigate random copolymers, expanding their systems to random polymers of three and four nucleotides, but the major problem of base sequence within a codon remained to be dealt with.

In addition to codon compositional analysis, the random copolymers also pointed to the degeneracy of the code; for example, proline was interpreted to have at least two codons. The development of two techniques in 1964 allowed the sequencing of bases within codons and led eventually to the complete elucidation of the genetic code.

Determination of the sequence of bases in a codon

In 1964, H. G. Khorana succeeded in synthesizing an mRNA that contained *not a random array* of ribonucleotides, but a regular *repeating sequence* of nucleotides; for example, ACACACACAC. . . . When used as a messenger, a polypeptide was synthesized containing a regularly repeating sequence of threonine-histidine-threonine-histidine-and so on. Similarly,

Copolymer	Amino acid incorporated	Codon assignment
CUC UCU CUC . . .	Leucine	5'CUC3'
	Serine	UCU
UGU GUG UGU . . .	Cysteine	UGU
	Valine	GUG
ACA CAC ACA . . .	Threonine	ACA
	Histidine	CAC
AGA GAG AGA . . .	Arginine	AGA
	Glutamine	GAG
UAU CUA UCU AUC UAU . . .	Tyrosine	UAU
	Leucine	CUA
	Serine	UCU
	Isoleucine	AUC
UUA CUU ACU UAC UUA . . .	Leucine	UUA
	Leucine	CUU
	Threonine	ACU
	Tyrosine	UAC

Copolymer	Codon recognized	Polypeptide made	Codon assignment
(AAG)$_n$	AAG AAG AAG . . .	Polylysine	AAG
	AGA AGA AGA . . .	Polyarginine	AGA
	GAA GAA GAA . . .	Polyglutamic acid	GAA
(UUC)$_n$	UUC UUC UUC . . .	Polyphenylalanine	UUC
	UCU UCU UCU . . .	Polyserine	UCU
	CUU CUU CUU . . .	Polyleucine	CUU
(UUG)$_n$	UUG UUG UUG . . .	Polyleucine	UUG
	UGU UGU UGU . . .	Polycysteine	UGU
	GUU GUU GUU . . .	Polyvaline	GUU

Table 9.5
Amino acid sequence of
polypeptides synthesized
from repeating
copolymers

CUCUCUCUCU . . . coded for a repeating sequence of leucine-serine-leucine-serine-and so on, and UGUGUGUG coded for a repeating sequence of cysteine-valine-cysteine-valine-and so on (Table 9.5). These copolymers substantiated the results of the copolymers synthesized randomly but were not unequivocal in their definition of the sequence of nucleotides within a codon. For example, ACA and CAC are the two possible sequences in the AC repeating copolymer, but it is not clear which triplet codes for threonine and which for histidine. The data from random copolymers suggest that both amino acids are coded for by codons containing 2C and 1A and that one is also coded for by a codon containing 1C and 2A. The data from the repeating-sequence copolymer reduce the possible codons specifying these two amino acids from six (AAC, ACA, CAA, CCA, CAC, and ACC) to two (ACA and CAC). Similarly, we can suggest that leucine or serine is coded for by either CUC or

UCU and cysteine or valine by UGU or GUG, but again we cannot be much more specific.

A copolymer containing the repeating sequence ACCACCACC . . . shows that translation into a sequence of amino acids can begin with any of the three possible triplets—ACC, CCA, or CAC—because from this copolymer three polypeptide species are produced: polythreonine, polyproline, and polyhistidine. Coupling these data with those from the regularly repeating AC copolymer, we can assign the CAC codon to histidine and, by the process of elimination, the ACA codon to threonine.

Thus the repeating-sequence copolymers can lead to additional codon assignments, but they frequently do not enable us to determine the sequence of bases in a codon that specifies a particular amino acid.

Concurrent with Khorana's work on the repeating nucleotide polymers was the development in Nirenberg's laboratory of a technique whereby defined sequences of trinucleotides were synthesized; each was shown to bind to only one species of tRNA. Under normal conditions, the binding of tRNA to the mRNA–ribosome complex is very specific. For example, only phenylalanine tRNA will bind to poly-U, and only lysine tRNA will bind to poly-A. This reaction is so specific that many trinucleotides can serve as mRNA and bind to both ribosomes and tRNA molecules.

Nirenberg was able to determine the sequence of the trinucleotide fragments, thus permitting him to determine the sequence of bases in many of the remaining codons. The trinucleotide technique involved adding a specific trinucleotide whose sequence was defined to a mixture of ribosomes and tRNA molecules charged with labeled amino acids. After allowing time for the ribosome–mRNA–tRNA complex to form, Nirenberg filtered the mixture and isolated the ribosome–mRNA–tRNA complex. The amino acid–tRNA that was bound to the trinucleotide mRNA was determined, and the trinucleotide was assumed to be the codon specifying that amino acid. Some of the codons determined by this method are shown in Table 9.6. However, a problem with this technique

Table 9.6
Trinucleotide binding results

Trinucleotide						tRNA bound
5'UUU3'	UUC					Phenylalanine
UUA	UUG	CUU	CUC	CUA	CUG	Leucine
AAU	AUC	AUA				Isoleucine
AUG						Methionine
GUU	GUC	GUA	GUG	UCU		Valine
UCU	UCC	UCA	UCG			Serine
CCU	CCC	CCA	CCG			Proline
AAA	AAG					Lysine
UGU	UGC					Cysteine
GAA	GAG					Glutamic acid

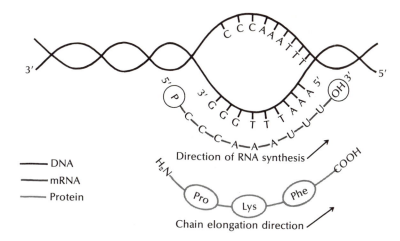

Figure 9.5
Prediction of the
orientation of codons and
the direction of
polypeptide synthesis

——— DNA

——— mRNA

——— Protein

Direction of RNA synthesis

Chain elongation direction

was that the specificity of tRNA binding *in vitro* was often low, giving some doubtful results—indeed, some that were wrong.

Before we examine the code itself, let us glance at the research that determined the direction of the reading of the code. We earlier learned that mRNA chains are synthesized in a 5′ to 3′ direction and that translation of the messenger took place in the same direction. Since polypeptides grow from their NH_2— end toward their —COOH end, it might be expected that in a tripeptide the amino acid with the free NH_2 group would be coded for by the 5′-end codon and the amino acid with the free COOH would be coded for by the 3′-end codon (Figure 9.5).

This prediction was experimentally verified by Ochoa and his colleagues, who synthesized AC copolymers using polynucleotide phosphorylase and an A/C ratio of 20/1. When treated with pancreatic ribonuclease, this copolymer was split between a purine (A) and a pyrimidine (C), leaving the phosphate attached to the 5′ end of A and the OH group attached to the 3′ end of C (Figure 9.6). The result of such hydrolysis was the generation of short RNA species that contained all A except for a terminal 3′C. This means that all codons will be 5′AAA3′ except one, which will be 5′AAC3′.

When the polypeptide synthesized from such a messenger was analyzed, it showed that the NH_2 terminal amino acid was a lysine, followed by several more lysine molecules and ending at the COOH end with asparagine. We conclude that AAA codes for lysine and that AAC

Figure 9.6
Ochoa experiment
designed to determine the
direction of code
translation (the arrows
indicate where the bonds
were split by pancreatic
ribonuclease)

UUU	Phenylalanine	UCU	Serine	UAU	Tyrosine	UGU	Cysteine
UUC		UCC		UAC		UGC	
UUA	Leucine	UCA		UAA	Terminate	UGA	Terminate
UUG		UCG		UAG		UGG	Tryptophan
CUU	Leucine	CCU	Proline	CAU	Histidine	CGU	Arginine
CUC		CCC		CAC		CGC	
CUA		CCA		CAA	Glutamine	CGA	
CUG		CCG		CAG		CGG	
AUU	Isoleucine	ACU	Threonine	AAU	Asparagine	AGU	Serine
AUC		ACC		AAC		AGC	
AUA		ACA		AAA	Lysine	AGA	Arginine
AUG^{init}	Methionine	ACG		AAG		AGG	
GUU	Valine	GCU	Alanine	GAU	Aspartic acid	GGU	Glycine
GUC		GCC		GAC		GGC	
GUA		GCA		GAA	Glutamic acid	GGA	
GUG		GCG		GAG		GGG	

Table 9.7
The RNA genetic code

codes for asparagine. We also conclude that the 5′ end of the mRNA contained AAA, and the 3′ end AAC. Therefore the direction of translation must have been 5′ to 3′, and the orientation of a codon must also be 5′ to 3′.

The genetic code   The results of the research into the genetic code are given in Table 9.7. The codons are ordered in a 5′ to 3′ direction. An examination of the genetic code reveals a number of interesting relationships.

1. The genetic code is highly degenerate, almost totally so. Three amino acids (Arg, Ser, and Leu) have six synonymous codons each; five (Val, Pro, Thr, Ala, and Gly) have four each; one (Leu) has three; and nine (Phe, Tyr, Glun, His, Aspn, Lys, Asp, Glu, and Cys) have two each. Met and Try are the only amino acids represented by one codon each.

2. Not all codons specify amino acids. The codons UAA, UAG, and UGA are nonsense codons and signify polypeptide chain termination, as noted earlier.

3. The genetic code has a regular scheme of degeneracy. With three exceptions (Ser, Leu, and Arg) the codons for any one amino acid resemble each other in their first two bases.

4. If an amino acid has a codon terminating in U, it also has one terminating in C; it is almost as consistently true that amino acids with a codon ending in A also have one ending in G.

5. In general, amino acids that occur frequently in proteins (Leu, Ala, Ser, and so on) have more codons coding for them than have the rarer amino acids (such as Met).

6. The genetic code is not ambiguous. Any one codon will code for only one amino acid, not two or three (a minor exception to this is the initiating codon AUG, which codes for F-met if initiating and for met if not initiating).

Perhaps the most interesting feature of the genetic code words is the existence of three triplets that code for no amino acid, the nonsense co-

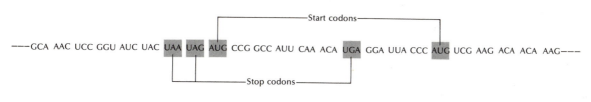

124   129

Figure 9.7
Partial RNA sequence of the MS2 bacterial virus, an RNA virus (Adapted with permission of the author from J. D. Watson, *Molecular Biology of the Gene,* 3rd edition. Copyright © 1976 by James D. Watson)

dons UAA, UAG, and UGA. They have been given quaint names. UAG is "amber," reportedly named for the mother of one of the graduate students (Bernstein) who found it. UAA is "ochre," a name apparently having no particular significance, and UGA is "opal," also a name of no particular significance. These three codons, acting in conjunction with specific protein release factors (Chapter 10), cause the termination of a polypeptide chain. Therefore, they are better referred to as terminators than as nonsense codons. Normally, only one terminator codon specifies the end of a gene, but an analysis of part of the RNA sequence of the phage MS2 reveals that this organism contains *two* stop codons at the end of its gene for coat protein (Figure 9.7).

Codons can change by the process of mutation so that they code for a different amino acid (*missense*) or for no amino acid (*nonsense*). Missense mutations are much more common, as might be expected, and often give rise to a polypeptide that retains partial function (a "leaky" mutant). A nonsense mutation produces incomplete polypeptides and hence does not normally produce "leaky" mutants.

Corroboration of the genetic code by *in vivo* studies

Independent *in vivo* confirmation of the codons determined by *in vitro* studies was essential, because some of the *in vitro* determinations were ambiguous. An elegant verification of the genetic code came from the work of Yanofsky on the *A* polypeptide chain of the tryptophan synthetase enzyme of *Escherichia coli.* As mentioned earlier, Yanofsky established the principle of colinearity with this polypeptide by demonstrating that the sequence of amino acid alterations in the *A* chain is the same as the sequence of mutation sites in the *A* cistron that caused them.

Of particular relevance to the verification of the genetic code was the observation that one amino acid in the normal *A* chain could be replaced by at least two different amino acids as the result of separate mutational

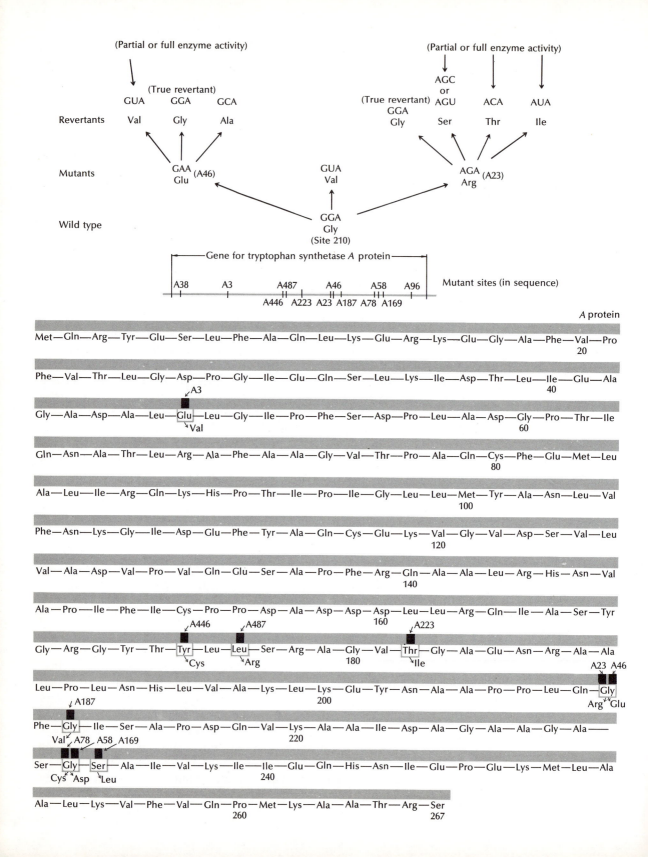

Figure 9.8 (opposite)
In vivo proof of the
genetic code, using
colinearity and the
tryptophan synthetase α
protein

events. For example, in mutant A23, glycine at position 210 in the normal *A* chain is replaced by arginine. In mutant A46, the glycine is replaced by glutamic acid. In a recombination analysis involving these two mutants, wild-type bacteria were recovered that contained the normal glycine at position 210 (Figure 9.8), suggesting that the two mutations involved different mutant sites *within a single codon* and furthermore that recombination might have occurred between *adjacent base pairs.*

Using the codons determined by *in vitro* studies, the most likely codons for arginine and glutamic acid in this case are GAA and AGA, respectively. Both these codons can be derived by a single base substitution of the glycine codon GGA.

The A23 and A46 mutants are revertible, either spontaneously or with chemicals, either to true revertants (those that have glycine at position 210) or to pseudorevertants with either full or partial enzyme activity (see Figure 9.8). These reversions allow further *in vivo* codon assignments, and they all corroborate the *in vitro* assignments. For example, the A23 mutant (AGA) reverts to strains carrying isoleucine (AUA), threonine (ACA), serine (AGC or AGU), or the normal glycine (GGA) at position 210 in the *A* chain. The A46 mutant (GAA) reverts to strains carrying alanine (GCA), valine (GUA), or the normal glycine (GGA) at position 210. One can conclude from these observations that the mutational event resulted from single base-pair substitutions within a codon.

These results also point out that no mutations, either forward or reverse, involved two or more simultaneous base-pair substitutions. That is, the A46 mutant (GAA) did not revert to strains carrying serine (AGC or AGU), threonine (ACA), or isoleucine (AUA) at position 210. Likewise, the A23 mutant (AGA) did not revert to strains carrying valine (GUA) or alanine (GCA) at position 210.

This work provides strong *in vivo* confirmation of the genetic code determined *in vitro.*

Interaction between mRNA codon and tRNA anticodon

Having traced the determination and verification of the genetic code and the elucidation of the base sequences of several tRNA molecules, we can now consider the base pairing between a codon on the mRNA molecule and the complementary triplet, or *anticodon,* on the tRNA molecule.

Examination of the codons in Table 9.7 reveals that many of the codons that specify a particular amino acid are equivalent in the first two bases but vary in the third. For example, the codons specifying proline all begin with CC, but the third base can be U, C, A, or G. In 1966, Crick proposed the *wobble hypothesis* to explain this observation. He suggested that the base at the 5' end of the *anticodon* has a flexibility that will allow it to form hydrogen bonds with more than one kind of base at the 3' end of

Figure 9.9
Diagram of the wobble hypothesis (note the equivalence of A and G in the codon)

Leu	5′ U—U—G 3′	mRNA (codon)
tRNA$_{leu}$	3′ A—A—C 5′	tRNA (anticodon)
Leu	5′ U—U—A 3′	mRNA (codon)
tRNA$_{leu}$	3′ A—A—C 5′	tRNA (anticodon)

the *codon* (Figure 9.9). The following observations led him to propose the wobble hypothesis.

1. Highly purified tRNA of one species (such as tRNA$_{alanine}$) having only one anticodon sequence can recognize *more than one* codon (Figure 9.9).

2. Several anticodons contain the base inosine at the 5′ end (this base is derived from the deamination of adenine to form a 6-keto purine; see Figure 9.10); inosine is a base that has more pairing flexibility.

3. The degenerate nature of the genetic code points out in many instances the lack of specificity of the third base in the codon.

Under the wobble concept, Crick visualized the pairing combinations shown in Table 9.8 and Figure 9.11. At the third position, I can pair with U, C, or A, thus eliminating the necessity for one anticodon to recognize four different codons. Only those anticodons with I in the third position can recognize three codons. And U and G can each recognize two. Another feature of the hypothesis is its prediction that not every codon has a separate tRNA species, thus making the total number of tRNA species considerably less than 61 (remember that 3 are nonsense). All available evidence to date lends support to the hypothesis.

Figure 9.10
Formation of inosine

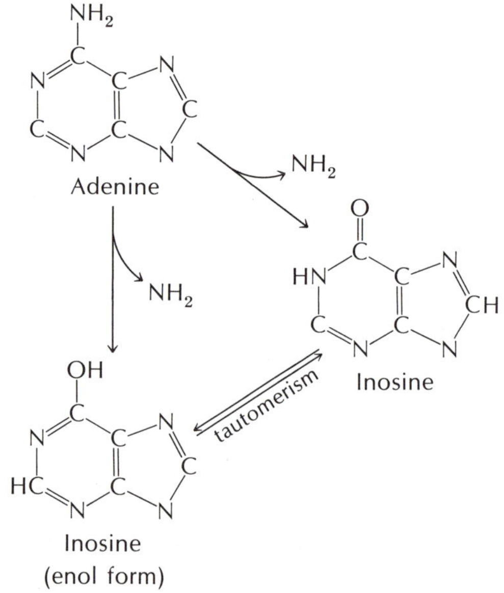

Adenine

Inosine

Inosine (enol form)

tautomerism

Table 9.8
Pairing combinations with
the wobble concept (third
position in the codon)

Base on the anticodon	Base recognized on the codon
U	A, G
C	G
A	U
G	U, C
I	U, C, A

The wobble hypothesis describes base-pairing equivalences at the third position in the codon–anticodon interaction, but in no way does it describe the *efficiency* of base pairing. The different wobble pairs exhibit much variation in pairing efficiencies in different organisms. In some organisms, one finds a preference for one of the codons specifying a particular amino acid, which probably reflects a greater pairing efficiency

Figure 9.11
Pairing possibilities under
the wobble concept (note
that inosine can pair with
C, A, and U)

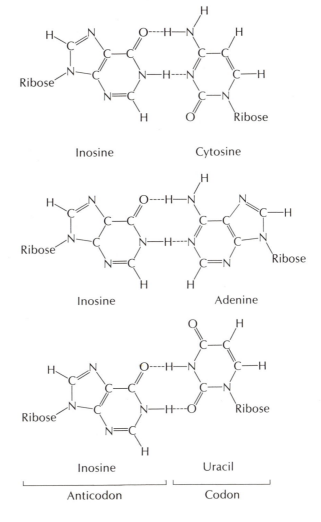

Inosine   Cytosine

Inosine   Adenine

Inosine   Uracil

Anticodon   Codon

between that particular codon and the tRNA anticodon. This preference may also account for variations in the DNA composition among organisms that have similar amino acid compositions.

The different binding efficiencies possible under the wobble hypothesis also suggest a means of controlling the rate of protein synthesis: Those codons with a low tRNA binding efficiency slow the process down, and those with a high tRNA binding efficiency keep the process going at a maximum rate.

Suppressor mutations

For many years, mutations have been known that reverse the effects of harmful mutations. Some are simply reverse mutations in which the original mutation reverted to normal. Others were more difficult to interpret, because they occurred at sites other than those at which the original mutation was located. This latter type of mutation, which restores full or partial activity to a mutant mapping at a different site, is called a *suppressor mutation*. There are two categories of suppressor mutations: those that map in the same gene affected by the original mutational event but at a different site (*intragenic suppressor*) and those that map in a different gene but restore activity lost by the original mutational event (*intergenic suppressor*).

Intragenic suppressor mutations restore partial or full enzyme activity to a mutation that originally had no enzyme activity, and this despite the presence of the original mutation. Such intragenic suppression can work in two ways. An addition (or deletion) of a base pair at point A can be corrected by the deletion (or addition) of a base pair at a different point B. This type of event restores the reading frame (see frameshift mutations), but is still mutant between the two sites. If the mutant amino acids between points A and B are not essential to the polypeptide's activity, full or partial restoration of activity can result. A second type of intragenic suppression occurs when a base substitution results in a missense mutational event (one amino acid substituted for another). A second missense mutation occurring at a different codon—and hence replacing a different amino acid—together with the first missense mutation can restore activity to the polypeptide. The original missense mutation may have resulted in the loss of polypeptide activity because of an altered three-dimensional configuration, but the second missense mutation could restore, at least partially, the original configuration and hence the activity.

Intergenic suppressor mutations are more complicated but still result in restoring the activity of a mutant gene. Such mutations involve genes other than the original mutant gene and fall into three main categories. First, there could be a mutation in the gene coding for the synthetase enzyme that joins an amino acid to a tRNA, and it could lead to the joining of an incorrect amino acid. This tRNA, with its incorrect amino

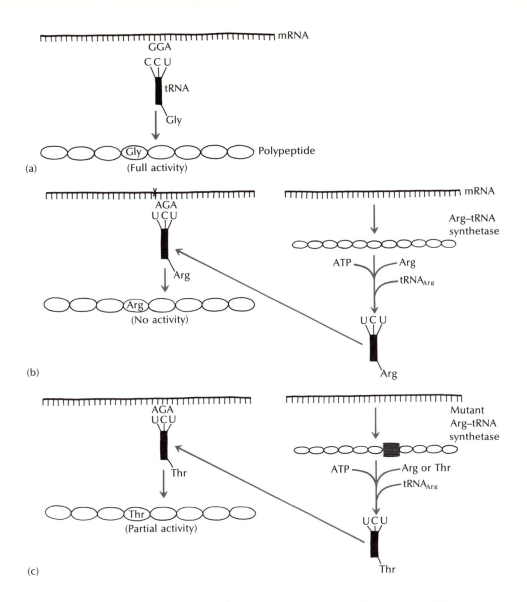

Figure 9.12

Diagram of a suppressor mutation in an aminoacyl-tRNA synthetase gene

acid attached, would insert that amino acid at the site specified by the tRNA, leading to amino acid replacement at that site in the polypeptide chain. If the mRNA codon at that site is a missense mutation and if the polypeptide resulting from the mutant gene is inactive, the insertion of this new amino acid could restore polypeptide activity without changing the mutant character of the gene (Figure 9.12).

A second type of intergenic suppressor mutation involves a mutant tRNA molecule. A mutation in the gene coding for a tRNA molecule can have two major consequences. It may alter the molecule's specificity for its proper amino acid, or it may alter the specificity of the anticodon for

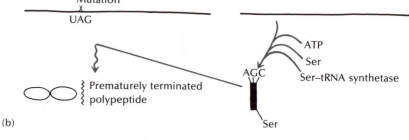

(a)

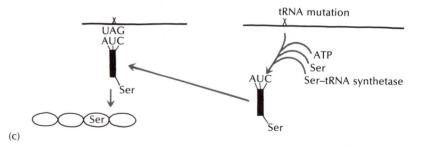

(b)

Figure 9.13
Model showing how a
nonsense mutation can be
suppressed by a tRNA
mutation

(c)

the mRNA codon. A change in the amino acid specificity has basically the
same consequences described in Figure 9.12: joining an incorrect amino
acid to a tRNA molecule. An altered tRNA anticodon results in the insertion of an amino acid at a site that specifies either no amino acid (a
nonsense codon) or a different amino acid—that is, a missense mutation
(Figure 9.13). It is also possible for a mutant tRNA to suppress a
frameshift mutation. It does this by having *four* bases in its anticodon, so
that addition mutations are corrected (Figure 9.14). Since no tRNA with
two bases in its anticodon are known, deletion mutants cannot be suppressed. Thus tRNA suppressor mutations can act by reading a chain
termination signal as if it were an amino acid, by suppressing a missense
mutation, or suppressing a frameshift mutation.

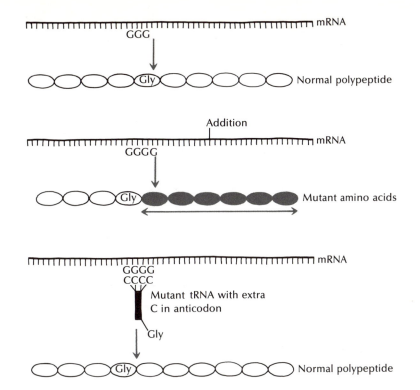

Figure 9.14
A model showing how an
addition frameshift
mutation is suppressed
by a mutant tRNA

The presence of nonsense suppressors raises an interesting problem. If a cell carries a suppressor of nonsense mutations, does this suppressor also suppress the normal chain-termination signal? If so, then nonsense suppressors would result in abnormally long polypeptide chains and, concurrently, cell lethality. One mechanism that has evolved to allow normal chain termination and suppression of nonsense mutations is double stop signals at the end of a gene (see Figure 9.7). One of the stop signals may be suppressed, but it would be rare indeed for *both* termination codons to be suppressed. The UAG and UGA codons are the most frequently suppressed, and the most frequently doubled up at the end of a gene. UAA is not so frequently suppressed, and is usually present alone as a normal chain-termination signal.

A third category of intergenic suppression involves mutant ribosomes. A mutation in any of the several ribosomal protein genes, or rRNA genes, can distort the structure of the ribosome so that it may be unable to choose a correct tRNA molecule. This could cause the insertion of incorrect amino acids into polypeptide chains, but if some of these incorrect insertions restore full or partial activity to a mutant polypeptide, the

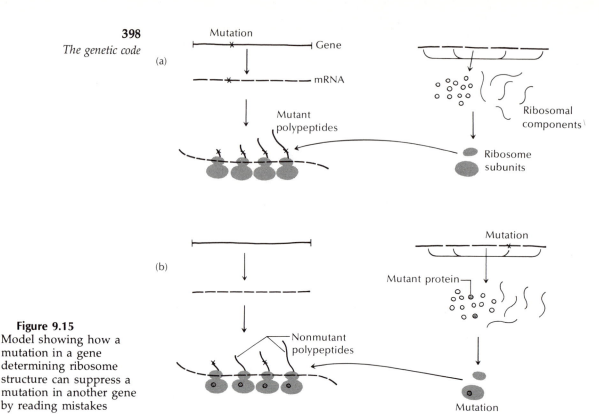

Figure 9.15
Model showing how a
mutation in a gene
determining ribosome
structure can suppress a
mutation in another gene
by reading mistakes

ribosomal mutation is described as a suppressor mutation of an intergenic variety (Figure 9.15).

Support for the idea that distorted ribosome structure may lead to misreading the genetic code comes from studies in which it has been shown that, in cells that are sensitive to the antibiotic streptomycin, extensive misreading occurs and results in cell death. Streptomycin is known to combine with the ribosome, and it is in this combination that the misreading takes place. The mutation to streptomycin resistance is accompanied by a failure of the drug to alter the ribosome structure to an appreciable extent, minimizing misreading and allowing most cells to survive (Figure 9.16).

A final note on intergenic suppression. We have discussed several mechanisms by which the amino acid sequence in a polypeptide chain has been altered, enabling the mutant polypeptide to function. However, we should also be aware of *environmental modifications* that allow mutants to function without polypeptide corrections. For example, suppressors may act to overcome the effects of a mutant gene by promoting synthesis of a necessary product by way of an alternative metabolic pathway. This type of intergenic suppression does not change amino acid sequences but rather by-passes a mutant effect.

Figure 9.16
Action of streptomycin on
protein synthesis in *E. coli*

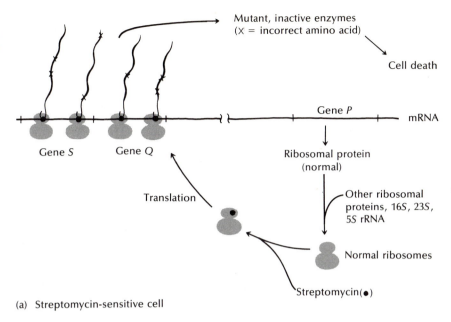

(a) Streptomycin-sensitive cell

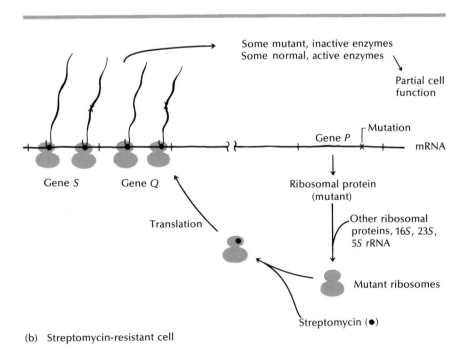

(b) Streptomycin-resistant cell

Problems concerning the evolution of the genetic code

The process of translating the genetic code into protein is a complicated but precise process. Proteins and nucleic acids interact in precise ways to produce specific proteins. Consider this tremendous complexity and precision. The aminoacyl-tRNA synthetase activates a specific amino acid with an accuracy of over 95 percent. This activated amino acid–enzyme complex must then recognize a specific tRNA and join the amino acid to it. This occurs with the same high frequency. Any error at the first step, rare as it may be, would almost certainly go no further, so the likelihood of an amino acid–tRNA mismatch is only about 1 in 10,000.

The tRNA molecules are highly specific macromolecules whose three-dimensional structure, just now beginning to be understood, is essential to their specificity. The tRNA molecules are perhaps the most complex of all molecules involved in protein synthesis, for they must have two specific sites for interaction with the aminoacyl-tRNA synthetase: an anticodon site to recognize the mRNA codon and a site to recognize the ribosome binding sites. The tRNA is actually more like a complex enzyme than a nucleic acid!

The ribosomes were thought by some to be straightforward sites for amino acid assembly. Now, however, they are emerging as highly complicated and highly efficient subcellular structures in the process of protein synthesis.

The complexity and efficiency of such a system leave no doubt about its long and extensive evolution. The development of the present genetic code must also have evolved simultaneously with the evolution of this translational apparatus. Thus, when we consider the evolution of the genetic code, we must also consider the evolution of the entire transcriptional and translational process.

This is precisely the area in which we lack information. It is possible to create a probable facsimile of the primitive, prebiotic environment, and it can be demonstrated that in this environment occurs the synthesis of amino acids, nucleotides, and even their polymers, as well as the synthesis of other compounds. It is also possible to examine the most primitive contemporary cells and viruses, study their mechanisms of protein biosynthesis, and find a very high level of complexity with respect to mechanisms of protein synthesis. Thus our problem in unraveling the evolution of the genetic code is similar to the problems faced in trying to reconstruct the origin of life: *All* the components have to be present in order for the system to work, but it is unlikely that they were all present in the beginning. Intermediary stages in the code's evolution undoubtedly existed, but the events that transpired between the abiotic synthesis of biological macromolecules and the development of a primitive cell capable of directing the synthesis of its own biological macromolecules are completely unknown.

How did protein synthesis come to be associated with nucleic acids?

What was the selective advantage of such an association? How did the genetic code evolve, and how is it that a particular codon specified one amino acid rather than another? The evidence to date suggests that the genetic code is universal—that the codons and their assigned amino acids (Table 9.7) apply to all organisms. But how does the code's universality shed light on its evolution? These and many other questions remain unanswered, for we are dealing with events that took place over three billion years ago.

To begin, let us evaluate some of the evidence that suggests the code is indeed universal, keeping in mind that we cannot *prove* this generalization but only establish evidence to support it. First, normal rabbit hemoglobin can be synthesized using aminoacyl-tRNA from *E. coli*, ribosomes from *E. coli*, and mRNA from rabbit red blood cells. Second, tobacco necrosis virus RNA will synthesize tobacco necrosis virus coat protein in a cell-free *E. coli* protein-synthesizing system. Third, a wide variety of organisms have shown relationships between codons and amino acids that are identical to those found in *E. coli*.

Although there is broad support for the universality of codons, there may well be considerable variation among organisms in tRNA species and their anticodons. The wobble hypothesis, which permits considerable flexibility in the third nucleotide position of a codon, predicts such variability. Different species may use different codons in different proportions. Recent studies indicate that not all of the 64 codons are used in every species. It is not at all uncommon to find a particular amino acid preferentially coded for by a specific codon in one species and by a different though synonymous codon in another species.

The universality of the code has generated a great deal of speculation. Were the code to evolve all over again, would it end up the way we find it today, or would it be entirely different? A number of investigators believe it would be the same. They reason that groups of codons and families of amino acids have enough stereochemical properties in common to ensure the continuation of the current correspondence between codons and amino acids, perhaps with some minor variations. Along these same lines, it has been suggested that there were no RNA intermediates in the very primitive cells and that amino acids associated directly with DNA codons. Since such associations would involve codons and *groups* of amino acids (basic amino acids, acidic amino acids, and so on), the polypeptides so synthesized from a given nucleotide sequence would probably be highly variable. Alternatively, it has been hypothesized that in primitive cells there was no DNA, only RNA.

The intervention of RNA in the process of protein synthesis, assuming the presence of DNA, would complicate the process but also make it more accurate, because the RNA could serve as adaptor. The step from an interaction between amino acids and DNA to one involving DNA, RNA, and amino acids is a major one, perhaps arising accidentally but afterwards being stringently selected for because of its increased specificity.

UUU	Phe 5.0	UCU		UAU	Tyr 5.4	UGU	Cys 4.8
UUC		UCC	Ser 7.5	UAC		UGC	
UUA	(Leu)	UCA		UAA[a]		UGA[a]	
UUG		UCG		UAG[a]		UGG	Trp 5.2
CUU		CCU		CAU	His 8.4	CGU	
CUC	Leu 4.9	CCC	Pro 6.6	CAC		CGC	
CUA		CCA		CAA		CGA	Arg 9.1
CUG		CCG		CAG	Gln 8.6	CGG	
AUU	Ile 4.9	ACU		AAU	Asn 10.0	AGU	
AUC		ACC	Thr 6.6	AAC		AGC	(Ser)
AUA	Ile	ACA		AAA		AGA	
AUG	Met 5.3	ACG		AAG	Lys 10.1	AGG	(Arg)
GUU		GCU		GAU	Asp 13.0	GGU	
GUC	Val 5.6	GCC	Ala 7.0	GAC		GGC	
GUA		GCA		GAA		GGA	Gly 7.9
GUG		GCG		GAG	Glu 12.5	GGG	

Table 9.9
Similar amino acids have similar codons (the numbers refer to amino acid "polar requirements"; they are derived from chromatography studies)

[a] Termination codon.

From C. R. Woese et al., "The molecular basis to the genetic code," *Proc. Natl. Acad. Sci. U.S.* 55:966 (1966).

An examination of the genetic code reveals two major features. First, it is highly degenerate but unambiguous, with one codon specifying starting points and only three codons specifying termination points (punctuation codons). Second, amino acids that have similar structural and functional properties have similar codons (Table 9.9), ensuring that a single-base alteration will either leave the same amino acid in place or replace it with a similar one.

It may be worthwhile to review some of the current ideas about these two features of the code, bearing in mind that these ideas are speculative. Sonneborn suggested that a degenerate and highly ordered code would be a great advantage to evolving populations, because it would minimize the possibilities for lethal mutations. He saw the forces of natural selection as leading to the minimization of nonsense codons, because these would be disadvantageous anywhere except at the end of the gene. He also saw selection pressure as leading to optimization of the frequency of mutations that involve a codon change but no concomitant amino acid change, along with optimization of the replacement of an amino acid by a *similar* amino acid in cases in which a mutation results in a changed amino acid.

Once a code such as this developed, selection would strongly favor its maintenance, because any variations would certainly result in a higher lethal mutation frequency. C. R. Woese proposed that translation was originally very imprecise so that a given sequence of nucleotides would rarely result in the same sequence of amino acids. A given nucleotide sequence would give rise to a population of polypeptides, perhaps related only by their sequence of amino acids according to family. Some of the proteins synthesized under such conditions might be quite advantageous to the cell and hence be selected for. So anything that might increase the probability of such a sequence being synthesized, such as more specific RNA species, would confer an adaptive advantage on that cell. Also, some codons might become more active and efficient than others, and some amino acids might become more influential in protein structure and function. Thus the intervention of a more specific tRNA and aminoacyl-tRNA synthetase would ensure greater specificity of amino acid sequence. Woese further noted that the present code increases the probability that a base change will result in no amino acid change, or if it does, that the changed amino acid will function similarly.

In 1965, T. H. Jukes proposed that primitive cells originally used only 15 amino acids, each coded by a nucleotide doublet ($4^2 = 16$ possible doublets), with one doublet serving as punctuation. Five additional amino acids came later, as did a third member of the codon. This proposal is flawed in that it neglects the lethality that would ensue in the transition from a doublet to a triplet code.

F. H. C. Crick offered a variation on the Jukes theme in his proposal of a small number of amino acids coded for by a small number of triplet codons. As time progressed, more codons and more amino acids associated, thus minimizing the existence of nonsense codons. A new amino acid could be incorporated into the code only if it first associated with a codon and then conferred a selective advantage on the cell.

These ideas are generated more out of a desire to *understand* the genetic code and protein synthesis than to merely determine the codons and describe the mechanics of polypeptide assembly. The broad implications are important, because they reflect on the origin of life itself. It is easy to say "God created," but this would hardly satisfy everyone. We want to know how and why. Unfortunately, though, our ability to re-create some three billion years of evolution is extremely limited, to say the least.

Overview

Our ideas about the genetic code have grown and matured during the past 25 years, and they will continue to do so. During the 1950s, we were immersed in theoretical approaches to the genetic code. Gamow suggested a diamond code, in which amino acids were assembled directly on a DNA template in a code that was overlapping. With the discovery of the intermediary role of RNA in protein synthesis, the diamond code was discarded. The idea of an overlapping code was also abandoned when it was shown that any one base is part of only one codon.

Crick proposed his adaptor hypothesis in 1957, suggesting that specialized adaptor molecules (unknown at the time, but now called tRNA) carried amino acids to the mRNA template. He also suggested a comma-free code, in which the 64 possible codons were split into two categories: 20 that specified amino acids, and 44 that did not. This was a nonoverlapping, nondegenerate code. When the first code word discovered proved not to be on Crick's list of 20 informational codons, the structure of the code required reevaluation. Based on studies of frameshift mutations, Crick and his colleagues summarized the basic qualities of the genetic code in four main points: The code is triplet; it is degenerate; it is not overlapping; and there is a fixed starting point to determine the correct sequence of triplets.

Experimental approaches to the genetic code began in 1961, when Nirenberg and Matthaei showed that a synthetic mRNA composed only of U coded for a polypeptide composed only of phenylalanine. Thus UUU coded for phenylalanine. By the same technique, CCC was shown to specify proline, and AAA lysine. Nirenberg and his associates next synthesized random copolymers of various RNA bases, and Khorana synthesized a repeating sequence of various copolymers. Both research efforts aided in uncovering more information about which codons specify which amino acids, but the most significant research in this area was Nirenberg's trinucleotide analysis. For the first time, codons could be sequenced (5' to 3') and accurately assessed.

When all 64 codons were eventually uncovered, the genetic code was shown to be degenerate. It has 3 nonsense codons. The codons for 1 amino acid commonly resemble each other in their first 2 bases. Common amino acids have more codons than rarer amino acids. And the genetic code is not ambiguous.

In vivo studies by Yanofsky and his colleagues confirmed the basic structure of the genetic code as it was established by the *in vitro* studies. They studied amino acid replacements in the α polypeptide chain of the tryptophan synthetase and showed that these replacements could be accounted for on the basis of the codons known from *in vitro* studies.

Crick evaluated pairing between codon and anticodon. He proposed the wobble hypothesis to account for one tRNA binding to more than one codon. The base in the third position of the anticodon has pairing flexibility, so it can pair with more than one base.

Suppressor mutations restore full or partial activity to a mutation that maps at a different site. Intragenic suppressions come about in compensating frameshift events or through missense mutations. Intergenic suppressors can involve mutant aminoacyl-tRNA synthetases, or they can involve mutant tRNA molecules. Mutant synthetases connect incorrect amino acids to tRNA molecules; mutant tRNA molecules can have altered affinities for amino acids or altered specificities for the codon. In this latter category, nonsense mutations are suppressed, as are missense mutations and some frameshift mutations. Ribosomes can also suppress mutations if they are themselves mutant. The mutation could be in any one of the ribosomal proteins or in rRNA.

Finally, we discussed some of the problems concerned with the evolution of the code. Many questions were raised, but few answers are available. How did protein synthesis come to be associated with nucleic acids? What selective forces acted on this association? How did the genetic code evolve? How did specific codons become associated with specific amino acids?

Questions and problems

9.1 Discuss the lines of evidence that lead to the conclusion that the genetic code is triplet.

9.2 A mutation is induced by a chemical whose mode of action is unknown. Using the T4 system, show how you would discriminate between the induction of a nonsense codon, a missense codon, and a frameshift mutation.

9.3 Discuss the concept of suppressor mutations. Include in your discussion the lines of experimental evidence that support their existence.

9.4 How would you explain the fact that a single species of tRNA (alanine tRNA) can recognize GCU, GCC, and GCA codons?

9.5 Explain the fact that a specific polypeptide, composed of 150 amino acids, has 15 known variants, all of which occur at 3 amino acid sites (6 variant forms at one site, 4 at a second, and 5 at a third). Does this argue for directed mutation?

9.6 A normal λ protein has the amino acid tyrosine at position 26, and a mutant strain has serine at that same position. Other than position 26, the polypeptide chains are identical. A reverse mutation, which maps at a site different from the original mutation, results in the insertion of tyrosine at position 26. Develop a model to explain this.

9.7 How was the intermediary role of RNA an important contributor to our understanding of the genetic code?

9.8 In the evolution of biological macromolecules—more specifically, the DNA–RNA–protein relationship—what do you think evolved first: DNA to protein, DNA to RNA, or RNA to protein? Explain.

9.9 Criticize the following statement: The genetic code is degenerate but not ambiguous.

9.10 We have discussed the gene from a structural and functional perspective from Mendel's time to the present. Discuss the major structural and functional interpretations of the gene during the past 75 years.

9.11 A polypeptide is isolated from a nonmutant *E. coli* strain. The same polypeptide is isolated from a mutant strain. The amino acid sequences are as follows.

H_2N-Ser-Pro-Ser-Leu-Asn-Ala-COOH   Nonmutant
H_2N-Val-His-His-Leu-Met-Ala-COOH   Mutant

Assuming a deletion in the NH_2 terminal codon, and an addition in the COOH terminal codon, are the amino acid sequences consistent with our understanding of the genetic code?

9.12 What is the minimum number of base substitutions that would be required to change codons from:
  (a) Ileu to Ser?
  (b) Thr to Pro?
  (c) Ser to Gly?
  (d) Thr to Tyr?

9.13 If nucleic acids were composed of only two kinds of bases, how large would codons have to be to specify 20 amino acids and termination and initiation signals?

9.14 How many and what kind of amino acid substitutions can there be for tryptophan, assuming only single base substitutions in the tryptophan codon?

9.15 Study the following nucleotide sequence for the coat-protein gene region of the RNA phage MS2. The coat-protein amino acid sequence appears below it. Knowing the nucleotide sequence and the amino acid sequence, line the amino acids up under their respective codons.

AUA	GAG	CCC	UCA	ACC	GGA	GUU	UGA	AGC	AUG·	GCU	UCU	AAC	UUU	ACU	CAG	UUC	GUU	CUC	GUC	GAC	AAU	GGC	GGA	ACU

GGC	GAC	GUG	ACU	GUC	GCC	CCA	AGC	AAC	UUC·	GCU	AAC	GGG	GUC	GCU	GAA	UGG	AUC	AGC	UCU	AAC	UCG	CGU	UCA	CAG	GCU	UAC	AAA	GUA

ACC	UGU	AGC	GUU	CGU	CAG·	AGC	UCU	GCG	CAG	AAU	CGC	AAA	UAC	ACC	AUC	AAA	GUC	GAG	GUG	CCU	AAA	GUG	GCA	ACC	CAG	ACU	GUU	GGU	GGU	GUA·

GAG	CUU	CCU	GUA	GCC	GCA	UGG	CGU	UCG	UAC	UUA	AAU	AUG	GAA	CUA	ACC	AUU	CCA	AUU	UUC	GCU	ACG	AAU	UCC	GAC·

UGC	GAG	CUU	AUU	GUU	AAG	GCA	AUG	CAA	GGU	CUC	CUA	AAA	GAU	GGA	AAC	CCG	AUU	CCC	UCA	GCA	AUC	GCA	GCA	AAC·

UCC	GGC	AUC	UAC	UAA	UAG	ACG	CCG	GCC	AUU	CAA	ACA	UGA	GGA	UUA	CCC	AUG	UCG	AAG	ACA	ACA	AAG	AAG	(U)

Nucleotide sequence for coat protein gene in MS2, an RNA phage. Part of the replicase gene is also included in the sequence.

Ala	Ser	Asn	Phe	Thr	Gln	Phe	Val	Leu	Val	Asp	Asn	Gly	Gly	Thr	Gly	Asp	Val	Thr	Val	Ala	Pro	Ser	Asn	Phe

Ala	Asn	Gly	Val	Ala	Glu	Trp	Ile	Ser	Ser	Asn	Ser	Arg	Ser	Gln	Ala	Tyr	Lys	Val	Thr	Cys	Ser	Val	Arg	Gln

Ser	Ser	Ala	Gln	Asn	Arg	Lys	Tyr	Thr	Ile	Lys	Val	Glu	Val	Pro	Lys	Val	Ala	Thr	Gln	Thr	Val	Gly	Gly	Val

Glu	Leu	Pro	Val	Ala	Ala	Trp	Arg	Ser	Tyr	Leu	Asn	Met	Glu	Leu	Thr	Ile	Pro	Ile	Phe	Ala	Thr	Asn	Ser	Asp

Cys	Glu	Leu	Ile	Val	Lys	Ala	Met	Gln	Gly	Leu	Leu	Lys	Asp	Gly	Asn	Pro	Ile	Pro	Ser	Ala	Ile	Ala	Ala	Asn

Ser	Gly	Ile	Tyr

Amino acid sequence of MS2 coat protein

After Min Jou, *et al.* 1972. *Nature* 237:82–88

The replicase enzyme begins with the amino acids Ser-Lys-Thr-Thr-Lys-Lys. Where does the replicase gene begin?

Brenner, S. 1957. On the impossibility of all overlapping triplet codes in information transfer from nucleic acids to proteins. *Proc. Natl. Acad. Sci. U.S.* 43:687.

Corwin, H. O., and J. B. Jenkins. 1976. *Conceptual Foundations of Genetics.* Boston, Houghton Mifflin.

Crick, F. H. C. 1958. On protein synthesis. *Symp. Soc. Exp. Biol.* 12:138.

Crick, F. H. C. 1966. Codon–anticodon pairing: The wobble hypothesis. *J. Molec. Biol.* 19:548–555.

Crick, F. H. C. 1967. The Croonian Lecture, 1966: The genetic code. *Proc. Roy. Soc. B.* 167:331.

Crick, F. H. C. 1968. The origin of the genetic code. *J. Molec. Biol.* 38:367–379.

Crick, F. H. C., L. Barnett, S. Brenner, and R. J. Watts-Tobin. 1961. General nature of the genetic code for proteins. *Nature* 192:1227.

Crick, F. H. C., J. S. Griffin, and L. F. Orgel. 1957. Codes without commas. *Proc. Natl. Acad. Sci. U.S.* 43:416–421.

Dounce, A. L. 1952. Duplicating mechanism for peptide chain and nucleic acid synthesis. *Enzymologia* 15:251–258.

Fitch, W. M. 1973. Aspects of molecular evolution. *Ann. Rev. Genetics* 7:343–380.

Gamow, G. 1954. Possible relation between DNA and protein structures. *Nature* 173:318.

Garen, A. 1968. Sense and nonsense in the genetic code. *Science* 160:149–159.

The Genetic Code. 1966. Cold Spring Harbor Symp. Quant. Biol. 31. Cold Spring Harbor, N.Y., Cold Spring Harbor Laboratory of Quantitative Biology.

Hartman, P. E., and J. R. Roth. 1973. Mechanisms of suppression. *Adv. Genetics* 17:1–105.

Jukes, T. H., and L. Gatlin. 1971. Recent studies concerning the coding mechanism. *Prog. Nuc. Acid Res. Molec. Biol.* 11:303–350.

Khorana, H. G. 1966–1967. Polynucleotide synthesis and the genetic code. *Harvey Lectures* 62:79–105.

Laycock, D. G., and J. A. Hunt. 1969. Synthesis of rabbit globin by a bacterial cell-free system. *Nature* 221:1118–1122.

Nirenberg, M., and P. Leder. 1964. RNA codewords and protein synthesis. *Science* 145:1399–1407.

Nirenberg, M. W., and J. H. Matthaei. 1961. The dependence of cell-free protein synthesis in *E. coli* upon naturally occurring or synthetic polyribonucleotides. *Proc. Natl. Acad. Sci. U.S.* 47:1588–1602.

Okada, Y., Y. Nozu, and T. Ohno. 1969. Demonstration of the universality of the genetic code *in vivo* by comparison of the coat proteins synthesized in different plants by tobacco mosaic virus RNA. *Proc. Natl. Acad. Sci. U.S.* 63:1189–1195.

Orgel, L. E. 1968. Evolution of the genetic apparatus. *J. Molec. Biol.* 38:381–393.

Sadgopal, A. 1968. The genetic code after the excitement. *Adv. Genetics* 14:325–404.

Smith, J. D. 1972. Genetics of tRNA. *Ann. Rev. Genetics* 6:235–256.

Woese, C. R. 1967. *The Genetic Code: The Molecular Basis for Genetic Expression.* New York, Harper and Row.

Ten

Mutation and the gene concept

The detection of mutations
Detection of mutations in *Drosophila*
Techniques of mutation study in *Neurospora*
Techniques of mutation study in bacteria
Study of mutation in viruses
Study of mutation in mammals

Rate and frequency of mutations
Spontaneous mutations
Mutator and antimutator genes

Position effect

Preadaptive and postadaptive mutations

Induced mutations of genes
The mutagenic properties of ionizing radiation
The mechanisms of UV-induced mutations
Chemically induced mutations
Base-pair substitutions
Frameshift mutations

Mutations and the environment

The gene concept

Overview

Questions and problems

References

In a broad sense, *mutation* refers to any change in the genetic information, including the addition of whole genomes (polyploidy) and individual chromosomes (aneuploidy), structural alterations of chromosomes (inversions), and qualitative changes *within* the gene itself. This chapter will concern itself with nuclear mutations and, more specifically, *intragenic* nuclear mutations. We can define these mutations as genotypic alterations that replicate as such, are passed on from generation to generation, and exhibit segregation and reassortment. Each mutation reflects a change in the genetic information, that is, in the DNA.

Mutation is the backbone of the evolutionary process. It is also a fundamental genetic process essential to our understanding of genetic phenomena. The expression of alternative phenotypes based on gene mutations has allowed investigators to uncover the physical and chemical basis of heredity. Mutations and the mutation process have enabled us to gain insight into the structure of individual genes. After discussing the mutation process, the concept of the gene will be reexamined.

The detection of mutations

It should first be helpful to discuss the techniques employed to detect mutations. In studying the mutation process, it is important to choose the system that will answer the questions being asked. For example, for questions about the mechanism of mutation at the DNA or RNA level, prokaryotes are the most suitable. Equally important is the study of mutations in eukaryotes, such as *Drosophila*, even though these organisms are less suitable for determining mutagenic mechanisms at the molecular level.

Detection of mutations in *Drosophila*

Using strains of *Drosophila* carrying special genetic markers, it is possible to detect sex-linked and autosomal recessive lethal mutations, dominant lethal mutations, mutations of genes that produce visible effects (such as eye color), and mutant genes that occur in only part of the organism's body (*mosaic mutants*).

Commonly employed to detect mutations in *Drosophila* is the *sex-linked recessive lethal system*. Recessive sex-linked lethal mutations are detected in males, because they have only one **X** chromosome and hence no normal dominant genes to mask the expression of the lethal gene. It has been estimated that there are around 1000 gene loci on the **X** chromosome that will, if mutated, result in a lethal effect when either homozygous or hemizygous. These are loci that are absolutely essential to the survival of the organism. Figure 10.1 outlines the technique used to uncover this kind of mutation. A wild-type male is treated with a mutagenic agent (the control is a nontreated male) and mated with a female that possesses a complex series of inversions in the chromosome to inhibit crossing over and is homozygous for the sex-linked traits w^a (white-apricot eye color)

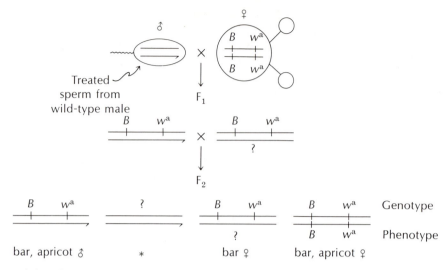

Figure 10.1
Sex-linked recessive lethal
test

bar, apricot ♂   *   bar ♀   bar, apricot ♀

*If this class is missing, ? = recessive lethal in all chromosomes.
  If this class is present as normal ♂, then
  (a) ? = normal, nonmutated chromosome in all cells.
  (b) ? = recessive lethal in some F_1 female chromosomes, normal in others
       (see text for explanation).

and *B* (bar eyes). This genetically marked fly was originally developed by
H. J. Muller and is referred to as Muller-5, or sometimes Basc. The F_1
females are heterozygous for the treated **X** chromosome in question. They
are mated singly with F_1 males (one female and one male per vial), and
the F_2 is analyzed for the presence or absence of wild-type males. Each F_1
female, then, represents a single treated chromosome obtained from the
treated parental male (Figure 10.1). If F_2 wild-type males are absent, all
the "?" chromosomes contained at least one recessive lethal mutation. If
wild-type males are *present*, lethal mutations may not have been induced;
or lethal mutations were induced, but only a portion of the F_1 fly's sex
cells contained mutant chromosomes (a *mosaic*).

Mosaics commonly arise when only one strand of a DNA molecule is
mutated. Upon replication, two DNA molecules are generated, one mutant and one nonmutant (Figure 10.2). If a *Drosophila* chromosome is
visualized as composed of one molecule of DNA and a mutation is induced in only one strand, a mechanism for mosaic origin can be visualized in which one chromosome is mutant after replication and one is
normal. Cells having the mutant chromosome are mutant, and cells having the nonmutant chromosome are normal.

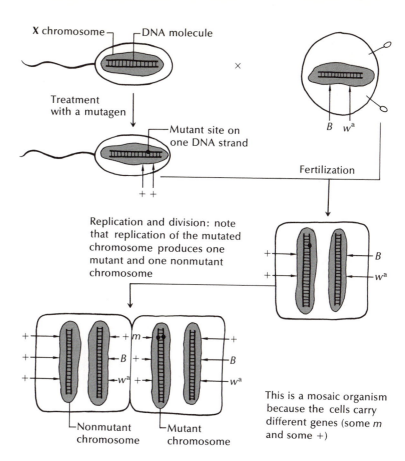

X chromosome — — DNA molecule

Treatment with a mutagen

Mutant site on one DNA strand

+ +

Replication and division: note that replication of the mutated chromosome produces one mutant and one nonmutant chromosome

Fertilization

B w^a

+ — — B

+ — — w^a

+ — — + m — — +

+ — — B + — — B

+ — — w^a + — — w^a

Nonmutant chromosome   Mutant chromosome

This is a mosaic organism because the cells carry different genes (some *m* and some +)

Figure 10.2
Origin of mosaics in *Drosophila*

The detection of sex-linked recessive lethal mosaics requires analyzing F$_2$ *Bw^a//?* females from apparently nonmutant vials, mating them individually to *Bw^a//* males, and scoring the F$_3$ for the presence of wild-type males. If any F$_2$ female gives rise to an F$_3$ devoid of wild-type males, the F$_1$ female from which she came would be classified as a mosaic (Figure 10.3).

Autosomal recessive lethals, recessive lethal mutations on the autosomes, can also be detected in *Drosophila,* but the system is more complicated because the mutation is expressed only if it is homozygous (Figure 10.4). For example, to make a second-chromosome analysis, investigators mate treated, or control, males to females heterozygous for *Cy,* a second-chromosome dominant gene that is lethal when homozygous and produces curly wings when heterozygous. Each F$_1$ male represents a single paternal second chromosome. The F$_1$ *Cy//?* males are mated individually to *Cy, L* females (*L* is a second-chromosome dominant gene, also lethal

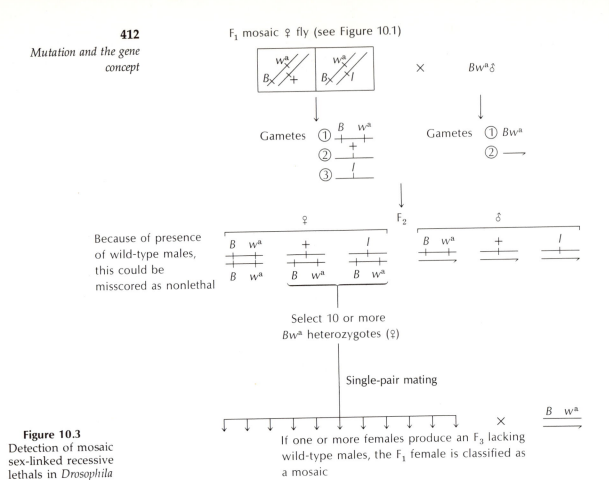

Figure 10.3
Detection of mosaic sex-linked recessive lethals in *Drosophila*

when homozygous, that produces lobe eyes when heterozygous). The females also possess inversions to inhibit crossing-over. In the F₂ generation, three classes are found: *Cy*//*L*, *Cy*//?, and *L*//?. Either the *Cy* or *L* flies could be used to establish an F₃ generation. Figure 10.4 shows *Cy*//? males mated to *Cy*//? virgin females in single-pair matings. The presence of all *Cy* flies in the F₃ establishes that a lethal mutation was induced in the second chromosome. The presence of non-*Cy* flies indicates either that no lethal mutations were induced or that a mosaic was produced (the latter possibility would require further testing).

Dominant lethal mutations, those mutations that are lethal in a single dose, are easy to detect and do not require specially marked genetic stocks. Wild-type males are treated with a mutagenic agent and then mated to wild-type females. Since a dominant lethal mutation is by definition lethal in a single dose, it is expressed in the first generation. However, because a dominant lethal mutation can result in death at any stage of development—including the adult stage—it is difficult to obtain accu-

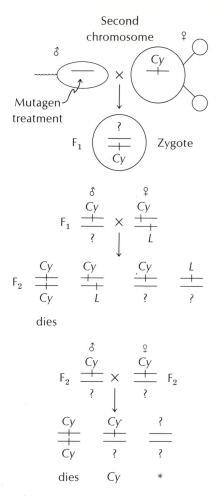

Figure 10.4
Detection of autosomal
recessive lethal mutations
in *Drosophila*

*If ? = a lethal mutation, this class will be missing.
If ? = no mutation, this class will be wild type.

rate values for mutation frequency. Normally, the following criteria are
considered indicative of the presence of dominant lethals:

Number of F$_1$ eggs laid compared with

↓ (lethal gene kills embryo)

Number of eggs that hatch (nonhatched = lethal) compared with

↓ (lethal gene kills larva)

Number of larvae that pupate (nonpupated larvae = lethal) compared with

↓ (lethal gene kills pupa)

Number of adults that emerge (nonemergent pupae = lethal)

Since dominant lethal mutations may not be expressed until well into adulthood, using these criteria could cause the mutations to be overlooked. So we are screening only a portion of the possible mutations induced. The same margin of error holds for autosomal and sex-linked recessive lethal mutations. The lethality could be expressed in mature adults and hence be missed.

Mutational processes occurring at specific gene loci that produce visible phenotypic alterations (autosomal or sex-linked visible mutations) are also easy to detect (Figure 10.5). For example, we can study mutation at three specific second-chromosome genes (*ed* = echinoid eyes, *dp* = truncated wings and thoracic tissue disturbances, and *cl* = clot-colored eyes). Wild-type males are treated with a mutagenic agent and mated to homozygous *ed dp cl* females (Figure 10.5a). All three genes are recessive; so, if no mutations are induced at these three loci, the F_1 will be phenotypically normal. If a mutation is induced at any one of the three loci, a mutant could appear in the F_1. Using mutagenic agents such as ethylmethanesulfonate (EMS) and ICR 170 Jenkins, Carlson, Corwin, and others found the mutation frequency at the dumpy region to be in excess of 1 percent.

Figure 10.5
Detection of mutations at specific loci

(a) Autosomal visible mutations

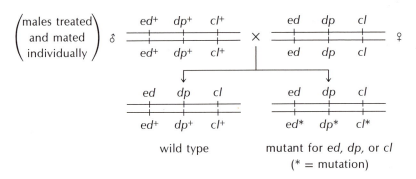

wild type

mutant for *ed*, *dp*, or *cl*
(* = mutation)

(b) Sex-linked visible mutations

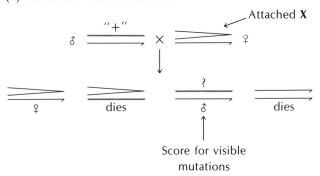

Score for visible
mutations

For mutations at specific visible loci on the **X** chromosome, attached-**X** females are used, because a treated male **X** chromosome can be analyzed in the first generation. Treated males are mated with attached-**X** females, and all F$_1$ males, having received a treated **X** from the male and a **Y** chromosome from the female, are scored for mutant phenotypes (Figure 10.5b).

Mosaic mutants are easily detected in specific visible mutations. Usually, a fly appears with part of its body normal and part mutant—for example, the mosaic dumpy (*dp*) fly in Figure 10.6.

Techniques of mutation study in *Neurospora*

Figure 10.6
Dumpy mosaic

The common bread mold *Neurospora* is haploid throughout most of its life cycle (Figure 1.30) and grows normally on a minimal medium containing only a single carbon and nitrogen source, inorganic salts, and the vitamin biotin. *Neurospora* is therefore highly useful in exploring nutritional mutations. Haploidy eliminates problems of dominance, and growth on a minimal medium provides an efficient way to detect nutritional mutations such as dependency on external sources of certain amino acids. The standard mutation-screening technique for nutritional mutations in *Neurospora* was established in 1945 by Beadle and Tatum (Figure 10.7).

Wild-type conidia (asexual spores) are treated with a mutagenic agent and then allowed to germinate on a medium containing all nutrients (called a complex medium). Each *Neurospora* colony so grown represents a single conidium. Samples of each colony are then transferred to a minimal medium. If a sample fails to grow, it is suspected that a nutritional mutation has been induced and that the new strain requires a particular nutritional supplement. The nutritional requirement of the mutant strain can be ascertained by transferring samples of the mutant colony from the complex medium to minimal media each of which is supplemented with one nutrient. If growth occurs on one of the single-nutrient supplemented minimal media, the mutational event can be described as one that interferes with the biosynthesis of that particular nutrient.

The genetic basis of the mutation can be verified by crossing the mutant strain with a wild-type strain. A one-to-one mutant-to-nonmutant ratio in the offspring confirms a genetic mutational event.

Techniques of mutation study in bacteria

Bacteria offer two of the same advantages for mutation study that *Neurospora* do—haploidy and growth on minimal, or simple, medium. A culture of prototrophic bacteria (which can grow on minimal medium) is treated with a mutagenic agent, diluted in order to allow separate growth of single bacterial cells, and plated on a complex medium supplemented with vitamins, amino acids, nucleic acids, minerals, and sugars. Mutants are detected when samples of each colony growing on the complex medium are transferred first to minimal medium and then to media each of which is supplemented with one nutrient. The transfer is accomplished by means of a cylinder of the same diameter as the inside of a petri dish

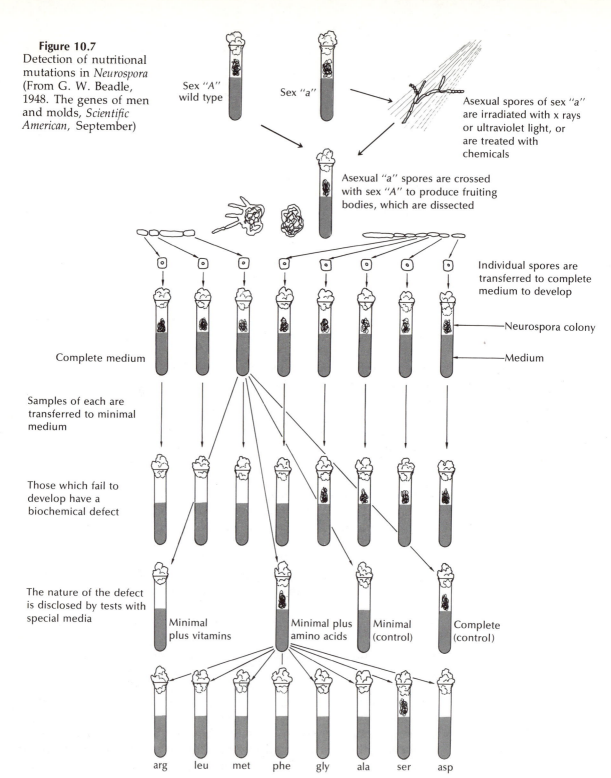

Figure 10.7
Detection of nutritional mutations in *Neurospora* (From G. W. Beadle, 1948. The genes of men and molds, *Scientific American,* September)

Sex "*A*" wild type

Sex "*a*"

Asexual spores of sex "*a*" are irradiated with x rays or ultraviolet light, or are treated with chemicals

Asexual "*a*" spores are crossed with sex "*A*" to produce fruiting bodies, which are dissected

Individual spores are transferred to complete medium to develop

Neurospora colony

Medium

Complete medium

Samples of each are transferred to minimal medium

Those which fail to develop have a biochemical defect

The nature of the defect is disclosed by tests with special media

Minimal plus vitamins

Minimal plus amino acids

Minimal (control)

Complete (control)

arg   leu   met   phe   gly   ala   ser   asp

Growth occurs only in minimal medium supplemented with serine; therefore, the mutation is in the gene that functions in the metabolism of serine

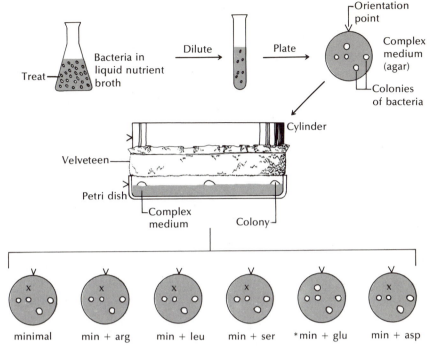

Figure 10.8
Detection of nutritional mutations in bacteria, using the replica-plating technique

*Growth on minimal medium plus glutamic acid indicates that a mutation has occurred that prevents the bacteria in the mutant colony from synthesizing glutamic acid

x Missing colony (mutant)

(Figure 10.8). The end of the cylinder is covered with velveteen, a fabric that has long fibers perpendicular to the surface of the medium (parallel to the axis of the cylinder). When the cylinder is set lightly against a petri dish covered with colonies of bacteria, samples of each colony adhere to the velveteen fibers. Its proper orientation maintained relative to the parent plate, the cylinder is then gently placed on new sterile surfaces. Sample cells from each parent colony have now been deposited on the new surface in the same location as on the parent plate. A single touch on the parent plate can be used to set up several replica transfers. (This technique, called *replica plating*, was developed by Lederberg in 1946.) If there is no growth on the minimal medium but there is growth on a supplemented medium, a nutritional mutation is indicated. Such bacteria are described as *auxotrophic*.

Variations of this detection technique can be developed to screen for other types of mutations. For example, bacteria can be plated on an agar plate covered with a strain of virulent bacteriophage. Only bacteria that are resistant to the phages will survive and grow into colonies. Antibiotics can also be used as a screening agent.

Study of mutation in viruses

The bacterial virus, or bacteriophage, has been critically important in the development of molecular biology and in studies of mutation. The reasons for the value of the bacteriophage as a genetic tool are as follows. Bacteriophage are easy to grow. They have an extremely short life cycle (the T4 bacteriophage can produce hundreds of progeny phage in just 30 minutes at 37°C). They can be stored safely in a refrigerator for years. And enormous populations can be generated in a short period of time, making it easy to select for rare mutational events. The bacteriophage has in many areas proved to be the most valuable genetic tool available to geneticists. This is especially true in the area of mutation study, in which we need to know actual mechanisms of mutation.

Some bacteriophage genes, such as the *r*II gene in T4, have been particularly valuable in mutation studies. Investigators detect mutations in the *r*II gene by screening large numbers of virus particles for their ability to cause large, sharp-edged plaques in a carpet of bacteria growing on a petri dish. (A *plaque* is a round, clear area on an otherwise opaque culture plate of bacteria. It signifies areas in which bacteria have been destroyed by bacteriophage.) The normal plaque is small and fuzzy-edged. Among the many advantages of the T4*r*II system is the ease with which the *r*II mutants are detected. Enormous numbers of wild-type T4 can be treated with a mutagen, plated out on a lawn of *Escherichia coli*, then studied for plaque morphology (Figure 10.9). To ascertain a mutation frequency, researchers compare the frequency of *r*II plaques from treated T4 with the frequency from an untreated T4 sample, or control.

The *r*II system has been employed to study dozens of compounds thought to be mutagenic. Because of the astronomical number of T4 progeny produced, even an extremely low rate of mutation can be easily detected. (One plate can have 200 plaques or more, and it takes only a few seconds to examine that plate. At 10 plates per minute, one person can examine almost 10^6 plaques in an 8-hour day.) All comparative studies so far have shown that compounds mutagenic in viruses are also mutagenic in bacteria and fungi and suggest that they are mutagenic in mammals.

Study of mutation in mammals

Mammals are not convenient organisms to use in genetic studies for a variety of reasons. Their reproductive cycle is long, they give birth to relatively few offspring, they are expensive to maintain, and they take up a great deal of laboratory space. Yet in spite of these and other drawbacks, the mammal, and more specifically the human, is of primary interest to us. To what extent can we extrapolate studies of T4 or *Drosophila* mutation to humans or to nonhuman mammals? In the case of humans, it is clear that we cannot subject ourselves to the types of analysis discussed so far. Yet we need to know the mutagenic potential of chemicals in humans.

One of the most common means of detecting mutagenic activity in mammals, including humans, involves indirect indicators. For example, chemicals are administered to experimental animals or to humans. Body fluids or tissues are then obtained from the experimental subjects. In most cases, blood or urine is extracted. For rodents, blood is withdrawn from the retroorbital sinus, and urine is collected in a special urine-collecting

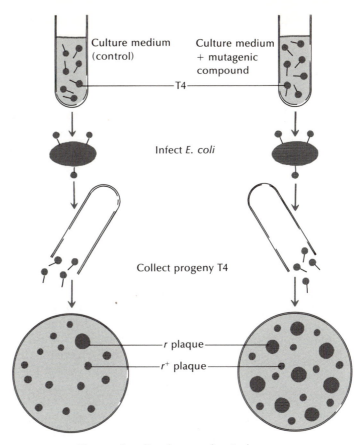

Figure 10.9
A technique for studying mutation frequency in T4 viruses

Plate on *E. coli* and screen for *r*II plaques

unit (Figure 10.10). The fluids are analyzed for mutagenic agents and/or used to treat specific test organisms such as viruses, *Drosophila*, *Salmonella*, or yeast. If the body fluids or tissues are carrying mutagenic

Figure 10.10
Urine-collecting unit

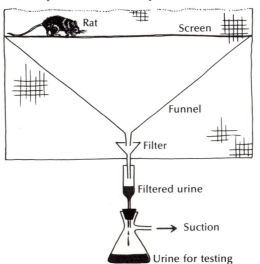

compounds, these indirect indicators carrying specific gene markers will produce mutants. In tests similar to these, the flame retardant called TRIS (used on children's clothing between 1973 and 1977, when it was banned) was shown to be absorbed by mammalian skin and excreted in the urine. It was also shown to be mutagenic and cancer-inducing.

We must, however, be cautious in our assessment of data from indirect indicators. A compound that is not mutagenic in viruses or bacteria may well be mutagenic in humans because that compound is metabolized to a substance that induces mutations. Conversely, a compound mutagenic in viruses may be inactivated in mammals. These and other factors make mutational analysis in mammals very complex.

Rate and frequency of mutations

The terms *mutation rate* and *mutation frequency* are often incorrectly thought of as interchangeable. Actually, they are quite distinct. *Mutation rate* involves a measure of recurrent events; that is, it is the probability of a mutation occurring per round of DNA replication. *Mutation frequency* is the proportion of mutant individuals in the population. An investigator scoring flies for the presence of a particular mutation would describe the mutation frequency as the percentage of mutant flies in the total population, and this may or may not reflect the mutation rate. A case in point is the occurrence of a mutation in an immature sex cell in *Drosophila*. Several rounds of DNA replication may take place before the cell is ready to participate in fertilization, which means that the original mutation has been replicated several times. If all the gametes derived from that single sex cell are viable and if they participate in the formation of zygotes, they could be all scored as *separate mutational events*. In this instance the mutation frequency would be the total number of mutants (x) over the total number of progeny (y): x/y, where x is greater than 1. The mutation rate, in this case *a single event*, would be $1/y$.

**Spontaneous
mutations**
The rate at which mutations occur in a species in its normal environment (or as simulated in a laboratory) is its *spontaneous mutation rate*. Spontaneous mutations may occur as a reaction to the physical–chemical components of an organism's external environment, they may result from certain metabolites, they may reflect inaccuracies in DNA replication or repair, or they may result from recombination mistakes.

Spontaneous mutation rates occurring at a single DNA base pair, and expressed per base pair per round of DNA replication, differ considerably among various types of organisms (Table 10.1). The mutations tabulated represent qualitative base-pair substitutions as well as the addition or deletion of base pairs. Whether one type of mutation predominates over another depends on the species involved. In phage T4, about 20 percent of the spontaneous mutations are base-pair changes. *Salmonella*, on the other hand, has about a 60–80 percent base-pair change rate. And for

Organism	Base pairs per genome	Mutation rate per base-pair replication	Total mutation rate
Bacteriophage λ	4.8×10^4	2.4×10^{-8}	1.2×10^{-3}
Bacteriophage T4	1.8×10^5	1.7×10^{-8}	3.0×10^{-3}
Salmonella typhimurium	4.5×10^6	2.0×10^{-10}	0.9×10^{-3}
Escherichia coli	4.5×10^6	2.0×10^{-10}	0.9×10^{-3}
Neurospora crassa	4.5×10^7	0.7×10^{-11}	2.9×10^{-4}
Drosophila melanogaster	2.0×10^5	7.0×10^{-11}	0.8
Human	2.0×10^9	?	?

Table 10.1
Comparative forward mutation rates

From J. W. Drake, "Comparative rates of spontaneous mutation," *Nature* 221:1132 (1969).

Neurospora, it is about 85 percent. Even the spontaneous mutation rates of different genes within the same organism differ, though this may be a reflection of genes differing in size and hence having more base pairs to mutate. It may also indicate what are known as *hot spots*—regions of a genome that are unstable and mutate with a high frequency.

Mutator and antimutator genes

The spontaneous mutation rate for any strain is usually constant, and although individual genes may exhibit different mutation rates, the rate for any one gene is fairly constant. Occasionally, however, we encounter organisms that exhibit a markedly different spontaneous mutation rate for all genes or perhaps for particular loci—either greatly increased or greatly decreased. Frequently, the cause for this deviation can be traced to a single gene and the location of the gene even mapped.

A *mutator gene*, a gene that causes another nonallelic gene to mutate more frequently, was discovered in maize by Marcus Rhoades (1941). In maize, the production of purple color in the kernels requires the interaction of the nonallelic genes *A*, *C*, and *R*. When these three genes are present as dominants, purple pigmentation results, because of the biosynthesis of anthocyanin (the upper-case and lower-case letters indicate a dominant or a recessive allele, respectively).

Genotype   Phenotype

A-C-R- = Purple kernel
aaC-R- = White kernel
A-ccR- = White kernel
A-C-rr = White kernel

The *A* locus is situated on the third chromosome. A ninth-chromosome dominant gene, *Dt* (dotted), when present, causes the *a* allele to mutate to *A* at a very high rate.

a/a, C/C, R/R, dt/dt = White kernel
a/a, C/C, R/R, Dt/Dt = White kernel with purple spots

The purple spots are clones of cells with the genotype *AaCCRRDtDt*. The presence of at least one *Dt* gene causes the *a* allele to mutate at a very high rate to *A*, giving rise to clones of purple cells.

The aleurone layer of a maize kernel is part of the endosperm, which arises from the fusion of two haploid female nuclei with a haploid sperm nucleus; hence it is triploid. The effects of 0, 1, 2, and 3 doses of *Dt* on the mutability of *a* to *A* in the aleurone are as follows.

aaa	*dt dt dt*	:	0.0 mutations per kernel
aaa	*Dt dt dt*	:	7.2 mutations per kernel
aaa	*Dt Dt dt*	:	22.2 mutations per kernel
aaa	*Dt Dt Dt*	:	121.9 mutations per kernel

Thus a mutator gene in maize, operating by unknown mechanisms, is able to increase the spontaneous mutation rate of another nonallelic gene.

Similar examples of mutator activity have been studied in microorganisms, and advances have been made in understanding the mechanisms involved. One of the most widely investigated mutator genes in bacteria is the *mutT* gene in *E. coli* (*T* stands for H. P. Treffers, who discovered the mutant). The most striking attribute of this mutation is its specificity. It acts predominantly on A-T base pairs and converts them to C-G pairs. This *mutT* gene can increase the mutation rate from A-T to C-G by 1000 times! The mechanism for *mutT* action is not thoroughly understood, but speculations center on a mutant form of the replicating enzyme DNA polymerase that increases the probability of mispairing at A-T sites.

Gene 43 in phage T4 codes for T4 DNA polymerase, and several gene 43 mutations exhibit mutator activity. The mutants induce a variety of base-pair substitutions as well as base-pair additions or deletions. Another series of mutations in gene 43 have an *antimutator* activity. That is, they specifically depress the frequency of base-pair substitutions, additions, and deletions and hence lower the spontaneous mutation rate of T4.

These discoveries indicate a likely role for DNA polymerase in determining spontaneous mutation rates, probably by altering its editing function or its synthesizing function (see Chapter 7).

Position effect

Certain alterations of the genetic material may appear to be mutational, but in fact they are more positional than mutational. Consider the phenomenon known as *position-effect variegation*. If, in *Drosophila*, the w^+ allele located near the tip of the **X** chromosome is inverted so that it comes to lie close to the centromere in the heterochromatic region* of the chromosome, the eye becomes mottled with red and white cells.

* A chromosome consists of an active, uncoiled, lightly staining euchromatin and a highly coiled, inactive, deeply staining heterochromatin.

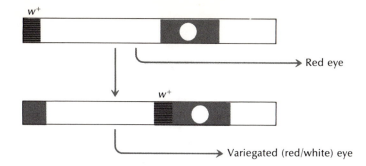

The cells are all genotypically w or w^+, simply because of a positional change in w^+ relative to the heterochromatin. When w^+ is in the vicinity of the heterochromatin, it is unstable, mutating from w^+ to w in the somatic cells.

To test whether the w^+ allele is defective, investigators cross it out so that it occupies its normal position on the end of the **X** chromosome. When this happens, the eye is normal and red. When the w^+ is put back into the heterochromatin, the variegated position effect is observed. Thus the gene is not defective, only unstable (highly mutable) in the vicinity of heterochromatin.

Preadaptive and postadaptive mutations

An issue often raised by evolutionists is whether specific mutations arise in response to a specific environmental stimulus or occur spontaneously and randomly and are then subject to environmental selection. Although there is no clear proof of adaptive mutations occurring *after* exposure to a selective environment (*postadaptive mutations*), there have been many observed instances of spontaneous mutations that were subsequently selected (*preadaptive mutations*). Two particularly good experiments will demonstrate that in microorganisms, at least, mutations are preadaptive.

Perhaps the most elegant technique used to distinguish between preadaptive and postadaptive mutations was Lederberg's replica-plating technique, in which drug-resistant mutants were shown to have existed *before* exposure to the drug (Figure 10.11). Bacteria from a drug-sensitive strain were plated on a drug-free medium, then replica-plated onto a drugged medium. If mutations arose as a specific response to the drug (that is, if the drug were directively mutagenic), the colonies on the drug-containing medium would be distributed at random. If the mutants already existed on the master plate *before* exposure to the drug, their distribution on the drug-containing medium would be nonrandom. That is, resistant colonies on each replica plate would occupy the same positions as the colonies on the other plates. Since the latter was the case, it was clear that these mutations had occurred prior to exposure to the drug

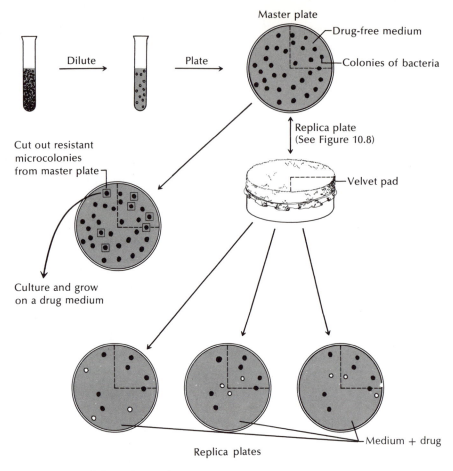

Dilute → Plate →

Master plate
Drug-free medium
Colonies of bacteria

Replica plate
(See Figure 10.8)

Velvet pad

Cut out resistant
microcolonies
from master plate

Culture and grow
on a drug medium

Replica plates

Medium + drug

Figure 10.11
Replica-plating technique
as a means of
distinguishing
preadaptive from
postadaptive mutations

● Colony almost always appears at this location
○ Colony appears at this location only once

and hence were preadaptive. For further verification of preadaptation, the presumed mutant colonies on the master plate were located and cut out. The cells from these colonies were grown in a drug-free medium and then plated out on a drug-containing medium. If the cells from this presumably mutant colony had actually been nonmutant, very few colonies would have appeared on the drug medium. However, if the colony had been mutant, large numbers of colonies would have appeared on the drug medium. The latter was found to be true, confirming the preadaptive nature of the drug-resistant mutation.

Another technique used to distinguish between preadaptive and postadaptive mutations is the *Newcombe spreading technique* (Figure 10.12). Bacteria from a sensitive strain are plated out on a drug-free medium, and colonies are allowed to develop. Half the plates are respread by dragging a sterile wire loop across the medium in various directions. Mutant col-

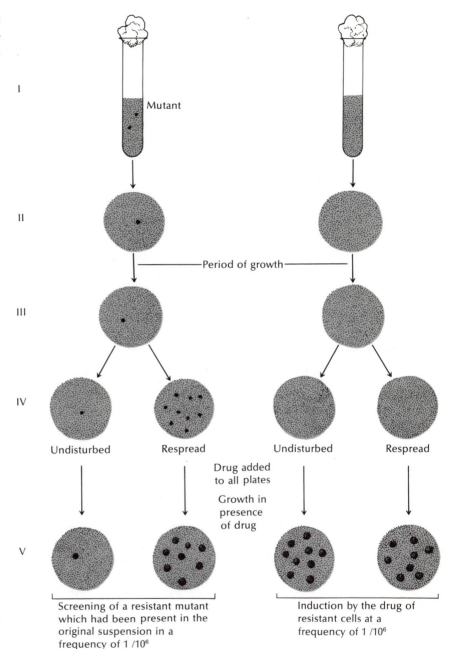

Figure 10.12
Newcombe spreading technique, which tests whether mutations occur as a response to a changing environment (right) or preexist and are subsequently selected by the environment (left).

I

II

— Period of growth —

III

IV

Undisturbed   Respread   Undisturbed   Respread

Drug added
to all plates

Growth in
presence
of drug

V

Screening of a resistant mutant
which had been present in the
original suspension in a
frequency of 1 /10⁶

Induction by the drug of
resistant cells at a
frequency of 1 /10⁶

onies growing on this drug-free medium are broken up by the spreading technique and redistributed over the surface. Separating the cells of the mutant colonies this way establishes many more colonies, all descended from the original parent colony. At this point, a drug is added to the plate and the resistant colonies counted. If a mutation is preadaptive, the number of resistant colonies on the spread plates greatly outnumbers those on the unspread plates because of the separation and redistribution of cells in the original mutant clone. If, on the other hand, mutations are postadaptive, spreading has no effect, because mutations have not arisen until *after* the drug has been added. The results of the experiment support the theory that adaptive mutations occur *before* an organism encounters a particular environment and not as a response to that environment.

Induced mutations of genes

The identification of DNA as the genetic material and the elucidation of its structure established a foundation for a well-defined model of the mutation process. The linear sequence of base pairs in the DNA duplex presents a reasonable basis for understanding the induction of mutations: Change the linear sequence and a mutation occurs. The first agent discovered that resulted in a high mutation frequency was the x ray, first employed by Muller in 1927. About 20 years later came the discovery that chemicals could also induce mutations. And by 1947, geneticists were using the techniques of mutation study to analyze the structure and function of genes.

**The mutagenic properties
of ionizing radiation**

Until 1927, the study of mutants was restricted to those that arose through spontaneous mutation, an occurrence about as common as the appearance of a $10 bill on a city street. In that year, Muller found that heavy doses of x rays produced a high rate of mutation in *Drosophila*. While Muller's experiments were in progress, similar investigations were being carried out to test the effects of x rays and radium on barley. In 1928, L. J. Stadler announced that gamma radiation from x rays and radium did indeed induce a high frequency of mutations.

X rays, high-speed noncharged particles like neutrons, and charged particles like helium nuclei (alpha particles) are all included in the term *ionizing radiation,* which means that when they traverse matter they can generate fast-charged particles, usually through collision with other atoms. Nonionizing radiation such as UV light does not give rise to charged particles when it traverses matter. Figure 10.13 places ionizing and nonionizing radiation in their respective positions on the scale of electromagnetic wavelengths.

Ionizing radiation is generally inadequate as a tool for probing gene structure and function, primarily because the type of mutational event it induces is often impossible to characterize. Radiation genetics has, however, been very useful in the study of mutation rates, chromosome structure, and genetic repair mechanisms. X rays can act directly or indirectly on the genetic material, but either way, the types of mutational events remain the same. Chromosome breaks are induced, and reconstitution produces new structures such as inversions, translocations, deletions, and duplications. Or intragenic mutations are induced that involve base-pair substitutions or the addition and deletion of base pairs. When x rays

Wavelength, cm (log scale)

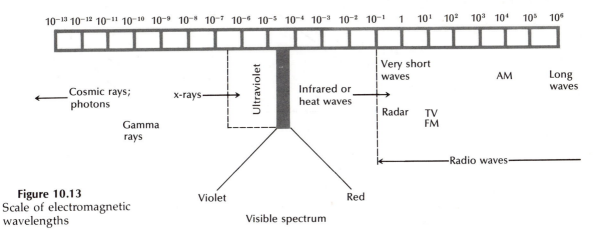

Figure 10.13
Scale of electromagnetic wavelengths

act directly on the genetic material, it is postulated that the radiation disrupts covalent linkages, causing a variety of structural alterations. Acting indirectly on the genetic material, x rays generate highly reactive free radicals from water, and these free radicals react with DNA to alter its structure.

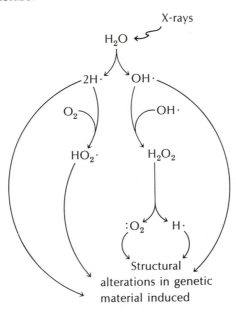

Dots indicate free radicals

There is a linear relationship between the mutation frequency of sex-linked recessive lethals in *Drosophila* and radiation dosage (Figure 10.14a). To explain the relationship, it has been suggested that one hit induces one mutation. If this assumption holds, it should be possible to estimate the

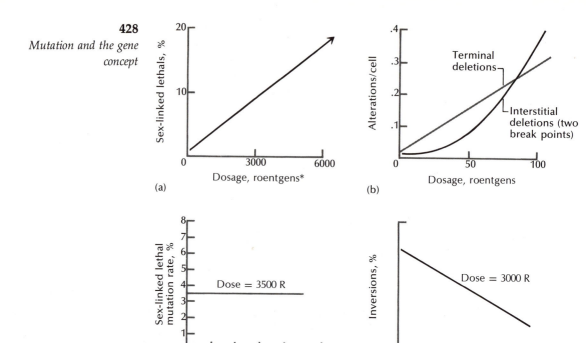

Figure 10.14
Ionizing radiation and
mutations

*1 roentgen (R) = amount of ionizing radiation required to produce one electrostatic unit of charge per cubic centimeter of air

average *size* of the gene, since larger genes should mutate more frequently than smaller ones. The *target theory* was an attempt to use x-ray–induced mutation frequency as a means of estimating size of genes. To illustrate by analogy, assume that you are blindfolded and standing before a wall of unknown size that is hung with balloons of known size and number. You could estimate the size of the wall if you could throw darts randomly at the wall. Knowing the size and number of the balloons, the number of darts thrown, and the number of balloons popped, you would be able to make a fairly good estimate of the size of the wall.

The target theory stimulated much debate, but was ultimately discarded as a means of estimating size of genes when it was shown that a medium irradiated with x rays was mutagenic to bacteria subsequently placed on it, thus establishing the indirect action of x rays. That is, one unit of x radiation could generate several highly mutagenic free radicals. This effect of x rays precluded a simple interpretation of the effect of x radiation on the genetic material.

The linearity between dose and the frequency of induced sex-linked recessive lethal mutations suggests that these mutations are one-hit events, that the mutation is caused by a single event, such as a base-pair substitution or a base-pair deletion. For two-hit events, such as inversions, which require two separate breaks, an exponential curve should, and indeed does, occur (Figure 10.14b). Mutational events involving two breaks are rare at low doses of radiation, but as dosage increases, an exponential increase in the occurrence of two-break events becomes evident.

The relationship between radiation intensity (the amount of radiation delivered per unit of time) and one-hit mutational events, such as sex-linked recessive lethals, points out the very great danger of x-ray radiation to biological systems. A dose of 3500 roentgens (denoted by R), whether administered in one dose or over a period of several hours, days, or weeks produces the same mutation rate (Figure 10.14c). So as far as mutations are concerned, radiation has a cumulative effect. This same curve does not hold for two-hit events, such as inversions, because the time interval preceding the administration of the second dose of radiation allows for repair of damage induced by the first exposure (Figure 10.14d).

The discovery of the indirect effects of x rays raises many questions about the use of x rays for sterilizing foods consumed by humans. Ionizing radiation is commonly used to break the dormancy of seeds, to inhibit the sprouting of stored potatoes and onions, and to sterilize food. Ionizing radiation is preferable to heat, because it does not break down the foods, as heat does, and it can be employed on food already packaged. But is this irradiated food mutagenic to us? The answers are not clear.

If a bacterial medium is irradiated heavily and then allowed to support bacterial growth, the bacteria experience a high kill rate and a high mutation rate. The basis of this appears to be a sugar broken down by the radiation, a substance that then produces mutagenic compounds. Mammalian cells exposed to irradiated blood exhibit an increased level of chromosome breakage.

Although it may be possible for irradiated foods to produce mutations through indirect action, we should not leap to conclusions. The free radicals generated by irradiating water molecules are extremely short-lived (a fraction of a millisecond) and of no danger to us in stored, irradiated food. The only way for irradiated food to be harmful would be through the generation of a stable mutagenic compound. Studies of *Drosophila, Habrobracon* (a parasitic wasp), and other organisms failed to show the production of mutagenic substances. But studies using irradiated glucose produced chromosome breaks in plant cells and in cultured human lymphocytes.

Nevertheless, ingesting irradiated sugar for 8-week periods had no deleterious effect on rats. When *Drosophila* ingested irradiated sugar, Rinehart found a slight increase in frequency of mutations. And Bugyaki and his colleagues found that rats fed on heavily irradiated flour throughout their entire life showed an increased rate of chromosome breakage. So ingesting large quantities of heavily irradiated food over a long period of time may be dangerous. But the jury is still out on whether small quantities of lightly irradiated foods are dangerous.

As mentioned earlier, radiation genetics suffers from a lack of information about what happens at the molecular level. No one doubts that ionizing radiation induces a wide range of mutational events in living systems, by either direct or indirect means. However, the mechanisms by which the radiation interacts with nucleic acids to induce specific types of events continue to inspire investigation.

The mechanisms of UV-induced mutations

In Chapter 7, we discussed how ultraviolet light, a nonionizing form of radiation, interacts with nucleic acids to produce pyrimidine dimers. If not removed, these dimers are usually lethal for a cell. But UV irradiation does more to a cell than induce dimerization; it induces mutations. These mutations can arise as a consequence of dimerization or as a consequence of the hydration of cytosine and, to a lesser extent, thymine.

Cytosine hydrate   Dihydro-thymine

Many UV-induced mutations occur at G-C sites, suggesting that hydrated cytosine may be an important source of UV-induced mutations. The hydration of cytosine can cause changes in the hydrogen-bonding capabilities of cytosine, leading it to mispair with A during replication or repair. The result of this would be a GC-to-AT mutational event. Evelyn Witkin and her colleagues at Rutgers University have studied UV-induced mutations for many years. They conclude that almost all UV-induced mutations in *E. coli* and many of its phages are the result of the inaccurate repair of UV-induced damage (dimers and single-stranded breaks). The repair systems discussed in Chapter 7 (photoreactivation, dark repair) are relatively accurate and only rarely lead to mutations. But another repair system, the "SOS" system (named for the international distress signal), is an error-prone system that can lead to a high mutation rate.

The SOS system is activated when a cell experiences damage such as single-stranded breaks or even dimers that block replication. The SOS enzyme system promotes cell and phage survival by, among other things, promoting DNA synthesis past pyrimidine dimers. However, in passing over these dimers, incorrect bases are inserted, increasing the mutation rate. The SOS repair activities are distinguishable from the activities of DNA polymerases I, II, III, and III*.

Dimerization can directly or indirectly result in mutations. Directly, the dimer can remain in the molecule through one round of DNA replication. The resulting distortion of the DNA molecule can influence hydrogen bonding and cause the insertion of incorrect bases in the newly synthesized DNA strand. Indirectly, the excision and repair of DNA carrying a dimer can result in a mutation, because the repair enzyme may not be as precise as the replicating enzyme. The two genes that appear to be crucial to this SOS repair system in *E. coli* are the *recA*+ and *lexA*+ genes. Figure 10.15 summarizes the possible pathways of error-prone SOS repair of UV-induced damage in *E. coli*.

Figure 10.15
Error-prone "SOS" repair of DNA after ultraviolet irradiation

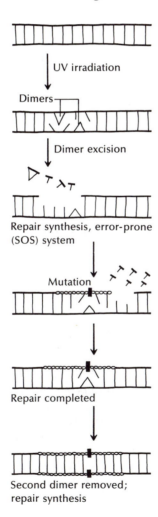

UV irradiation

Dimers

Dimer excision

Repair synthesis, error-prone (SOS) system

Mutation

Repair completed

Second dimer removed; repair synthesis

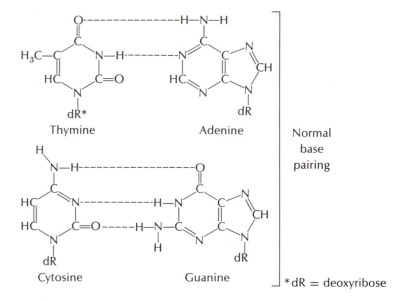

Figure 10.16
Tautomeric shifts of the DNA bases. Abnormal base pairing leading to base-pair substitution

Thymine   Adenine

Cytosine   Guanine

Normal base pairing

*dR = deoxyribose

Chemically induced mutations

In 1942, C. Auerbach and J. M. Robson in England discovered that a potent World War I weapon, mustard gas, was extremely mutagenic in *Drosophila*. Whether for reasons of politics or security, the discovery was not made public until 1946. Up to that time, ionizing radiation had been the most widely used mutagenic agent, but its poorly defined effects limited its value. The discovery of chemical mutagens, which permit a more predictable type of mutational event and thus a more refined approach to the structure of a gene, opened up an entirely new era of mutation study.

Chemical mutagens can be loosely categorized by their effects on the genetic material. Some mimic the effect of ionizing radiation (*radiomimetic* chemicals), which brings about poorly defined genetic damage. Some chiefly cause *base-pair substitutions*. Others result mainly in the *addition and deletion of base pairs*. Our discussion will focus on the latter two categories, because they have contributed the most to our understanding of mutation mechanisms, gene structure, and gene function.

Base-pair substitutions   The model of DNA proposed by Watson and Crick implied a mechanism for mutation by base-pair substitution. Under normal physiological conditions, the DNA bases exist as *tautomeric* molecules. That is, they are in equilibrium between two alternative chemical configurations termed the keto and enol forms, but the equilibrium is such that these bases are usually keto and only rarely enol (Figure 10.16).

Tautomeric shifts in the four DNA bases

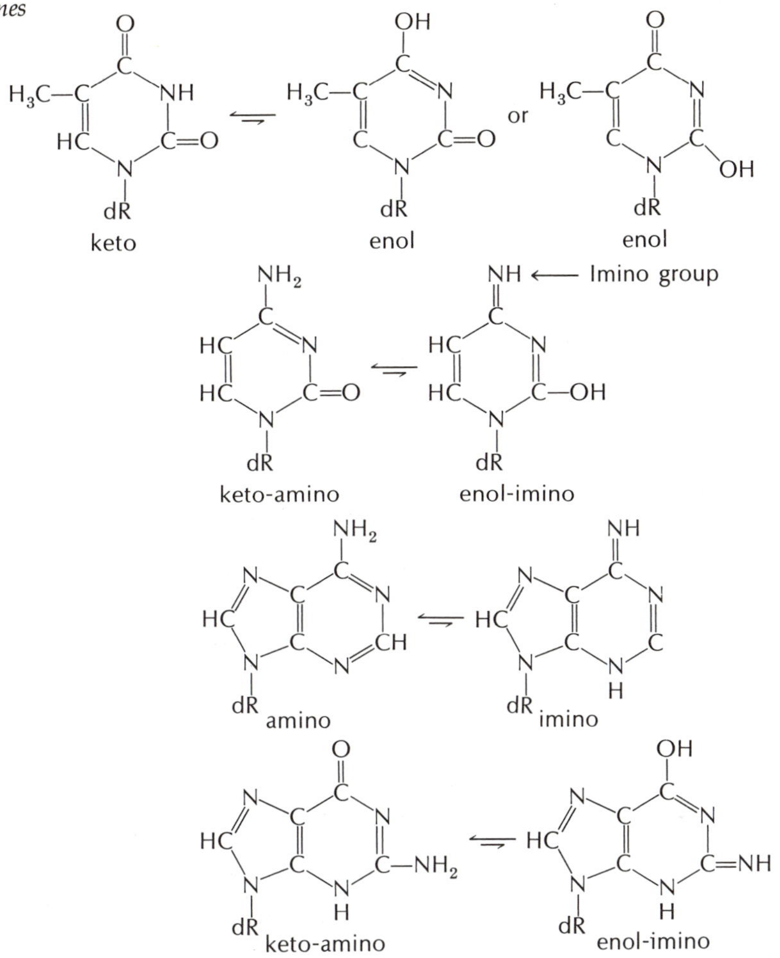
keto
enol
or
enol
NH ⟵ Imino group
keto-amino
enol-imino
amino
imino
keto-amino
enol-imino

Thymine

Cytosine

Adenine

Guanine

Figure 10.16 (cont.)

Abnormal base pairing leading to base-pair substitution

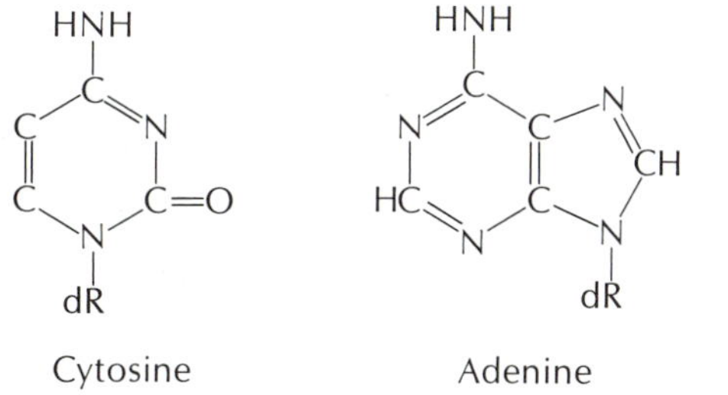
Cytosine
Adenine

Hydrogen bonding highly improbable

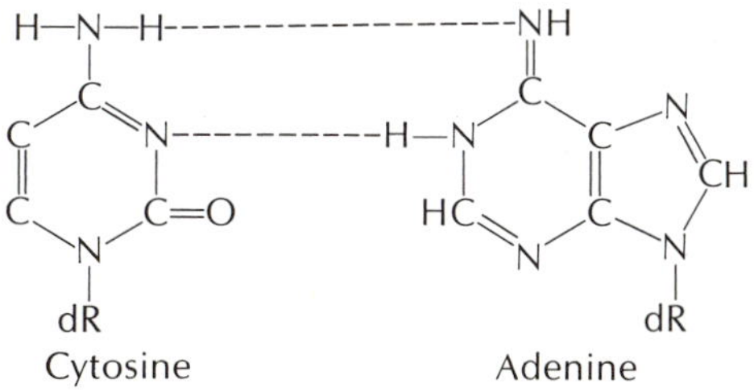
Cytosine
Adenine

Hydrogen bonding possible between adenine and cytosine when adenine is shifted into imino form

Mispairing becomes more likely in the enol state, leading to the following sequence of events.

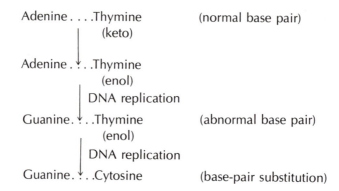

Adenine . . . .Thymine   (normal base pair)
          (keto)

Adenine . ⅄ . .Thymine
          (enol)
          DNA replication

Guanine. ⅄ . .Thymine   (abnormal base pair)
          (enol)
          DNA replication

Guanine. ⅄ . .Cytosine   (base-pair substitution)

A G-C base pair is substituted for an A-T base pair, because a tautomeric shift makes mispairing more probable.

This proposed mechanism for mutation has led many investigators to seek out chemicals that would facilitate substitution. A number of chemical agents have been found to induce a base-pair substitution. For convenience, they can be grouped into two categories. *Base analogs* are nitrogenous bases that resemble the normal DNA bases but are sufficiently different in structure that when *incorporated into* replicating DNA in place of the regular bases, the analogs increase the probability of incorrect base pairing. Such base analogs include 5-bromouracil and 2-aminopurine (Figure 10.17a). *Agents that act on nonreplicating DNA*—among which are nitrous acid (HNO_2), an alkylating agent called ethyl methanesulfonate (EMS), and hydroxylamine (HA)—directly alter resting, nonreplicating DNA (Figure 10.17b). The DNA so altered is likely to undergo base-pairing mistakes during subsequent replication.

Let us consider the ways in which these two classes of mutagenic agents interact with DNA and then construct models that predict how the agents may cause mutational events (Figure 10.18). For the base analogs (Figure 10.18a and Figure 10.18b), two events must take place before a mutation occurs. First, the analog must be *incorporated* in DNA. Second, it must *mispair* during replication in such a way that a *transition* mutation is brought about—a base-pair substitution in which a pyrimidine base on one strand is exchanged for a different pyrimidine base and in which the purine base on the complementary strand is exchanged for a different purine base. For example, an A-T pair is substituted for a G-C pair. A *transversion* occurs when the purine–pyrimidine orientation is reversed (that is, when an A-T pair substitutes for a T-A pair or an A-T pair for a C-G pair).

(a) Base analogs

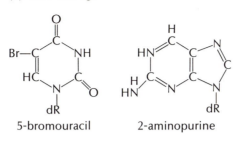

5-bromouracil   2-aminopurine

(b) Agents which act on resting DNA

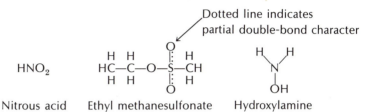

Figure 10.17
Common chemical
mutagens

Nitrous acid   Ethyl methanesulfonate   Hydroxylamine

There are actually two mechanisms by which base analogs can induce mutations, and the thymine analog *5-bromouracil* (BU) demonstrates both. In an *error of replication*, BU is incorporated into replicating DNA in place of T and opposite A so that we have an A-BU pair. When BU is in its keto form, this is the most probable event. After one or more rounds of replication, the BU, in its much less common enol form, can mispair with G, which on further replication gives rise to a G-C pair at the site where an A-T was originally. In an *error of incorporation*, BU, temporarily in its rarer enol state, acts as a C and is incorporated opposite a G *during replication*, thereby forming a G-BU pair. Subsequent replication has a high probability of fixing an A-T pair at what was originally a G-C site. These two mechanisms, presumably characteristic for all base analogs, are summarized in Figure 10.19.

Nitrous acid (HNO$_2$) acts by removing an —NH$_2$ group from adenine to form hypoxanthine, an —NH$_2$ from cytosine to form uracil, and an —NH$_2$ from guanine to form xanthine (Figure 10.18c). So altered, these bases are more likely to mispair (except for xanthine, which normally pairs with cytosine) and hence lead to transition mutations of both the A-T to G-C variety and the G-C to A-T variety.

Hydroxylamine (HA) is specific for cytosine. By adding an —OH to the —NH$_2$ group of cytosine, the modified cytosine can then mispair with an adenine, resulting in a G-C to A-T transition mutation (Figure 10.18d).

Ethyl methanesulfonate (EMS), one of the most powerful mutagenic agents known, acts primarily by adding an ethyl group (—CH$_2$CH$_3$) to

Figure 10.18
Models of mechanisms by which some chemical mutagens induce base-pair substitutions (Figure continues on pages 437 and 438.)

(a) 2-aminopurine mutagenesis

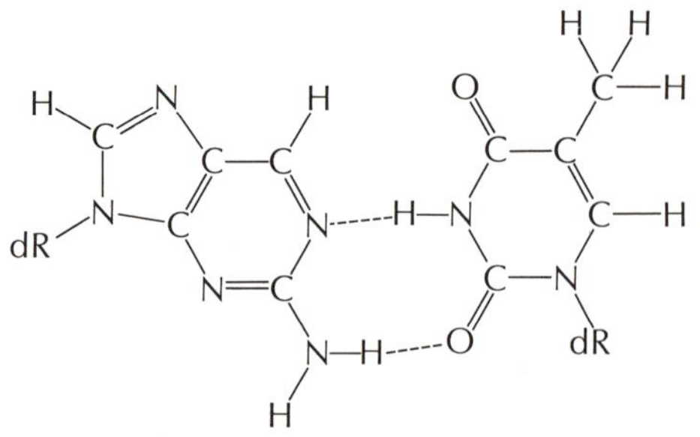

2-aminopurine (normal state)　　　Thymine

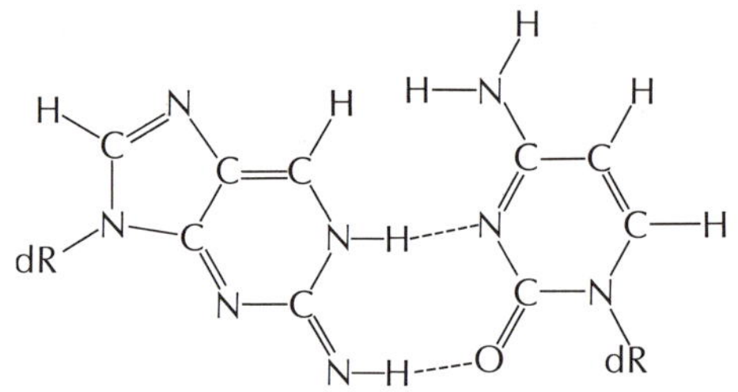

2-aminopurine (rare imino state)　　　Cytosine

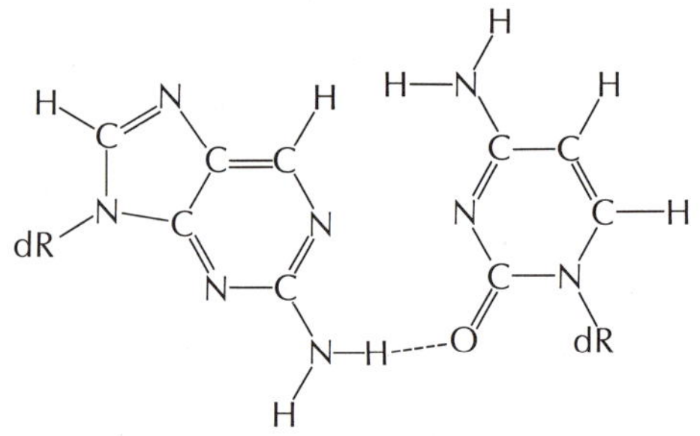

2-aminopurine (normal state)　　　Cytosine

(b) 5-bromouracil mutagenesis

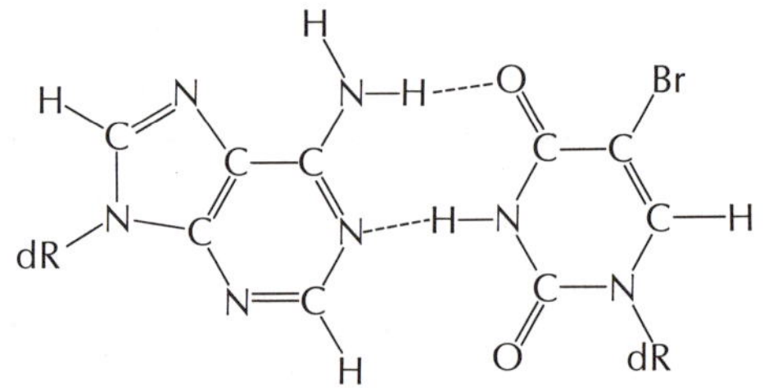

Adenine　　　5-bromouracil (normal keto state)

Figure 10.18 (cont.)

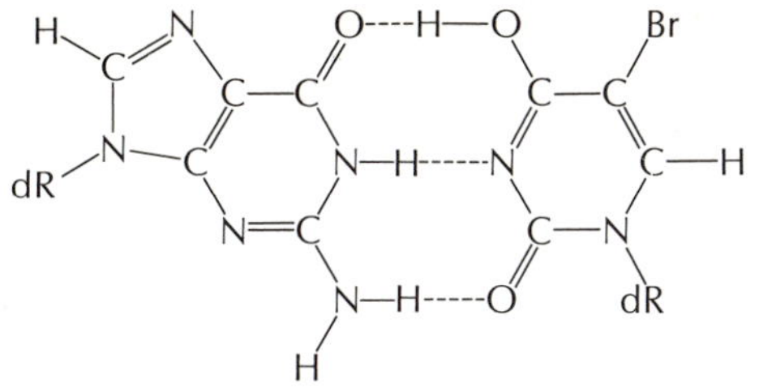

Guanine   5-bromouracil
                 (rare enol state)

(c) Nitrous acid mutagenesis

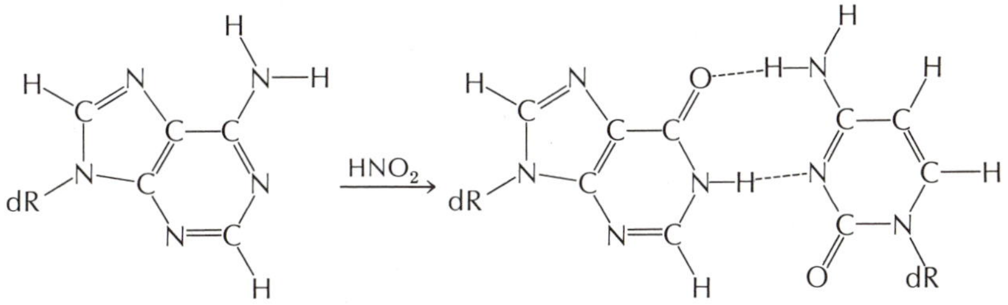

Adenine   Hypoxanthine   Cytosine

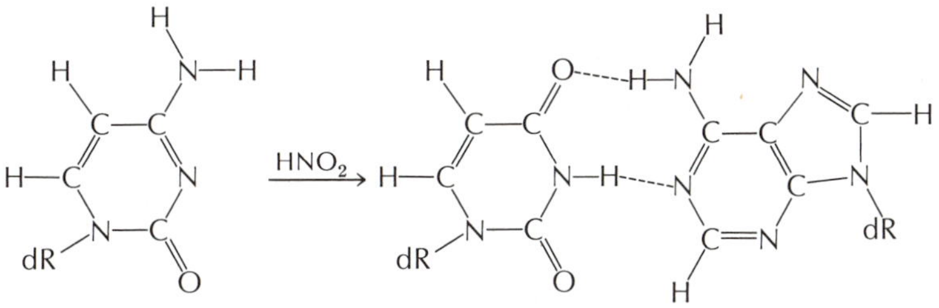

Cytosine   Uracil   Adenine

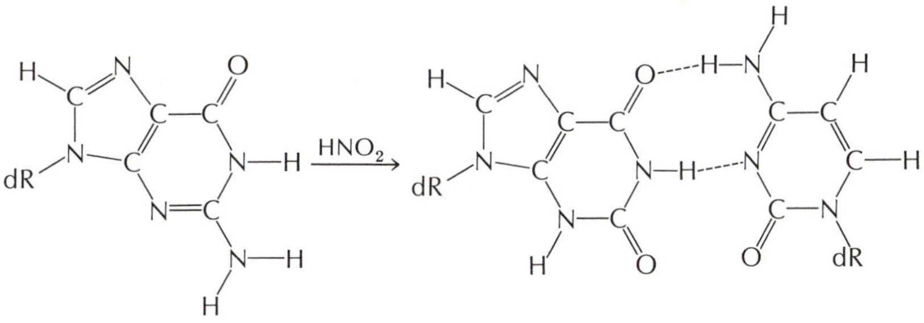

Guanine   Xanthine   Cytosine

(d) Hydroxylamine mutagenesis

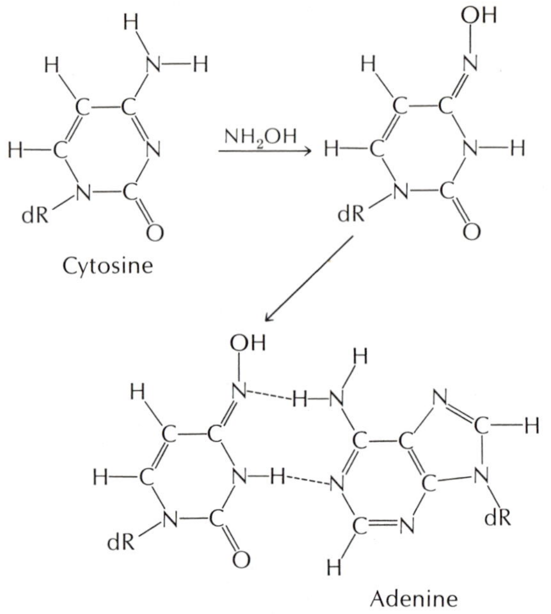
NH2OH
Cytosine
Adenine

Figure 10.18 (cont.)

(e) Ethyl methanesulfonate mutagenesis

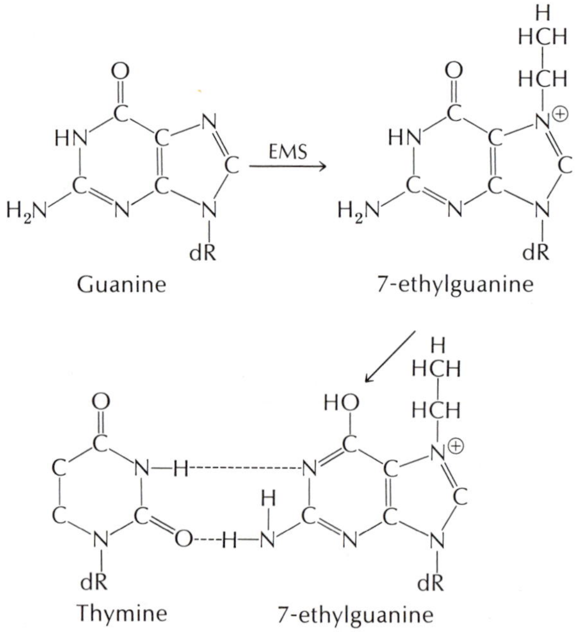
EMS
Guanine
7-ethylguanine
Thymine
7-ethylguanine

(a) Mutagenesis by 5-bromouracil (BU)

Erroneous pairing with guanine during:

(b) Mutagenesis by nitrous acid

Deamination of:

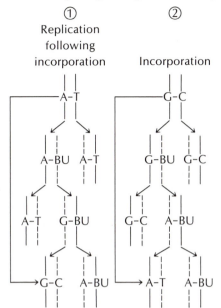

① Replication following incorporation

② Incorporation

③ Adenine (A) to hypoxanthine (HX)

④ Cytosine (C) to uracil (U)

Figure 10.19
Proposed mechanisms for bromouracil mutagenesis

guanine and, though to a much lesser extent, to adenine. The ethylation changes the hydrogen-bonding properties of G and A and often leads to a mispairing involving the alkylated base, which results in transition mutations. EMS also induces transversions by removing purines from DNA and inserting any one of the four bases into the empty gap. A summary of the proposed mutagenic results of these chemical mutagens appears in Table 10.2.

Presently, the best way to verify the various models of chemical mutagenesis is through *reverse-mutation analysis*. If a particular mutagen induced a G-C to A-T transition mutation, a mutagen that induces A-T to

Table 10.2
The base-pair substitutions assumed to be induced by various chemical mutagens

Mutagen	Substitutions	Type of substitutions
5-bromouracil (BU)	A-T ⟷ G-C	Two-way transitions
2-aminopurine (AP)	A-T ⟷ G-C	Two-way transitions
Nitrous acid (NA)	A-T ⟷ G-C	Two-way transitions
Hydroxylamine (HA)	G-C ⟶ A-T	One-way transition
Ethyl methanesulfonate (EMS)	G-C ⟶ A-T	One-way transition
	G-C ⟷ C-G	Transversion
	A-T ⟷ T-A	Transversion
	G-C ⟷ T-A	Transversion
	A-T ⟷ C-G	Transversion

Table 10.3
Summary of AP- or
BU-induced reversion of
mutants produced by
various mutagens

Mutagen-inducing mutation	Total number	Reversible with AP or BU (percent)	Nonreversible with AP or BU (percent)
2-aminopurine	98	98	2
5-bromouracil	64[a]	95	5
Nitrous acid	47	87	13
Hydroxylamine	36	94	6

[a]This number includes untested mutants which have the same genetic location and spontaneous reversion index as a mutant that was tested for reversion induction. They are assumed to be reversion-inducible or noninducible according to the mutant tested.

Data from E. Freese, *Proc. Vth Int. Cong. Biochemistry*, pp. 204–226 (1961).

G-C transitions should be able to revert it to normal. Table 10.3 summarizes reversion studies of the various chemically induced mutations. (These reversion studies are usually done with T4 bacteriophage, because they are easy to handle and screen.) The data accumulated in studies of reverse mutation support the basic models of mutagenesis already discussed. For example, a BU-induced mutation in T4 is weakly revertible by BU but strongly revertible by aminopurine (AP) and by HNO_2 and not at all revertible by HA. This suggests that BU induces primarily G-C to A-T transitions. An AP mutation is weakly revertible by AP but is strongly revertible by BU and HNO_2, suggesting that AP induces both A-T to G-C and G-C to A-T transitions.

Reversion studies do not always agree with the predictions, however. Note, for example, that 5 percent of the BU-induced mutations, which presumably arose by A-T to G-C or G-C to A-T transitions, are *not* revertible with either BU *or* AP. This does not conform to the model of BU mutagenesis but rather serves to remind us that a precise and predictable pattern of chemical mutagenesis has not been entirely realizable. Data on chemical mutagenesis should be interpreted very cautiously.

Frameshift mutations   In 1959, Freese suggested that mutations induced in T4 by *proflavin* (Figure 10.20a) and *acridine* (Figure 10.20b) were

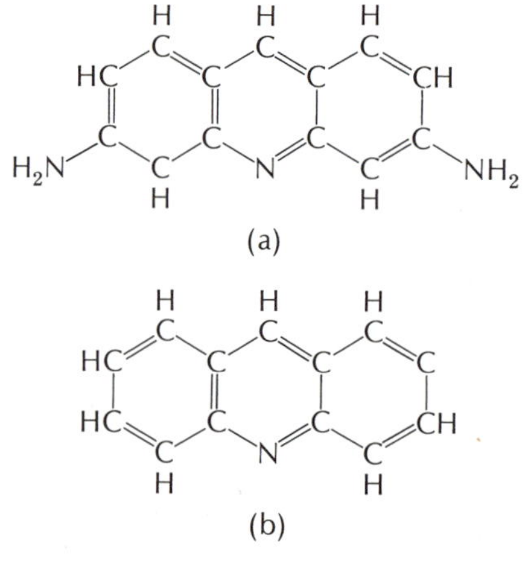

Figure 10.20
Structure of (a) proflavin
and (b) acridine

(a)

(b)

transversions. But in 1961, Crick and his colleagues suggested that base-pair substitutions, either transitions or transversions, were not induced by these two mutagens. They proposed instead that proflavin and acridine caused the *addition or deletion* of a base pair, an event called a *frameshift mutation*. (See Chapter 9 for a review of frameshift mutations.) Frameshift mutations alter from one to several amino acids in a polypeptide chain, in direct contrast to base-pair substitution mutations, which alter only one amino acid.

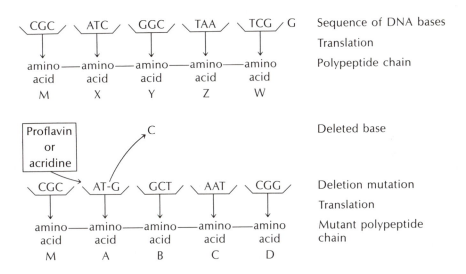

The suggestion that an addition–deletion type of mechanism accounted for proflavin and acridine mutagenesis gained considerable support in 1963, when L. S. Lerman showed that acridines bound to DNA in such a way that the ringed molecule *intercalated* between the adjacent base pairs (Figure 10.21). This stretched the internucleotide distance between them from 3.4 Å to 6.8 Å, thus making it possible for a base pair to be added or deleted by reason of the instability created in the distorted DNA backbone.

In 1966, Streisinger and his colleagues substantiated an addition–deletion mechanism. They analyzed the amino acid sequence of the T4 enzyme lysozyme, which causes the bacterial host cell to lyse. They showed that lysozyme mutants induced by proflavin carry abnormal *sequences* of amino acids compared to the normal enzyme (Figure 10.22). Base-substitution mutations carry single amino acid alterations.

Figure 10.21
Schematic model for the intercalation of acridines (shaded) into DNA [From L. S. Lerman, *Proc. Natl. Acad. Sci. U.S.* 49:94 (1964)]

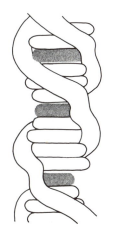

Mutations and the environment

Our environment contains an increasingly large number of chemicals that may induce deleterious mutations with long-term effects. We should be aware of the potential hazards we face as a consequence of these chemi-

Figure 10.22
(a) Amino acid sequence of areas of lysozyme (continued on page 443)

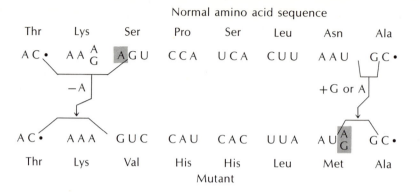

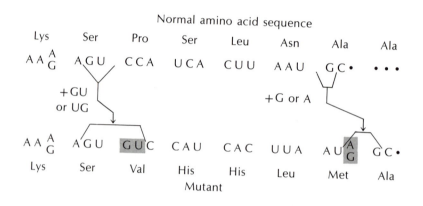

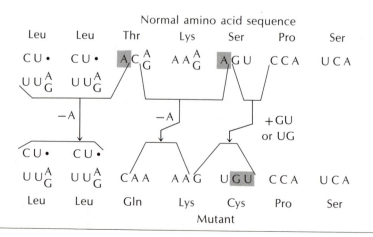

Figure 10.22(a) (cont.)

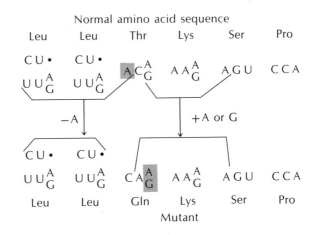

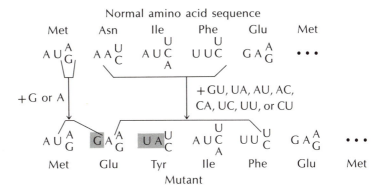

cals. Should these chemicals induce mutations in sex cells, we may be condemning future generations to lives of misery and disease. Such things as mercury, toxic at modest levels, is now known to be mutagenic at low levels at which it is not toxic.

H. J. Muller and L. J. Stadler were the first to show that a humanly controlled agent, x rays, was mutagenic and a danger to all of us. Up to that time, x rays were used indiscriminately. I personally remember shoving my feet into a special x-ray machine in a shoe store when I was a child, so that I could watch my toes wiggle. We know now that x rays increase the burden of genetic defects in humans, but what about the other environmental agents?

Geneticists are increasingly vocal about their concerns for the potential genetic hazards created by the chemicals that we encounter almost daily. New techniques are being developed that permit detection of low-level mutations (Figure 10.23 shows a technique developed by Bruce Ames to detect mutagenic agents). We are discovering that agents that are carcinogenic are also mutagenic.

Figure 10.22 (cont.)
(b) A model for the
production of frameshift
mutations (after
Streisinger)

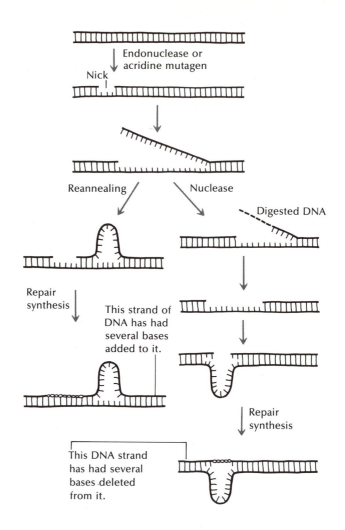

Endonuclease or
acridine mutagen

Nick

Reannealing   Nuclease

Digested DNA

Repair
synthesis

This strand of
DNA has had
several bases
added to it.

Repair
synthesis

This DNA strand
has had several
bases deleted
from it.

Many common components of our environment are mutagens or potential mutagens. X rays and mercury have already been mentioned. Cigarette smoke and hair dyes also contain mutagenic chemicals. The artificial sweetener cyclamate is mutagenic, and its use has been prohibited in many countries. The study of saccharin, another artificial sweetener, has not yet yielded conclusive results, though some studies show it to be carcinogenic and mildly mutagenic. Nitrites, used as color enhancers in bacon, ham, hot dogs, and other meats, are mutagenic. Vinyl chloride, used in plastics, phonograph records, floor tiles, and many other products, is mutagenic when used in Ames's test system.

These are but a small sample of the thousands of chemicals we are exposed to in our increasingly polluted world. It is clear that if we are to survive as a healthy, vigorous species, we must be aware of the potential mutagenic hazards of our environment.

Figure 10.23
A screen for detecting mutagenicity. The indicator organism, a *Salmonella* histidine-requiring mutant induced as a transition mutant, is spread on a histidine-deficient medium. Only reverse mutations (his⁻ → his⁺) survive and grow on this medium. A disc impregnated with the test compound is placed on the center of the plate, and the compound diffuses from the disc outward, creating a decreasing concentration gradient. In this figure, we see that MNNG is highly mutagenic, TEM less so. The control plate shows the frequency of spontaneous mutation.

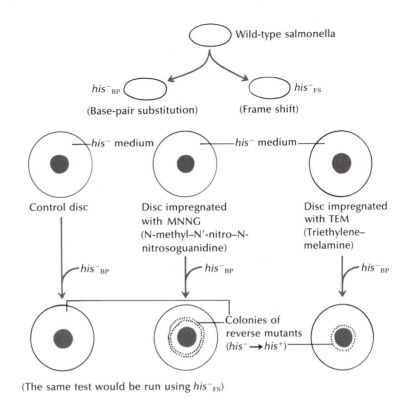

The gene concept

Ionizing radiation and chemical mutagens enabled geneticists to accumulate mutants in large numbers. Mutants became so numerous and so easy to induce that, in many cases, individual genes were represented by several independently arising mutant alleles. The abundance of mutants stimulated investigations into the *functional composition* of the gene, and this in turn led to our current functional interpretation of the gene.

The gene concept began with what Mendel called "factors," units of unknown chemical nature that were passed from generation to generation in a specific fashion and that determined specific characters. Genes were later regarded as indivisible units of a chromosome, within which mutation, but not crossing-over, could occur. However, the discovery of multiple alleles soon showed that this view was too simple, for it was possible to demonstrate that recombination could occur between different mutants of the same gene. Multiple allelism developed into a concept called *pseudoallelism*, which proposed that alleles that were separable by recombination were at the same time functionally related. Pseudoallelism stimulated extensive research into the functional interaction between different alleles in a pseudoallelic series.

445

Primarily as a result of the analysis of pseudoallelic series, the gene emerged conceptually as a *unit of function*, or *cistron*. The cistron has been defined by a genetic test designed to show the functional relatedness of genes. The test was first developed for higher organisms and subsequently applied to phages. The dumpy (*dp*) pseudoallelic series in *Drosophila* offers an opportunity to explore the functional relatedness of alleles and to use this property as a basis for defining the gene. If one member of a homologous pair of chromosomes contains the dp^{ov} allele and the other contains the dp^v allele, the phenotype is mutant. The two pseudoalleles are said to be *noncomplementing* because in a *trans* position (opposite each other on separate chromosomes), they produce a mutant phenotype. Both alleles on the same chromosome (*cis* position) give a normal phenotype. However, if another dumpy pseudoallele, dp^l, is on one chromosome and dp^{ov} on the other (putting them in a *trans* position), the phenotype is normal, or wild-type. In the latter case, the pseudoalleles are said to *complement* each other because in a *trans* configuration, they give rise to a nonmutant phenotype.

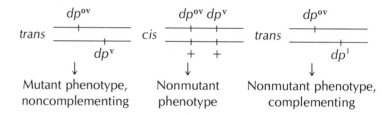

The *cis-trans position effect* forms the basis of the cistron concept: Two mutations that in the *trans* position produce a mutant phenotype are said to belong to the same functional unit, or cistron. If the two mutations complement each other in the *trans* position, they are said to belong to different functional units, or cistrons.

The cistron, defined by complementation or noncomplementation, has come to be regarded as synonymous with the gene. However, this interpretation of the gene in higher organisms suffers from a failure to show a relationship between the DNA molecule itself and the protein product of the gene. Pseudoallelic series have been viewed in two ways. E. B. Lewis and M. M. Green suggested that each mutation separable by recombination was actually a distinct gene with a distinct function. They explained the effect of the *cis-trans* position by arguing that each gene in a pseudoallelic series produced a product that was passed on to the gene next to it. There the product acted as a substrate for a new reaction, the product of which went over to the next gene, and so forth. This idea, referred to by some as the "bucket brigade hypothesis," explained a mutant phenotype in the *trans* configuration with the hypothesis that a gene product could not be passed from one homologous chromosome to another, either because the distance was too great or because the product was produced in too small a quantity or in too unstable a form to survive

Figure 10.24
Heterokaryon formation in *Neurospora*

• Nucleus, mating type +
o Nucleus, mating type −

such a distance. This was a purely speculative model; no gene products were analyzed.

An alternative interpretation was presented by G. Pontecorvo, who proposed that pseudoalleles represent mutations at different and separable sites within a single unit of function, or cistron. On this basis, a gene functions normally if it is nonmutant, but a mutation at any point within the gene will result in either a changed gene product or no product at all. Pontecorvo added that a single gene (cistron) is divisible through recombination (in contrast to the suggestions by Lewis and Green), although it may occasionally appear to be indivisible because of some difficulties encountered in obtaining recombinants.

The Lewis–Green interpretation of pseudoallelism is probably applicable to some situations. Their ideas may be appropriate to situations in which it can be demonstrated that gene products actually interact. However, great care should be exercised in assessing the "bucket brigade hypothesis." The idea that recombination can occur *within* a cistron, as proposed by Pontecorvo, has gained much wider support.

Genetic complementation in higher organisms has in general proved difficult to fully grasp. It has, however, been successfully examined in *Neurospora*. A general molecular model can be developed from a complementation study of the *arg*-1 mutation (mutants that are unable to synthesize arginine) and several mutants of the *am* locus (codes for glutamic dehydrogenase production).

Neurospora is haploid, and complementation studies require a homologous pair of chromosomes. But a state of pseudodiploidy is sometimes observed in the formation of a *heterokaryon* (Figure 10.24), a binucleate situation in which two different nuclei occupy the same cytoplasm. The genotype of each nucleus is defined. Nuclei of differing genotypes are used, and complementation can be determined, provided that the gene products leave the nucleus and interact in the common cytoplasm.

Using heterokaryons, investigators carried out complementation tests among *arg*-1, *am*-1, *am*-3, *am*-4, and *am*-14 mutants by having each nucleus carry a different mutant. The results are shown in Figure 10.25.

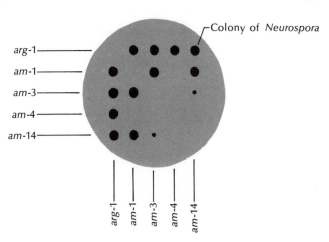

Figure 10.25
Complementation study in *Neurospora*

Figure 10.26
Model for complementation between the *arg*-1 and *am* loci

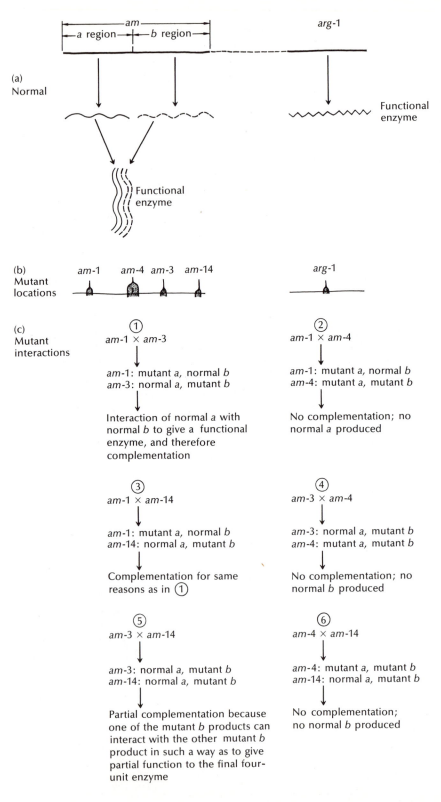

(a) Normal

(b) Mutant locations

(c) Mutant interactions

① *am*-1 × *am*-3

am-1: mutant *a*, normal *b*
am-3: normal *a*, mutant *b*

Interaction of normal *a* with normal *b* to give a functional enzyme, and therefore complementation

② *am*-1 × *am*-4

am-1: mutant *a*, normal *b*
am-4: mutant *a*, mutant *b*

No complementation; no normal *a* produced

③ *am*-1 × *am*-14

am-1: mutant *a*, normal *b*
am-14: normal *a*, mutant *b*

Complementation for same reasons as in ①

④ *am*-3 × *am*-4

am-3: normal *a*, mutant *b*
am-4: mutant *a*, mutant *b*

No complementation; no normal *b* produced

⑤ *am*-3 × *am*-14

am-3: normal *a*, mutant *b*
am-14: normal *a*, mutant *b*

Partial complementation because one of the mutant *b* products can interact with the other mutant *b* product in such a way as to give partial function to the final four-unit enzyme

⑥ *am*-4 × *am*-14

am-4: mutant *a*, mutant *b*
am-14: normal *a*, mutant *b*

No complementation; no normal *b* produced

Arg-1 complements everything except *arg*-1, which is expected because the *arg*-1 and *am* loci produce totally different enzymes. It is more notable that *am*-1 and *am*-3 complement each other, as do *am*-1 and *am*-14. Even more interesting is the interaction between *am*-3 and *am*-14, where a "mini-colony" of *Neurospora* is observed, which suggests *partial* complementation. A model that attempts to resolve these results appears in Figure 10.26. The *am*-1 mutant could be a point mutant in the *A* region, whereas *am*-3 and *am*-14 could be point mutants in the *B* region. Mutant *am*-4 could be a deletion or frameshift mutation involving both the *A* and *B* regions.

The *am*-4 mutant fails to complement any of the other *am* mutants, because it is mutant for both *A* and *B* products, so there is either no normal *A* or no normal *B* produced in each cross. Complementation is observed in the crosses *am*-1 × *am*-3 and *am*-1 × *am*-14 because, in each case, normal *A* and normal *B* product are present. Weak complementation between *am*-3 and *am*-14 could be explained as the result of interaction between the two different mutant *B* products, giving a product with partial function. Isolation of glutamic dehydrogenase from the various complementation situations confirms the observation. Its enzymatic activity is zero in noncomplementing mutants, full in complementing mutants, and weak in the *am*-3, *am*-14 heterokaryon.

Complementation studies can be diagrammed as a series of interacting cistrons in a *complementation map*. For example, the various dumpy pseudoalleles interact as shown in Table 10.4. This pseudoallelic interaction can be transcribed onto a map (Figure 10.27), although the biochemical significance of the map is not yet understood. The map shows that where there are overlapping lines, the *trans* heterozygote has a mutant phenotype.

The mapping of mutant sites within a gene achieved the ultimate in refinement when S. Benzer (1955) mapped literally thousands of inde-

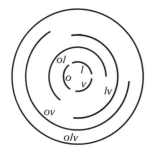

Figure 10.27
Complementation map of dumpy pseudoalleles

Table 10.4
Complementation at the dumpy locus

Genotypes	dp^l	dp^{ol}	dp^{olv}	dp^o	dp^{lv}	dp^{ov}	dp^v	+	$dp^o dp^v$
dp^l	l	l	l	+	l	+	+	+	+
dp^{ol}	l	l	l	o	l	o	+	+	o
dp^{olv}	l	l	l	o	l	ov	v	+	l
dp^o	+	o	o	o	+	o	+	+	o
dp^{lv}	l	l	l	+	l	v	v	+	l
dp^{ov}	+	o	**ov**	o	v	**ov**	v	+	**ov**
dp^v	+	+	v	+	v	v	v	+	v
+	+	+	+	+	+	+	+	+	+
$dp^o dp^v$	+	o	l	o	l	**ov**	v	+	**ov**
$dp^o dp^{lv}$	l	l	l	o	l	**ov**	v	+	l
$dp^{ol} dp^v$	l	l	l	o	l	**ov**	v	+	**ov**

o, oblique wings; l, lethal; v, vortex. Boldface type indicates greater intensity of expression.

Adapted from E. A. Carlson, "Allelism, complementation, and pseudoallelism at the dumpy locus in *Drosophila melanogaster*," *Genetics* 44:347–373 (1959).

pendently arising *r*II mutations in bacteriophage T4. His high-resolution analysis of intragenic recombination down to the level of the nucleotide is called *fine-structure genetic mapping.* It served to bridge the gap between the concept of the gene as viewed in higher organisms and its more chemical interpretation.

The *r*II mutations, in addition to being plaque morphology mutants, are also host range mutants; they cannot grow on *E. coli* K strains that harbor the λ (lambda) phage in a repressed state. Benzer mapped the *r*II mutations by crossing pairs of *different* *r*II mutants on wild-type *E. coli* B cells, collecting the progeny phages, and then infecting *E. coli* K (λ). Only *r*II⁺ (wild-type T4) will grow on strain K (λ). So any *r*II⁺ recombinants generated during the lytic cycle on *E. coli* B produce plaques on *E. coli* K (λ) (Figure 10.28). By this technique, Benzer was able to arrange the first

Figure 10.28
Mapping of T4 *R*II mutants; crossing of point mutants

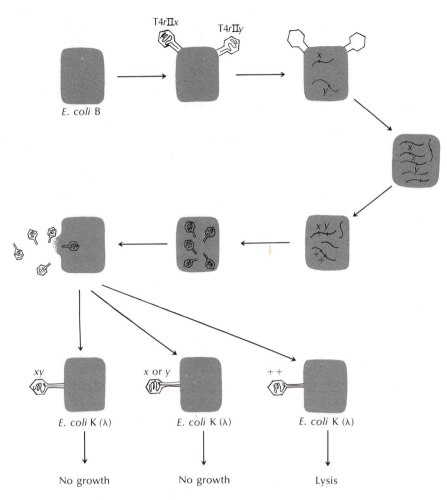

Note: *r*IIx and *r*IIy are independently arising mutations in the *r*II region.

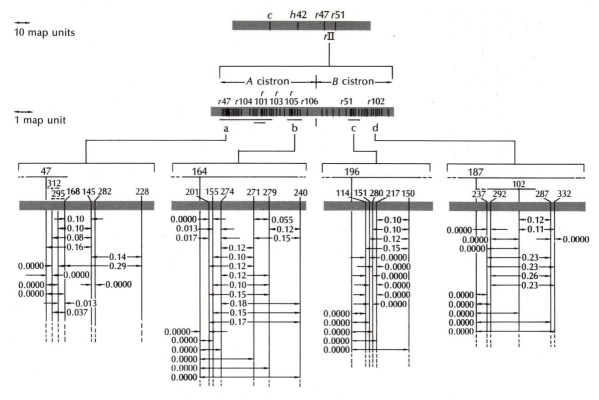

Successive levels in this figure correspond to progressively greater magnifications of the viral chromosome. At the lowest level, numbered vertical lines represent individual *r*II mutants, and the decimals indicate the percentage of recombination found in crosses between two mutants connected by an arrow. The horizontal bars shown in the middle and lowest levels represent the extent of deletions of portions of the *r*II region.

Figure 10.29
First fine-structure genetic map of the *r*II region of T4, based on the recombination frequency produced in pairwise crosses between *r*II mutants of independent origin.
[From S. Benzer, *Proc. Natl. Acad. Sci. U.S.* 41:344 (1955)]

60 *r*II mutants in linear order (Figure 10.29). He estimated that the sensitivity of this cross would enable him to detect recombination frequencies as low as 10^{-6}, although the lowest he found was 10^{-4}. Thus it appeared that he had discovered the smallest distance between two genetic mutant sites that were still separable by recombination.

The lower limit of recombination detection can be used to estimate the size of the *r*II region and to relate this to the DNA molecule. The circular genetic map of T4 is 1500 map units, and a recombination frequency of 0.01 percent (10^{-4}) would be equivalent to 0.02 units. [Remember that a recombinational event that gives an *r*II⁺ recombinant occurs about as frequently as the reciprocal event that gives the double mutant. The double mutant does not grow on *E. coli* K (λ), so to get a reliable estimate of recombinants, the frequency of *r*II⁺ recombinants must be multiplied by 2.] A map distance of 0.02 map units amounts to 1.3 × 10^{-5} of the entire phage chromosome. The T4 chromosome contains

451

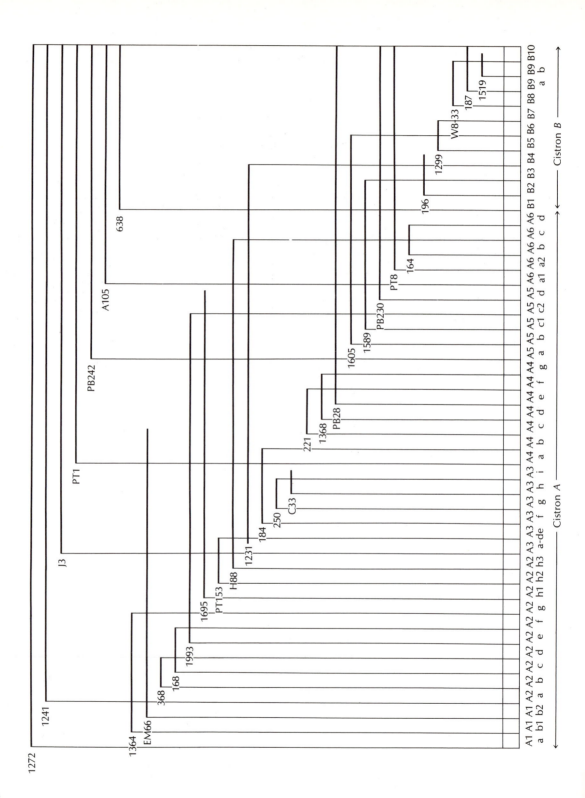

Figure 10.30 (opposite)
Map of *r*II deletions that
serve to divide the *r*II
region into 47 ordered
segments (some ends are
not used to define a
segment; they are shown
with no vertical lines
extending from them)
[From S. Benzer, *Proc.
Natl. Acad. Sci. U.S.*
47:403 (1961)]

Figure 10.31
Deletion mapping. An *r*II
mutation (X) whose map
position is unknown is
crossed with a selected
group of reference
deletion mutations. The
*r*II deletions are of known
length. When mutant X is
crossed with deletion
mutants N, O, P, and R,
no standard recombinants
are observed because
both copies of the DNA
molecule are defective at
the same place. However,
mutant X crossed with
deletion mutants Q and
M produces standard
recombinants because the
DNA molecules are intact
at that point. This tells us
that X is to the right of Q
and to the left of the right
limit of the O deletion.
One can further locate X
by using additional test
mutants with appropriate
deletions, then point
mutants that map in the
restricted region being
investigated. (From "The
Fine Structure of the
Gene" by Seymour
Benzer, *Scientific
American*, January 1961.
Copyright © 1962 by
Scientific American, Inc. All
rights reserved)

about 2×10^5 nucleotide pairs. Therefore, assuming recombination is
equally probable at all points in the genome, the minimal recombinational
distance is $(1.3 \times 10^{-5}) \times (2 \times 10^5)$, or about three nucleotide pairs.
Subsequent work has shown that this figure may be too high and that
recombination may occur between *contiguous* nucleotides in phage DNA.

As the number of *r*II mutants grew, the pairwise mating scheme used
for mapping them became inefficient. Fortunately, a second group of *r*II
mutants occurs that makes it unnecessary to cross each mutant with every
other, thus saving thousands of hours of labor. These are *deletion mutants*;
that is, they lack part of the *r*II region. A group of deletion mutants has
been collected and classified according to the length and position of the
deletion. The criterion for describing a mutant as due to a deletion is that
it does not revert to wild type (in contrast to point mutations, which are
largely base-pair substitutions or frame-shifts and do revert to wild type).
The basis of *classifying* deletion mutants is that recombination does not
occur within the region of the deletion. Thus an *r*II mutant crossed with a
deletion mutant shows no recombination if the *r*II mutant maps in the
region deleted on the other chromosome. A map of *r*II deletions is shown
in Figure 10.30.

Once characterized, the deletion mutants are used to localize new,
unmapped *r*II mutations. Crossing a mutant with a series of deletions
allows its position in the *r*II region to be ascertained by noting whether or
not recombination occurs (Figure 10.31). Once the general segment of a
mutant is fixed, it can be crossed to other mutants that map in the same
segment, and its relative position can be determined. Thus mutants that
fall outside the specific segment defined by deletion mapping need not be

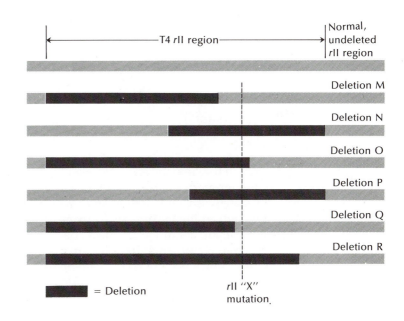

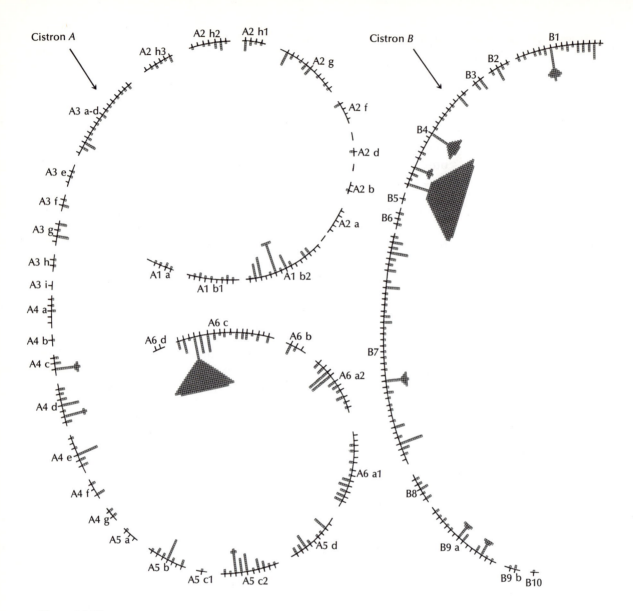

Figure 10.32
Map of spontaneous mutations arising at different loci of the two cistrons in the *r*II region of bacteriophage T4 (each square represents a mutation that has been observed at a particular locus)

used in the final process of localizing the mutant. Figure 10.32 depicts the contemporary view of the genetic fine structure of the *r*II region.

The complementation test, which Lewis developed to probe gene function in higher organisms, was adapted by Benzer to provide an experimental definition of the gene. He infected *E. coli* K (λ) with two different *r*II mutants. If such mutants complement each other, they grow and produce plaques; if they are noncomplementing, no plaques appear. This idea of complementation for *r*II mutants developed when it was

Table 10.5
Complementation test of
*r*II mutants

*r*IIA + *r*IIA + *E. coli* K (λ) ⟶ no growth	
*r*IIB + *r*IIB + *E. coli* K (λ) ⟶ no growth	
*r*IIA + *r*IIB + *E. coli* K (λ) ⟶ lysis	

observed that an *E. coli* K (λ) infected with both a normal T4 and an *r*II mutant lysed and yielded both normal and *r*II progeny. The normal T4 evidently supplied an essential component that enabled the *r*II mutant to develop. If two *r*II mutants can grow on strain K cells, each must supply an essential factor not produced by the other.

Benzer found that the *r*II mutants fell into two groups that could be distinguished by function. Mutants that map in one part of the *r*II region (the *A* region) do not complement each other, but they do complement mutants in the other *r*II region (the *B* region). Similarly, mutants with a defect in the *B* region do not complement each other, but they do complement *A* region mutants (Table 10.5). Thus we can see that the *r*II region is composed of two functional segments, which Benzer referred to as *cistrons*. A mutation in the *A* cistron means that no functional product coded by *A* is produced; a mutation in the *B* cistron means that no functional product coded by *B* is produced. For normal phage growth, both products are required in a nonmutant form. If both the phages infecting an *E. coli* K (λ) cell carry mutations in the *A* cistron, no growth occurs because no normal *A* product is produced; if they both carry mutations in the *B* cistron, no normal *B* product is produced. If one mutant carries a mutation in the *B* cistron, it produces normal *A*. If the other mutant carries a mutation in the *A* cistron, it produces normal *B*. Phages grow, therefore, because normal *A* and *B* products are present (Figure 10.33).

The length of the DNA in the *r*IIA and *r*IIB cistrons can be estimated. The length of the *r*IIA cistron is 6 map units, and that of the *r*IIB cistron 4 map units. Recalling that the T4 genome is 1500 map units and contains 2×10^5 nucleotide pairs, we set

$$r\text{IIA} = \tfrac{6}{1500} \times 2 \times 10^5 \quad \text{(or about 800 base pairs)}$$
$$r\text{IIB} = \tfrac{4}{1500} \times 2 \times 10^5 \quad \text{(or about 500 base pairs)}$$

Benzer's work with the *r*II region in T4 emphasized the concept that the gene is a unit of function, or cistron. Gene and cistron are equivalent terms and will be used interchangeably. It is common to interpret a cistron as a unit of function that produces a single polypeptide chain, but as we shall see, there are exceptions to this generalization.

More recently, techniques have been developed for sequencing extensive regions of DNA, and from these sequence analyses have emerged some interesting insights into our concept of the gene. For example, in the single-stranded DNA bacteriophage φX174, Barrell, Air, and Hutchison

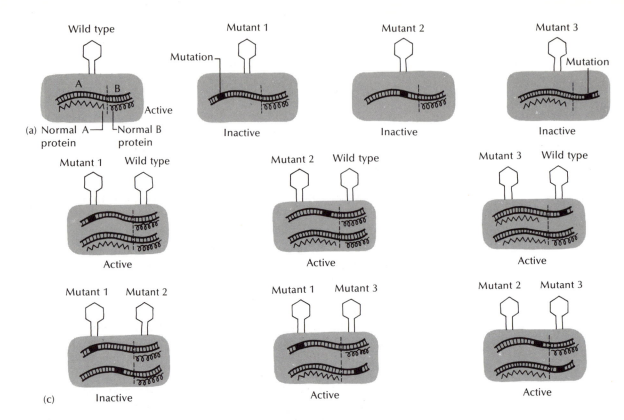

(a) *A* and *B* genes are intact in the wild-type *r*⁺ phage genome so that both *A*-gene and *B*-gene polypeptides are formed. Mutants 1 and 2 have defects at two different sites of their *A*-gene and can only form the *B*-gene polypeptide; mutant 3 has a defect in its *B*-gene and can form only the *A*-gene polypeptide. Since the presence of both polypeptides is required for growth in bacteria of strain K, all three mutants are inactive on this strain. (b) In mixed infection with any of the three defective mutants, the wild type can supply both *A*-gene and *B*-gene polypeptides; therefore, growth takes place. (c) Since neither mutant 1 nor mutant 2 can form the *A*-gene polypeptide, no growth can take place in their mixed infection. However, in mixed infection of mutant 1 or mutant 2 with mutant 3, mutant 1 or 2 supplies the *B*-gene polypeptide, and mutant 3 supplies the *A*-gene polypeptide; hence growth proceeds.

Figure 10.33
Complementation test of *r*II mutants in *E. coli*, strain K.

have found that one DNA segment codes for two polypeptides (Figure 10.34), but that the genes overlap! How to interpret this strip of DNA, given our prior characterization of a gene as a specific nucleotide sequence that specifies a specific polypeptide? Here one sequence specifies *two* polypeptides, so that some of the nucleotides are involved in two distinct genes. Granted that one of the polypeptides (A) is read out of phase compared to the other (A*), and that a different start sequence is involved. But how do we now characterize a gene in light of this new information? Perhaps we should change nothing at all and simply recognize these intriguing diversions from the norm for what they are: excep-

Figure 10.34
Overlapping genes. Structure of the ϕX174 genome in the A-B region and the proteins it codes for. MW = molecular weight. (From Weisbeek, et al. 1977. *Proc. Nat. Acad. Sci. U.S.* 74:2504–2508)

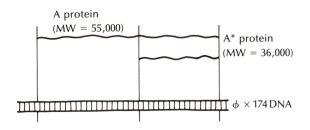

tions to the rule, but not contradictory to the rule. After all, a polypeptide is still specified by a unique DNA sequence, even if part of that sequence is involved in coding for another polypeptide.

Overview

The study of mutagenic agents resulted in the characterization of mutational events at the molecular level. The genetic damage induced by ionizing radiation is difficult to characterize, but it consists of breakage events and base-pair alterations. UV irradiation causes dimerization and hydration, both of which can lead to mutation, primarily as a consequence of error-prone repair systems.

Spontaneous mutations arise as a consequence of tautomeric shifts, or imperfections in the replication process. When mutations occur in genes that code for replication enzymes, the spontaneous mutation rate can increase (mutator genes) or decrease (antimutator genes).

Numerous techniques exist for the detection of mutations in organisms. We discussed techniques for mutation detection in *Drosophila, Neurospora*, bacteria, viruses, and mammals.

We also discussed how the position of a gene relative to the chromosome's heterochromatic region can alter its stability and increase its spontaneous mutation rate (position effects).

Chemical mutagens have a more predictable effect on the genetic material, so they have been a great help to investigators characterizing the mutational events. Base-pair substitutions are induced by mutagenic chemicals such as BU, AP, HA, HNO_2, and EMS. Frameshift mutations are induced by proflavine and acridine.

Several mutagenic agents exist as part of our daily environment, and they endanger our immediate health and long-term health as a species. X rays were the first component of our environment discovered to be dangerous. Cyclamates, saccharin, nitrites, vinyl chloride, mercury, cigarette smoke, and hair dyes have all been shown to be actual or potential mutagens and/or carcinogens.

Our ability to induce specific mutational events in large numbers has enabled us to more effectively study the gene. Complementation studies in *Drosophila* and *Neurospora* led to the gene/cistron concept. It was refined in Benzer's analysis of the *r*II cistrons in bacteriophage T4, in which he examined mutants no more than three base pairs apart. Recent studies on DNA sequences have shown overlapping genes, but they should not detract from our concept of the gene as the basic unit of function.

Questions and problems

10.1 Using the technique of replica plating, design a test for the detection of penicillin-resistant mutations.

10.2 For the detection of autosomal recessive lethal mutations in *Drosophila*, the test shown in Figure 10.4 can be employed. An F_3 generation is produced by single-pair matings of F_2 *Cy/l?* females with *Cy/l?* males. Why is it necessary to use single-pair matings?

10.3 Why are forward mutation rates generally higher than reverse mutation rates?

10.4 How would you distinguish a base-pair substitution mutation from a minute deletion?

10.5 Discuss the concept of the gene as it has been so far developed.

10.6 What is the purpose of the *cis-trans* complementation test?

10.7 An alkylating agent like EMS (see p. 435) is known to induce mutational events in single bases, which upon replication become base-pair substitutions. If EMS acts on DNA after replication (in a mature sperm cell, for example), one expects to find only mosaic mutants. Yet complete, or whole-body, mutants are found. How would you explain this?

10.8 Since direct analysis of the nucleotide sequence within a DNA molecule is not usually feasible, any analysis used to detect mutations must be indirect, such as examining viral RNA's, gene products, or organismal phenotypes. Using indirect analysis, introduce ways in which the original mutation may go undetected. Using indirect screening analysis, discuss the major reasons for loss of mutational events.

10.9 Five histidine mutations are isolated in *Neurospora*, and their complementation patterns are studied through heterokaryon formation. The following table summarizes the data.

		Histidine mutants				
		a	b	c	d	e
Histidine mutants	a	x	x	x	x	x
	b	x	x	√	√	√
	c	x	√	x	√	√
	d	x	√	√	x	x
	e	x	√	√	x	x

x = No growth
√ = Growth

Using these data, draw a complementation map and determine the number of functional units involved.

10.10 Using the *r*II deletion mutants from Figure 10.31, it was discovered that a newly arisen mutant produced recombinants only with test mutant D. Where is the new mutant located?

10.11 Can attached-**X** *Drosophila* females be used to detect low frequencies of sex-linked recessive lethal mutations in *Drosophila* males?

10.12 The following are data from a complementation study done for different mutants of a particular gene region in bacteriophage T4. From these data, how many cistrons comprise the region?

	1	2	3	4	5	6
1	−	+	−	+	−	−
2	+	−	+	−	+	+
3	−	+	−	+	−	−
4	+	−	+	−	+	+
5	−	+	−	+	−	−
6	−	+	−	+	−	−

\+ = Complementation
− = No complementation

10.13 When a son is born to a man who works in a plant that manufactures chemical pesticides, the infant is found to be suffering from the **X**-linked disease Duchenne muscular dystrophy. Neither the man's family nor his wife's family has had any occurrence of this rare disease for at least five generations. Claiming that the disease is a consequence of a mutation caused by the chemicals he worked around, the man sues the company for failing to protect him from the mutagenic hazards of his working environment. Is this man's claim justified?

10.14 In testing the mutagenicity of a chemical, you have induced 100 *r*II mutations in bacteriophage T4. Of these, 49 were revertible by AP and BU but not by HA, and 7 were revertible by all three.
(a) How would you classify this group of 56 mutations?
(b) The remaining 44 mutants were for the most part revertible with proflavine but not base analogs. How would you classify these mutations?
(c) Four mutants were not revertible by any chemical. How would you explain these mutations?

References

Auerbach, C. 1962. *Mutation: Part 1. Methods.* Edinburgh, England, Oliver and Boyd.

Auerbach, C. 1967. Changes in the concept of mutation and the aims of mutation research. In R. A. Brink (ed.), *Heritage from Mendel.* Madison, Wis., University of Wisconsin Press.

Auerbach, C. 1976. *Mutation Research.* London, Chapman and Hall.

Auerbach, C., and B. J. Kilbey. 1971. Mutation in eukaryotes. *Ann. Rev. Genetics* 5:163–218.

Baer, A. S. 1977. *Heredity and Society*. New York, Macmillan. (See the section on "Genetic Aspects of Environmental Hazards.")

Barrell, B. G., G. M. Air, and C. A. Hutchison, III. 1976. Overlapping genes in bacteriophage ϕX174. *Nature* 264:34.

Beadle, G. W., and E. L. Tatum. 1945. *Neurospora:* II. Methods of producing and detecting mutations concerned with nutritional requirements. *Amer. J. Bot.* 32:678–686.

Benzer, S. 1955. Fine structure of a genetic region in bacteriophage. *Proc. Natl. Acad. Sci. U.S.* 41:344–354.

Brenner, S., L. Barnett, F. H. C. Crick, and A. Orgel. 1961. The theory of mutagenesis. *J. Molec. Biol.* 3:121–124.

Carlson, E. A. 1966. *The Gene: A Critical History*. Philadelphia, Saunders.

Committee 17. 1975. Environmental mutagenic hazards. *Science* 187:503–514.

Corwin, H. O., and J. B. Jenkins. 1976. *Conceptual Foundations of Genetics*. Boston, Houghton Mifflin.

Cox, E. C. 1976. Bacterial mutator genes and the control of spontaneous mutation. *Ann. Rev. Genetics* 10:135–156.

Drake, J. W. 1970. *The Molecular Basis of Mutation*. San Francisco: Holden-Day.

Drake, J. W., and R. H. Baltz. 1976. The biochemistry of mutagenesis. *Ann. Rev. Biochemistry* 45:11–38.

Fincham, J. R. S. 1966. *Genetic Complementation*. New York, Benjamin.

Freese, E. 1959. The difference between spontaneous and base analogue induced mutation of phage T4. *Proc. Natl. Acad. Sci. U.S.* 45:622–633.

Genetic hazards to man from environmental agents (a symposium). 1975. *Mutation Research* 33:1–105.

Green, M. M. 1963. Pseudoalleles and recombination in *Drosophila*. In W. J. Burdette (ed.), *Methodology in Basic Genetics*. San Francisco, Holden-Day.

Hollaender, A. 1971–1976. *Chemical Mutations: Principles and Methods for Their Detection*, Vols. 1, 2, 3, and 4. New York, Plenum.

Jenkins, J. B. 1967. Mutagenesis at a complex locus in *Drosophila* with the monofunctional alkylating agent, ethyl methanesulfonate. *Genetics* 57:783–793.

Lederberg, J., and E. M. Lederberg. 1952. Replica plating and indirect selection of bacterial mutants. *J. Bact.* 63:399.

Lerman, L. S. 1963. The structure of the DNA–acridine complex. *Proc. Natl. Acad. Sci. U.S.* 49:94.

Lewis, E. B. 1955. Some aspects of position pseudoallelism. *Amer. Nat.* 89:73–89.

Lewis, E. B. 1967. Genes and gene complexes. In R. A. Brink (ed.), *Heritage from Mendel*. Madison, Wis., University of Wisconsin Press.

Muller, H. J. 1927. Artificial transmutation of the gene. *Science* 66:84–87.

Newcombe, H. B. 1949. Origin of bacterial variants. *Nature* 164:150.

Pontecorvo, G. 1958. *Trends in Genetic Analysis*. New York, Columbia University Press.

461
References

Rhoades, M. M. 1946. Plastid mutations. *Cold Spring Harbor Symp. Quant. Biol.* 11:202–207.

Rinehart, R. R., and F. J. Ratty. 1965. Mutation in Drosophila melanogaster cultured on irradiated food. *Genetics* 52:1119–1126.

Streisinger, G., Y. Okada, J. Emrich, J. Newton, A. Tsugita, E. Terzaghi, and M. Inouye. 1966. Frameshift mutations and the genetic code. *Cold Spring Harbor Symp. Quant. Biol.* 31:77–84.

Sutton, H. E., and R. P. Wagner. 1975. Mutation and enzyme function in humans. *Ann. Rev. Genetics* 9:187–212.

Weisbeck, P. J., W. E. Borrias, S. A. Langveld, P. D. Baas, and G. A. van Arkel. 1977. Bacteriophage ϕX174: Gene A overlaps gene B. *Proc. Natl. Acad. Sci. U.S.* 74:2504–2508.

Witkin, E. M. 1976. Ultraviolet mutagenesis and inducible DNA repair in *E. coli*. *Bact. Rev.* 40:869–907.

Eleven

The genetics of viruses

The virus is the simplest of all living things, and its simplicity makes it an ideal research tool in genetics. Questions we are hard pressed to answer using eukaryote systems can often be answered using a viral system—questions dealing with gene structure, gene function, gene organization, and recombination, to mention a few.

We want to focus in this chapter on the transmission and recombination of the genetic material in viruses, directing our attention more specifically to the bacterial viruses (*bacteriophage* or simply *phage*). We direct our attention to bacterial viruses because they are easier to study and have provided us with most of our knowledge of virus genetics.

Through our analysis of the bacteriophage, we shall be able to examine in detail the process of genetic recombination discussed more extensively in earlier chapters on eukaryotes. But we also want to compare genetic recombination in eukaryotes with that observed in prokaryotes, specifically viruses. We shall also examine the role of animal viruses in cancer.

Events leading to the use of phages as genetic tools

The term *virus* has had a confusing, often contradictory history. The ancient Romans called any "poison" of animal origin a virus. Thus any disease caused by an animal, such as the bacterially caused bubonic plague, was considered viral. During the Middle Ages, it became clear that there were noncontagious diseases caused by poisons and contagious diseases caused by infectious particles. The term *virus* came to be attached only to infectious particles, even though some of these particles were bacteria. With the discovery of bacteria, the term *virus* became even more vague.

Viruses were recognized as a distinct class of organisms in 1899 when M. W. Beijerinck described microbes that could pass through ceramic filters designed to stop bacteria. In addition, they could not be detected with a microscope and would not grow on an artificial medium designed for growing bacteria. Most scientists at the time viewed Beijerinck's organisms merely as aberrant forms of bacteria and wrote off the possibility that organisms smaller than bacteria existed.

Frederick Twort, one of the few microbiologists who believed that viruses were indeed unique, thought that disease-causing viruses were descended from non-disease-causing ancestors and tried to detect these ancestors in smallpox vaccine fluid. Assuming that the ancestors could grow on artificial media, Twort inoculated nutrient agar with samples of smallpox vaccine fluid and looked for the growth of organisms. But the only growth he observed came from bacteria that were found in the fluid. After a time, though, some of the bacterial colonies underwent a transformation, manifested by a "glassy" appearance. The bacteria appeared to be dissolving.

When other colonies were inoculated with samples of fluid from the glassy colonies, they took on the same glassy look. Twort suspected some kind of bacterial disease and suggested three possible causes: a peculiar stage in the life cycle of the microorganism, a stage that influenced neighboring cells; a dissolving enzyme secreted by the bacteria; or a virus that infects and kills the cells. Published in 1915, Twort's observations on this "bacterial disease" were the first records of bacterial viruses in action, but his work was largely ignored.

Two years later, Felix d'Hérelle published an account of an agent—which he called a *bacteriophage*—that destroyed the bacillus bacteria causing, of all things, diarrhea in locusts. D'Hérelle's article did not refer to Twort's published observations, so it is difficult to ascertain whether d'Hérelle was familiar with them. In 1920, Gratia severely criticized d'Hérelle for not citing Twort's work, but d'Hérelle countered by arguing that the phenomena described by Twort bore no resemblance to his bacteriophage. Gratia, however, demonstrated that d'Hérelle's bacteriophage could indeed induce the same glassy transformation observed by Twort.

D'Hérelle suggested that the bacterial disease he observed was due to bacterial viruses, but there were still those who believed it was caused by bacterial enzymes. For example, Jules Bordet, director of the Pasteur Institute in Paris, proposed that d'Hérelle's observations could be accounted for by just such a bacterial enzyme. But d'Hérelle was undissuaded, and he stuck by his original interpretations, which have since been verified a number of times.

It is interesting to note that the primary motivation for continuing the research with bacteriophages was based not on genetics but medicine, because d'Hérelle and others believed that such bacterial diseases as typhus, cholera, anthrax, and scarlet fever might be cured by treating diseased patients with appropriate bacteriophages that would destroy the virulent bacteria. Unfortunately, because of immune reactions and inactivating enzymes in the host, viruses did not perform *in vivo* the way they did *in vitro*.

The rise of modern phage genetics began in the 1930s through the work of three men—F. M. Burnet, Martin Schlesinger, and Max Delbrück—who independently investigated phages because of their simplicity rather than their practical applicability to medicine. Burnet instituted phage genetics in 1936 when he discovered a mutant virus that had lost its capacity for living harmoniously with bacteria. This harmonious capacity, called *lysogeny*, was an inherited trait, as was its loss. D'Hérelle had asserted that the bacteriophage was a self-reproducing and always destructive virus that was parasitic on bacteria. In 1921, however, *lysogenic* bacteria were discovered that harbored bacteriophages with *po-*

Frederick Twort and (far right) Felix d'Hérelle

F. M. Burnet and (far right) Max Delbrück

Salvatore Luria

465

tential lytic, or destructive, capabilities. Burnet's discovery of a phage that had lost its ability to enter a lysogenic state was the first record of a mutation in a phage.

In 1936, Schlesinger analyzed the biochemical composition of viruses and concluded that they consisted mainly of DNA and protein. This was the first indication that the chemical composition of viruses resembled that of chromosomes, but it was certainly not the first attempt to associate bacteriophages with genetics. In 1921, H. J. Muller (discussed earlier in connection with Mendelian genetics) speculated that d'Hérelle's bacteriophages might be nothing more than naked genes. Muller's hypothesis apparently fell on unreceptive ears, but if Burnet, Schlesinger, or Delbrück had heard his prophetic remarks, bacteriophage genetics might have been born earlier.

Max Delbrück may rightly be referred to as the father of modern phage genetics. Trained as a physicist, he became interested in the phenomenon of inheritance through informal discussions with other physicists who were concerned with physical interpretations of biological phenomena. In 1935, he published a paper dealing with a physical model of the gene. In 1937, as the Nazi movement became more threatening, he moved from Germany to the California Institute of Technology, where he was introduced to bacteriophages. Soon he was deeply involved in phage research. One of Delbrück's most important contributions to phage genetics came in 1939. He and Emory Ellis designed the one-step growth experiment, which showed that a bacterium infected with a single phage liberates over 100 progeny phages within half an hour of infection. Thus by 1940 a major question in phage genetics had taken shape: What are the mechanisms involved in replicating and assembling these progeny phages in such a short span of time?

The structure and life cycle of bacteriophages

During the 1940s, Delbrück joined with two other scientists interested in phage genetics—Salvatore Luria and A. D. Hershey. The work of this trio is the intellectual precursor of most modern phage research.

In the late 1940s, studies by Delbrück, Hershey, and R. Rotman revealed a parasexual process by which phage chromosomes may exchange genetic information, a phenomenon encountered in eukaryotes earlier and called *genetic recombination*. This discovery led to some of the most detailed genetic studies ever undertaken, and from it emerged a body of information on gene structure, mechanisms of recombination, and gene function. However, before discussing these topics, we should consider the structure and life cycle of bacteriophages, especially within the context of DNA structure and replication.

There is no universally used scheme to designate varieties, or strains, of bacteriophages, although several rational ones have been proposed. In general, phage strains are designated by numbers, Roman or Greek letters, or combinations thereof, and the designations usually include the bacterial strain the bacteriophages infect. For example, coliphage T4 is a strain of bacteriophage that infects *Escherichia coli*; Salmonella phage P22 is a bacteriophage that infects *Salmonella*. Figure 11.1 gives a classification scheme for bacterial viruses suggested by Bradley in 1967. It is widely accepted.

Figure 11.1
Morphological types of bacteriophage [After Bradley, *Bact. Rev.* 31:230 (1967)]

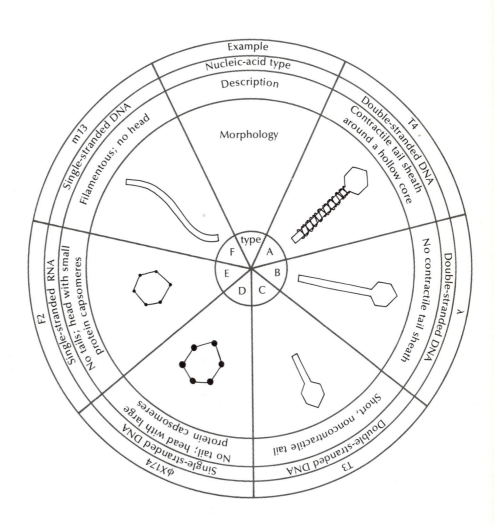

Structurally, viruses are the simplest of all organisms. Bacteriophages are composed basically of protein and nucleic acid (animal viruses, such as those causing chickenpox, may be more complex, containing some carbohydrate and lipid). Phage diversity and structural complexity result from the number and organization of the protein species that make up the various outer coats of phages, as well as the type and structure of the chromosome.

The phage types most extensively used to explore genetic phenomena are the *E. coli* T-phages, numbered T1 through T7 (Figure 11.2). Variations exist in the physical characteristics of the seven types of T-phages, but they are in general tadpole-like in appearance. The

Figure 11.2
Structure of T-phage
(Photos courtesy of Lee D.
Simon)

Head —
DNA —
Collar —
Core —
Tail sheath —
Base plate —
Tail pin
Tail fibers

(a) Diagram of T2

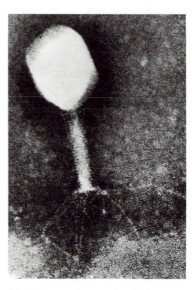

(b) Electron micrograph of T4

(c) Electron micrograph of T4 baseplate

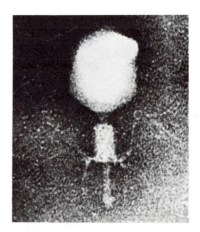

(d) Electron micrograph of T4 tail with contracted sheath

T-phage *head* is a hexagonal prism of protein containing within it the phage chromosome, a double-stranded DNA molecule. Attached to the head is a *tail*, usually composed of two coaxial hollow tubes—an inner *needle* and an outer *sheath*—terminating in a *baseplate*. The baseplate usually consists of *tail fibers* and *tail pins* and is the point of initial contact between the E. coli host and the T-phage. On attachment of the T-phage to *E. coli*, the phage chromosome passes through the hollow tail into the host, where it is replicated and subsequently packaged into progeny virus particles.

Other bacteriophages commonly used in genetic research are φX174, F2, M13, and lambda (λ). The φX174 phage is much less of a structural enigma than it once was. Its claim to fame is its chromosome, a single-stranded molecule of DNA, used as a template for the complete *in vitro* synthesis of biologically active DNA. The φX174 protein coat is icosahe-

Figure 11.3
Electron micrographs and model of φX174 (Courtesy of R. W. Horne)

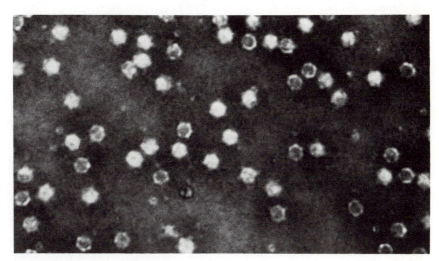

(a) φX174 phage particles.

(b) Enlargement of a single φX174.

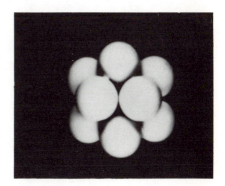

(c) Modular interpretation based on the electron micrograph in (b).

dral (having 20 plane surfaces) and probably lacks a tail (Figure 11.3). The spikes on the protein surface may play a role in φX174 attachment specificity (this phage infects only strains C and 15 of *E. coli*). However, the precise mechanism by which φX174 attaches to a host and transfers its genetic information to the host's interior remains obscure. The F2 phage is a small, tailless, single-stranded DNA virus, and M13 is a headless, filamentous phage.

Lambda is referred to as a *temperate* phage, because it can enter into a lysogenic relationship with the host cell in which it does not always destroy the cell, but can be replicated concurrently with the host cell's chromosome. Structurally, lambda resembles the T5 bacteriophage in that it has a long, thin tail, one tail pin, and no tail fibers (Figure 11.4).

Figure 11.4
Electron micrograph of bacteriophage lambda (λ) (Courtesy of E. Kellenberger)

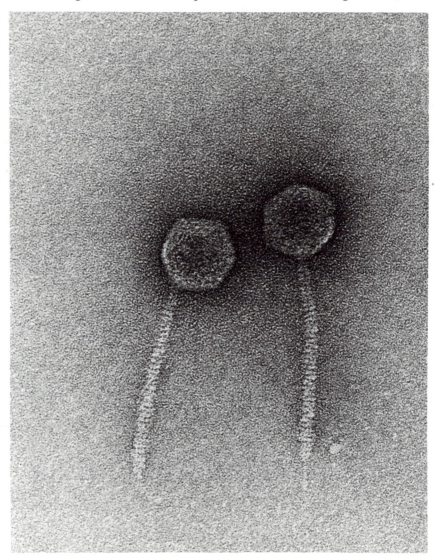

Our ability to grasp the genetics of bacterial viruses depends on our knowledge of the stages in their life cycle. Understanding life cycles is crucial to a delineation of genetic principles, as was evident in the relationship between Nägeli and Mendel (Chapter 1). Remember that when he completed his pea experiments, Mendel sought the advice of Nägeli, who encouraged Mendel to study the hawkweed. Mendel could not find his numerical ratios in the offspring of this plant, because it did not go through a sexual life cycle. Had Mendel and Nägeli been aware of the hawkweed's asexual life cycle, Mendelism might have followed a different historical course.

E. L. Ellis and Delbrück demonstrated in 1939 that within 30 minutes of infection, a single virus can give rise to over 100 progeny viruses. The two techniques used to demonstrate this were the one-step growth experiment and the single-burst experiment.

The *one-step growth experiment* was designed to study the kinetics of phage growth in a *population* of bacteria. A bacteria population growing in a nutrient broth were infected by T4 bacteriophages in a ratio of 1 phage to 1 bacterium. At regular intervals after infection, samples of infected cells were drawn and analyzed to see how many infected units there were by screening for plaque-forming units. A *plaque* is a clear, circular area that develops in a dense bacterial lawn on a solid agar surface. Each plaque may be thought of as the eventual result of either a single phage or a single infected bacterium containing many phages. The phages go through several rounds of infection and, because the bacteria and viruses are confined by the agar, destroy bacteria in a limited area. A plaque, then, is an area of destroyed bacteria on an agar surface.

For the first 24 minutes after the initial infection, the number of plaque-forming units remained at a constant level for all samples drawn (Figure 11.5). Ellis and Delbrück proposed that phages were developing

Figure 11.5
One-step growth experiment

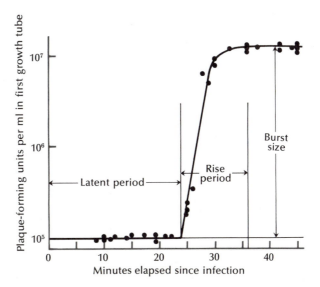

inside the bacteria and that each plaque resulted from a single infected bacterium that was plated onto the medium. After 24 minutes of infection, the number of plaque-forming units being produced began to increase exponentially, and this was interpreted as the bursting of bacteria releasing progeny phages, plaques now being the result of single phage or a single infected bacterium. By 36 minutes after infection, the number of plaque-forming units stabilized, which means that all infected bacteria had released progeny phages, and each plaque was now the result of a single phage. Ellis and Delbrück characterized this growth pattern as having an initial *latent* period in which intracellular growth of phages occurred, then a *rise* period in which infected cells were bursting open, and finally a *plateau* period in which the cycle reached completion.

To study the process of infection in *single* bacteria, Ellis and Delbrück appropriately diluted infected populations so that a test tube contained only one infected cell. The progeny released by the infected cell were counted by plaque formation analysis. This study, the single-burst experiment, revealed that single cells exhibit wide variability with respect to the numbers of progeny phages they release. The variability is due to minor perturbations, such as physiological differences in the host cells, that occurred in various stages of the intracellular growth phase.

The report by Ellis and Delbrück was important for a variety of reasons. It stimulated the investigation of intracellular events leading to the production of progeny phages. Among those investigations was Hershey and Chase's study proving that only DNA entered the host cell, thereby establishing DNA as the genetic material (see pages 249–253). It also focused attention on bacteriophages as useful tools to explore basic genetic phenomena of a biochemical nature in that phages were the simplest organisms known and had very short life cycles. Since 1939, we have learned much about the life cycle of bacteriophages, and before we explore the phage as a genetic system, we should summarize its life cycle as it is currently understood.

The life cycle The initial step in the life cycle of a bacteriophage—and here we shall be referring to T2, T4, and T6 phages (T-even phages) unless otherwise stated—is the attachment of the phage to the bacterial host. This is a two-step process including first a collision between bacterium and phage and then the attachment of phage to bacterium. The attachment process is a highly specific one. The surface of the bacterium appears to possess distinct areas that the phage must "recognize" before attachment occurs. In *E. coli*, two structures could conceivably be attachment sites—the cell wall and the cell membrane. It has been shown that if the cell wall is removed, T-phages can no longer attach. Conversely, if only *E. coli* cell walls are present in a solution, T-phages will attach to them. The cell wall, then, appears to contain the specificity for phage attachment. The polysaccharides that make up *E. coli*'s cell wall define, in essence, which phage types will be able to attach to it.

Figure 11.6
Attachment of phage to
bacterium (Courtesy of
Lee D. Simon)

(a) Tail fiber attachment.

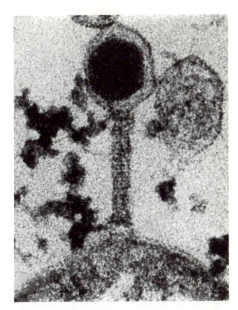

(b) Tail pin attachment to *E. coli.*

The attachment process itself appears rather complicated. The tail fibers, the attachment structures of the phage, are usually coiled around the baseplate until an appropriate collision occurs. They then uncoil and attach to the cell wall (Figure 11.6a). It has been speculated that a chemical component of the cell wall triggers the uncoiling. After attachment, the main body of the phage moves closer to the bacterial surface until it makes contact with its tail pins (Figure 11.6b), which may serve to attach the tail firmly to the cell wall. For phages without tail fibers, such as lambda, the attachment mechanism is unknown.

Following attachment of the baseplate to the cell wall, a series of events takes place that involves the contraction of the tail sheath (Figure 11.7). The baseplate remains attached to the distal end of the sheath, but releases itself from the tip of the needle so that it looks as if it is sliding up the needle and away from the cell wall as the sheath contracts. However, short tail fibers, originating perhaps from the tail pins (Figure 11.7), and the long tail fibers keep the baseplate in contact with the cell wall.

On contraction of the sheath, the needle penetrates the cell wall, perhaps aided by a digestive enzyme contained in the structure of the tail. The needle does not appear to penetrate the cell membrane. Then DNA from the phage head passes through the needle and enters the host cell by active transport across the cell membrane (Figure 11.8). There is some speculation that the release of DNA is triggered by the enzymatic mechanisms that aid the needle's penetration of the cell wall. Figure 11.9 summarizes these events.

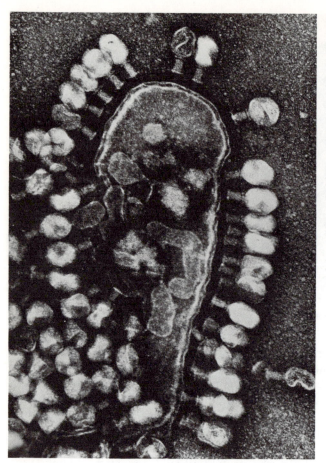

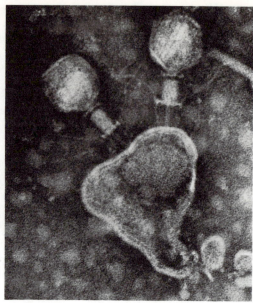

Once inside the cell, the phage DNA first directs the synthesis of enzymes that stop bacterial DNA synthesis, and then it begins to direct the synthesis of phage DNA. As phage DNA is synthesized, so too are protein components of the phage's coat. About halfway through the latent period, the DNA begins to be incorporated into protein coats, and this continues at a constant rate until the cell undergoes *lysis* (the breaking open, or dissolution, of the cell membrane). Figure 11.10 is an electron micrograph of *E. coli* 30 minutes after infection with phage T2. It shows some mature phage particles just before they are released.

The climax of the phage life cycle is the bursting forth from the host cell of the progeny phages. Contrary to what might be expected, lysis is not a function of internal pressures caused by the accumulation of intracellular phages. Rather, the phage DNA directs the synthesis of an enzyme, *lysozyme,* that digests away the cell wall and membrane and thus releases the progeny phages.

Virulent phage such as T4 always destroy the bacterial host cell, but not all virally infected bacterial cells terminate in lysis. A gentler relation-

Figure 11.8
Injection of T4 DNA into *E. coli* B. Ribosomes and DNA are visible (arrow points to a T4 needle that has penetrated the cell wall). (Courtesy of Lee D. Simon)

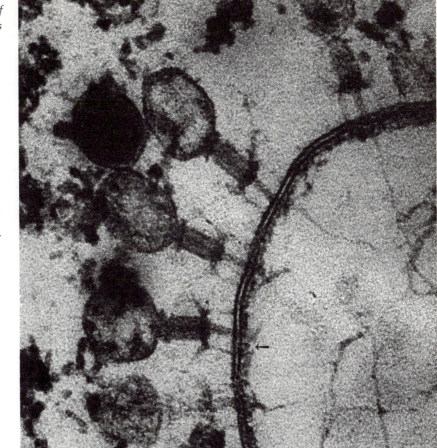

Figure 11.9
Steps in attachment of T2 and T4 phages to *E. coli* cell wall

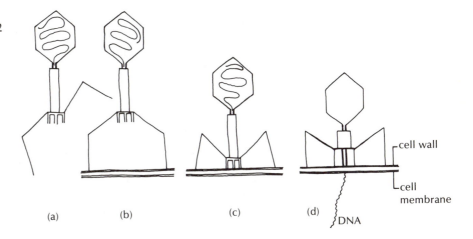

(a)   (b)   (c)   (d)

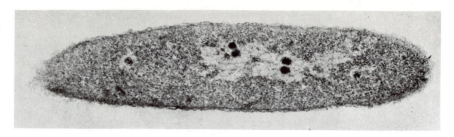

ship exists between the bacterial cell and certain bacteriophage, a relationship known as *lysogeny*. In lysogeny, the DNA of the *temperate* phage, rather than commandeering the cell's resources for the cell's ultimate destruction, is actually incorporated into the cell's own genetic material. This means that the phage DNA, called a *prophage*, behaves as if it were a bacterial gene. The bacterial cell carrying the prophage functions and reproduces normally, producing daughter cells that all carry the prophage. On occasion, the prophage leaves the confines of the bacterial chromosome to become an independent phage chromosome. It now enters the lytic cycle, culminating in lysis and the release of progeny phage. We call the process by which the prophage enters a lytic cycle *induction*, and it can be caused by a variety of external stimuli, such as chemicals, x rays, and UV irradiation. The lambda (λ) phage is a well studied example of a temperate phage. Lysogeny has proven to be a valuable tool in the mapping of certain bacterial genomes, as we shall see in Chapter 12. It has also provided some interesting models of gene control, as Chapter 13 will show.

Genetic recombination in phages

The study of genetic recombination in any organism requires an ability to discriminate between alternative phenotypes. In our discussion of Mendelian phenomena, we dealt with phenotypes that were easily discernible: green seeds versus yellow seeds, tall plants versus short plants, white eyes versus red eyes, and so forth. In these cases, no special observational techniques are required to distinguish phenotypes. Bacteriophages, however, do not have flowers or seeds or eyes, and they can be observed only with an electron microscope. So the basic approach to studying phage phenotypes focuses on the phage–bacterium interaction. From this interaction can be discerned specific inherited traits, such as the size or shape of the plaque or the type of bacterial cell that can be infected.

Plaque size and shape are very characteristic of the viral strain employed and of different mutants within a strain. A plaque may be large or small, for example, and its edges may be fuzzy or sharp. The size and shape of a plaque are functions of the physiological interaction between phage and bacterium on the one hand *and* phage genotype on the other. The most extensively investigated plaque mutant is the *r* mutant (*r* standing for rapid lysis) of T-phage. A normal T-phage produces a plaque that is small and has a fuzzy edge—an *r+* plaque. The *r* mutant produces a plaque about twice as large and has a clear, sharp edge (Figure 11.11).

Phage mutants are known that are unable to infect certain bacterial strains. These are termed *host range* mutants, and they generally affect the attachment phase of the phage growth cycle. Consider, for example, *E. coli* B. Under normal circumstances, *E. coli* B will serve as host for T2 phages. Occasionally, however, a mutation occurs in the bacterium that renders it resistant to T2; such a bacterial strain is referred to as *E. coli* B/2. A mutation designated *h* (*h* for host range) can occur in T2 that permits the phage to use both *E. coli* B and *E. coli* B/2 as host. *E. coli* B/2 can in turn acquire immunity to T2*h* through another mutation, the mutant bacterial strain referred to as *E. coli* B/2/*h*. As you may already have guessed, T2*h* can mutate to form (T2*h'*), which can infect B, B/2, and B/2/*h*. Such host range mutants are found in a number of different phages.

Figure 11.11
Morphology of T2*r+* and T2*r* plaques on a lawn of *E. coli* (the plate also contains mottled plaques, caused by growth of both *r* and *r+* phages in the same spot)

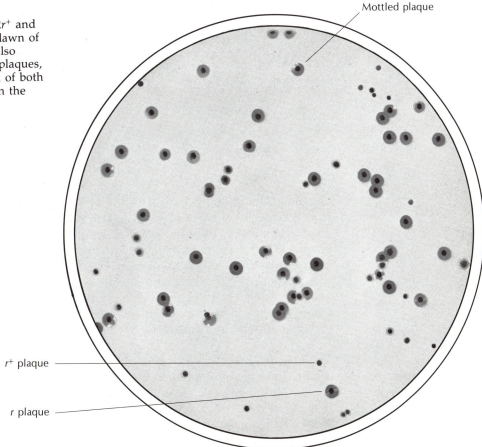

Mottled plaque

r+ plaque

r plaque

Figure 11.12
Plaques formed by T2*h*
and T2*h*⁺ phages growing
on a mixed lawn of *E. coli*
B and *E. coli* B/2 cells

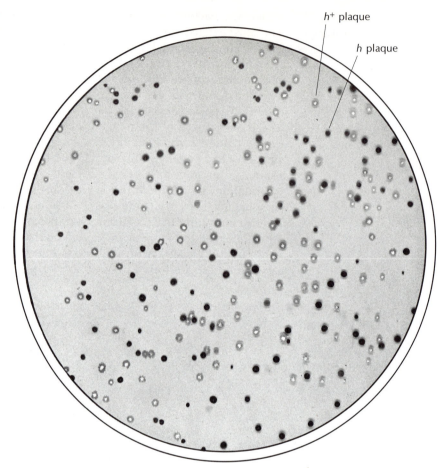

h⁺ plaque

h plaque

Since all *h* mutants have similar plaques, they can be distinguished from each other by infecting with T2 a mixture of B and B/2 cells growing on an agar surface. The *h* mutants, infecting both B and B/2, will give clear plaques. The *h*⁺ nonmutant phage, able to infect only B cells, will produce plaques that are identical to *h* in size but *turbid*, because only B cells are lysed (Figure 11.12). The *h* and *h'* mutants can be distinguished in like manner, using a mixed lawn of B/2 and B/2/*h'*. A basic understanding of plaque morphology and host range mutants permits us to analyze the early genetic studies of phages.

Hershey and Rotman demonstrate genetic recombination in phages

The first detailed study of genetic recombination in phages was carried out by Hershey and Rotman in 1948 and 1949. Using three independently arising T2*r* mutants (*r1*, *r7*, and *r13*) and a T2*h* mutant, they demonstrated that phage genes could recombine as genes do in higher organisms. However, it was soon shown that recombination in phages and higher organisms differ in important ways.

A typical cross carried out by Hershey and Rotman showed that when *E. coli* B cells were mixedly infected with h^+r and hr^+ strains of T2, recombinant hr and h^+r^+ progeny were recovered. Single cells were infected by both types of parental phages. The genotypes of the progeny phages from this cross were determined by plating them on a mixed lawn of *E. coli* B and *E. coli* B/2 cells and scoring the plaque phenotypes for recombinants. The nonrecombinant parental genotypes would produce turbid r plaques (h^+r) and clear r^+ plaques (hr^+), whereas the recombinant progeny (h^+r^+ and hr) would produce turbid r^+ plaques and clear r plaques, respectively (Figure 11.13). The results of this cross and others appear in Table 11.1.

The data collected by Hershey and Rotman demonstrate two important things. First, recombinant genotypes appear in all crosses, and reciprocal recombinants are present in roughly equal proportions when the data for *all crosses* are averaged. However, the data from individual single-burst experiments show that one recombinant type usually predominates numerically over the other. This suggests that *reciprocal* recombination, analogous to that found in higher organisms and resulting from meiosis, probably does not occur or occurs rarely in phages. The reason

Figure 11.13
Types of plaques formed by T2*rh*, T2*r*+*h*+, T2*rh*+, and T2*r*+*h* on a mixed lawn of *E. coli* B and *E. coli* B/2 cells

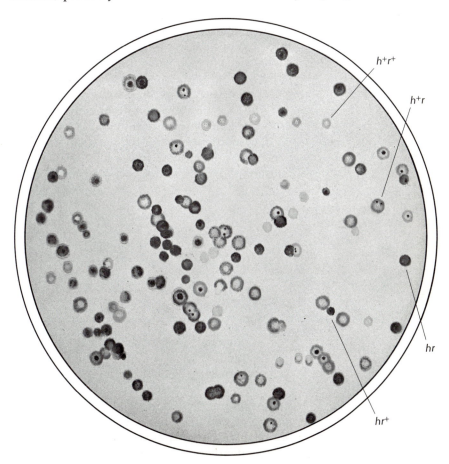

Cross	No. of experiments		h^+r^+	hr^+	h^+r	hr	Burst size
$r^+h \times r1h^+$	5	Input	0	53	47	0	
		Yield	12	42	34	12	630
$hr1 \times h^+r^+$	3	Input	57	0	0	43	
		Yield	44	14	13	29	680
$hr^+ \times h^+r7$	4	Input	0	49	51	0	
		Yield	5.9	56	32	6.4	650
$hr7 \times h^+r^+$	2	Input	49	0	0	51	
		Yield	42	7.8	7.1	43	690
$hr^+ \times h^+r13$	3	Input	0	49	51	0	
		Yield	0.74	59	39	0.94	510
$hr13 \times h^+r^+$	4	Input	52	0	0	48	
		Yield	50	0.83	0.76	48	590

Table 11.1
Average percent distribution of phage genotypes before and after genetic crosses are performed

Adapted from A. D. Hershey and R. Rotman, "Genetic recombination between host-range and plaque-type mutants of bacteriophage in single bacterial cells," *Genetics* 34:44–71 (1949).

for this will be explored later. Second, the frequencies of the various recombinant types involving *r1*, *r7*, and *r13* relative to *h* suggest that the three *r* genes are found on different parts of the T2 chromosome. If the probability of recombination between two genes is a function of the distance separating them, then *r1* must be the farthest relative to *h*, ((24 + 27)/2 or 25.1 percent recombination), *r13* the closest ((0.74 + 0.94 + 0.831 + 0.76)/2 or 1.6 percent recombination), and *r7* intermediate ((12.3 + 14.9)/2 or 13.6 percent recombination).

These crosses made by Hershey and Rotman were insufficient to permit mapping of the four genes relative to each other. The genes in question (*h*, *r1*, *r7*, and *r13*) could be arranged in four possible ways:

1.	*h*	*r13*		*r7*	*r1*
2.	*r13*	*h*		*r7*	*r1*
3.	*r1*		*r13*	*h*	*r7*
4.	*r1*		*h*	*r13*	*r7*

Additional crosses using these four genes and others established the gene sequence as *h-r13-r7-r1*. However, an interesting fact emerged from T2 phage crosses. The maximum amount of recombination between any two genes was 30 percent, even though they might actually be separated by several hundred map units. In a Mendelian cross, the maximum detectable recombination was 50 percent, because multiple crossover events between widely separated genes made the genes appear to be assorting independently of each other. The 30 percent value found in T2 phages is best understood by examining the recombination mechanism employed by these phages.

A phage cross has so far been presented simply as a reciprocal exchange of genetic information between two genotypically different molecules of viral DNA within the confines of a sensitive host cell. This is far from a complete description of the actual events of genetic recombination. Recombination in phages involves repeated interactions among the intracellular population of phage DNA molecules replicated from the parental chromosome. Any one DNA molecule can undergo repeated recombinational events up to the time it is either incorporated into a mature virus particle or released unincorporated through lysis. In higher organisms, recombination is restricted to the prophase stage of the first meiotic division.

A cross of three genetically distinct phages illustrates the multiple rounds of exchange a phage genome may engage in. It is possible for *E. coli* to be simultaneously infected with three phages: T2(ab^+c^+), T2(a^+bc^+), and T2(a^+b^+c). Some progeny phages from such a triply infected bacterium have the genotype *abc*, a combination possible only when the phage obtains one gene from each parent. A possible mechanism for this kind of recombination is diagrammed in Figure 11.14. It is evident that the generation of *abc* genotypes could occur by two recombinational events involving three distinct genomes. Note also that recombination is not depicted as involving chromosome pairing. In the next section of this chapter, the experimental basis for such a breakage-fusion type of recombination will be presented.

Figure 11.14
Multiple recombination events occurring in phage chromosomes

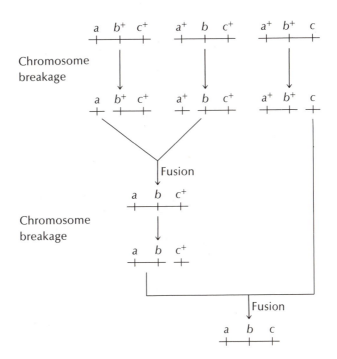

Viewing phage recombination as a population phenomenon with hundreds of chromosomes undergoing multiple rounds of exchange within a host cell, we can understand a maximum recombination frequency of less than 50 percent. Consider a phage cross involving xy and x^+y^+ in a 1:1 ratio. The only way a recombinant genome (x^+y or xy^+) is produced is through the interaction of xy with x^+y^+, and such an interaction has a probability of $\frac{1}{2}$, assuming random interaction (the probability of xy interacting with x^+y^+ is $\frac{1}{2} \times \frac{1}{2} = \frac{1}{4}$, and that of x^+y^+ interacting with xy is $\frac{1}{2} \times \frac{1}{2} = \frac{1}{4}$; $\frac{1}{4} + \frac{1}{4} = \frac{1}{2}$). If each such heterozygous mating were to result in a recombinant, the maximum percentage of recombinants would be 50 percent, assuming unlimited rounds of mating. However, not all heterozygous "matings" (xy by x^+y^+) generate recombinants. When x and y are widely separated, they tend to assort independently so that xy, x^+y^+, x^+y, and xy^+ all have identical frequencies. This means that only half of all heterozygous matings—assuming x and y are widely separated—produce recombinants, an observed recombination frequency of $(\frac{1}{2})^2$, or 25 percent (the frequency of xy interacting with x^+y^+ is $\frac{1}{2}$, and the frequency of an xy and x^+y^+ mating yielding a recombinant is $\frac{1}{2}$). As more matings occur within the confines of the host cell, recombinant frequencies increase and approach 50 percent, but matings are not unlimited. After a certain length of time, phage chromosomes are withdrawn from the chromosome pool and incorporated into mature phage particles. This continuous removal of chromosomes from the chromosome pool means that the recombinant frequency for widely separated genes will be less than 50 percent. It can be generally assumed that the smaller the percentage of recombination, the more restricted are the opportunities for chromosome exchange.

It is evident from this discussion that there are major differences between recombination in phages and recombination in a Mendelian system, even though the end products are recombinant chromosomes in both cases. Phage recombination must be considered more of a population phenomenon and treated accordingly. With this in mind, N. Visconti and Delbrück in 1953 devised a statistical model to analyze recombination frequencies as a function of the number of chromosome exchanges in a host cell. The T2 and T4 chromosomes undergo several exchanges up until they are incorporated into mature phages. This is reflected in their recombination frequency of 30 percent for widely separated genes. Lambda undergoes perhaps one exchange round prior to incorporation. This is reflected in a 15-percent recombination frequency for widely separated genes. It is not well understood what determines the number of exchanges a phage chromosome undergoes inside a host cell.

In addition to rounds of chromosome exchange, recombination frequencies in phages are influenced by the number of DNA molecules in an infected cell, by the proportions of different parental genotypes, by the replication rates of the different genotypes, and by the environment of the host cell. In spite of these complexities, phage maps can be constructed that show linear relationships between genes.

In 1960, G. Streisinger and V. Bruce developed the three-point test-cross in phages to demonstrate that all the T2 and T4 genes were located on a single chromosome—in other words, that each phage had one linkage group. Before this time, the data had suggested that a phage contained more than one linkage group.

The three-point testcross involves three genes, two of which are known to be linked and one whose linkage is uncertain. In a phage cross involving linked loci ($ab \times a^+b^+$), the recombinants (a^+b and ab^+) are examined for the frequency with which they include one of the two alleles (c and c^+) at a third locus whose linkage to ab is in question.

If the c locus is not linked to ab, then c and c^+ should assort independently in the a^+b and ab^+ recombinant classes.

$$\text{Frequency of } a^+bc = \text{Frequency of } a^+bc^+$$

$$\text{Frequency of } ab^+c = \text{Frequency of } ab^+c^+$$

If the c locus is linked to ab, then c and c^+ will not assort independently, and their linkage could be either to the right or left of ab.

$$a\ b\ c \quad \text{or} \quad c\ a\ b$$

The cross to test which of these two possibilities is actually the case puts abc on one chromosome and $a^+b^+c^+$ on the other.

$$a\ b\ c \times a^+\ b^+\ c^+ \quad \text{or} \quad c\ a\ b \times c^+\ a^+\ b^+$$

In the cross $abc \times a^+b^+c^+$, recombination between a and b should lead to a greater number of ab^+c^+ than ab^+c progeny, because the latter would require two recombination events and the former only one. Also, and for the same reason, a^+bc should be more frequent than a^+bc^+.

$$\longrightarrow a^+bc \quad \text{and} \quad ab^+c^+$$

$$\longrightarrow a^+bc^+ \quad \text{and} \quad ab^+c$$

In the cross $cab \times c^+a^+b^+$, the same reasoning would lead to cab^+ being more frequent than c^+ab^+ and c^+a^+b being more frequent than ca^+b.

$$\longrightarrow cab^+ \quad \text{and} \quad c^+a^+b$$

$$\longrightarrow c^+ab^+ \quad \text{and} \quad ca^+b$$

483

Figure 11.15
Map of the bacteriophage
T4 genome (From W. B.
Wood and H. R. Revel,
Bact. Rev. 40:847–868,
1976)

This technique of comparing recombination frequencies allowed Streisinger and Bruce to conclude that all the genes known in T2 and T4 were contained in one linkage group. In 1961, Streisinger suggested that the T4 genetic map was circular. This would explain crosses in T4 in which genes mapped to both the left *and* right of known markers. Since then, all the available data seem to support a circular linkage map for T2 and T4; the latter is shown in Figure 11.15.

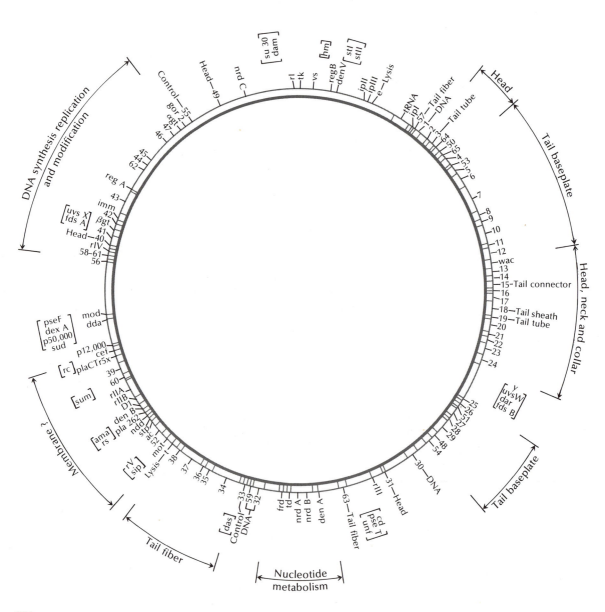

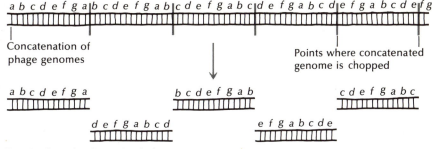

Concatenation of
phage genomes

Points where concatenated
genome is chopped

Figure 11.16
Formation of terminally
redundant, circularly
permutated phage
genomes

Terminally redundant, circularly
permuted phage chromosomes

The circular genetic map raises the issue of whether the chromosomes of T2 and T4 are *physically* circular. Evidence suggests that the chromosome is linear while in the phage head. The circularity of the genetic maps could be explained by *terminal redundancy* (each chromosome begins and ends with the same nucleotide sequences) and *circular permutations* (the redundant extremities are not the same for each chromosome but rather appear to be random). Each linear chromosome with any pair of genes at the extremities will map as a circular chromosome, because any pair of genes will be found linked on the majority of chromosomes. These facts suggest that T2 and T4 chromosomes are formed from fixed lengths of DNA—longer than the complete genome—being chopped off from a long chain of genomes (*concatenation*) linked together (Figure 11.16). Terminal redundancy is a feature common to nearly all DNA phages that have linear chromosomes, and this may reflect the mechanism by which such phages replicate their chromosomes—perhaps a rolling-circle scheme or a Cairns scheme—because evidence suggests that the linear molecules circularize while in the host cell (Figure 11.17).

Constructing a genetic map in lambda

In 1955, A. D. Kaiser performed a series of two-point and three-point crosses in lambda, and from his studies a genetic map of lambda emerged. It would be valuable for us to look closely at Kaiser's studies in order to appreciate gene mapping in a bacteriophage. The five genes analyzed are

s = Small plaque
mi = Minute plaque
c = Clear plaque
co_1 = Clear plaque with a few bacterial colonies
in the center
co_2 = Like co_1 but with more colonies in the center

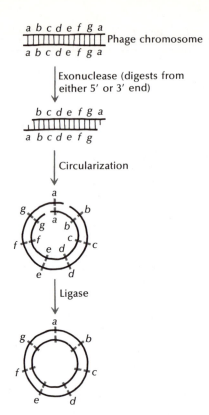

Figure 11.17
Circularization of a
terminally redundant
DNA molecule (phage
chromosome)

In a series of 10 two-point crosses, Kaiser accumulated the data shown in Table 11.2. From these data, we can draw certain conclusions. For example, the co_1–mi distance is about 5.4 map units, the co_1–s distance is about 2.6, and the s–mi distance about 8.5. From this, we can suggest that s and mi are two end points with co_1 in the middle:

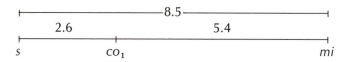

When we look at c relative to s and mi, we find that it is very close to co_1. The s–c distance is 2.8 map units, and the mi–c distance is 6.2 map units. The c–co_1 distance is only 0.1 map unit, confirming the close association of these two genes. But from the information given so far, we cannot tell whether the correct gene sequence is

$$s–co_1–c–mi \quad \text{or} \quad s–c–co_1–mi$$

Table 11.2
Two-factor crosses involving λ

Parents	Progeny				% Recombinants
$co_1+ \times +mi$	6341 co_1+	7580 $+mi$	372 $++$	412 co_1mi	5.3
$co_1mi \times ++$	111 co_1+	86 $+mi$	1800 $++$	1576 co_1mi	5.5
$s+ \times +co_1$	7101 $s+$	5851 $+co_1$	145 $++$	169 sco_1	2.4
$sco_1 \times ++$	46 $s+$	53 $+co_1$	1615 $++$	1774 sco_1	2.8
$s+ \times +mi$	647 $s+$	502 $+mi$	65 $++$	56 smi	9.5
$smi \times ++$	1024 $s+$	1155 $+mi$	13083 $++$	13253 smi	7.6
$s+ \times +c$	808 $s+$	566 $+c$	19 $++$	20 sc	2.8
$c+ \times +mi$	1213 $c+$	1205 $+mi$	84 $++$	75 cmi	6.2
$c+ \times +co_1$	6000 $c+$	6000 $+co_1$	14 $++$	$-cco_1$[a]	0.1
$co_2+ \times +mi$	1477 co_2+	1949 $+mi$	109 $++$	131 co_2mi	6.6

[a] $c\,co_1$ indistinguishable from $c+$

From A. D. Kaiser, 1955. *Virology* 1:424.

Nor can we tell where co_2 is located. We know that the co_2–mi distance is 6.6 map units, but we do not know whether co_2 maps to the right or left of mi:

or

We can extricate ourselves from this dilemma by performing a three-point cross, and this is indeed what Kaiser did (see Table 11.3). In cross 1, classes G and H are the double-crossover classes, suggesting that co_1 is the middle gene in the sequences s–co_1–mi, a fact we already deduced from the two-point cross. In cross 2, classes A and B are the double-crossover classes, again confirming the position of co_1 relative to s and mi. In cross 3, we assume class A is a double crossover (class F being perhaps a spontaneous mutation), meaning that co_1 is central relative to c and mi. In cross 4, F is the double-crossover class, meaning that c is the middle gene relative to co_2 and mi.

Table 11.3
Three-point cross with λ

Parents	$+++$	$scomi$	$s++$	$+comi$	$sco+$	$++mi$	$s+mi$	$+co+$	Total
1. $sco_1mi \times +++$	975	924	30	32	61	51	5	13	2091
2. $s+mi \times +co_1+$	38	23	273	318	112	121	6389	5050	12324
3. $c++ \times +co_1mi$	8	(—)[a]	(—)	(—)	(—)	1	(—)	(—)	6600
4. $co_2++ \times +cmi$	28	(—)	(—)	(—)	(—)	13	(—)	(—)	5800
	A	B	C	D	E	F	G (Mutation?)	H	

[a] (—) means no data given.

Now we can combine our data from the two-point and three-point crosses to make a reasonable map of these five genes. From the three-point cross, crosses 1 and 2 establish the gene sequence as being as follows.

$$s-co_1-mi$$

From cross 3, the sequence is

$$c-co_1-mi$$

And from cross 4, the sequence is

$$co_2-c-mi$$

Integrating crosses 3 and 4, we get

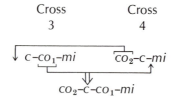

From the two-point cross data, we know that s maps to the left of co_2, so the final sequence, including map distances, is

$$s \quad \overset{1.5}{} \quad co_2 \quad \overset{1}{} \quad c \quad \overset{0.1}{} \quad co_1 \quad \overset{5.4}{} \quad mi$$

Interference and coefficient of coincidence in phage crosses

In our study of gene mapping in eukaryotes, we noted that interference was usually positive. That is, a crossover in one region usually suppressed a crossover in an adjacent region, causing the observed double crossovers to be fewer than expected. The coefficient of coincidence was normally less than 1.0. But in phage, the situation is often quite different. Consider, for example, the gene sequence we just calculated in lambda.

$$s \overset{2.6}{\rule{2cm}{0.4pt}} co_1 \overset{5.4}{\rule{3cm}{0.4pt}} mi$$

In this three-point cross, we expect the double exchanges to appear with the frequency

$0.026 \times 0.054 = 0.0014$, or 14/10,000

However (see Table 11.3), we observed

18/2091, or 86/10,000

and this is more than six times our expectation. The coefficient of coincidence in this case is 6.14, a value showing *negative interference* (observed double exchanges *exceed* expected double exchanges).

Part of the reason for this negative interference is the multiple rounds of mating between different chromosomes, which we discussed earlier. Another contributing factor is that recombination in the same geographic region is more likely because of the concentration of enzymes assisting in the recombination process. This negative interference, which also occurs in eukaryotes, can be extremely high in an extremely small map interval (Figure 11.18). We often call this *high negative interference*, as opposed to *low negative interference*. High negative interference is probably caused by the recombinational enzymes concentrated in a small region, whereas only multiple exchanges produce the low values. All of this leads us now to consider the mechanisms of recombination.

Heterozygosity in phages

Phages are haploid organisms; that is, except for terminally redundant genes, each gene is present only once. However, in 1951 Hershey and Chase made some observations indicating that perhaps the organism is on occasion partially diploid for some genes.

If *E. coli* is simultaneously infected with T2r and T2r⁺, and if the infected bacteria are plated on a lawn of *E. coli* before lysis, many of the resultant plaques that form are "mottled" in appearance (see Figure 11.11) because they contain a mixture of *r* and *r⁺* phages. However, if the progeny phages from the simultaneous infection are collected first and then plated out on a lawn of *E. coli*, about 2 percent of the plaques that form are mottled and can be shown to contain both *r* and *r⁺* phages. This means that a *single phage* gave rise to both *r* and *r⁺* progeny. How could this

Figure 11.18
The relationship between coefficient of coincidence and the distance between three points (RI × RII) (After Amati and Meselson, 1965, *Genetics* 51:369)

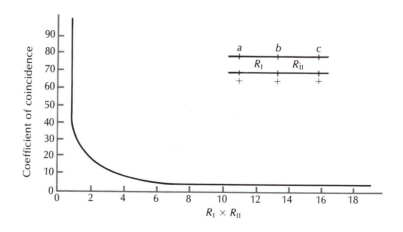

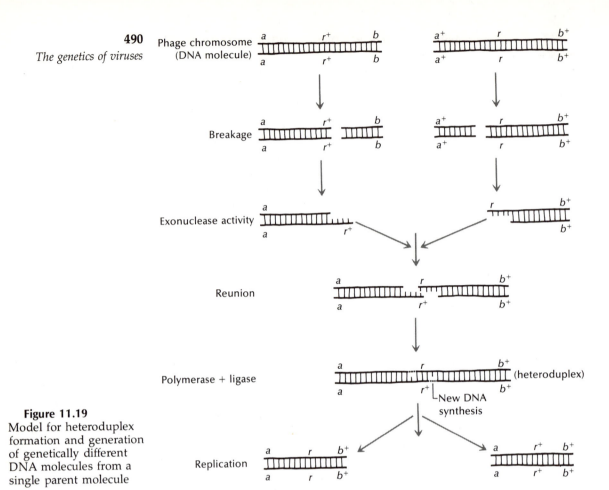

Figure 11.19
Model for heteroduplex
formation and generation
of genetically different
DNA molecules from a
single parent molecule

occur? One possible mechanism is the formation of a *heteroduplex*, a DNA molecule composed of one strand with an *r* gene sequence and one strand with an *r*⁺ gene sequence. On replication of such a heteroduplex, one daughter molecule would be *r* and one *r*⁺ (Figure 11.19). The heteroduplex forms as a consequence of recombination; it then replicates, giving *r* and *r*⁺ chromosomes.

Molecular models of recombination

Recombination in bacteriophages involves interaction between homologous DNA molecules and results in the generation of recombinant structures containing nucleotide sequences from both parental molecules. A number of studies deal with the mechanism of recombination at the molecular level, and the majority of them favor a *breakage–fusion* type of mechanism as opposed to a *copy-choice* mechanism; the latter mechanism explains recombination as a result of DNA synthesis.

The copy-choice model   According to the copy-choice model, recombination occurs in conjunction with DNA synthesis (Figure 11.20a). During chromosome replication, homologous sections of two or more chromosomes come to lie opposite each other and, in the process of replication, copy part of each other. This model, first proposed for higher organisms in 1931 by Belling, was later revived by Hershey and Rotman, who suggested it in connection with their observation that recombination is not always a reciprocal event. That is, in a phage cross of $ab \times a^+b^+$, some single bursts produce only one recombinant type (say, a^+b and not ab^+), and some produce large numbers of one type relative to the other.

In general, the copy-choice model presents many problems and is not well supported by experimental evidence. For example, if DNA replication is semiconservative, only one strand of the DNA molecule would switch over to copy a strand of another molecule. This means that recombinant molecules could exist with large regions of noncomplementary DNA sequences (Figure 11.20a). But even if such sequences did not represent an insurmountable problem, the uncoiling of DNA strands at the completion of synthesis would constitute a very major physical difficulty (Figure 11.20b).

Despite these drawbacks, however, the copy-choice model, or at least a modified version of it, may best explain some recombinational events. It was suggested that the formation of phage heterozygotes involved recombination through breakage-reunion *and* limited DNA synthesis. Gene conversion in *Neurospora* (see pages 303–306) also involves DNA synthesis.

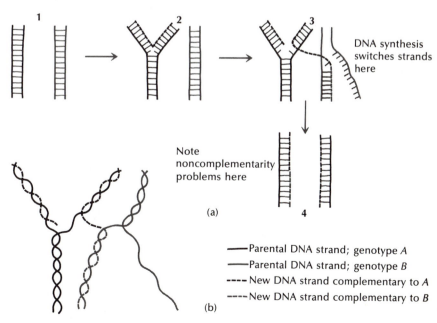

Figure 11.20
(a) Copy-choice model of genetic recombination: a diagrammatic view
(b) Amplification of step 3, demonstrating coiling in the copy-choice model (uncoiling here would present a major problem)

DNA synthesis switches strands here

Note noncomplementarity problems here

(a)

(b)

———Parental DNA strand; genotype *A*
———Parental DNA strand; genotype *B*
---- New DNA strand complementary to *A*
---- New DNA strand complementary to *B*

491

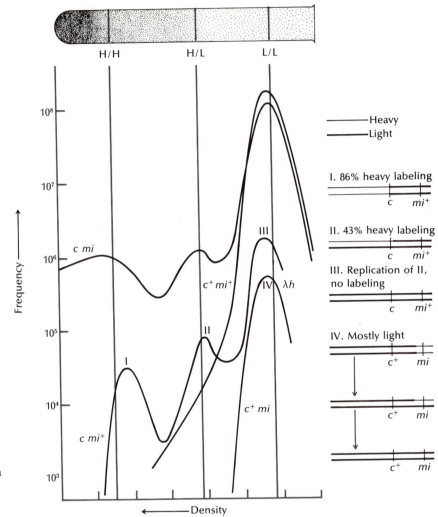

Figure 11.21
Cross between λ *c mi* labeled with ^{13}C and ^{15}N and unlabeled λ *c$^+$ mi$^+$*; the progeny phages were collected and centrifuged, with λ*h* added as a reference for density and band shape [Adapted from M. Meselson and J. J. Weigle, *Proc. Natl. Acad. Sci. U.S.* 47:857 (1961)]

The breakage–fusion model

This model, based on the breaking and rejoining of DNA molecules, was introduced in conjunction with our discussion of phage heterozygotes and heteroduplex formation. It is the most widely accepted general model of recombination. It does not require that recombination and DNA synthesis go hand in hand, but it does suggest that limited DNA synthesis accompanies the breakage–fusion event. Results of studies by Meselson and J. J. Weigle in 1961 and Meselson in 1964 strongly support the breakage–fusion hypothesis.

In the Meselson–Weigle experiments, two mutant strains of the bacteriophage lambda were employed, one labeled with the density labels ^{13}C and ^{15}N and one without density labels (^{12}C and ^{14}N). Lambda *c mi* (*c* = clear plaque, *mi* = minute plaque) labeled with ^{13}C and ^{15}N and lambda *c$^+$ mi$^+$* unlabeled (^{12}C and ^{14}N) simultaneously infected a sensitive *E. coli* host cell growing in an unlabeled (^{12}C and ^{14}N) medium. The progeny phages were collected and centrifuged in a CsCl gradient. Progeny from different density levels were sampled and their genotypes determined. The results (Figure 11.21) suggested that breakage and reunion accounted for the *c mi$^+$* recombinant class, which was in some cases 86 percent heavy. The *c$^+$ mi* recombinant class was too light to indicate anything positive, because it could be interpreted as either copy-choice or breakage–fusion. A copy-choice mechanism would produce recombinants that were at most 50 percent heavy (a DNA molecule having one parental strand, labeled with ^{13}C and ^{15}N, and one unlabeled recombinant strand).

However, the fact that some unlabeled DNA was present in the *c mi$^+$* recombinant class implies that some DNA synthesis may have accompanied the recombinational event—recombination perhaps similar to that shown in Figure 11.19. To determine whether any DNA synthesis was required for recombination, Meselson (1964) crossed ^{13}C^{15}N-labeled lambda *hc* (*h* = host range, *c* = clear plaque) with ^{13}C^{15}N-labeled lambda *h$^+$c$^+$* in a fashion similar to that described for the Meselson–Weigle experiment. On centrifugation of the progeny phages in a CsCl density gradient, recombinants (*h$^+$c*) were found with chromosomes formed entirely, or almost entirely, of parental DNA (Figure 11.22). This demonstrated that genetic recombination occurred by breakage and fusion of double-

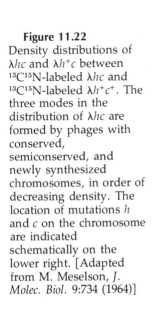

Figure 11.22
Density distributions of λ*hc* and λ*h$^+$c* between ^{13}C^{15}N-labeled λ*hc* and ^{13}C^{15}N-labeled λ*h$^+$c$^+$*. The three modes in the distribution of λ*hc* are formed by phages with conserved, semiconserved, and newly synthesized chromosomes, in order of decreasing density. The location of mutations *h* and *c* on the chromosome are indicated schematically on the lower right. [Adapted from M. Meselson, *J. Molec. Biol.* 9:734 (1964)]

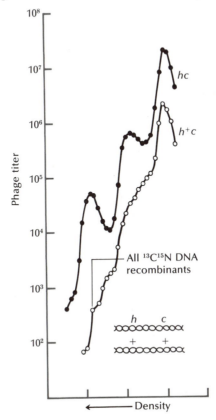

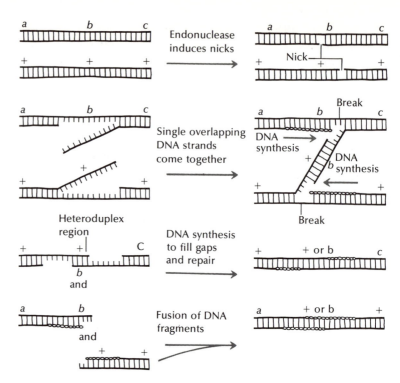

Figure 11.23
A molecular model for
the generation of
reciprocal recombination

stranded DNA, but there is also some indication that a small amount of DNA synthesis occurs in the formation of recombinant molecules.

To date, our most reasonable models of recombination involve breakage and fusion along with limited synthesis. In Figure 11.23, we see a proposal to explain the generation of a reciprocal recombinational event, perhaps applicable to eukaryotic recombination. The key to this model is the base-pairing that occurs between complementary single DNA strands. A somewhat modified version of this model appears in Figure 11.24. Here both reciprocal and nonreciprocal recombination occur through a more complex series of breaks. But as attractive as these models are, they do not explain everything about recombination. There is still a great deal we do not understand.

Viruses and cancer

A cancerous cell is one that has lost its ability to regulate its own growth and division and has acquired new surface characteristics. The transformation of a normal cell into a cancerous cell may be the result of one or both of the following phenomena.

1. Mutations in somatic cells that lead to dysfunctions in cell regulatory mechanisms; such mutations may arise spontaneously or they may be induced by irradiation (UV or x-ray) or chemical carcinogens.

2. Animal viruses that infect and transform cells into cancerous growths; the addition of viral genetic material to the host cell genome is seen as the causative agent in this cancer transformation.

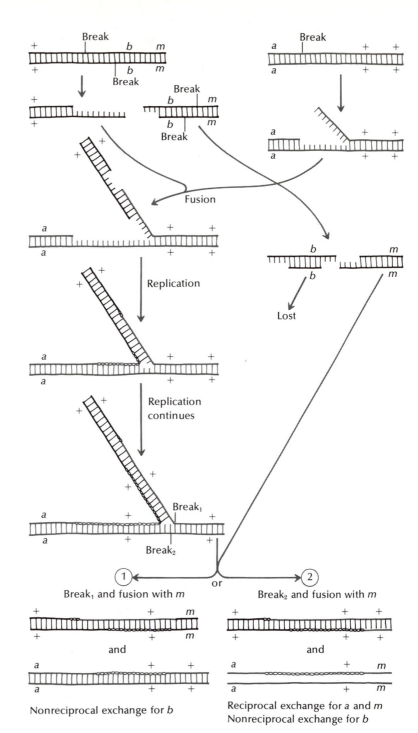

Figure 11.24
A model of recombination based on a complex series of breaks and joins and DNA synthesis (after Boon and Zinder, *J. Mol. Biol.* 58:133, 1971)

Viruses have been implicated in cancer for well over 60 years, when in 1912 the Rous Sarcoma Virus (RSV) was shown to be the cause of connective tissue tumors (Sarcomas) in chickens. But how extensive is the viral connection to cancer? Recent research suggests that virus genomes, or segments of the genomes, are perhaps the major cause of all cancers. We are now beginning to understand how viral genes transform normal cells into cancerous cells. This understanding comes from exciting and extensive research into the two classes of tumor viruses: DNA tumor viruses and RNA tumor viruses.

DNA tumor viruses

The DNA tumor viruses are better understood than their RNA counterparts. These DNA tumor viruses include SV40, a monkey tumor virus; herpes [a group of viruses that causes cold sores, mononucleosis, probably Burkitt lymphoma (a rare cancer found in certain African regions), and perhaps other cancers]; and adenoviruses, a group of viruses that induce common coldlike symptoms and may also cause tumors. To get some idea of how a DNA virus may cause cancer, let us focus on the SV40 virus.

The SV40 virus is a small organism composed of a protein coat surrounding a double-stranded DNA molecule (Figure 11.25). Recent studies indicate that the SV40 chromosome may contain only three genes:

VP 1   (major coat protein)
VP 2   (minor coat protein)
T-antigen

When the SV40 virus infects a monkey cell, it follows one of two pathways. Either it is inactivated and ultimately destroyed by the host cell's defense mechanisms, or it replicates in the cell's nucleus, destroying the cell while it produces hundreds of thousands of progeny. It is when the virus enters a foreign host, such as a mouse, that we observe the transformation from normal to cancer.

Studies of the SV40 life cycle help us gain insight into its transforming capabilities. When the SV40 virus enters a cell, the first gene to be activated is the T-antigen gene. The T-antigen is a protein that moves quickly to the nucleus. Its function appears to be to induce the synthesis of DNA-replicating enzymes in the host nucleus, and these enzymes begin replicating both host and SV40 DNA. At about the same time as replication begins, the mRNA species that code for the coat proteins make their appearance. We do not know how the coat protein genes are specifically regulated.

Figure 11.25
An SV40 virus showing its protein capsomeres, or structural subunits

In a cell being transformed, such as an SV40-infected mouse cell, the T-antigen mRNA, T-antigen, and SV40 DNA are made. But no coat proteins are synthesized, even though coat protein mRNA is synthesized. We cannot explain how the coat protein translation process is defective. When the cell has become transformed into a cancerous cell, SV40 DNA is detected in the host cell's chromosomal material. It appears that the SV40 DNA has physically inserted itself into the host's DNA and is thus behaving as a chromosomal gene.

Transformed cells do not have any visible SV40 particles. Nor can these viruses be induced to appear (that is, to enter a lytic phase and form progeny). But if a transformed mouse cell is fused (using Sendai virus, see page 80) with a normal monkey cell free of SV40 infection, the hybrid cell produces mature SV40 particles. Clearly something in the normal host cell enabled the SV40 transformed mouse cell to by-pass whatever blocked its normal life cycle. We still do not know what the monkey cell provided to enable the transformed cell to proceed through a lytic cycle.

When SV40 DNA is integrated into the host's DNA, the T-antigen gene remains active. The T-antigen is evidently the culprit in SV40-induced cancer, but precisely how it works is not clear. It does cause the synthesis of DNA-replicating enzymes essentially on a continuous basis, leading to unchecked growth. But it may also activate other host genes, genes normally inactive, and these genes may cause the cancer. There are many questions yet to be answered, but we are much closer today to those answers than we were five years ago.

RNA tumor viruses

The RNA tumor viruses are more structurally complex than their DNA counterparts. Their RNA chromosome containing about 10 genes is surrounded by a protein coat, which is enclosed in a cell-type membrane composed of lipids and proteins. Glycoprotein knobs are embedded in the surface of the membrane (Figure 11.26). The RNA tumor viruses are structurally similar, but since most research has focused on the Rous Sarcoma Virus (RSV), we shall discuss it in more detail.

RSV is a normal parasite of chickens. The life cycle of RSV is similar to that of all RNA tumor viruses (Figure 11.27). The virus penetrates the cell by membrane fusion. The protein coat is shed and the RNA is transcribed into a complementary DNA strand by virally coded *reverse transcriptase*, an enzyme specified by the virus and discovered by H. Temin and D. Baltimore. This RNA–DNA hybrid then is transcribed into double-stranded DNA, which circularizes, enters the nucleus, and integrates into the host DNA. Once integrated, RSV mRNA is transcribed and RSV protein synthesized. This protein encapsulates RSV RNA, moves to the cell membrane, and buds out. The budding process results in cell membrane covering the protein coat, but the membrane is covered with RSV-specific glycoproteins. Under normal conditions, this life cycle does not result in cell death.

How does the RNA virus cause cancer? Some have suggested that the budding process disrupts the cell membrane and causes the cell to lose control of its growth. This idea seems unlikely in light of the fact that RSV strains exist that cannot complete their life cycle but can cause cancer. Strains also exist that can proceed through their normal life cycle but are unable to cause cancer.

This latter mutant strain is particularly interesting. This strain, which reproduces normally but is unable to transform cells into cancerous cells, is usually a strain carrying a deletion. The deleted genetic information is called the *oncogene*, or cancer gene. Since it is not essential to the RSV life

Figure 11.26
A schematic diagram of an RNA tumor virus, such as the Rous Sarcoma Virus (RSV)

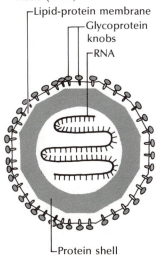

┌Lipid-protein membrane
┌Glycoprotein knobs
┌RNA

└Protein shell

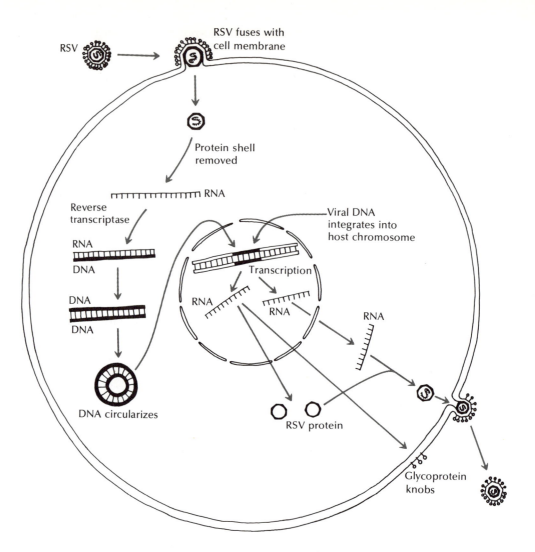

RSV

RSV fuses with cell membrane

Protein shell removed

RNA

Reverse transcriptase

RNA
DNA

DNA
DNA

DNA circularizes

Viral DNA integrates into host chromosome

Transcription

RNA

RNA

RNA

RSV protein

Glycoprotein knobs

Figure 11.27
Life cycle of RSV

cycle, the oncogene may be a host gene inserted into the RSV genome. When in its normal location in the host genome, the host gene is under normal gene-regulatory influences. But when inserted into RSV DNA, it is no longer regulated by the cell and is activated at times when it should be repressed. In RSV, we suspect that the ''oncogene'' is a block of genes that control embryonic development. If it continues to function or is induced to function through RSV infection, cells multiply as if they were embryonic and generate tumors as a consequence.

What is intriguing about the oncogene idea is that *all* cells in the body have the oncogene. In normal cells, the oncogene is repressed; in cancer cells, it is activated.

If the RNA virus studies can boast of a ubiquitous oncogene, the DNA tumor virus studies can boast of a ubiquitous *provirus* in all cells, both normal and cancerous. It has been suggested that all normal cells carry proviruses in their genome and that they are inherited as Mendelian units. Activation of the provirus, perhaps through carcinogenic chemicals or irradiation, leads to tumor formation.

Phenotypic mixing

The phenomenon of phenotypic mixing has been saved for the last portion of the chapter, because it is recombination in an entirely different sense. The recombinational events discussed so far have involved the interaction of genetically distinguishable DNA molecules. Phenotypic mixing entails the interaction between phage chromosomes and phage protein in such a way that DNA from one phage strain is packaged in the protein coat of another (Figure 11.28).

The phenomenon was first observed by Delbrück and W. T. Bailey in 1946 and elaborated on by A. Novick and L. Szilard in 1951. *E. coli* B was simultaneously infected with T2 and T4 phages and the progeny tested on either *E. coli* B/2 (T2-resistant) or *E. coli* B/4 (T4-resistant). About half the total progeny phages formed plaques on B/4 (the T2 phage), and about half formed plaques on B/2 (the T4 phage). However, about half the progeny phages released by B/2 bacteria are incapable of forming plaques on B/2 but can form plaques on B/4. The B/4 bacteria also released progeny phages half of which could not reinfect B/4 but could infect B/2. These observations suggest that the T2 and T4 chromosomes can be packaged into T2 or T4 protein coats with equal probability. For example, a T2 chromosome packaged into a T4 protein coat will be injected into a B/2 bacterium but not into a B/4, which is resistant to T4 attachment (attachment being a function of the protein tail structure). The T2 chromosome generates T2 protein, and mature T2 progeny are produced that can no longer infect B/2 cells.

Phenotypic mixing is a unique form of recombination, and it also provides a strong argument in support of the notion that DNA is the genetic material. The progeny of a T2 chromosome and a T4 protein parent phage are all T2, the T4 protein playing no role in the composition of the offspring.

Figure 11.28
Phenotypic mixing

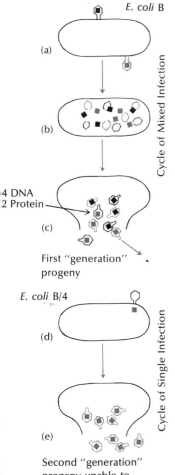

E. coli B

(a)

(b)

4 DNA
2 Protein

(c)

First "generation" progeny

E. coli B/4

(d)

(e)

Second "generation" progeny unable to further infect *E. coli* B/4

Cycle of Mixed Infection

Cycle of Single Infection

◀

Cycle of Mixed Infection (a, b, c) The two types of genetic material become randomly associated with the two types of phage coat. Thus in half of the progeny particles the specificity of the protein coat is different from that of the genome it carries.

Cycle of Single Infection (d, e) When a particle in which the protein coat has a different specificity from the genome singly infects a new bacterium, as shown in (d), phage protein synthesis is determined solely by the injected genetic material. Consequently, the phage progeny particles (e) are pure, and their phenotype indicates the genotype of the infecting particle.

Viruses were recognized as a distinct class of organisms after a vigorous debate over the cause of the "glassy" transformation, ultimately shown to be due to bacteriophage infection. The bacteriophage, focal point of most viral genetic studies, is a relatively simple organism composed of protein and nucleic acid (usually DNA). Its life cycle involves attachment to a susceptible host, penetration, reproduction inside the host cell using host cell machinery, and finally lysis. Occasionally, however, a gentler phage–bacterium relationship called lysogeny exists, in which the phage DNA is integrated into the host cell DNA and its functions are for the most part repressed.

Hershey and Rotman discovered that viral genes recombine and can be mapped. But they also discovered, in single-burst experiments, that this recombination is not necessarily reciprocal. This led to much speculation about mechanisms of recombination.

Gene mapping in bacteriophage T4 and λ is best carried out using the three-point cross. That is, bacteria are simultaneously infected by bacteriophage that differ at three genetic loci. One interesting recombinant showed that a single phage produced progeny of two different genotypes. This led to the concept of the heteroduplex, a model of recombination involving the reunion of two DNA fragments. The ends of the two fragments have regions of single-strandedness, and when they overlap during reunion, one strand differs from the other, thus leading to differing progeny on replication.

The two main models of recombination are copy-choice and breakage–fusion. When it was originally conceived, copy-choice suggested that when DNA molecules were replicating, strands switched templates, yielding recombinants as a consequence of synthesis. Breakage–fusion suggested that recombination came about when DNA fragments fused. This model did not require any DNA synthesis. Our most widely supported models of recombination today show recombination occurring by breakage–fusion with limited DNA synthesis to fill the gaps.

Viruses have been implicated in cancer for over 60 years, but only recently has substantial evidence accumulated, allowing us to understand how they transform normal cells into cancerous cells. A DNA tumor virus such as SV40 integrates into a host cell's genome, where it continues to synthesize a protein (T-antigen) that appears to regulate the synthesis of DNA-replicating enzymes. An RNA tumor virus, such as RSV, carries an oncogene, which in turn induces cancers.

Questions and problems

11.1 A series of crosses is made using T4:

    a. $m^+nr \times mn^+r^+$
    b. $mn^+r \times m^+nr^+$
    c. $mnr^+ \times m^+n^+r$

Wild-type recombinants in the first cross are much less frequent than in the other two crosses. Using this observation, how would you sequence the genes?

11.2 Discuss the similarities and differences in recombination as it occurs in viruses on the one hand and eukaryotes on the other.

11.3 How has the study of viral genetics contributed to the chromosome theory of inheritance?

11.4 André Lwoff, in a poetic simile reminiscent of Gertrude Stein, stated that "the virus is a virus is a virus." If you disagree with Lwoff, how would you characterize the virus? If you agree with Lwoff's frustration, why can you not define the virus in any other way?

11.5 Two strains of T4 infect *E. coli* cells: strain 1 is abc, and strain 2 is $a^+b^+c^+$. The recombination data are as follows.

Class	Frequency
abc	0.34
$a^+b^+c^+$	0.36
ab^+c^+	0.05
a^+bc	0.05
abc^+	0.08
a^+b^+c	0.09
ab^+c	0.02
a^+bc^+	0.02

Are the genes a, b, and c linked or not? If so, determine the sequence and map the distances separating them. What is the coefficient of coincidence, and what does it mean?

11.6 Hershey and Chase crossed three different T2 strains and obtained the following results.

	$h^+m^+ri^+$	h^+m^+ri	hm^+ri^+	h^+mri^+	hm^+ri	$hmri^+$	h^+mri	$hmri$
$hm^+ri^+ \times h^+mri^+ \times h^+m^+ri$	25	17	18	16	7	8	6	3
$hmri^+ \times h^+mri \times hm^+ri$	3	4	7	10	16	22	15	23

What conclusions can you draw from this experiment pertaining to recombination?

11.7 The following T4 crosses are performed.

	rh^+	r^+h	r^+h^+	rh
$r_xh^+ \times r^+h$	0.38	0.44	0.09	0.09
$r_yh^+ \times r^+h$	0.43	0.48	0.04	0.05
$r_zh^+ \times r^+h$	0.47	0.51	0.01	0.01

(a) Construct a map for each cross.
(b) Construct the possible genetic map(s) for all four genes.
(c) The cross $r_x^+r_z \times r_yr_z^+$ yields 13% recombinants. What does this tell you about the h-r_x-r_z gene sequence?
(d) From what you know about the T4 genetic map, can you resolve the remaining gene maps so that they are all compatible?

11.8 Study the following map in phage.

From 80,000 progeny scored in the cross $x^+m^+n^+ \times xmn$, 76 are x^+mn^+ or xm^+n^+. Is this consistent with our concept of negative interference in phage recombination?

11.9 Three independently arising *rIIB* mutants are crossed in T4, and the following results are obtained.

	Percent recombinants
$rIIB_1 \times rIIB_2$	0
$rIIB_1 \times rIIB_3$	0
$rIIB_2 \times rIIB_3$	0.9

Interpret these results.

11.10 How do you determine whether a particular mutation in phage is a deletion?

11.11 Two T4 *rII* mutants (*p* and *q*) simultaneously infect *E. coli* B cells and produce *r* plaques. When these same two mutants infect *E. coli* K cells, no plaques are produced. If, however, the progeny are collected from *rIIp* × *rIIq* infected *E. coli* B cells and used to infect *E. coli* K cells, a few wild-type plaques are produced. How do you interpret this? (*Hint*: Review material in Chapter 10.)

11.12 Compare and contrast the meanings of the term *heterozygous* as it is used in describing viruses and *Drosophila*.

11.13 How can a physically linear chromosome produce a circular map?

11.14 How does interference in phage recombination compare with interference in eukaryote recombination?

11.15 The lambda chromosome is linear except when it is replicating inside a host cell. Then it is circular. Yet even though it circularizes and recombines in its circular form, it produces a linear map. How would you explain this?

References

Baltimore, D. 1976. Viruses, polymerases, and cancer. *Science* 192:632–636.

Becker, F. F. (ed.). 1975. *Cancer*. New York, Plenum.

Beijerinck, M. W. 1899. Über ein contagium vivum fluidum als Ursache der Fleckenkrankheit der Tabaksblätter. *Zentr. Bakteriol. Parasitenk.* (II) 5:27.

Belling, J. 1933. Crossing over and gene rearrangement in flowering plants. *Genetics* 18:388.

Bordet, J. 1922. Concerning the theories of the so-called "bacteriophage." *Brit. Med. J.* (19 August, 1922):296.

Burnet, F. M. 1934. The bacteriophages. *Biol. Rev. Cambridge Phil. Soc.* 9:332.

Cairns, J., G. S. Stent, and J. D. Watson (eds.). 1966. *Phage and the Origins of Molecular Biology*. Cold Spring Harbor, New York, Cold Spring Harbor Laboratory of Quantitative Biology.

Corwin, H. O., and J. B. Jenkins. 1976. *Conceptual Foundations of Genetics: Selected Readings*. Boston, Houghton Mifflin.

Delbrück, M., and W. T. Bailey. 1946. Induced mutations in bacterial viruses. *Cold Spring Harbor Symp. Quant. Biol.* 11:33.

Doermann, A. H. 1973. T4 and the rolling circle model of replication. *Ann. Rev. Genetics* 7:325–342.

Duckworth, D. H. 1976. Who discovered the bacteriophage? *Bact. Rev.* 40:793–802.

Dulbecco, R. 1976. From the molecular biology of oncogenic RNA viruses to cancer. *Science* 192:437–440.

Ellis, E. L., and M. Delbrück. 1939. The growth of bacteriophage. *J. Gen. Physiol.* 22:365–384.

Gratia, A. 1922. Concerning the theories of the so-called "bacteriophage." *Brit. Med. J.* (19 August, 1922):296.

Hausmann, R. 1973. The genetics of T-odd phages. *Ann. Rev. Microbiol.* 27:51–68.

Hayes, W. 1968. *The Genetics of Bacteria and their Viruses.* New York, Wiley.

d'Hérelle, F. 1917. Sur un Microbe invisible antagoniste des bacilles dysentériques. *C. R. Acad. Sci. Paris* 165:373.

Hershey, A. D., and M. Chase. 1951. Genetic recombination and heterozygosis in bacteriophage. *Cold Spring Harbor Symp. Quant. Biol.* 16:471.

Hershey, A. D., and R. Rotman. 1948. Linkage among genes controlling inhibition of lyses in bacterial viruses. *Proc. Natl. Acad. Sci. U.S.* 34:89.

Hershey, A. D., and R. Rotman. 1949. Genetic recombination between host-range and plaque-type mutants of bacteriophage in single bacterial cells. *Genetics* 34:44.

Hotchkiss, R. D. 1974. Models of genetic recombination. *Ann. Rev. Microbiology* 28:445–468.

Meselson, M. 1964. On the mechanism of recombination between DNA molecules. *J. Molec. Biol.* 9:734.

Meselson, M. 1967. The molecular bases of recombination. In R. A. Brink (ed.), *Heritage from Mendel.* Madison, Wis., University of Wisconsin Press.

Meselson, M., and J. J. Weigle. 1961. Chromosome breakage accompanying genetic recombination in bacteriophage. *Proc. Natl. Acad. Sci. U.S.* 47:857.

Mosig, G. 1970. Recombination in bacteriophage T4. *Adv. Genetics* 15:1–53.

Mosig, G., W. Berquist, and S. Bock. 1977. Multiple interactions of a DNA binding protein *in vivo.* III. Phage T4 gene—32 mutations differentially affect insertion-type recombination and membrane properties. *Genetics* 86:5–23.

Muller, H. J. 1922. Variation due to change in the individual gene. *Amer. Nat.* 56:32.

Novick, A., and L. Szilard. 1951. Virus strains of identical phenotype but different genotype. *Science* 113:34.

Radding, C. M. 1973. Molecular mechanisms in genetic recombination. *Ann. Rev. Genetics* 7:87–112.

Schlesinger, M. 1936. The feulgen reaction of the bacteriophage substance. *Nature* 138:508.

Stent, G. S. 1963. *Molecular Biology of Bacterial Viruses.* San Francisco, Freeman.

Stent, G. S. 1965. *Papers on Bacterial Viruses.* Boston, Little, Brown.

Streisinger, G., and V. Bruce. 1960. Linkage of genetic markers in phages T2 and T4. *Genetics* 45:1289.

Studier, F. W. 1972. Bacteriophage T7. *Science* 176:367–376.

Temin, H. M. 1976. The RNA provirus hypothesis. *Science* 192:1075–1080.

Tumor Viruses. 1974. *Cold Spring Harbor Symp. Quant. Biol.* Vol. XXXIX.

Twort, F. W. 1915. An investigation on the nature of the ultramicroscopic viruses. *Lancet* 189:1241.

Visconti, N., and M. Delbrück. 1953. The mechanism of recombination in phage. *Genetics* 38:5.

Watson, J. D. 1976. *The Molecular Biology of the Gene.* Menlo Park, Calif., Benjamin.

Wood, W. B., and H. R. Revel. 1976. The genome of bacteriophage T4. *Bact. Rev.* 40:847–868.

Twelve

Transmission of the genetic material in bacteria

Transformation
Initial interaction between donor DNA and recipient cells
The uptake and incorporation of transforming DNA
Transformation and gene mapping

Conjugation
The discovery of conjugation
Conjugation is a one-way genetic transfer
The sex factor
Chromosome mobilization in *E. coli*
Sexduction
Function of the sex factor
The process of conjugation

Transduction and lysogeny
The discovery of transduction
Generalized transduction
Lysogeny
The prophage
Zygotic induction
Genetic control of lysogeny
Prophage integration
DNA insertion elements
Specialized transduction
Abortive transduction

Plasmids and recombinant DNA

Overview

Questions and problems

References

We have already noted that the cellular process of meiosis, along with crossing-over, accounts for segregation, independent assortment, and complete and incomplete linkage. Meiosis, however, is known to occur only in eukaryotic organisms. One of the more significant outcomes of meiosis is the reshuffling of genetic material and the consequent increase in genetic variability. Prokaryotic organisms achieve variability by means of processes other than meiosis. We saw in Chapter 11 how viruses can attain this end, and in this chapter we will examine mechanisms evolved by bacteria for recombination and transmission of genetic information.

Bacteria reproduce by asexual fission, but apart from mutation, this does not lead to any change in genetic information from one generation to the next. There are, however, three *parasexual* mechanisms whereby chromosomes of different organisms exchange genetic information without gamete formation. The first, *transformation*, comes about by the addition of foreign DNA to a culture. In the second, *conjugation*, bacteria of different mating types participate in a one-way transfer of genetic information. The third, *transduction*, is a viral-mediated mechanism of recombination. The importance of genetic recombination in the successful evolution of species can be reckoned from the diverse mechanisms employed by organisms to achieve that recombination.

Transformation

Basically, genetic *transformation* is the transfer of a fragment of DNA from a donor genome through a cell membrane of a recipient cell and the incorporation of that donor fragment into the recipient cell's genome. It usually occurs only when the donor DNA and the recipient are of the same species, but this may not always be so. Incorporation occurs when, through recombination, some of the recipient cell's DNA is replaced by the donor DNA. Transformation can be detected only when the newly incorporated DNA fragment gives rise to a new cell phenotype. For instance, a formerly drug-sensitive strain may develop genetically determined resistance to the drug. Transformation was first described by Griffith in 1928 using *Diplococcus pneumoniae* (also called *Pneumococcus*) bacteria, but it has since been shown to occur in other bacteria (such as *Haemophilus influenzae, Bacillus subtilis,* and *Escherichia coli*). Although the precise mechanisms of transformation differ somewhat for each species, we shall utilize information from several systems to present a general picture of the process.

Transformation is not simply a laboratory phenomenon. It also occurs in natural populations of microorganisms. For example, when two *Pneumococcus* populations are mixed, each carrying a different drug-resistance mutation, cells resistant to both drugs are recovered. Studies reveal that when a cell dies, it often lyses, sending its DNA into the medium. This DNA, or a piece of it, is picked up by another cell and transforms that cell. Even without dying, *Bacillus* cells can extrude pieces of DNA into the medium—DNA that can be taken up by other cells. So transformation appears to be a natural phenomenon, of particular importance, perhaps, when different bacterial strains infect a single host organism.

Initial interaction between donor DNA and recipient cells

The first step in transformation is the successful interaction between transforming DNA and the recipient bacterium. The size of the transforming fragments, their concentration, and the physiological state of the recipient cell are critical factors.

To successfully transform a *Pneumococcus* cell, a transforming DNA fragment must have a minimal size of about 800 nucleotide pairs. But for *Bacillus*, the minimal size is around 16,000 nucleotide pairs. Although there is apparently no upper size limit for transforming DNA, with high concentrations of high-molecular-weight DNA, only portions of each molecule get into recipient cells.

For any particular gene, a correlation exists between the number of donor DNA molecules present and the successful transformation of cells. For example, when DNA extracted from a streptomycin-resistant strain of bacteria is used to transform streptomycin-sensitive cells, the number of resistant transformants is directly proportional to the number of DNA molecules present, up to a maximum of about 10 DNA molecules per bacterium (Figure 12.1). The presence of more than 10 DNA molecules per recipient bacterium has no effect on the number of transformants generated, as shown by the plateau in Figure 12.1. How can this figure be interpreted?

Figure 12.1
Relation between the concentration of transforming DNA and the number of transformants obtained

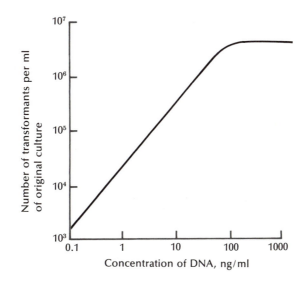

It has been estimated that when DNA is extracted from donor bacteria, the genome of each bacterium fragments into anywhere from 200 to 500 molecules. Thus, for any one gene, the maximum probability of transformation is between $\frac{1}{200}$ and $\frac{1}{500}$ if each recipient bacterium takes up one donor DNA molecule. Stated another way, if a recipient bacterium takes up between 200 and 500 DNA molecules, the probability that transformation will occur with any one particular gene is high. The data presented in Figure 12.1 suggest that between 100 and 200 DNA molecules are taken up by the recipient bacteria population for every streptomycin-resistant transformant.

The leveling off of DNA uptake in bacteria at concentrations above 10 molecules per bacterium may indicate that each recipient cell has a fixed number of DNA receptor sites on the cell membrane or cell wall. Once those receptor sites are saturated, additional DNA has no effect on the number of transformants generated. To test this possibility, transforming DNA from streptomycin-resistant cells was mixed with nontransforming DNA from streptomycin-sensitive cells, and this mixture was used to transform streptomycin-sensitive cells into streptomycin-resistant cells. The donor streptomycin-sensitive DNA is "nontransforming," because donor and recipient genes are identical; hence no transformation is detected. However, nontransforming DNA can compete with the streptomycin-resistant DNA for receptor sites on the recipient cell wall or cell membrane. If the nontransforming DNA competes successfully, the plateau shown in Figure 12.1 should be lowered—assuming that the total *number* of DNA molecules per unit volume remains the same as in the figure, but the *proportion* of transforming DNA molecules is decreased. A lowered plateau was indeed observed, suggesting that competition for receptor sites did occur. Alternative interpretations of this competitive inhibition are, of course, possible. For example, the internal volume of the recipient cell may be limiting, just as the enzymes that aid the transport of the donor DNA fragment across the cell wall and cell membrane may be limiting. However, data from several sources support the presence of cell-wall receptor sites.

Transformation of a cell from one genotype to another requires more than a physical interaction between transforming donor DNA and a recipient cell. The recipient cell must be physiologically *competent*. That is, it must be in a physiological state such that it makes not only possible but likely the movement of a donor DNA fragment across the cell membrane and the incorporation of that fragment into the recipient cell genome. Studies show that transformation can occur only within a certain interval of the bacterial growth cycle. A small number of bacteria placed in an enriched nutrient broth increase exponentially until the concentration of cells puts a severe strain on the available oxygen and food resources, at which time the population ceases to increase. Sampling cells at regular intervals in this growth cycle and testing them for competence demonstrated that maximum competence was achieved just prior to the plateau

Figure 12.2
Rise and fall of
competence during the
growth cycle of a culture
of recipient bacteria

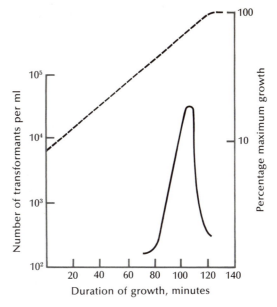

Dashed Curve   Growth of recipient population as a percentage of maximum.

Solid Curve   Number of transformants obtained when samples of the population are
removed at intervals during the growth cycle, exposed for a few minutes to an excess of
transforming DNA, and then treated with DNase.

phase (Figure 12.2). Competence, then, is observed during the very brief
period when the recipient cells are dividing at a maximal rate.

The specific physiological state that makes a recipient cell competent
has proved difficult to define. The key to competence may lie in the
biochemical configuration of the cell wall, because *protoplasts* (bacteria
with cell walls removed) are completely refractory to transformation. Apparently cells classified as competent have just ceased DNA synthesis and
yet continue active protein synthesis, which suggests that protein synthesis is an important requirement in achieving competence. It has been
proposed that the active protein synthesis occurring at the competence
phase of the growth cycle somehow modifies the bacterial cell wall, making it receptive to transforming DNA. It has been further suggested that
competence is induced in a population by a protein released into the
medium by a few cells that have achieved competence. This protein,
called a *competence factor,* is assumed to induce competence in noncompe-
tent cells.

Let us summarize our discussion of transformation to this point.
Transforming DNA, in a size range of about 800 to 16,000 nucleotide
pairs, interacts with competent bacterial cell walls as a prelude to movement to the cell's interior. Apparently the presence of more than 10 DNA
molecules per recipient cell does not result in increased numbers of trans-

509

formants; but for fewer than 10 molecules, there is a linear relationship between the number of DNA molecules and number of cells transformed. The movement of transforming DNA into a recipient cell's interior, followed by its incorporation into the recipient cell's genome, will be discussed next.

The uptake and incorporation of transforming DNA

The uptake of DNA by a recipient cell is a two-step process including first a loose, reversible association of DNA with the recipient cell wall and then a fixed, irreversible association of DNA with an enzyme system that transports it to the cell's interior. Furthermore, the DNA must be double-stranded in its initial contact with recipient cells; single-stranded DNA is incapable of transforming cells.

Once inside the cell, the donor DNA is altered so that it loses its potential as transforming DNA, at least for a time. The reason for this appears to be that the double-stranded donor DNA becomes single-stranded on entering the cell's interior. Single-stranded donor DNA in the cell's interior has been tested for its transforming ability and found to be unable to transform cells. This short period after uptake, during which donor DNA loses its transforming capacity, is the *eclipse period.* It is found in *Pneumococcus* and *Bacillus subtilis* but not in *Haemophilus.*

The integration of a single-stranded molecule of donor DNA into a double-stranded recipient chromosome is not completely understood, but two things are apparent. Donor DNA *replaces* a segment of recipient DNA, and the donor DNA is integrated into and covalently bonded with the recipient DNA. That transforming DNA actually replaces a segment of recipient DNA was demonstrated by H. E. Taylor (Figure 12.3). Recipient cells (*A*) and donor DNA (*a*) were mixed, and *a* transformants were recovered. If these *a* transformants were due to the *addition* of an *a* gene, they would be "heterozygous" (*aA*), with *a* dominant over *A*. If the *a* transformants were due to the *replacement* of *A* by *a*, the cells would be simply *a*. To decide between these alternatives, Taylor mixed the *a* transformants with *A* DNA and obtained *A* transformants. If transformation were an addition mechanism, the *A* transformants would have been *AAa* genotypically and *A* phenotypically, a difficult situation since *Aa* would have been phenotypically *a*. A replacement mechanism would offer a simpler explanation, because the *A* transformant could arise the same way the *a* transformant arose—by replacement.

Taylor thus presented a strong argument for a replacement mechanism, but the details of the process remained elusive. In 1964, M. S. Fox and M. K. Allen presented evidence that the integration of donor DNA into recipient DNA occurred through a single-stranded intermediate. They used transforming DNA density-labeled with ^{15}N and radioactively labeled with ^{32}P. After mixing *Pneumococcus* cells with this labeled DNA and allowing enough time for the DNA to be transported to the cell's interior, they extracted the DNA from recipient cells at regular intervals

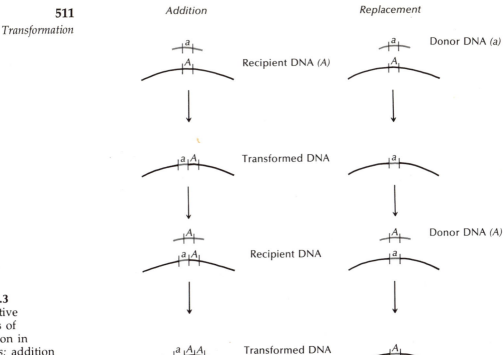

Addition   Replacement

Donor DNA *(a)*

Recipient DNA *(A)*

Transformed DNA

Recipient DNA

Donor DNA *(A)*

Transformed DNA

Figure 12.3
Two alternative
explanations of
transformation in
Pneumococcus: addition
and replacement

and analyzed it. During the eclipse period, they found heavy (^{15}N), ^{32}P-labeled single-stranded DNA, ^{15}N-^{32}P nucleotides, and nonlabeled double-stranded DNA. As time passed, the single-stranded DNA disappeared and the ^{32}P in double-stranded DNA increased. This suggests that the ^{15}N-^{32}P-labeled donor DNA was degraded intracellularly to single-stranded DNA and then incorporated into recipient double-stranded DNA.

In order to prove that the donor DNA was covalently incorporated into recipient DNA—that is, that 5' to 3' phosphodiester linkages were established—the recipient DNA was fragmented by means of ultrasonic waves. This was necessary because the total labeled area in a recipient chromosome would be small, and fragmenting the DNA would make the density label in any one fragment a more significant proportion of the whole fragment. When the fragments were centrifuged at this point, a band appeared that was intermediate in density between ^{15}N DNA and ^{14}N DNA and was ^{32}P-labeled. Further analysis of this system made it appear likely that a single strand of transforming DNA (available evidence indicates that the strand is selected randomly) is integrated into a double-stranded recipient DNA molecule. Furthermore, the integration appears to involve a replacement reaction, resulting in the donor frag-

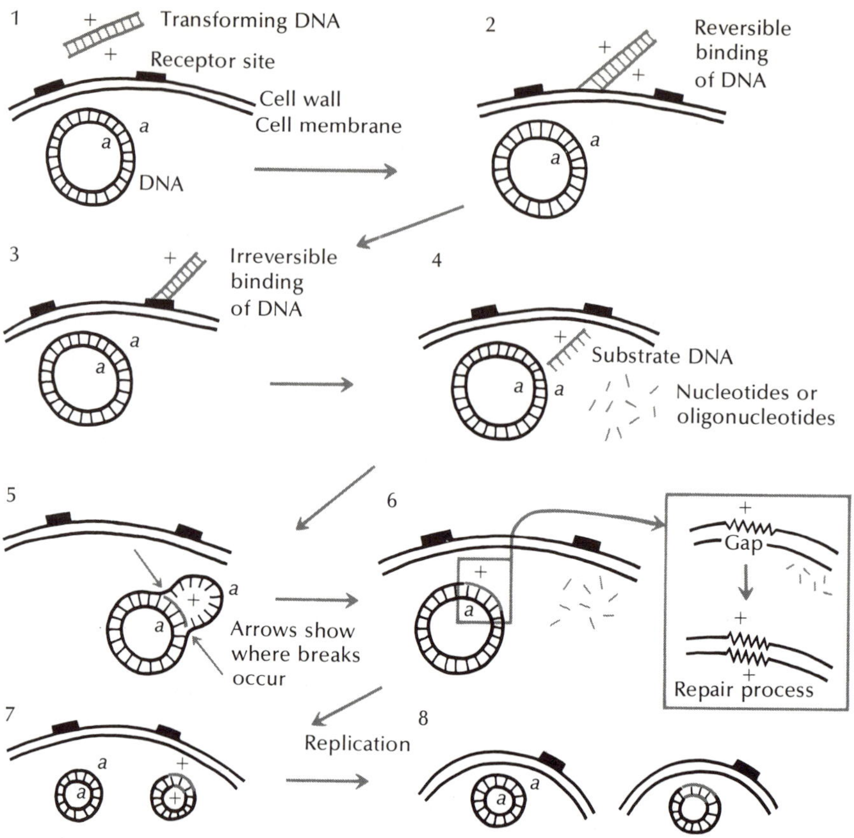

Figure 12.4
Schematic view of transformation

ment being covalently integrated into the recipient. Figure 12.4 summarizes the probable mechanism of DNA uptake and incorporation.

Transformation and gene mapping

From a purely genetic perspective, transformation—indeed almost any parasexual system in bacteria—involves three components. First, there is pairing of homologous regions of donor and recipient chromosomes. Second, the donor alleles replace the recipient alleles. Third, alleles that are close together on the same fragment have a greater probability of being transformed together than do widely separated genes. The first two points have been either discussed or diagrammed (see Figure 12.4) and need not be considered further. The third point applies classical genetic analysis to a recombination system that is markedly different from those examined in connection with Mendelian principles. In discussing linkage and mapping frequencies in transformation, we want to ascertain that groups of donor genes have actually entered a recipient cell and that some have not been selectively excluded. We also want to ascertain that the donor genes entered on the same DNA molecule.

Linkage in transformation generally means that two or more genes are located on the *same donor DNA fragment*. This idea of linkage differs considerably from a Mendelian interpretation, in which all genes on one chromosome are said to be linked. In haploid prokaryotes such as bacteria, all genes are now generally assumed to be on one chromosome. In our earlier discussion, in which the number of transformation events was considered a function of the number of DNA molecules, we saw that a competent cell could take up more than one DNA molecule. Therefore, if two genes were carried on different molecules, the probability of double transformation would be the product of each independent event. For example, if a population of *ab* cells were exposed to donor *Ab* and *aB* DNA in equal proportions and if each separate transformation (*ab* to *Ab* or to *aB*) had a probability of 1 percent, the probability of a double transformation (*ab* to *AB*) is 0.01 × 0.01, or 1 in 10,000. If *AB* donor DNA were used with *ab* recipient cells and if *A* and *B* were on different molecules (unlinked), the *AB* double transformants would appear with a frequency of 10^{-4}. If the double transformant frequency was actually 10^{-3}, we might conclude that *A* and *B* were linked. This would not necessarily be true, though, because transformation depends not only on the *number of DNA molecules* but also on the *proportion of competent recipient bacteria*.

The concentration factor could distort our interpretation of linkage, because at low DNA concentrations, only one molecule is taken up by a competent cell. At high DNA concentrations, however, 10 molecules may be taken up (see Figure 12.1), so there is a higher probability of double transformation by unlinked genes, which may suggest linkage. The other complicating factor, competence, is critical because transformation frequency is expressed as a proportion of the *total population* of recipient cells. If a small number of cells are competent, then even for linked genes the probability of double transformants will be low, which suggests no linkage. Conversely, unlinked genes may actually appear to be linked if the initial proportion of competent cells in the population is low. For example, *AB* donor DNA is mixed with *ab* recipient cells, and *Ab* and *aB* transformants are found in equal frequencies of 1 percent. We therefore expect double transformants to appear with a frequency of 10^{-4}. But if competent cells accounted for only 10 percent of the original population, the original *Ab* and *aB* transformation frequency of 1 percent must actually have been 10 percent. In that case, the calculated expected frequency of double transformation would be 0.10 × 0.10, or 10^{-2}—a hundred times greater than our original estimation, which assumed 100 percent competence. Thus, if we used the 1-percent value in a population with, say, 50 percent competent cells, we might infer linkage where in fact none existed.

To establish linkage, while avoiding the possible pitfalls, it is easier to deal with individual DNA molecules, which can be labeled with genes *A*, *B*, and *AB*. Figure 12.1 showed that below the plateau phase, transformation and DNA concentration revealed a linear relationship. If we use steadily decreasing concentrations of transforming DNA and measure

Figure 12.5

Effect of decreasing concentration of transforming DNA on the relative number of single and double transformants for the two markers *A* and *B* [Adapted from S. H. Goodgal, *J. Gen. Physiol.* 45:205 (1961)]

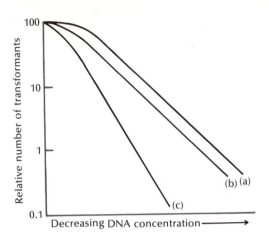

Curve (a) shows the number of transformations from *ab* to *Ab* or *aB*. Curve (b) shows the number of transformations from *ab* to *AB* when *A* and *B* are linked. Curve (c) shows the number of transformations from *ab* to *AB* when *A* and *B* are not linked.

single transformants (*ab* to *Ab* or to *aB*), the frequency of single transformants should decrease linearly (Figure 12.5a). If *A* and *B* are closely linked, a curve with the *same slope* is expected for *AB* double transformants (Figure 12.5b). If *A* and *B* are not linked, however, double transformation occurs only if *two* molecules are taken up. This means that if the DNA concentration is reduced 10-fold, double transformation drops 100-fold, giving a curve with a steeper slope (Figure 12.5c). Thus, by analyzing single and double transformation solely as a function of DNA concentration, we can unambiguously determine linkage.

Using the same general principles of gene mapping that we discussed earlier, Nester, Schafer, and Lederberg mapped genetic loci in *Bacillus* by transformation. They set up a two-point cross by employing a recipient bacterial strain carrying two mutations (*trp⁻* and *tyr⁻*, both amino-acid-synthesizing mutations), and donor DNA from three strains. Two of the three carried only one mutation (*trp⁺ tyr⁻* and *trp⁻ tyr⁺*) and one carried none (*trp⁺ tyr⁺*). When the transformant classes were isolated and analyzed, the following results were obtained.

Experiment	Donor DNA	Recipient DNA	Transformant classes trp	Transformant classes tyr	Number	%
1	*trp⁺ tyr⁻* *trp⁻ tyr⁺*	*trp⁻ tyr⁻*	A + B − C +	− + +	190 256 2	42.4 57.1 .4
2	*trp⁺ tyr⁺*	*trp⁻ tyr⁻*	D + E − F +	− + +	196 328 367	22.0 36.8 41.2

One of the first things we can suggest is that the two loci are carried on the same transforming DNA fragment (that is, they are linked). We can suggest this by comparing the double transformation frequency in experiment 1 (0.004) with that in experiment 2 (0.412). There is more than a 100-fold difference. In experiment 1, double transformation was the result of two DNA fragments. If the same were true for experiment 2, assuming all other things equal, then we should have had a comparably low frequency of double transformants. That the difference is so large suggests linkage. But how do we calculate the linkage distance between the two points?

In order to calculate the map-unit distance between *trp* and *tyr*, we need to know the number of recombinations that occurred between them. Transformant classes A and B could be recombinations either between the two loci or outside the two loci. We cannot use them in our calculation, because we cannot tell which of the following events produced the transformant class.

Transformant class C occurred as a result of two events, but again we cannot discriminate between two different possibilities.

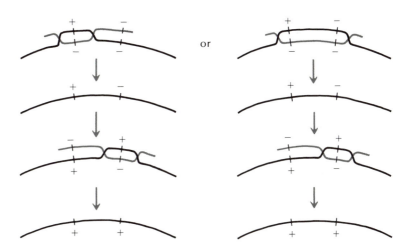

Because of the ambiguity, we cannot use these data.

However, we can use the data from transformant classes D, E, and F. Classes D and E can only have arisen in one way, and as a result of a recombination between *trp* and *tyr*.

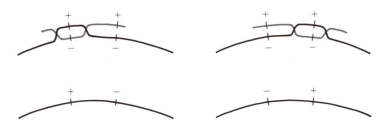

Transformant class F is a recombination outside the two gene markers.

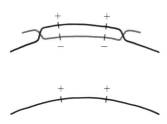

To construct a map, we use classes D, E, and F. The distance between *trp* and *tyr* is the percentage recombination between them: (D + E)/(D + E + F). The *trp*-to-*tyr* distance is thus (196 + 328)/(196 + 328 + 367), or 59 map units.

This two-point transformation cross provides a good estimate of distance between genes. But there is considerable variation from one experiment to another, making the precision of the map units questionable. Figure 12.6 is a linkage map of *Bacillus* constructed in part from transformation mapping.

Conjugation

In 1946, J. Lederberg and E. L. Tatum established for the first time that *E. coli* bacteria engage in a form of sexual reproduction that involves mating between sexually differentiated strains of bacteria and that there is transfer of chromosomal material from one strain to the other. This process, called *conjugation*, make the bacteria useful for rigorous genetic analysis and also establishes their chromosomal organization. Since chromosomes are only partially transferred and move in only one direction, conjugation is actually a *parasexual* process that has enabled investigators to deal with gene-map distances in physical–chemical terms. In this section, we will discuss the discovery of conjugation, analyze the process, and explain its application to genetic analysis.

The discovery of a sexual process in bacteria was a direct outgrowth of G. W. Beadle and E. L. Tatum's biochemical investigations in *Neurospora*. These two investigators had used x rays to induce mutations in biochemical pathways in *Neurospora*, and Lederberg and Tatum attempted to extend the same techniques to bacteria. In 1946, Lederberg and Tatum published a paper in the journal *Nature*, which began:

Figure 12.6
Linkage map of the *Bacillus subtilis* genome (the map is constructed in part from data obtained using transformation) [Adapted from D. Dubnau et al., *J. Molec. Biol.* 27:163 (1967)]

> Analysis of mixed cultures of nutritional mutants has revealed the presence of new types which strongly suggest the occurrence of a sexual process in the bacterium *Escherichia coli*.

They had mixed together two strains of *E. coli*, each strain requiring two different nutritional supplements in the media. After plating the mixture on minimal medium, which was unable by itself to support the growth of

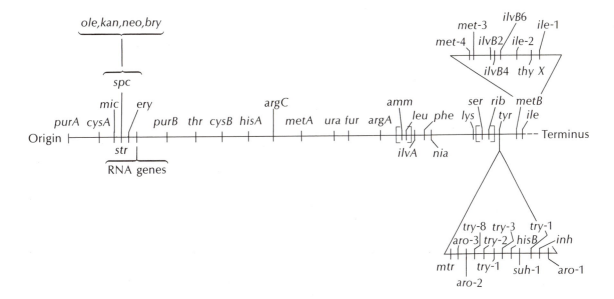

amm = inability to assimilate NH₄⁺	mic = micrococcin resistance
arg = arginine requirement	neo = neomycin resistance
aro = aromatic amino acid requirement	nia = niacin requirement
bry = bryamycin resistance	ole = oleandomycin resistance
cys = cysteine requirement	phe = phenylalanine requirement
ery = erythromycin resistance	pur = adenine or guanine requirement
fur = fluorouracil resistance	rib = riboflavin
his = histidine requirement	ser = serine
ile = isoleucine requirement	spc = spectinomycin resistance
ilv = isoleucine and valine requirement	str = streptomycin resistance
kan = kanamycin resistance	thr = threonine requirement
leu = leucine requirement	thy = thymine requirement
lys = lysine requirement	try = tryptophan requirement
met = methionine requirement	tyr = tyrosine requirement
	ura = uracil requirement

either strain, they discovered colonies that could grow. Normal, or wild-type, *E. coli* will grow on a minimal medium containing only a sugar (usually glucose) and some inorganic salts. From this minimal medium, the cells synthesize all their proteins, nucleic acids, lipids, and carbohydrates as needed. One strain used by Lederberg and Tatum had mutations that prevented it from synthesizing two amino acids—threonine (*thr*) and leucine (*leu*). This strain is symbolized as *thr⁻ leu⁻*. A second strain was unable to synthesize the amino acids phenylalanine (*phe*) and cystine (*cys*). This strain is symbolized as *phe⁻ cys⁻*. These two strains are called *auxotrophs*, because they cannot grow on minimal medium. *Prototrophs* are cells that can grow on minimal medium.

When Lederberg and Tatum mixed the two auxotrophic *E. coli* strains on minimal medium, prototrophic colonies were observed: *phe⁺ cys⁺ thr⁺ leu⁺*. These colonies could be propagated indefinitely on minimal medium and arose about once per every million bacteria plated. Mutation of the auxotrophs to prototrophs was eliminated as a mechanism, because mutations of two mutant genes to normal would have occurred at a much lower frequency than the frequency observed. The appearance of nutritionally independent bacteria at such a high frequency could be interpreted as indicating either that some sort of transformation was occurring or that the two bacterial strains were actually mating and exchanging genes.

Transformation was eliminated when it was demonstrated that DNA extracted from either strain could not transform the reciprocal strain. The conclusion that the cells were actually mating (that is, that physical contact was essential) was the only possibility remaining, and intensive research was begun to describe the sexual process. Figure 12.7 summarizes this experiment.

**Conjugation is a one-way
genetic transfer**

In conjugation, *E. coli* cells do not *reciprocally* exchange genetic information. Genes pass *in one direction* only, from a donor to a recipient cell. This discovery brought the first clear indication that the sexual process of *E. coli* differs significantly from that of higher organisms. Evidence supporting a one-way genetic transfer emerged from reciprocal crosses of two auxotrophic *E. coli* strains on a minimal medium. One of the mutant strains was streptomycin-sensitive (denoted by *strˢ*) and could not survive on a medium containing this antibiotic drug. The other mutant strain had a mutation that made it streptomycin-resistant (*str^r*).

Strain A (*strˢ*) cells that were mixed with strain B (*str^r*) cells and plated on a minimal medium containing streptomycin yielded prototrophic colonies. However, the reciprocal cross, strain A (*str^r*) × strain B (*strˢ*), did not produce prototrophs. This was believed to indicate that strain B cells must be viable at all times in order for prototrophic colonies to grow, but that strain A cells need not be. It appeared that B cells were functioning as recipients for genetic information donated by A cells (Figure 12.8). And once the A cells donated their genetic information, their presence was no

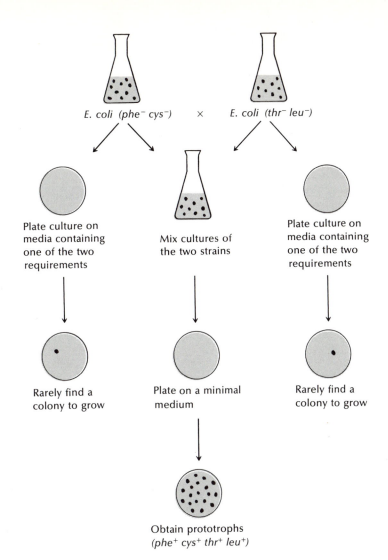

Figure 12.7
Lederberg–Tatum
experiment leading to the
discovery of sexuality in
bacteria

longer required. That is to say that strain B cells were "zygotes" in which recombination was occurring, and if they were killed, no recombinants would be isolated. This interpretation was confirmed in another experiment, in which the two strains were treated separately with heavy doses of streptomycin and then washed and mated with nontreated cells on a minimal medium without streptomycin. Treating donor strain A (str^s) did not impair its ability to donate genes to B, so prototrophs formed. But treating strain B (str^s) destroyed its ability to function as a recipient "zygote" and form prototrophic colonies (Figure 12.8c and d).

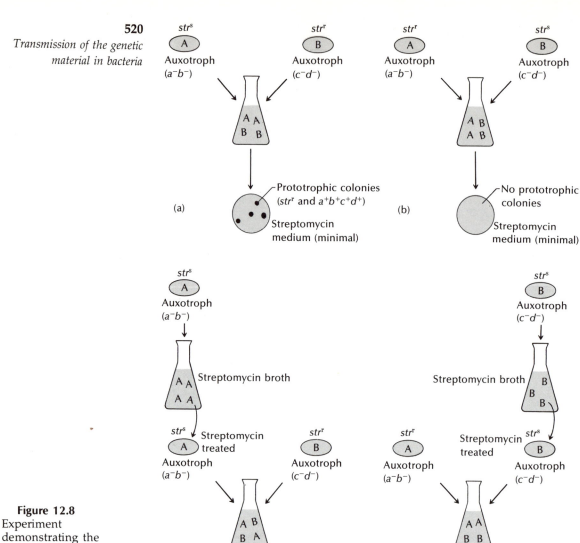

Figure 12.8
Experiment
demonstrating the
one-way transfer of the
genetic material in
conjugation and the
required viability of
recipient cells at all times.
The reciprocal crosses in
(a) and (b) show that
strain A must be a donor.

The sex factor Cells that serve as donors of genetic information are denoted by F⁺, those
that serve as recipients by F⁻ (F = fertility). The next problem was to
identify the determinant that makes a cell either a donor (F⁺) or a recipient
(F⁻). Among the first possibilities considered for conferring F⁺ or F⁻ properties on cells was a single gene, with F⁺ and F⁻ being alleles. This idea

was rejected, however, because it was shown that when F⁺ cells genotypically $a^+b^+c^-d^-$ were mated with F⁻ ($a^-b^-c^+d^+$) cells, there was a recombination frequency of about 10^{-6}, but after about an hour, over 50 percent of the F⁻ recipient cells were converted into F⁺ cells! It thus appeared that the donor state was determined by a factor F, the sex factor, that is transmitted to recipient cells *independently* of the bacterial chromosome.

Without belaboring the experiments that led to its characterization, the F factor is known to be a circular molecule of DNA about a hundredth the size of the *E. coli* chromosome. It is usually *autonomous*, meaning that it replicates at its own rate, and is *autotransmissible*, which means that it can be independently passed to recipient cells lacking the F factor (that is, F⁻ cells). It is normally attached to the inside of the cell membrane.

Chromosome mobilization in *E. coli* In 1950, an F⁺ strain (one having the sex factor) was isolated that generated about 1000 times more recombinants than the standard F⁺ strains. This strain, called a *high-frequency recombinant* (Hfr), and others subsequently isolated transfer bacterial genes at a high frequency but only *rarely* transfer the F (sex) factor. Hence they contrast with F⁺ cells, which rarely transmit bacterial genes but which almost always transmit the sex factor to F⁻ cells (cells lacking the sex factor).

The various Hfr strains are also unique in that each strain transfers only *some* of the bacterial genes at a high frequency. Other genes are transferred with frequencies approaching those for F⁺ cells. For each independently arising Hfr strain, the genes showing a high frequency of recombination are the same, but strains differ from each other in that the genes transferred at a high frequency differ for each strain.

Another notable property of Hfr is that its frequency in F⁺ populations can be increased by at least two to three times by treating F⁺ cells with agents that increase recombination (ultraviolet light and some chemicals). This led to the proposal that the conversion of an F⁺ cell to an Hfr cell involved an association, perhaps a recombination, between the sex factor and the bacterial chromosome. This association determines that a particular segment of the donor chromosome is passed on to recipient cells at a high probability; other segments, including the sex factor itself, are passed on at lower probabilities.

The transfer of an Hfr donor chromosome to F⁻ recipient cells became clearer in 1955 when E. L. Wollman and F. Jacob developed the *interrupted-mating technique*. They used an Hfr strain designated Hfr H (H = Hayes) and carrying genes we shall symbolize as *A B C D E*. They mixed this strain with F⁻ cells (*a b c d e*) and selected *A* recombinants at different time intervals. They found that genes *B, C, D,* and *E* appeared in the recipients with *A* at frequencies of about 90, 75, 40, and 25 percent, respectively. This suggested a sequential transfer of genes from donor to recipient. To test this, they interrupted conjugation at different time intervals by placing mating pairs in a high-speed mixer to break apart the mating cells. During the first 8 minutes of mixing, no recombinants were found. This indicated that the first 8 minutes of conjugation were occupied by activities other than chromosome transfer. After 8 minutes, *A* recombinants were found but no *B*, *C*, *D*, or *E*. As time passed, *A* recombinants were found with first *B* and then successively *C*, *D*, and *E*.

Wollman and Jacob inferred that the genes in the Hfr donor could be mapped in sequence according to the time taken to transfer from Hfr to F⁻. The distance between genes would be expressed in time units rather than in percentage of recombination units. The F factor was the last to

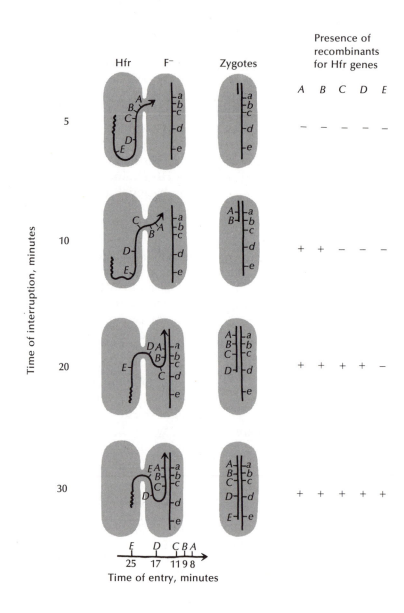

Figure 12.9
Diagram of the
interrupted-mating
experiments in *E. coli*
[Adapted from W. Hayes,
*The Genetics of Bacteria and
their Viruses,* copyright ©
1968 by John Wiley &
Sons, Inc., by permission
of Blackwell Scientific
Publications]

enter the recipient cell and did so only rarely, because the slender conjugation bridge that connected the Hfr and F⁻ cells usually broke before chromosome transfer was complete. Figure 12.9 summarizes the process of conjugation as interpreted by Wollman and Jacob.

The low frequency of gene transfer in F⁺ populations is now thought to be due to the low spontaneous conversion of F⁺ to Hfr, which in turn transfers chromosomal genes. In other words, F⁺ cells do not transfer chromosomal genes. This raises the following question: How do Hfr cells form from F⁺? Wollman and Jacob suggested that conversion from F⁺ to Hfr involved recombination between the F factor and the F⁺ chromosome. However, the actual mechanism of F integration into the F⁺ chromosome was not understood until the organization of these two molecules was elucidated.

Wollman and Jacob were the first to suggest that the *E. coli* chromosome was circular. Using several different Hfr strains, they carried out interrupted-mating experiments with each and determined the gene sequence for each strain. The results, given in Table 12.1, at first imply that the gene sequences in all the strains differ. It appears that any gene can enter first after the origin (*O*, which stands for "origin of oriented chromosome transfer"). However, closer scrutiny of the data reveals that any one gene always has the same neighbors and that the differences between the various Hfr strains can be accounted for by the different positions *O* assumes and by the *direction* of gene transfer. Note, for example, that Hfr H differs from Hfr 3 in *O* position and in direction of transfer. If *O* in Hfr H were moved between the *pur* and *gal* genes and if the direction of transfer were reversed (*pur-lac-pro,* etc.), then Hfr 3 and Hfr H would be identical. In 1957, Wollman and Jacob postulated, therefore, that the *E. coli* chromosome was circular and that the F factor, which determined *O*, could integrate at any one of a number of chromosomal positions. The idea of a circular chromosome in *E. coli* was confirmed by Cairns in 1962, as we noted in Chapter 7.

Table 12.1
Gene order in conjugational transfer by several different Hfr strains

Hfr strain	Order of gene transfer
Hayes	*O-thr-leu-azi-ton-pro-lac-pur-gal-trp-his-gly-str-mal-xyl-mtl-ile-met-thi*
1	*O-leu-thr-thi-met-ile-mtl-xyl-mal-str-gly-his-trp-gal-pur-lac-pro-ton-azi*
2	*O-pro-ton-azi-leu-thr-thi-met-ile-mtl-xyl-mal-str-gly-his-trp-gal-pur-lac*
3	*O-pur-lac-pro-ton-azi-leu-thr-thi-met-ile-mtl-xyl-mal-str-gly-his-trp-gal*
4	*O-thi-met-ile-mtl-xyl-mal-str-gly-his-trp-gal-pur-lac-pro-ton-azi-leu-thr*
5	*O-met-thi-thr-leu-azi-ton-pro-lac-pur-gal-trp-his-gly-str-mal-xyl-mtl-ile*
6	*O-ile-met-thi-thr-leu-azi-ton-pro-lac-pur-gal-trp-his-gly-str-mal-xyl-mtl*
7	*O-ton-azi-leu-thr-thi-met-ile-mtl-xyl-mal-str-gly-his-trp-gal-pur-lac-pro*
AB311	*O-his-trp-gal-pur-lac-pro-ton-azi-leu-thr-thi-met-ile-mtl-xyl-mal-str-gly*
AB312	*O-str-mal-xyl-mtl-ile-met-thi-thr-leu-azi-ton-pro-lac-pur-gal-trp-his-gly*
AB313	*O-mtl-xyl-mal-str-gly-his-trp-gal-pur-lac-pro-ton-azi-leu-thr-thi-met-ile*

From F. Jacob and E. L. Wollman, *Sexuality and the Genetics of Bacteria,* New York: Academic, 1961.

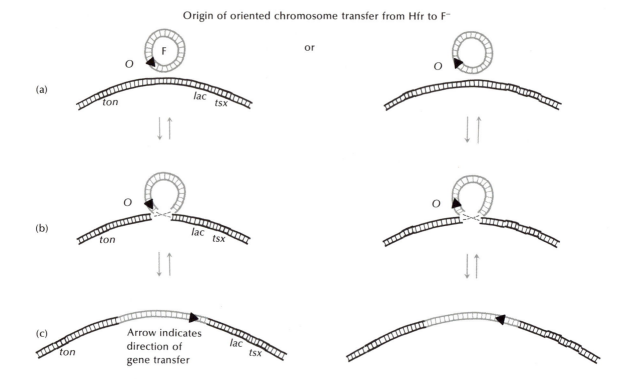

Origin of oriented chromosome transfer from Hfr to F⁻

(a)

(b)

(c)

Arrow indicates direction of gene transfer

Figure 12.10
Conversion of an F⁺ bacterium to the Hfr state. This sequence of events is reversible, so the Hfr chromosome can loop out to reconstitute the circular F factor and revert to the F⁺ state by another reciprocal crossover.

A. Campbell suggested that the generation of Hfr from F⁺ involved the integration of two circular molecules of DNA (Figure 12.10). The circular F factor, containing the gene *O*, pairs with a homologous region on the chromosome. A reciprocal crossover occurs, integrating the F factor into the chromosome in either a clockwise or a counterclockwise direction. The position and direction of integration determine the sequence of gene transfer.

The linkage map of *E. coli* is shown in Figure 12.11, and the functions of all genes depicted are listed in Table 12.2. The map is set off in time units in the manner originated by Wollman and Jacob. The overall length of the genetic map—the time it takes for the transfer of the entire *E. coli* chromosome, including F—is 90 minutes.

Summarizing our discussion to this point, the F factor is a self-reproducing, circular piece of DNA that can exist in either an *autonomous* or an *integrated* state. When autonomous, it replicates independently of the bacterial chromosome. It also confers on its carriers the properties that make possible its transfer to an F⁻ cell. When integrated, the F factor is replicated with the bacterial chromosome. It cannot be transferred independently to F⁻ cells and can be transferred only *after* the entire chromosome into which it is integrated has been transferred.

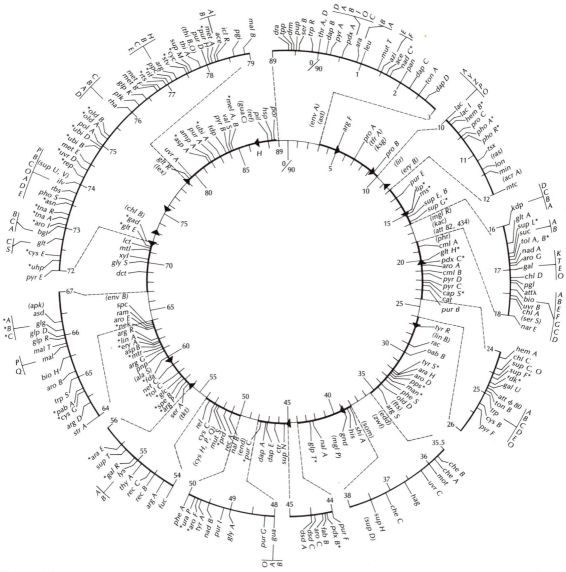

Figure 12.11
Circular linkage map of *E. coli* (the arrows on the inner circle indicate some of the different Hfr strains and the direction of chromosome transfer) [From A. L. Taylor, *Bact. Rev.* 34:155 (1970), courtesy of the American Society for Microbiology]

The sex factor is one of a new class of genetic elements that are added to a genome. The added elements are called *episomes* and can be acquired through an *external* source only—not by mutation, or rearrangement, of the existing genome. Episomes can be autonomous or integrated; when autonomous, they are commonly attached to cell membranes. Temperate bacteriophages, those capable of entering into a lysogenic state, represent another type of episome, as we shall soon point out.

Table 12.2
List of genetic markers of
E. coli (continued on pp.
527–534)

Gene symbol	Mnemonic	Map position	Alternate gene symbols; phenotypic trait affected
aceA	Acetate	78	*icl;* utilization of acetate: isocitrate lyase
aceB	Acetate	78	*mas;* utilization of acetate: malate synthetase A
aceE	Acetate	2	*aceE1;* acetate requirement; pyruvate dehydrogenase (decarboxylase component)
aceF	Acetate	2	*aceE2;* acetate requirement; pyruvate dehydrogenase (lipoic reductase-transacetylase component)
acrA	Acridine	(11)[a]	Sensitivity to acriflavine, phenethyl alcohol, Na dodecyl sulfate
alaS	Alanine	(60)	*ala-act;* alanyl-tRNA synthetase
ampA	Ampicillin	82	Resistance or sensitivity to penicillin
apk		(66)	Lysine-sensitive aspartokinase
araA	Arabinose	1	L-Arabinose isomerase
araB	Arabinose	1	L-Ribulokinase
araC	Arabinose	1	Regulatory gene
araD	Arabinose	1	L-Ribulose 5-phosphate 4-epimerase
araE	Arabinose	56	L-Arabinose permease
araI	Arabinose	1	Initiator locus
araO	Arabinose	1	Operator locus
argA	Arginine	54	*argB, Arg1, Arg₂;* N-acetylglutamate synthetase
argB	Arginine	77	*argC;* α-N-acetyl-L-glutamate-5-phosphotransferase
argC	Arginine	77	*argH, arg2;* N-acetylglutamic-γ-semialdehyde dehydrogenase
argD	Arginine	64	*argG, Arg₁;* acetylornithine-δ-transaminase
argE	Arginine	77	*argA, arg4;* L-ornithine-N-acetylornithine lyase
argF	Arginine	5	*argD, Arg5;* ornithine transcarbamylase
argG	Arginine	61	*argE, Arg6;* argininosuccinic acid synthetase
argH	Arginine	77	*argF, Arg7;* L-argininosuccinate arginine lyase
argP	Arginine	57	Arginine permease
argR	Arginine	62	*Rarg;* regulatory gene
argS	Arginine	35	Arginyl-tRNA synthetase
aroA	Aromatic	21	3-enolpyruvylshikimate-5-phosphate synthetase
aroB	Aromatic	65	Dehydroquinate synthetase
aroC	Aromatic	44	Chorismic acid synthetase
aroD	Aromatic	32	Dehydroquinase
aroE	Aromatic	64	Dehydroshikimate reductase
aroF	Aromatic	50	3-deoxy-D-arabinoheptulosonic acid-7-phosphate (DHAP) synthetase (tyrosine-repressible isoenzyme)
aroG	Aromatic	17	DHAP synthetase (phenylalanine-repressible isoenzyme)
aroH	Aromatic	32	DHAP synthetase (tryptophan-repressible isoenzyme)

[a] Parentheses indicate approximate locations.

From A. L. Taylor, "Current linkage map of *E. coli,*" *Bact. Rev.* 34:155–175 (1970). Additional genetic markers are listed in Taylor and Trotter (1972).

Table 12.2 (cont.)

Gene symbol	Mnemonic	Map position	Alternate gene symbols; phenotypic trait affected
arol	Aromatic	73	Function unknown
asd	—	66	*dap + hom;* aspartic semialdehyde dehydrogenase
asn	—	73	Asparagine synthetase
aspA	—	82	Aspartase
aspB	Aspartate	62	*asp;* aspartate requirement
ast	Astasia	(4)	Generalized high mutability
attλ	Attachment	17	Integration site for prophage λ
attφ80	Attachment	25	Integration site for prophage φ80
att82	Attachment	(17)	Integration site for prophage 82
att434	Attachment	(17)	Integration site for prophage 434
azi	Azide	2	*pea, fts;* resistance or sensitivity to Na azide or phenethyl alcohol; filament formation at 42C
bglA	β-glucoside	73	*β-glA;* aryl β-glucosidase
bglB	β-glucoside	73	*β-glB;* β-glucoside permease
bglC	β-glucoside	73	*β-glC;* regulatory gene
bioA	Biotin	17	Group II; 7-oxo-8-aminopelargonic acid (7KAP) $\longrightarrow$ 7,8-diaminopelargonic acid (DAPA)
bioB	Biotin	17	Conversion of dethiobiotin to biotin
bioC	Biotin	17	Unknown block prior to 7KAP synthetase
bioD	Biotin	17	Dethiobiotin synthetase
bioE	Biotin	17	Unknown block prior to 7KAP synthetase
bioF,G	Biotin	17	7KAP synthetase
bioH	Biotin	66	*bioB;* early block prior to 7KAP synthetase
capS	Capsule	22	Regulatory gene for capsular polysaccharide synthesis
cat	—	23	*CR;* catabolite repression
cheA	Chemotaxis	36	*motA;* chemotactic motility
cheB	Chemotaxis	36	*motB;* chemotactic motility
cheC	Chemotaxis	37	Chemotactic motility
chlA	Chlorate	18	*narA;* pleiotropic mutations affecting nitrate-chlorate reductase and hydrogen lyase activity
chlB	Chlorate	(71)	*narB;* pleiotropic mutations affecting nitrate-chlorate reductase and hydrogen lyase activity
chlC	Chlorate	25	*narC;* structural gene for nitrate reductase
chlD	Chlorate	17	*narD, narF;* nitrate-chlorate reductase
cmlA	Chloramphenicol	19	Resistance or sensitivity to chloramphenicol
cmlB	Chloramphenicol	21	Resistance or sensitivity to chloramphenicol
ctr	—	46	Mutations affecting the uptake of diverse carbohydrates
cyc	Cycloserine	78	Resistance or sensitivity to D-cycloserine
cysB	Cysteine	25	Pleiotropic mutations affecting cysteine biosynthesis
cysC	Cysteine	53	Adenosine 5′-sulfatophosphate kinase
cysE	Cysteine	72	Apparently pleiotropic
cysG	Cysteine	65	Sulfite reductase

Gene symbol	Mnemonic	Map position	Alternate gene symbols; phenotypic trait affected
cysH	Cysteine	(53)	Adenosine 3'-phosphate 5'-sulfatophosphate reductase
cysP	Cysteine	(53)	Sulfate permease and sulfite reductase
cysQ	Cysteine	(53)	Sulfite reductase
dapA	Diaminopimelate	47	Dihydrodipicolinic acid synthetase
dapB	Diaminopimelate	0	Dihydrodipicolinic acid reductase
dapC	Diaminopimelate	2	Tetrahydrodipicolinic acid $\longrightarrow$ N-succinyl-diaminopimelate
dapD	Diaminopimelate	3	Tetrahydrodipicolinic acid $\longrightarrow$ N-succinyl diaminopimelate
dapE	Diaminopimelate	47	*dapB;* N-succinyl-diaminopimelic acid deacylase
darA	—	—	*See uvrD*
dct		69	Uptake of C_4-dicarboxylic acids
deo	Deoxythymidine	—	*See dra, drm, pup* and *tpp*
dra	—	89	*deoC, thyR;* deoxyriboaldolase
drm	—	89	*deoB, thyR;* deoxyribomutase
dsdA	D-Serine	45	D-Serine deaminase
dsdC	D-Serine	45	Regulatory gene
edd	—	(35)	Entner-Doudoroff dehydrase (gluconate-6-phosphate dehydrase)
end	—	(50)	*endoI;* endonuclease 1
envA	Envelope	(3)	Anomalous cell division involving chain formation
envB	Envelope	(65)	Anomalous spheroid cell formation
eryA	Erythromycin	62	Resistance or sensitivity to erythromycin
eryB	Erythromycin	(11)	High level resistance to erythromycin
exr	—	—	*See lex*
fabB	—	44	Fatty acid biosynthesis
fda	—	60	*ald;* fructose-1, 6-diphosphate aldolase
fdp	—	84	Fructose diphosphatase
ftsA	—	—	*See azi*
fts	—	(35)	*fts-9;* filamentous growth and inhibition of nucleic acid synthesis at 42 C
fuc	Fucose	54	Utilization of L-fucose
gad	—	72	Glutamic acid decarboxylase
galE	Galactose	17	*galD;* uridinediphosphogalactose 4-epimerase
galK	Galactose	17	*galA;* galactokinase
galO	Galactose	17	*galC;* operator locus
galT	Galactose	17	*galB;* galactose 1-phosphate uridyl transferase
galR	Galactose	55	*Rgal;* regulatory gene
galU	Galactose	25	*UPDG;* uridine diphosphoglucose pyrophosphorylase
glc	Glycolate	58	Utilization of glycolate; malate synthetase G
glgA	Glycogen	66	Glycogen synthetase
glgB	Glycogen	66	α-1, 4-Glucan: α-1, 4-glucan 6-glucosyltransferase
glgC	Glycogen	66	Adenosine diphosphate glucose pyrophosphorylase

Table 12.2 (cont.)

Gene symbol	Mnemonic	Map position	Alternate gene symbols; phenotypic trait affected
glpD	Glycerol phosphate	66	*glyD;* L-α-glycerophosphate dehydrogenase
glpK	Glycerol phosphate	76	Glycerol kinase
glpT	Glycerol phosphate	43	L-α-Glycerophosphate transport system
glpR	Glycerol phosphate	66	Regulatory gene
gltA	Glutamate	16	*glut;* requirement for glutamate; citrate synthase
gltC	Glutamate	73	Operator locus
gltE	Glutamate	72	Glutamyl-tRNA synthetase
gltH	Glutamate	20	Requirement
gltR	Glutamate	79	Regulatory gene for glutamate permease
gltS	Glutamate	73	Glutamate permease
glyA	Glycine	49	Serine hydroxymethyl transferase[b]
glyS	Glycine	70	*gly-act;* glycyl-tRNA synthetase
gnd	—	39	Gluconate-6-phosphate dehydrogenase
guaA	Guanine	48	*gua$_b$;* xanthosine-5′-monophosphate aminase
guaB	Guanine	48	*gua$_a$;* inosine-5′-monophosphate dehydrogenase
guaC	Guanine	(88)	Guanosine-5′-monophosphate reductase
guaO	Guanine	48	Operator locus
hag	H antigen	37	*H;* flagellar antigens (flagellin)
hemA	Hemin	24	Synthesis of δ-aminolevulinic acid
hemB	Hemin	10	*ncf;* synthesis of catalase and cytochromes
his	Histidine	39	Requirement
hsp	Host specificity	89	*hs, rm;* host restriction and modification of DNA
icl	—	—	See *aceA*
iclR	—	78	Regulation of the glyoxylate cycle
ilvA	Isoleucine-valine	74	*ile;* threonine deminase
ilvB	Isoleucine-valine	74	Condensing enzyme (pyruvate + α-ketobutyrate)
ilvC	Isoleucine-valine	74	*ilvA;* α-hydroxy-β-keto acid reducto-isomerase
ilvD	Isoleucine-valine	74	*ilvB;* dehydrase
ilvE	Isoleucine-valine	74	*ilvC;* transminase B
ilvO	Isoleucine-valine	74	Operator locus for genes *ilvA, D, E*
ilvP	Isoleucine-valine	74	Operator locus for gene *ilvB*
kac	K-accumulation	17	Defect in potassium ion uptake
kdpA-D	K-dependent	16	Requirement for a high concentration of potassium
ksg	Kasugamycin	(8)	Resistance or sensitivity to kasugamycin (30S ribosomal subunit)
lacA	Lactose	10	*a, lacAc;* thiogalactoside transacetylase
lacI	Lactose	10	*i;* regulator gene
lacO	Lactose	10	*o;* operator locus
lacP	Lactose	10	*p;* promoter locus
lacY	Lactose	10	*y;* galactoside permease (M protein)
lacZ	Lactose	10	*z;* β-galactosidase
lct	Lactate	71	L-Lactate dehydrogenase

[b] Enzymatic defect is inferred from studies on the homologous mutant in *S. typhimurium.*

Table 12.2 (cont.)

Gene symbol	Mnemonic	Map position	Alternate gene symbols; phenotypic trait affected
leuA	Leucine	1	α-Isopropylmalate synthetase
leuB	Leucine	1	β-Isopropylmalate dehydrogenase
lex	—	(79)	Resistance or sensitivity to x-rays and UV light
linA	Lincomycin	62	Resistance or sensitivity to lincomycin
linB	Lincomycin	(28)	High-level resistance to lincomycin
lip	Lipoic acid	15	Requirement
lir	—	(11)	Increased sensitivity to lincomycin or erythromycin, or both
lon	Long form	11	*capR*, *dir*, *muc*; filamentous growth, radiation sensitivity, and regulation of capsular polysaccharide synthesis
lysA	Lysine	55	Diaminopimelic acid decarboxylase
lysB	Lysine	55	Lysine or pyridoxine requirement
malB	Maltose	79	*mal-5*; maltose permease and phage λ receptor site
malP	Maltose	66	*malA*; maltodextrin phosphorylase
malQ	Maltose	66	*malA*; amylomaltase
malT	Maltose	66	*malA*; probably a positive regulatory gene
man	Mannose	33	Phosphomannose isomerase
melA	Melibiose	84	*mel-7*; α-galactosidase
melB	Melibiose	84	*mel-4*; thiomethylgalactoside permease II
metA	Methionine	78	met_3; homoserine O-transsuccinylase
metB	Methionine	77	*met-1*, met_1; cystathionine synthetase
metC	Methionine	59	Cystathionase
metE	Methionine	75	$met\text{-}B_{12}$; N^5-methyltetrahydropteroyl triglutamate-homocysteine methylase[c]
metF	Methionine	77	*met-2*, met_2; N^8, N^{10}-methyltetrahydrofolate reductase[c]
mglP	Methyl-galactoside	(40)	*P-MG*; methyl-galactoside permease
mglR	Methyl-galactoside	(17)	*R-MG*; regulatory gene
min	Minicell	11	Formation of minute cells containing no DNA
mot	Motility	36	Flagellar paralysis
mtc	Mitomycin C	12	*Mb*, *mbl*; sensitivity to acridines, methylene blue and mitomycin C
mtl	Mannitol	71	Utilization of D-mannitol
mtr	Methyl tryptophan	61	Resistance to 5-methyltryptophan
mutS	Mutator	53	Generalized high mutability
mutT	Mutator	1	Generalized high mutability; specifically induces AT ⟶ CG transversions
nadA	Nicotinamideadenine dinucleotide	17	*nicA*; nicotinic acid requirement
nadB	Nicotinamideadenine dinucleotide	49	*nicB*; nicotinic acid requirement
nadC	Nicotinamideadenine dinucleotide	2	Quinolinate phosphoribosyl transferase
nalA	Nalidixic acid	42	Resistance or sensitivity to nalidixic acid
nalB	Nalidixic acid	51	Resistance or sensitivity to nalidixic acid
nar	Nitrate reductase	—	*See chl*
narE	—	18	Nitrate reductase (see also *chl*)

Table 12.2 (cont.)

Gene symbol	Mnemonic	Map position	Alternate gene symbols; phenotypic trait affected
nek	—	63	Resistance to neomycin and kanamycin (30S ribosomal protein)
nic	—	63	See nad
oldA	Oleate degradation	75	old-30; thiolase
oldB	Oleate degradation	75	old-64; hydroxyacyl-coenzyme A dehydrogenase
oldD	Oleate degradation	34	old-88; acyl-coenzyme A synthetase
pabA	p-Aminobenzoate	65	Requirement
pabB	p-Aminobenzoate	30	Requirement
pan	Pantothenic acid	2	Requirement
pdxA	Pyridoxine	1	Requirement
pdxB	Pyridoxine	44	Requirement
pdxC	Pyridoxine	20	Requirement
pfk	—	76	Structural or regulatory gene for fructose 6-phosphate kinase
pgi	—	79	Phosphoglucoisomerase
pgl	—	17	6-Phosphogluconolactonase
pheA	Phenylalanine	50	Prephenic acid dehydratase
pheS	Phenylalanine	33	phe-act; phenylalanyl tRNA synthetase
phoA	Phosphatase	11	P; alkaline phosphatase
phoR	Phosphatase	11	R1 pho, R1; regulatory gene
phoS	Phosphatase	74	R2 pho, R2; regulatory gene
phr	Photoreactivation	(17)	Photoreactivation of UV-damaged DNA (K12-B hybrids)
pil	Pili	88	fim; presence or absence of pili (fimbriae)
pnp	—	61	Polynucleotide phosphorylase
polA	Polymerase	75	DNA polymerase
por	P1 restriction	89	Restriction of phage P1 DNA
ppc	—	77	glu, asp; succinate, aspartate, or glutamate requirement; phosphoenolpyruvate carboxylase
pps		33	Utilization of pyruvate or lactate; phosphopyruvate synthetase
prd	Propanediol	53	1,2-Propanediol dehydrogenase
proA	Proline	7	pro_1; block prior to L-glutamate semialdehyde
proB	Proline	9	pro_2; block prior to L-glutamate semialdehyde
proC	Proline	10	pro_3; Pro2; probably Δ-pyrroline-5-carboxylate reductase
pup	—	89	Purine nucleoside phosphorylase
purA	Purine	82	ade_k, Ad_4; adenylosuccinic acid synthetase
purB	Purine	23	ade_h, adenylosuccinase
purC	Purine	48	ade_g; phosphoribosyl-aminoimidazole-succinocarboxamide synthetase
purD	Purine	78	$adth_a$; phosphoribosylglycineamide synthetase[c]
purE	Purine	13	ade_3; ade_f; Pur_2; phosphoribosyl-aminoimidazole carboxylase
purF	Purine	44	purC, $ade_{u,b}$; phosphoribosyl-pyrophosphate amidotransferase[c]

Table 12.2 (cont.)

Gene symbol	Mnemonic	Map position	Alternate gene symbols; phenotypic trait affected
purG	Purine	48	*adth$_b$;* phosphoribosylformylglycine-amidine synthetase[c]
purH	Purine	78	*ade$_i$;* phosphoribosyl-aminoimidazole-carboxamide formyltransferase
purI	Purine	49	Aminoimidazole ribotide synthetase[c]
pyrA	Pyrimidine	0	*cap, arg + ura;* glutamino-carbamoyl-phosphate synthetase
pyrB	Pyrimidine	84	Aspartate transcarbamylase
pyrC	Pyrimidine	22	Dihydroorotase
pyrD	Pyrimidine	21	Dihydroorotic acid dehydrogenase
pyrE	Pyrimidine	72	Orotidylic acid pyrophosphorylase
pyrF	Pyrimidine	25	Orotidylic acid decarboxylase
rac	Recombination activation	29	Suppressor of *recB* and *recC* mutant phenotype
ram	Ribosomal ambiguity	64	Nonspecific suppression of all nonsense codons
ras	Radiation sensitivity	(11)	Sensitivity to UV and x-ray irradiation
rbs	Ribose	74	Utilization of D-ribose
recA	Recombination	52	Ultraviolet sensitivity and competence for genetic recombination
recB	Recombination	55	Ultraviolet sensitivity and competence for genetic recombination
recC	Recombination	55	Ultraviolet sensitivity and competence for genetic recombination
ref	Refractory	(88)	*ref II;* specific tolerance to colicin E2
rel	Relaxed	54	*RC;* regulation of RNA synthesis
rep	Replication	74	Inhibition of lytic replication of temperate phages
rhaA	Rhamnose	76	L-Rhamnose isomerase
rhaB	Rhamnose	76	L-Rhamnulokinase
rhaC	Rhamnose	76	Regulatory gene
rhaD	Rhamnose	76	L-Rhamnulose-1-phosphate aldolase
rif	Rifampicin	77	DNA-dependent RNA polymerase sensitivity to rifampicin
rns	Ribonuclease	15	Ribonuclease I
rts	—	77	*ts-9;* altered electrophoretic mobility of 50S ribosomal subunit
serA	Serine	57	3-phosphoglyceric acid dehydrogenase
serB	Serine	89	Phosphoserine phosphatase
serS	Serine	(18)	Seryl-tRNA synthetase
shiA	Shikimic acid	38	Shikimate and dehydroshikimate permease
som	Somatic	(37)	*O;* somatic (O) antigens
spc	Spectinomycin	64	Resistance or sensitivity to spectinomycin
speB	Spermidine	57	Putrescine (or spermidine) requirement; agmatine ureohydrolase
strA	Streptomycin	64	Resistance, dependence, or sensitivity; "K-character" of the 30S ribosomal subunit
stv	Streptovaricin	77	DNA-dependent RNA polymerase sensitivity to streptovaricin

Table 12.2 (cont.)

Gene symbol	Mnemonic	Map position	Alternate gene symbols; phenotypic trait affected
sucA	Succinate	17	*suc, lys + met;* succinate requirement; α-ketoglutarate dehydrogenase (decarboxylase component)
sucB	Succinate	17	*suc, lys + met;* succinate requirement; α-ketoglutarate dehydrogenase (dihydrolipoyltranssuccinylase component)
supB	Suppressor	16	su_B; suppressor of *ochre* mutation (not identical to *supL*)
supC	Suppressor	25	su_C; suppressor of *ochre* mutation (possibly identical to *supO*)
supD	Suppressor	(38)	su_I, *Su-1;* suppressor of *amber* mutations
supE	Suppressor	16	su_{II}; suppressor of *amber* mutations
supF	Suppressor	25	su_{III}, *Su-3;* suppressor of *amber* mutations
supG	Suppressor	16	*Su-5;* suppressor of *ochre* mutations
supH	Suppressor	38	
supL	Suppressor	17	Suppressor of *ochre* mutations
supM	Suppressor	78	Suppressor of *ochre* mutations
supN	Suppressor	45	Suppressor of *ochre* mutations
supO	Suppressor	25	Suppressor of *ochre* mutations (possibly identical to *supC*)
supT	Suppressor	55	
supU	Suppressor	74	*su7;* suppressor of *amber* mutations
supV	Suppressor	74	*su8;* suppressor of *ochre* mutations
tdk	—	25	Deoxythymidine kinase
tfrA	T-four	(8)	ϕ^r; resistance or sensitivity to phages T4, T3, T7, and λ
thiA	Thiamine	78	*thi;* synthesis of thiazole
thiB	Thiamine	(78)	Thiamine phosphate pyrophosphorylase
thiO	Thiamine	(78)	Probable operator locus for *thiA* and *thiB* genes
thrA	Threonine	0	Block between homoserine and threonine
thrD	Threonine	0	*HS;* aspartokinase I-homoserine dehydrogenase I complex
thyA	Thymine	55	Thymidylate synthetase
tkt	—	(55)	Transketolase
tnaA	—	73	*ind;* tryptophanase
tnaR	—	73	R_{tna}; regulatory gene
tolA	Tolerance	17	*cim; tol-2;* tolerance to colicins E2, E3, A, and K
tolB	Tolerance	17	*tol-3;* tolerance to colicins E1, E2, E3, A, and K
tolC	Tolerance	58	*colE1-i, tol-8, ref1;* specific tolerance to colicin E1
tonA	T-one	2	*T1, T5 rec;* resistance or sensitivity to phages T1 and T5
tonB	T-one	25	*T1 rec;* resistance to phages T1, π80 and colicins B, I, V; active transport of Fe
tpp	—	89	*deoA, TP;* thymidine phosphorylase
trpA	Tryptophan	25	*tryp-2;* tryptophan synthetase, A protein
trpB	Tryptophan	25	*tryp-1;* tryptophan synthetase, B protein

Table 12.2 (cont.)

Gene symbol	Mnemonic	Map position	Alternate gene symbols; phenotypic trait affected
trpC	Tryptophan	25	*tryp-3;* indole-3-glycerol phosphate synthetase
trpD	Tryptophan	25	*tryE;* phosphoribosyl anthranilate transferase
trpE	Tryptophan	25	*tryD, anth, tryp-4;* anthranilate synthetase
trpO	Tryptophan	25	Operator locus
trpR	Tryptophan	90	*Rtry;* regulatory gene
trpS	Tryptophan	65	Tryptophanyl-tRNA synthetase
tsx	T-six	11	*T6 rec;* resistance or sensitivity to phage T6 and colicin K
tyrA	Tyrosine	50	Prephenic acid dehydrogenase
tyrR	Tyrosine	27	Regulatory gene for *tyrA* and *aroF* genes
tyrS	Tyrosine	32	Tyrosyl-tRNA synthetase
ubiA	Ubiquinone	83	4-Hydroxybenzoate $\longrightarrow$ 3-octaprenyl 4-hydroxybenzoate (OPHB)
ubiB	Ubiquinone	75	2-Octaprenylphenol $\longrightarrow$ ubiquinone
ubiD	Ubiquinone	75	OPHB decarboxylase
uhp	—	72	Uptake of hexose phosphates
uraP	Uracil	50	Uracil permease
uvrA	Ultraviolet	80	*dar-3;* repair of ultraviolet radiation damage to DNA
uvrB	Ultraviolet	18	*dar-1,6;* repair of ultraviolet radiation damage to DNA
uvrC	Ultraviolet	37	*dar-4,5;* repair of ultraviolet radiation damage to DNA
uvrD	Ultraviolet	74	*dar-2, rad;* repair of UV radiation damage to DNA
valS	Valine	84	*val-act;* valyl-tRNA synthetase
xyl	Xylose	70	Utilization of D-xylose
zwf	Zwischenferment	(35)	Glucose-6-phosphate dehydrogenase

Sexduction　Careful examination of Figure 12.10 reveals that the process of Hfr formation should be a reversible one, and indeed it is. The Hfr chromosome can loop out and regenerate a circular F factor. In 1959, E. A. Adelberg noted that one of his Hfr strains had reverted to F$^+$ but that it had some unusual properties. The reverted F$^+$ sex factor formed Hfr cells at a very high frequency; furthermore, it always reintegrated at the same chromosomal site at which the original Hfr strain had carried its F sex factor. This new F$^+$ strain of *E. coli* was called *sex factor affinity* (sfa). Probably sfa originates when the F factor incorrectly loops out from an Hfr cell, leaving part of the F factor behind on the bacterial chromosome. This F factor remnant in the chromosome is a site for homologous pairing of F factors, and hence reintegration of F factors at that site is greatly enhanced (Figure 12.12).

Encouraged by the discovery that defective F factors can form from Hfr cells, Jacob and Adelberg studied F⁺ revertants to see whether chromosomal genes could be incorporated into F factors. Studying an F⁺ revertant of an Hfr strain whose F factor was integrated between the *ton* and *lac* loci, they noted that when this F⁺ revertant was mated with an F⁻ strain carrying many mutant genes, genetic recombinants for the mutant characters arose only in the low frequency expected for an F⁺ × F⁻ cross. There was one exception, however. The *lac⁺* gene was transferred to the F⁻ *lac⁻* recipient at a frequency at least as high as that expected for an Hfr × F⁻ mating. Closer scrutiny of the cross revealed that the F⁺ *lac⁺* recombinants actually contained not one but two *lac* genes: *lac⁺* and *lac⁻*! The recipient cell was partially diploid for the *lac* locus and was phenotypically *lac⁺*, showing that *lac⁺* was *dominant* over *lac⁻*.

Figure 12.12
Formation of sex factor affinity (sfa) strains of *E. coli*

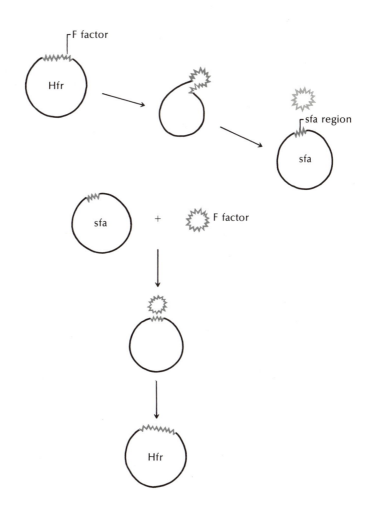

This observation led to the proposal that the F⁺ revertant underwent inexact excision from the Hfr chromosome (as in the case of sfa) and in so doing incorporated the lac^+ gene into the F factor (Figure 12.13). The F⁺ revertant carried the F factor along with its incorporated lac^+ gene, and, in mating with F⁻ lac^-, the revertant transferred this modified F factor to the recipient. The recipient, converted to F⁺ and partially diploid for the *lac* locus, we symbolize as $lac^-/F\text{-}lac^+$. This F factor carrying chromosomal genes is symbolized F′ and has been useful in studying dominance relations in normally haploid cells. The process of F′ gene transferral is called *sexduction*.

Episomes are common to many bacteria. In addition to the sex factor, there are episomes that contain genes for resistance to antibiotics, mercury ions, cobalt and nickel ions, and a host of drugs. Many of these episomes are passed not only among individuals of the same species but also among individuals of different species, which makes the study of episomes vital to our programs to control bacteria-caused diseases.

Figure 12.13
Generation of an F factor carrying bacterial genes (an F′ factor); improper looping out of the Hfr chromosome results in lac^+ genes being incorporated into the F factor

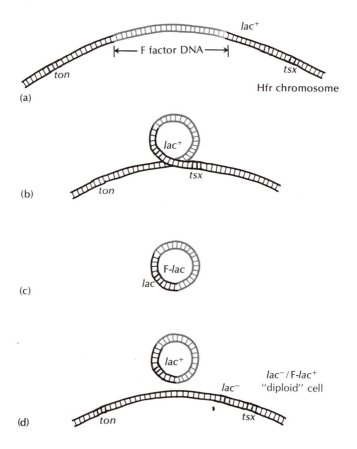

The origin of episomes is an intriguing one. Many see them as phage DNA that has entered into a symbiotic relationship with a bacterial host. P. Fredericq, referring to the origin of the F factor, stated in 1963

> A phage is a particle that can establish effective contact with a cell and inject into it genetic material.
> What else is an F^+ cell?

Function of the sex factor

Before analyzing the process of conjugation, we will explore the role of the F factor in differentiating F^+ and Hfr cells from F^- cells. Donor bacteria (F^+ or Hfr) have the ability to form mating pairs with F^- recipients and occasionally with other F^+ cells. This is the single most important feature of the conjugation process. Donor cells have surface components that are missing in recipient cells. Donor bacteria cell surfaces, when compared to those of recipient cells, have different antigenic properties and different electrophoretic mobilities, and they are susceptible to attack by certain phages that do not attack F^- cells. These phages have been shown to attach to specific cell surface structures called *F-pili*—long, filamentous appendages (Figure 12.14). These F-pili are not present in cells lacking the F factor, suggesting that genes in the F factor are responsible for these structures. The F-pili differ from the pili common to all bacteria in the same class as *E. coli*, the class Enterobacteriacae, in that they are few in number (perhaps one for each sex factor the donor contains), they are longer, and they appear to contain different molecular components. These F-pili are essential to conjugation, but their exact role is poorly understood. It has been suggested that they may serve as a "conjugation tube" through which donor DNA passes to recipient cells.

A summary of the sexual types of *E. coli* appears in Figure 12.15 on page 539, which depicts F^- cells, F^+ cells, Hfr and sfa strains, and the F' cells.

The process of conjugation

The *general* mechanism of conjugation is well understood and supported by experimental data, though many details of the process remain obscure. What follows is a provisional model based largely on experimental data and partly on educated speculation. The discussion will trace the events shown in Figure 12.16, on page 540.

The first step in conjugation is *specific pair formation* (Figure 12.16a). It involves the F-pilus of an Hfr cell and the cell surface of an F^- cell. This initial attraction is based on cell surface differences in the two strains.

The next step is *effective pair formation* (Figure 12.16b and c). It involves the formation of a conjugation tube through which the donor DNA will pass. The retraction of the F-pilus and the contact of the two cell surfaces are the key events here. As depicted, the conjugation tube forms at the site of the F-pilus (though this is not known with certainty).

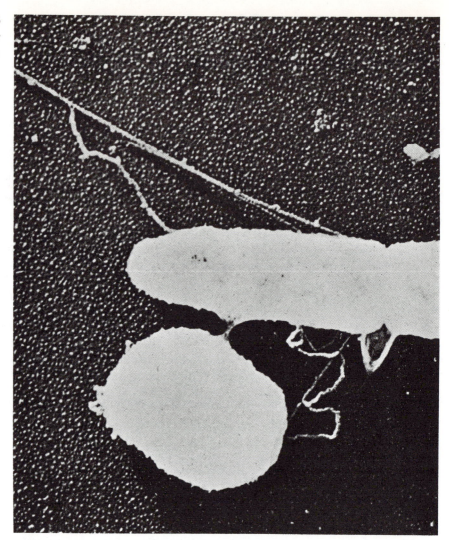

Figure 12.14
Electron micrograph of
conjugating bacteria.
(Courtesy of W. Hayes,
by permission of
Blackwell Scientific
Publications, Ltd.)

Chromosome mobilization (Figure 12.16c) probably begins as soon as
contact is made between the two cells. This contact could trigger the
release in Hfr of an enzyme (endonuclease) that induces a single-strand
break at the site of F integration. The exposed 5′ end of a single strand of
DNA is the origin of chromosome transfer.

Chromosome transfer of the Hfr single strand of DNA to an F⁻ cell is
seen in Figure 12.16d, e, and f. First, DNA polymerase that is attached to
or part of the cell membrane replicates the Hfr chromosome at the site of

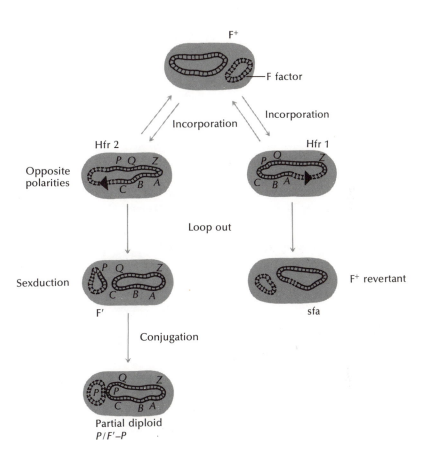

Figure 12.15
Sexual types of *E. coli*

the enzyme-induced break. The single strand of DNA is "pushed" through the conjugation tube and into the recipient cell. This step, then, requires DNA synthesis in the donor cells. Then DNA replication in the recipient cell is shortly completed. The third step in chromosome transfer is the synapsis and pairing of homologous regions of the donor and the recipient DNA. The donor strand displaces the comparable strand of the recipient molecule, thus forming the usual Watson–Crick base pairs. The last step in chromosome transfer is a cooperative effort in which the

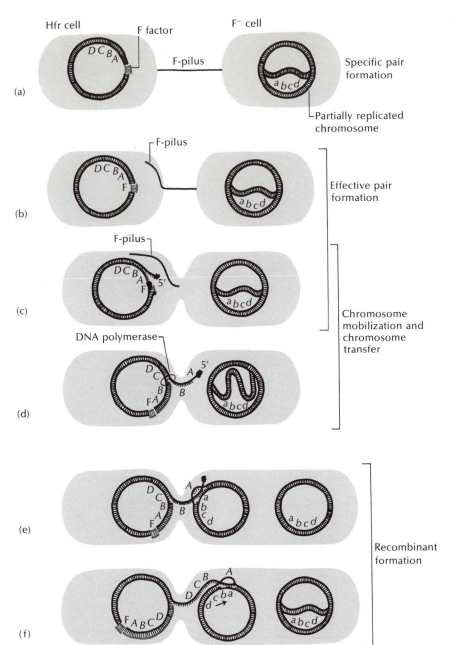

(a)

(b)

(c)

(d)

(e)

(f)

Figure 12.16
The stages of bacterial conjugation: provisional model

recipient cell is "pulling" the chromosome in and the donor is "pushing" it out.

The formation of *recombinants* involves first the pairing of homologous DNA regions (Figure 12.16e and f). Regions of single-strandedness in the recipient chromosome, occurring perhaps because of single-stranded breaks in the recipient chromosome, facilitate homologous pairing and hydrogen bonding between complementary bases. The replacement of the recipient strand by the donor strand is shown in Figure 12.17. After effective pairing of donor and recipient strands of DNA (Figure 12.17a and b), single-stranded breaks occur in the recipient strand and it is displaced (Figure 12.17c). Similar single-stranded breaks occur in the other strand of the recipient chromosome, and that strand is also displaced, resulting in the structure shown in Figure 12.17d. Using the donor strand as a template, DNA synthesis occurs simultaneously with strand displacement so that the structure shown in Figure 12.17d probably does not exist. On completion of DNA synthesis, the structure shown in Figure 12.17e is

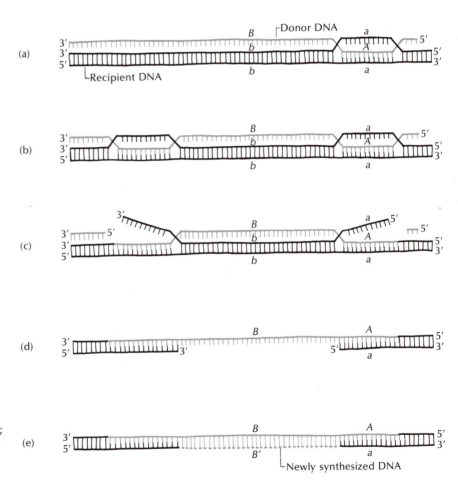

Figure 12.17
Provisional model of genetic recombination following chromosome transfer in matings between Hfr and F⁻ cells; the stage diagrammed in part (a) is equivalent to stage (e) in Figure 12.16

formed—a recombinant chromosome. Close examination reveals the basis for heterozygosity for some loci, a matter already discussed in connection with recombination in phages.

A somewhat modified view of DNA transfer during conjugation appears in Figure 12.18. It suggests that the single DNA strand entering the recipient cell is replicated by recipient DNA-polymerizing enzymes. Recombination in this instance would involve interaction between two double-helical molecules and would differ from the model presented in Figures 12.16 and 12.17.

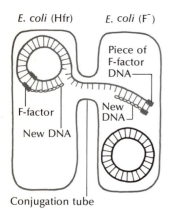

E. coli (Hfr)   *E. coli* (F⁻)

Piece of F-factor DNA

F-factor

New DNA

New DNA

Conjugation tube

Figure 12.18
An alternative model of DNA transfer during conjugation

Transduction and lysogeny

Transduction is a third mechanism of genetic exchange in bacteria, and it differs significantly from the mechanisms of transformation and conjugation in that it is mediated by a bacterial virus. Transduction is the transfer of bacterial genetic information from one bacterium to another using a phage as the vector. There are two main types of transduction: *general transduction,* in which the phage can transfer *any* segment of the bacterial genome to another bacterium, and *specialized transduction,* in which only *restricted* segments of the bacterial genome can be transduced. The two types of transduction usually occur by mechanisms that differ fundamentally from each other, as we shall see. Transduction, like transformation and conjugation, is extremely important in the mapping of bacterial genes and is widely used for fine-structure mapping—the mapping of sites *within* a gene.

The discovery of transduction

The first instance of phage-mediated transfer of genetic information from one bacterium to another was uncovered in 1951 by Lederberg and his graduate student, N. Zinder, who were looking for a process in *Salmonella typhimurium* analogous to *E. coli* conjugation. They mixed two auxotrophic strains of *Salmonella*—one unable to synthesize phenylalanine, tryptophan, and tyrosine (*phe⁻ trp⁻ tyr⁻*) and one unable to synthesize methionine and histidine (*met⁻ his⁻*)—and found prototrophic colonies that grew on unsupplemented medium. Because the frequency of appearance of prototrophs (10^{-5}) was too high to be attributed to reverse mutation, they hypothesized that *Salmonella,* like *E. coli,* was capable of genetic exchange.

In order to determine whether *Salmonella* engaged in conjugation, Lederberg and Zinder grew the two strains in a common medium in a Davis U-tube, but separated from each other by a glass filter. The filter prevented physical contact between the two strains but allowed material smaller than bacteria to pass through (Figure 12.19). Recombinants formed under these conditions, indicating that contact between the par-

Figure 12.19
Davis U-tube. The two arms of the U are separated by a sintered glass filter that is impervious to bacterial cells, but allows the growth medium to pass through freely. One parent strain is placed in each arm, and alternating pressure and suction are applied to one arm to stimulate movement of material through the filter. (After Davis, *J. Bact.* 60:507, 1950.)

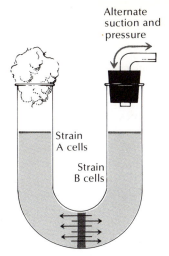

Alternate suction and pressure

Strain A cells

Strain B cells

The two arms of the U are separated by a sintered glass filter that is impervious to bacterial cells but allows the growth medium to pass through freely. One parent strain is placed in each arm of the tube, and alternating pressure and suction are applied to one arm to stimulate movement of material through the filter.

ent strains was not required for recombination to occur. Thus a process analogous to conjugation was ruled out. Genetic exchange in *Salmonella* was apparently mediated by a *filterable agent* (FA) that could pass through the filter, bringing genes from donor to recipient cells.

Transformation was eliminated from consideration as a possible mechanism when it was demonstrated that the FA was not affected by DNase, an enzyme that degrades free DNA and therefore destroys the transforming principle. Careful analysis of the FA revealed that it was a bacteriophage, called P22, for which one of the parent strains was lysogenic; that is, the viral genetic information was integrated into the host's genetic information in a manner analogous to Hfr formation. One strong observation supporting the viral-mediated basis of recombination in *Salmonella* is the inactivation of FA by anti-P22 serum.

With a virus implicated in bacterial genetic recombination, the next task was to describe the transduction process. The following general picture of transduction emerged. The lysogenic *Salmonella* is somehow changed so that the repressed P22 phage enters a lytic growth cycle. The donor cell's DNA is broken down into fragments, and as mature P22 form, some bacterial DNA is incorporated into a few mature P22. On release from the donor bacteria, these unusual P22 phages infect other cells and inject the bacterial genes. If the recipient bacterial cell survives the infection, the transduced bacterial DNA may then undergo genetic recombination with its homologous alleles on the recipient host chromosome and produce recombinant genotypes.

Generalized transduction   The P22 phage is one of several phage types that are capable of *generalized transduction*; that is, they can transduce any bacterial gene. The question then arises as to just how the phage particles pick up the fragments of bacterial chromosome. Two models of generalized transduction have been proposed. In the first, a fragment of phage DNA recombines with a

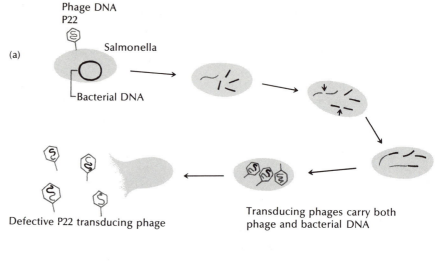

(a)

Phage DNA
P22

Salmonella

Bacterial DNA

Transducing phages carry both
phage and bacterial DNA

Defective P22 transducing phage

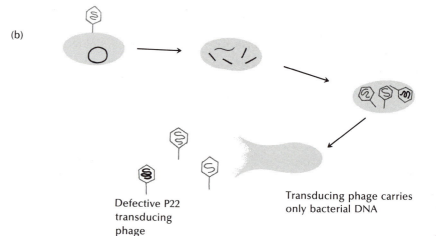

(b)

Defective P22
transducing
phage

Transducing phage carries
only bacterial DNA

Figure 12.20
Two models of the
formation of generalized
transducing phage

fragment of bacterial DNA to produce a hybrid DNA molecule (Figure
12.20a). This recombinant chromosome is deficient in phage genetic material, so it cannot direct the synthesis of progeny phage. But once injected,
it can recombine with the host-cell DNA to yield recombinants. This
model of transducing phage formation is not well supported by experimental evidence. However, it may be applicable in certain transducing
situations.

A second model of transducing phage formation for generalized
transduction, and one that has supportive experimental evidence, depicts
fragments of bacterial chromosomes incorporated into phage heads *in lieu
of* phage chromosomes. This situation is in many respects similar to
phenotypic mixing. According to this model (Figure 12.20b), there is no

physical association between phage and bacterial DNA in the formation of transducing phage.

Generalized transduction can be used to map gene loci via a technique whereby some genes are selected for and others are not selected. Let us use a study of the *E. coli* phage P1 to illustrate our point. The P1 phage is a general transducing phage. We use it to infect *E. coli* cells that carry three genes the sequence of which we wish to determine: *leu*⁺ *thr*⁺ *azi*^r (*leu* = leucine, *thr* = threonine, *azi*^r = azide-resistant). Transducing P1 phage from these donor cells are then used to infect *E. coli* cells carrying the genetic markers *leu*[−], *thr*[−], and *azi*^s.

The next step in this mapping exercise is to place the cells on media that will select for one or two of the genetic markers, leaving the remaining marker(s) unselected for. For example, if the recipient cells are placed on a minimal medium free of azide but supplemented with *thr*, then *leu* is the selected marker, because only *leu*⁺ cells will grow. *Thr* and *azi* are the unselected markers, because in the presence of *thr*, the genetic marker could be *thr*⁺ or *thr*[−]; in the absence of azide, the genetic marker could be *azi*^r or *azi*^s.

For each experiment with the selected marker, the frequency of the unselected marker(s) is determined. Using the foregoing example, those cells that are *leu*⁺ are tested by further plating to determine whether they are *azi*^r or *azi*^s, *thr*⁺ or *thr*[−]. This will give us relative linkage strengths, as we shall see. Three experiments were done, and the following data were collected.

Experiment	Selected marker(s)	Unselected marker(s)	
1	*leu*⁺	50% *azi*^r	2% *thr*⁺
2	*thr*⁺	3% *leu*⁺	0% *azi*^r
3	*leu*⁺, *thr*⁺	0% *azi*^r	

We interpret this experiment in terms of the linkage relationships between the three loci. From experiment 1, we conclude that the *leu* and *azi* loci are close and that *leu* and *thr* are not close. We conclude this because half the time that the *leu*⁺ gene was carried from the donor, it was accompanied by *azi*^r; in only 2% of the cases did *thr*⁺ accompany *leu*⁺. From this analysis, we can suggest the models

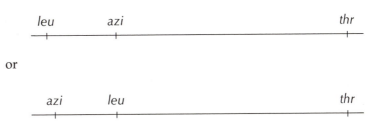

or

From experiment 2, we conclude that *leu* is closer to *thr* than is *azi*. We can do this because 3% of the time *leu*⁺ and *thr*⁺ are cotransduced, but never did *thr*⁺ and *azi*⁺ get cotransduced. So we can now suggest the gene sequence to be

Experiment 3 is consistent with this sequence, since the *thr*⁺ *leu*⁺ transducing fragment never carries *azi*ʳ.

The size of the transducing fragment is also evident from this analysis. The length of DNA between *thr* and *leu* must be close to the size limit, because they are cotransduced only 2–3% of the time. The *azi* locus, which is close to *leu*, is never cotransduced with *thr*. So the maximum size of the transducing fragment runs from the *thr* locus to between *leu* and *azi*.

Lysogeny

In order to fully understand specialized transduction, we must understand lysogeny. Soon after the discovery of bacterial viruses, strains of bacteria were isolated that seemed to carry phages permanently; that is, the bacteria were *lysogenic*. At that time, such strains were in marked contrast to the particularly brutal and lethal interaction between bacterial viruses and bacteria, in which viruses infected and always destroyed the host cells. During the course of extensive investigations into these lysogenic strains, André Lwoff, in describing *lysogeny*, suggested that each lysogenic bacterium harbors and maintains a noninfective viral chromosome (called a *prophage*) that has the ability to produce mature, infective phage particles *without* the intervention of exogenous phage particles. When a prophage begins to produce progeny phage particles, it is said to be *induced*, and this induction leads to the ultimate death and lysis of the host cell. Induction, Lwoff added, is under the control of external factors, and although induction does occur spontaneously at a low frequency, ultraviolet light can artificially induce it in almost 100 percent of a population of lysogenic bacteria. Once induction begins, the phages go through the pattern of development identical to that discussed in Chapter 11: a *latent* period, followed by a *rise* period, and ending with a *plateau* period.

When used to infect nonlysogenic bacteria, the phages released from a lysogenic strain of bacteria generally proceed through a lytic cycle of development; that is, they infect and immediately destroy the host cells. However, some of the nonlysogenic bacteria survive the attack, and when these survivors are examined, they are found to have been converted to lysogenic cells. Thus is lysogeny induced in bacteria by the viruses. Some phages may therefore enter one of two phases on infection of nonlysogenic bacteria: the *lytic* phase, in which the phage infects and destroys a cell, or the *lysogenic* phase, in which the infecting phage does not multiply but instead becomes a prophage, allowing the host cell to survive as a lysogenic bacterium. Phages *capable* of entering the lysogenic relationship are called *temperate phages,* whereas those that cannot enter a lysogenic state (and hence can never form a prophage) are referred to as *virulent phages*.

Lysogenic bacteria are *immune* to infection by other phage particles of the same type as the prophage. If it were any other way, spontaneous induction of a single lysogenic bacterium in a population of lysogenic bacteria would result in the destruction of the entire population. This does not occur. The lysogenic bacteria are, however, susceptible to attack by all other viruses to which the bacterial strain is normally sensitive.

A lysogenic bacterium, in addition to experiencing induction, can also be *cured* of the prophage. The prophage simply disappears from the cell, and the cured cell becomes nonlysogenic and susceptible to attack by temperate viruses like the lost prophage. Figure 12.21 summarizes lysogeny.

Figure 12.21
Summary of lysogeny [After A. Lwoff, *Bact. Rev.* 17:269 (1953)]

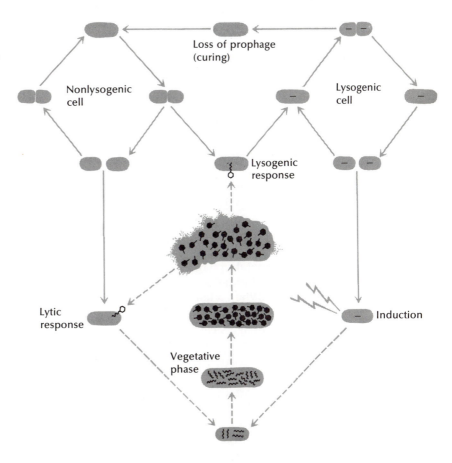

The relationship between the prophage (the temperate virus chromosome, which is a molecule of DNA) and the bacterial chromosome became intelligible in 1951, when Esther Lederberg found that the *E. coli* strain used by J. Lederberg in describing conjugation (*E. coli* K12) was lysogenic for the λ (*lambda*) phage. The discovery of lysogenic and nonlysogenic strains of sexually fertile *E. coli* K12 paved the way for bacterial crosses in which the distribution of the λ prophage among recombinant bacteria could be ascertained and mapped as genes are. Studies soon revealed that the λ prophage behaved like a chromosomal gene or segment of genes and, further, that it was linked to the *gal* region of *E. coli*, the region responsible for galactose fermentation.

In a cross in which Hfr and F⁻ parents were lysogenic for λ prophage and carried different genetic markers, it was shown that the prophage enters the F⁻ recipient after the *gal* genes and before the *trp* genes. Other crosses established that the *E. coli* chromosome contained a gene (*hfl*) causing a high frequency of lysogeny. It can therefore be concluded that the λ prophage becomes part of the *E. coli* chromosome between the *gal* and *trp* loci and that lysogeny is influenced by bacterial genes. In contrast to λ, some temperate phage, such as P2, can have the prophage insert into more than one site.

Zygotic induction When an Hfr strain that is lysogenic for λ mates with an F⁻ strain that is nonlysogenic for λ, the F⁻ strain that receives the Hfr chromosome will lyse shortly after the prophage enters. This is called *zygotic induction*, and it occurs only when the donor is lysogenic and the recipient is not. If the transfer of the chromosome is terminated prior to the entrance of the prophage, normal recombination occurs. It is only when the prophage passes from its lysogenic environment to a nonlysogenic environment that the prophage is induced.

Zygotic induction argues rather conclusively that the F⁻ cytoplasm is crucial to the induction of the prophage. A cytoplasmic factor, the product of a phage gene, represses the intracellular prophage, making possible the lysogenic condition. When this factor, called a *repressor* (Chapter 13), is missing or inactivated (as by UV light), prophages are induced and the temperate phages go through a lytic cycle. Figure 12.22 summarizes zygotic induction.

When a temperate phage infects a nonlysogenic bacterium, there is a "race" between synthesis of the cytoplasmic repressor and synthesis of the phage's early proteins. If environmental conditions favor the synthesis of the repressor, a lysogenic response results; if phage protein synthesis is favored, lysis results.

Genetic control of lysogeny There are three genes in the λ genome that determine whether an infecting λ can become a prophage. In a nonmutant state, these genes (*cI, cII,* and *cIII*) allow λ to become a prophage. However, a *cI* mutation greatly reduces or eliminates the λ potential for becoming a prophage. The *cI* gene is responsible for synthesizing the cytoplasmic repressor that con-

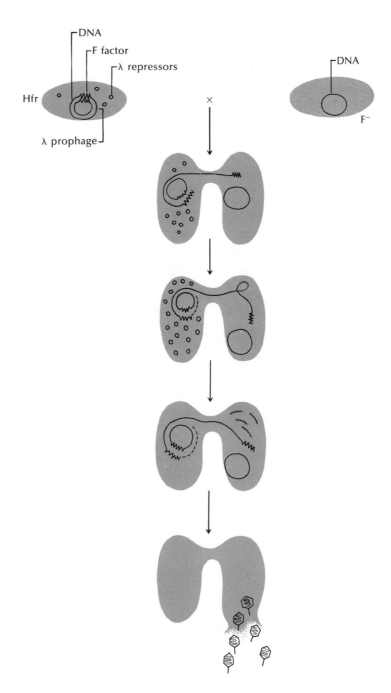

Figure 12.22
Zygotic induction

verts λ into a prophage. The *c*II and *c*III genes are involved in the activation of certain gene functions. In Chapter 13 we shall examine in more detail the genetic control of lysogeny.

Lambda-c mutants are detected by analyzing the plaques formed when λ infects nonlysogenic *E. coli* cells. Wild-type (*c*⁺) plaques are

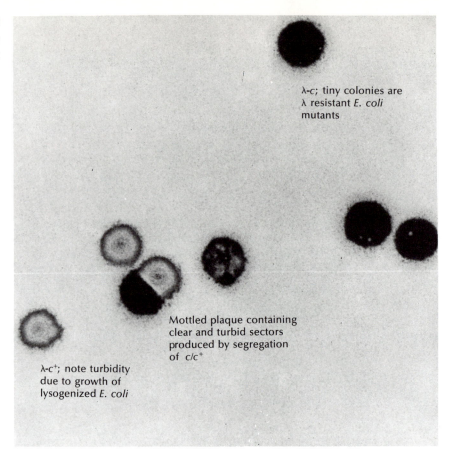

λ-*c*; tiny colonies are λ resistant *E. coli* mutants

Mottled plaque containing clear and turbid sectors produced by segregation of *c*/*c*⁺

λ-*c*⁺; note turbidity due to growth of lysogenized *E. coli*

Figure 12.23
Plaques formed by λ phage and its *c* mutants on nonlysogenic *E. coli* [From J. J. Weigle, *Proc. Natl. Acad. Sci. U.S.* (1953), courtesy of M. Meselson]

somewhat turbid because of the growth of cell colonies that have been converted from nonlysogenic to lysogenic (Figure 12.23). The plaques formed by λ-*c* mutants are clear, because the potential for prophage formation has been greatly reduced or eliminated. (The λ-*c* mutant served as a genetic marker in experiments by Meselson and Weigle on the mechanisms of recombination in viruses discussed in Chapter 11.)

Prophage integration   The integration of a prophage such as λ requires a genetic exchange between homologous regions of the λ and *E. coli* chromosomes. This was suggested when a λ strain deficient for the *b2* region (Figure 12.24) was found to be incapable of integration into the host chromosome but was perfectly capable of entering the lytic cycle. Thus it appeared that *b2* and a region on the *E. coli* chromosome between the *gal* and *trp* genes called *lambda attachment* (λ*att*) were homologous.

In 1962, Campbell suggested that the actual process of λ integration was exactly analogous to the integration of the F factor into the *E. coli* chromosome forming an Hfr strain. The chromosome of mature λ is *linear*,

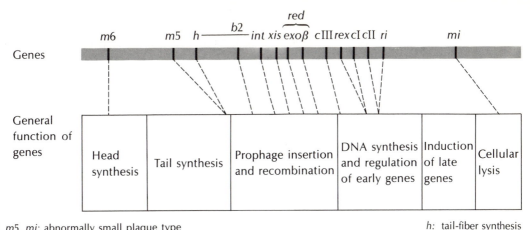

Genes

red

m6   m5   h————b2——int xis exoβ cIII rex cI cII ri   mi

General
function of
genes

Head synthesis	Tail synthesis	Prophage insertion and recombination	DNA synthesis and regulation of early genes	Induction of late genes	Cellular lysis

m6, m5, mi: abnormally small plaque type
b2: attachment region to bacterial chromosome
xis: excision protein
int: integration protein
red (exo and β): genetic recombination proteins
cI, cII, cIII: immunity (and clear plaques)
rex: gene controlling exclusion of T-even rII mutant phage growth

h: tail-fiber synthesis
ri: DNA synthesis

Figure 12.24
Genetic map of λ phage

but on injection into nonlysogenic strains of *E. coli*, it circularizes (Figure 12.25). The *b2* and λ*att* regions synapse, and then follows the insertion of the prophage into the host chromosome (Figure 12.26). The insertion of the prophage is achieved by an *integration protein* that is coded for by a λ gene (*int*). Prophage induction is the reverse of integration, except that a second protein works in conjunction with the integration protein—the *excision protein*, coded for by the *xis* λ gene.

This model of prophage formation predicts that the *mi* and *m6* loci, which map at the ends of the λ linear chromosome, will be next to each other after insertion (Figures 12.24 and 12.26). Conjugation experiments have verified that *c* and *h* are the end markers of the prophage, with *m6* and *mi* neighbors in the middle.

Temperate phages may therefore be considered episomes. Indeed, episomes such as the F factor may once have been temperate phages that lost their ability to be induced but not their ability to produce phage proteins. The F-pili may be a manifestation of a temperate phage turned F factor, in which the chromosome is interior and the protein coat exterior.

DNA insertion elements   Recent research into mechanisms of inserting DNA into chromosomes has uncovered the fascinating phenomenon of genes that move around on the chromosome, and indeed move from chromosome to episome and back. This movement of genes ("jumping genes," as they are called) is the result of small segments of DNA with specific sequences called *insertion elements* (IS). There are four such insertion elements, ranging in size from about 700 to 1400 base pairs and called IS1, IS2, IS3, and IS4. These

551

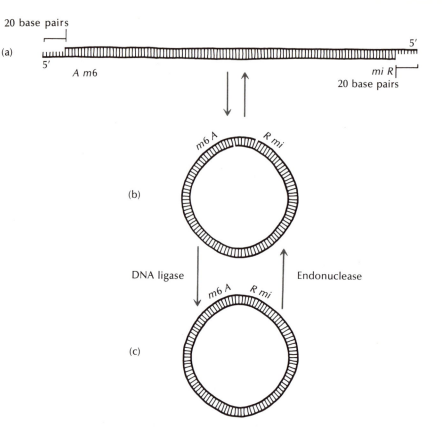

20 base pairs

(a)

5′

A m6

mi R

20 base pairs

(b)

DNA ligase   Endonuclease

(c)

Figure 12.25
Circularization and cohesive ends of λ phage DNA. [Adapted from A. D. Kaiser, in F. O. Schmitt (ed.), *The Neurosciences: Second Study Program*, New York: Rockefeller University Press, 1971]

(a) The chromosome of the infective λ phage is a linear DNA molecule carrying gene *A* at one end (the "left" end) and gene *R* at the other (the "right" end). At both ends, the polynucleotide chain carrying the 5′ phosphate terminus is longer than that carrying the 3′ hydroxyl terminus. (b) The base sequence of the two terminal-producing ends is complementary, thus providing the entire DNA molecule with "cohesive ends," making it possible to convert the linear DNA molecule into a circle. (c) In the cohesive end joint, the 5′ phosphate termini of both polynucleotide chains are adjacent to their own 3′ hydroxyl termini, thus allowing for the action of the DNA ligase enzyme, which seals these interruptions; this gives rise to a fully covalently bonded, circular double helix without any interruptions. Within five minutes of the injection of λ DNA into an *E. coli* host cell, the linear DNA molecule of the mature phage particle has been converted into a covalently linked circular molecule, in which form it replicates for the generation of progeny DNA molecules. Upon the encapsulation of these progeny DNA circles into the heads of progeny phage particles, the circular DNA is nicked to give rise to linear molecules like (a) above.

insertion elements are found in temperate phage, such as λ, and in episomes, such as the F factor. They are also found in the *E. coli* chromosome. Their existence seems to facilitate an integrative type of recombination, such as we see occurring in λ. The IS elements, when homologous and properly aligned, specify location of insertion. The IS elements

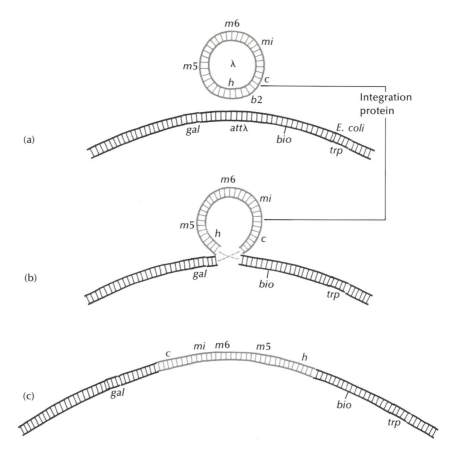

Figure 12.26
Insertion of λ prophage into the bacterial chromosome. (a) The circular phage chromosome synapses with the λ attachment site of the bacterial chromosome. (b) The phage chromosome breaks between *h* and *c* genes in the *b2* region, the bacterial chromosome breaks between *gal* and *bio* genes, and the different DNA pieces rejoin. (c) The crossover generates one continuous genetic structure containing the λ genome interposed between the bacterial *gal* and *bio* genes.

specify where the F factor integrates and where λ integrates. But other extrachromosomal segments of DNA also have IS elements, in addition to their other genes. These frequently insert into sites specified by similar IS elements in the chromosome, but they can later leave the site and reinsert at a different site. Thus genes seem to be jumping around on the chromosome.

The IS elements, in addition to facilitating the integration of extrachromosomal DNA into chromosomes, may also be an important force in the evolution of primary chromosome structure. In Chapter 5, we mentioned that the duplications generated by unequal crossing-over provide a reasonable basis for increasing the amount of genetic material in the genome. IS elements may provide another way. Foreign DNA, such as viral DNA, may insert into chromosomes as a result of the IS elements, thus increasing the amount of genetic material in the genome. This may be especially pertinent to the integration of tumor virus DNA in human genomes (see Chapter 11). Meanwhile, research into the IS elements continues, increasing our understanding of these remarkable entities.

In contrast to generalized transduction, which can involve any bacterial gene, specialized transduction is limited to a select segment of the bacterial genome. Specialized transduction was discovered in 1956 by Lederberg when he induced lysis in an *E. coli* K12 strain lysogenic for λ. The mature phages released after induction, when used to transduce nonlysogenic strains of *E. coli* carrying several mutant genes, transduced *only* the *gal* and *bio* loci. Thus λ transduced only those genes in the immediate vicinity of λ*att*, which in this case happened to be the *gal* and *bio* genes.

The *gal*⁺ transductants displayed some interesting properties. First, they could be induced to liberate infective λ, and they were also immune to further λ infection, which verified that the λ chromosome carrying the *gal* genes entered the cell. Second, the *gal*⁺ transductants were *genetically unstable;* that is, they reverted to *gal*⁻ at a frequency of between 1 and 10 percent. This latter characteristic of the *gal*⁺ transductant suggested that it was a partial heterozygote (*gal*⁺/*gal*⁻) and that the *gal*⁺ gene brought in from the donor was actually *added to* the recipient chromosome and did not replace the *gal*⁻ genes. The *gal*⁻ revertants from the *gal*⁺ transductants were cells that had lost the *gal*⁺ donor fragment.

Induction of the *gal*⁺/*gal*⁻ heterozygote produced infective λ particles with an extraordinarily high transducing potential (about half of the new particles could transduce *gal*⁺ genes into *gal*⁻ recipients). A lysate that produces such high percentages of transducing phages is called a *high-frequency transductant* (HFT). Further analysis of the phages found in the HFT lysate revealed that they were of two types: normal, nontransducing λ and *defective* λ. Defective λ can infect a cell and transduce it, but they are incapable of generating progeny phages (that is, they cannot be induced). These defective λ phages carry the *gal* gene in place of a segment of λ genes and are called *lambda dg* (λ*dg*, where *d* = defective and *g* = gal). Defective λ carrying the *bio* gene are called λ*db*. The only way λ*dg* can be induced to produce progeny phages is if another nondefective λ particle is in the same cell. This nondefective particle is called a *helper phage,* because it supplies the missing genes needed for complete λ development.

The generation of λ*dg* particles is exactly analogous to F′ factor formation. Improper looping out and excision of the λ prophage (Figure 12.27) can lead to a λ particle that is defective for some of its own genes and carries in their place the bacterial *gal* or *bio* genes. On infection of nonlysogenic *gal* cells by λ*dg*, the λ *b2* region synapses with the bacterial λ*att* region and is integrated, thereby producing the *gal*⁺/*gal*⁻ partial heterozygote (Figure 12.28). Excision and loss of the prophage again generates *gal*⁻ cells.

Specialized transduction requires the integration of the transducing viral chromosome into the host's genome, its excision accompanied by the inclusion of some bacterial genes, infection of new recipient cells, and the integration of the transducing prophage into the recipient chromosome. It may happen, however, that the transducing prophage is not integrated but simply passed on, without replication, to one of two daughter cells.

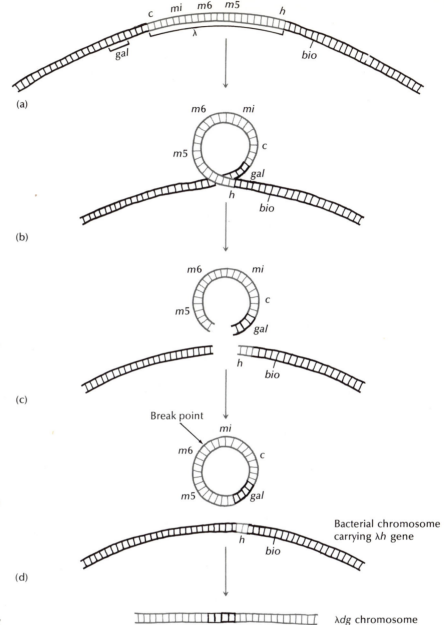

Figure 12.27
Generation of a λ*dg* transducing phage. (a) The chromosome of the lysogenic K12 (λ)*gal*⁺ bacterium (b) Improper looping out of the prophage (c) Excision of the improper loop gives rise to a circular phage chromosome that includes the bacterial *gal* genes but leaves behind in the bacterial chromosome the *h* gene region of the phage. (d) Normal opening of the excised circle between *m6* and *mi* by the nicking enzyme gives rise to the linear λ*dg* chromosome.

This was discovered in *Salmonella* when P22 phages were tested for their ability to transduce motility genes (*fla*⁺ *Salmonella* strains have flagella and hence are motile; *fla*⁻ mutations result in the absence of flagella). In one experiment, a number of "trails" were observed leading out from regions

556
Transmission of the genetic
material in bacteria

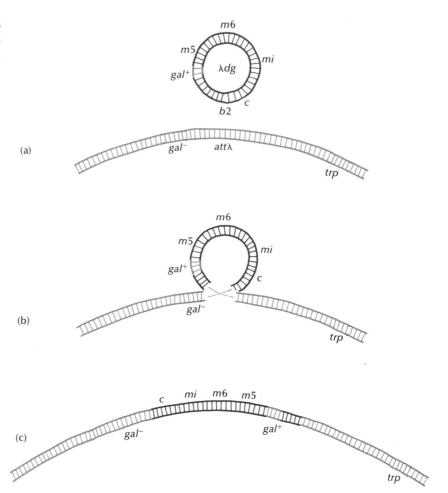

(a)

(b)

(c)

Figure 12.28
Transduction of a nonlysogenic *K12 gal⁻* recipient cell by a *gal⁺ λdg* phage into a *K12 (λdg)* *gal⁺/gal⁻* partially heterozygous transductant. (a) Synapsis of the recipient chromosome and the λ*dg* phage chromosome in the homologous λ*att* and *b2* regions. (b) Crossover of the two chromosomes in the region of synapsis. (c) The continuous genetic structure resulting from the crossover. This structure contains a defective lambda prophage between two bacterial *gal* genes, one a *gal⁺* donor and the other a *gal⁻* recipient.

of colonial growth (Figure 12.29). These trails were explained by the suggestion that the *fla⁺* gene brought into the *fla⁻* cell by the transducing P22 phage neither recombined with the host nor was replicated. On cell division, only one of the two daughter cells received *fla⁺*, and this cell moved off, leaving the *fla⁻* cell immobile. When *fla⁺* divided again, one motile and one nonmotile cell were produced, the motile cell moving off and the nonmotile one forming a colony of nonmotile cells (Figure 12.30). This unilinear inheritance of a transferred gene is called *abortive transduction*, and it continues until the gene is somehow lost.

Plasmids and recombinant DNA

Plasmids differ from episomes in that they are extrachromosomal genetic elements that *do not* integrate into the bacterial chromosome. In *E. coli*,

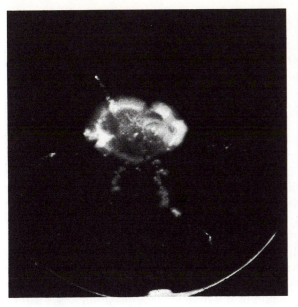
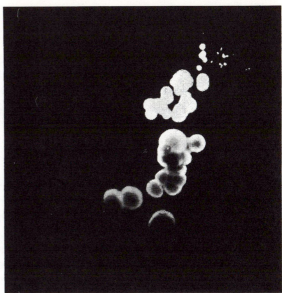

Figure 12.29
The "trail phenomenon" in *Salmonella*. Nonmotile *Salmonella* (*fla⁻*) are infected with transducing P22 (*fla⁺*) phages. Trails of small, isolated colonies are seen emanating from the central, original colony. [Reproduced from B. Stocker, *J. Gen. Microbiol.* 15:525 (1956), by permission of B. Stocker and Cambridge University Press]

Figure 12.30
Underlined cells are nonmotile, or, in the case of abortively transduced nutritional characters, unable to divide. Only daughter cells that inherit the abortively transduced donor fragment are shown as capable of motility or division.

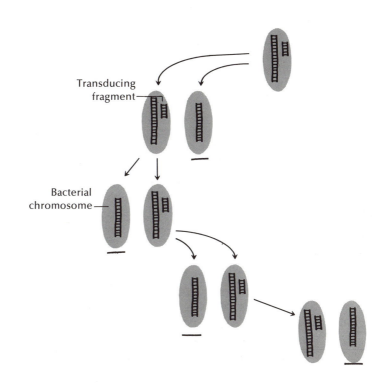

Transducing fragment

Bacterial chromosome

there is a wide variety of these plasmids. Though plasmids, like episomes, are small, they do carry genes that often confer a selective advantage on the cell. For example, colicin factors or plasmids excrete colicin, a protein that destroys other bacteria, and R plasmids make cells resistant to various antibiotics. Many plasmids can be passed from cell to cell just as the F factors are, so that a single cell resistant to an antibiotic can transmit that antibiotic-resistance plasmid through an entire population in a very short time.

Plasmids are part of a new genetic technology that has the potential to lead to new and important discoveries about gene structure and function. The new technology is called recombinant DNA technology. It was made possible by the discovery of DNA restriction enzymes that cleave DNA molecules at specific sites in the base sequence (Table 12.3). When a plasmid is treated with a restriction enzyme, it opens up to form a linear molecule with complementary single-stranded ends. DNA from other sources can be similarly treated and the fragments spliced together to form a new, hybrid, recombinant DNA molecule (Figure 12.31). This recombinant DNA molecule can be introduced, intact, into viable *E. coli* cells, and the genetic function of the inserted (nonplasmid) DNA can be studied.

Many interesting combinations of plasmid DNA–"foreign" DNA have been made and studied. Recombinant plasmids carrying *Drosophila* DNA or *Xenopus* (African clawed toad) DNA have been inserted into *E. coli* cells. Monkey DNA has also been used in recombinant DNA studies. In an interesting experiment, globin mRNA from humans can be isolated, converted into double-stranded DNA using reverse transcriptase (see Chapter 11), then spliced into a bacterial plasmid, inserted into *E. coli*, and studied. The possibilities for inserting eukaryote genes into the easily studied *E. coli* cell are enormous. And the technology is just beginning to open up these exciting possibilities that have before eluded investigators.

But recombinant DNA is also potentially hazardous. Tumor virus genes incorporated into a recombinant plasmid could spread cancer through a population if the host bacterium were a natural inhabitant of the human body. The complexities of this new genetic technology will be further explored in the last chapter.

Table 12.3
DNA restriction enzymes, their source, and the base sequence they cleave

Enzyme	Bacterial source	Sequence ($\downarrow$ indicates cleavage site)
		$5'$ $\qquad\qquad$ $3'$
Eco RI	E. coli	T A G$\downarrow$p A A T T C A T
Hind III	Haemophilus influenzae	A$\downarrow$p A G C T T
Hae II	Haemophilus aegyptius	Pu^{a}G C G C$\downarrow$p Py^a
Hpa II	Haemophilus parainfluenzae	C$\downarrow$p C G G
Hha I	Haemophilus haemolyticus	G C G$\downarrow$p C
Bam HI	Bacillus amyloliquefaciens	G$\downarrow$p G A T C C

^{a}Pu = purine, Py = pyrimidine.

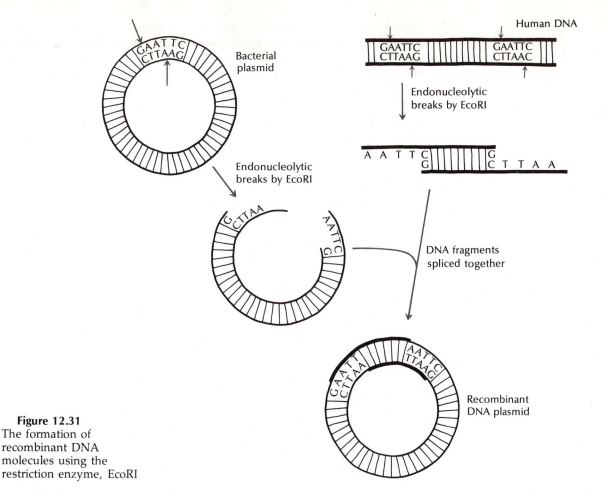

Figure 12.31
The formation of
recombinant DNA
molecules using the
restriction enzyme, EcoRI

Overview

A variety of schemes have evolved in bacteria that function to recombine the genetic material. These schemes are considerably more varied than recombination in sexually reproducing organisms, but this is to be expected, because bacteria lack a true sexual cycle. What is important to keep in mind is that, though the means may vary, recombination of the genetic material is of such evolutionary importance that different means have evolved for achieving it in organisms ranging from viruses to humans.

Transformation is one mechanism by which bacteria can undergo genetic recombination. It is the transfer of DNA fragments from one bacterial strain to another. DNA fragments from a donor strain are transported into a physiologically competent recipient cell where, as single-stranded DNA, they recombine with the recipient cell's DNA to produce a recombinant. Studies have shown that some genes are cotransformed, and the frequency of cotransformation is an indication of the map units separating them. Thus transformation has been useful for mapping genes in bacterial species refractive to other forms of genetic analysis.

Conjugation is another mechanism by which bacteria can undergo genetic recombination. It is a parasexual process in *E. coli* that involves the fusion of two cells (F^+ or Hfr with F^-, usually) and the unidirectional transfer of genetic material from one cell (F^+ or Hfr) to the other (F^-). An episome called a *sex factor* or *F factor* distinguishes F^+ and Hfr from F^- cells: The former two have the F factor, and the latter does not. If the F factor is integrated into the host cell's chromosome, then during conjugation it facilitates the transfer of chromosomal genes into recipient F^- cells. A donor cell with an integrated F factor is designated Hfr. If the F factor is not integrated, it is the only thing transferred during conjugation. A donor cell with an unintegrated F factor is designated F^+.

Conjugation mapping is done by measuring the length of time it takes for a donor gene to be transferred from an Hfr cell to a recipient cell. Map units in this case are expressed as time intervals. Occasionally an integrated F factor is released from the chromosome, creating an F^+ cell, and in the process taking some bacterial chromosome genes with it. This new F factor is called F', because it carries genes in addition to its F genes. During conjugation, bacterial genes incorporated into F' are transferred to recipient cells with a high frequency, but the other chromosomal genes are not transferred at all. This F'-mediated gene transfer is called *sexduction*. Conjugation appears to involve recombination by single-stranded DNA from the donor Hfr cell with double-stranded recipient (F^-) DNA.

Transduction is a third scheme found in bacteria that results in recombination. It involves the transfer of genetic material from one bacterium to another using a phage as a vector. There are two types of phage-mediated transduction: generalized and specialized. In generalized transduction, a phage infects a cell, but during phage reproduction, a piece of bacterial DNA is incorporated into a phage protein coat. This piece of DNA is carried by the phage to a recipient cell where it is injected. Once injected, it can recombine with the recipient cell's DNA. Generalized transduction can be used to map genes by determining the frequencies with which genes are cotransduced.

Specialized transduction involves lysogeny and is similar in many ways to F and F' formation in conjugation. In this case, a temperate phage such as λ infects a cell, but instead of lysing the cell, the phage is repressed, and the phage DNA (prophage) becomes integrated into a specific region of the bacterial chromosome by a process similar to Hfr formation. When the prophage is induced to enter a lytic cycle, it occasionally takes bacterial DNA with it when it leaves the bacterial chromosome, a process similar to F' formation. When this phage DNA–bacterial

DNA is injected into a recipient cell, recombination can occur, though it is usually recombination by addition, not replacement.

In generalized transduction, any bacterial DNA piece can be incorporated into the phage protein coat. In specialized transduction, only those bacterial genes closely linked to the specified site of the prophage integration can be incorporated into the phage protein coat (therefore it is specialized). A generalized transducing phage has no phage DNA; a specialized transducing phage has a hybrid phage–bacterium DNA molecule.

Bacterial plasmids differ from episomes in that they are not integratable into the bacterial chromosome. But with specific restriction enzymes, hybrid or recombinant plasmids can be constructed using bacterial DNA and DNA from such sources as *Drosophila*, toads, or humans. When the recombinant plasmids are reinserted into host *E. coli* cells, the gene function of the "foreign" DNA can be conveniently studied. But this new technology introduces many potential hazards as well as benefits.

Questions and problems

12.1 In Figure 12.1, we presented a correlation between the concentration of transforming DNA and the number of transformants obtained. It was suggested that the curve was the result of saturation of active sites. Suggest an alternative basis for the curve and design an experiment to test your proposal, including clear predictions.

12.2 If an *E. coli* cell has multiple copies of its chromosome, it is still considered haploid. Can this be justified?

12.3 Design an experimental procedure by which you could differentiate between F+, F−, Hfr, and F' *E. coli* cells.

12.4 An F' cell is in many ways analogous to a diploid cell. Describe the similarities and differences, and argue *against* inclusion of an F' cell and an exconjugant F− (containing Hfr donor DNA) cell in the class of diploid cells.

12.5 In the bacteria and viruses, circular DNA molecules are commonly encountered. Discuss the significance of circular DNA.

12.6 How would you prove that a bacterial chromosome is physically circular? What genetic tests would you use to demonstrate circularity?

12.7 A new species of bacteria is discovered, and two different auxotrophic strains are isolated. When the two auxotrophic strains are mixed together, prototrophic cells are recovered. Among the possible mechanisms to account for this, you consider conjugation, transduction, transformation, and mutation. Design a study that would distinguish among these possibilities.

12.8 What would you expect the relationship to be between the size of a recipient cell's genome and the efficiency of transformation per gene per unit of DNA? Justify your expectation.

12.9 If an *E. coli* cell does not lyse when placed in a medium containing λ phage, can we conclude that the *E. coli* cell is lysogenic?

12.10 Compare the genome of *E. coli* with the genome of *Drosophila*, and do so from both a functional and a structural perspective.

12.11 Two *Bacillus* genes, *a* and *b*, are studied in a transformation mapping experiment. The data are recorded in the following table.

Experiment	Donor DNA	Recipient DNA	Transformant classes		Number
			a	*b*	
A	$a^+ b^-$	$a^- b^-$	+	−	232
	$a^- b^+$		−	+	341
			+	+	7
B	$a^+ b^+$	$a^- b^-$	+	−	130
			−	+	96
			+	+	247

(a) Are the two genes linked, according to our definition of *linkage* in transformation analysis?

(b) If they are linked, how far apart are they?

12.12 A three-point cross is done in *Bacillus*, using transformation, and the following data are collected.

Donor: $a^+ b^+ c^+$
Recipient: $a^- b^- c^-$

	(1)	(2)	(3)	(4)	(5)	(6)	(7)
a:	−	−	+	−	+	+	+
b:	−	+	−	+	−	+	+
c:	+	−	−	+	+	−	+
	700	400	2600	3600	100	1200	12000

What is the sequence and distance between these three points?

12.13 In an $a^+ b^+ c^- d^+ \times a^- b^- c^+ d^-$ cross using conjugating *E. coli* cells, the $b^+ c^+$ genes were selected for among the recombinants, and the *a* and *d* alleles were not selected. When the $b^+ c^+$ recombinants were checked, most were $a^- d^-$.

(a) Which strain was the donor?

(b) What conclusions can you draw from this experiment?

12.14 In a conjugation experiment, you are confronted with two genes that lie very close to each other, so close that you cannot tell the sequence by time mapping. A colleague suggests a way to resolve the problem. Given the map

you screen for *a* to make sure that *b* and *c* are in the recipient. You set up a cross so that the donor is $+ + c$ and the recipient is $a b +$ in one experiment and another cross with the donor $a b +$ and the recipient $+ + c$. When you screen for the $+ + +$ recombinants and compare the frequencies of the $+ + +$ recombinants for both crosses, you can tell if the gene sequence is *abc* or *acb*. How can you tell what the gene sequence is?

12.15 In 1965, Ikeda and Tomizawa (*J. Mol. Biol.* 14:85) studied the origin of the bacterial genetic markers carried by the general transducing phage P1. They were considering two models of generalized transducing phage (see Figure 12.20). They grew thymine-requiring *E. coli* cells (*thy⁻*) in three different ways.

(a) *E. coli* was grown in thymine, then infected with P1, and progeny P1 were collected.

(b) *E. coli* was grown on BU, a thymine analog that serves as a density marker. The cells were then washed, infected with P1, and placed in a thymine medium. Progeny P1 were collected.

(c) *E. coli* was grown on BU medium and then infected with P1. Progeny P1 were collected.

The progeny from each experiment were centrifuged and analyzed for being either nontransducing (infective) or transducing (noninfective). The results are as follows. From these data, what do you conclude about the two models, and why?

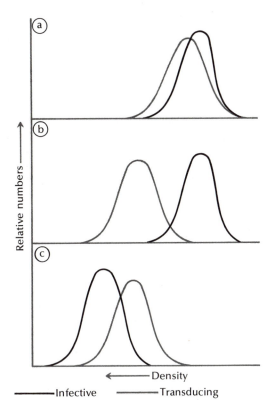

References

Adelberg, E. A. 1966. *Papers on Bacterial Genetics.* Boston, Little, Brown.

Adelberg, E. A., and S. N. Burns. 1959. A variant sex factor in *E. coli. Genetics* 44:497.

Anderson, T. F., E. L. Wollman, and F. Jacob. 1957. Sur les processus de conjugaison et de recombinaison chez *E. coli:* III. Aspects morphologiques en microscopie électronique. *Ann. Instit. Pasteur* 93:450.

Baer, A. 1977. *Heredity and Society,* New York, Macmillan.

Beadle, G. W., and E. L. Tatum. 1941. Genetic control of biochemical reactions in *Neurospora. Proc. Natl. Acad. Sci. U.S.* 27:499.

Borek, E., and A. Ryan. 1973. Lysogenic induction. *Prog. Nuc. Acid Res. Molec. Biol.* 13:249–300.

Braun, W. 1965. *Bacterial Genetics.* Philadelphia, Saunders.

Bukhari, A. I., J. A. Shapiro, and S. Adhya (ed.) 1977. *DNA Insertion Elements, Plasmids, and Episomes.* Cold Spring Harbor, N. Y., Cold Spring Harbor Laboratory.

Campbell, A. 1962. Episomes. *Adv. Genetics* 11:101.

Campbell, A. 1969. *Episomes.* New York: Harper and Row.

Clark, A. J. 1973. Recombination deficient mutants of *E. coli* and other bacteria. *Ann. Rev. Genetics* 7:67–86.

Corwin, H. O., and J. B. Jenkins. 1976. *Conceptual Foundations of Genetics: Selected Readings.* Boston, Houghton Mifflin.

Curtiss, R. 1969. Bacterial conjugation. *Ann. Rev. Microbiol.* 23:69–136.

Curtiss, R. 1976. Genetic manipulation of microorganisms: Potential benefits and biohazards. *Ann. Rev. Microbiology* 30:507–534.

Dubnau, D. 1976. Genetic transformation of *Bacillus subtilus:* A review with emphasis on the recombination mechanism. In *Microbiology–1976.* D. Schlessinger (ed.). Washington, D.C., American Society for Microbiology.

Dubnau, D., I. S. Goldthwaite, and J. Marmur. 1967. Genetic mapping in *Bacillus subtilis. J. Molec. Biol.* 27:163.

Eisenstark, A. 1977. Genetic recombination in bacteria. *Ann. Rev. Genetics* 11:369–396.

Fox, M. S., and M. K. Allen. 1964. On the mechanism of DNA integration in pneumococcal transformation. *Proc. Natl. Acad. Sci. U.S.* 48:1043.

Fredericq, P. 1963. On the nature of colicinogenic factors: A review. *J. Theor. Biol.* 4:159.

Goodgal, S. H. 1961. Studies on transformation of *Hemophilis influenzae:* IV. Linked and unlinked transformations. *J. Gen. Physiol.* 45:205.

Hayes, W. 1968. *The Genetics of Bacteria and Their Viruses.* New York, Wiley.

Helinski, D. R. 1973. Plasmid determined resistance to antibiotics: Molecular properties of R factors. *Ann. Rev. Microbiol.* 27:437–470.

Hotchkiss, R. O., and M. Gabor. 1970. Bacterial transformation with special reference to recombination processes. *Ann. Rev. Genetics* 4:183–224.

Kotewicz, M., S. Chung, Y. Takeda, and H. Echols. 1977. Characterization of the integration protein of bacteriophage λ as a site-specific DNA-binding protein. *Proc. Natl. Acad. Sci. U.S.* 74:1511–1515.

Lederberg, E. M. 1951. Lysogenicity in *E. coli* K-12. *Genetics* 36:560.

Lederberg, J. 1956. Genetic transduction. *Amer. Scientist* 44:264.

Lederberg, J., E. M. Lederberg, N. D. Zinder, and E. R. Lively. 1951. Recombination analysis of bacterial heredity. *Cold Spring Harbor Symp. Quant. Biol.* 16:413.

Lederberg, J., and E. L. Tatum. 1946. Gene recombination in *E. coli. Nature* 158:558.

Lederberg, J., and E. L. Tatum. 1946. Novel genotypes in mixed cultures of biochemical mutants of bacteria. *Cold Spring Harbor Symp. Quant. Biol.* 11:113.

Low, K. B. 1972. *Escherichia coli* K-12 F-prime factors, old and new. *Bact. Rev.* 36:587–607.

Lwoff, A. 1959. Bacteriophage as a model of host–virus relationship. In F. M. Burnet and W. M. Stanley (eds.), *The Viruses.* New York, Academic.

Nester, E. W., M. Shafer, and J. Lederberg. 1963. Gene linkage in DNA transfer: A cluster of genes concerned with aromatic biosynthesis in *Bacillus subtilus. Genetics* 48:529–551.

Notani, N. K., and J. K. Setlow. 1974. Mechanism of bacterial transformation and transfection. *Prog. Nuc. Acid Res. Molec. Biol.* 14:39–100.

Oishi, M., and S. D. Cosloy. 1974. Specialized transformation in *E. coli* K12. *Nature* 248:112–116.

Science, 8 April 1977 (Vol. 196, No. 4286). A collection of articles and essays pertinent to recombinant DNA.

Shapiro, J. A. 1977. DNA insertion elements and the evolution of chromosome primary structure. *Trends in Biochemical Sci.* 2:176–180.

Sinsheimer, R. L. 1977. Recombinant DNA. *Ann. Rev. Biochemistry* 46:415–438.

Stent, G. S. 1971. *Molecular Genetics: An Introductory Narrative.* San Francisco, Freeman.

Susman, M. 1970. General bacterial genetics. *Ann. Rev. Genetics* 4:135–176.

Taylor, A. L., and C. D. Trotter. 1972. Linkage map of *E. coli* K12. *Bact. Rev.* 36:504–524.

Taylor, H. E. 1949. Transformations réciproques des formes R et ER chez le pneumocoque. *Compt. Rend. Acad. Sci.* 228:1258.

Taylor, J. H. 1965. *Selected Papers on Molecular Genetics.* New York, Academic.

Weisberg, R., and S. Adhya. 1977. Illegitimate recombination in bacteria and bacteriophage. *Ann. Rev. Genetics* 11:451–473.

Willetts, N. 1972. The genetics of transmissible plasmids. *Ann. Rev. Genetics* 6:257–268.

Wollman, E. L., and F. Jacob. 1955. Sur le méchanisme du transfer de matériel génétique au cours de la recombination chez *E. coli* K-12. *Compt. Rend. Acad. Sci.* 24:24449.

Wollman, E. L., and F. Jacob. 1957. Sur les processus de conjugaison et de recombinaison chez *E. coli:* II. La localisation chromosomique du prophage λ et les conséquences génétiques de l'induction zygotique. *Ann. Instit. Pasteur* 93:323.

Thirteen

Developmental genetics

Nucleocytoplasmic interaction
Equivalence of nuclei during early development
Regional differentiation in eggs and early embryos

Patterns of protein and RNA synthesis
Evidence for differential gene activity
Hormones as regulators of gene activity
Patterns of RNA synthesis during early development

Phage T4 morphogenesis: A model of sequential gene action

Regulation of gene activity in prokaryotes
Development of the operon concept
Induction and repression
The promoter
Cyclic AMP and the lac *operon*
The histidine operon
The arabinose operon
Control of lysogeny in lambda
Regulation of rRNA and tRNA synthesis

Regulation of gene activity in eukaryotes
Cell determination versus cell differentiation
Gene regulation: Histone and nonhistone proteins
Gene amplification
mRNA in eukaryotes
Repeated DNA and the structure and organization of
    eukaryote chromosomes
A model of eukaryotic gene regulation
The genetics of aging

Overview

Questions and problems

References

Our study of genetics would be totally inadequate without an investigation of the role of genes in the course of an organism's development and their function in various environments. Problems of gene activity during development can be explored at three levels: in the virus, in the single-celled organism, and in the multicellular organism.

In most multicellular organisms, a single cell, the zygote, gives rise to an array of daughter cells that exhibit wide functional diversity—nerve cells, muscle cells, epidermal cells, secretory cells, ciliated cells, and others. All derive from a progression of mitotic divisions of the original zygote; hence, for the most part, they contain the same genetic information. Because this is so, the observed differences among cell types are due either to differential gene activity—that is, certain genes must function in some cells but not in others—or to the loss of some genetic information in the course of cell division.

By contrast, a single-celled organism such as *Escherichia coli* contains about 3000 different types of protein molecules, which engage in a wide range of structural and enzymatic functions. Many proteins vary greatly in their respective concentrations according to the cellular environment. So, given a specific set of environmental conditions, some genes are more active than others; indeed, some may not be active at all.

Viruses, the simplest of all living systems, depend on living cells for their reproduction. Yet they too exhibit a precise and sequential order of gene activity in the course of their development. Some viral genes function early, while others function late in the viral life cycle.

In a fundamental sense, development does not differ among viral, unicellular, and multicellular organisms. In each, we are concerned with the regulation of gene activity primarily through the interaction between genes and their surrounding environment. Since information is more complete for *E. coli* than for any other organism, our molecular models of gene regulation have developed largely from research on this bacterium. This is not to say that regulation in *E. coli* is completely understood or that it is the same as in a eukaryote. As a matter of fact, recent evidence suggests many fundamental differences, some of which we shall explore in detail.

Nucleocytoplasmic interaction

In the period before the rediscovery of Mendel's work in 1900, a number of ideas were advanced to explain how a single cell, the zygote, differentiated into cells with new and different functions. Perhaps the most widely accepted concept was one devised by Weismann, who in 1892 suggested that as a cell divides, its hereditary determinants (genes) assort into different cells in such a way that each cell contains a different complement of genetic information. That this is not the case, at least in the early embryonic stages, was shown by H. Driesch, H. Spemann, R. Briggs, T. J. King, and others.

Equivalence of nuclei during early development

In 1893, Driesch used sea urchins to demonstrate that even if a nucleus is in a cytoplasm other than the one it would normally occupy, development of the resulting embryo still proceeds normally. This suggests that the nuclei in the early embryo (eight-cell stage) are equivalent and not differentiated (Figure 13.1). Two vertical cleavages divide the sea urchin egg into four cells; the third cleavage plane is normally horizontal, at a right angle to the first two planes. Between two slides, Driesch gently compressed the four-cell embryo, which had the effect of making the third cleavage vertical too. This altered plane of cleavage at the third division cuts the cytoplasm differently, causing a different cytoplasm to contact the nuclei. In other words, the nucleus is in a different cytoplasm than it would normally be in if cleavage had not been modified. If not tampered with, the fourth cleavage corrects the missegregation of cytoplasm by cutting all eight cells equatorially, which restores the cells to their correct cytoplasmic lineage, but there is still a different nucleus (in terms of normal nuclear lineage) in each cell. That the embryo develops normally suggests that nuclear reorientation fails to disrupt development and, further, that the nuclei—at the eight-cell stage at least—are equivalent.

In 1903, Spemann demonstrated that a nucleus from a 16-cell newt embryo retains the potential for stimulating full and normal development from a single cell (Figure 13.2). Spemann constricted a fertilized egg in such a way that one part contained the nucleus and some cytoplasm and the other part contained only cytoplasm. A small cytoplasmic bridge connected the two halves. At the 16-cell stage, a nucleus was allowed to pass into the nonnucleated segment. The result was the formation of two completely normal embryos. This would not be expected to occur if each nucleus did not contain the full amount of genetic information.

More substantial proof for nuclear equivalence at later stages of embryogenesis was presented by Briggs and King in 1952. They showed that nuclei from more differentiated cells could support the development of normal embryos from single enucleated eggs (Figure 13.3). First they pricked a frog egg (*Rana pipiens*) with a needle, inducing it to begin protein

Figure 13.1
Driesch experiment: (a) normal cleavage; (b) the result of compressing the egg during the first three cleavages

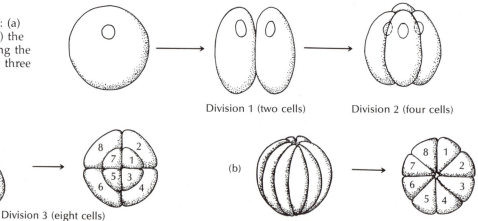

Division 1 (two cells)   Division 2 (four cells)

(a)

Division 3 (eight cells)

(b)

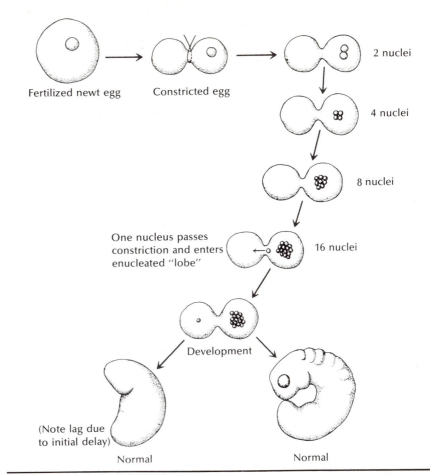

Figure 13.2
Spemann's experiment
demonstrating that a
nucleus from the 16-cell
stage can support normal
development in the newt

Fertilized newt egg   Constricted egg

2 nuclei

4 nuclei

8 nuclei

One nucleus passes
constriction and enters
enucleated "lobe"   16 nuclei

Development

(Note lag due
to initial delay)

Normal   Normal

Figure 13.3
Briggs–King experiment
demonstrating that nuclei
from differentiated cells
can support the
development of normal
embryos from single
enucleated eggs.
[Adapted from R. Briggs
and T. G. King, *Biological
Specificity and Growth*
(Elmer G. Butler, ed.),
copyright 1955, by
permission of Princeton
University Press]

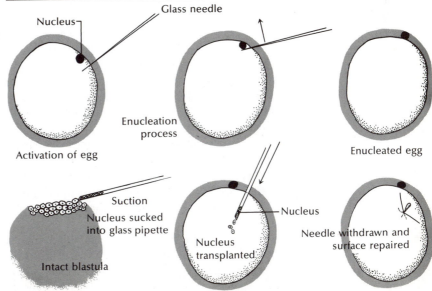

Glass needle

Nucleus

Enucleation
process

Activation of egg   Enucleated egg

Suction

Nucleus sucked
into glass pipette

Intact blastula

Nucleus

Nucleus
transplanted

Needle withdrawn and
surface repaired

synthesis. Then they removed the nucleus. A nucleus from a donor cell was inserted into the enucleated recipient egg, and the development of the embryo was followed. Their studies showed that transplanted nuclei from cells advanced even as far as the late blastula stage allowed the zygote to develop into normal embryos, but that transplanted nuclei from more advanced embryonic cells progressively lost their potential for supporting such development. This suggests that as cell specialization progresses, changes take place in the nucleus that may prevent it from supporting complete development. J. B. Gurdon, however, has demonstrated that even nuclei from cells as specialized as epithelial cells from the gut of the tadpole stage of the African clawed toad (*Xenopus laevis*) can support complete development.

These experiments of Driesch, Spemann, Briggs, King, and Gurdon argue against models such as Weismann's, in which genetic determinants assort during cell division. Rather they suggest that nuclei are genetically equivalent during early embryogenesis, but that changes must occur in later stages of development, because the potential of a nucleus for supporting complete development diminishes.

A number of conclusions can be drawn from the nuclear transplantation experiments. For example, it could be argued that as cells become more highly differentiated, they become more sensitive to the transplantation operation, making for a lowered probability of success. Or it could be argued that actual nuclear changes occur as development proceeds. Present evidence supports both arguments. Removal of nuclei from differentiated cells is more traumatic than their removal from undifferentiated cells, but nuclei also undergo what frequently appear to be irreversible changes during the process of differentiation.

Regional differentiation in eggs and early embryos

Although there is considerable evidence to support the idea that nuclei are equivalent, at least in early developmental stages, the same is not true for the cytoplasm in different regions of the egg. If regional cytoplasmic differences exist, a model of cell differentiation can be developed that is based on genetically equivalent nuclei in cytoplasmically different environments. The model asserts that the environment differentially affects gene activity. But we are getting ahead of our story; we must first establish the basis for our statement that an egg shows regional cytoplasmic differentiation.

The most extreme example of cytoplasmic differences is found among the annelids and mollusks. In organisms from these phyla, egg cytoplasm is very highly structured. Indeed, after the first cleavage, the cytoplasmic contents of the two cells are so unlike that when the two-cell stage is separated into blastomeres, they often can't develop into normal embryos.

The phylum Echinodermata, which includes the sea urchin, exhibits no extreme cytoplasmic differentiation in the egg, but it does show some. It was originally proposed that the four-cell embryo of a sea urchin could be disaggregated into single blastomeres and that each blastomere would develop into smaller but normal embryos (Figure 13.4). But it now

 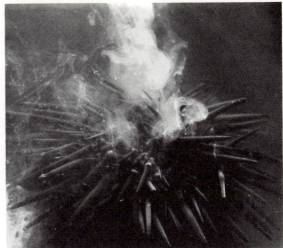

Figure 13.4
Photographs showing
emission of sea urchin
eggs (left) and sperm
(right). The diagram
illustrates disaggregation
of a normal eight-cell
embryo and subsequent
development of isolated
blastomeres. (Photos by
Walter Dawn)

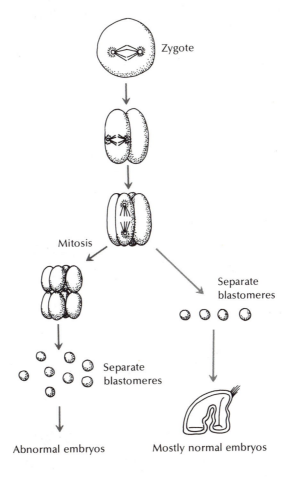

Zygote

Mitosis

Separate
blastomeres

Separate
blastomeres

Abnormal embryos

Mostly normal embryos

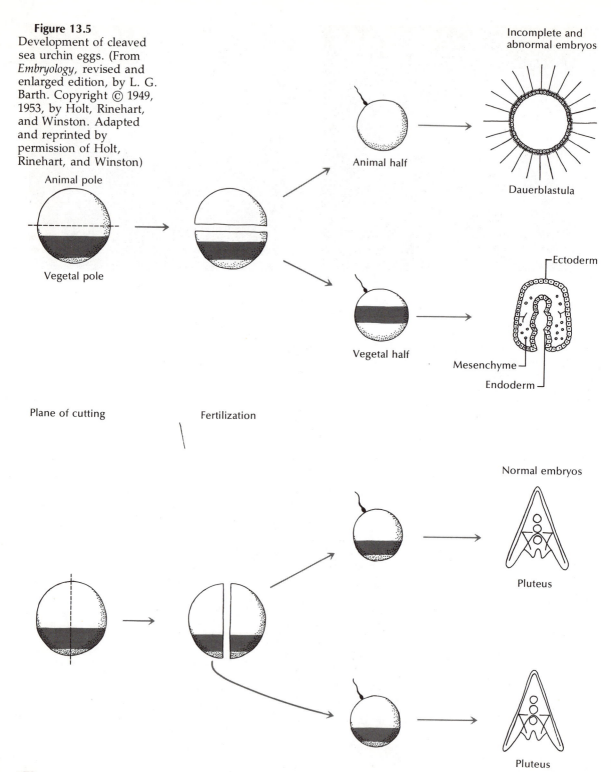

Figure 13.5
Development of cleaved sea urchin eggs. (From *Embryology*, revised and enlarged edition, by L. G. Barth. Copyright © 1949, 1953, by Holt, Rinehart, and Winston. Adapted and reprinted by permission of Holt, Rinehart, and Winston)

Animal pole

Vegetal pole

Plane of cutting

Fertilization

Incomplete and abnormal embryos

Animal half

Dauerblastula

Vegetal half

Ectoderm

Mesenchyme

Endoderm

Normal embryos

Pluteus

Pluteus

appears that when the two or four cells are isolated and observed *separately*, some fail to give rise to complete embryos. Some embryos are decidedly abnormal, suggesting that the cells in early embryos are perhaps not all equivalent and that the sea urchin egg may be regionally more differentiated than was once thought. The regional differentiation in early echinoderm embryos is seen when sea urchin eggs are cut in half along different planes, fertilized, and allowed to develop (Figure 13.5). When a *vertical* separation is followed by fertilization, each half gives rise to a normal embryo. With a horizontal separation between the animal and vegetal poles, incomplete and abnormal development results. Thus the cytoplasms from these two regions of the egg have dissimilar properties.

The normal early development of a sea urchin embryo (Figure 13.4) is, as we noted, characterized by two vertical cleavages followed by a series of horizontal cleavages. An eight-cell embryo, since it results from a horizontal cleavage, should, if the cells are disaggregated, give us the same abnormal embryos as would an egg cut horizontally in half—and this is indeed the case. These observations support the concept of regional cytoplasmic differences in the egg and early embryo of the sea urchin. Cytoplasm from *both* the animal and vegetal poles is required for the normal development of a zygote. If either part is omitted, development fails.

Figure 13.6
Cleavage in an amphibian egg, showing the importance of the gray crescent. (From *Embryology*, revised and enlarged edition, by L. G. Barth. Copyright © 1949, 1953, by Holt, Rinehart, and Winston. Adapted and reprinted by permission of Holt, Rinehart, and Winston) See next page for text discussion.

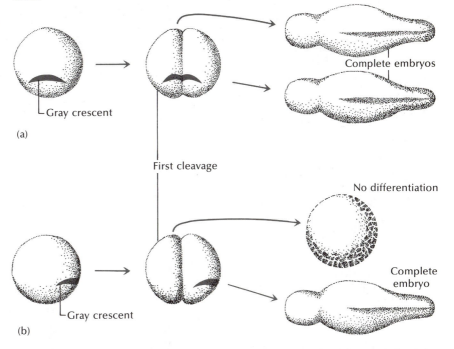

(a) Gray crescent

First cleavage

Complete embryos

(b) Gray crescent

No differentiation

Complete embryo

(a) When the egg divides, half the gray crescent passes into each of the two cells. If the cells are separated from each other, each will form a complete embryo. (b) A side view of the gray crescent. In the cleavage shown here, one cell contains all of the gray crescent, the other none of it. If the cells are separated from each other, only the one with the gray crescent will develop into a complete embryo.

In amphibians, a particular region of the zygote cytoplasm called the *gray crescent* (Figure 13.6) is crucial to future development. This region forms opposite the point of entry of the sperm. If the first cleavage bisects the gray crescent and if the two cells are separated, two normal embryos develop. But if at the first cleavage only one cell gets the gray crescent, that cell develops into a normal embryo and the other does not.

We conclude from these observations that cellular development requires interaction between nucleus and cytoplasm; that is, the activity of nuclear genes during development is limited by the properties of the cytoplasm. It should be emphasized, however, that the phenotype can be *regulated* but not *determined* by the cytoplasm. The genes determine a cell's potential; the cytoplasm determines whether that potential will be reached. In 1943, Hammerling performed a classic experiment, demonstrating that point with the unicellular alga *Acetabularia* (Figure 13.7).

Figure 13.7
Acetabularia life cycle
(From Brachet, *Biochemical Cytology,* Academic Press, 1957)

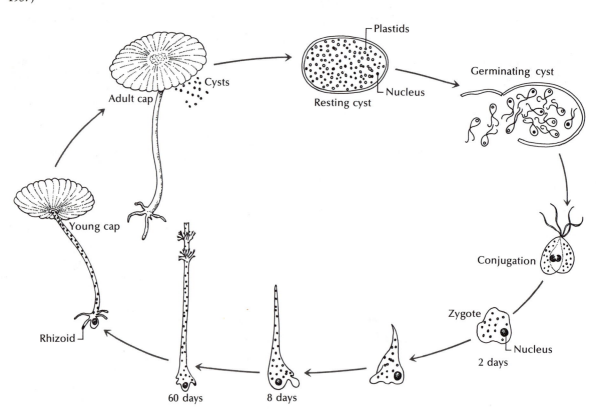

Figure 13.8
The Hammerling experiment with *Acetabularia* (based on Hammerling, *Z. Abstg. Vererb.* 81:114–180, 1943)

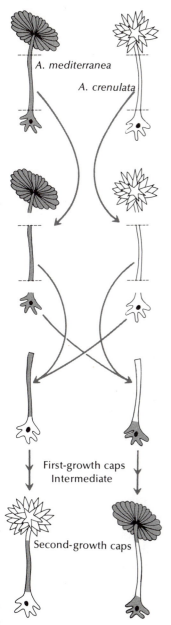

A. mediterranea

A. crenulata

First-growth caps
Intermediate

Second-growth caps

The adult *Acetabularia* is composed of a rhizoidal holdfast, a stalk, and a cap that is characteristic for each species. The nucleus of this alga is located in a lobe of the holdfast. If the rhizoid and cap are removed, leaving only the enucleate stalk, a new cap will be regenerated; but this will not occur a second time if *that* new cap is removed. After removal of the nucleus, gene products remain in the cytoplasm of the stalk and a new cap can be regenerated; but the gene products present in the stalk are sufficient to regenerate only one cap.

If a stalk of *A. crenulata* from which the cap, rhizoidal holdfast, and nucleus have been removed is grafted onto a rhizoid base and nucleus of *A. mediterranea,* the regenerated cap will be intermediate between *A. crenulata* and *A. mediterranea* (Figure 13.8). If this intermediate cap is removed, a second cap is regenerated that is identical to *A. mediterranea.* Moreover, in the reciprocal experiment, in which an *A. mediterranea* stalk is grafted onto an *A. crenulata* rhizoid base and nucleus, the second regenerated cap is that of *A. crenulata.*

We can interpret these results as indicating that after removal of the nucleus, certain of its gene products remain in the stalk and mingle with the gene products of the transplanted nucleus. The first cap shows the interaction of the gene products of *two nuclei;* but the second cap reflects gene products of the transplanted nucleus only, the gene products of the original nucleus having been used up. Thus the nucleus determines the phenotype.

This experiment brings out a further important point: The nucleus controls the synthesis of products at points far removed from its own location. Today we would interpret Hammerling's results in terms of mRNA being transported from nucleus to cytoplasm.

Patterns of protein and RNA synthesis

Our discussion thus far has established the probability that nuclei, at least in the early developmental stages, are equivalent. But changes also occur in nuclei as they become more highly differentiated, so they are often unable to support a complete developmental sequence from zygote to adult. Cell differentiation is characterized as resulting from differential gene activity, which is stimulated by gene–cytoplasmic interactions. Our concern in this section is to examine the evidence for gene activity and to establish that different genes are active at different stages of their development.

In the African clawed toad *Xenopus* (Figure 13.9), the biosynthesis and accumulation of rRNA occurs at a specific chromosomal area called the *nucleolar organizer* (NO) *region.* The accumulation of rRNA and protein at this genetic region is cytologically visible as the *nucleolus* (we have already discussed the presence of the nucleolus during cell division). In the course of development of *Xenopus*, rRNA is synthesized in the developing oocyte but not again until the gastrula stage. Until gastrulation, then, all the ribosomes necessary to protein synthesis are present in the unfertilized ovum.

Figure 13.9
Xenopus laevis (African
clawed toad) (Photo by
Walter Dawn)

If a nucleus from a postgastrulation cell (synthesizing rRNA) is placed in an enucleated egg (with no rRNA synthesis occurring), will rRNA synthesis cease as a result of cytoplasmic inhibitory factors, or will it continue? In 1965, Gurdon and Brown showed that in such an environment, the nucleus responds by ceasing rRNA synthesis (Figure 13.10). This demonstrated that the genes for rRNA are controlled by their cytoplasmic environment.

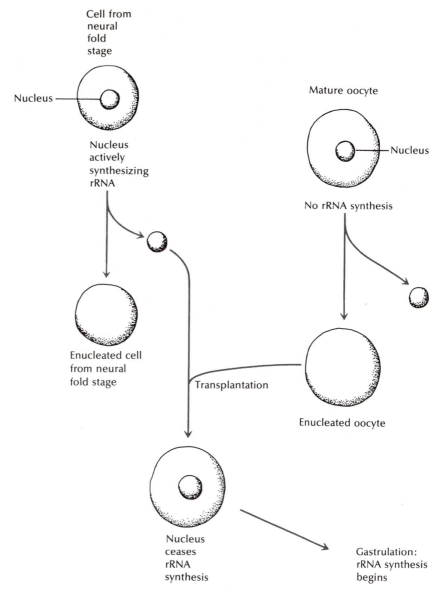

Cell from neural fold stage

Nucleus

Nucleus actively synthesizing rRNA

Mature oocyte

Nucleus

No rRNA synthesis

Enucleated cell from neural fold stage

Transplantation

Enucleated oocyte

Nucleus ceases rRNA synthesis

Gastrulation: rRNA synthesis begins

Figure 13.10
Gurdon–Brown experiment

Another example of differential gene activity under environmental control concerns the *in vitro* culture of embryonic chick muscle cells. If these cells are grown on a medium containing collagen (a proteinaceous component of supportive connective tissue), they synthesize the contractile muscle proteins actin and myosin. If these same cells are grown on the identical medium minus the collagen, however, they do not synthesize the contractile proteins and assume an undifferentiated fibroblastic morphology. Collagen thus induces the activity of the contractile muscle protein genes.

A final example of differential gene activity is the synthesis of hemoglobin during the course of human development. The hemoglobin molecule is composed of four polypeptide chains. In the developing fetus, the molecule is composed of two α chains and two γ chains ($\alpha_2\gamma_2$). In the adult, the hemoglobin is composed largely of two α chains and two β chains ($\alpha_2\beta_2$), though some of the adult hemoglobin is composed of two α and two δ chains. The α, β, and δ polypeptide chains are encoded by different genes, and each complete hemoglobin molecule has properties that enable it to function efficiently under either fetal or adult conditions. Thus the α gene is active throughout the life of the individual, the γ gene functions only during fetal development, and the β and δ genes, which are inactive during fetal development, are active during postnatal development.

These examples support the assertion that the synthesis of cell-specific macromolecules such as rRNA, actin, myosin, and hemoglobin is variable and dependent on the cellular milieu.

Hormones as regulators of gene activity

Hormones have often been said to be regulators of gene activity. There are many indications that hormones promote the biosynthesis of protein. For example, the administration of glucocorticoid hormones has been observed to stimulate the liver tissue to synthesize a wide range of different enzymes. Recent studies show that the hormone is stimulating mRNA transcription followed by translation, supporting the contention that gene activity can be controlled by supplying appropriate environmental stimuli.

Among the most illuminating research on the role of hormones in activating genes is U. Clever's work with *Diptera*. The salivary glands of this insect group contain large, banded, polytene chromosomes. The banding pattern in these chromosomes can easily be distinguished with a microscope, and it can be seen that the patterns change as the physiological environment changes. Some bands "puff" at a particular developmental stage, with other bands showing *puffs* (or *balbiani rings* as they are sometimes called) at other stages. Puffing is indicative of gene activity, and Figure 13.11 shows examples of chromosome puffs in *Sciara* and *Chironomus*.

These puffs are the result of the uncoiling of the highly condensed

Figure 13.11
Chromosome puffs in
Sciara (courtesy of
Hewson Swift, Whitman
Laboratory, Chicago) and
(below) *Chironomus*
(Courtesy of Dr. Claus
Pelling)

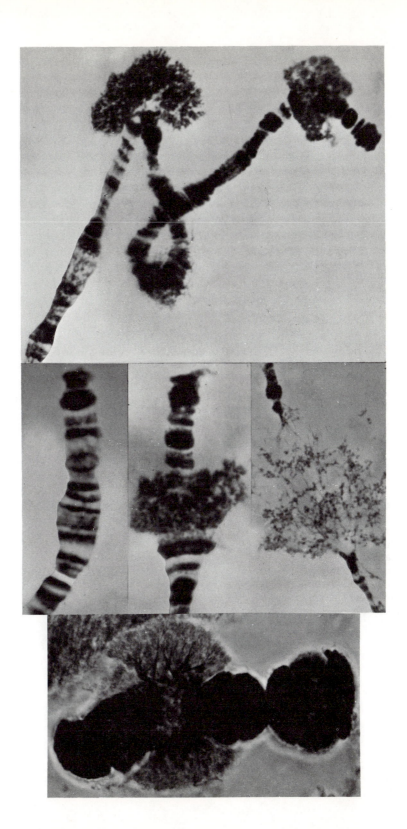

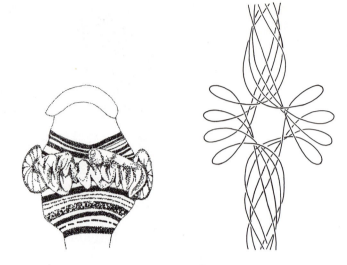

Figure 13.12
Diagram of a
chromosome puff

chromatin material (Figure 13.12). In 1966, Clever and Romball showed that injection of the molting hormone *ecdysone* into *Chironomus* larvae brought about a specific sequence of chromosome puffs. If these puffs are sites of gene activity, they should also be sites of active RNA synthesis. This has indeed been established using ³H-uridine and autoradiography (Figure 13.13). The chromosome puffs have been shown to be engaged in the rapid synthesis of RNA after exposure to ecdysone. *Actinomycin D,* an antibiotic that inhibits RNA synthesis and thus transcription, prevented the formation of puffs and blocked the uptake of ³H-uridine.

Figure 13.13
RNA synthesis in
chromosome puffs (the
synthesis of RNA from
the radioactive precursor
occurs mainly in the
heavily labeled bands)
(Courtesy of Dr. Claus
Pelling)

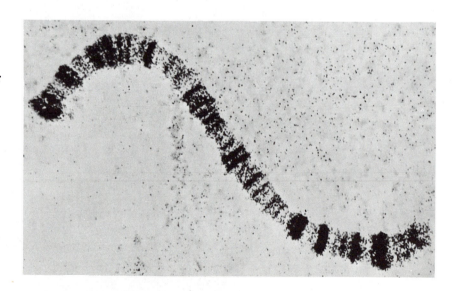

Cycloheximide, a drug that inhibits translation but not transcription of RNA, has no effect on puffs that form early but does prevent the appearance of puffs that form later. This suggests the highly coordinated series of reactions summarized in Figure 13.14. The actinomycin D inhibits any puff formation, because it effectively blocks RNA synthesis. Cycloheximide allows the first puffs to form, but if subsequent puffs depend on the gene products of the second or third puffs, inhibiting the formation of these gene products prevents further puffs from forming.

In short, there is ample experimental evidence to support the contention that environmental components such as hormones and drugs play a direct role in the activation of genes and, further, that the gene products so formed can influence future patterns of gene activity.

Figure 13.14
Model for the coordination of puff formation

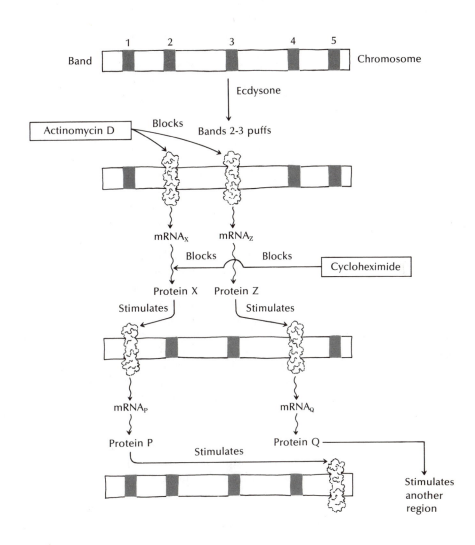

We have established that the synthesis of specific proteins is frequently controlled by the cellular environment, but this can mean two different levels of control. The transcription of mRNA can be regulated, and the translation of preexisting mRNA can be regulated. In our discussion of chromosome puffing, we saw that RNA synthesis was controlled by hormones and drugs, but that the synthesis of the contractile proteins actin and myosin and also the liver enzymes could be explained either by the initiation of mRNA synthesis or by the translation of preexisting mRNA molecules.

Originally, mRNA was defined by its physical properties in a living cell. It is heterogenous in size and has a very short half-life—approximately three or four minutes in prokaryotes—and is translated into polypeptides. The short half-life, while true of the prokaryotes, is not necessarily true of the eukaryotes. Eukaryotes sometimes contain mRNA that is considerably more stable. For our purposes, we can classify these long-lived mRNA species of eukaryotes either as *stable mRNA* (a species that has a long half-life, usually several hours, and is translated during much of its existence) or as *masked mRNA* (usually present in unfertilized eggs and early embryos, bound to protein, and incapable of being translated until activated).

In a number of laboratories, stable mRNA species have been verified as a means of controlling cell differentiation. These stable mRNA molecules are apparently essential to highly differentiated cells, such as erythrocytes, or red blood cells (RBC's). The RBC represents the extreme in differentiation, because it often has no nucleus but continues to synthesize hemoglobin. The absence of a nucleus dictates that the mRNA must be stable in order for the cell to continue to produce hemoglobin. Even in differentiated, nucleated cells, however, stable mRNA species are commonly found. For example, HeLa cells (a strain of cancer cells maintained *in vitro* and originally obtained from a patient named Henrietta Lacks) turn over 50 percent of the mRNA in three hours compared with three or four minutes in bacteria.

The rapid turnover of mRNA in bacteria is understandable, because the organisms must adjust very rapidly to any changes in the environment. And because they are unicellular, each cell must be self-sufficient. The more highly differentiated cells of multicellular organisms exist in a more stable environment and would be wasting energy if they synthesized and degraded mRNA at such a rate. It is tempting to speculate on this stable mRNA; perhaps to have stable, differentiated cells, the mRNA must first be stabilized. Possibly it is stabilized by an association with proteins. Some experiments indicate that masked mRNA in unfertilized eggs is complexed with protein, and a comparable situation may exist for stable mRNA. We do not yet know.

The discovery of masked mRNA established the fact that messages may be transcribed some time before they are translated. During oogenesis, there is active synthesis of RNA of all types; but on maturation of the egg, the synthesis of RNA may cease, not to begin again until later in development. So all the protein synthesized before gastrulation is translated from *preexisting* mRNA. Actinomycin D, an inhibitor of RNA synthesis, has no effect on development until gastrulation, again arguing that active RNA synthesis is unimportant during early embryo development.

For sea urchin eggs, it appears that mRNA is masked by some proteinaceous material and that the mask is later removed to allow translation. There is some speculation that the entire mRNA–ribosome complex is held inactive until needed (this view is losing support). A hypothetical scheme for the formation of masked mRNA and its subsequent activation is shown in Figure 13.15. The mRNA synthesized in the nucleus can follow either of two possible pathways. If the mRNA is being actively translated, it associates with 40S ribosomal subunits (comparable to the

Figure 13.15
Hypothetical scheme of
mRNA stabilization

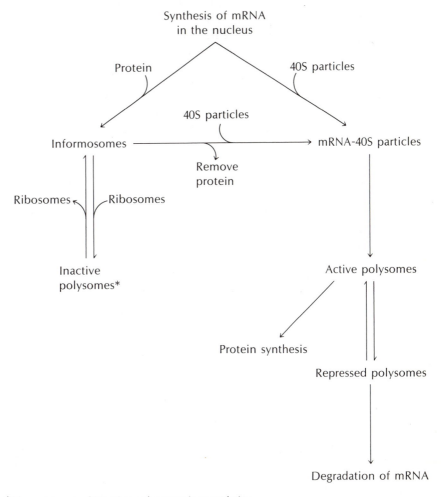

*The existence of inactive polysomes is speculative.

30S subunit in prokaryotes), then forms active, protein-synthesizing *polysomes* (several ribosomes on a single messenger). The active polysomes can be repressed, which subsequently leads to the degradation of mRNA. Alternatively, if the mRNA is not to be used immediately, it is complexed with protein to form *informosomes*, which then may complex with ribosomes to form inactive polysomes. The inactive polysomes can be activated to form protein-synthesizing polysomes. Tyler has summarized the process of differentiation, saying that it involves the progressive, controlled unmasking and translation of mRNA. The tRNA molecules do not exist in masked form and probably play only a minor role in differentiation, though they can influence the rate of protein synthesis. The tRNA molecules are synthesized from the moment of fertilization, so they are probably not a determining factor in differentiation. However, the relative concentrations of tRNA or amino-acid-activating enzymes in a cell could be a significant force in differentiation.

Ribosomes and rRNA, on the other hand, do influence cell differentiation through the control of protein-synthesis rate. We have already been introduced to the nucleolar organizer (NO) region in *Xenopus*. This is the site of rRNA synthesis, and it is inactive until gastrulation (pregastrula protein synthesis is carried out on ribosomes synthesized during oogenesis). A mutation in *Xenopus* has been found in which the NO region is lacking. Since each homologous member of the chromosome pair has an NO region (and, therefore, two nucleoli), this trait follows a Mendelian pattern of inheritance. The homozygous mutant lacking an NO region can synthesize no ribosomes and survives only until the swimming stage, using ribosomes that were present in the egg. Thus the synthesis of rRNA can be instrumental in the differentiation of cells, as exemplified by the extreme case of embryo death.

We know now that rRNA and tRNA can influence *protein-synthesis rate*. But mRNA is the key to differentiation, because it controls the *protein quality* through sequential translation, aided by such devices as the masking and stabilization of mRNA.

Phage T4 morphogenesis: A model of sequential gene action

One of the more intriguing problems in developmental genetics concerns the coordinated genetic activity seen in the T4 phage. When T4 infects a sensitive *E. coli* cell, some of the T4 genes are active immediately, while others are active later in the life cycle. The gene products, which also appear at different points in the life cycle, undergo a process of sequential self-assembly, which is controlled in large part by viral genes.

There is an evolutionary logic to this coordination. The early genes are involved primarily in DNA synthesis, and the late genes in the synthesis of T4 coat proteins. Blocking the genes involved in DNA synthesis results in no late gene activity. This is sound from an energy standpoint, because making T4 proteins without T4 DNA to package is a waste of energy. Also the T4 enzyme lysozyme, which is responsible for lysing the host cell and liberating the progeny T4, must be synthesized late in the T4 life cycle. If it were synthesized early, then the cell would lyse prematurely, releasing incomplete T4 particles. Thus the sequence of gene activity is important to the success of the T4 life cycle.

How is this sequential gene activity determined? The type of control seen in T4 is called *positive control,* because it relies on the interaction of a protein factor with RNA polymerase in the synthesis of a specific mRNA. When T4 first infects an *E. coli* cell, *E. coli* σ factors (see page 343) and *E. coli* RNA polymerase interact to transcribe T4 early genes. About five minutes into the infection, the early genes are shut off and the middle and late genes turned on. This comes about in the following way. One of the early genes produces a protein that blocks the association of *E. coli* σ with RNA polymerase. Another series of early genes specifies factors that associate with the RNA polymerase and transcribe middle and late genes.

The middle genes assist many of the early genes in the continuation of DNA synthesis. The late genes are concerned primarily with the synthesis of T4 structural proteins.

From the late genes come about 40 different structural proteins that interact in a precise sequence to form mature T4 phage. There are also proteins that are not structural, but rather enzymatic, insofar as they facilitate the formation of various phage parts. Figure 13.16 summarizes T4 morphogenesis as it proceeds along three different pathways: head, tail, and tail fibers. It is important to note that the components assemble in a precise sequence in this morphogenetic pathway. A block in one step prevents all subsequent steps from occurring.

Recent work from the laboratories of R. L. Wood and J. T. King has explored the function of four genes—48, 54, 19, and 3—that engage in the assembly of the phage tail. It has been shown that genes 48, 54, and 19 are involved in the formation of the baseplate (see Figure 13.16) and that the products of these genes are utilized in a specific order, 48-54-19. The product of gene 3 reacts with the tail core to catalyze the formation of the core–sheath–baseplate complex. The product of gene 15 links them all together once the complex has formed.

Such a highly sequential pathway has some obvious selective advantages. T4 heads cannot attach to tails unless they are complete; heads, therefore, cannot be wasted on incomplete and uninfectious particles. Likewise, sheaths cannot attach to baseplates until baseplates have complexed with the tail core. Each step in the morphogenesis of T4 is dependent on all the preceding steps, creating an efficient assembly line for the production of complete, infectious phage particles.

Gene activity during the development of all organisms is characterized by a specific temporal program of gene activity and gene product interaction. The mechanisms controlling such regulation are beginning to be understood, at least in microorganisms, but much more work remains. In the next section, we shall explore some of the mechanisms by which gene activity is coordinated.

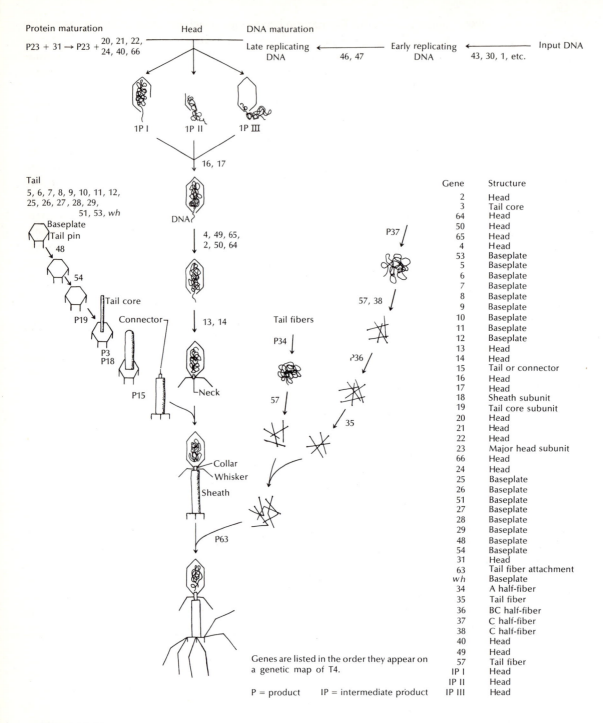

Protein maturation
P23 + 31 → P23 + 20, 21, 22, 24, 40, 66

Head

DNA maturation

Late replicating DNA ← 46, 47 ← Early replicating DNA ← 43, 30, 1, etc. ← Input DNA

1P I 1P II 1P III

16, 17

Tail
5, 6, 7, 8, 9, 10, 11, 12,
25, 26, 27, 28, 29,
51, 53, wh

DNA

Baseplate
Tail pin
48

54

P19

Tail core

Connector

P3
P18

P15

4, 49, 65,
2, 50, 64

13, 14

Neck

P37

57, 38

Tail fibers

P34

P36

57

35

Collar
Whisker
Sheath

P63

Genes are listed in the order they appear on a genetic map of T4.

P = product   IP = intermediate product

Gene	Structure
2	Head
3	Tail core
64	Head
50	Head
65	Head
4	Head
53	Baseplate
5	Baseplate
6	Baseplate
7	Baseplate
8	Baseplate
9	Baseplate
10	Baseplate
11	Baseplate
12	Baseplate
13	Head
14	Head
15	Tail or connector
16	Head
17	Head
18	Sheath subunit
19	Tail core subunit
20	Head
21	Head
22	Head
23	Major head subunit
66	Head
24	Head
25	Baseplate
26	Baseplate
51	Baseplate
27	Baseplate
28	Baseplate
29	Baseplate
48	Baseplate
54	Baseplate
31	Head
63	Tail fiber attachment
wh	Baseplate
34	A half-fiber
35	Tail fiber
36	BC half-fiber
37	C half-fiber
38	C half-fiber
40	Head
49	Head
57	Tail fiber
IP I	Head
IP II	Head
IP III	Head

Figure 13.16
T4 morphogenesis

586

Regulation of gene activity in prokaryotes

What we know of the mechanisms of protein synthesis and of gene activity at different developmental stages prepares us to ask how genes are regulated by the environment. To answer this question at the molecular level—analyzing the DNA–RNA–protein sequence—we must rely heavily on microbial systems, because they provide us with most of the information. The research on eukaryotes is more difficult and far from complete. We do know that the basic pattern of controlling gene activity in prokaryotes is in some respects applicable to eukaryotes, though there are fundamental differences.

Development of the operon concept

Induction and repression An *E. coli* cell contains some 3000 genes, and such a cell, growing on a minimal medium, produces about 700 different enzymes. The total number of protein molecules in this cell is around 10,000,000. If we ignore the structural proteins for a moment, we can estimate that each enzyme is represented by roughly 14,000 copies. But analysis of an *E. coli* cell shows this is not at all the case. Some enzymes are present in as few as 10 copies, others in as many as 500,000! In other words, *E. coli* does not synthesize all its enzymes in equal amounts. This is reasonable; it would be most efficient for the cell to synthesize only those enzymes it needs and in the needed proportions. We shall now explore how regulatory mechanisms operate in *E. coli* to control enzyme synthesis.

One of the most intensely studied proteins in *E. coli* is the enzyme *β-galactosidase,* which hydrolyzes the disaccharide lactose to form two monosaccharides, galactose and glucose (Figure 13.17). In a cell growing

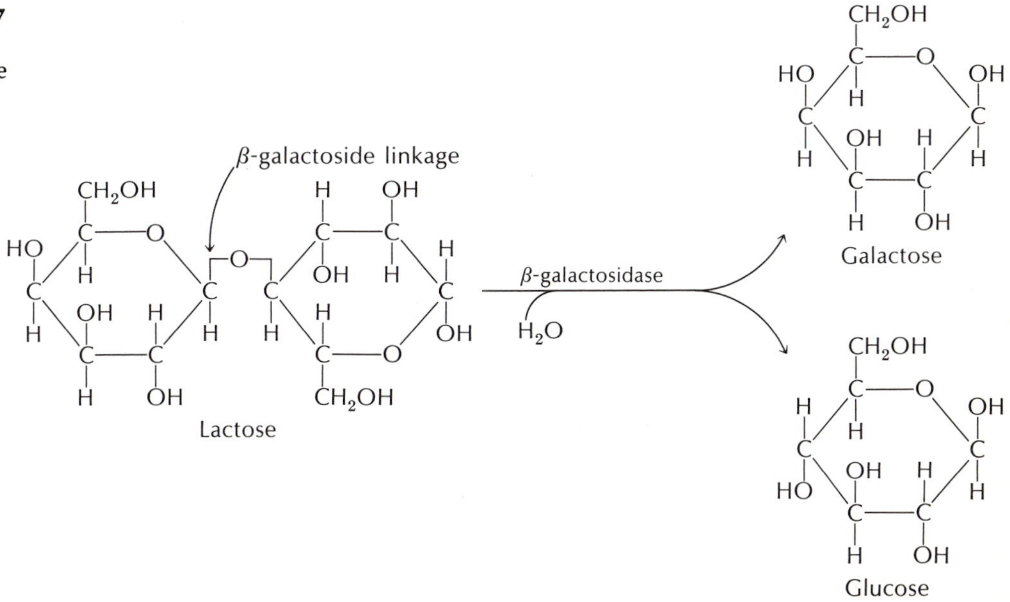

Figure 13.17
The action of β-galactosidase

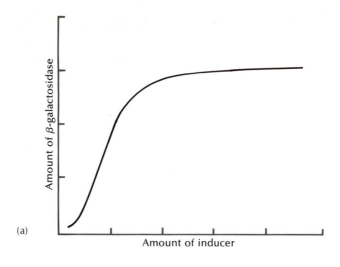

(a)

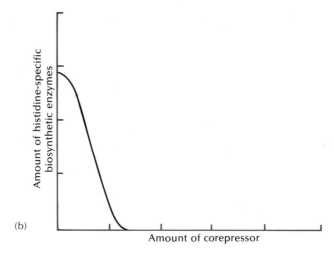

(b)

Figure 13.18
(a) Induction and (b) repression

on a lactose medium, there are some 3000 copies of this enzyme per cell. But in a medium free of lactose (a glucose medium), there are only about two copies per cell. As a result of this phenomenon, β-galactosidase was originally termed an *adaptive enzyme*, because its presence or absence depended on the environment in which the cell was growing. *Constitutive enzymes* are produced at a constant rate, irrespective of the cell's environment. In 1953, Monod and his collaborators replaced the term *adaptive enzyme* with *inducible enzyme*. In the presence of a specific substrate called an *inducer* (in this case lactose or, more correctly, a chemical derivative of lactose), the corresponding enzyme (inducible β-galactosidase) is synthesized (Figure 13.18). Another type of enzyme system—one that is quite common for amino acid biosynthetic pathways—is termed *repressible*, because the end product (called a *corepressor*) of a particular amino acid biosynthetic pathway (for example, histidine) inhibits genes that code for the corresponding enzymes. The value of both inducible and repressible enzyme systems to bacteria is apparent. In the presence of a given substrate, enzymes are induced that degrade the substrate; in the absence of the substrate, the cell does not waste energy producing enzymes if they are unneeded. For a repressible enzyme system, in the presence of a metabolite that the cell would normally synthesize, the cell shuts down manufacture of synthesizing enzymes and utilizes the exogenous source, again conserving energy and achieving greater efficiency.

An inducible enzyme, such as β-galactosidase, is due to *de novo* protein synthesis and not to the conversion of already existing inactive enzyme precursors into active enzymes (Figure 13.19). This was amply demonstrated when cells were grown for several generations in a medium radioactively labeled with ^{35}S (but without lactose) so that all proteins synthesized were labeled (including any precursor β-galactosidase). The cells were then transferred to a nonlabeled medium containing the inducer lactose and allowed to grow for a time. The β-galactosidase was isolated and analyzed for ^{35}S label. If the enzyme had initially been present as an inactive precursor, it would be ^{35}S-labeled, but if it resulted from *de novo* synthesis on exposure to lactose, it would be unlabeled. The latter was found to be the case, so *de novo* synthesis of inducible enzymes was inferred.

It was soon found that in addition to β-galactosidase, two other enzymes were involved in lactose metabolism and were coordinately induced with β-galactosidase: *β-galactoside permease* (functions in the transport and concentration of lactose inside the cell) and *β-galactoside transacetylase* (acetylates lactose, but what function this might serve is not well understood).

A conjugational analysis of the genes responsible for these three enzymes showed them to be contiguous and in the following sequence: galactosidase (Z), permease (Y), transacetylase (A). As studies progressed on this system, a number of mutants were isolated:

$Z^-Y^+A^+$ (no galactosidase, others normal)
$Z^+Y^-A^+$ (no permease, others normal)
$Z^+Y^+A^-$ (no transacetylase, others normal)

Perhaps the most intriguing mutant, however, was one that had lost its ability to respond to the inducer. In the presence or absence of lactose, genes Z, Y, and A constitutively synthesized their respective enzymes; that is, these genes were no longer inducible. This mutant was designated I (*lacI⁻*) and mapped at a site that was not contiguous with Z, Y, and A.

The *lacI⁻* mutant stimulated much research into the question of how a mutation in *lacI* could control the expression of three other distal genes.

β-galactosidase synthesized in an inactive form; activated by presence of inducer

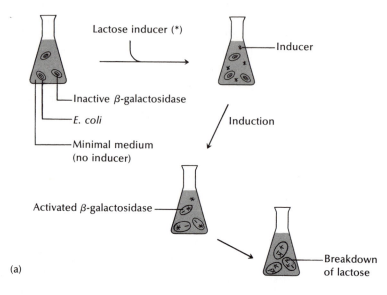

(a)

β-galactosidase not synthesized until an inducer is present

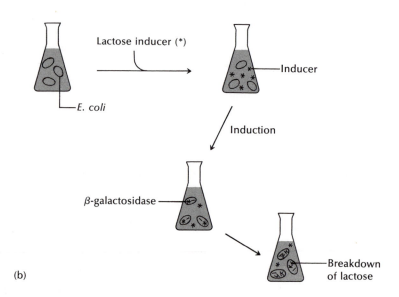

Figure 13.19
Two possibilities for the
origin of β-galactosidase   (b)

One idea proposed was that the mutant *I* gene synthesized an "internal inducer" that maintained a constant supply of inducer and kept the genes *Z, Y,* and *A* in a state of constant operation. This was disproved and replaced by another concept proposed by A. B. Pardee, F. Jacob, and J.

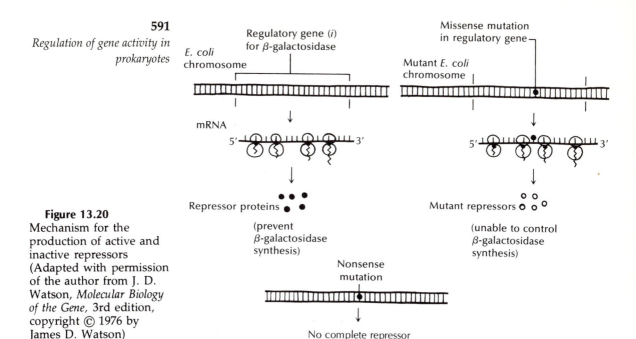

Figure 13.20
Mechanism for the production of active and inactive repressors (Adapted with permission of the author from J. D. Watson, *Molecular Biology of the Gene,* 3rd edition, copyright © 1976 by James D. Watson)

Monod. These three investigators suggested that the *I* gene be termed a *regulatory* gene, which constitutively produced a *repressor* molecule that could diffuse to genes *Z, Y,* and *A* and effectively inhibit the transcription of mRNA. The *lacI⁻* mutants either lacked this repressor or produced a mutant form of it, hence the constitutive synthesis of the three enzymes (Figure 13.20).

The fact that many *lacI⁻* mutations were suppressible by *E. coli* strains known to have suppressor tRNA molecules suggested that the *lacI* gene product might be a protein. This was verified in 1967 by W. Gilbert and B. Müller-Hill, who isolated the *lac* repressor by binding it to a lactose analog isopropylthiogalactoside (IPTG) and then isolating the complex. The *lac* repressor consists of four identical polypeptide chains, each with a molecular weight of 40,000, to give a total molecular weight of 160,000. (There are only about 15 copies of *lac* repressor present per cell.)

Our next concern is the matter of how the protein repressor inhibits the transcription of three contiguous yet distal genes. Some six years before the isolation of the repressor protein, Jacob and Monod had proposed the *operon theory,* which suggested that the target of the repressor was another regulatory gene called the *operator* (*O*). The operator gene is contiguous with genes *Z, Y,* and *A* (*structural genes*) and controls their action. The theory further proposes that the *Z, Y,* and *A* genes are transcribed as a unit into a single polycistronic molecule of mRNA. And the total unit of operator and its associated structural genes is called an *operon.*

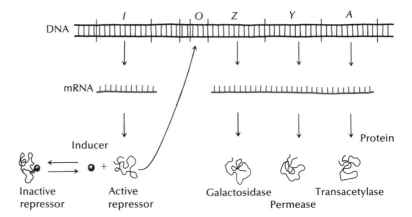

Figure 13.21
The original
Jacob–Monod operon
model

Figure 13.21 summarizes the theory as it applies to an inducible enzyme system. A repressor protein binds to the operator gene and in essence prohibits the transcription of mRNA from the structural genes. An inducer can inactivate the repressor by binding to it and rendering it incapable of acting on the operator. The repressor protein has two active sites, one for inducer recognition and one for operator recognition. If the inducer site is occupied, it distorts the molecule so that the operator site no longer recognizes the operator (*allosteric effect*). The association of inducer (also known as a *corepressor* in repressible enzyme systems) with repressor is a weak one involving hydrogen bonds, salt linkage, or van der Waals forces, so it is easily reversible.

A repressible enzyme system works on the same basic principle but with some rearranging. In this case, the corepressor binds to the repressor protein to activate it and allow it to recognize the operator. Without a corepressor, the repressor cannot bind to the operator and block the synthesis of the structural gene products. The interaction of inducer or corepressor with the repressor is summarized in Figure 13.22.

The model of the operon proposed in 1961 by Jacob and Monod developed from a study of several *lac* mutants, especially the *lacI⁻* mutant. Since 1961, genetic and biochemical evidence has supported it. In 1965 Jacob, Monod, and Lwoff were awarded the Nobel Prize for their research on the regulation of enzyme synthesis by the cell. Their model has been altered little in the past decade, a remarkable tribute to their research and creative interpretation of data.

Genetic proof of the operon model was developed from experimentation with partially diploid strains of *E. coli*—that is, F′ cells—carrying *lac* genes on the chromosomes and *lac* genes on the F′ factor. Table 13.1 and Figure 13.23 (pages 594–595) show some of these studies. The normal genotype ($I^+Z^+A^+$) is essentially inactive in the absence of the inducer and fully active in its presence. The $I^-Z^+A^+$ genotype is constitutive, because the repressor is absent or nonfunctional; that is, the inducer has no effect

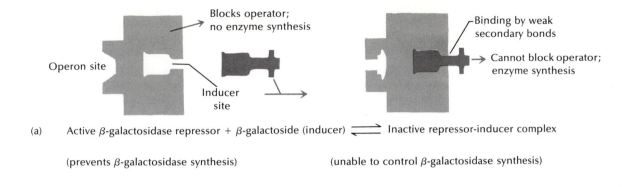

(a)   Active β-galactosidase repressor + β-galactoside (inducer) ⇌ Inactive repressor-inducer complex

      (prevents β-galactosidase synthesis)   (unable to control β-galactosidase synthesis)

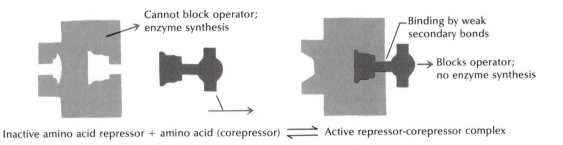

(b)   Inactive amino acid repressor + amino acid (corepressor) ⇌ Active repressor-corepressor complex

      (unable to control synthesis   (controls rate of synthesis of enzymes
of enzymes for amino acid   for amino acid synthesis)
synthesis)

Figure 13.22
Illustration of the opposite effects of corepressors and inducers on the activity of repressors (whether the free repressors are active or inactive depends on whether the enzymes are inducible or repressible)

on the activity of the structural genes. (The lower level of A^+ activity is due to the unequal production of Z and A enzymes by a *translational control* system, to be discussed later.)

The partially diploid strain $I^-Z^+A^+/I^+Z^+A^+$ behaves just like the normal strain, so we can conclude that the repressor synthesized by the single I^+ gene is present in sufficient quantity to repress both sets of structural genes. Genetically, the I^+ gene is dominant over I^-. A *lac* mutant called I^s (s = superrepressor) has provided an interesting insight into the activity of the repressor. The repressor from this mutant has a very weak affinity for the inducer and a strong affinity for binding to the operator. Hence, in an $I^sZ^+A^+$ strain, the inducer is unable to inactivate the repressor (except when present in very large quantities). The strain is constantly repressed.

The partially diploid strain $I^sZ^+A^+/I^+Z^+A^+$ does not synthesize an appreciable amount of these enzymes because of the presence of the superrepressor binding to both operons, so we regard I^s as dominant over

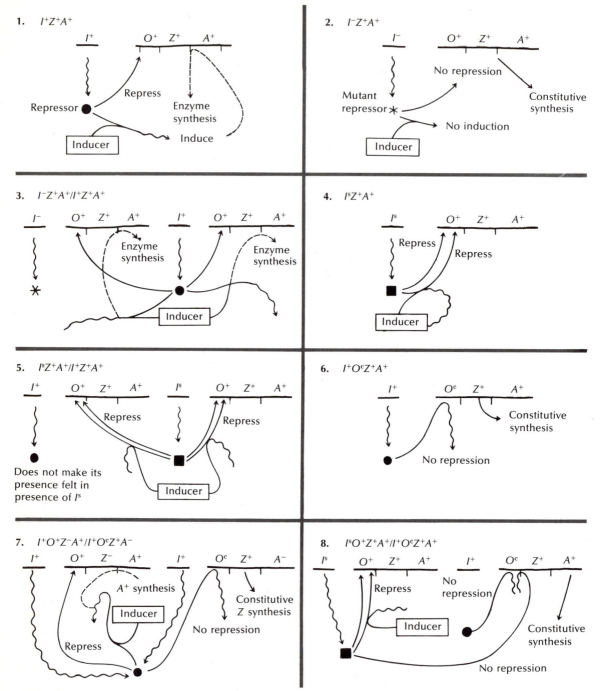

Figure 13.23
Interpretation of data in Table 13.1

Chromosome genes	F' factor genes	Galactosidase (lacZ)		Galactoside-transacetylase (lacA)	
		Noninduced	Induced	Noninduced	Induced
1. $lacI^+,Z^+,A^+$		0.1	100	1	100
2. $lacI^-,Z^+,A^+$		100	100	90	90
3. $lacI^-,Z^+,A^+$	$I^+Z^+A^+$	1	240	1	270
4. $lacI^s,Z^+,A^+$		0.1	1	1	1
5. $lacI^s,Z^+,A^+$	$I^+Z^+A^+$	0.1	2	1	3
6. $lacI^+O^c,Z^+,A^+$		25	95	15	100
7. $lacI^+O^+,Z^-,A^+$	$I^+O^cZ^+A^-$	180	440	1	220
8. $lacI^s,O^+,Z^+,A^+$	$I^+O^cZ^+A^+$	190	219	150	200

Table 13.1
Genetic analysis of *lac* mutants using partially diploid strains of *E. coli*

Adapted from F. Jacob and J. Monod, "Genetic regulatory mechanisms in the synthesis of proteins," *J. Molec. Biol.* 3:318–356 (1961).

I^+. Another mutation, this one in the operator itself, gives rise to constitutive enzyme synthesis. The mutant O^c (c = constitutive) has a weak affinity for the repressor, so the strain $I^+O^cZ^+A^+$ shows considerable activity even in the absence of an inducer.

In the partially diploid strain $I^+O^cZ^+A^-/I^+O^+Z^-A^+$, the Z^+ gene product is constitutively produced, but the A^+ gene product is inducible. This shows that the genes linked to O^c are behaving constitutively and that those linked to O^+ are inducible; thus we can infer that O does not produce a gene product that diffuses from one chromosome to another.

The final genotype we shall consider is the partially diploid strain $I^sO^+Z^+A^+/I^+O^cZ^+A^+$. This strain is constitutive for Z and A gene products. The superrepressor probably blocks the activity of the genes linked to O^+ but not of those linked to the O^c mutation, because the superrepressor does not recognize it—hence the constitutive synthesis of Z^+ and A^+ gene products from the genes linked to O^c.

Corroborative evidence for the operon model is the demonstration that not only do repressors inhibit the synthesis of the enzymes controlled by these structural genes, but they also block the synthesis of $Z\ Y\ A$ mRNA, thus establishing the level of control at RNA transcription (Figure 13.24). Induced and noninduced *E. coli lac*$^+$ cultures were labeled with ^{14}C-uracil, and the mRNA was isolated. This labeled mRNA was then used in a DNA–RNA hybridization experiment using denatured *lac*$^+$ DNA. If an induced culture is producing more mRNA (because the Z, Y, and A genes are being transcribed) than a noninduced culture, then, compared to the noninduced culture, more mRNA hybridizes with *lac*$^+$ DNA. This was found to be the case, supporting the operon model.

Another important confirmation of the operon is worth noting. If the O gene is the site of action of the repressor, it should be possible to demonstrate the interaction of the two. This became possible in 1967 with

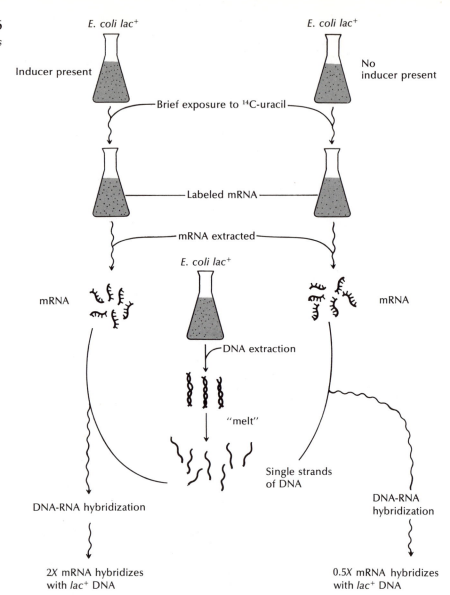

Figure 13.24
Repressors inhibit
synthesis of mRNA by
structural genes

the isolation of the *lac* repressor protein (Figure 13.25). *E. coli* was grown
over several generations in the presence of ^{35}S radioactive label, and the
[35]S-labeled repressor was isolated.* The repressor molecules were mixed
in various combinations, with or without IPTG (isopropylthiogalactoside,
an analog of lactose that binds firmly to the repressor protein, rendering it
a stronger inducer than lactose) and with DNA from various *lac* strains of

* A mutant strain of *E. coli* that produces abnormally large quantities of the repressor
protein was used in these studies. The mutation is designated *i*[Q].

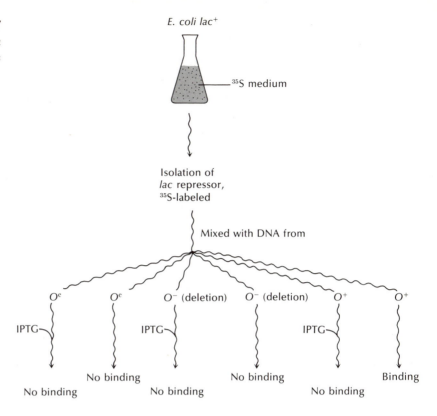

Figure 13.25
Interaction of repressor
protein with operator
region

E. coli. The repressor was bound to the DNA only if the nonmutant
operator gene was present *and* if IPTG was absent, confirming the predictions of the operon model.

   The promoter   The one major addition to the original Jacob–Monod
operon model is the assignment of a *promoter* region to the left of the
operator. Promoter regions are sites for the attachment of RNA
polymerase and hence starting points for RNA transcription (Figure
13.26). If the operator gene is bound by a repressor, the RNA polymerase
either cannot interact with the promoter or cannot pass it; either way, no
RNA synthesis occurs.

   The first suggestion of the existence of the promoter region came
from observations in *E. coli* in which the *lac* operon was transcribed very
poorly in the presence of an I^- mutation (deletion) and in the presence of
an O^c mutation. In both cases, one would expect normal but constitutive
transcription (Figure 13.27).

   Although the promoter (*P* gene) appears not to overlap with or run
into the *O* gene in the *lac* operon, it may do so in other operons. Its
position to the left of the operator also suggests that the *O* gene is transcribed, though perhaps not translated, and there are recent data to support this. The *I* gene, which produces the *lac* repressor, appears to have its

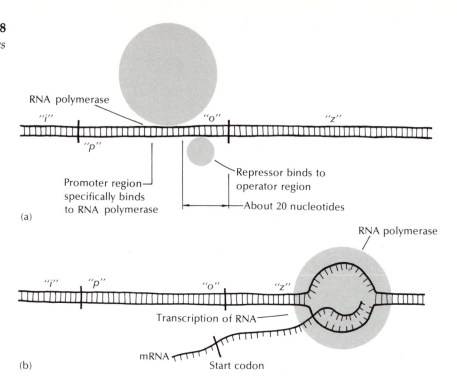

Figure 13.26
Binding of RNA polymerase to the promoter for the *lac* operon: (a) with repressor; (b) without repressor

Figure 13.27
Demonstration of the existence of the promoter region for the *lac* operon. As shown in (a) and (b), in the absence of an active repressor or the presence of a mutant operator, synthesis of *lac* mRNA takes place. Parts (c) and (d) show a promoter mutation that blocks synthesis of *lac* mRNA even when no functional repressor or nonfunctional operator is present. (Adapted with permission of the author from J. D. Watson, *Molecular Biology of the Gene*, 3rd edition, copyright © 1976 by James D. Watson.)

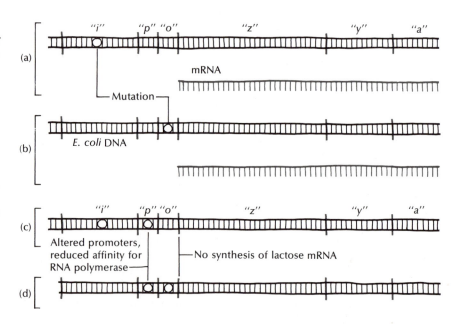

own promoter region, but the rate of RNA transcription from the *I* gene is constitutive, is not subject to inducers or repressors, and is slow. The promoter gene in this case must have a lower affinity for RNA polymerase attachment than does the promoter gene to the left of the *lacO* gene, because the *I* gene is transcribed at the approximate rate of 1 mRNA molecule per cell generation, in contrast with the dozens transcribed from genes *Z*, *Y*, and *A*. (The *laci*^{Q} mutation noted in the footnote on page 596 is apparently a promoter mutation that increases *i* transcription.) This suggests that either the *P* genes themselves are variable or perhaps that different sigma-type factors are involved in RNA polymerase recognition of the *P* gene.

Cyclic AMP and the lac operon   When *E. coli* is grown on a medium containing both glucose and lactose, only the glucose is utilized. The *lac* operon is inactive. The same situation holds in the presence of both glucose and galactose: The galactose operon is inactive. Somehow glucose blocks the activity of other operons, but how? A clue to this problem emerged when cyclic AMP (cAMP, Figure 13.28) was found to occur in *E. coli* cells.

Before this discovery, cAMP was thought to occur only in higher animals, in which it acted as a messenger in hormonal action. In normal *E. coli* cells growing on a glucose medium, the cellular level of cAMP is very low. Adding cAMP along with lactose to the medium resulted in the *lac* operon's being active. It was soon clear that cAMP promoted mRNA transcription from the glucose-repressed *lac* operon. It freed the cell from *catabolite repression*, the phenomenon whereby a glucose breakdown product (catabolite) blocks the cAMP required for glucose-repressed operon transcription.

Pastan and Perlman found that cAMP-stimulated operons required another protein called *cAMP receptor protein* (CRP), also known as *catabolite gene-activator protein* (CAP). When both CAP and cAMP are present, transcription of the *lac* operon is normal. If either cAMP or CAP is absent, transcription is very slow.

Recent studies show that cAMP and CAP first complex and then, in the presence of RNA polymerase, promote the initiation of transcription

Figure 13.28
Cyclic AMP

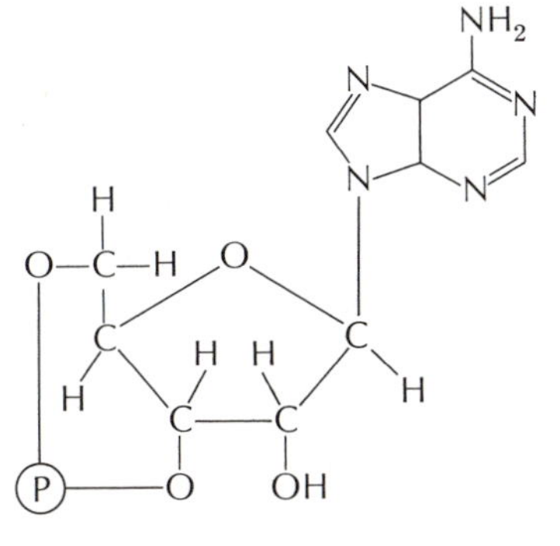

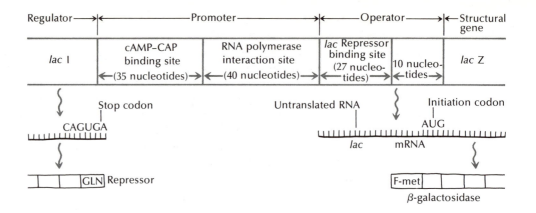

Figure 13.29
The structure of the promoter–operator region of the *lac* operon

from the *lac* operon. Because CAP and cAMP promote transcription, they are referred to as *positive* controlling elements in contrast to the *lacI* repressor, which is a *negative* controlling element. Recently Dickson and his colleagues determined the DNA sequence of the promoter–operator region of the *lac* operon. From this sequence, they ascertained that the promoter region consisted of a cAMP–CAP site and an RNA polymerase interaction site (Figure 13.29).

A model for the initiation of transcription of the *lac* operon is seen in Figure 13.30. The cAMP–CAP complex binds to its specific site and moves to the initiation site, where RNA transcription begins.

Why should glucose be preferred over other sugars? There is no clear answer to this question. It may be that, in the early evolution of genetic systems, glucose was the primary source of carbon, but this does not really explain catabolite repression. We can say that when glucose is available as an energy source, cAMP levels are low, which prevents the functioning of other operons involved in the metabolism of sugars like lactose, arabinose, galactose, or maltose. When glucose levels fall, cAMP levels rise, and the other sources of sugar can be utilized. We still do not know how cAMP levels are depressed by glucose, nor do we fully understand the evolutionary significance of the relationship.

The histidine operon   The histidine operon, a repressible rather than an inducible system, provides some interesting contrasts to the *lac* operon. First, there appear to be five regulatory genes (*hisR, hisU, hisS, hisT, hisW*) regulating the histidine operator (*hisO*), rather than one. A mutation in any one of the five results in constitutive enzyme synthesis. All five regulatory genes appear to be involved with the synthesis of tRNA$_{his}$ and histidyl-tRNA synthetase. Second, neither tRNA$_{his}$ nor histidine is by itself capable of turning off the histidine operon. For the histidine operon, the repressor appears to be the complex of his-tRNA$_{his}$ bound to histidyl-tRNA synthetase, as shown in Figure 13.31.

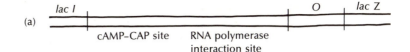

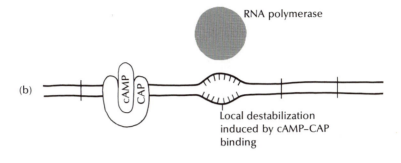

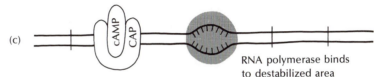

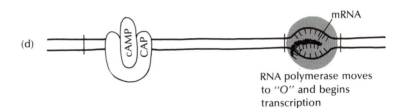

Figure 13.30
A model for initiating transcription at the *lac* operon (modified from Dickson, *et al. Science* 187:27, 1975)

In a number of other repressible enzyme systems (valine, isoleucine, and leucine), it has been shown that the amino acid–tRNA complex, and not simply the amino acid, is the corepressor of the operons. But how this complex interacts with the operon is not yet known.

The arabinose operon   The inducible and repressible enzyme systems discussed so far are *negative control systems* in that the repressor inhibits the operons. In 1965, E. Englesberg and his colleagues presented evidence in *E. coli* that suggested a *positive control system* in that the product of the repressor gene was *required* for structural gene activity. They were studying the arabinose operon, which consists of four structural genes and a regulatory gene (Figure 13.32). The four structural genes (*A, B, D,* and *E*) code for enzymes that convert the sugar arabinose to xylulose-5-phosphate. The product of the regulatory gene (*ara-C,* which is comparable to the *I* gene in the *lac* operon) is required for the induction of the operon. Nonsense mutations or deletions in the regulatory gene have been found, and these render the operon noninducible so that no tran-

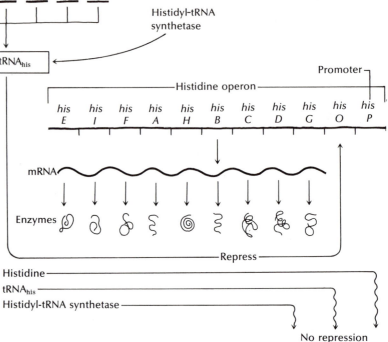

Figure 13.31
The histidine operon: a
repressible enzyme
system

scription occurs. This is inconsistent with a negative control operon. A
nonsense mutation or deletion in the *O* or *I* gene in an inducible or
repressible system would result in constitutive enzyme synthesis. Constitutive mutants (*araC*^{c}) of the *araC* gene are known, and in partial diploids these are dominant over noninducible *araC*$^-$. Recently Englesberg
suggested that the product of the normal *araC* gene is a repressor that is
converted to an activator by an inducer. Using this suggestion, we can
hypothesize that the product of the *araC*^{c} gene is a mutant protein that
stimulates transcription even in the absence of an inducer. Thus the *araC*^{c}
mutant is a constitutive mutation. Having lost its ability to recognize the
inducer, the *araC*$^-$ gene product results in a noninducible phenotype. So
the normal *C* gene product has a positive and a negative role in the
arabinose operon. But how this product, in combination with an inducer,
can *activate* a sequence of structural genes is not known. Possibly the
product of the repressor gene interacts with the inducer, and this complex
then binds to the promoter, allowing the RNA polymerase molecule to
transcribe the arabinose structural genes. The *C* protein–inducer complex
in this case would be comparable to the cAMP–CAP complex.

Some interesting evidence in support of the dual function of the *C*
protein emerged from studies of partially diploid *E. coli.* In cells that were

C^c/C^+, we might expect constitutive synthesis of the arabinose mRNA, but in fact the operon is not constitutive. It is inducible. Evidently the C^+ protein is repressing the operon so that the C^c protein cannot activate it. If arabinose (inducer) is present, the operon is active. This emphasizes the repressing qualities of the C protein.

Control of lysogeny in λ

When the bacteriophage λ infects an *E. coli* cell, it can proceed through a lytic cycle or it can enter into a lysogenic relationship with that cell. The maintenance of lysogeny requires the repression of almost all of the λ genome—all except the repressor gene. Leaving lysogeny to enter the lytic cycle involves first repressing the repressor and then following a specific sequence of gene transcription.

The key to lysogeny with λ is the λ repressor, a protein coded for by the C1 gene. In a lysogenized bacterium, the λ repressor binds to the operator region of two operons that function early in the λ life cycle, thus repressing them. These operons function in λ DNA replication and in the excision of the λ chromosome from its integrated state in the *E. coli* chromosome.

When the λ repressor is inactivated (by UV light or certain chemicals), a specific sequence of events begins (Figure 13.33). The two early operons are transcribed, and the gene product from the right operon's *tof* gene shuts off transcription from the C1 repressor gene.

At this point another key λ gene, *N*, comes into play. If the *N* gene is mutant or deleted, transcription proceeds only part way through the early operons. This occurs because there are stop signals in these λ operons that are read by *E. coli* ρ termination factors. But in the presence of the *N* protein, these termination signals are not read by the ρ factors, and transcription proceeds through them. The *N* gene product somehow inactivates the host cell ρ factors. Interestingly, these ρ factors may have evolved to protect the cell from lytic infection by λ or λ-like phage. But the phage evolved a mechanism to circumvent the host's defense.

Figure 13.32
A model of the arabinose operon of *E. coli*, a positively controlled enzyme system. Note that *araE* maps to a different part of the chromosome; it is, however, coordinately controlled with B, A, and D.

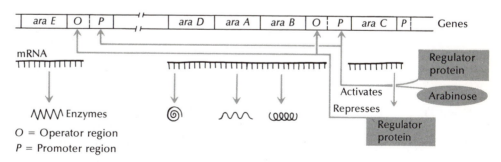

O = Operator region
P = Promoter region

A model of the arabinose operon of *E. coli*, a positively controlled enzyme system. Note that *ara E* maps to a different part of the chromosome; it is, however, controlled coordinately with *B*, *A*, and *D*.

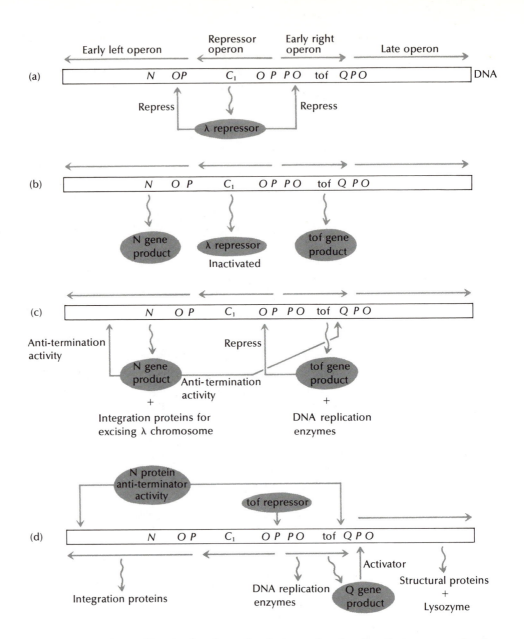

Figure 13.33
A diagrammatic summary of some of the steps involved in λ development; The arrows for each operon indicate transcription direction.

About 13 minutes into the lytic cycle, the fourth λ operon begins transcription. This is a late operon that codes primarily for the structural proteins of λ. The Q gene, which is in the early right operon, codes for a product that activates the late operon. Lambda development is thus an elegant example of precisely coordinated regulation of genes.

The operon model of Jacob and Monod presents a mechanism of regulation at the transcription level, but is there evidence of regulation at the translation level? The first indication that translation is regulated was the observation that in a normal *lac*⁺ cell ($Z^+Y^+A^+$), the enzymes β-galactosidase, β-galactoside permease, and β-galactoside transacetylase are produced in unequal amounts (the ratio is 10:5:2). There are a number of possible explanations for this. The attachment of ribosomes to different points along the mRNA molecule may vary; and since the mRNA from the *lac* operon is *polycistronic* (that is, one messenger codes for more than one cistron), some of the cistrons may be translated more efficiently than others. (Here ribosomes may encounter a polypeptide termination signal, fall off, and fail to reattach as efficiently at the next cistron.)

Another possibility for translational control is the presence of limited amounts of certain amino acid–tRNA species such that the translational process is slowed up at codons specifying them. These and other possibilities may all be applicable to different operons. All we can say with certainty is that there is unequal translation along a polycistronic mRNA molecule. This does appear to have adaptive significance, because those enzymes needed in greatest quantity (β-galactosidase) are produced in the greatest quantity and those needed least (for example, β-galactoside transacetylase) are present in the smallest amounts.

There are other means of controlling enzyme production at a nontranscriptional level. For example, the life span of the *Z, Y, A* polycistronic mRNA is only a few minutes on removal of an inducer (Figure 13.34), so any one mRNA is limited in the quantity of enzymes it can generate. Another common means of nontranscriptional regulation is *feedback inhibi-*

Figure 13.34
Life span of polycistronic
mRNA transcribed from
the *lac* region (*ZYA*)

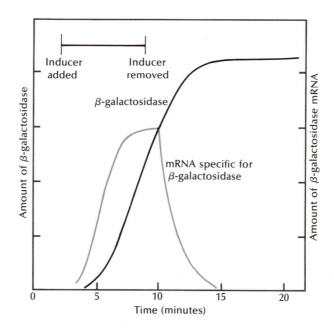

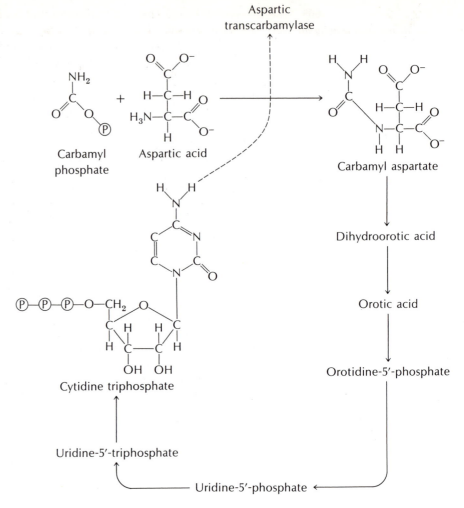

Figure 13.35
Feedback inhibition
controls the biosynthesis
of pyrimidines in *E. coli.*

tion (Figure 13.35), a system in which the end product of a particular biosynthetic pathway inhibits the function of the enzyme involved in the initial reaction.

Nevertheless, transcriptional control is more common than translational control, and this is as it should be. Regulating metabolism at the beginning of a process conserves more energy than regulating it at the end. It spares the cell the necessity of synthesizing mRNA that will not be used.

Regulation of rRNA
and tRNA synthesis When an *E. coli* cell is starved for amino acids and thus cannot synthesize proteins, the synthesis of rRNA and tRNA stops. Without rRNA, the number of ribosomes drops by almost half. The cessation of rRNA and tRNA synthesis under conditions of amino acid starvation is called the

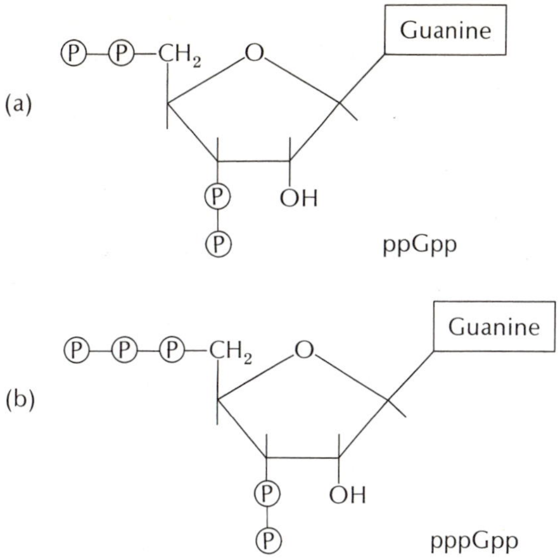

(a)

ppGpp

(b)

pppGpp

Figure 13.36
Unusual nucleotides responsible for the control of rRNA and tRNA synthesis

stringent response. Mutants are known, however, in which rRNA and tRNA synthesis continues under conditions of amino acid starvation. These mutants are said to manifest a *relaxed response.*

In studies of the stringent response, a nucleotide was discovered that was absent in cells expressing the relaxed response. This unusual nucleotide is called guanosine-5′-diphosphate 2′ (or 3′)-diphosphate, or ppGpp (Figure 13.36a). Another unusual nucleotide was also discovered in the company of ppGpp. This one is guanosine-5′-triphosphate 2′ (or 3′)-diphosphate, or pppGpp (Figure 13.36b). Both these unusual bases are synthesized by ribosomes in the presence of ATP, GDP, mRNA, and free tRNA (tRNA with no amino acid bound to it).

In some as yet unknown way, ppGpp and pppGpp block the synthesis of rRNA and tRNA when amino acids are in short supply. Clearly, when protein synthesis is blocked by amino acid starvation, there is no need for ribosomes or tRNA, so their synthesis is halted. Thus the stringent response conserves energy for the cell.

A protein called the *stringent factor* appears to be involved in the stringent response. It binds to ribosomes and aids in the conversion of ppG to ppGpp. Relaxed mutants lack this factor.

Regulation of gene activity in eukaryotes

While we are able to describe prokaryotic gene regulation from several different perspectives, our understanding of eukaryotic gene regulation is relatively skimpy. There are many reasons for this. Eukaryotes have more DNA than prokaryotes, and that DNA is usually distributed over several chromosomes. In eukaryotes, we have to contend with diploidy and dominance. Furthermore, eukaryotic chromosome structure is not well understood, but it is certainly more complex than the *E. coli* or λ chromosome. The eukaryotic nuclear membrane separates the transcriptional from the translational operations. Unlike prokaryotic organisms, eukaryotic cells become irreversibly differentiated. Thus differentiated cells have irreversibly repressed genes, unlike prokaryotic organisms, in which repression is easily reversible. (Even though they are differentiated, it is increasingly clear that differentiated cells, such as muscle and kidney, carry the same genetic information.)

Gene regulation in prokaryotes occurs predominantly at the transcriptional level. In eukaryotes we find—in addition to transcriptional regulation—mRNA processing, translational control, and posttranslational control. In this section of the chapter, we shall explore some of these regulatory mechanisms.

Cell determination versus cell differentiation

There is only a narrow area of separation between *cell differentiation* and *cell determination*. Nevertheless, many developmental geneticists make such a distinction. At some time during embryogenesis, a cell becomes committed or determined to a limited range of future phenotypes, and this determination is passed on to its daughter cells. That cell, referred to at this stage as a *determined cell*, may divide for several more generations until it is triggered into expressing those predetermined potentialities. The expression of these potentialities is termed *cell differentiation*.

The difference between determination and differentiation may best be described with an example or two. In the early gastrula of the newt, the cells that will eventually form the neural tube are located in the animal hemisphere adjacent to the dorsal lip (Figure 13.37). The cells that will eventually form the epidermis are also in the animal hemisphere but on the side opposite the presumptive neural-tube cells. If a small piece of presumptive neural-tube tissue of the *early gastrula* is transplanted to the region of the presumptive epidermis, the presumptive neural-tube tissue will develop into epidermis tissue. The reciprocal experiment, transplanting presumptive epidermis tissue to the region of the presumptive neural tube, causes the presumptive epidermis to develop into neural-tube tissue. At the early gastrula stage, then, the presumptive epidermis and presumptive neural tube are not yet determined. That is, they are not committed to a specific future phenotype but rather conform to the dictates of the cellular environment into which they are transplanted.

However, if this same transplantation experiment is carried out at *late gastrulation*, the results are markedly different. Presumptive epidermis tissue transplanted into presumptive neural-tube tissue develops only into epidermis tissue, and vice versa. By late gastrulation, the presumptive neural tube and presumptive epidermis *have been* determined, or committed to a specific future phenotype. When these cells express that future phenotype (neural tube or epidermis), they are said to be differ-

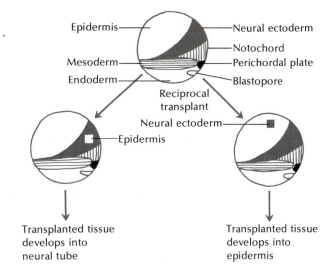

Figure 13.37
Reciprocal tissue transplantation in early newt gastrula

entiated, but this may not occur until well after they have been determined.

Another example that may serve to distinguish determination from differentiation is the formation of the limb in a developing chick embryo. If the cells from the limb-forming region of the chick embryo are isolated and transplanted to a different, abnormal site before the limb bud actually appears, the transplanted tissue will develop into a complete limb—regardless of where it is transplanted. This is understood to mean that the presumptive limb-bud cells were *determined* to be limb cells *before* they actually *differentiated* into a limb structure.

An interesting example of a cell type determined to differentiate into more than one phenotype is Wolffian lens regeneration in the newt. If the lens of the newt eye is removed (the lens being made up of highly differentiated cells), the iris cells (also highly differentiated) surrounding it will redifferentiate into new lens cells. In other words, iris cells are already determined for the lens phenotype as one of their possible modes of expression. It has even been demonstrated that iris cells contain some lens proteins, which suggests that these cells are determined for more than one differentiated state.

Alteration of the onset of cell determination has been attempted but has generally been unsuccessful, which emphasizes that cell determination is a very stable condition once it begins. The remarkable stability of the determined state has advantages in a developing embryo, because a cell's environment is constantly changing. If a cell's potential phenotype were to change with each environmental change, the intricate pattern of organization required for successful development would be disrupted and chaotic.

Hadorn and Gehring explored this stability of determined cells in *Drosophila* and found that under extreme conditions it could be altered. The primordia of various adult structures—such as genitalia, wings, antennae, and legs—are present in the larvae as clusters of cells called *imaginal discs*. These discs can be dissected out of the larva and transplanted into either a larval or an adult abdomen, but differentiation of the disc occurs only if it is situated in the abdomen of a metamorphosing larva.

The stability of determined cells was tested by transplanting a disc or a disc fragment into an adult abdomen and continuing to subculture the disc over several generations (Figure 13.38). At regular intervals, the discs were tested for their determined properties by transplanting them into a metamorphosing larva. Initially the transplanted discs differentiated into structures in accordance with their original determination (wing discs into wing tissue, leg discs into leg tissue, and so on). However, after a prolonged growth period in adult hosts, *transdetermination* occurred; that is, an original leg disc differentiated into antennae, or genitalia discs differentiated into legs. A further series of experiments established that one differentiated state was being converted into another. One possible explanation of transdetermination is similar to that used to explain Wolffian lens regeneration—the determined cells were committed to more than one differentiated state, but one of the states was preferred to others.

Figure 13.38
The phenomenon of transdetermination as it occurs in *Drosophila*

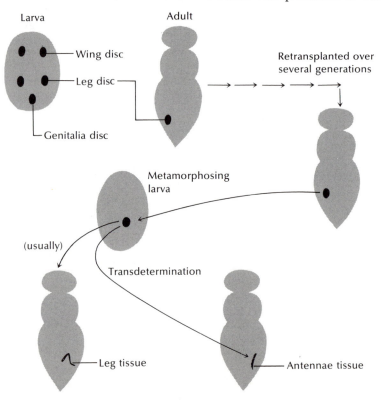

Alternatively, the treatment the cells received could have changed the determined state. Which of these possibilities, if either, applies to trans-determination is unknown.

Determination and transdetermination are poorly understood phenomena. In both instances, however, the key lies in the effect of the cellular environment on the regulation of genetic activity.

Gene regulation: Histone and nonhistone proteins

Histones are small (m.w. 10,000–20,000), basic (positively charged at pH 7.0) proteins that are complexed with DNA in eukaryotic chromosomes. The histone proteins, of which there are 6 classes (H1, H2A, H2B, H3, H4, and H5), may serve a structural function, but they also prevent transcription of DNA. However, though histones may prevent transcription, many question whether they are in fact specific repressors analogous to the repressors we have just discussed.

There are problems with the idea that histones are specific repressors. For example, with only 6 different histone species, it is difficult to visualize them as the sole repressors of the complex eukaryotic genome. Also, the 6 histones are almost equimolar in all cells and all tissues, though gene activity is extremely varied in different tissues. However, it has been suggested that histone specificity may occur when histones first aggregate to form dimers, tetramers, and oligomers. Like the 4 nucleotides read as triplets to produce 64 possibilities, 5 histone classes clumped in pairs, tetramers, and oligomers can produce hundreds of different possibilities. But even the most ardent histone researchers now believe that histones are probably not specific repressors. Rather they bind DNA and serve as the structural framework for eukaryotic structural gene transcription.

Eukaryotic chromatin also contains nonhistone proteins. These are usually acidic and far more varied in their structure than the histones. These nonhistone proteins include RNA polymerase, DNA polymerase, and regulatory proteins. But we still do not know what these regulatory proteins are or how they work, leaving us with the problem of identifying eukaryotic repressor molecules. Recently, however, it has been found that the acidic proteins may be complexing with steroid hormones and serving as activators of specific genes.

The fine structure of chromatin (that is, the structural relationship between DNA and the histone proteins) has recently been clarified. There are particles called *nucleosomes*, which are histones (H2A, H2B, H3, H4) complexed with coiled DNA and regularly spaced along the DNA molecule. These nucleosomes are separated by *spacer DNA*, which has low amounts of histones H1 and H5 associated with it.

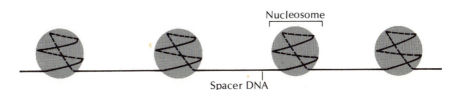

Nucleosome

Spacer DNA

An analysis of these nucleosomes shows that the DNA is wound around the outside of the histone complex. We do not yet understand how this intriguing structure relates to gene function.

Gene amplification   The switching on and off of clusters of functionally related genes by means such as those found in inducible and repressible enzyme systems is one way cells become differentiated from each other during the course of development, but it is certainly not the only way. It has been discovered in the oocytes of amphibians and some insects that rRNA genes whose products are required in large quantities at specific stages of development are often differentially replicated so that massive RNA synthesis can occur. This differential replication of genes in certain cells is called *gene amplification.*

One of the most thoroughly studied examples of gene amplification is the nucleolar organizer region (NO) in *Xenopus.* The NO region contains about 500 copies each of 28S and 18S rRNA. A homozygous mutant with the NO region deleted does not synthesize these rRNA species, so it does not survive beyond the early swimming stage. Until that time, protein synthesis occurs on ribosomes made in the developing oocyte. In a normal embryo (homozygous or heterozygous for the NO region), rRNA—and hence ribosomes—are synthesized beginning at gastrulation.

In the *Xenopus* oocyte, there are about 1000 times more 28S and 18S rRNA genes than in a somatic cell. This massive proliferation of rRNA genes is undoubtedly tied to the extraordinarily high rate of protein synthesis in the very early embryo, but what controls this differential gene proliferation is unknown. Visual evidence of these extra rRNA genes appears in Figure 13.39, where hundreds of nucleoli are in evidence.

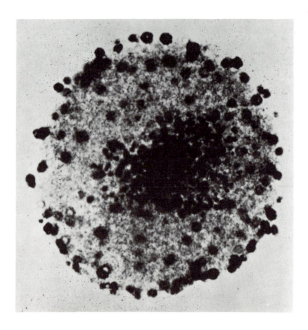

Figure 13.39
Photomicrograph of an isolated germinal vesicle of *Xenopus laevis* treated with a violet stain (the deeply stained spots are some of the hundreds of nucleoli). (From D. D. Brown and J. B. David, *Science* 160:272; reprinted by permission of the authors and the American Association for the Advancement of Science; copyright © by the American Association for the Advancement of Science)

612

(Remember that each nucleolus is indicative of one NO region—a diploid cell has two nucleoli, a tetraploid cell four, and so on.)

The *Xenopus* oocyte nucleoli contain rRNA and rDNA (the rRNA genes) and also protein. O. Miller and his colleagues actually isolated these rRNA genes (Figure 13.40) and examined them using the electron microscope. The electron micrographs have been interpreted as indicating the transcription of rRNA genes and also the presence of DNA regions that are not transcribed.

Gene amplification contributes to the differentiation process by differentially altering the gene content. But it now appears that differential

Figure 13.40
Electron micrographs of gene transcription in the newt nucleolus. [From O. L. Miller and B. R. Beatty, "Visualization of nucleolar genes," *Science* 164:956 (1969); reprinted by permission of O. L. Miller and the American Association for the Advancement of Science; copyright © by the American Association for the Advancement of Science]

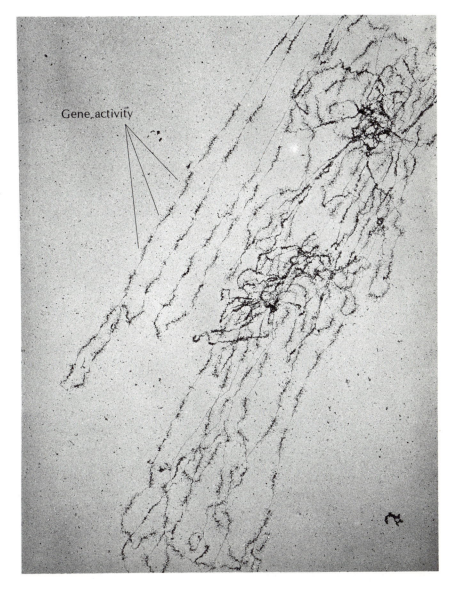

Gene activity

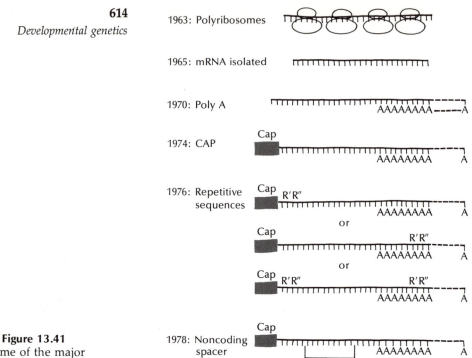

1963: Polyribosomes

1965: mRNA isolated

1970: Poly A

1974: CAP

1976: Repetitive sequences

or

or

1978: Noncoding spacer sequences

Noncoding sequence

Figure 13.41
Some of the major
discoveries about
eukaryotic mRNA

gene amplification is not likely to be a general mechanism for differentiation. It has been confirmed only for rRNA genes.

mRNA in eukaryotes   Only within the last 10 years or so have the structure, function, and regulation of eukaryotic mRNA been successfully examined. Figure 13.41 summarizes some of the major discoveries about eukaryotic mRNA.

Mammalian mRNA was first isolated as a *polyribosome*, ribosomes attached to mRNA. Shortly after this, mRNA was isolated and characterized as about 1500 nucleotides long, low in G+C, and coding for only one polypeptide (that is, eukaryotic mRNA is monocistronic). But before the mRNA is translated in the cytoplasm, it undergoes posttranscriptional modification.

When first transcribed, nuclear RNA is heterogenous in size and is therefore called *heterogenous nuclear RNA* (hnRNA). It is generally much larger than the functional mRNA, tRNA, rRNA species. Some of this RNA never leaves the nucleus. It appears that the hnRNA is tailored by a series of posttranscriptional events into functional mRNA. One event is the addition of poly-A to the 3' end of mRNA in the nucleus after tran-

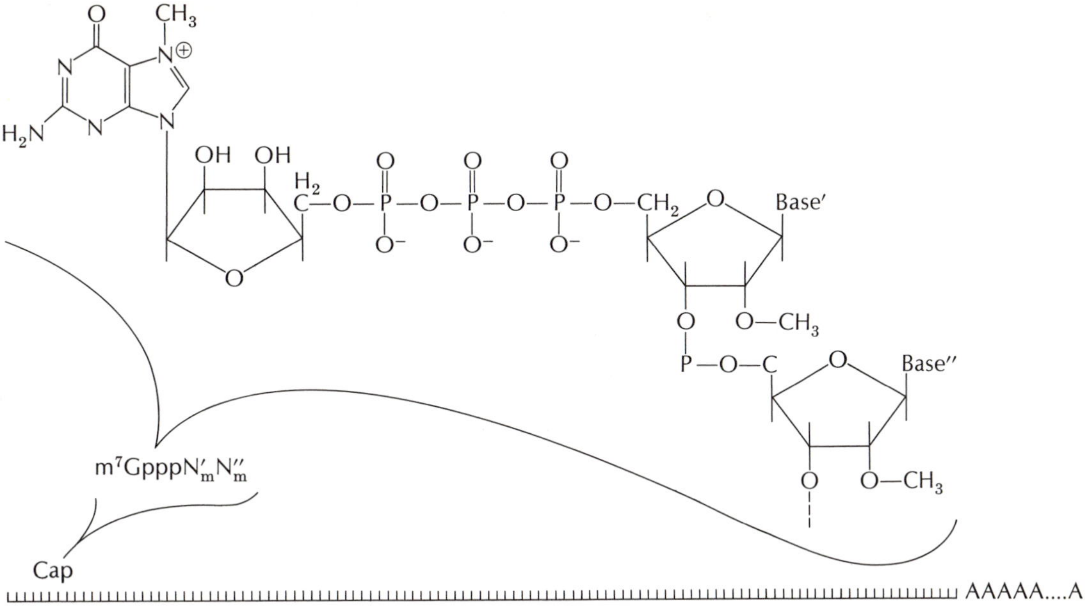

CH₃ → CH_3

Structure labels (Figure 13.42): H_2N, $N^\oplus$, CH_3, OH OH, H_2 C—O—P—O—P—O—P—O—CH_2, O⁻ O⁻ O⁻, Base′, O—CH_3, P—O—C, Base″, O—CH_3

$m^7GpppN'_mN''_m$

Cap

AAAAA....A

Figure 13.42
Eukaryotic mRNA, showing structure of methylated "cap"

scription. This may function in the transport of mRNA out of the nucleus or in the ultimate breakdown of the mRNA. Another event is the cleavage of hnRNA into smaller molecules. The sequences that are cleaved may be transcriptional control sequences of no further importance in translation. Finally, eukaryotic mRNA has a methylated "cap" added to its 5′ end. This cap (Figure 13.42) probably plays a role in binding mRNA to ribosomes. It is apparently added in the nucleus.

Other details of the structure of eukaryotic mRNA may provide important insights into translational control. The mRNA molecules whose sequences have been determined contain short repetitive regions, perhaps at both ends of the molecule. These repetitive sequences may enable mRNA to fold back on itself and form double-stranded regions that would be protected against degradation (Figure 13.43).

Another structural oddity has emerged from studies of mRNA: mRNA molecules contain extensive noncoding nucleotide sequences, as

Figure 13.43
Repetitive sequences in hnRNA and their possible role in protecting the molecule from degradation. (R = repeated region)

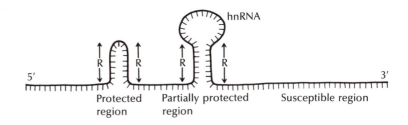

hnRNA

5′   R   R   R   R   3′

Protected region   Partially protected region   Susceptible region

well as coding sequences. For example, in rabbit α-globin mRNA, we have the following structure.

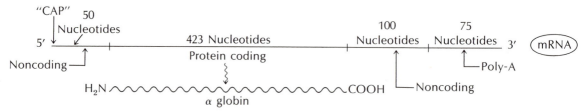

What are these noncoding regions doing? We do not know whether they are involved in initiation or termination, cleavage, poly-A addition, or some other function. But they do appear to be a regular feature of eukaryotic mRNA molecules, and they are important in either transcriptional or translational control.

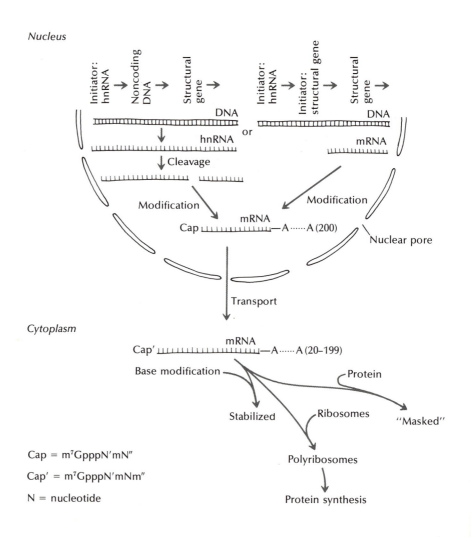

Figure 13.44
Eukaryotic mRNA formation. (After J. E. Darnell, 1976. *Prog. Nuc. Acid Res.* and *Mol. Biol.* 19:493)

Cap = m⁷GpppN′mN″

Cap' = m⁷GpppN′mNm″

N = nucleotide

More recent studies of the structure of eukaryotic mRNA have added a new dimension to the field. Some of the mRNA sequences have noncoding sequences located in the *middle* of the sequence. These are called *intervening* or *spacer* sequences. This suggests that mRNA of nucleated cells is synthesized as hnRNA, which may then be tailored by clipping out *interior* rather than end regions. The spacer sequences may aid in regulating protein synthesis after mRNA synthesis, but this is only speculation. There is also the distinct possibility that the noncoding sequences in eukaryotic mRNA are crucial to masking and stabilizing (see page 582).

We summarize some of our thoughts on eukaryotic mRNA in Figure 13.44. The hnRNA is transcribed in the nucleus and either cleaved and modified, or just modified. Modifications involve adding about 200 A units and a cap. This mRNA is transported out of the nucleus, where some of the A units are removed and the cap is further modified. The mRNA so formed can now follow one of three paths: It can be translated, stabilized, or masked.

Repeated DNA and the structure and organization of eukaryote chromosomes

Much of our understanding of chromosome structure comes from the giant chromosomes (polytene chromosomes) of the *Drosophila* salivary gland. We can view this chromosome as extended DNA–protein fibers separated by dark-staining bands (heterochromatin) in which the DNA is highly coiled and supercoiled (Figure 13.45). The puffs we discussed ear-

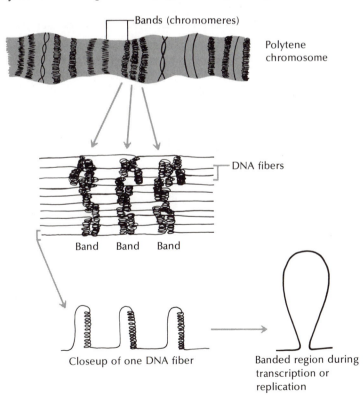

Bands (chromomeres)

Polytene chromosome

DNA fibers

Band   Band   Band

Closeup of one DNA fiber

Banded region during transcription or replication

Figure 13.45
Organization of a eukaryote chromosome (*Drosophila*)

lier are seen as loops of uncoiled DNA ready for transcription.

The DNA of higher organisms is often arranged in a manner that distinguishes it from the DNA of prokaryotes. Higher-organism DNA has sequences that are repeated several times as well as sequences that are present only once (*unique sequences*). Bacterial DNA, by contrast, is a linear array of unique sequences except for the rRNA genes, which may be repeated. Questions perplexing researchers today concern the function and organization of repeating and unique DNA sequences. Do the repeated sequences code for polypeptides? And how are these sequences organized in chromosomes? Many biologists believe that the organization of repeated DNA sequences is related to gene control.

The most recent investigations suggest that at least 95 percent of the mRNA in a sea urchin cell is transcribed from *unique* DNA sequences— genes of which there is only one copy each. If it is generally true that only unique sequences are transcribed into mRNA, what function does the repeated DNA serve? As mentioned, many believe that repeated DNA can be built into models of eukaryotic gene control, but there must first be an understanding of DNA sequence organization.

Britten and E. H. Davidson recently speculated on the organization of DNA in animal chromosomes. They suggest that about 50 percent of animal chromosomal DNA consists of closely interspersed sequences of repetitive (200–400 nucleotides) and unique (650–900 nucleotides) sequences. The remaining 50 percent consists of long unique sequences (about 4000 nucleotides) interspersed with short repetitive sequences and other poorly understood sequences.

C. Thomas, elaborating on the work of Britten and Davidson, has proposed an interpretation of DNA organization in the *Drosophila* chromosome. He suggests that the *Drosophila* chromosome is composed of DNA segments 5 μm long. Each segment is organized as repeated DNA bracketed by equal amounts of nonrepeated, unique DNA sequences. However, the following question arises. How does the organization of *Drosophila* DNA relate to the cytological organization of the *Drosophila* chromosome and, more specifically, to the chromosome banding pattern? Recall that the *Drosophila* chromosome is composed of a series of bands, or chromomeres. Most of the DNA (95 percent) is in the dark bands, and each of these bands is proposed to have one genetic function. But if a band produces one type of protein, there is enough DNA in a band to carry about 30 copies of a gene. There are two ways of interpreting this DNA-band relationship. Either the gene coding for the protein is repeated several times, or most of the DNA in a band does not code for protein. Thomas favors the first interpretation, and although he cannot yet prove that protein-coding genes are multiply represented in a band, his data do not contradict the hypothesis.

J. Bonner and J. R. Wu analyzed the problem of DNA organization in *Drosophila* chromosome bands and developed a model that is consistent with the data of Britten and Davidson and of Thomas. Their model presents each *Drosophila* chromomere as having 30 to 35 unique sequences that are separated by short, repeated sequences. The repeated sequences in any one band are all alike. The unique sequences could each code for a specific protein, or they might be functionally related so that we could coin the phrase "one band, one function." If all the proteins were the same, we could say "one band, one gene, one protein" (assuming protein and polypeptide are equivalent).

Much of the hnRNA discussed in the previous section may relate to DNA organization in the chromomeres. The repeated DNA sequences may be transcribed into large hnRNA molecules and then tailored down to mRNA molecules that are ultimately translated.

A final aspect of repeated DNA should be considered. If, as Thomas suggests, a polypeptide-coding gene is multiply represented in a band, how can mutations be induced? Mutating one gene would still leave other normal copies. The *master–slave hypothesis,* suggested by Callan in 1963, provides an interpretation of the function of redundant genes. This hypothesis states that there is a single "master" gene and multiple copies of "slave" genes. During meiotic prophase, each slave gene pairs with the master gene and is corrected according to the DNA sequence of the master. In this way, mutations in the slave genes are corrected. However, a mutation in the master gene will be passed on to all slave genes, because they are not corrected against it. Though not established experimentally, the hypothesis is an interesting one and certainly pertinent to the problems outlined in this section.

A model of eukaryotic gene regulation

Regulation of gene activity in eukaryotes requires concepts that differ from the classic operons of prokaryotes. In eukaryotes, structural genes are not arranged contiguously; coordinately regulated genes are often on different chromosomes. With a relatively small number of chromosomal protein species, it may be that our concept of the repressor requires alteration. The necessity for this is apparent when you consider the circuitous route a protein repressor must traverse: synthesized in the cytoplasm from a processed mRNA, then transported back to the nucleus for its repressing duties.

One model for the regulation of gene activity that takes into account the complexities of the eukaryotic genome is the Britten–Davidson model (Figure 13.46). They propose the existence of *sensor regions* that respond to chemical messengers such as cAMP, hormones, or specific enzymes. When activated, the sensor region promotes the transcription of an *integrator gene.* The product of this integrator gene is an *activator RNA,* which is recognized by one or more *receptor sites* located at other regions of the genome. The activator RNA stimulates transcription at adjacent structural gene(s) and the eventual production of a protein product.

Building on this simplified scheme, Britten and Davidson suggest that by employing multiple sensors, integrators, and receptors, complex

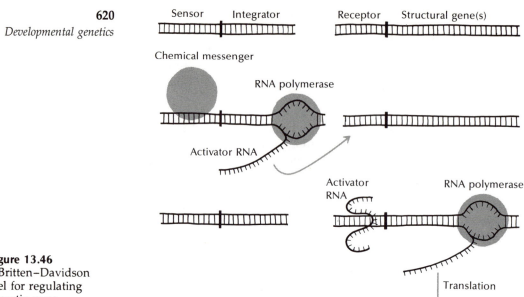

Sensor   Integrator   Receptor   Structural gene(s)

Chemical messenger

RNA polymerase

Activator RNA

Activator RNA   RNA polymerase

Translation

Polypeptide

Figure 13.46
The Britten–Davidson
model for regulating
eukaryotic gene
expression

genetic coordination can be achieved (Figure 13.47). From a model such as this, we can realize considerable flexibility in gene regulatory patterns.

The Britten–Davidson model has some attractive features. By employing an RNA species as an activator, we need not be concerned with proteins diffusing in from the cytoplasm. We also see that regulatory regions (sensors, integrators, and receptors) can make up a large part of the genome relative to structural genes, and recent studies indicate that regulatory DNA regions are extensive. But this is still a model, and others do exist. Several years will probably be required before this and other models can be thoroughly tested.

The genetics of aging   It is appropriate that we conclude this chapter with a discussion of that most visible of developmental phenomena: aging, a process referred to by many as programmed death. Aging is characterized by two general features. First, there is a gradual decline in the adaptability of the organism to its normal environment following the onset of reproductive maturity. Second, there is a more or less fixed species-specific life span.

Research in aging suffers from the lack of a unifying paradigm. Instead, we have an abundance of pet theories and hypotheses. In all likelihood, aging will eventually be understood as the result of a combination of mechanisms that interact to cause cell dysfunction, cell death, tissue dysfunction, organismal decline, and finally organismal death. Basic to all the aging mechanisms is the gene and its regulation. Species-specific life spans certainly argue for genetic control of aging.

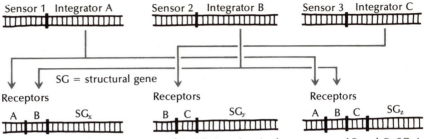

Figure 13.47
The Britten–Davidson model of differential gene activity as a consequence of differing patterns of activator RNA species. In the presence of activators A and B, SG_x is active; SG_y is active in the presence of B and C; all three SG's are active in the presence of A, B, and C. (SG = structural gene)

In the presence of integrators A and B, SG_x is active. In the presence of B and C, SG_y is active. In the presence of A, B, and C, all three SG's are active.

In viewing aging, we can suggest a general model (Figure 13.48). Following birth there is a period of growth and development, culminating in the attainment of reproductive maturity. During this process, a process programmed in the genetic material, cellular environmental changes occur. As we pass through reproductive maturity, deteriorating functions usher us into senescence and ultimately death. The Hayflick limit, to be

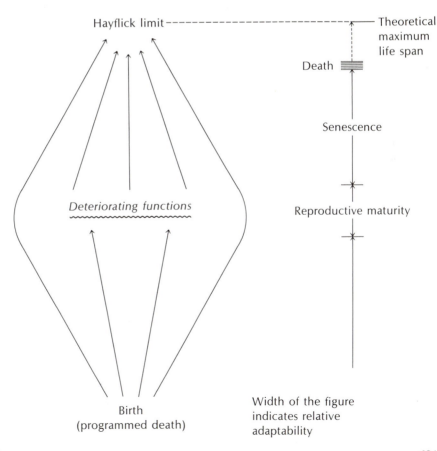

Figure 13.48
A general model of aging

discussed shortly, is the absolute maximum life span attainable by a cell. As such, it reflects the theoretical maximum life span of an individual. The Hayflick limit could be attained in humans, for example, by eliminating the premature causes of death, such as cancer and heart disease.

Let us examine some of the ideas that fit our general model of aging. First we shall look at the idea of programmed death, or a developmental program for aging. Leonard Hayflick found that diploid cells in culture have a clearly defined life span, which is dependent on the number of cell doublings rather than chronological age. Human embryo fibroblasts in culture divide about 50 times before they deteriorate, become senescent, lose their capacity to divide, and finally die. Human fibroblasts taken after an infant is born divide only 20 to 30 times. In general, investigators have found that the greater the life span of the species, the greater the number of times its cells divide in culture. This limit to the number of divisions is called the *Hayflick limit,* and many believe it is related to aging. Senescence is seen as a deterioration in cell function before the Hayflick limit is reached.

Many investigators have suggested that a developmental clock determines the Hayflick limit. One such model has been suggested by Strehler (Figure 13.49). He argues, in his *codon-restriction theory of aging,* that the kinds of proteins synthesized by a cell are controlled by the set of code words that a cell can decode. He finds that tRNA and aminoacyl-tRNA synthetase species change in different tissues during development and aging. To explain these changes, Strehler postulates sequential gene

Figure 13.49
Codon-restriction theory of aging (see text)

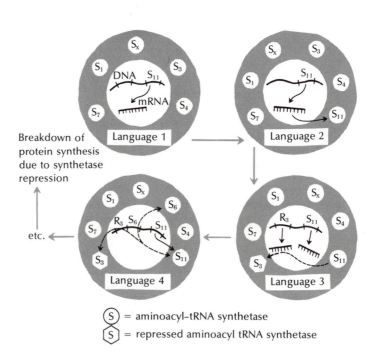

$\textcircled{S}$ = aminoacyl–tRNA synthetase

$\boxed{S}$ = repressed aminoacyl tRNA synthetase

expression controlled by changes in codon-reading capacities. A given set of codon-reading capacities leads automatically to the addition of new and/or the subtraction of old codon usages as a consequence of the translatability of synthetase mRNA or synthetase-repressor mRNA encoded in the prior language set. Accordingly, aging is the result of sequential activation and repression of gene activity.

On the assumption that cells have a specific life span, and that it is defined by intrinsic genetic programming, what other factors may contribute to deteriorating cell functions and senescence? Mutation and repair may be crucial to the aging process, as we see in Figure 13.50 (an expanded version of Figure 13.48).

Under some circumstances, x rays can shorten a life span; an alkylating agent such as EMS does not. The fact that a powerful chemical mutagen such as EMS does not shorten life spans is seen as an argument against the *somatic mutation theory*, which suggests that aging occurs as the consequence of a constant somatic cell mutation rate. This in turn causes cellular dysfunction and ultimately death. But we need to temper the EMS and x-ray studies by pointing out that x-ray-induced aging is similar

Figure 13.50
A summary of the
contributing causes of
aging (see text)

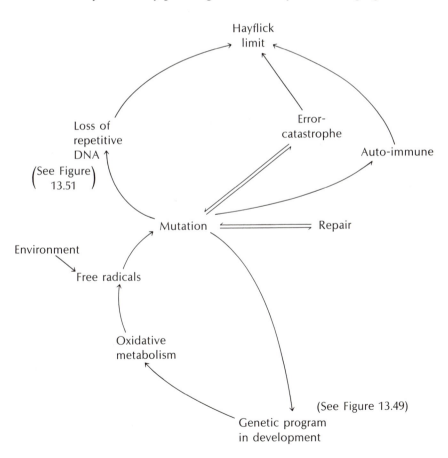

but not identical to natural aging. Point mutations such as those induced by EMS are repairable in younger cells but rarely in older cells, and chromosome breaks are generally not repairable. Support for the mutation theory of aging comes from studies that show that mutations accumulate in older tissues and that DNA repair is more effective in longer-lived species.

Mutations in dominant control genes could influence the genetic program, speeding up the deterioration (aging process). How might these mutations occur? In a replicating cell, errors in DNA replication may take place. In a postreplicating cell, it has been suggested that highly reactive free radicals may damage DNA and other cellular organelles. Free radicals, such as superoxide ($\cdot O_2^-$), can be generated during the course of normal cell metabolism or by environmental pollution (ionizing radiation or radiomimetic chemicals).

Mutations may have different consequences, depending on where in the genome they occur. For example, mutation may lead to auto-immune diseases, such as pernicious anemia and arthritis, in which antibodies to the self are generated. Auto-immunity may occur when a repressed virus in the genome is derepressed and synthesizes an antigen that elicits an antibody response. Or another protein may become so altered by mutation that it stimulates an antibody response.

But aging may not involve programmed genetic death or somatic mutations. Orgel suggests that an *error-catastrophe model* may apply. According to this model, errors accumulate in the amino acid sequence of proteins and result in other metabolic mistakes. Eventually a snowballing effect takes over, leading to cell deterioration and death. Evidence exists on both sides of this issue. On the pro side, abnormal enzyme molecules accumulate with age. On the con side, viruses grown on aged cells are no different from viruses grown on young cells. Since viruses use cell protein-synthesizing machinery, we would expect errors to occur in viruses grown on aged cells, on the assumption that aged cells had accumulated errors in their translational apparatus.

Repetitive or redundant DNA may be one way of determining the life span of species. It has been argued that as mutations accumulate in functioning genes, reserve sequences containing the same genetic information take over. When the redundancy is exhausted, senescence and death occur. Those species with the most redundancy would have the longest life span.

Redundancy can also be eliminated by the actual excision of redundant DNA sequences (Figure 13.51). Here we see slippage occurring in repeated DNA sequences, with the subsequent excision of repeated segments.

Aging can probably not be explained by any one theory, but rather by a complex of mechanisms such as those discussed here. There is currently a great deal of research on mechanisms of aging, and we may soon see a major breakthrough.

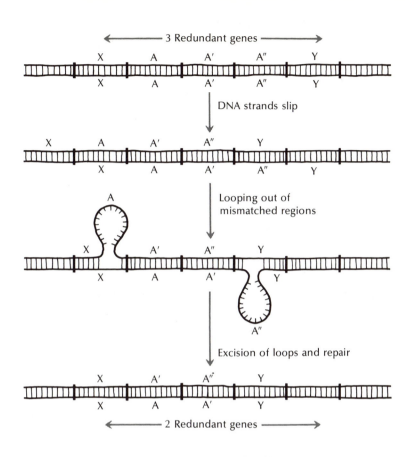

Figure 13.51
A model for the loss of
genetic redundancy

Overview

In this chapter, we examined the regulation of gene activity in prokaryotes and eukaryotes. We began by looking at some important studies showing that nuclei are genetically equivalent, at least during early embryogenesis. Driesch demonstrated that even if a nucleus is in a cytoplasm other than the one it would normally occupy, development of the resulting embryo still proceeds normally. Spemann showed that a nucleus from a 16-cell embryo retains the potential for stimulating normal development from a single cell. Briggs and King, using nuclear transplantation studies, proved that even nuclei from differentiated cells can support normal development in enucleated eggs.

We next examined evidence for regional differentiation in eggs and early embryos. The cytoplasm in different regions of the egg is differentiated, and it is clear that the activity of the nuclear genes during development is limited by the properties of the cytoplasm. The cytoplasm thus regulates a cell's development, but the nucleus determines the properties of the cell.

What evidence do we have that different genes are active at different developmental stages? In *Xenopus,* it was shown that rRNA synthesis is cytoplasmically regulated; rRNA-synthesizing nuclei transplanted into an oocyte in which no rRNA are active stop synthesizing rRNA. Hemoglobin synthesis during human development also shows that different genes are active at different times. Chromosome puffs provide cytological evidence for differential gene activity.

The mRNA of prokaryotes is short-lived, but the mRNA of eukaryotes is often long-lived. This longevity may be the result of stable mRNA species, such as hemoglobin mRNA. Or the longevity may be the result of masked mRNA, in which the mRNA is not translated until long after it is transcribed. A protein complex usually maintains the masked mRNA condition. The mRNA species are, of course, essential to eukaryote development.

In microorganisms, we also see that gene activity during development is characterized by a specific temporal program of gene activity and gene product interaction. To demonstrate this point, we looked at the morphogenesis of the T4 bacteriophage.

The regulation of gene activity is more thoroughly understood in prokaryotes than it is in eukaryotes. The classic model of prokaryotic gene regulation is the operon, developed by Jacob and Monod. The *lac* operon consists of three structural genes, an operator, and a promoter; there is also a regulatory gene. The regulatory gene codes for a repressor protein that binds to the operator and prevents the RNA polymerase from transcribing the structural genes. But when the repressor is complexed with an inducer, it no longer can bind to the operator and RNA transcription occurs. This is an inducible enzyme system.

In a repressible operon, a corepressor binds to the repressor; this complex then binds to the operator, preventing transcription.

Cyclic AMP plays an important role in the function of many operons. It is now known that cAMP and catabolite gene-activator protein (CAP) first complex with the promoter and facilitate the binding of RNA polymerase. When cAMP levels are depressed, for example during glucose metabolism, the *lac* operon is poorly transcribed.

The arabinose operon is an example of positive and negative genetic control. In this instance, the product of the regulatory gene binds to and represses the operator. But in the presence of the inducer, the repressor protein becomes an inducer by binding to the promoter and facilitating mRNA transcription by RNA polymerase.

The lysogenic behavior of the λ bacteriophage is another example of prokaryotic gene regulation. A λ repressor shuts off almost all λ gene activity in the establishment of lysogeny. In the lytic phase, however, the repressor gene is shut down, and the two early operons are active and code for DNA synthesis and excision. Early gene products induce the late operon to function, producing structural proteins.

Though transcriptional regulation is the most common means of regulation, translational control is also important. All of the *lac* structural genes, for example, are not translated with equal efficiency.

When *E. coli* is starved for amino acids, the synthesis of rRNA and tRNA stops. This type of regulation is called the *stringent response.* The unusual nucleotides ppGpp and pppGpp are involved in this regulatory mechanism.

We next turned our attention to the regulation of gene activity in eukaryotes. The processes of cell differentiation and cell determination are distinct. During embryogenesis a cell becomes committed or determined to a specific phenotype; it does not become differentiated until that phenotype is expressed.

Histones, small proteins, complex with DNA and prevent transcription. But because they lack specificity and because there are only six classes of them, they are not considered specific repressors like those present in prokaryotic systems. Acidic proteins may be more specific.

Gene amplification of rDNA to transcribe the enormous amount of rRNA required by cells is an effective regulatory device in eukaryotes. But gene amplification is probably not a model for how cells differentiate.

In eukaryotes, we have recently discovered a considerable degree of control in mRNA transcription; hnRNA is transcribed in the nucleus and either cleaved and modified, or just modified. These modifications involve adding poly-A and a cap. Following transport out of the nucleus, there is further modification before the mRNA is translated, stabilized, or masked.

The genome of eukaryotes is organized differently from that of prokaryotes. For one thing, there are extensive repeated DNA sequences that probably have a control function. The banding pattern we observe in polytene chromosomes may reflect a one gene–one band type of organization in which each band has a specific function with multiply repeated genes. Histones are viewed as organizing the structure here.

We examined a model for gene regulation in eukaryotes. The genome in this model was organized into sensor regions that respond to chemical messengers: the integrator gene that transcribes an activator RNA, which acts at a receptor site to promote transcription of structural genes.

Finally, we looked at aging and considered some of the ideas that have been advanced about this developmental phenomenon. We view the process as a playing out of a genetic tape, leading to deteriorating functions and finally death. The Hayflick limit may reflect the theoretical maximum life span attainable by a species.

Questions and problems

13.1 Discuss the experimental findings that argue against a preformationist concept and support an epigenetic point of view.

13.2 Evaluate evidence that supports the idea that a gene's activity is limited to certain periods during development.

13.3 How have nuclear transplantation studies contributed to the concept of the gene?

13.4 How might the information contained in this chapter contribute to our understanding of mosaicism?

13.5 Transcription does not usually occur during mitosis. Describe a reasonable basis for this observation.

13.6 Review the key differences between inducible, constitutive, and repressible enzyme systems.

13.7 Using all sources of information, discuss the concept of the gene.

13.8 In what ways can the concept of induction and repression, as developed in microbial systems, be applied to eukaryotes?

13.9 A student of developmental biology may approach the experimental analysis of development either with or without the use of genetic tools. Compare and contrast the kinds of information that each experimental approach may yield, including the limitations and advantages of each approach.

13.10 Discuss instances of positive and negative control in the life cycle of the λ bacteriophage (both lytic and lysogenic pathways).

13.11 An *E. coli* strain is isolated that does not synthesize any β-galactosidase, even when an inducer is present. What genetic defects might this strain have?

13.12 The *E. coli* strain mentioned in Question 13.11 has a nonsense mutation suppressor introduced into it. The strain now makes β-galactosidase, but the enzyme has only partial activity. How would you now interpret this mutant strain?

13.13 Another *E. coli* strain constitutively synthesizes β-galactosidase. When this strain is made partially diploid by the insertion of F' I^+Z^-, it remains a constitutive phenotype. What is the basis of the genetic defect in this strain?

13.14 How would your answer to Question 13.13 change if, when the F' I^+Z^- plasmid was introduced, the strain became inducible?

13.15 The following strains of *E. coli* are examined for arabinose operon activity. What phenotypes do you expect for each of the following four genotypes? Explain each answer.
   (a) *araA⁻ araC^c/F':araA⁺ araC⁻*
   (b) *araA⁻ araC^c/F':araA⁺ araC⁺*
   (c) *araA⁺ araC⁺/F':araA⁻ araC⁻*
   (d) *araA⁺ araC⁻/F':araA⁻ araC⁺*

13.16 Describe the phenotype of an *araC* mutant with only its operator region deleted.

13.17 The galactose operon in *E. coli* has the following properties. The *galR* gene synthesizes a regulatory protein; *galR⁻* cells exhibit constitutive enzyme synthesis; *galRˢ* cells are not inducible; *galRˢ/F':galR⁺* cells are not inducible; and *galR⁻/F':galR⁺* cells are inducible. How would you characterize this operon?

13.18 If the *C1* gene in λ had a nonsense mutation, what phenotype would the phage express? What if the *tof* gene were deleted?

13.19 Actinomycin D, an inhibitor of RNA synthesis, does not seriously affect protein synthesis in newly fertilized eggs. It does not seriously affect protein synthesis in embryos until the blastula or even gastrula stage. After this, protein synthesis declines and the embryo's further development ceases. How would you interpret this?

13.20 Some investigators have pointed out that the Hayflick limit does not pertain to all cells. Assuming this is true, must we reject the pertinence of the Hayflick limit to the aging process?

References

Barth, L. J. 1964. *Development: Selected Topics.* Reading, Mass., Addison-Wesley.

Beckwith, J., and D. Zipser (eds.). 1970. *The Lactose Operon.* Cold Spring Harbor, N.Y., Cold Spring Harbor Laboratory of Quantitative Biology.

Beerman, W. (ed.). 1972. *Developmental Studies on Giant Chromosomes.* New York, Springer Verlag.

Bertrand, K., L. Korn, F. Lee, T. Platt, C. L. Squires, C. Squires, and C. Yanofsky. 1975. New features of the regulation of the tryptophan operon. *Science* 189:22–26.

Black, L. W., and L. M. Gold. 1971. Pre-replicative development of the bacteriophage T4: RNA and protein synthesis *in vivo* and *in vitro. J. Molec. Biol.* 60:365–388.

Bonner, J., and J. R. Wu. 1973. A proposal for the structure of the *Drosophila* genome. *Proc. Natl. Acad. Sci. U.S.* 70:535–537.

Brenchley, J. E., and L. S. Williams. 1975. Transfer RNA involvement in the regulation of enzyme synthesis. *Ann. Rev. Microbiology* 29:251–274.

Brenner, M., and B. N. Ames. 1971. The histidine operon and its regulation. In H. Vogel (ed.), *Metabolic Regulation.* New York, Academic.

Briggs, R., and T. J. King. 1953. Factors affecting the transplantability of nuclei of frog embryonic cells. *J. Expl. Zool.* 122:485–506.

Briggs, R., and T. J. King. 1960. Nuclear transplantation studies on the early gastrula (*Rana pipiens*): I. Nuclei of presumptive endoderm. *Dev. Biol.* 2:252–270.

Britten, R. J., and E. H. Davidson, 1969. Gene regulation for higher cells: A theory. *Science* 165:349–357.

Brown, D. D., and I. B. Dawid. 1968. Specific gene amplification in oocytes. *Science* 160:272–280.

Brown, S. W. 1966. Heterochromatin. *Science* 151:417–425.

Burnet, M. 1974. *Intrinsic Mutagenesis: A Genetic Approach to Aging.* New York, Wiley.

Calendar, R. 1970. The regulation of phage development. *Ann. Rev. Microbiology* 24:241–296.

Calhoun, D. H., and G. W. Hatfield. 1975. Autoregulation of gene expression. *Ann. Rev. Microbiology* 29:275–301.

Callan, H. G. 1967. The organization of genetic units in chromosomes. *J. Cell Sci.* 2:1–7.

Calvo, J. M., and G. R. Fink. 1971. Regulation of biosynthetic pathways in bacteria and fungi. *Ann. Rev. Biochem.* 40:943–968.

Campbell, A. M. 1969. *Episomes.* New York, Harper and Row.

Cashel, M. 1975. Regulation of bacterial ppGpp and pppGpp. *Ann. Rev. Microbiology* 29:301–318.

Casjens, S., and J. T. King. 1975. Virus assembly. *Ann. Rev. Biochem.* 44:555–611.

Chromosome structure and function. 1974. *Cold Spring Harbor Symp. Quant. Biol.* 38.

Clever, U. 1968. Regulation of chromosome function. *Ann. Rev. Genetics* 2:11–30.

Corwin, H. O., and J. B. Jenkins. 1976. *Conceptual Foundations of Genetics: Selected Readings.* Boston, Houghton Mifflin.

Crick, F. H. C. 1971. General model for the chromosomes of higher organisms. *Nature* 234:25–27.

Davidson, E. H. 1976. *Gene Activity in Early Development.* New York, Academic.

Davidson, E. H., and R. J. Britten. 1973. Organization, transcription, and regulation in the animal genome. *Quart. Rev. Biol.* 48:565–613.

Davies, J., and F. Jacob. 1968. Genetic mapping of the regulator and operator genes of the *lac* operon. *J. Molec. Biol.* 36:413–417.

Dickson, R. C., J. Abelson, W. M. Barnes, and W. S. Reznikoff, 1975. Genetic regulation: The *lac* control region. *Science* 187:27–35.

Dove, W. F. 1968. The genetics of the lambdoid phages. *Ann. Rev. Genetics* 2:305–340.

Driesch, H. 1892. The potency of the first two cleavage cells in echinoderm development. Experimental production of partial and double formations. Reprinted in B. H. Willer and J. M. Oppenheimer (eds.), *Foundations in Experimental Embryology,* 1964. Englewood Cliffs, N.J., Prentice-Hall.

Ebert, J. W., and I. M. Sussex. 1970. *Interacting Systems in Development.* New York, Holt, Rinehart, Winston.

Echols, H. 1972. Developmental pathways for the temperate phage: Lysis vs. lysogeny. *Ann. Rev. Genetics* 6:157–190.

Englesberg, E., C. Squires, and F. Meronk. 1969. The L-arabinose operon in *E. coli* B/r: A genetic demonstration of two functional states of the product of a regulator gene. *Proc. Natl. Acad. Sci. U.S.* 62:1100–1107.

Epstein, W., and J. Beckwith. 1968. Regulation of gene expression. *Ann. Rev. Biochem.* 37:411–436.

Finch, C. E., and L. Hayflick (eds.). 1977. *Handbook of the Biology of Aging.* New York, Van Nostrand Reinhold.

Gehring, W. 1967. Clonal analysis of determination dynamics in cultures of imaginal discs in *Drosophila melanogaster*. *Dev. Biol.* 16:438–456.

Gehring, W. J. 1976. Developmental genetics of *Drosophila*. *Ann. Rev. Genetics* 10:209–252.

Gelderman, A. H., A. V. Rake, and R. J. Britten. 1971. Transcription of non-repeated DNA in neonatal and fetal mice. *Proc. Natl. Acad. Sci. U.S.* 68:172–176.

Gene regulation and development in *Drosophila* (a symposium). 1977. *Amer. Zoologist,* Summer (Volume 17).

Gilbert, W., and B. Müller-Hill. 1966. Isolation of the *lac* repressor. *Proc. Natl. Acad. Sci. U.S.* 56:1891–1898.

Gilbert, W., and B. Müller-Hill. 1967. The *lac* operator is DNA. *Proc. Natl. Acad. Sci. U.S.* 58:2415–2421.

Gurdon, J. B. 1963. Nuclear transplantation in amphibia and the importance of stable nuclear changes in promoting cellular differentiation. *Quart. Rev. Biol.* 38:54–78.

Gurdon, J. B. 1968. Transplanted nuclei and cell differentiation. *Sci. Amer.* 219:24–35.

Gurdon, J. B., and D. D. Brown. 1965. Cytoplasmic regulation of RNA synthesis and nucleolus formation in developing embryos of *Xenopus laevis. J. Molec. Biol.* 12:27–35.

Gross, P. R. 1967. The control of protein synthesis in embryonic development and differentiation. *Curr. Topics Dev. Biol.* 2:1–46.

Gross, S. R. 1969. Genetic regulatory mechanisms in the fungi. *Ann. Rev. Genetics* 3:395–424.

Hadorn, E. 1968. Transdetermination in cells. *Sci. Amer.* 219:110–120.

Hammerling, J. 1943. Ein- und zweikernige Transplantate zwischen *Acetabularia mediterranea* and *A. crenulata. Z. Induktive Abstammungs- und Vererbungslehre* 81:114–180.

Hartwell, L. H. 1970. Biochemical genetics of yeast. *Ann. Rev. Genetics* 4:373–396.

Hershey, A. D. (ed.). 1971. *The Bacteriophage Lambda.* Cold Spring Harbor, N.Y., Cold Spring Harbor Laboratory of Quantitative Biology.

Hess, O., and G. F. Meyer. 1968. Genetic activities of the **Y** chromosome in *Drosophila* during spermatogenesis. *Adv. Genetics* 14:171–223.

Holliday, R., and J. E. Pugh. 1975. DNA modification mechanisms and gene activity during development. *Science* 187:226–232.

Irr, J., and E. Englesberg. 1970. Nonsense mutants in the regulator gene ara C of the L-arabinose system of *E. coli* B/r. *Genetics* 65:27–39.

Jacob, F., and J. Monod. 1961. Genetic regulatory mechanisms in the synthesis of proteins. *J. Molec. Biol.* 3:318–356.

Jacob, F., and J. Monod. 1965. Genetic mapping of the elements of the lactose region in *E. coli. Biochem. Biophys. Res. Comm.* 18:693–701.

Jobe, A., and S. Bourgeois. 1972. *lac* repressor–operator interaction: VI. The natural inducer of the *lac* operon. *J. Molec. Biol.* 69:397–408.

Johns, E. W., and T. A. Hoare. 1970. Histones and gene control. *Nature* 226:650–651.

Kasai, T. 1974. Regulation of the expression of the histidine operon in *Salmonella typimurium. Nature* 249:523–527.

Kauffman, S. 1973. Control circuits for determination and transdetermination. *Science* 181:310–318.

Kornberg, R. D. 1977. Structure of chromatin. *Ann. Rev. Biochem.* 46:931–954.

Lewin, B. 1975. Units of transcription and translation: The sequence components of heterogenous nuclear RNA and messenger RNA. *Cell* 4:77–93.

Lewin, B. 1975. Units of transcription and translation: The relationship between heterogenous nuclear RNA and messenger RNA. *Cell* 4:11–20.

Lodish, H. F. 1976. Translational control of protein synthesis. *Ann. Rev. Biochem.* 45:39–72.

MacLean, N. 1976. *Control of Gene Expression.* New York, Academic.

Markert, C., and H. Ursprung. 1971. *Developmental Genetics.* Englewood Cliffs, N.J., Prentice-Hall.

McMahon, D. 1974. Chemical messengers in development: A hypothesis. *Science* 185:1012–1021.

Metzenberg, R. L. 1972. Genetic regulatory mechanisms in *Neurospora*. *Ann. Rev. Genetics* 6:111–132.

O'Malley, B. W., H. C. Towle, and R. J. Schwartz. 1977. Regulation of gene expression in eukaryotes. *Ann. Rev. Genetics* 11:239–276.

Pardee, A. B., F. Jacob, and J. Monod. 1959. The genetic control and cytoplasmic expression of inducibility in the synthesis of β-galactosidase by *E. coli*. *J. Molec. Biol.* 1:165–178.

Pastan, I. 1972. Cyclic AMP. *Sci. Amer.* (August)

Pastan, I., and R. Perlman. 1969. Cyclic AMP in bacteria. *Nature* 169:339–344.

Perry, R. P. 1976. Processing of RNA. *Ann. Rev. Biochem.* 45:605–630.

Postlethwait, J. H., and H. A. Schniederman. 1973. Developmental genetics of *Drosophila* imaginal discs. *Ann. Rev. Genetics* 7:381–434.

Ptashne, M. 1967. The isolation of the λ phage repressor. *Proc. Natl. Acad. Sci. U.S.* 57:306–313.

Ptashne, M. 1967. Specific binding of the λ phage repressor to λ DNA. *Nature* 214:232–234.

Ptashne, M., K. Backman, M. Z. Humayun, A. Jeffrey, R. Maurer, B. Meyer, R. T. Saver. 1976. Autoregulation and function of a repressor in bacteriophage lambda. *Science* 194:156–161.

Radding, C. M. 1969. The genetic control of phage-induced enzymes. *Ann. Rev. Genetics* 3:363–394.

Reichardt, L., and A. D. Kaiser. 1971. Control of λ repressor synthesis. *Proc. Natl. Acad. Sci. U.S.* 68:2185–2189.

Reznikoff, W. S. 1972. The operon revisited. *Ann. Rev. Genetics* 6:133–156.

Ruddle, F. H. (ed.). 1973. *Genetic Mechanisms of Development*. New York, Academic.

Rutter, W. J., R. L. Pictet, and P. W. Morris. 1973. Toward molecular mechanisms of developmental processes. *Ann. Rev. Biochem.* 42:601–646.

Ryan, A. M., and E. Borek. 1971. The relaxed control phenomenon. *Prog. Nuc. Acid Res. Molec. Biol.* 11:193–228.

Salser, W., R. F. Gesteland, and A. Bolle. 1970. Transcription during bacteriophage T4 development: A demonstration that distinct subclasses of the "early" RNA appear at different times and that some are "turned off" at late times. *J. Molec. Biol.* 49:271–295.

Saunders, J. W. 1970. *Patterns and Principles of Animal Development*. New York, Holt, Rinehart, Winston.

Scherrer, K., et al. 1970. Nuclear and cytoplasmic messenger-like RNA and their relation to the active mRNA in polyribosomes of HeLa cells. *Cold Spring Harbor Symp. Quant. Biol.* 35:539–554.

Sonneborn, T. M., 1977. Genetics of cellular differentiation: Stable nuclear differentiation in eukaryotic unicells. *Ann. Rev. Genetics* 11:349–368.

Spemann, H. 1938. *Embryonic Development and Induction*. New Haven, Conn., Yale University Press.

Strehler, B. 1971. Codon-restriction theory of aging and development. *J. Theor. Biol.* 33:429–474.

Studier, F. W. 1972. Bacteriophage T7. *Science* 176:367–376.

Sueoka, N., and T. Kano-Sueoka. 1970. Transfer RNA and cell differentiation. *Prog. Nuc. Acid Res. Molec. Biol.* 10:23–55.

Summers, W. C. 1972. Regulation of RNA metabolism of T7 and related phages. *Ann. Rev. Genetics* 6:191–202.

Sussman, M. 1970. Model for quantitative and qualitative control of mRNA translation in eukaryotes. *Nature* 225:1245–1246.

Tartof, K. D. 1975. Redundant genes. *Ann. Rev. Genetics* 9:355–386.

Thomas, C. 1974. The rolling helix: A model for the eukaryote gene. *Cold Spring Harbor Symp. Quant. Biol.* 38:347–352.

Tompkins, G. M., T. D. Gelehrter, D. Granner, D. Martin, H. H. Samuels, and E. B. Thompson. 1969. Control of specific gene expression in higher organisms. *Science* 166:1474–1480.

Tompkins, G. M., and D. W. Martin. 1970. Hormones and gene expression. *Ann. Rev. Genetics* 4:91–106.

Travers, A. 1970. Positive control of transcription by a bacteriophage sigma factor. *Nature* 225:1009–1012.

Tyler, A. 1967. Masked mRNA and cytoplasmic DNA in relation to protein synthesis and processes of fertilization and determination in embryonic development. In M. Locke (ed.), *Control Mechanisms in Developmental Processes.* New York, Academic.

Wallace, H., J. Morray, and W. H. R. Langridge. 1971. Alternative model for gene amplification. *Nature (New Biology)* 290:201–203.

Watson, J. D. 1976. *Molecular Biology of the Gene,* 3d ed. New York, Benjamin.

Wood, W. B. 1973. Genetic control of bacteriophage T4 morphogenesis. In F. H. Ruddle (ed.), *Genetic Mechanisms in Development.* New York, Academic.

Wright, T. R. F. 1970. The genetics of embryogenesis in *Drosophila. Adv. Genetics* 15:262–395.

Zubay, G., D. Schwartz, and J. Beckwith. 1970. Mechanism of activation of catabolite-sensitive genes: A positive control system. *Proc. Natl. Acad. Sci. U.S.* 66:104–110.

Fourteen

Genes in populations, I

Darwin's theory of evolution, as proposed in 1859, ranks among the most fundamental and profound of all the intellectual revolutions that have occurred in human history. It eliminated anthropocentrism by placing the human species alongside all other species, and showed that the natural forces operating in the process of speciation operate on all species. The Darwinian theory has affected nearly all of our metaphysical and ethical concepts, and it has had an enormous impact on our history, literature, and society.

Yet interestingly, Darwin's theory of evolution by natural selection was weakened by the lack of a sound mechanism to explain the origin of variation. Recall from earlier chapters that Darwin, unaware of Mendel's research, suggested a Lamarckian type of mechanism called *pangenesis* (see pages 5 and 6) which was rooted in the notion of acquired characteristics. This idea was discarded and eventually replaced by Mendelism. But before evolution could be fully appreciated, the fusion of Darwinism with Mendelism required a new dimension. Principles had to be formulated so that we could evaluate and predict changes in gene frequency in natural populations of sexually reproducing organisms. In other words, we had to learn how to monitor the flow of genetic material in a freely interbreeding population, and to assess the forces that act to alter gene frequencies and thus lead to the formation of species.

The study of population genetics enables us to evaluate mechanisms of evolution. In this chapter and the next, we shall examine those mechanisms. This chapter will focus on what we might call *microevolutionary changes:* Those changes which occur in populations that lead to the formation of races and new species. In the next chapter we shall examine the process of species formation, as well as such *macroevolutionary changes* as the formation of higher taxa, extinction, specialization, and other larger aspects of the evolutionary process.

Brief history of evolutionary theory

Before we discuss the mechanisms by which changes in gene frequency accrue in populations, let us trace the development of the theory of evolution by natural selection. Any explanatory system developed to explain the living world had to deal with three of its general features.

1. Living organisms exhibit great structural complexity.
2. Living organisms possess adaptive features that may appear to have been specially designed.
3. There is great diversity in the living world.

From these general observations, we can pose a series of questions that generate diverse responses.

1. How have complex organisms and structures come into existence?
2. What forces operated to mold adaptive characteristics?
3. How did diversity originate in the living world, and how is this diversity maintained?

Darwin's theory of evolution provided a reasonable interpretation of the accumulated data, but it was not the only interpretation offered.

The idea of evolution—of descent with modification—had been around for well over 100 years before Darwin's *Origin of Species* was first published in 1859. Yet ideas of evolution were for the most part rejected and ridiculed. Why was this so? Why did such prominent scientists as Cuvier, Lyell, and Agassiz oppose evolutionism and accept the idea— theologically rooted—of special creation? Indeed, most scientists opposed Darwinism. The evidence in support of evolution was available long before Darwin's *Origin of Species*, yet there was a remarkable lag in evolutionary thinking.

The most reasonable explanation for the failure to accept evolutionism is the enormous power of the opposing ideas. Theology and special creation had a firm grip on people's minds. Powerful figures in the world of science opposed evolution; their authority alone was a sufficient persuasive force for many.

Furthermore, we should not overlook the importance of *catastrophism*, a doctine that attempted to explain the fossil discoveries made with ever-increasing frequency during the eighteenth and nineteenth centuries. Fossils and geological strata were interpreted as evidence of the Great Flood described in the Old Testament. As geologists discovered more and more strata in the Earth's crust, it was necessary to argue for more and more floods, until the idea of progressive flooding became absurd.

Another factor that undoubtedly contributed to the lag in evolutionary thinking was *typological thinking*, or *essentialism*. The living world was viewed as containing basic organismal types, such as dog, cat, monkey, spider, and fly. There was variation within each type, but there was no organic link between the basic types. The acceptance of such an idea is incompatible with populational thinking so characteristic of evolutionism.

Long before Darwin, Lamarck suggested a theory for the origin of species. He reasoned that there was a gradual transformation of one species into another by the accumulation of acquired characteristics. This was an important idea, both because it was evolutionist and because it was in marked contrast to typological thinking. One interesting feature of Lamarck's thesis was his denial of extinction. Organisms were transformed via acquired characteristics, but they did not become extinct. They were modified according to environmental dictates.

Lyell's contributions to evolutionary theory were enormous. His book, *Principles of Geology*, took the rather radical position that natural forces operating on Earth have been fairly constant throughout Earth's history. These forces gradually changed the surface of the Earth, creating mountains, canyons, desert, grasslands, and other features of Earth's outer crust. Darwin read and admired Lyell's geology and was profoundly affected by it. In addition, he and Lyell were close friends.

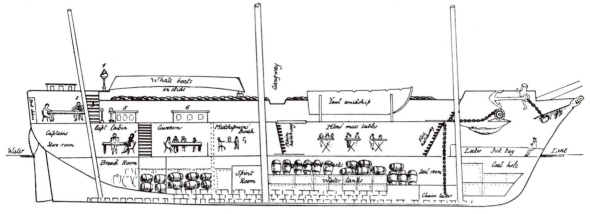

H.M.S. Beagle 1832

Figure 14.1
The H.M.S. *Beagle*
(Courtesy of the Royal
College of Surgeons of
England)

What factors led Darwin to his theory of evolution? The single most important event in the series of events leading to the *Origin of Species* was Darwin's voyage aboard H.M.S. *Beagle* (Figure 14.1). This five-year journey (1831–1836) afforded the observations that led Darwin to conclude that there was a natural explanation for the origin of adapted species. Without attempting to discuss the voluminous amount of data Darwin collected while aboard the *Beagle*, we can summarize some of the more important observations.

1. In South America, Darwin collected fossils that were similar but not identical to existing species. He puzzled over this. If the fossils were remnants of a great flood, and if a new creation had followed the flood, why were entirely new organisms not created? What is the point of creating similar but not identical species? Could it be that there was a genetic link between the fossil organisms and those currently occupying the region?

2. Darwin observed the full effects of an earthquake in Chile and noted the alterations in the landscape that resulted. Land had moved and been rearranged. This and other geological observations convinced Darwin of the truth of Lyell's gradualistic geology.

3. In the Galapagos Islands, Darwin became firmly convinced of the evolutionary origin of species. No longer could he entertain even the possibility of special creation. The key observation here involved 13 rather dull-looking species of finches. All of them were similar to South American finches, and all 13 of the Galapagos Island finch species differed from each other in body size and in beak morphology and function. They were all morphologically similar to each other—yet they differed as a result of occupying different ecological niches. It was reasonable to suggest that a finch species from South America successfully traveled to the Galapagos Islands and diversified throughout the island chain, becoming differentially adapted to different life styles dictated by slightly differing conditions (available food and nest sites, competitors, and predators) on the individual islands.

Darwin spent several months after his return to England thinking about his data and trying to organize them into a coherent theory. He had the idea of evolution, but not the mechanism. But 15 months after his return, and just after reading Malthus' essay on the principle of population, Darwin had his theory. Malthus' essay provided the key: populational thinking. The observations and deductions that culminate in the theory of evolution are outlined in the accompanying diagram.

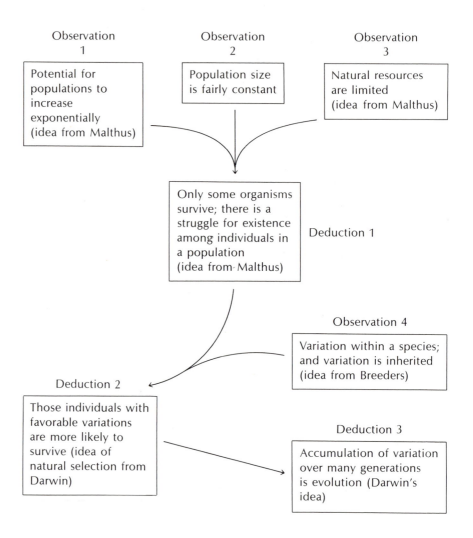

The Darwinian revolution—and it *was* a revolution—went on for many years after Darwin published his *Origin of Species.* It was a complex revolution, because it involved much more than the biological sciences. Ernst Mayr characterized this revolution as having six main elements.

1. Recognition of the vast age of the Earth (not just 4000–6000 years, as suggested by biblical scholars)
2. Rejection of catastrophism
3. Rejection of a striving for "perfection," or a goal-oriented process
4. Rejection of special creation
5. Replacement of typological thinking with populational thinking
6. Abolition of anthropocentrism

The complexity of these elements explains why the Darwinian concept of evolution was slow in gaining a foothold.

Darwin's theory suffered from one major weakness. It offered no mechanism to explain the origin of variation. In Chapter 1 we examined this problem and traced its ultimate resolution. The fusion of Darwinism and Mendelism solved most of the problems inherent in Darwin's theory. But it was not until population genetics developed a strong base that we were able to fully appreciate evolution and its mechanisms.

Population: Structure and concept

Neo-Darwinian evolution, a fusion of nineteenth-century Darwinism with twentieth-century genetics, offers explanations of the changes in the frequencies of different alleles and in the associations between alleles at different gene loci. It also explains changes in populations, not in individuals, for the population is the basic evolutionary unit.

The term *population* is ambiguous, meaning different things to different people. When we use the term, we refer specifically to a *Mendelian population* or *reproductive community*. A Mendelian population consists of a community of individuals of a sexually reproducing species within which matings take place. When mating in a population is random, we refer to its mating scheme as *panmictic*. Often the term *deme* is used to describe a localized population or subgroup of a larger population, whether or not the members of the subgroup reproduce sexually.

Populations can be described in terms of their structure. This structure consists of characteristics, such as spatial organization and breeding system. Spatially, populations can be large and continuous; small and colonial, or insular; or linear (such as a narrow but continuous population along a riverbank). With respect to breeding systems, populations range from the extreme of complete self-fertilization to the other extreme of complete and random outcrossing. There are many intermediate mating systems, such as preferential outcrossing with close neighbors only.

Through extensive studies of natural populations, investigators have found that a population's *gene pool* (the sum total of all genotypes in a Mendelian population) consists of extensive heterozygosity at most loci. This *genetic polymorphism* is maintained by mutation pressure and by the forces of natural selection, as we shall see shortly. The questions that geneticists must confront in dealing with evolution concern this genetic polymorphism (the existence of two or more genetically different classes in the same Mendelian population). For example, how much genetic variation is present in a population? What is its source? What is its function? How is it maintained? And how does it change over time? Seeking answers to these questions has been the goal of population geneticists for the past several decades.

Development of the Hardy–Weinberg principle of genetic equilibrium

Two major lines of thought led to the discipline of population genetics, the discipline that studies gene frequency changes in populations. One line began with Galton and ended with Nilsson-Ehle (see Chapters 1 and 4). This was the biometrical school of thought that culminated in the demonstration of multiple-gene or quantitative inheritance—that is, several genes influence a single trait—and provided a basis for the inheritance of small phenotypic differences. The second line began with Mendel and progressed to the analysis of the effects of inbreeding and crossbreeding in randomly mating populations. This latter approach, coupled with the principles of quantitative genetics, led first to the establishment of population genetics and ultimately to the formulation of mechanisms that could account for the processes of organic evolution.

The biometricians argued that the study of quantitative characters and the use of statistical tools would eventually produce a more general theory of inheritance. The Mendelians argued that Mendel's principles for the inheritance of contrasting traits would form the basis of a general theory of inheritance. The reconciliation of these two views came about only after a lengthy and heated debate that raged throughout the first decade of this century. Studies revealed that the two apparently contradictory positions were in fact complementary.

The application of Mendelian principles to a freely interbreeding population of individuals began in 1902. In this year, G. U. Yule pointed out that when the members of an F_2 population, segregating for a single pair of alleles (A and a), interbreed at random, the three possible genotypes (AA, Aa, aa) are represented in the same proportions in the F_3 and all succeeding generations. This idea was expanded in 1903 by W. E. Castle, who showed that the genotypic ratios would change in each generation if the aa class were either kept from mating or removed from the parental generation. It is worth noting that Mendel's 1866 paper actually contained the seeds of modern population genetics, for Mendel maintained that constant self-fertilization in pea populations caused a progressive decline in the heterozygous frequency and an increase in the frequency of the homozygous classes. Consider, for example, inbreeding in a population that switches from a cross-fertilizing to a self-fertilizing mode of reproduction.

	Genotype frequencies		
Generation	AA	Aa	aa
0	$\frac{1}{4}$	$\frac{1}{2}$	$\frac{1}{4}$
1	$\frac{3}{8}$	$\frac{1}{4}$	$\frac{3}{8}$
2	$\frac{7}{16}$	$\frac{1}{8}$	$\frac{7}{16}$
3	$\frac{15}{32}$	$\frac{1}{16}$	$\frac{15}{32}$
4	$\frac{31}{64}$	$\frac{1}{32}$	$\frac{31}{64}$
n	$(1 - \frac{1}{2}^n)/2$	$\frac{1}{2}^n$	$(1 - \frac{1}{2}^n)/2$

In this mating scheme, it is clear that we are reducing heterozygosis and increasing homozygosis. The basic principle of population genetics applies to the same situation, except that it is based on randomly crossbreeding populations.

The Hardy–Weinberg principle

G. H. Hardy, a mathematician at Cambridge University, was encouraged by Punnett (Bateson's colleague) to make public his views on genetic equilibrium in populations of organisms. Punnett and Hardy had discussed Yule's work, which promoted the hypothesis that if a dominant gene entered a human population—regardless of its initial frequency—it would eventually result in a distribution of $\frac{3}{4}$ dominant phenotypes and $\frac{1}{4}$ recessive. This seemed reasonable, because it was known that a monohybrid cross between two heterozygous individuals would yield the F_2 3:1 phenotypic ratio. However, Hardy pointed out that in order for a 3:1 phenotypic ratio to occur, the frequency of the dominant allele had to be precisely 50 percent, as it is in a monohybrid cross. And in 1908, he proposed that under a set of standard conditions the gene frequencies would be constant, as would the genotypic ratios $AA:Aa:aa$. Hardy reasoned that the gene frequencies would determine the genotypic ratios.

That same year W. Weinberg, a German physician, had applied Mendelian principles to human populations. Working independently, Hardy and Weinberg both discovered the basic principle of population genetics: that an equilibrium is established for allele frequencies in a freely interbreeding, sexually reproducing population and that the frequency of each allele remains constant generation after generation. The constancy is maintained as long as the following conditions are met.

Figure 14.2
G. H. Hardy

1. The population must be large enough so that sampling errors can be disregarded.

2. There can be no mutations, either A to a or a to A. If there are, the A to a mutation frequency must be the same as the a to A frequency. If this is not the case, the A and a gene frequencies will change over time.

3. Mating must be random; that is, all mating combinations must be equally probable.

4. All the genotypes must be equally viable and equally fertile.

5. All gametes must have equal probabilities of forming zygotes.

6. The population must be isolated to preclude individuals, and hence genes, from migrating into and out of the population.

If these conditions are met, genetic equilibrium will be established; if not, gene frequencies will change. The changing of gene frequencies over time is the entire basis of evolution, which may be defined as *the change in the genetic composition of populations through time, and the conversion of intrapopulation variation to interpopulation variation in the process of speciation.*

The principle of genetic equilibrium states that if the aforementioned conditions are met, gene frequencies and phenotypic ratios in a population will remain stable. A hypothetical population that has achieved equilibrium, composed of 490 individuals of genotype *AA*, 420 *Aa*, and 90 *aa* may be used to illustrate this principle.

The frequency of allele *A* is the ratio of the actual number of *A* alleles to the total number of *A* + *a* alleles; the frequency of *a* is the ratio of the actual number of *a* alleles to the total number of *A* + *a* alleles. If we let

p = Frequency of *A*
q = Frequency of *a*

(noting that $p + q = 1$), then, for a population in genetic equilibrium, the frequency of *AA* is expected to be the likelihood of 2*AA* gametes fusing, or $p \times p$ (p^2); the expected frequency of *Aa* is $p \times q$; the expected frequency of *aA* is $q \times p$; and the expected frequency of *aa* is $q \times q$ (q^2). In summary:

Frequency of *AA* genotypic class = p^2
Frequency of *Aa* genotypic class = pq $\left.\right\}$
Frequency of a*A* genotypic class = qp $\left.\right\}$ = $2pq$
Frequency of aa genotypic class = q^2

and the genotype frequencies of *AA*, *Aa*, and *aa* can be summarized as

$$p^2 + 2pq + q^2 = 1$$

which is the expansion of the binomial $(p + q)^2$. Returning to our hypothetical population, we see that the *genotype frequencies* are

$$AA = \frac{490}{1000} = 0.490$$

$$Aa = \frac{420}{1000} = 0.420$$

$$aa = \frac{90}{1000} = 0.090$$

643

Development of the
Hardy–Weinberg principle of
genetic equilibrium

The *A* and *a* allele frequencies can be determined from the genotype frequencies.

Frequency of *A* in population =
    Frequency of *A* in *AA* + Frequency of *A* in *Aa* + Frequency of *A* in *aa*

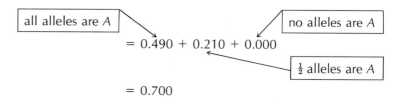

    = 0.490 + 0.210 + 0.000

    = 0.700

Frequency of *a* in population =
    Frequency of *a* in *AA* + Frequency of *a* in *Aa* + Frequency of *a* in *aa*

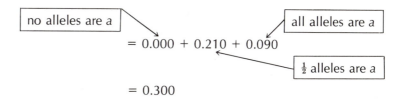

    = 0.000 + 0.210 + 0.090

    = 0.300

    Having calculated gene frequencies for *A* and *a* in our hypothetical population, we should consider whether these frequencies change from one generation to the next if conditions 1 through 6 hold. To answer this, we must consider the flow of genetic information in a dynamic population. Such a flow obeys precise statistical laws, as we shall soon see. Assuming random mating in our population, we develop the probabilities of *mating pair formation* as follows.

| | Males Genotypes and genotype frequencies | | |
Females Genotypes and genotype frequencies	*AA* 0.49	*Aa* 0.42	*aa* 0.09
AA (0.49)	*AA* × *AA* 0.240	*AA* × *Aa* 0.206	*AA* × *aa* 0.044
Aa (0.42)	*Aa* × *AA* 0.206	*Aa* × *Aa* 0.176	*Aa* × *aa* 0.038
aa (0.09)	*aa* × *AA* 0.044	*aa* × *Aa* 0.038	*aa* × *aa* 0.008

For each mating pair, the fraction of offspring having *A* alleles and the fraction having *a* alleles can be calculated.

Mating	Mating probability	Offspring ratio	Allele ratio in offspring	
			A	*a*
AA × AA	0.240	1 AA	0.240	0.000
AA × Aa	0.206	1 AA:1 Aa	0.155	0.051
AA × aa	0.044	1 Aa	0.022	0.022
Aa × AA	0.206	1 AA:1 Aa	0.155	0.051
Aa × Aa	0.176	1 AA:2 Aa:1 aa	0.088	0.088
Aa × aa	0.038	1 Aa:1 aa	0.009	0.029
aa × AA	0.044	1 Aa	0.022	0.022
aa × Aa	0.038	1 Aa:1 aa	0.009	0.029
aa × aa	0.008	1 aa	0.000	0.008
Total	1.000		0.700	0.300

Note that in these operations, the gene frequencies among the first generation are identical to those among the parents. This is a characteristic feature of an equilibrium population and would hold true no matter how many generations were studied. (This assumes an equilibrium population and conditions 1–6 listed on pages 641–642.)

The foregoing example illustrates genetic equilibrium as proposed by Hardy and Weinberg, but let us return now to an issue raised earlier by Yule, who proposed that a dominant gene introduced into a population would eventually be expressed in three-fourths of the population. Using the principle of genetic equilibrium, we can demonstrate that Yule was wrong. Assume a hypothetical population composed of 18 individuals, all genotypically homozygous recessive *aa*. Into this population we introduce two homozygous dominant *AA* individuals. How will these *A* alleles be distributed and what will their final frequency be?

To begin with, we can calculate the genotype frequencies for the new population.

$$AA = \frac{2}{20} \quad Aa = \frac{0}{20} \quad \text{and} \quad aa = \frac{18}{20}$$

that is,

$$0.100 + 0.000 + 0.900 = 1.000$$

From the genotype frequencies, the *A* and *a* gene frequencies will be

$$AA + \tfrac{1}{2}Aa = 0.100 + 0 = 0.100 \quad (A)$$
$$aa + \tfrac{1}{2}aa = 0.900 + 0 = 0.900 \quad (a)$$

If we assume random mating among the 20 individuals in the population, the mating frequencies will be

645

Development of the
Hardy–Weinberg principle of
genetic equilibrium

	$\vardelta$ AA (0.1)	aa (0.9)
♀ AA (0.1)	0.01	0.09
aa (0.9)	0.09	0.81

The ratio of allele A to a in the offspring is calculated as follows.

Mating	Probability	Offspring ratio	Allele ratio in offspring A	a
$AA \times AA$	0.01	all AA	0.01	0.00
$AA \times aa$	$0.09 + 0.09 = 0.18$	Aa	0.09	0.09
$aa \times aa$	0.81	all aa	0.00	0.81
Total	1.00		0.10	0.90

Hence the A and a allele frequencies in the first generation following migration will be 0.100 and 0.900, respectively, the same as in the initial population. But the genotypic ratios will be different.

$$AA = (0.10)^2 = 0.01 \qquad (AA)$$
$$Aa = 2(0.10)(0.90) = 0.18 \qquad (Aa)$$
$$aa = (0.90)^2 = 0.81 \qquad (aa)$$

With further random mating, the composition of the subsequent generations remains unchanged. In other words, there will be no change in the genotypic ratio or gene frequencies in the second or the third generation or in later generations. This is an equilibrium population. The dominant allele accounts for 10 percent of the total, and dominant phenotypic classes account for 19 percent of the total—nowhere near the 75 percent predicted by Yule. This will continue unchanged unless another migration occurs. Then, after one generation of random mating, a new equilibrium will be established.

One final point should be made about calculating gene frequencies. For a population that has achieved equilibrium, the frequencies of A and a can be calculated directly from the genotype frequencies. If

$$p^2 = 0.021$$
$$2pq = 0.248$$
$$q^2 = 0.731$$

then

$$q = \sqrt{0.731} = 0.855$$

And since $p + q = 1$, then

$$p = 1 - q = 1 - 0.855 = 0.145$$

In cases in which one allele is completely dominant over the other (*A* is dominant over *a*), two of the genotypic classes (*AA* and *Aa*) are phenotypically identical. So in determining gene frequencies we can use the square root of the homozygous recessive class if we assume Hardy–Weinberg equilibrium. For example, the ability to taste the chemical phenylthiocarbamide (PTC) is based on the presence of a dominant gene (*T*). All *Tt* and *TT* individuals find PTC very bitter and are classed as tasters, whereas all *tt* individuals find PTC tasteless and are considered nontasters. In classroom samples taken at Swarthmore College, 218 tasters and 102 nontasters were found. The 218 tasters could have been either *TT* or *Tt*, but the nontasters could only have been *tt*. Assuming that mating had been random with respect to this trait (that is, the students' parents did not form mating pairs based on ability to taste PTC) and that this is a random sample of the larger population, we can estimate the *T* and *t* gene frequencies in the population. Since the genotypes *TT* and *Tt* account for the 218 tasters and *tt* represents the 102 nontasters, we have

$$\text{Frequency of } tt = \tfrac{102}{320} = 0.32 = q^2$$
$$q = \sqrt{0.32} = 0.56 \qquad \text{(frequency of } t\text{)}$$
$$p = 1 - q = 1 - 0.56 = 0.44 \qquad \text{(frequency of } T\text{)}$$

Furthermore, from the gene frequencies, we can estimate the genotype frequencies of the original sample.

$$p^2 = (0.44)^2 = 0.19 \qquad \text{(that is, 61 } TT \text{ students)}$$
$$2pq = 2(0.44)(0.56) = 0.49 \qquad \text{(that is, 157 } Tt \text{ students)}$$
$$q^2 = (0.56)^2 = 0.32 \qquad \text{(that is, 102 } tt \text{ students)}$$

Bear in mind that, in order to make valid estimates for the homozygous recessive class, we must have a population that resulted from random mating. For example, if we merge two arbitrarily selected populations that have been isolated from each other, even if each is in a state of genetic equilibrium, determinations of gene frequency of the newly formed population will not necessarily reflect the genotype frequencies. Suppose the genotypes are distributed as follows.

Population A			*Population B*		
AA	Aa	aa	AA	Aa	aa
640	320	40	250	500	250

Population C (A and B merged)		
AA	Aa	aa
640	320	40
250	500	250
Total 890	820	290
(0.445)	(0.410)	(0.145)

We can compute

Frequency of a = $0.145 + \frac{1}{2}(0.410) = 0.35$
Frequency of A = $1 - q = 1 - 0.35 = 0.65$

Using these gene frequencies, we can construct our expected genotypic classes.

$p^2 = (0.65)^2 = 0.42$	(that is, 840 *AA* individuals)
$2pq = 2(0.65)(0.35) = 0.46$	(that is, 920 *Aa* individuals)
$q^2 = (0.35)^2 = 0.12$	(that is, 240 *aa* individuals)

The observed and expected classes do not agree, because population C is not a product of random mating; it is not in equilibrium. Genetic equilibrium will be reached, however, after one generation of random mating.

Gene frequency determinations for codominant alleles do not differ in principle from the determinations already discussed. The M and N blood antigens in humans are genetically determined, are inherited in a Mendelian fashion, and are codominant, so that the genotype *MM* results in the M blood type, *MN* results in the MN blood type, and *NN* results in the N blood type.

A study of M and N blood types was made in a small Mideastern community in order to estimate the frequencies of the *M* and *N* alleles. The following data were obtained.

Frequency of *MM* = 0.512
Frequency of *MN* = 0.400
Frequency of *NN* = $\underline{0.088}$
    1.000

The gene frequencies are determined to be

$M = MM + \frac{1}{2}MN = 0.512 + \frac{1}{2}(0.400) = 0.71$
$N = NN + \frac{1}{2}NN = 0.088 + \frac{1}{2}(0.400) = 0.29$

The expected genotype frequencies, assuming the population to be randomly mating, are

$p^2 = (0.71)^2 = 0.505$	(that is, 505 *MM* individuals)
$2pq = 2(0.71)(0.29) = 0.412$	(that is, 412 *MN* individuals)
$q^2 = (0.29)^2 = 0.083$	(that is, 83 *NN* individuals)

Because the expected class size agrees so well with the observed class size, we can conclude that the population is an equilibrium population.

In comparing observed and expected genotype frequencies for the *M* and *N* alleles, we pointed out that the agreement between the observed and expected classes was excellent. In the previous example, in which we (artificially) merged two populations (see page 646), we concluded that observations and expectations did not agree. We should now look at the basis for making such judgments, the chi-square analysis (see pages 134–136 for a review of this procedure).

In our M and N example, we determined that $MM = 0.512$, $MN = 0.400$, and $NN = 0.088$. From this, we said that $M = 0.71$ and $N = 0.29$. Our χ^2 analysis looks like this.

	O	E	$O - E$	$(O - E)^2$	$(O - E)^2/E$
MM	512	505	7	49	0.097
MN	400	412	−12	144	0.350
NN	88	83	5	25	0.301
	1000	1000			$\chi^2 = 0.748$

We use only one degree of freedom in an analysis of this type, because we have no theoretical basis for obtaining expected gene frequencies. We estimate those frequencies from the data sampled. We lose one degree of freedom for each parameter (each allele frequency) that is estimated from the data. In this case, we lose two degrees of freedom. For a χ^2 value of 0.748 and one degree of freedom, $p \cong 0.4$, which means that the probability of getting deviations as large as or larger than those observed by chance alone is about 0.4. This is within our limits of acceptability, so we conclude that observations and expectations are in agreement.

Our hypothetical population gives us a different result. We determined that $AA = 890$ (0.445), $Aa = 820$ (0.410), and $aa = 290$ (0.145). From this, we said that $A = 0.65$ and $a = 0.35$. Our χ^2 analysis of these data looks like this.

	O	E	$O - E$	$(O - E)^2$	$(O - E)^2/E$
AA	890	840	50	2500	2.976
Aa	820	920	−100	10000	10.870
aa	290	240	50	2500	10.417
	2000	2000			$\chi^2 = 24.263$

With one degree of freedom, the probability of getting deviations as large as or larger than those observed is essentially 0, so we conclude that observations and expectations are not in agreement.

Gene frequency determinations for a multiple allelic series

The ABO blood group series in humans is determined by a multiple allelic series with three alleles (I^A, I^B, i) producing a total of six possible genotypes ($I^A I^A$, $I^A i$, $I^A I^B$, $I^B I^B$, $I^B i$, ii). (Some writers use the symbols A, B, and O instead of I^A, I^B, and i.) The i allele, which determines blood type O, is recessive, so there are only four observable phenotypes.

649

Development of the
Hardy–Weinberg principle of
genetic equilibrium

Blood type	Genotype
A	$I^A I^A, I^A i$
B	$I^B I^B, I^B i$
AB	$I^A I^B$
O	ii

In a sample of 500 people, the distribution of blood types was found to be

A: 195   B: 70   AB: 25   O: 210

To estimate gene frequencies here, we assume this is a randomly mating population, and add the additional symbol r to represent the frequency of i, so that

p = Frequency of I^A
q = Frequency of I^B
r = Frequency of i

The genotype frequencies can be represented by the expression

$$p^2 \underset{I^A I^A}{} + \underset{I^A i}{2pr} + \underset{ii}{r^2} + \underset{I^A I^B}{2pq} + \underset{I^B i}{2qr} + \underset{I^B I^B}{q^2} = 1$$

	A	O	AB	B
	$\dfrac{195}{500}$	$\dfrac{210}{500}$	$\dfrac{25}{500}$	$\dfrac{70}{500}$
	(0.39)	(0.42)	(0.05)	(0.14)

To calculate the gene frequencies, we begin with

$r^2 = 0.42$ (frequency of ii)
$r = \sqrt{0.42} = 0.65$ (frequency of i)

The frequencies of the A and O phenotypes are given by

$$p^2 + 2pr + r^2 = (p + r)^2$$
$$(p + r)^2 = 0.39 + 0.42$$
$$= 0.81$$

Then $p + r = \sqrt{0.81} = 0.90$. We can estimate p by substituting the estimated value of r. We obtain

$$p + 0.65 = 0.90$$
$$p = 0.25$$

Since $p + q + r = 1$, the value of q can be determined as

$$0.65 + 0.25 + q = 1$$
$$q = 0.10$$

Calculating the expected genotype frequencies from the gene frequencies, we find that the population is an equilibrium population.

$p^2 = (0.25)^2 = 0.06$	(that is, 30 $I^A I^A$ individuals) ⎫ Phenotype A
$2pr = 2(0.65)(0.25) = 0.33$	(that is, 165 $I^A i$ individuals) ⎭
$2pq = 2(0.25)(0.10) = 0.05$	(that is, 25 $I^A I^B$ individuals) Phenotype AB
$r^2 = (0.65)^2 = 0.42$	(that is, 210 ii individuals) Phenotype O
$2qr = 2(0.10)(0.65) = 0.13$	(that is, 65 $I^B i$ individuals) ⎫ Phenotype B
$q^2 = (0.10)^2 = 0.01$	(that is, 5 $I^B I^B$ individuals) ⎭

Gene frequency determinations for sex-linked alleles

In organisms such as humans, males have only one **X** chromosome, but females have two, so determinations of gene frequency for sex-linked alleles must take into account the fact that females carry twice as many **X**-linked genes as males. At equilibrium, therefore, assuming that $p + q = 1$, the genotype frequencies are

$$p + q \qquad \text{for males (two genotypic classes)}$$
$$p^2 + 2pq + q^2 \qquad \text{for females (three genotypic classes)}$$

Colorblindness is a trait caused by a sex-linked recessive gene (d). In a sample of 2000 individuals (1000 men and 1000 women), 90 were colorblind—all male! Therefore, with p as the frequency of the normal allele D and q as the frequency of the colorblind allele d,

$$p + q = 1$$
$$p + \frac{90}{1000} = 1$$
$$p = 1 - 0.09 = 0.91$$

With $p = 0.91$ and $q = 0.09$, and assuming a population with random mating proportions, the expected numbers of females expressing the trait and carrying it in heterozygous fashion can be estimated as follows.

$$p^2 + 2pq + q^2 = 1$$
$$(0.91)^2 + 2(0.91)(0.09) + (0.09)^2 = 1$$
$$0.83 + 0.16 + 0.01 = 1$$

Thus 830 females will have the genotype DD, 160 the genotype Dd, and 10 the genotype dd. In other words, we can expect 99 percent of the women to be phenotypically normal, compared with 91 percent for males. An

651

*Development of the
Hardy–Weinberg principle of
genetic equilibrium*

equilibrium population for sex-linked alleles requires males and females to have the same gene frequencies, but not the same genotype or phenotype frequencies.

In summary, as long as genotypes can be distinguished, gene frequencies can be determined. However, determination of gene frequency is not necessarily the most important aspect of population genetics. Most important in the discipline is the usefulness of determining a shift in gene frequencies that results from such forces as mutation, selection, and genetic drift.

Inbreeding, panmixis, and heterosis

A basic assumption of the Hardy–Weinberg equilibrium is random mating among all individuals in a population. A population expressing such a mating structure is called a *panmictic* population, and it is said to exhibit *panmixis*. For many traits, we might conclude that mating is random. In the human species, for example, mates are not selected on the basis of M and N blood antigens or the ability to taste phenylthiocarbamide (PTC). On the other hand, mates *are* selected on the basis of many physical and behavioral qualities, making the mating process in humans nonrandom.

The situation in which male–female pairing in a Mendelian population is not random, but rather involves a tendency for a particular type of female to mate with a particular type of male, is called *assortative mating. Positive assortative mating* involves pairs that are more alike than expected from random couplings; *negative assortative mating* involves pairs that are less alike than expected.

Assortative mating is the rule, not the exception, in natural populations. A consequence of this is a restriction in the flow of genetic material through a population. The overall gene pool of the population may in fact consist of a series of smaller subpools, each—because of assortative mating—being somewhat different from the average of all the gene pools that comprise the population.

One of the consequences of restricted gene movement in a population is increased *inbreeding*. Individuals may tend to mate with those individuals that are close by, and this may result in matings between relatives. An extreme form of inbreeding is self-fertilization, the consequences of which were outlined earlier (see page 641). Even though gene frequencies did not change, a strict program of self-fertilization would eventually reduce the genotypic classes, and possibly the phenotypic classes, from three to two. Inbreeding reduces the genetic variability in a population. When a population's variability is reduced, its adaptiveness is also reduced. The potential for evolutionary success is quite limited under these conditions. Another consequence of inbreeding is that a rare recessive gene has an increased likelihood of becoming homozygous and thus being acted on by the forces of natural selection. Inbreeding obviously has important consequences in a population.

If a normally outcrossing population is forced into a strict inbreeding mating scheme, it usually causes an *inbreeding depression,* characterized by (among other things) a general decline in vigor. One reason for the depression is the increased homozygosity of recessive alleles that may be harmful when homozygous.

Inbreeding also increases the variability in different inbred lines in a population. Different families have different combinations of homozygous genes, and this accentuates the differences between families. Ironically, then, in a population with several inbred lines, phenotypic variability may actually increase, but genotypic variability decreases because of the loss of heterozygotes.

In striking contrast to inbreeding depression is the phenomenon called *heterosis*, a phenomenon associated with increased heterozygosity. When two highly inbred lines (each perhaps suffering from inbreeding depression) are crossed, the hybrid offspring usually show greater vigor as measured by such traits as growth, survival, and fertility. We often refer to the increased fitness of the hybrids as *hybrid vigor*.

The classic example of the importance of heterosis is corn, or maize. Corn has been a crop plant for more than 5000 years. For most of that time, improvement in yield was achieved by careful selection. Kernels were collected from the largest and most vigorous plants and used for subsequent plantings. By this type of selection, corn showed considerable improvement. But this improvement soon leveled off. Corn is almost 100 percent cross-fertilizing, so most corn varieties are highly heterozygous. This high degree of heterozygosity maintained by outcrossing makes selection difficult beyond a certain point because of the wide range of variation caused by genetic recombination.

The problem of heterozygosity was overcome through a program of systematic inbreeding followed by outcrossing. Pure lines of corn were established through self-fertilization over several generations. Each of the inbred lines suffered from inbreeding depression, but each line was also selected for some desirable traits (such as blight resistance). When two inbred lines were crossed to each other, the hybrid offspring were taller and more vigorous and had larger ears than either inbred line. The overall yield and quality of the hybrid line were superior to the yield and quality of either inbred line. The hybrid line of corn was exhibiting hybrid vigor, or heterosis.

There is, however, a problem associated with this technique. The inbred strains used as seed parents produced a vigorous F_1 that had small ears with small numbers of kernels. To overcome this problem, the *double-cross technique* was developed (Figure 14.3).

The double-cross technique involves four inbred strains and three separate crosses. Two of the inbred strains are crossed to produce a single-cross hybrid; the other two inbred strains are also crossed to produce a single-cross hybrid. The two single-cross hybrids are then crossed to produce a double-cross hybrid, which yields the seeds that are planted by the farmer. Though two seasons are required before the seed that the farmer plants is produced, the desired hybrid seed is produced in great

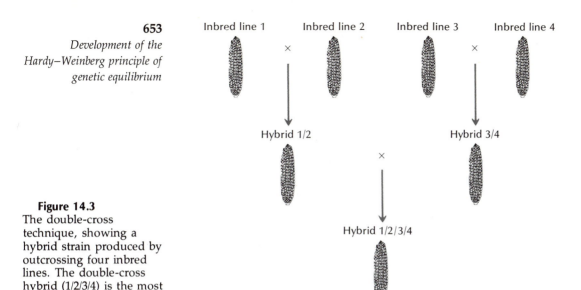

Inbred line 1 × Inbred line 2   Inbred line 3 × Inbred line 4

Hybrid 1/2   Hybrid 3/4

×

Hybrid 1/2/3/4

Figure 14.3
The double-cross technique, showing a hybrid strain produced by outcrossing four inbred lines. The double-cross hybrid (1/2/3/4) is the most vigorous.

abundance by this technique. In addition, the crop yield in certain areas in which this technique has been used has risen from 25 bushels of corn per acre to more than 50.

At first, the double-cross technique was very time-consuming, because plants had to be detasseled to prevent self-fertilization (tassels are the pollen-bearing flowers). But a technique was soon developed that eliminated the need for detasseling. Cytoplasmic factors were found that rendered the male flowers sterile. A nuclear gene, called a *restorer*, restores fertility to the male flower (Figure 14.4). Unfortunately, the cytoplasmic male-sterility factors also seem to make the plants more susceptible to blight, so until that connection can be severed, the use of these cytoplasmic factors is limited.

How does heterosis work? What makes the hybrids more vigorous than their inbred parents? There are two main schools of thought on this problem, but so far the available evidence does not favor either idea over the other. The *overdominance hypothesis,* proposed originally by G. H. Schull and E. M. East in 1908, argues that the heterozygous genotypes are superior to the homozygotes. It might be, for example, that an enzyme composed of two polypeptide chains coded for by a single gene locus is functionally superior when the chains are different (*Aa*) rather than when they are the same (*AA* or *aa*). In other words, the heterozygote produces a superior gene product. The example of sickle-cell anemia is a case of heterozygote superiority.

A second idea, proposed by K. Mather and others, is called the *dominance hypothesis.* Supporters of this hypothesis argue that inbreeding leads to increased homozygosis for deleterious recessive genes. Outcross-

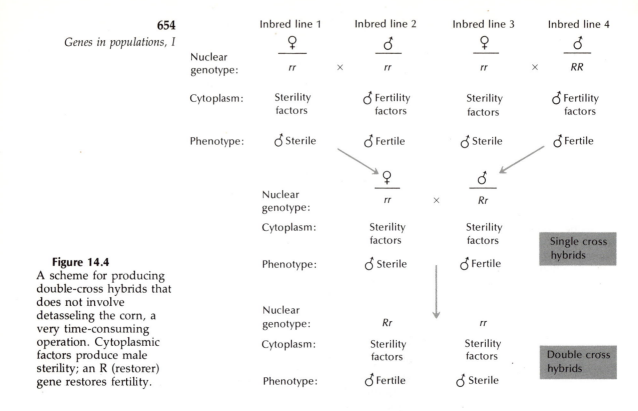

Figure 14.4
A scheme for producing double-cross hybrids that does not involve detasseling the corn, a very time-consuming operation. Cytoplasmic factors produce male sterility; an R (restorer) gene restores fertility.

ing or hybridizing masks the deleterious recessives behind the dominant alleles of the different inbred lines. Here it is the masking effects of dominant genes that leads to heterosis. According to the overdominance hypothesis, heterozygous loci or masked recessives—not just the presence of superior dominant alleles—are the root cause of heterosis.

Most geneticists believe that there is some truth in both the dominance and the overdominance hypotheses. Nevertheless, the dominance hypothesis appears to account for most examples of heterosis.

Microevolutionary changes in populations

The Hardy–Weinberg equilibrium population is in a steady state (all alleles are maintained at a constant frequency), and this is precisely what Hardy and Weinberg intended to demonstrate. However, populations generally do not remain fixed. Gene frequencies change over time, thus forming the basis for evolution. The Hardy–Weinberg equilibrium model must therefore be expanded to take into account the forces that change or shift the equilibrium point—forces such as mutation, natural selection, genetic drift, meiotic drive, and migration.

Mutation In the broadest sense, a mutation is any change in the genotype, and this might include such events as translocations and inversions. In current usage, however, mutation usually refers to a qualitative change within a gene, and it is in this context that the term will be used in this section. Mutation is important in the genetics of populations, and therefore in evolution, because it alters gene frequencies and also because it provides the raw material by which natural selection too can alter gene frequencies. (For a review of the mutation process, see Chapter 10.)

Take gene A, which can mutate at a constant rate (u) to its allele a:

$$A \xrightarrow{u} a$$

If this mutation were indeed a one-way event, as depicted, given enough time, all the A alleles would mutate to a. Mutations, however, rarely operate in only one direction. Usually, a reverse rate (v) accompanies the forward rate.

$$A \underset{v}{\overset{u}{\rightleftharpoons}} a$$

Considering the mutation rates u and v and their effects on the A and a frequencies in a population, we may set

Initial frequency of $A = p_0$
Initial frequency of $a = q_0$
$A \longrightarrow a$ mutation rate $= u$
$a \longrightarrow A$ mutation rate $= v$

At generation 1, where the proportion of a alleles mutated to A is vq_0 and where the proportion of A alleles mutated to a is up_0,

Frequency of $A = p_0 + vq_0 - up_0$
Frequency of $a = q_0 + up_0 - vq_0$

These relationships are convincing evidence that if $u = v$ and if $p_0 = q_0$, there will be no change in the frequency of A or a. However, if $u \neq v$ or if $p_0 \neq q_0$, such a change will occur as a function of time. Where the loss from $a \longrightarrow A$ is vq_0 and where the gain from $A \longrightarrow a$ is up_0,

Change in frequency of $a = up_0 - vq_0$

and where the loss from $A \longrightarrow a$ is up_0 and where the gain from $a \longrightarrow A$ is vq_0,

Change in frequency of $A = vq_0 - up_0$

If the frequency of A is much greater than the frequency of a, then the change in the frequency of a will be large and there will be a steady increase of a at the expense of A. At some time, a will have increased to such a degree that its reverse rate to A will balance the forward mutation rate $A \longrightarrow a$. This point of steady-state equilibrium is called *mutational equilibrium* (q) and is expressed as follows:

$$up - vq = 0$$
$$up = vq$$

Since $p = 1 - q$,

$$u(1 - q) = vq$$
$$u - uq = vq$$
$$u = uq + vq = q(u + v)$$
$$q = \frac{u}{u + v}$$

Similarly, the mutational equilibrium for A (p) is expressed as follows:

$$vq - up = 0$$
$$up = vq$$

And since $q = 1 - p$,

$$up = v(1 - p)$$
$$up = v - vp$$
$$v = vp + up = p(v + u)$$
$$p = \frac{v}{u + v}$$

Attaining mutational equilibrium is rare in nature, because it requires thousands of generations at a normal, spontaneous, mutation frequency of 10^{-5} to 10^{-6}. This is so slow that other evolutionary forces will have otherwise influenced gene frequencies long before this equilibrium is achieved. Let us now examine one of these forces: migration.

Migration The movement of individuals or groups of individuals into or out of isolated populations can have a marked effect on gene frequencies. A model giving quantitative treatment to the effects of migration on an isolated population is shown in Figure 14.5. A number of individuals from population 2 have migrated to population 1, and they now make up 20 percent of it. Since the gene frequencies of the two populations differ, the equilibrium of the new population will shift accordingly after the migration. In each generation of this population, the value of q is the average of

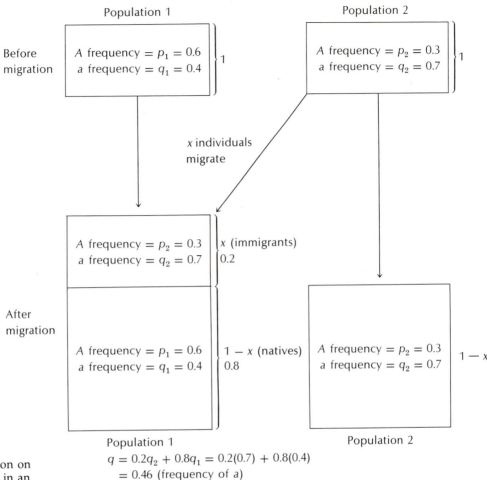

Figure 14.5
Effects of migration on
gene frequencies in an
isolated population

$q = 0.2q_2 + 0.8q_1 = 0.2(0.7) + 0.8(0.4)$
   $= 0.46$ (frequency of a)
$p = 1 - q = 0.54$ (frequency of A)

the gene frequencies q_1 and q_2 weighted by the relative fractions of natives and migrants.

In summary, the Hardy–Weinberg equilibrium law serves as the basis for estimating the effect of some of the forces responsible for changing gene frequencies—changes that ultimately lead to the formation of new species, the topic we shall consider next.

Natural selection   *Basic concepts*   Natural selection is the principal and perhaps the only significant force in altering gene frequencies and hence in bringing about evolution. Darwin's major premise was that populations of organisms undergo modifications as responses to environmental change, a phenomenon we refer to as natural selection. In general, all organisms produce more progeny than the environment can support, so some progeny are eliminated. Since genetic differences exist among all individuals

657

in a population (except identical twins), some progeny are better adapted to survive than others. Therefore, the better adapted organisms produce more offspring than the less-fit members of the population. In this way, the genes of the better adapted individuals increase in frequency in the population at the expense of the genes of the more poorly adapted individuals. Natural selection, then, is the differential and nonrandom perpetuation of dissimilar genotypes in violation of the assumption that all genotypes have the equal viability required to maintain an equilibrium population.

An important aspect of natural selection is the concept of fitness. It's hard to define fitness fully, but the term is so widely used that a working definition is essential. To think of natural selection as the "survival of the fittest" is a serious oversimplification (it argues in a circular fashion that the fittest are those that survive, hence those that survive are the fittest). For our purposes, *fitness* might best be described as the degree of a phenotype's capacity to donate genes to a subsequent generation. This implies that fitness can be quantified. We shall estimate fitness coefficients when we discuss the ways in which natural selection modifies gene frequencies.

The effect of natural selection on the gene frequencies in a population can be illustrated by a simplified model. Given the allelic pair A and a, with A completely dominant over a, and the genotypes AA, Aa, and aa, let us say that the homozygous recessive class aa is adaptively inferior to Aa and AA, or that aa is selected against by the selection coefficient s. The selection coefficient s has a range from 0 to 1. The fitness of the three genotypes can be designated as w_0 for AA and Aa and w_1 for aa.

Genotype	AA	Aa	aa
Fitness value (w)	w_0	w_0	w_1
Selection (s)	0	0	s

Fitness value and selection are related as follows:

$$w = 1 - s$$

The AA and Aa genotypes are the most fit, and therefore they are assigned a fitness value of $w = 1.00$ ($s = 0$). At generation 0 (before selection begins), the distribution of the genotypes in the population is a function of the binomial expansion $(p + q)^2$, where $p = 1 - q$. Substituting this value for p into the expression $p^2 + 2pq + q^2$ yields an equation containing only q:

$$\begin{array}{ccc} AA & Aa & aa \\ (1 - q)^2 + & 2q(1 - q) + & q^2 \end{array}$$

Each genotype contributes to the next generation in proportion to its fitness. Multiplying the frequency of each genotype by its fitness yields, at generation 1,

$$(1 - q)^2 w_0 + 2q(1 - q)w_0 + q^2 w_1$$

Since $w_0 = 1$ and $w_1 = 1 - s$, we can rewrite this expression as follows:

$$\begin{array}{ccc} AA & Aa & aa \end{array}$$
$$(1 - q)^2(1) + 2q(1 - q)(1) + q^2(1 - s)$$
$$\text{or}$$
$$(1 - q)^2 \quad + 2q(1 - q) \quad + q^2(1 - s)$$

Since $p^2 + 2pq + q^2 = 1$,

$$(1 - q)^2 + 2q(1 - q) + q^2(1 - s) = 1 - sq^2$$

At generation 0, the frequency of A is

$$\frac{(1 - q)^2 + (1 - q)q}{1} = 1 - q$$

and the frequency of a is

$$\frac{q^2 + (1 - q)q}{1} = q$$

At generation 1, the frequency of A is

$$\frac{(1 - q)^2 + (1 - q)q}{1 - sq^2} = \frac{1 - q}{1 - sq^2}$$

and the frequency of a is

$$\frac{q^2(1 - s) + (1 - q)q}{1 - sq^2} = \frac{q(1 - sq)}{1 - sq^2}$$

The change in the frequency of A from generation 0 to generation 1 is

$$\frac{1 - q}{1 - sq^2} - (1 - q)$$

and the change in the frequency of a from generation 0 to generation 1 is

$$\frac{q(1 - sq)}{1 - sq^2} - q = \frac{-sq^2(1 - q)}{1 - sq^2}$$

Careful scrutiny of these expressions reveals that the change in frequency of A is positive (that is, the frequency of A has increased) and that the change in frequency of a will be negative (that is, the frequency of a has decreased).

The most extreme form of selection pressure occurs when one of the genotypes is lethal. If aa is lethal ($s = 1$), what will happen to the frequencies of A and a after one generation of random mating, where the frequencies of A and a are both 0.5 initially? At generation 0,

Frequency of A = 0.5
Frequency of a = 0.5

At generation 1,

$$\text{Frequency of } A = \frac{1 - q}{1 - sq^2} = \frac{0.50}{0.75} = 0.67$$

$$\text{Frequency of } a = \frac{q(1 - sq)}{1 - sq^2} = \frac{0.25}{0.75} = 0.33$$

Then at generation 2,

$$\text{Frequency of } A = \frac{0.67}{0.89} = 0.75$$

$$\text{Frequency of } a = \frac{0.33(1 - 0.33)}{0.89} = 0.25$$

Under extreme selection pressure, then, the nonselected allele continues to decrease in frequency and is eventually lost from the population (assuming a finite population) except when it is replaced through mutation or migration. This is a dramatic case, but it does occur in nature. Cystic fibrosis, for example, is caused by an autosomal recessive gene. Even when properly treated, it is generally fatal, especially in childhood. Selection pressures range from 0 to 1, but in genotypes with s near zero, changes in gene frequency caused by selection would be difficult to verify. The weaker the selection pressure, the slower the rate of loss of an allele; the stronger the pressure, the faster the rate of loss (Figure 14.6).

It is important to bear in mind that selection operates on entire phenotypes, and therefore on entire genotypes. Gene complexes or linkage groups are often a unit of selection. So, although a single gene pair might have a detrimental effect on the carrier, it may be preserved because other genes in the complex confer a strong beneficial effect. It is also common for a single gene to have more than one phenotypic effect (such as the white eye allele in *Drosophila*). We call this condition *pleiotropy*. This suggests that one of the effects may have a negative adaptive value but that, because the other effects have strong positive selective value, the gene could nevertheless be favored. One of the phenotypic effects of the

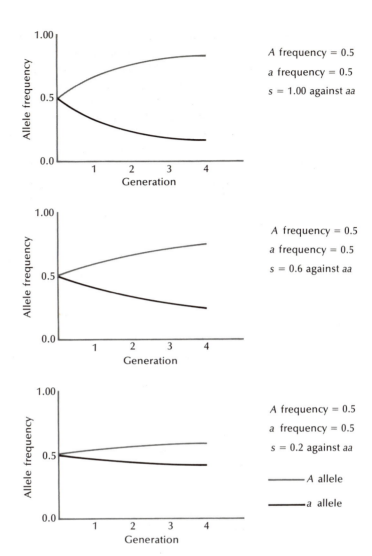

Figure 14.6
Effects of three different selection pressures on the frequencies of a pair of alleles over four generations

gene for sickle-cell anemia is to equip the carrier with resistance to malaria when the carrier is heterozygous. When the individual is homozygous for the sickle-cell gene, the phenotypic effect is commonly painful death from hemolytic anemia. Nevertheless, the sickle-cell gene is sometimes maintained in a population, because the heterozygous phenotype (in this case, resistance to malaria) is selected for.

Examples   It is difficult to acquire data to support the operation of natural selection in natural populations, because, first, experimenters cannot adequately control the conditions and, second, they must have observations that go back several years to give them an indication of what the population was like in the past. Nevertheless, we do have examples of well-documented gene frequency changes brought about by the selection of adaptive variants or gene combinations. For example, scale insects that infect the citrus orchards of southern California have acquired an increasing resistance to insecticides; and the housefly has become resistant to DDT in the 20 odd years that DDT has been in wide use. In both instances, the insecticide has been the environmental selecting agent, selecting insect strains resistant to the insecticides.

One of the most complete studies of the effects of natural selection on a natural population is the study of industrial melanism in moths of the industrialized regions of Europe. Over the past 100 years or so, several moth populations have changed color from whitish gray to charcoal gray, a change that has accompanied industrial expansion in areas such as the Ruhr Valley in Germany and the English Midlands, where factory smoke and soot have discolored the trees and buildings.

The dark color in the moths was the result of a single dominant mutation. The whitish gray genotype is *cc*, and the charcoal gray is *Cc* or *CC*. Tests have shown that these dominant mutations were not induced by the pollutants in the atmosphere but rather appeared spontaneously at a constant rate. (In the less industrialized regions, the whitish gray moth is still the predominant one.) These observations suggested that the change in color from whitish gray to charcoal gray was the result of the increased fitness of the dark variety in the dark environment. Sitting on a darkened tree limb, the dark variety was less conspicuous to its bird predators (such as thrushes and robins that clearly hunt by sight) than if it were on a nonpolluted, light-colored limb. In the countryside, the situation was reversed: The dark variety was conspicuous against the light background, whereas the light variety blended very well with the environment (Figure 14.7). Current observations support the hypothesis that the change in color was a result of natural selection favoring moths that blend best with their habitat.

H. B. D. Kettlewell demonstrated that natural selection was responsible for this change in moth color when he showed that, in the industrial regions, predator birds ate darker moths less frequently than lighter moths. In the countryside, the opposite was true. Kettlewell bred both the light and dark varieties of moths, marked them with a dot of paint, and released them in the country or in an industrialized area. In the countryside, he later recaptured 15 percent of the light moths and 5 percent of the dark moths. In the industrialized city of Birmingham, he recaptured 13 percent of the pale moths and 28 percent of the dark moths. He also filmed birds attacking the more conspicuous moth, while the nearby moth that blended with its environment went largely unnoticed.

Another example of the operation of natural selection in natural populations involves the European land snail *Cepaea nemoralis*. Shell color in this organism is determined by a multiple allelic series.

y = Yellow shell and is recessive
y^1 = Pink shell and is dominant over y
y^2 = Brown shell and is dominant over both y and y^1

(a)

Figure 14.7
(a) Two individuals of the moth *Biston betularia*, one typical and one melanic, resting on an unpolluted, lichen-covered tree trunk. (b) The same two moths resting on a polluted, soot-covered tree trunk. (From Ehrlich, Holm, and Parnell, *The Process of Evolution*, 1958; redrawn after H. B. D. Kettlewell, *Proc. X Int. Cong. Ent. 2,* by A. Ehrlich; used by permission of McGraw-Hill.)

(b)

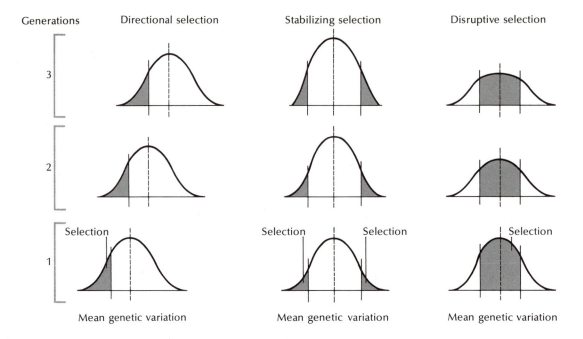

Figure 14.8
Three selection regimes
and their consequences

Shell color is crucial to the snail's ability to remain obscure and thus hidden from its predators, thrushes and other birds that hunt for the snails by sight. In woods composed of beech trees, in which the forest floor is red-brown, the snail populations are mainly brown and pink. This protective coloration enhances the snail's survival. In open green fields, the yellow snail is the dominant type. It blends in with its background more effectively than the brown or pink snails would.

In forests that undergo changes with the seasons, the snail populations also change. In the spring, snails with brown and pink have a selective advantage. In the summer when the forest floor has a green covering, yellow shells offer the best protection, and indeed yellow snails are predominant. Shell coloration in *Capaea nemoralis* is thus a trait responding to strong selective pressures.

Selection regimes Selection operates in populations in many different ways and with different consequences. We shall discuss three selection regimes and their consequences: directional selection, stabilizing selection, and disruptive selection (Figure 14.8).

Industrial melanism in the moth *Biston betularia* is a good example of *directional selection*. With only one of several variants being selected against, there is a progressive, unidirectional change in the genetic composition of the population. In the industrialized regions of Europe, the lighter moth was selected against, resulting in a progressive change in the population toward the darker varieties. We commonly observe directional selection when the environment is steadily changing and the population is changing with it.

In *stabilizing selection,* the variants at both extremes are selected against so that the mean in the population remains the same. For a population that is well adapted to its environment, the genetic variants that arise as a consequence of recombination, mutation, or migration are eliminated, thus stabilizing the population. Stabilizing selection is the most common selection regime in nature. For example, as urban areas of Europe became more polluted, directional selection toward more melanism in *Biston betularia* was quite evident. But in the areas that were not subjected to industrial pollution, the moth population was experiencing stabilizing selection. The dark variants were selected against, keeping the mean pigment quotient at an optimum for survival on the lighter backgrounds.

Disruptive selection acts in such a way as to favor the extreme variants and eliminate those organisms in the average range. The consequences of this type of selection are to fragment one large population into two smaller ones, each quite different from the other with respect to the trait in question. In an interesting experiment to test the effects of disruptive selection, Thoday and Boam established a *Drosophila* population and then began selecting as parents those flies with high and low numbers of body bristles. In other words, those flies with an average number of bristles were not permitted to contribute their genetic information to the following generation. But even though the investigators selected flies with high and low numbers of bristles, they counteracted this selection pressure by intercrossing these two lines to produce additional generations. The results of this selection regime over 36 generations (about a year and a half) appear in Figure 14.9. There was no evidence of any real disruption until

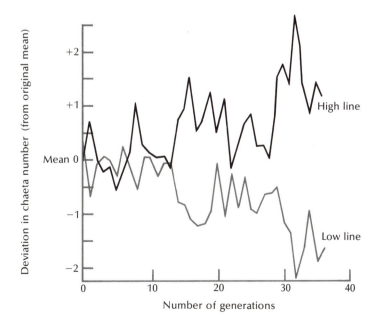

Figure 14.9
The results of a disruptive selection regime for bristle number in *Drosophila;* mating is random after the mean is selected against.

the fourteenth generation, when the population diverged into "high" and "low" lines—and this in spite of the crossbreeding opportunities.

It is difficult to establish instances of disruptive selection in nature, though there are several likely candidates. For instance, sexual selection may favor those individuals with the most prominent sexual displays and thus tend to eliminate the intermediate levels. Selection for wing color in some butterfly species appears to be disruptive, and there are cases of disruptive selection in plants (see Remington and Antonovics in the References at the end of this chapter).

In disruptive selection, if we had *AA*, *Aa*, and *aa* genotypes, the homozygotes would be favored. But it is also common to see the heterozygote selected for at the expense of the homozygote. Heterozygote superiority was discussed in connection with carriers of sickle-cell anemia (see page 661). When heterozygotes have the advantage, we have a balancing type of selection.

A well-studied instance of heterozygote superiority involves a series of inversions in chromosome 3 of *Drosophila pseudoobscura* (see pages 220–226 for a review of inversions). If we start with a *Drosophila* population composed of 10 percent flies homozygous for the standard (*ST*) inversion and 90 percent flies homozygous for the chiricahua (*CH*) inversion, after several generations the inversion frequencies change drastically (Figure 14.10). But it is interesting to note that the inversions stabilize at *ST* = 70 percent and *CH* = 30 percent. If one inversion were superior to the other, we might expect the inferior inversion to be eliminated. This did not happen, because the *ST/CH* heterozygote was superior to both of the inversion homozygotes. Thus a low but constant

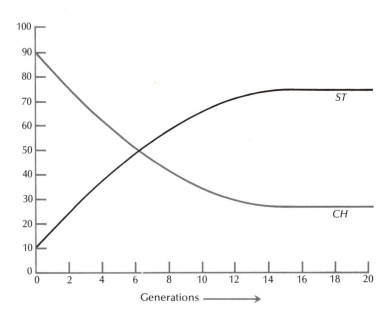

Figure 14.10
Selection against *CH* and *ST* inversions favors the heterozygote.

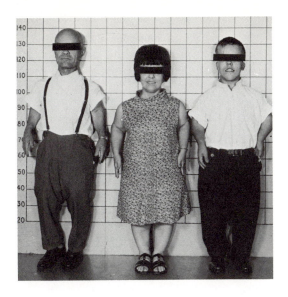

Figure 14.11
A father, mother, and son with achondroplasia. (These parents have another son who is of average height.) (Photograph courtesy of Dr. Victor A. McKusick, Johns Hopkins Hospital)

number of decidedly inferior *CH/CH* homozygotes were produced in order to maintain an optimal level of *ST/CH* heterozygotes. If we assign fitness values to the three genotypes and let *ST/CH* = 1.0, then *ST/ST* = 0.90 and *CH/CH* = 0.41. In Chapter 15, we shall see how these inversions contribute to population diversification and speciation processes.

Cost Natural selection has both positive and negative aspects. Some genotypes are successful at reproduction and contribute effectively to the following generation. But some genotypes, those selected against, do not contribute to the gene pool of the next generation (or they contribute very little). These genetic dead ends reduce the overall fitness of a population, since evolutionary fitness is quantified in terms of reproductive success. Thus populations carry a genetic load in that segment of the population that is less fit. Genetic load in a population is defined as 1.0 − mean population fitness.

Balance between selection and mutation We mentioned earlier that mutational equilibrium is rarely if ever achieved in natural populations, because it is so slow that other forces acting more quickly come to dominate. A case in point is the equilibrium reached in the human population between mutation pressure increasing the incidence of a dominant gene and selection pressure decreasing its incidence.

Achondroplasia is a form of dwarfism (Figure 14.11) caused by a dominant allele that is lethal in the homozygous state. The frequency of this trait in the human population is 1 per 10,000 births, and about 80 percent of these individuals are the offspring of normal parents with no history of the trait. This 80 percent, then, arose from gametes carrying a new mutation. The mutation rate for this dominant allele has been calculated to be about 1 per 20,000 gametes.

Natural selection operates on achondroplasia in two ways. The homozygous dominant is lethal, and the heterozygote has only a 20-

Figure 14.12
A hypothetical situation
in which the population
begins with no
achondroplastic dwarfs.
There is a mutation rate
of 5×10^{-5} at each
generation. In (a), there is
no selection against
dwarfs; in (b), selection is
0.80 against dwarfs, so
that selection and
mutation balance out.

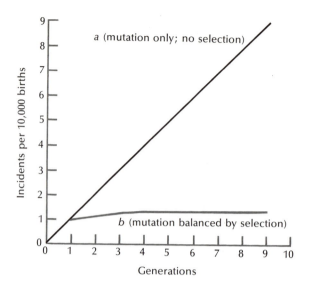

percent chance of leaving offspring ($s = 0.80$; $w = 0.20$). With no selection
pressure operating, and with a mutation rate of 1 per 20,000, the frequency of the trait shows a linear increase over several generations (until
eventually balanced by reverse mutations). But with a selection pressure
of 0.80, the incidence of the trait in the human population reaches an
equilibrium after four generations (Figure 14.12).

This stabilizing of the incidence of achondroplasia is the result of a
balance between mutation pressure and selection pressure. The equilibrium point is usually higher than the mutation rate. For achondroplasia,
the equilibrium point is an incidence of 1.25 per 10,000.

We can predict the frequency of individuals expressing the dominant
trait at the equilibrium point by the formula

$$\text{Frequency at equilibrium } (b) \text{ approaches } \frac{2\mu}{1 - w}$$

where

μ = Mutation rate
w = Fitness of a dominant trait

In our analysis of achondroplasia,

$$\mu = \frac{1}{20,000} = 0.00005, \quad 2\mu = 0.0001$$

$$w = 0.2, \quad 1 - w = 0.8$$

$$b = \frac{2\mu}{1 - w} = \frac{0.0001}{0.8} = 0.000125$$

From this formula, you can see that the higher the fitness and the higher the mutation rate, the greater the frequency of the dominant trait.

We can use this formula to estimate mutation frequency. For example, a dominant trait has a fitness of 0.6 and a frequency in the population of 1 per 8000. The mutation rate is

$$b = \frac{2\mu}{1 - w}$$

$$\mu = \frac{b(1 - w)}{2}$$

$$= \left(\frac{1}{8000}\right)(0.4)\left(\frac{1}{2}\right) = 2.5 \text{ per } 100,000$$

These estimates must be used cautiously, however, because the fitness factor is subject to many sources of error.

Random genetic drift   Each individual in a Mendelian population generates millions upon millions of gametes. From those millions, however, only one is involved in the formation of each offspring. If the number of offspring in a population is small, the number of gametes that go to form those offspring will also be small. Concomitantly, a large number of offspring means that a large number of gametes are involved. The smaller number of offspring represents a smaller sample of parental gametes; and the smaller the sample of parental gametes, the greater the chance of sampling errors. The following example will illustrate this concept (Figure 14.13).

Figure 14.13
Genetic drift: the smaller the sample size, the greater the chance of sampling error

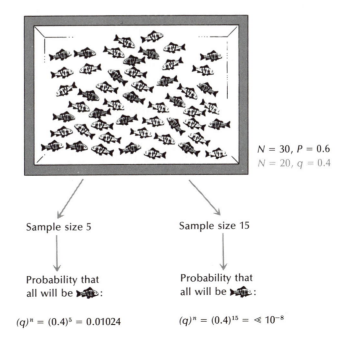

$N = 30, P = 0.6$
$N = 20, q = 0.4$

Sample size 5

Sample size 15

Probability that all will be 🐟:

Probability that all will be 🐟:

$(q)^n = (0.4)^5 = 0.01024$

$(q)^n = (0.4)^{15} = \ll 10^{-8}$

From a bowl containing 30 black fish and 20 orange fish, 1 fish at a time is drawn, then recorded and replaced. If only 5 fish are recorded, the probability that all 5 will be orange is a little over 1 percent. If the sample size is increased to 15, the odds against their all being orange are more than 100,000,000 to 1!

This example demonstrates that in small populations, the offspring may all be the same type simply by chance. *Random genetic drift*, then, is a random fluctuation in allele frequencies. Its effect on large populations is negligible. We have already encountered genetic drift in our discussion of Mendelian ratios, when we noted deviations from the expected ratios. We can expand on this point by analyzing a large population in which the frequency of *A* is $p = 0.4$ and the frequency of *a* is $q = 0.6$. If we take a random sample from this population, we expect the allele frequencies to remain 0.4 and 0.6, respectively, with some expected deviation occurring because of sampling error. How far observed data depart from the average expected value is expressed in terms of the *standard deviation*. The standard deviation (σ) expected in a sample drawn from a large population is

$$\sigma = \sqrt{pq/2N}$$

where N = number of diploid individuals in the sample. In a sample of 10,000 individuals, the standard deviation is

$$\sigma = \sqrt{0.6 \times 0.4/20,000} = 0.0035$$

A standard deviation means that for a normal distribution curve (Figure 4.11) the mean, or average, value, plus or minus the standard deviation, takes in about two-thirds of the total area under the curve. The mean, plus or minus 1.96 multiplied by the standard deviation, accounts for 95 percent of the area. For a sample size of 20,000 with $p = 0.4$ and $q = 0.6$, two-thirds of the time the frequency of *A* will lie within $0.4 \pm \sigma$, or between 0.3965 and 0.40035.

However, if the sample is made up of only 20 individuals, the standard deviation is

$$\sigma = \sqrt{0.6 \times 0.4/40} = 0.077$$

In this case, two-thirds of the time the frequency of *A* will lie within $0.4 \pm \sigma$, or between 0.323 and 0.477.

If a new population is being established from a sample of a large parent population, a sample size of 20,000 ensures that the gene frequencies—at least at the beginning—in both the old and the new populations are essentially the same. However, if the new population is founded with a sample made up of only 20 individuals, chances are that the gene frequencies in the two populations will exhibit significant differences simply because of sampling error.

The role of genetic drift in evolution has been actively debated, but no solid conclusions have resulted. In the human population, there are instances of what appears to be genetic drift in operation. For example, the uneven distribution of the ABO blood groups in different parts of the world (Figure 14.14) could be the result of some areas having been settled by small migrant groups that were atypical of their populations. The high frequency of the O blood type among American Indians could stem from a small founding population of Asians crossing the Bering Strait and subsequently founding a series of populations throughout North and South America. The O blood type is uncommon in Asia, but the original migrants to North America may have been an atypical sample. Instances of drift are difficult to verify, though, because of the following restrictions.

1. The traits in question must be well understood genetically.

2. The traits must be adaptively neutral, something very difficult to establish (otherwise, natural selection might be the determining factor in gene frequency distributions).

3. The population must be small, either at the time of origin or at some later time in its existence, and reproductively isolated.

4. The population's parent population and the population that now surrounds it must be genetically well defined with respect to the traits in question.

5. Migration of genotypes into the population must either not have occurred or be amenable to precise quantification so that its effects can be taken into account.

Such populations are difficult to find, as may be imagined, but they do exist. Bentley Glass has studied a population of "Dunkers" (German Baptist Brethren), numbering about 350, in south-central Pennsylvania. The population fits nearly all the necessary criteria, and the data collected from it support an argument for genetic drift. A model of the population and its analysis are given in Figure 14.15.

The Dunkers migrated from Germany early in the eighteenth century, and their communities have in large part remained reproductively isolated from each other and from the surrounding American population. Marriages between the Dunkers and individuals from other communities have been so monitored that compensations in the gene frequencies can be made. The frequency of blood group A is 0.6 in the Dunkers but only 0.40 in the United States and 0.45 in the parent German population. The B and AB groups are almost entirely absent from the Dunkers (0.05), but they comprise 0.15 of the United States and German populations. The same fluctuations hold for the M and N alleles and also for other presumably nonadaptive traits, such as attached ear lobes, "hitchhiker's thumb," and mid-digital hair (Figure 14.16). In all these traits, the genes in the

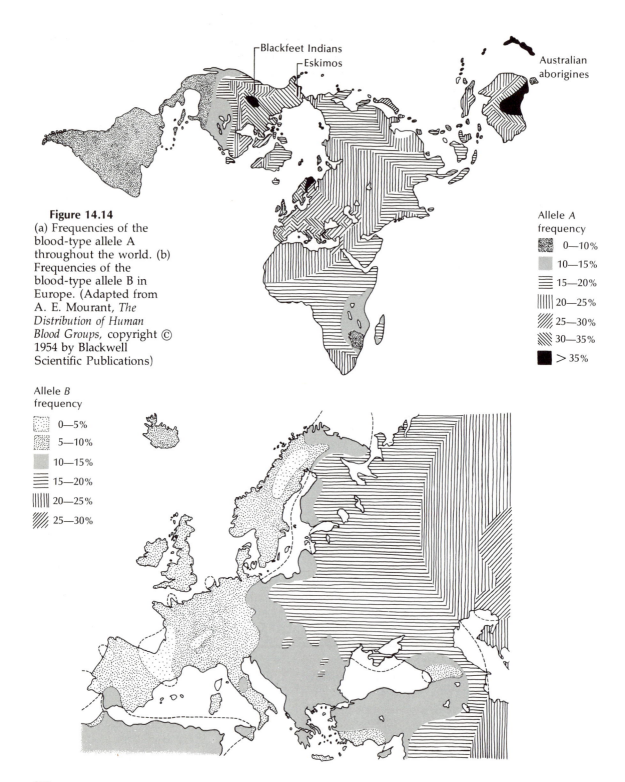

Figure 14.14
(a) Frequencies of the blood-type allele A throughout the world. (b) Frequencies of the blood-type allele B in Europe. (Adapted from A. E. Mourant, *The Distribution of Human Blood Groups,* copyright © 1954 by Blackwell Scientific Publications)

Blackfeet Indians
Eskimos
Australian aborigines

Allele *A* frequency
0—10%
10—15%
15—20%
20—25%
25—30%
30—35%
> 35%

Allele *B* frequency
0—5%
5—10%
10—15%
15—20%
20—25%
25—30%

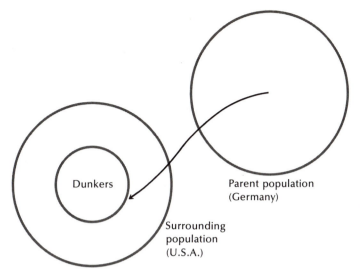

Dunkers

Parent population
(Germany)

Surrounding
population
(U.S.A.)

Figure 14.15
Model of a Dunker community, comparing some of its genotype frequencies with the frequencies in the parent and surrounding populations. [Adapted by permission of the University of Chicago Press from B. Glass et al., *The American Naturalist* 86:145 (1952), published for the American Society of Naturalists by the Science Press, Lancaster, Pa., 1952]

Results

	Trait	Class frequency		
		U.S.A.	Dunker	Germany
Blood groups	A	0.40	0.60	0.45
	B—AB	0.15	0.05	0.15
	M	0.30	0.45	0.30
	MN	0.50	0.41	0.50
	N	0.20	0.14	0.20

Dunkers have frequencies like neither the Germans nor the Americans surrounding them, nor like anything in between. Instead, the frequencies have fluctuated widely from one extreme to the other, suggesting that genetic drift has been at work.

In conclusion, the mathematical theory of genetic drift has received support from data such as the Dunker community afforded, but concern over the evolutionary significance of genetic drift continues. The controversy will not be resolved until more work is done.

"Non-Darwinian" evolution: The controversy continues

The neo-Darwinian view of evolution proposes that natural selection is the predominant evolutionary force and that new mutations are subject to a rigorous screening process. A mutation that increases the fitness of a population persists and increases in frequency, while a mutation that decreases the fitness of a population does not persist, except as it is introduced by mutation pressure. The key assumption of this viewpoint is that mutations have positive or negative adaptive value.

But do all mutations that occur have positive or negative adaptive value? It is clear from earlier discussions that some codon changes do not lead to amino acid changes in a polypeptide. And some amino acid changes result in no perceptible change in the function of a polypeptide.

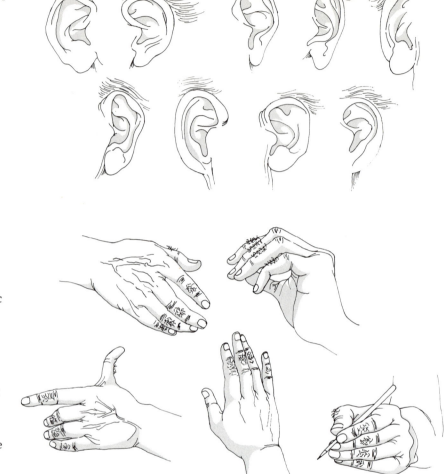

Figure 14.16
Some genetically determined human traits used in studies of genetic drift: attached ear lobes, mid-digital hair, and hitchhiker's thumb. [Adapted by permission of the University of Chicago Press from B. Glass et al., *The American Naturalist* 86:145 (1952), published for the American Society of Naturalists by the Science Press, Lancaster, Pa., 1952]

These and other observations have led some evolutionary geneticists to propose that some mutations are neutral and become fixed in populations as a consequence of random genetic drift, not selection pressure. This position is called the *neutralist hypothesis* or, incorrectly, *non-Darwinian evolution* (Darwin actually considered neutral mutations). Let us examine this position more carefully.

Motoo Kimura (Figure 14.17) at the National Institute of Genetics in Japan, is probably the main spokesman for the neutralist position. He has methodically evaluated rates of substitution of amino acids in a variety of proteins in a wide range of organisms. Studies of this type have shown a high degree of enzyme polymorphism. On the basis of their analysis of rates of substitution of amino acids, investigators have estimated that in

Figure 14.17
Motoo Kimura

mammals there are at least six nucleotide substitutions per year per genome. Kimura suggests that such an enormously high rate of molecular evolution cannot be explained by natural selection acting independently at each genetic locus. It would require too high a cost of selection.

For example, to maintain polymorphism at 1000 separate genetic loci, and assuming only a slight selective advantage at each locus, a female would have to produce about 40,000 offspring to ensure replacement and thus avoid extinction. To circumvent such a high cost of selection, Kimura and his co-workers propose instead that most polymorphic protein systems are composed of neutral mutations.

Kimura and his co-workers also cite other evidence that they contend supports their position. They show that functionally less important molecules or parts of molecules evolve faster than more important ones with respect to amino acid substitutions. Also, amino acid substitutions that do not seriously change the function of a protein occur more frequently than those that are more disruptive. From these observations, the neutralists conclude that evolutionary rates are higher in unimportant molecules or parts of molecules only if we assume a greater likelihood of random neutral mutations in these regions.

How do the selectionists interpret these observations? They suggest that the astronomically high cost of selection required to maintain extensive polymorphism is calculated on the basis of a mistaken assumption. Selection does not operate on individual genes. It operates on the entire phenotype determined by the combined effects of genotype and environment. Granting this, the cost of selection becomes far more reasonable, because it does not require selection at each locus. Another point that advocates of natural selection make is that in many cases the extensive polymorphism that exists in natural populations actually exhibits a pattern. In other words, there is a correlation between certain protein polymorphisms and ecological distributions. This suggests that at least some polymorphisms are maintained by selection.

R. Ambler studied the cytochrome-*c* enzyme from different and diverse populations of the protozoan *Pseudomonas aeruginosa*. The enzyme from nine diverse populations had identical amino acid sequences in eight of those populations, and it differed by only one amino acid in the ninth. This enzyme, which does show extensive polymorphisms, does not show the variation the neutralists expect in these nine diverse populations. Selection pressures, not at all understood, may account for this stability.

Whether neutral mutations in fact occur is still an open issue. Nobody has yet disproved the neutralist hypothesis. Indeed, many observations are consistent with the hypothesis. But caution must be exercised in viewing neutralism. Just because we cannot perceive an advantage or disadvantage of an amino acid substitution does not mean that one does not exist. A future environmental shift may make a mutation "neutral" at one time, advantageous at another. Furthermore, while some codon changes may not result in an amino acid substitution (review degeneracy in the genetic code, page 388), translation efficiency may be altered by such mutations. So these mutations may not qualify as "neutral."

T. Ohta has presented a somewhat modified view of the neutralist theory. She suggests that "neutral mutations" may have very slight negative selective values, around $S = 0.0001$. She argues that in small populations, these nearly neutral alleles behave essentially as if they were completely neutral. As the population increases, the effectiveness of negative selection increases according to the formula

Selection effectiveness = NeS

where Ne = effective size of the population (counting only those individuals participating in the mating process).

The concept of acquired characters

Among the many controversies spawned by evolutionary theory, none has been more intense or heated than the idea of the inheritance of acquired characters. It is thus appropriate at this point to consider the issues surrounding this controversy. All too often the idea of acquired characters is ignored by students of evolution, or dismissed simply as a historical curiosity and nothing more. But the idea is more important than many care to admit, and it should be carefully evaluated.

The first aspect we need to deal with is the meaning of the term *acquired character*. There are two levels that interest us when we use this term. At one level, we have environmental modifications of the phenotype that are not in any way transmitted back to the genotype. In other words, the genotype interacts with a range of environments to produce a range of phenotypes.

At another level, phenotypes are acquired through the acquisition of new genetic information. It is this level that has been of such great concern to evolutionists. Can phenotypic modifications induced by some component of the environment be incorporated into the genetic material and faithfully transmitted to subsequent generations?

In one respect, we must answer in the affirmative. The acquisition of some microorganisms by some eukaryotic organisms can change the phenotype of the eukaryote in a permanent, genetically transmissible way. Recall that cytoplasmic viral particles, transmitted from generation to generation, made *Drosophila* sensitive to CO_2 (see page 170). Many people believe that organelles such as mitochondria and chloroplasts are remnants of bacteria that set up a symbiotic relationship with an early eukaryotic cell. The eukaryote phenotype was dramatically and permanently altered by this acquisition. The kappa particle of *Paramecium* is a genetically stable environmental acquisition (see page 169). Other environmentally induced phenotypic modifications that are genetically stable occur during the process of transformation and transduction in bacteria (see Chapter 12).

But the key feature of these examples is the randomness or purposelessness of the acquisitions. *Drosophila* had no plan to become CO_2-sensitive; there was no design to acquire chloroplasts or mitochondria (but there certainly was selection pressure to retain them once they were acquired). These examples are analogous to the studies we discussed earlier on the randomness of mutations (see page 423).

There are, however, a couple of studies that can be interpreted as supporting *directed* acquired characteristics. One study can be explained, but the other presents us with some difficulties. The explainable study involves bleaching (loss of chloroplasts) in the green flagellate *Euglena*. If *Euglena* is raised in the dark, the chlorophyll fails to develop and consequently the chloroplasts fail to develop. Eventually the chloroplasts disappear from these dark-raised *Euglena* cells and they appear "bleached." This is a permanent genetic trait in these cells (recall that chloroplasts have their own DNA), such that new chloroplasts do not appear in the presence of light. It thus may appear that Euglena perceived no need for chlorophyll or chloroplasts in a dark environment and so got rid of this photosynthesizing material. This should remind you of the Lamarckian idea of use and disuse (see page 636). But though there is a specific genetic response (change) to a specific environment, need does not enter into the picture. Cytoplasmic DNA was lost because chloroplasts required light in order to develop. But in one sense, we have here a case of inherited acquired characteristics. Bear in mind, however, that the *Euglena* study does not contradict any known biological principles.

A study involving flax is more difficult to interpret. When fertilized with N, P, and K, flax plants grew larger and appeared more vigorous. But what is not easy to explain is that the offspring of these fertilized plants were also large, even though they received no fertilizer. The large size was a permanent trait that persisted through several generations. It appears that the fertilizer-induced largeness became a genetically transmitted trait. The cytoplasm does not appear to be involved in this trait, but it does appear that nuclear DNA increased in the large plants.

We do not yet understand the connection between this fertilizer and increased nuclear DNA, but some argue that the study supports environmentally directed mutation (see Evans, Durrant, and Rees in the References). A reasonable possibility consistent with our knowledge of cell structure and function can be presented, and it does not invoke a Lamarckian explanation. The fertilizer may have interfered with a cellular mitosis so that a cell line came to have a higher DNA content. These cells, for some reason, divided more rapidly and eventually replaced the unaltered cells. We have here, then, a combination of cellular and individual selection.

Acquired characters may be operating at some levels in the living world, but it is extremely unlikely that they are operating in a truly

Lamarckian sense. All available evidence shows that nuclear gene muta-tions are random. Cytoplasmic genes offer the only plausible arena for discussion of acquired characters. At a more basic level, for phenotypic modifications to be incorporated into nuclear genes, the following infor-mation flow would have to hold:

$$\text{Environmental stress} \longrightarrow \text{Phenotypic modification} \longrightarrow \text{Protein} \longrightarrow \text{RNA} \longrightarrow \text{DNA (gene)}$$

There is no evidence anywhere supporting such a scheme of information flow. We can only conclude that acquired characters, except in a limited sense, do not occur.

Overview

In this chapter, we examined the transmission of genetic material through a population by the application of Mendelism to populations of sexually reproducing organisms.

After tracing the roots of Darwinism, we looked at the structure and concept of populations. Our discussion was limited to Mendelian popula-tions. Within a population, mating can range from self-fertilization to random and complete outcrossing. Analysis of the genetic structure of populations has revealed a high degree of genetic polymorphism (a high degree of heterozygosity at most gene loci).

If a Mendelian population is randomly mating, is large enough to avoid sampling errors, experiences no mutations, is composed of genotypes that are equally viable and fertile, produces gametes that all have equal probabilities of forming zygotes, and is isolated, then gene frequencies do not change from one generation to the next. This condition is called the Hardy–Weinberg equilibrium.

But evolution occurs, and it involves changes in gene frequency. So we proceeded to show why each of the foregoing assumptions does not necessarily hold. After discussing how gene frequencies are calculated using the formula $p^2 + 2pq + q^2 = 1$, we examined some forces that disrupt that equilibrium. Mutation pressure and migration patterns both operate to alter gene frequencies.

The most significant and certainly the most creative of all evolution-ary forces is natural selection. Natural selection is the differential repro-ductive success of different genotypes. We looked at some quantitative models that assessed changes in gene frequency as a consequence of some selective regimes. The regimes we studied were directional, stabiliz-ing, and disruptive selection.

Natural selection has both positive and negative aspects. Populations contain individuals that may not contribute to the gene pool; thus they reduce the overall fitness of the population. These individuals constitute the population's genetic load.

Mutation and natural selection often counteract each other and reach an equilibrium point. Selection acts to remove undesirable genes from the gene pool, but mutation pressure continues to feed the undesirable genes into the gene pool.

Genetic drift is another force that tends to disrupt genetic equilibrium. It is particularly important in small populations, because they are more susceptible to random fluctuations in gene frequency. Drift is based simply on sampling error.

The mating structure of populations, if it involves a high degree of inbreeding, can lead to inbreeding depression. This depression, characterized by a decline in vigor, is caused in part by increased homozygosis. Often, however, when two different inbred lines are crossed, a vigorous hybrid is produced. This is called hybrid vigor, or heterosis. Two hypotheses that attempt to explain heterosis are the overdominance hypothesis (heterozygote superiority) and the dominance hypothesis (superiority of dominant alleles).

Are all mutations subject to selection forces, or can mutations be neutral? We looked at both sides of this controversy and can only suggest that the issue is not yet resolved.

Finally, we looked at the notion of acquired characteristics. Some observations appear to be consistent with the idea, but these observations largely concern the addition or deletion of cytoplasmic material. No study supports the idea that the environment can direct or orient a specific mutation in the nuclear DNA, and there is much evidence to discredit the idea.

This chapter looked at how evolution occurs—how microevolutionary changes occur in a population. In Chapter 15, we shall look at the evidence for macroevolutionary changes.

Questions and problems

14.1 In estimating the number of colorblind females in a population of 1000, we arrived at the number 10. What factors could account for the fact that all the colorblind individuals were males, with no observed colorblind females?

14.2 How would you criticize the idea that the frequency of a mutant gene in a population is an inverse function of the selection pressure against it?

14.3 Using population genetics theory, how would you account for the lack of eyes in some cave-dwelling fish?

14.4 What are the *A* and *a* gene frequencies in a human population in which the *Aa* heterozygote is 50 percent? (Assume the population is in equilibrium.)

14.5 A population is composed of *AA*, *Aa*, and *aa* genotypes in a 1:1:1 ratio. The offspring produced by this population number 996, 996, and 224 respectively. What are the selection pressures on the three genotypes?

14.6 Bruce Wallace studied the elimination of an autosomal recessive lethal mutation from an experimental population of *Drosophila melanogaster*. He obtained the data shown in the table.

Generation	n	Observed frequency of a	Expected frequency of a
0	—	0.500	—
1	454	0.284	0.333
2	194	0.232	0.250
3	212	0.189	0.200
4	260	0.188	0.167
5	290	0.090	0.143
6	398	0.085	0.125
7	366	0.082	0.111
8	382	0.065	0.100
9	388	0.054	0.091
10	394	0.041	0.083

Are these data perfectly consistent with the hypothesis that $S_{AA} = 0$, $S_{Aa} = 0$, and $S_{aa} = 1.0$? Explain. (S = selection pressure.)

14.7 How will gene frequencies change in three generations if $S_{AA} = 0$, $S_{Aa} = 1.0$, $S_{aa} = 0$, and $p_0 = q_0 = 0.50$?

14.8 How will gene frequencies change in three generations if $S_{AA} = 0$, $S_{Aa} = 1.0$, $S_{aa} = 0$, $p_0 = 0.4$, and $q_0 = 0.6$?

14.9 In 1954, Kerr and Wright (*Evolution* 8:172–177) studied 96 lines of *Drosophila*, each line starting with the same 8 parent flies (f = forked bristles, a recessive sex-linked trait).

$$♀ = 1 \ f/f \quad 2 \ f/+ \quad 1 \ +/+$$
$$♂ = 2 \ f/Y \quad 2 \ f/Y$$

At each generation, 8 flies were chosen at random from the progeny to be parents for the next generation. The phenotypes of the 8 flies were recorded. If all 8 parental flies chosen were forked, then f became fixed in that line and + was lost. If all 8 flies were +/+ or +/Y, then + was fixed and f was lost. The data collected are given in the table, which lists, for each generation, the number of populations that either contained only wild-type or mutant (f) alleles or were still segregating. Offer an interpretation of this experiment.

Generation	Wild	Unfixed	Forked
0	0	96	0
1	1	94	1
2	1	92	3
3	2	87	7
4	7	79	10
5	10	70	16
6	11	66	19
7	16	59	21
8	17	56	23
9	20	52	24
10	24	47	25
11	29	39	28
12	31	37	28
13	34	34	28
14	37	30	29
15	38	29	29
16	41	26	29

14.10 Gordon (*Am. Nat.* 69:381, 1935) released a large number of ebony heterozygous *Drosophila* (*e* is an autosomal mutant) in a part of England normally devoid of this species. After six generations, the frequency of *e* was 0.11. How would you interpret this?

14.11 In a natural population of field mice, you find the following distribution of genotypes for the sex-linked alleles *s* and *t* (*s* = striped). What are the *s* and *t* gene frequencies, and is this population in equilibrium?

	s/s	s/+	+/+	s/⋏	+/⋏
♂	—	—	—	32	40
♀	17	22	26	—	—

14.12 What frequencies of *A* and *a* in a population produce the greatest number of heterozygotes?

14.13 Consider two large, isolated populations, each carrying a different recessive lethal allele. In population 1, the recessive lethal (*a*) has a frequency of 0.06. In population 2, the recessive lethal (*b*) has a frequency of 0.03. The *b* allele is not found in population 1, and *a* is not found in population 2. Following an environmental change, the two populations merge and exhibit random mating. What is the frequency of the lethal phenotype in population 1, population 2, and the new population (once it reaches equilibrium)?

14.14 Studies of 108 chondrodystrophic dwarfs showed that they produced 27 offspring. These dwarfs had 457 normal siblings who produced 582 children. Calculate a fitness value for this dominant trait.

14.15 Of 94,075 children born to normal (non-chondrodystrophic dwarf) parents, 8 were chondrodystrophic dwarfs. These would all be heterozygous (homozygotes are lethal) and the result of new mutations. Use this information to estimate the mutation frequency at this locus.

14.16 In a class of 25 students, 14 found the taste of PTC bitter. They were *tt*. How many students were carriers?

14.17 Dobzhansky and Pavlovsky studied two autosomal chromosomal inversions (*A* and *D*) in *Drosophila tropicalis*. Individuals can be either *AA*, *AD*, or *DD* for the two inversions. In a sample of 140 flies, they found 3 *AA*, 134 *AD*, and 3 *DD*. On the basis of the Hardy–Weinberg principle, is this population an equilibrium population for the observed genotypic classes? If you decide that the answer is no, how would you interpret the data?

References

Antonovics, J. 1971. The effects of a heterogeneous environment on the genetics of natural populations. *Amer. Sci.* 59:593–599.

Castle, W. E. 1903. The law of heredity of Galton and Mendel and some laws governing race improvement by selection. *Proc. Amer. Acad. Arts Sci.* 39:233–242.

Crow, J. F., and M. Kimura. 1970. *An Introduction to Population Genetics Theory*. New York, Harper and Row.

Darwin, C. 1859. *On the Origin of Species by Means of Natural Selection*. (The original and later editions have been republished by many publishers.)

Denell, R. E., B. H. Judd, and R. H. Richardson. 1969. Distorted sex ratios due to segregation distorter in *Drosophila melanogaster*. *Genetics* 61:129–139.

Dobzhansky, Th. 1970. *Genetics of the Evolutionary Process.* New York, Columbia University Press.

Dobzhansky, Th., F. Ayala, G. L. Stebbins, and J. W. Valentine. 1977. *Evolution.* San Francisco, Freeman.

Dobzhansky, Th., M. K. Hecht, and W. C. Steere (eds.). 1967–1978. *Evolutionary Biology* (in 10 volumes). New York, Plenum.

Ehrlich, P. R., R. W. Holm, and D. R. Parnell. 1974. *The Process of Evolution.* New York, McGraw-Hill.

Evans, G. M., A. Durrant, and H. Rees. 1966. Associated nuclear changes in the introduction of flax genotypes. *Nature* 212:697–699.

Felsenstein, J. 1976. The theoretical population genetics of variable selection and migration. *Ann. Rev. Genetics* 10:253–280.

Ford, E. B. 1964. *Ecological Genetics.* New York, Wiley.

Glass, B. 1954. Genetic changes in human populations, especially those due to gene flow and genetic drift. *Adv. Genetics* 6:95–139.

Grant, V. 1977. *Organismic Evolution.* San Francisco, Freeman.

Hardy, G. H. 1908. Mendelian proportions in a mixed population. *Science* 28:49–50.

Hecht, M. K., and W. C. Steere (eds.). 1970. *Essays in Evolution and Genetics* (in honor of Theodosius Dobzhansky). New York, Appleton-Century-Crofts.

Hedrick, P. W., M. E. Ginevan, and E. P. Ewing. 1976. Genetic polymorphism in heterogeneous environments. *Ann. Rev. Ecol. and Syst.* 7:1–32.

Jain, S. K. 1976. The evolution of inbreeding in plants. *Ann. Rev. Ecol. and Syst.* 7:469–495.

Kerr, W. E., and S. Wright. 1954. Experimental studies of the distribution of gene frequencies in very small populations of *Drosophila melanogaster:* I. Forked. *Evolution* 8:172–177.

Kimura, M. 1970. The length of time required for a selectively neutral mutant to reach fixation through random frequency drift in a finite population. *Genet. Res.* 15:131–133.

Levine, L. 1971. *Papers on Genetics.* St. Louis, Mo., Mosby.

Lewontin, R. C. 1973. Population Genetics. *Ann. Rev. Genetics* 7:1–18.

Lewontin, R. C. 1974. *The Genetic Basis of Evolutionary Change.* New York, Columbia University Press.

Li, W-H., and M. Nei. 1977. Persistence of common alleles in two related populations or species. *Genetics* 86:901–914.

MacIntyre, R. J. 1976. Evolution and ecological value of duplicate genes. *Ann. Rev. Ecol. and Syst.* 7:393–420.

Mayr, E. 1972. The nature of the Darwinian revolution. *Science* 176:981–989.

Mettler, L. E., and T. G. Gregg. 1969. *Population Genetics and Evolution.* Englewood Cliffs, N.J., Prentice-Hall.

Milkman, R., and M. Kimura. 1976. Molecular evolution. *Trends in Biochem. Sci.,* July.

Molecular Evolution: See responses to the Milkman and Kimura debate in *Trends in Biochem. Sci.,* August, September, and November, 1976.

683

References

Ohno, S. 1970. *Evolution by Gene Duplication.* New York, Springer.

Ohta, T. 1974. Mutational pressure as a main cause of molecular evolution. *Nature* 252:351–354.

Peters, J. A. 1959. *Classic Papers in Genetics.* Englewood Cliffs, N.J., Prentice-Hall.

Selander, R. K., and W. E. Johnson. 1973. Genetic variation among vertebrate species. *Ann. Rev. Ecol. and Systematics* 4:75–92.

Spiess, E. B. 1962. *Papers on Animal Population Genetics.* Boston, Little, Brown.

Wallace, B. 1968. *Topics in Population Genetics.* New York, Norton.

Weinberg, W. 1908. Über den Nachweis der Vererbung beim Menschen. *Jahresh. Verein f. vaterl. Naturk. Württem.* 64:368–382.

Wright, S. 1948. On the roles of directed and random changes in gene frequency in the genetics of populations. *Evolution* 2:279–295.

Yule, G. U. 1902. Mendel's laws and their probable relation to intraracial heredity. *New Phytol.* 1:192–207, 222–238.

Zimmering, S., L. Sandler, and B. Nicoletti. 1970. Mechanisms of meiotic drive. *Ann. Rev. Genetics* 4:409–436.

Fifteen

Genes in populations, II

In Chapter 14, we discussed the way genetic material is transmitted in a population from one generation to another. The genetic constitution of a population does change over time as a consequence of the forces that disrupt genetic equilibrium. We discussed those forces and their effects on a population. But though we assessed the changes in gene frequency in populations, we have yet to connect these changes to the process of speciation. That is what this chapter is all about. We shall approach the process of speciation by examining the way populations differentiate into species and races, the way isolating mechanisms operate, and the way various models of speciation apply in the natural world. We shall also examine the relationship between speciation and transspecific evolution, the emergence of taxonomic groups above the species level. Finally, we shall investigate the relationship between specialization and extinction, and look at the process of evolution at the molecular level.

The process of speciation

In discussing the evolutionary mechanisms that lead ultimately to the emergence of new species, we must examine the concepts of race and species. Let us see how populations become differentiated into races, or subspecies, and how isolating mechanisms foster the conversion of races into new species.

The differentiation of populations

A Mendelian population is a breeding unit occupying a particular space and time. That space is probably quite heterogeneous, and when the time factor is added, the heterogeneity becomes even more complex. The heterogeneity of the physical environment operates on the population as a differentially selective force and creates a population system composed of differentiated local populations linked by breeding relationships. We call these local breeding units—these subgroups of a larger Mendelian population—*races* or *subspecies*.

Subspecies exhibit quantitative differences in certain gene frequencies as a result of selection or drift, but they are all reproductively connected to each other. They can and do interbreed.

A *species*, on the other hand, is in general characterized as a population of interbreeding individuals that is reproductively isolated from other populations of interbreeding organisms. We shall soon see that these rather inflexible definitions create problems for us in certain situations, so be prepared to rethink the concepts of race and species.

What kind of evidence shows that one population is differentiated from another? One of the classic studies pointing out regional differentiation in a species was performed by J. Clausen, D. D. Keck, and W. M. Hiesey in 1940. They studied regional differentiation in the plant species known as *Achillea millefolium* (the Western yarrow), a species that is widely distributed in the Western United States. They collected seeds

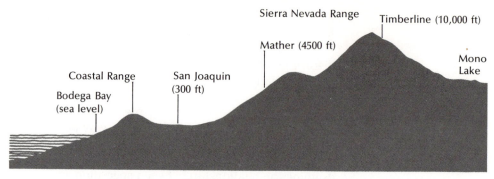

Transect across central California

	Stem length, cm			Stem number (relative vigor)			First flowers		
Planting stations	Stanford	Mather	Timberline	Stanford	Mather	Timberline	Stanford	Mather	Timberline
Seeds collected from									
Bodega Bay	49	31	—	30	29	—	5/21	7/12	—
San Joaquin	79	37	—	25	9	—	4/15	6/15	—
Mather	80	82	34	7	28	1	5/17	6/30	9/20
Timberline	21	32	24	4	6	3	5/13	6/10	8/18

Figure 15.1
Seeds from various ecological races of *Achillea lanulosa*, collected at the three locations indicated, were planted at the various stations shown. The data clearly show that the populations are highly differentiated. (From Clausen, Keck, and Hiesey, 1940)

from *Achillea* in four California locales and planted them at Stanford University laboratory. The plant from each seed was separated into three clumps, so that each clump was genetically identical (a clone).

They transplanted the clumps from each plant to three stations: one at Stanford (sea level), one at Mather (elevation 4500 feet), and one at Timberline (elevation 10,000 feet). The results are shown in Figure 15.1.

It is clear from the data that the plants, though they are of the same species, respond in a dramatically different way to the three environments. We see, for example, that plants that normally inhabit the coast do not survive at 10,000 feet, a region in which other populations of the same species survive quite well.

Figure 15.2 is a drawing that emphasizes the same point. In this experiment, two populations of *Achillea borealis* (one from Seward, Alaska, and one from Berkeley, California) were tested in the same fashion. Seeds were collected, plants grown, and the plants broken into groups of three and transplanted to Stanford, Mather, and Timberline. The Alaskan plant—a coastal plant, but separated from the Berkeley plant by 22° of latitude—grew well at 4600 feet and less well at sea level and at 10,000 feet. The conditions at 4600 feet in California are probably the closest to those that obtain in its native environment on the Alaskan coast. The Berkeley plant does best at Stanford, which is essentially its native habitat, and less well at 4600 feet. It does not survive at 10,000 feet. Again

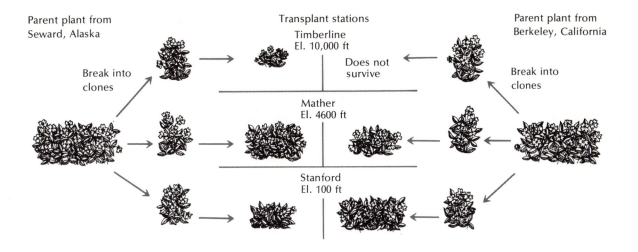

Figure 15.2
Clones of two latitudinal races of *Achillea borealis* transplanted at three different altitudes. (From Clausen, Keck, and Hiesey, 1940)

we see that populations of the same species respond differentially to similar environmental conditions. This argues strongly that these *Achillea* species are genetically different. They are said to consist of several races, or subspecies.

In another study, Dobzhansky and his colleagues demonstrated that *Drosophila pseudoobscura* consists of a series of ecological races based on the frequencies of three inversions found in different populations. They sampled flies from different altitudes in a transect across California. Then they examined each population for the frequency of three characteristic inversions: Standard (*ST*), Arrowhead (*AR*), and Chiricahua (*CH*). The results are shown in Figure 15.3.

The flies sampled from each of the three elevations shown had strikingly different frequencies of the three types of inversions. The observations suggest that, for some unknown reason, the specific inversion frequencies increase the fitness of the populations at the different locales. The differences between the three populations must be physiologically

Figure 15.3
Frequencies of three common third-chromosome inversions in populations of *Drosophila pseudoobscura* sampled at different altitudes. (From Dobzhansky, *Evolution, Genetics, and Man*, Wiley, 1955)

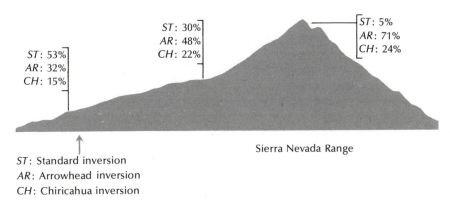

ST: 53%
AR: 32%
CH: 15%

ST: 30%
AR: 48%
CH: 22%

ST: 5%
AR: 71%
CH: 24%

Sierra Nevada Range

ST: Standard inversion
AR: Arrowhead inversion
CH: Chiricahua inversion

687

Figure 15.4

Changes in frequency of chromosomes carrying two inversion types in natural populations of *Drosophila pseudoobscura* in the San Jacinto Mountains during the advance of the season from March to October (From Grant, *The Origin of Adaptations*, Columbia University, 1963)

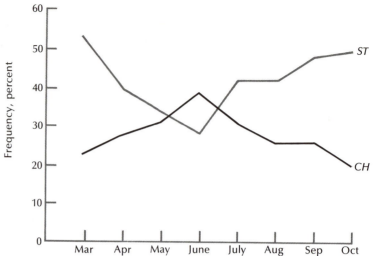

subtle, because we can detect no morphological differences among the three populations. To make matters even more complex, these inversions shift in frequency with respect to the season (Figure 15.4).

On a somewhat broader geographical scale, Dobzhansky and his colleagues showed that populations of *Drosophila pseudoobscura* from various regions in the Southwest varied markedly for five different inversions (Figure 15.5). The *Drosophila* populations, like the *Achillea* populations, exhibited a great deal of genetic differentiation, and may be considered ecological races.

Though the human species is subject to the same evolutionary forces as other species, the idea of human races is generally an emotionally charged issue. It should not be, for the human species is clearly a genetically diverse species, and its diversity indicates evolutionary success. Even a casual observer of the peoples of the world sees that individuals from Black Africa, Northern Europe, and Asia exhibit obvious differences in physical structure.

These differences, however, become more subtle when we look at people not quite so geographically disparate. For example, Asians from Thailand, Cambodia, Laos, Vietnam, China, and Japan all have unique features that generally make them distinguishable from each other. But the differences are really continuous, and involve traits that are not clearly defined genetically. Though they may have a high heritability, skin color, hair color, hair texture, head shape, nose shape, and other external features are not well understood genetically. We can suggest, however, that these external differences indicate that there are internal differences as well.

Recall that the concept of a race involves interbreeding populations that differ in certain gene frequencies. A pioneering effort to examine race formation in the human species was made by W. Boyd in 1950 when he studied the distribution of the A-B-O alleles in different human popula-

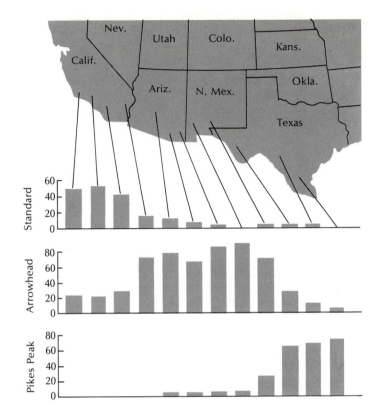

Figure 15.5
The frequencies of three different third-chromosome inversions in different populations of *Drosophila pseudoobscura*. (From Dobzhansky, 1958)

tions. The A-B-O blood-group alleles are well defined genetically and show specific differences in frequency in various parts of the world. Since this important study, the human population has been analyzed for a number of genetic traits. All the studies show that human populations vary extensively in the frequencies of different genes or alleles.

Be sure to notice something very important in what we have said here: The different races consist of *populations* that show differences in gene frequency. We are *not* dealing with the individual and trying to assign that individual to a specific race. It is unsound to attempt to do such a thing, because the individual may or may not have an allele in question. In any case, the individual does not have an allelic frequency to display. *Race* is a meaningful term only when we use it in the context of populations.

We study racial variation in the human species by coupling external morphology with internal factors that are well defined genetically. An example may serve to demonstrate what we mean. Suppose that a building housed 400 people: 100 from Nigeria, 100 from Japan, 100 from Sweden, and 100 from Sri Lanka (formerly Ceylon). Your partner instructs these 400 people to assemble in four different rooms according to their race. They do this according to external appearance alone. Now you enter the hall and are asked to determine which group is in each of the four rooms without opening the doors to those rooms. You could make the determination by ascertaining the frequencies of certain alleles in each of the four groups. Knowing these frequencies, you could tell which was the Black African group, the Northern European group, the Asian group, and the Sri Lanka group.

Some genetically determined traits that are often used to assess racial divergence in the human species are given in Tables 15.1, 15.2, 15.3, 15.4. Note that differences in gene frequency are quite distinct in the populations shown. According to the criteria that we set forth earlier for a race, we would have to say that the human species, like most other species, is composed of many races.

Table 15.1
Variation in blood group (percentages) among populations

Population	A_1	A_2	B	O	Rh negative	Duffy factor
Caucasians	5–40	1–37	4–18	45–75	25–46	37–82
Negroes	8–30	1–8	10–20	52–70	4–29	0–6
East Asians	0–45	0–5	16–25	39–68	0–5	90–100
American Indians	0–20	about 0	0–4	68–100	about 0	22–99

Adapted from Carleton S. Coon, *The Living Races of Man* (New York: Alfred A. Knopf, 1965), p. 286.

Table 15.2
Percentage of PTC tasters in different populations

Group	Percentage of tasters
Europeans	60–80
Negroes	90–97
East Asians (Chinese, Japanese, and related populations)	83–100
Australian Aborigines	50–70
Micronesians	70–80

Adapted from Carleton S. Coon, *The Living Races of Man* (New York: Alfred A. Knopf, 1965), p. 264.

Table 15.3
Percentage of population bearing the dry ear wax gene

Population examined	Percentage bearing gene
Northern Chinese	98
Southern Chinese	86
Japanese	92
Melanesians	53
Micronesians	61
Germans	18
American Whites	16
American Blacks	7

Adapted from E. Matsunaga, *Annals of Human Genetics*, 25(1962) pp. 277–286.

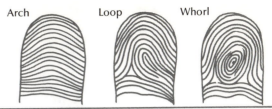

Group	Arches	Loops	Whorls
European	0–9	63–76	20–42
Negroes	3–12	53–73	20–40
Bushmen	13–16	66–68	15–21
East Asians (Chinese, Japanese, and related populations)	1–5	43–56	44–54
American Indians	2–8	46–61	35–57
Australian Aborigines	0–1	28–46	52–73
Micronesians	2	49	49–50

Table 15.4
Percentage range of types of fingerprints among different populations

The species concept

At several points in this chapter we have used the term *species* without really defining it. What we have done is to say that reproductive isolation is the basis of our modern concept of species. But the concept of species is more complex; it includes four main features.

1. A biological species occurs only among sexually reproducing organisms.
2. Populations of interrelated individuals comprise a species.
3. All members of a species are able to share in a common gene pool; that is, all members of a species are able to interbreed.
4. Populations that comprise a species are reproductively isolated from other species.

Ernst Mayr summarized the concept of biological species by saying that a species consists of groups of actually or potentially interbreeding natural populations that are reproductively isolated from other such groups.

But we should be aware of the dangers of rigidly defining a concept that is dynamic. The concept of species is not meant to be rigid. Some examples may serve to illustrate the complexities involved in characterizing a species.

Drosophila pseudoobscura has a habitat that covers most of the Western half of the United States. *Drosophila persimilis* has a range covering most of the West coast of the United States and the Sierra Nevada mountains (Figure 15.6). The two species overlap in the West-coast states and in the mountains. Yet no hybrids are formed in the areas in which they overlap, even though the two species are morphologically almost identical. If,

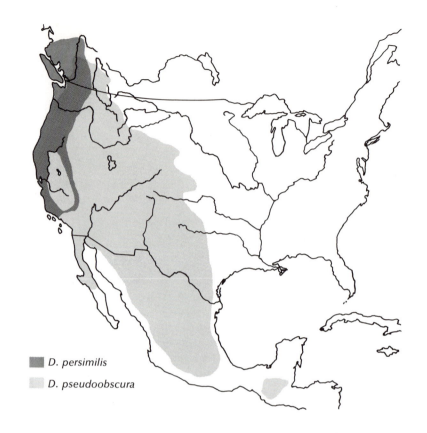

Figure 15.6
Geographic distribution of two sibling species of flies, *Drosophila pseudoobscura* and *D. persimilis*. The two species are sympatric over a large area from British Columbia to the mountain regions of California. (From Dobzhansky, *Evolution, Genetics, and Man*, Wiley, 1955, page 170)

■ *D. persimilis*

■ *D. pseudoobscura*

however, the temperature in a laboratory population cage is lowered to 16°C, or if the females are lightly etherized, the two species will mate and produce vigorous F_1 offspring. The male offspring are usually sterile, and the females, when backcrossed, produce offspring that have a low viability.

For all practical purposes, these two species are reproductively isolated from each other, but the fact that they produce vigorous F_1 interspecific hybrids tells us that they are closely related and only recently diverged from each other. Cytological sudies confirm this. We see here the process of reproductive isolation nearly but not quite complete.

Another study of natural *Drosophila* populations points out dramatically the futility of a rigid definition of species. *D. arizonensis* occupies a territory from Southeastern Arizona to Northwestern Mexico (Figure 15.7). *D. mojavensis* exists as three disjunct populations: one in South-central California; one along the Eastern coast of Baja California; and one in the Northwestern part of Mexico (Figure 15.7). The species overlap in Sonora, Mexico. All three disjunct populations of *D. mojavensis* are able to interbreed. When they do, they produce vigorous, fertile offspring. They thus share a common gene pool. Where the two species overlap, in Sonora, Mexico, they remain reproductively isolated from each other. But *D. arizonensis* hybridizes with the disjunct populations of *D. mojavensis* in the

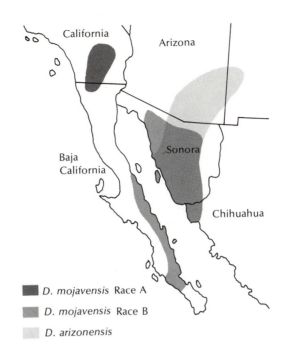

Figure 15.7
The distribution of the closely related flies, *Drosophila mojavensis* and *D. arizonensis*. *Mojavensis* consists of three major disjunct populations; the one in southern California is a morphologically distinct geographic race (semispecies). Although members of the two species will effectively hybridize in the laboratory, they are biotically sympatric and reproductively isolated in Sonora, Mexico.

laboratory and produce viable and fertile offspring. These two species are in a state of evolutionary intermediacy and are not yet fully isolated from each other reproductively.

What these and other studies tell us is that evolution is a continuous, dynamic process, and that speciation cannot be constrained by inflexible limits. The four main features of the concept of species that we listed earlier are meant to be flexible guidelines that describe the process, rather than define it.

Isolating mechanisms The single most important feature distinguishing two different species is reproductive isolation, and the process of reproductive isolation usually begins with some sort of physical separation of populations. Separation restricts the exchange of genes between the two groups and allows the independent accumulation of genetic differences. Perhaps the most common type of extrinsic barrier is geographical. The formation of a river, a mountain range, or a desert, and the rise and fall of land masses can all cause *geographical isolation*.

One of the most spectacular examples of the role of geographical isolation in evolution is the unique fauna of Australia. During the Cretaceous period of the Mesozoic Era, when marsupials and placental mammals were radiating, a land bridge existed between Australia and the Asian mainland. Over this land bridge passed representative marsupials, but no significant number of placental mammals. When the land bridge disappeared at the end of the Mesozoic Era, the marsupials were isolated in Australia and were able to radiate into the available ecological niches without competition from placental mammals. On the mainlands of Eurasia, North America, South America, and Africa, the placental mammals survived at the expense of the marsupials, whereas in Australia the marsupials thrived. The same is true, though to a lesser degree, of South America, where marsupials were long isolated by a continental separation then existing at the Isthmus of Panama. The link, broken in the early Tertiary period of the Cenozoic Era, was reestablished by the end of the period.

Ecological isolation is another mechanism that can bring about reproductive isolation. It may actually be the first of a series of isolating events, because within a genetically diverse population existing in an ecologically diverse habitat, some groups become better adapted to certain conditions than others, and thus the population becomes partitioned into subgroups. An example of ecologically separated species comes from the desert region of the Southwestern United States and Northern Mexico. *Drosophila pachea* lives with several other *Drosophila* species in this region, but is ecologically isolated from them, because it reproduces only in the stems of a particular species of cactus (*Lophocereus schotti*). This cactus contains material required for *D. pachea* reproduction *and* materials that inhibit other *Drosophila* species from reproducing in it. Only a few genetic loci are involved in these qualities. This provides a simple genetic basis for the establishment of complete reproductive isolation.

A summary of the most important isolating mechanisms appears in Table 15.5. The two main groups of mechanisms include those that prevent fertilization and the formation of zygotes (*prezygotic mechanisms*) and those in which zygotes are created, but in lethal, semilethal, or sterile form (*postzygotic mechanisms*). Two species of toads, *Bufo americanus* and *Bufo fowleri*, offer an example of *seasonal isolation*. The two species occupy the same region (though not exactly the same ecological niche), but they mate at different times of the year, which keeps them reproductively separate.

Behavioral isolation, which occurs only in animals, is a major isolating mechanism. Matings that include complex courtship patterns do not occur if the pattern is not followed precisely. For example, *Drosophila pseudoobscura* and *Drosophila persimilis* follow complex but different courtship patterns in which the male flicks his wings, taps and licks the female, and dances. Under normal conditions, the females of *D. pseudoobscura* will not mate with *D. persimilis* males, because their courtship rituals differ. However, if the female is made drowsy with ether or 16°C temperatures, or if her antennae are removed, she loses her power to discriminate and will then accept a *D. persimilis* male.

Mechanical isolation, in which mating could occur but the genitalia are incompatible, is rare in animals but moderately significant in plants. Flower structure can encourage cross-pollination or self-pollination by attracting or not attracting specific types of pollinators. The structure of the flower can even physically prevent some pollinators from entering it.

Prezygotic Mechanisms (Prevent fertilization and zygote formation)

1. Geographical or ecological: The populations live in the same regions but occupy different habitats.

2. Seasonal or temporal: The populations live in the same regions but are sexually mature at different times.

3. Behavioral (only in animals): The populations are isolated by different and incompatible behavior before mating.

4. Mechanical: Cross-fertilization is prevented or restricted by differences in reproductive structures (genitalia in animals, flowers in plants).

5. Physiological: Gametes fail to survive in alien reproductive tracts (sometimes called immunological).

Postzygotic Mechanisms (Fertilization takes place and hybrid zygotes are formed, but these are nonviable or give rise to weak or sterile hybrids)

1. Hybrid nonviability or weakness.

2. Developmental hybrid sterility: Hybrids are sterile because gonads develop abnormally or meiosis breaks down before completion.

3. Segregational hybrid sterility: Hybrids are sterile because of abnormal segregation to the gametes of whole chromosomes, chromosome segments, or combinations of genes.

4. F_2 breakdown: F_1 hybrids are normal, vigorous, and fertile, but F_2 contains many weak or sterile individuals.

Table 15.5
Summary of the most important isolating mechanisms

From G. L. Stebbins, *Processes of Organic Evolution,* 2nd edition, © 1971, p. 100. Reprinted by permission of the author and Prentice-Hall, Inc., Englewood Cliffs, N.J.

More notable than mechanical isolation is *physiological isolation,* in which, for example, male gametes of one species are rendered nonfunctional by the physiological conditions in the reproductive tract of a female of another species. This is an important isolating mechanism in *Drosophila,* in which interspecific mating barriers of a behavioral nature occasionally break down.

Postzygotic mechanisms are inefficient and uneconomical because energy and resources are wasted, especially if a weak and sterile F_1 is produced. Occasionally, however, postzygotic mechanisms fail to keep species reproductively separate. An example is the formation of polyploids, but often this failure rapidly gives rise to new varieties, and ultimately, new species. The basic cause of postzygotic isolation mechanisms is chromosome incompatibility, which arises when chromosomes from different species come to lie in the same nucleus. This causes problems in development and cell division, as discussed in Chapter 5.

In conclusion, several mechanisms are important in the step-by-step process leading to final reproductive isolation, the major criterion separating one species from another. Often two or more mechanisms may be operating together, such as seasonal and ecological isolation, to strengthen separation between species. Even in cases of successful interspecific breeding, the offspring are usually nonviable or sterile (witness the mule, a sterile hybrid of a horse and a donkey).

Modes of speciation

Several models have been suggested for the origin of new species. We shall examine some of them in this section. The key to all the models is the ultimate acquisition of reproductive isolation.

Phyletic speciation   From extensive analysis of the fossil record for some lineages, it does not appear that a population has split into two distinct species. Rather it seems that the population has continued to accumulate variation in such a way that the morphological change is continuous. It is impossible, in this scheme, to pinpoint any kind of discontinuity, except at the beginning and end of the continuum. If we had samples of populations at the two ends, we would see that they are reproductively isolated from each other. But because the populations are separated by so much time, this comparison is not possible. Reproductive isolation in this case is inferred, not demonstrable. The fossil record for some conservative animal groups such as horseshoe crabs suggests this scheme of speciation. Figure 15.8 is a schematic diagram of phyletic speciation.

Allopatric speciation   Also known as geographic speciation, this type of speciation may be the most common mechanism for the formation of species in the living world. Figure 15.9 summarizes the process. A large outcrossing population is genetically polymorphic and thus phenotypically polymorphic. The space it inhabits is ecologically diverse, which means that the large population is composed of smaller demes that are particularly well suited to the local environments. This type of regional differentiation leads to the formation of races, or subspecies, that are all potentially interbreeding, but in fact practice assortative mating. At least they interbreed within the deme rather than within the population as a whole. Barriers thus exist that restrict—but do not prohibit—the flow of genes between the races.

The key element of the model is the next step, which shows the formation of an extrinsic barrier. This barrier prevents the exchange of genetic material between the two populations, so that new mutants and new combinations of alleles are not swamped out or diffused throughout the greater population. The barrier permits the development of unique genotypes and distinct gene frequencies to the extent that the two populations are eventually reproductively isolated from each other. If the barrier is later removed, the populations may remain disjunct, in which case they are referred to as *allopatric species*. Or the populations may merge into the same habitat, in which case they are referred to as *sympatric species* (still reproductively isolated from each other).

There is another possibility in this model of allopatric speciation. The

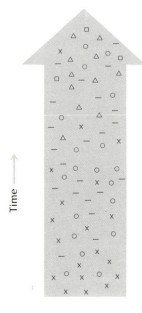

Figure 15.8
A scheme of phyletic evolution. The symbols inside the arrow indicate genotypes. They change over time, but the number of species at any one point in time is constant.

Time

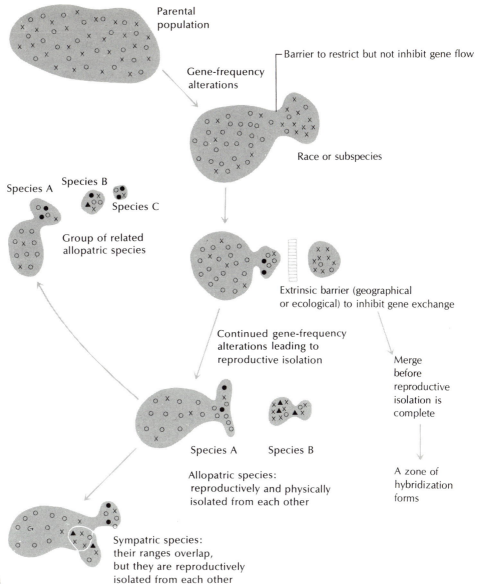

Figure 15.9
Schematic view of allopatric speciation

Parental population

Gene-frequency alterations

Barrier to restrict but not inhibit gene flow

Race or subspecies

Extrinsic barrier (geographical or ecological) to inhibit gene exchange

Species A

Species B

Species C

Group of related allopatric species

Continued gene-frequency alterations leading to reproductive isolation

Merge before reproductive isolation is complete

Species A   Species B

Allopatric species: reproductively and physically isolated from each other

A zone of hybridization forms

Sympatric species: their ranges overlap, but they are reproductively isolated from each other

reproductive isolation may not be complete when the extrinsic barrier breaks down. This could have a number of consequences. For example, the two populations may once again merge and hybridize, so that there is an area in which phenotypes are changing rapidly (the hybridization zone). If the hybridization occurs at the juncture of several distinct races, and if the hybrids are successfully adapted, then species swarm may form. A *species swarm* is a group of closely related species occupying a restricted geographical region far removed from other related species. We see some evidence for such species swarms in certain fish populations in the older freshwater lakes of East Africa.

Sympatric speciation Whether sympatric speciation is possible is a debatable point. This model requires the splitting of a population into two species *without* the intrusion of an extrinsic barrier first separating the population into two isolated groups. But can speciation occur without isolating mechanisms? Some think yes and some think no.

Under some circumstances, sympatric speciation can be said to occur without doubt. The examples of allopolyploidy we discussed in an earlier chapter (see page 185) serve as models for sympatric speciation, because the populations were sympatric when new species formed. In these cases, speciation resulted from the rapid increase in chromosome number of a hybrid. Second, sympatric speciation could easily occur in self-fertilizing organisms. Here each organism is of necessity reproductively isolated from all other organisms in a sympatric field. In cases such as these, there really is not much debate. The debate begins when it is suggested that sympatric speciation occurs in an outcrossing population—that primary divergence can occur in an outcrossing population that is not experiencing hybridization.

There is some evidence that sympatric speciation may occur under certain circumstances that do not involve self-fertilization or hybridization. We refer to these circumstances as speciation in *neighboringly sympatric zones.* These are different zones that border on each other and are inhabited by a single population. Selection occurs as a disruptive force in the bordering areas, which serves to partially isolate the populations occupying the different zones. But the populations occupying the different zones are part of one species and are at least potentially interbreeding. Under conditions such as these, where the zones are ecologically differentiated even though they are contiguous, we have the possibility for formation of races and eventually speciation.

An example of sympatric speciation in neighboring zones involves the fruit fly *Rhagoletis pomonella*. One group of these flies carries out its life cycle on the fruit of the apple tree, and another group carries out its life cycle on the fruit of the native hawthorn tree. In many areas these two trees are sympatric. The two fly populations are racially distinct, each being restricted to its specific tree for maintenance and reproductive purposes. It has been suggested that the original population of *Rhagoletis pomonella* lived on the hawthorn. When apple trees were introduced to the region, some of the flies switched to the apple trees. What would cause such a shift?

We know that in a species related to *Rhagoletis,* host-plant preference and recognition is a trait determined by a single gene pair. So it is reasonable to suggest that a mutation occurred in the original population that caused a shift of a female from hawthorn to apple, where she laid her eggs and started a new race. Genetic diversification occurred in this incipient race, so that it eventually became racially distinct from the hawthorn population.

Though the *Rhagoletis* example supports a sympatric model of race formation (a primary divergence), do we have evidence to support sympatric speciation? It is far from conclusive, but there are four species of *Rhagoletis* that live on four different plant families. The suggestion here is that speciation occurred as a consequence of a series of host-plant shifts in a sympatric field. This extrapolation is entirely speculative, however.

Can speciation occur in a population that is *biotically sympatric*? In such a population, the organisms come into contact with each other regularly, including times during which they are in the reproductive phase of their life cycles. As new mutations or new recombinants come into existence, it is hard to see how they would not be reabsorbed into the population and thus swamped. How could reproductive isolation ever be achieved under conditions of outcrossing in a biotically sympatric population? You might argue that a newly arisen form could be kept intact (not swamped out) by disruptive selection against the hybrid. But this would involve a tremendous selection cost that could probably not be tolerated.

One way around this dilemma is to suggest that when a new group arises, there is positive assortative mating. This reduces the cost of selection to the point at which a new group might be a primary divergence and succeed even though it overlaps biotically with the rest of the population.

An experiment by Thoday and Gibson showed that, in a very limited sense, this type of model may work. They took *Drosophila* and selected for lines with high bristle number and lines with low bristle number. They mixed the high and low lines and allowed them to freely intermate with each other. However, the offspring were constantly selected for high and low before the next generation of mating took place. The results are shown in Table 15.6. After 19 generations of selection, positive assortative mating was very obvious, involving 82 percent of all matings. In other words, after initial selection pressures against the hybrid, a positive assortative mating scheme developed that helped cut down the high cost of selection. But this study, suggestive as it may be, is a long way from proving that speciation can occur by a similar mechanism.

Quantum speciation The allopatric model of speciation required race formation followed by the physical isolation of the race from the parent population and the ultimate formation of a new reproductively isolated species. Quantum speciation occurs more rapidly and commonly along

Table 15.6
Results of tests for mating preferences among *Drosophila melanogaster* that have been selected for high bristle number (*H*) and low bristle number (*L*), and in which males and females are given a free choice of mates

Generation of selection	Number of matings			
	H × *H*	*H* × *L*	*L* × *H*	*L* × *L*
7	12	3	4	12
8	14	2	6	10
9	10	4	6	7
10	8	4	3	13
19	27	2	8	20
	71	15	27	62

From Thoday, 1972, *Proc. Roy. Soc. Lond.* 182:109.

the margins of a population. It usually involves a small number of organisms that get pinched off from the main body of the population and found a new population. The key feature in this new, small population is that genetic drift and natural selection can operate together to produce rapid changes. In the allopatric model, genetic drift is not an important force. Within this small population, however, inbreeding results in new phenotypes that may lead to new species, though this outcome is not at all guaranteed.

This model of speciation may also result in species swarms. Several small groups, pinched off from a main and larger genetically polymorphic population, go on to form new species. The new species should exhibit gene frequency differences characteristic of genetic drift.

Some situations seem to be best interpreted by the quantum speciation model. The founder effect we discussed earlier in the context of genetic drift in humans is an example of an event that can lead to quantum speciation. The Galapagos finches are descendants of a small number of finches that were pinched off from the main South American finch population. Speciation of *Drosophila* in the Hawaiian Islands is also an example of quantum speciation. Small groups along the margins of larger island populations have founded new populations and species in neighboring islands.

A well-studied example of quantum speciation involves the annual flowering plant *Clarkia. C. amoena* and *C. rubicunda* have the distributions shown in Figure 15.10. *C. franciscana* is a predominantly self-fertilizing species restricted to an area of serpentine rock in San Francisco. It is well within the range of *C. rubicunda*. The three species differ from each other

Figure 15.10
The distribution of three *Clarkia* species in the San Francisco Bay area

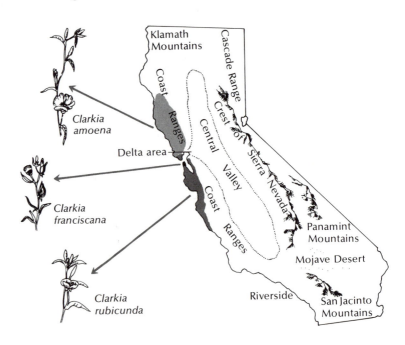

by specific series of inversions and translocations. On the basis of their geographical relation to each other and their cytological relationships, Harlan Lewis suggested that *C. franciscana* developed from *C. rubicunda* by the rapid accumulation of specific chromosomal rearrangements, perhaps by genetic drift. A similar event was postulated to have produced *C. rubicunda* from *C. amoena* earlier in Earth's history.

More recent studies have cast some doubt on that mechanism, however. Of eight enzyme systems analyzed in *Clarkia,* six have become fixed for an allele in *C. franciscana* that is not found in *C. rubicunda.* The genetic differences between the two species are much greater than one would expect if the two species had recently diverged from each other by the model of quantum speciation. It may be that the two species diverged from each other according to the quantum speciation model, but at a much earlier time. Alternatively, *C. franciscana* may be the surviving population of an earlier species divergence that produced both *C. franciscana* and *C. rubicunda.* This latter event could result from allopatric speciation. We cannot distinguish between these two alternatives, but neither can we eliminate quantum speciation as a model. Indeed, when we consider all the evidence, it seems most compatible with a model of quantum speciation.

Transspecific evolution

The development of intrapopulational variation and the ultimate formation of new species characterize what we might call *microevolution:* the evolutionary forces that operate in the formation of a species. *Transspecific evolution* is the evolution of taxonomic groups larger than a species. It is *macroevolution.* It concerns the emergence of genera, families, orders, classes, and phyla, and it involves time spans often several millions of years in length. In microevolutionary studies, we rely almost exclusively on genetic changes that accumulate in populations. In macroevolutionary studies, we rely heavily on paleontological records and comparative anatomy.

Are the evolutionary forces that lead to the emergence of the larger taxonomic groups different from those forces that lead to new species? A few evolutionary biologists seem to think so. They argue that the large differences between families, orders, and classes arise from unique processes not found in speciation processes. What appear to be sudden changes in morphology are explained by *saltation:* the sudden acquisition of large, genetically based morphological differences in a cluster of populations. These events are like mutation processes that occur throughout the population, not at the individual level. Proponents of this theory have also claimed that certain evolutionary trends—such as a general increase in body size, the loss of all but one toe in the horse family, and the rapid increase in the cerebral hemispheres in primates—suggest a type of *orthogenesis,* evolution guided by some directing mystic force with a particular goal in mind. We shall discuss orthogenesis shortly.

Most evolutionary biologists reject the notion of special evolutionary forces operating at the macroevolutionary level and not at the microevolutionary level. There is simply no evidence from any quarter to

support these ideas. On the other hand, we need to be aware of certain differences between microevolution and macroevolution. The greatest difference is time. There may also have been major differences in degree of habitat diversity, constancy of mutation rate, and population size so that simple extrapolation from microevolution to macroevolution is unreasonable. However, we feel that, in a general sense, the principles we have developed for the speciation problem also apply to the transspecific level.

Evolutionary trends

One of the best-studied examples of evolutionary trends is the horse. A complete fossil record points out clear trends in the horse's teeth and feet and in other body parts. (Figure 15.11 summarizes the evolution of the horse.)

In the foot of the horse, the trend was toward a reduction in the number of toes, from five to one (Figure 15.12). This change occurred only

Figure 15.11 (opposite) Radiation of forms within the horse family. The earliest horses, successful as three-toed browsers, gave rise to a number of separate lines. Evolution of the ability to graze and invasion of the continent of South America resulted in a further multiplication of equine genera. (From Stahl, *Vertebrate History: Problems in Evolution.* McGraw-Hill, 1974)

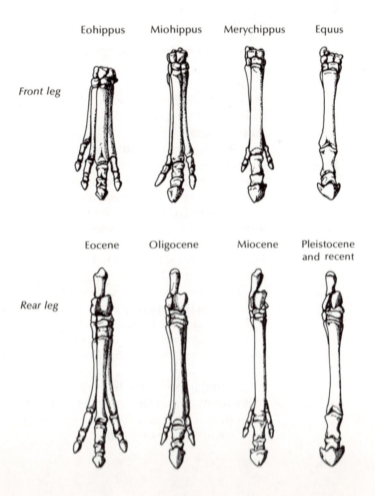

Figure 15.12
The evolution of the foot in horses

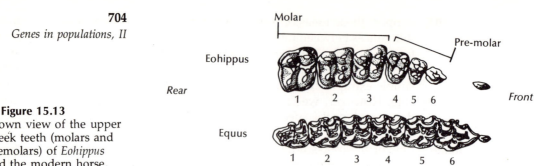

Figure 15.13
Crown view of the upper cheek teeth (molars and premolars) of *Eohippus* and the modern horse, *Equus*

in certain lines and was not very gradual. But the trend was clear.

In the horse's teeth, there is a gradual trend correlated with changes in diet. At the most primitive level of horse evolution, the teeth were suited for the browsing these animals engaged in. The teeth were smaller, had simpler patterns of crests and ridges, had no cement, and were less differentiated between front and rear. As the horse's habitat shifted from the forest to the grassy plains, its teeth became larger, showing more complicated crests and ridges, and hard cement in the valleys between the ridges for added strength. These teeth were admirably suited to a diet of tough field grasses. Figure 15.13 shows the teeth of a modern horse and those of a primitive horse. Another trend in the horse was an increase in body size. This gave it added strength and enhanced its survival on the grassy plains.

Though some might argue that the story of the horse's evolution supports a directed or orthogenetic type of evolution, the more reasonable explanation is that populations exploit new environments during climatic changes. The genetic potential for change was there all along but was not selected for until the environment changed. This type of evolution, *preadaptive evolution*, was discussed earlier (page 423) in connection with mutations to phage resistance or antibiotic resistance in bacteria. Without the genetic potential *and* the environmental stress, the modern horse would not have evolved.

Evolutionary trends such as those found in the horse result from natural selection operating in a specific direction for a long period of time. It is not necessary to invoke orthogenesis to explain such trends.

There are two classic cases in which the idea of orthogenesis has been invoked to explain the extinction of two well-known animals, the Irish elk (actually a deer) and the saber-toothed tiger (Figure 15.14 and Figure 15.15). In the case of the Irish elk, it has been argued that an evolutionary momentum caused this animal's antlers to reach absurd proportions and was ultimately responsible for its extinction. This idea is clearly not consistent with neo-Darwinism.

Figure 15.14
Irish elk (Courtesy
American Museum of
Natural History)

However, we do not need to invoke a mystic force creating an evolutionary "momentum" to explain the Irish elk's extinction. There is a much more reasonable explanation. There may have been a strong selection for rapid antler growth in that those young males with the most impressive head gear mated the most successfully. Once they mated and left their genes to the next generation, it mattered little whether the antlers continued growing. The older animals with the large antlers were no longer important to the reproductive fitness of the population. Only when the large antlers interfered with the reproduction of the males would a negative selective value appear. If this happened, the stage could be set for extinction.

Figure 15.15
Saber-toothed tiger
(Courtesy of American
Museum of Natural
History)

An alternative proposal is that selection really favored a large body size and the antlers were pulled along with the increased body size. This happens in modern deer. Larger males are reproductively more successful, and antler size is coordinated with body size. The problem is that when increased body size is selected for, antler size increases faster than body size. Selection continued to favor increased body size until the antler size was detrimental to the animal's reproductive success. At this point, the animal's body may have been smaller than optimum, and the antlers larger than optimum, and this may have been a key to the Irish elk's extinction.

In the saber-toothed tiger, it has been suggested that the animal's canines grew to such massive proportions that they eventually interfered with its eating and the animal starved itself to extinction. Considering the paleontological history of the saber-toothed tiger, it is not likely that this was the cause of its extinction. When it first appeared some 40,000,000 years ago, its canines were *larger* than those of the last saber-toothed tiger that roamed the Earth some 30,000 years ago. The canines showed considerable variation in length over the 40,000,000 years of its existence, so one can hardly say that they caused its extinction. The fact that they existed 40,000,000 years suggests that the large canines enhanced the fitness of the animal. The demise of the saber-toothed tiger was probably connected with the extinction of many of its natural food sources, such as the cave bear. The saber-toothed tiger was a slow animal that relied heavily on its strength and tremendous jaws to pull down large, slow animals. When those animals disappeared, so did the saber-toothed tiger.

In both the Irish elk and the saber-toothed tiger, specialization and extinction can be explained by natural processes. Orthogenetic reasoning is not required.

Regulatory mechanisms to explain transspecific shifts

The type of evolution that is based on the gradual substitution of one structural gene allele for another over time may not be completely adequate to explain the very large differences that exist among the higher taxonomic groups. It is becoming more and more apparent that the enormous anatomical, physiological, behavioral, and ecological differences between phyla such as Coelenterata and Chordata cannot be explained simply on the basis of gene substitution. The differences are too great and the time span not long enough. Is there a way within the framework of neo-Darwinism to explain the rapid evolution of major morphological differences? There may be if we look at alterations in regulatory genes whose functions affect entire batteries of structural genes.

It is reasonable to suggest that some of the major transspecific shifts have been the result of changes in development and differentiation. In Coelenterates, for example, there are 7 main cell types; in Chordates, there are over 200. With some minor exceptions, all 200 cells in a Chordate body have the same genotype. This is true for most other multicellular organisms. The differences between the cell types are the result of differential gene activity. To go from a level of organization with 7 cell types to a level with more than 200 may well require reorganization of the gene regulatory system. Alterations in the structural genes cannot by themselves provide a satisfactory explanation for the great differences that characterize the higher taxonomic groups.

We earlier considered the Britten–Davidson model of gene regulation in eukaryotes, and we feel that this model can provide a basis for understanding the evolution of transspecific differences. (You should review the elements of the Britten–Davidson model in Chapter 13.) Mutations in any of the regulatory genes (sensors, integrators, and receptors) can have profound effects on entire series of genes. For example, a mutation in the sensor gene may make that gene sensitive to a different chemical messenger. This means that the integrator and receptor regions will be activated at new times and under new circumstances. The result could be a radically new developmental phenotype.

The same general result could occur when an integrator gene mutates. Its activator product may now recognize a new receptor and thus activate a new gene series that would not normally be active at that time. Inversions may switch structural genes around so that they come under the control of new receptors and thus create new developmental phenotypes.

These alterations in regulatory genes—and we have mentioned only a few of the possibilities—can lead to major changes in organisms over a short period of time. But is there any evidence for this scheme? There is some indirect evidence suggesting that the differences between taxonomic groups above the species level cannot be explained by allelic substitution of structural genes. King and Wilson studied the genetic differences between humans and chimpanzees (two different families) and found that structural differences in genes are no greater than you would expect to find for sibling species. In other words, by comparing proteins (by amino acid sequencing, immunology, and electrophoresis) and nucleic acids (by hybridization), they found that the genetic distance between human and chimpanzee is too small to account for the tremendous anatomical, physiological, behavioral, and ecological differences.

Figure 15.16 summarizes the contrast between the biological and molecular evolution since the divergence of the human and chimpanzee lineages from a common ancestor. The two plots show that the biological, or organismal, change has been tremendous in humans, while the chimpanzees have remained fairly conservative. The molecular evolution, however, of those changes—manifested as alterations in amino acids or changes in nucleotide sequences—has been the same in both animals. As a matter of fact, studies show that human and chimpanzee proteins are 99 percent identical. This molecular similarity is characteristic of sibling species, not families. The implication of this study is that a relatively few changes in regulatory genes may account for the major organismal differences between humans and chimpanzees. Analysis of the banding pat-

Figure 15.16
The contrast between biological evolution and molecular evolution since the divergence of the human and chimpanzee lineages from a common ancestor. Left, zoological evidence indicates that far more biological change has taken place in human lineage (y) than in chimpanzee lineage ($y \gg x$). Right, evidence from both proteins and nucleic acids indicates that as much change has occurred in chimpanzee genes (w) as in human genes (z). (From M. C. King and A. C. Wilson, "Evolution at Two Levels in Humans and Chimpanzees," *Science* 188:107, 11 April 1975. Reprinted by permission of A. C. Wilson and the American Association for the Advancement of Science; © 1975, American Association for the Advancement of Science.)

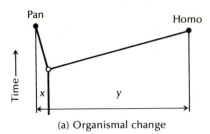

(a) Organismal change

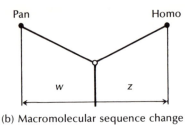

(b) Macromolecular sequence change

tern in both organisms' chromosomes suggests that the major cause of these regulatory changes is inversion.

But some other aspects of the human–chimpanzee relationship and the King and Wilson study require further discussion. For one thing, many of the important differences between the two organisms result from cultural evolution in the hominids. This has tended to accentuate the differences between humans and chimpanzees. On a purely biological level, it may not be appropriate to consider the two animals as belonging to different families.

Second, studies of interspecific hybridization in plants reveal rather striking genetic differences between the species. By extrapolation, we can expect similar degrees of difference in the animal world, which suggests that the structural similarities in genes observed for human and chimpanzee were not based on a representative sample of the gene pool. Third, though the Wilson and King paper is probably correct in concluding that regulatory genes play an important role in speciation, we must remember that the nature of gene regulation in eukaryotes is poorly understood at best.

All the cautionary notes that must apply to investigations such as King and Wilson's study suggest that, although regulatory genes may have been important in differentiating the two animals, the genetic differences in structural genes are probably greater than the molecular studies now indicate.

Overview

In this chapter, we examined two very complicated aspects of evolutionary genetics: the process of speciation and evolutionary processes above the species level.

The first step in the process of speciation is commonly the differentiation of a large Mendelian population into races. Using examples drawn from studies of *Achillea* and *Drosophila,* we showed that populations are actually genetically diverse. In humans, race classification is often based on external morphology, but recent studies combine characterizations of gene frequency with morphological considerations to arrive at a much more meaningful assessment of variation of human races.

A biological species is a population of sexually reproducing, interrelated individuals. The individuals comprising a species are able to interbreed and are reproductively isolated from individuals in other species. But a rigid species definition is fraught with difficulties. For example, we noted that some *Drosophila* species, under some circumstances, can and do interbreed.

The key event in the process of speciation is the acquisition of reproductive isolation. Mechanisms of isolation are either prezygotic or postzygotic.

We looked at four modes of speciation. Phyletic speciation is a steady, nonbranching type of speciation in which beginning and culminating forms are different species, but many of the intermediate forms are not. Allopatric or geographic speciation involves the formation of local races followed by physical isolation and then reproductive isolation. Sympatric speciation is a controversial mode that requires the acquisition of reproductive isolation without the intrusion of an extrinsic barrier first separating the population into isolated groups. Neighboringly sympatric zones may lead to the formation of new species, but the problems are much more acute with biotically sympatric populations. The last mode we looked at was quantum speciation. This is best seen in speciation along the margins of populations where small groups get pinched off or found new populations based to some extent on the combined forces of drift and selection. *Clarkia,* in the San Francisco Bay area, served as a model of quantum speciation.

Transspecific evolution is the evolution of genera, families, orders, classes, and phyla. Though the basic principles developed in our discussion of speciation apply to transspecific evolution, we did note some important differences.

In transspecific evolution, we are able to discuss trends over extended periods of time. We examined trends in the horse, Irish elk, and saber-toothed tiger and viewed these trends from the perspectives of orthogenesis (a process rooted in mysticism) and neo-Darwinism. It is clear that we do not need to invoke orthogenetic arguments to explain speciation, extinction, and evolutionary trends.

Finally, we looked at the rapid evolution of taxa above the species level by considering regulatory mutations rather than structural gene alterations. A comparison of structural gene products in humans and chimpanzees showed them to be very similar, which suggests basic developmental differences. We commented on the possible pitfalls of this study.

15.1 You discover two disjunct populations of *Drosophila*. How would you go about ascertaining whether the two populations are members of the same or of different species?

15.2 Compare allopatric speciation with sympatric speciation and phyletic speciation.

15.3 Can natural selection act on a clone?

15.4 Why is sympatric speciation such a controversial model?

15.5 How would you criticize the contention of a colleague who claims that he can look at an organism and tell you what race it belongs to?

15.6 Should race be considered a category of classification?

15.7 Suppose you cross two presumably different species and obtain interspecific hybrids that are partially or completely fertile. Would you reject the classification scheme that places the organisms into two distinct species?

15.8 Can a single mutational event give rise to a new species?

15.9 Working on a column of rock in a Wyoming dig site, a paleontologist uncovers the following sequence of fossils.

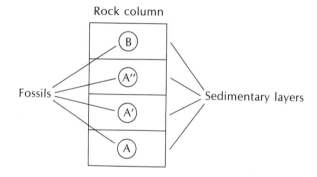

The fossils at the bottom level represent species A. At the next level up (more recent), species A' is only slightly different. One level further up is A'', which is slightly different from A'. Above the A'' level is a more radically different species that the paleontologist calls species B. It is more different from A'' than A'' is from A. The paleontologist suggests that this supports a model of phyletic gradualism. That is to say that A evolved gradually into A', which evolved gradually into A''. Between A'' and B, there was a long period where sediments were not laid down so fossils were not made. When fossils once more formed, evolution had progressed, by gradualism, to the B stage. Offer an alternative interpretation.

15.10 From the perspective of the modes of speciation discussed in this chapter, interpret the following two types of phylogenetic trees.

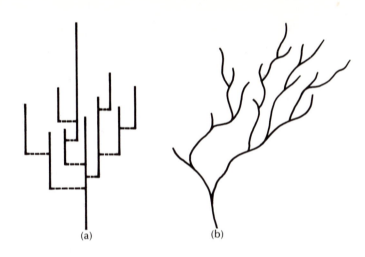

(a)   (b)

15.11 The human species is genetically diverse, yet we still consider it a single species. Why?

15.12 What is a Mendelian population, and why is it so important to our concept of evolution?

15.13 On a trip to the zoo, your friend sees a gorilla and remarks, "You mean we descended from that animal?" How would you clear up your friend's confusion about the relationship between gorilla and *Homo sapiens*?

15.14 You observe two morphologically similar populations that are geographically isolated from each other. How would you determine whether they are different races or distinct species?

References

Ayala, F. J., and Th. Dobzhansky (eds.). 1974. *Studies in the Philosophy of Biology.* University of California Press.

Ayala, F. J., M. L. Tracey, D. Hedgecock, and R. C. Raymond. 1974. Genetic differentiation during the speciation process in *Drosophila. Evolution* 28:576–592.

Betz, J. L., P. R. Brown, M. J. Smyth, and P. H. Clarke. 1974. Evolution in action. *Nature* 247:261–264.

Bock, W. J. 1970. Microevolutionary sequences as a fundamental concept in macroevolutionary models. *Evolution* 24:704–722.

Brues, A. M. 1964. The cost of natural selection vs. the cost of not evolving. *Evolution* 18:379–383.

Bush, G. L. 1975. Modes of animal speciation. *Ann. Rev. Ecol. and Syst.* 6:339–361.

Carson, H. L. 1975. The genetics of speciation at the diploid level. *Amer. Nat.* 109:83–92.

Carson, H. L., and K. Y. Kaneshiro. 1976. *Drosophila* of Hawaii: Systematics and ecological genetics. *Ann. Rev. Ecol. and Syst.* 7:311–346.

Clausen, J., D. D. Keck, and W. W. Hiesey. 1940. Experimental studies on the nature of species. *Carnegie Inst. Washington Publ. No. 520,* pages 1–542.

Dobzhansky, Th. 1958. Genetics of natural populations. XXVII. The genetic changes in populations of *Drosophila pseudoobscura* in the American southwest. *Evolution* 12:385–401.

Dobzhansky, Th. 1970. *Genetics of the Evolutionary Process.* New York, Columbia University Press.

Dobzhansky, Th., F. J. Ayala, G. L. Stebbins, and J. W. Valentine. 1977. *Evolution.* San Francisco, Freeman.

Dubinin, N. P. 1966. *Evolution of Populations and Radiation.* Moscow, Atomizdat.

Eldredge, N. 1971. The allopatric model and phylogeny in Paleozoic invertebrates. *Evolution* 25:156–167.

Eldredge, N., and S. J. Gould. 1972. Punctuated equilibria: An alternative to phyletic gradualism. In *Models in Paleobiology,* T. J. M. Schopf (ed.). San Francisco, Freeman.

Fitch, W. M. 1973. Aspects of molecular evolution. *Ann. Rev. Genetics* 7:343–380.

Fitch, W. M. 1976. Molecular evolutionary clocks. In *Molecular Evolution,* F. J. Ayala (ed.). Sunderland, Mass., Sinauer.

Ford, E. B. 1975. *Ecological Genetics.* 4th ed. London, Chapman and Hall.

Gould, S. J., and N. Eldredge. 1977. Punctuated equilibria: The tempo and mode of evolution reconsidered. *Paleobiology* 3:115–151.

Grant, V. 1971. *Plant Speciation.* New York, Columbia University Press.

Grant, V. 1977. *Organismic Evolution.* San Francisco, Freeman.

Hedrick, P. W., M. E. Ginevan, and E. P. Ewing. 1976. Genetic polymorphism in heterogeneous environments. *Ann. Rev. Ecol. and Syst.* 7:1–32.

Jukes, T. H., and R. Holmquist. 1972. Evolutionary clock: Nonconstancy of rate in different species. *Science* 177:530–532.

King, M. C., and A. C. Wilson. 1975. Evolution at two levels: Molecular similarities and biological differences between humans and chimpanzees. *Science* 188:107–116.

Koopman, K. F. 1950. Natural selection for reproductive isolation between *Drosophila pseudoobscura* and *Drosophila persimilis. Evolution* 4:135–148.

Lerner, I. M., and W. J. Libby. 1976. *Heredity, Evolution, and Society.* San Francisco, Freeman.

Lewis, H. 1966. Speciation in flowering plants. *Science* 152:167–172.

Lewontin, R. C. 1974. *The Genetic Basis of Evolutionary Change.* New York, Columbia University Press.

Maynard-Smith, J. 1970. Genetic polymorphism in a varied environment. *Amer. Nat.* 104:230–234.

Mayr, E. 1970. *Populations, Species, and Evolution.* Cambridge, Mass., The Belknap Press, Harvard University Press.

McDonald, J. F., and F. J. Ayala. 1974. Genetic response to environmental heterogeneity. *Nature* 250:572–574.

Mettler, L. E., and T. G. Gregg. 1969. *Population Genetics and Evolution*. Englewood Cliffs, N.J., Prentice-Hall.

Selander, R. K. 1976. Genetic variation in natural populations. In *Molecular Evolution*, F. J. Ayala (ed.). Sunderland, Mass., Sinauer.

Simpson, G. G. 1949. *The Meaning of Evolution*. New Haven, Conn., Yale University Press.

Simpson, G. G. 1953. *The Major Features of Evolution*. New York, Columbia University Press.

Soans, A. B., D. Pimentel, and J. S. Soans. 1974. Evolution of reproductive isolation in allopatric and sympatric populations. *Amer. Nat.* 108:117–124.

Stebbins, G. L. 1971. *Processes of Organic Evolution*. Englewood Cliffs, N.J., Prentice-Hall.

Stebbins, G. L. 1974. *Flowering Plants: Evolution Above the Species Level*. Cambridge, Mass., The Belknap Press, Harvard University Press.

Throckmorton, L. H. 1977. *Drosophila* systematics and biochemical evolution. *Ann. Rev. Ecol. and Syst.* 8:235–254.

Wallace, B. 1968. *Topics in Population Genetics*. New York, Norton.

Williams, G. C. 1966. *Adaptation and Natural Selection*. Princeton University Press.

Wilson, E. O. 1975. *Sociobiology, The New Synthesis*. Cambridge, Mass., The Belknap Press, Harvard University Press.

White, M. J. D. 1978. *Modes of Speciation*. San Francisco, Freeman.

Sixteen

Genetics: past, present, and future

Genetics has had and will continue to have a strong impact on our daily lives. No science really touches the core of human existence to the same extent that genetics does. Great social tragedies such as the Immigration Restriction Laws in the United States, the Nazi movement in Germany, and the Lysenko affair in the Soviet Union are all rooted in spurious concepts of genetics.

Today we are embroiled in several other controversies that have immediate impact on our social structure and social consciousness. These controversies are focused on genetics and geneticists. Are there genetically based racial differences in IQ scores? Should researchers continue using the techniques of recombinant DNA to study gene structure and function when these techniques are potentially hazardous to our well-being? How should we handle food additives, beauty aids, drugs, and other substances in our environment that are possible mutagenic agents? Our future as a species depends to a large extent on how we respond to these questions and others. And it depends on how we use the new genetic technology that we have created.

This chapter will raise many questions about genetics and society. I shall couch these questions in discussion of the major issues, but I shall refrain from offering glib solutions. These are complex issues, and they require careful analysis. A word of caution seems appropriate before we proceed with our discussion of the issues raised in this chapter. I have selected the problems raised, and I have chosen to evaluate them in a rather personal way, giving you, the reader, some insight into my own viewpoints. Whether you agree or disagree with them is not so important as your willingness to think about the crucial and complex issues raised.

The past

Eugenics in the United States: 1900–1924

Eugenics is that branch of biology that applies the principles of genetics to the human population with the goal of improving the human genotype. Positive eugenics seeks this improvement by encouraging the "most desirable" genotypes to reproduce more—to have more offspring. Negative eugenics seeks to improve the human genotype by discouraging the "less desirable" genotypes from having offspring.

Between 1905 and 1915, during the time that Mendelism was expanding, eugenics was developing into a popular discipline. It was included in the curricula of many colleges and universities around the country, reaching a high point in 1928, when 75 percent of all U.S. colleges offered courses devoted exclusively or in part to the study of eugenics. Interest in eugenics extended well beyond the realm of academic institutions. It reached the general population—and eventually the Congress of the United States.

The eugenics movement frequently attracted people who were uncritical thinkers and who harbored strong sentiments of Anglo-Saxon and

Nordic superiority. These racist sentiments were expressed in various ways. By 1930, 24 states had passed sterilization laws for various types of social "misfits": the mentally retarded, the insane, or criminals. Certain strong undercurrents were detectable in this type of legislation. One was the assumption that these very complex characteristics were hereditarily determined. Another was the assumption that the traits were generally associated with the lower economic classes in society. Both these assumptions are unwarranted and, for the most part, wrong.

Also by 1930, 30 states had passed miscegenation statutes restricting interracial marriages. The underlying assumption here was the genetic inferiority of nonwhites. Interracial marriages "diluted" the so-called superior white genes. But the crowning achievement for the eugenicists and their supporters came in 1924, when Congress passed the Johnson Act (Immigration Restriction Laws). This federal act legitimized racism by placing severe restrictions on immigration from Eastern European countries and countries around the Mediterranean. Immigration from Northern Europe was essentially unlimited. Asians were already restricted by legislation passed during the 1880s. The basis of the Johnson Act was the belief that immigrants from Eastern Europe and the Mediterranean region were genetically inferior to Northern Europeans.

How did eugenics, with its flagrantly inaccurate genetic doctrines, manage to get so far in the face of developing Mendelism?

Perhaps one reason for its success was the widespread belief among the white middle and upper classes that they were indeed superior. The eugenics movement gave them an outlet for their latent views, which for the most part were purely racist. These classes may also have felt threatened by the lower working class. But just as important, perhaps, was the support—tacit or explicit—for eugenics programs from within the ranks of geneticists and biologists. Such well-established geneticists as C. B. Davenport, Francis Galton, and W. E. Castle all openly supported the ideals of the eugenics movement—and, indirectly, the concept of the superiority of white Anglo-Saxon and Nordic peoples. They tried to demonstrate that traits such as alcoholism, seafaringness, degeneracy, feeblemindedness, delinquency, sexual promiscuity, and criminality were not only determined by the genotype, but in some cases assignable to a single pair of Mendelian factors. By extension, if these traits were found in high frequency in an urban ghetto, it must mean that the inhabitants of that ghetto were genetically inferior. One's economic condition was seen by many as a reflection of her or his genetic worth. Such people did not consider the influence of environment on these characteristics.

How did people justify the laws prohibiting interracial marriage? Some biologists and geneticists argued that interracial marriages would produce hybrid offspring that would not be fit because of an incompatible admixture of parts of the two races. This argument completely overlooked (or ignored) the concept of hybrid vigor and heterosis.

After 1915, geneticists began to withdraw their support for the eugenics movement. They did so for a number of reasons. First, it became clear that few if any traits, especially behavioral traits, showed a simple Mendelian inheritance pattern. The traits were much more complex than this. Second, studies such as Johannsen's study with the broad bean (see page 154) showed that identical genotypes expressed a variety of phenotypes in response to a range of environments. This clearly showed that the environment was important in the expression of a genotype, and that genotypic expression could be modified by the environment. Third, the principles of Hardy–Weinberg equilibrium established the enormous difficulty of using controlled breeding to select undesirable recessive genes out of a population. Finally, the data collected by eugenicists were often questionable, and their "experiments" not replicable. All these factors combined to make the eugenicists' position a weak one. But this did not cause their downfall as a major force in society.

As geneticists withdrew their support for the eugenics movement, the void they left behind was filled to a large extent by uneducated, uncritical, and bigoted people with no background in genetics. Two books published during this period reflect the attitude of many eugenicists, and stimulated the movement of lay people into the eugenics fold. These books were Madison Grant's *Passing of the Great White Race* (1916), and Lothrop Stoddard's *This Rising Tide of Color Against White Supremacy* (1920). Both writers warned of the serious consequences of unrestricted foreign immigration into the United States. Both warned of the declining Nordic "civilization" in the West. Both writers referred extensively to Davenport, Castle, and other geneticists to support their warped views. Grant took the position that racism and segregation were natural phenomena that served to maintain racial purity. It was his contention that we should not fight these natural human feelings. Both Grant and Stoddard presented what they called "evidence" to show that the Anglo-Saxon and Nordic groups were superior to all other racial groups.

These authors and others were very influential. They convinced Congress and many other people that disease, illiteracy, poverty, and crime were inherited, and that immigrants from Eastern Europe and the Mediterranean were more apt to have these traits, and hence were genetically inferior to the Anglo-Saxon and Nordic peoples. In 1924, Congress passed the Immigration Restriction Act; it remained on the books until it was repealed in 1965.

Geneticists were in a position to effectively counter the spurious claims of the eugenicists, but they did not. For the most part, they remained silent. Why? This is a difficult question, and we can only suggest some possible reasons. It may have been that many scientists felt that it was inappropriate to get involved in political matters; T. H. Morgan was typical of scientists who subscribed to this view of the scientist in society. Another factor may have been the scientists' tendency to avoid public controversy. Another reason why geneticists were silent in the face of mass acceptance of eugenicist positions may have been that they did not want to get involved in personal conflicts with colleagues who supported the movement. Finally, money may have bought some silence. Some of the major supporters of genetic research were also major supporters of the eugenics movement (the Rockefeller Foundation, the Carnegie Foundation, and J. H. Kellogg, founder of Kellogg Foods). It may be that some geneticists chose to remain silent rather than to speak out and risk losing their research support.

For our society, the Immigration Restriction Act was a tragedy. It was a triumph of blind ignorance over justice and wisdom.

Eugenics was also a big loser. Many serious and concerned geneticists, worried about the ever-increasing genetic load in the human species, were unable to attract a receptive audience once it became clear that the eugenics movement of the first two decades was an aberrancy. It was difficult to overcome the ignorance and confusion generated by that movement. It was even more difficult once the horrors of the Nazi eugenics movement became known.

Eugenics in Germany: 1919–1933

The rise of Nazism in Germany is intimately connected with perversions of genetics and eugenics, and with social Darwinism. By the time World War II had ended, more than 6 million people had been put to death by the Nazis in a systematic attempt to destroy what they perceived to be genetically inferior races. Most of those exterminated were Jews or Gypsies.

I think it is accurate to say that the roots of Nazism ran deep. The National Socialist German Worker's Party (Nazi) was organized in 1919, and it assumed the leadership of Germany in 1933 when Hitler came to power. But Nazi ideas and national racism were prevalent before 1919, most notably in the writing of the prominent biologist Ernst Haeckel and his followers. Haeckel was the leading spokesman for Darwinism in Germany during the latter part of the nineteenth century and first two decades of this century. Under Haeckel's guidance, Darwinism became more than a mechanistic interpretation of the origin and diversity of life. It became the total explanation of the world. Haeckel and his followers applied biological laws literally to society, and they urged a religious reformation in Germany based on Darwinistic natural laws. Haeckel's status as a scientist and a communicator of science was enormous, so his proto-Nazi ideas were highly influential.

Darwinism had major implications for traditional religion, philosophy, and social science. In Haeckel's own work, we can see how he developed his own strange brand of Social Darwinism. His view of human society strongly advocated struggle and competition; he considered them to be basic, natural laws of human society. This belief was important to the development of National Socialism. Haeckel also argued that there were "higher" and "lower" human races, and that racial conflict was natural. The Aryans were naturally the "higher" race, according to Haeckel. At the turn of the century, it was clear that the German people were beginning to make this Social Darwinism, with its racist overtones, into a religion. The primitive forces of nature were being deified. Biology and genetics in Germany—rather than providing counterarguments to this mystical kind of Darwinism—were actually supporting the growth of Nazism, and Haeckel was in large part responsible for this.

Haeckel's ideology provided a rallying point for loosely connected, poorly articulated, and vague sentiments of racism, anti-Semitism, and nationalism among various disgruntled segments of German society. This was especially true of the lower middle class, who felt that they were being threatened by the working class.

Haeckel cultivated the idea of race and gave it a substance and momentum that carried it to the more modern Nazis. He blended his Social Darwinism with the racist writings of the French Count Arthur de Gobineau. Gobineau's most influential piece of work was his book *The Inequality of the Human Races* (1853), in which he developed his theme that the Aryan was *the* superior race in all the world. He felt that the Aryans were doomed by racial interbreeding. Haeckel agreed with Gobineau's views, but argued that German racial purity could be maintained through the radical use of racial eugenics. Because of Haeckel's stature and influence, these ideas were widely discussed in Germany. They eventually formed the basis of Nazi doctrines.

Hitler assumed power in 1933, and though he never admitted his indebtedness to Haeckel, it is clear that his ideas on racial purity paralleled those of Haeckel. German eugenicists who worked closely with Heinrich Himmler in formulating German racial policy admitted that they were impressed by the writing of Haeckel.

In 1933, Germany passed into law the Sterilization Act, a radical program of negative eugenics. It required the sterilization of people with traits such as alcoholism, feeblemindedness, and schizophrenia. It was passed quickly and with little debate. Geneticists remained silent on the issue. Frick, who was Hitler's Interior Minister and was charged with administering these laws, said, "The fate of race-hygiene, of the Third Reich and the German people will in the future be indissolubly bound together."

It is only fair to say that not all German geneticists accepted the eugenics and racial doctrines of the Nazis. Many of these serious scientists underestimated the dangers of the Nazi movement until it was too late. In this tragic instance of misuse of genetics, guilt must be shared by those who supported the movement and those teachers who failed to point out the distinctions between true and false science.

When the extermination camps were discovered and the Nazi eugenics program became known to the world, people were horrified. Human genetics and anthropology were set back 10 years. The word eugenics became synonymous with Nazi atrocities. Even eminent geneticsts such as H. J. Muller, who were interested in the improvement of the genetic constitution of the human species without regard to race, creed, or ethnic origin, had a difficult time. People did not want to listen to eugenics arguments. To some extent, this is still true today.

The Lysenko affair   One of the most bizarre cases of politics and science forming an unholy alliance occurred in the Soviet Union between 1925 and 1965. T. D. Lysenko, a scientific hack, became a dictator of genetics for almost 20 years. During his reign, he stifled research into modern genetics (Mendelism, molecular biology, and so on), and caused the dismissal, exile, or execution of many Soviet geneticists. It was only when Khrushchev was deposed in 1965 that Lysenko finally lost his powerful grip on Soviet science.

In order to appreciate the absurdity of the Lysenko affair, we should note some of the incredible "discoveries" Lysenko claimed he made, and mention some of the "principles" he adhered to. Lysenkoists rejected the role of chromosomes in heredity, and denied that DNA played any important role in inheritance. They accepted the theory of acquired characteristics, and expressed a belief in the notion of spontaneous generation of a species from the acellular material of another species. Lysenkoists claimed that they could transform one species into another. For example, they claimed that they could transform wheat into barley, oats, or even cornflowers. They could transform beets into cabbage and pine trees into fir trees. They even claimed that they could produce mammalian cells from cereal grains! Lysenko rejected Mendelism because it was nothing more than abstract statistical ratios, and he found these ratios biologically irrelevant.

Lysenko's experimentation was scientifically appalling. He never bothered to run control experiments, and was interested in obtaining only the results he expected. He surrounded himself with people who would produce the results that he wanted. With these farcical ideas in mind, let us see how Lysenko came to assume complete control of Soviet genetics.

In 1925, Lysenko, a poorly educated peasant, began working as an agronomist at an agricultural experiment station in Azerbaijan. At this time, genetic research in Russia was in the extremely talented hands of N. I. Vavilov. Through the 1920s and 1930s, Vavilov directed Russian genetics to a position of world prominence. Certainly any discussion of the development of Mendelism during this time would be incomplete without mention of Soviet contributions to the field.

In 1928, Lysenko formulated a vague theory of plant development. He put this theory to a field test when he transferred to the Odessa Institute of Selection and Genetics. Some of this field work was successful in that he was able to show a steady improvement in agricultural yields. His work was appreciated by the farm workers, who formed a nucleus of peasant support. As Lysenko gained popular support, he began to make his move against Vavilov. Lysenko's goal was to gain political power, and he planned to do this at the expense of Vavilov and other true scientists.

Lysenko's next major move was to team up with I. I. Prezent, a philosopher with essentially no background in biology or science in general. These two launched an attack against Vavilov and the Mendelians. They denied that Mendelism contributed anything to the development of agriculture. At a 1936 conference in Moscow, Lysenko and Prezent entered into a public debate with Vavilov, who was then vice president of the Lenin All-Union Academy of Agricultural Sciences. The attack on Vavilov and Mendelism was a vicious, *ad hominem* assault. Lysenko was a master of avoiding the issue and directing attention away from areas in which he was ignorant—areas which included just about the entire field of genetics. The result of this debate was a victory for Lysenko. Shortly thereafter, Vavilov's duties became more restricted, and Lysenko's responsibilities expanded.

Stalin was a big help to Lysenko's career. He found Lysenko's ideas exciting, mainly because of their Lamarckian flavor. He felt that acquired characteristics could help to build a strong Soviet society. In 1938, Stalin's purges wiped out many of the top scientists in Russia. Lysenko made his move. Vavilov was arrested in 1940, convicted of espionage, and sentenced to death. He was not executed, but was shipped to a labor camp in Siberia, where he died three years later.

When the war ended, Lysenko rose to assume complete power over Soviet agriculture and most of biology. Genetics was outlawed, except for Lysenkoist genetics. Textbooks were completely rewritten, with Mendelism and DNA omitted, and acquired characteristics included. Scientists who did not agree with the Lysenkoist position were executed, imprisoned, or sent to Siberia. Or perhaps they changed fields altogether.

During Khrushchev's regime, the door to free inquiry opened somewhat. Khrushchev was a friend of Lysenko's, but he was also receptive to other points of view. Opposing ideas began to be heard as the level of fear subsided. Opposition to Lysenko began to build steadily until, in 1965, when Khrushchev was deposed, Lysenko was removed. Lysenko died an obscure figure in 1976.

Lysenkoists still remain in the Soviet Union, though they are few. Soviet genetics, repressed for 40 years, is once more moving ahead vigorously. The Fourteenth International Congress of Genetics met in Moscow in the summer of 1978. Only in a few isolated areas of agronomy—and, interestingly, in the People's Republic of China—do we see the remnants of Lysenkoism.

Lysenkoism was clearly an aberrancy. It was the result of a power struggle for political gains. The politics of Stalinist Russia was grasping for a way to improve the faltering Russian agricultural system and to inspire the people. Lysenkoism, with its theory of acquired characteristics, provided such a mechanism. Lysenko, a master at the art of manipulation, was able to realize his goal of political power by prostituting genetics. In the Lysenko affair, everyone eventually lost.

The IQ controversy

In 1969, Arthur R. Jensen, a professor of psychology at the University of California at Berkeley, published a paper that had the impact of a sledgehammer. In this paper, Jensen reports that there is about a 15-point spread between mean IQ scores for Blacks and Whites, with Whites scoring consistently higher. He argues that heredity is the basic reason for this IQ difference. More specifically, the argument that Jensen and others develop is as follows.

1. Since the IQ score has an 0.80 heritability factor (see page 165 for a review of heritability), Blacks are genetically incapable of scoring as well as Whites on an IQ test.

2. The Blacks' IQ score is not only lower than that of Whites, but it is also lower than that of American Indians, who are considered far more disadvantaged than American Blacks. In addition, on scholastic achievement tests and other similar tests, Blacks scored lower than Whites, Mexican Americans, Oriental Americans, Puerto Ricans, and American Indians.

3. When Black and White children were compared by socioeconomic level, Black children achieved lower IQ scores than White children in the same or even lower socioeconomic class.

Before we examine these points in detail, let us note that, even if these assertions are true, they do not in any way prove that the score differences are genetically determined. For one thing, the differences between Blacks and Whites go beyond simple racial categorization. Even when socioeconomic levels are the same, Blacks and Whites for the most part go to different schools in different neighborhoods, associate with different people, and generally lead culturally different lives. These variables make it virtually impossible to ascribe IQ differences to genes alone.

Does IQ performance have a high heritability? We earlier suggested that IQ scores may have a heritability as high as 0.80. This value was obtained primarily from twin and adoption studies in White children. Recently, however, many of these data have been carefully reevaluated and it has been found that there may have been some serious methodological errors. Indeed, it has even been suggested that some of the data were manufactured. Recent calculations indicate that the heritability of IQ score probably ranges between 0.45 and 0.60.

Another problem with IQ heritability is that the heritability values have been derived mainly from studies in the White population. Values may be different in the Black population. We simply cannot extrapolate from Whites to any other race with absolute confidence.

Can we state with reasonable certainty that the difference in IQ scores between Blacks and Whites is the result of genetic differences rather than environmental differences? Probably not. If we look at the other racial minorities, we see that for the most part their members went to predominantly White schools. Blacks went to predominantly Black schools. It is interesting to note that Blacks attending predominantly White schools scored higher than Blacks attending predominantly Black schools. All this suggests strong environmental differences in educational experience for the different racial groups. Yet children from all races and backgrounds take an IQ test that was developed and standardized with White middle-class populations in mind.

How can we interpret the difference in IQ scores for Blacks and Whites at the same socioeconomic level? There are many differences in background and life style between Blacks and Whites at the same socio-economic level, differences that serve to complicate the picture. Consider, for example, an upper-middle-class socioeconomic level. Proportionately fewer Blacks reach this level than Whites. In comparing Blacks and Whites at this level, we should be aware of the differences as well as the similarities. The Blacks have had to face discrimination in the housing market, which creates a housing shortage for them and which severely restricts the areas in which they have bought or rented homes. This means that a Black family at the upper-middle-class socioeconomic level may still live in a predominantly Black neighborhood, attend predominantly Black schools, and associate predominantly with Blacks. Since most Blacks are poor, the social and cultural environment for Blacks at this level is in many ways strikingly different from that of their White counterparts, who live in suburbia, attend White schools, and associate predominantly with other upper-middle-class Whites.

The environmental differences experienced by both groups can result in tremendous differences in attitudes, value systems, perception, and, finally, IQ scores. In some recent studies, in which these variables were taken into consideration, the differences in IQ between Blacks and Whites were only between 4 and 6 points. This is the range of difference one finds in monozygotic (genetically identical) twins.

It seems reasonable to conclude that a person's intellectual development is determined to a large extent by the genotype. But we cannot conclude that the 15-point IQ difference between Blacks and Whites is genetically determined. Complex cultural differences make it impossible to draw such a conclusion. The problem becomes even more acute when we consider the difficulties of labeling people Black or White.

There is a danger in having to choose between environmental and genetic hypotheses. If society elects to support the genetic argument, policies of segregation, racial discrimination, and even worse could be instituted.

On the other hand, if society recognizes the environment as the major cause of differences in IQ score, then it can eliminate the gap through concerted economic, educational, and social programs. Social policies predicated on the equal intellectual potential of all races will benefit us far more than policies based on racial inequality. Individual abilities, defects, and talents need to be dealt with without regard to race.

Recombinant DNA molecules

For the past three years or so, the public has been bombarded with warnings about the dangers that lie ahead if we continue pursuing research into recombinant DNA (review pages 656–658 for the techniques of constructing recombinant DNA molecules). But much of what has been written about recombinant DNA research is more fiction than fact, and it serves only to frighten the public into making unwise decisions. Let us examine recombinant DNA research here, and correct some of the misconceptions.

The techniques of recombinant DNA research are, by themselves, neither good nor evil. They are techniques that can be used for good or evil purposes. The same can be said of nuclear energy, lasers, or solar energy. But because a technique can be used in both a beneficial and a hazardous way, do we ban the technique entirely, or do we control it? In studies using recombinant DNA techniques, there are some clearly hazardous experiments that should not be done, such as constructing antibiotic-resistant pathogenic bacteria or bacteria capable of synthesizing the *botulinum* or diphtheria toxin. Even if these experiments were allowed, responsible scientists (alas, there are irresponsible scientists) would not do them. There are also experiments that have no hazards at all connected with them. For example, we can cut up an *E. coli* genome and resplice it into different sequences. This can tell us a great deal about gene function, and it presents no hazard.

The major difficulties with recombinant DNA research emerge from those experiments in which there *may* be a hazard. For such experiments, the National Institutes of Health (NIH) has tried to assess risk factors and prescribe levels of experimental containment commensurate with the risk involved. Yet how do you evaluate risk when you have no hard facts to guide you? The NIH guidelines are based on hypothetical situations and speculations. Though these risk factors are speculative, they are often presented to the general public as fact. It is interesting that the guidelines were generated by the scientists working in the area and are designed to protect people against hypothetical hazards not yet known to exist. Guidelines for the study of pathogenic bacteria and viruses, toxic substances, and radioactive material were set up on the basis of known hazards.

It has been argued that we should not do research using recombinant DNA because results that are not seen as hazardous today may prove to be hazardous in the future. This is true. But it is also true for many other things that we do. When we administer vaccinations, for example, we do so knowing that there is a clear benefit and essentially no hazard. But the vaccine *could* lead to cancer in 20 years. Does this mean that we should not administer vaccines because we cannot prove that in 20 years there will be no danger of cancer? We cannot guarantee that the vaccine will not cause cancer, but then few aspects of our lives come with these kinds of guarantees. Should we proceed through life never taking risks when there are clear benefits to be gained? The critics of DNA research seem to base many of their arguments on fear of the unknown rather than fact.

In connection with this discussion, we can ask whether scientists should be free to pursue new knowledge regardless of the consequences. Actually, scientists have not normally been free to experiment in areas that pose clear threats to the health and welfare of the public. Unfortunately, experiments that *do* threaten public safety have been carried out by scientific arms of governmental agencies, and by some private research institutes. But in research programs that are publicly funded, we can and do regulate research so that it presents no hazards to people. It is a rare person indeed who argues that scientists should be free to do whatever experiments they wish.

Critics also argue that recombinant DNA research may upset the delicate balance of nature that evolutionary processes have produced for billions of years. I find this argument unconvincing for a number of reasons. First, nature is not so delicately balanced that it cannot absorb change. It does so all the time. Second, humans have been modifying and controlling evolutionary processes for centuries without adversely affecting the balance of nature. Our domestication of animals and development of food crops are examples of how we have meddled in the evolutionary process in a positive way. The consequences of this controlled evolution have been crucial to the further evolution of the human species.

Our constant battle to preserve the lives of genetically disabled individuals is a form of tampering with the forces of evolution and natural selection, yet would anyone argue that we should stop such interference? Remember, too, that natural evolutionary processes have produced such disasters as the bubonic plague, polio, venereal disease, smallpox, cancer viruses, typhoid fever, and yellow fever, to mention only some of the scourges that have hit the human species.

And how should we view the parasitic infections that sap the strength and energy of millions of people in the third-world countries? These parasites are the product of natural evolutionary processes. Should we stop fighting these infections? It is clear that we should continue to try to control the hazardous components of our natural environment. Recombinant DNA research may help us do so.

Admittedly there are hypothetical risks involved in recombinant DNA research, but let us look at some of the benefits, which are not hypothetical. These benefits are of two types. First, recombinant DNA research enables us to understand gene structure and function more completely than ever before. Specifically, by constructing new plasmids we are able to learn a great deal about genes that control, for example, antibiotic resistance. And we can discover how these genes are themselves in turn controlled.

Viewing the problem from a more pragmatic level, we can see some spectacular benefits looming on the horizon. The techniques for inserting human hormone and antibody genes into bacterial plasmids are almost perfected. Soon we shall be able to construct bacteria that synthesize human hormones and antibodies such as gamma globulin. Insulin and somatostatin genes have already been successfully spliced into *E. coli* cells. These and other specialized human proteins are now rare and expensive to extract, but recombinant DNA research may soon provide us with a way to get these substances in quantity and inexpensively.

Vaccines are also being planned using the technique of recombinant DNA. One specific vaccine is nearly ready to go into production. It will protect cattle from a strain of *E. coli* that produces a lethal toxin. The gene producing the toxin has been cut out of the plasmid, and the plan is to cut a piece of the gene out and reinsert the defective gene into the *E. coli* plasmid. The bacterium carrying this plasmid would produce a protein that has no toxic properties but retains its antigenic properties so that the host cattle could build up antibodies against the toxin.

In a more speculative vein, we can suggest the insertion of nitrogenase genes into bacteria that inhabit nonleguminous crop plants so that these plants could fix nitrogen. This would create tremendous savings in nitrogen fertilizers, which are expensive and becoming scarce, and the overuse of which lends to environmental pollution. And there are many other potential benefits emerging from recombinant DNA research. These are real benefits, not imaginary, and they should be contrasted with the more hypothetical and speculative biohazards discussed earlier.

Recognizing the speculative nature of the biohazards involved in recombinant DNA research, the NIH has still set forth a catalog of guidelines to be followed in any such research. Some research, involving organisms that are obviously dangerous to society, is prohibited. Researchers using dangerous microorganisms such as Lassa fever virus, Marburg virus, and Zaire hemorrhagic fever virus (nearly 100% fatal) require a P4 containment facility. This is a specially built laboratory with filters, airlocks, special cabinets, and other safeguards. There are only four or five P4 labs in the country. P3 and P2 containment facilities are designed for research involving organisms not quite so dangerous as those mentioned in connection with the P4 facility. A P1 containment facility is designed for research that involves genes in bacteria that could be exchanged by normal biological processes such as conjugation, transduction, or transformation.

The guidelines set down by the NIH are far more than adequate to protect us against "leakage." The NIH has added to that protection by requiring the use of microorganisms that, even if set free in the population, could not survive to propagate. They are special laboratory strains that cannot survive outside the laboratory.

In spite of these safeguards, people still worry. There is, of course, reason to be concerned about the implications of any new scientific development, but in this case the alarmists are misguided. Some are also guilty of charlatanism. At a meeting in Washington, D.C., in April 1977, I heard educated people warning about cloning new Hitlers and Frankenstein monsters using the techniques of recombinant DNA. They were trying to intimidate through fear—not reason—and thus were only marginally successful (they did manage a few converts). Reasonable people who study the issues usually conclude that the way development has proceeded on recombinant DNA research represents one of science's finest hours.

And there is another, though not a strictly scientific, beneficial by-product of the recombinant DNA issue. Scientists are setting up lines of communication with nonscientists, saying "This is what we have discovered, these are the potential dangers and benefits, and this is what we think we should do about it." Scientists then ask others what they think and what they recommend. City governments all over the country have organized discussions about the new techniques. This type of open communication is an ideal we should all strive for.

Environmental mutagenic hazards

A dangerous consequence of modern industrialized society is the effects its products can have on the genetic material. In Chapter 10, we discussed the Ames system for detecting mutagenic chemicals in the environment, and we found that many commonly used products such as mercury (in paints), hair dyes, vinyl chloride, and cyclamates are mutagenic. But environmental mutagens are far more ubiquitous than we might imagine. For example, let us focus on some of the compounds found in our atmosphere that have mutagenic potential.

Under natural conditions, a number of gaseous compounds are produced, such as CO, CO_2, SO_2, NO, NO_2, NH_3, and H_2S. These are generated by volcanic processes, naturally caused forest fires, breakdown of rock material through erosion, sea spray, and the decay of biological material. But our consuming, technologically oriented society has greatly increased the amounts of these gases (primarily through the use and burning of fossil fuels), and they can be quite hazardous in higher concentrations.

The sulfurous gases that enter the atmosphere through industrial and automobile combustion can participate in reactions that are harmful. Consider the reactions of SO_2, for example.

$$SO_2 + H_2O \rightleftharpoons H_2SO_3 \quad \text{(hydrogen sulfite)}$$
$$H_2SO_3 \rightleftharpoons H^+ + HSO_3^- \quad \text{(bisulfite)}$$
$$HSO_3^- \rightleftharpoons H^+ + SO_3^{-2}$$
$$SO_3^{-2} + H_2O \longrightarrow H_2SO_4 \quad \text{(sulfuric acid)}$$

Bisulfite has been shown to be mutagenic in a number of organisms. So too has SO_2, though it is a somewhat weaker mutagen.

The oxides of nitrogen have also been shown to be mutagenic. They are a chief component of the smog that blankets the larger industrialized regions of the world. NO and NO_2 are produced during the combustion of fossil fuels and can react with water or light to produce dangerous chemicals. NO_2 reacts with water to form nitrous acid (HNO_2), which is a potent mutagenic agent (see page 435). NO_2 also reacts with light to form NO.

$$NO_2 \xrightarrow{h\nu} NO + O$$

All these compounds are mutagenic.

Combustion of fossil fuels in industry and automobiles produces another class of hazardous chemicals known as *polynuclear aromatic hydrocarbons*. The most dangerous of these are illustrated in the diagram.

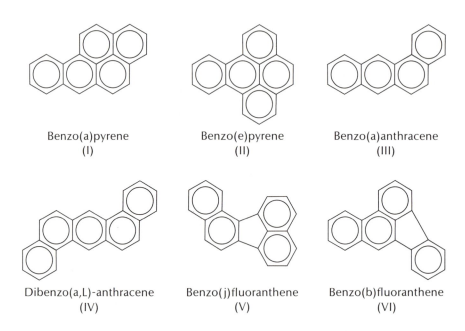

Benzo(a)pyrene
(I)

Benzo(e)pyrene
(II)

Benzo(a)anthracene
(III)

Dibenzo(a,L)-anthracene
(IV)

Benzo(j)fluoranthene
(V)

Benzo(b)fluoranthene
(VI)

Some of these compounds are directly mutagenic, while others are metabolically converted into mutagens that induce frameshift mutations.

Ozone is another important compound that is found in the atmosphere and that has mutagenic capabilities. It is normally produced in the upper atmosphere by the reaction of O_2 with light (wavelengths less than 240 nm).

$$O_2 \xrightarrow{h\nu} 2O$$
$$O + O_2 \longrightarrow O_3$$

Ozone is essential to life on Earth, because it absorbs lethal UV radiation. But although it plays a protective role on the one hand, it also presents us with a very real hazard when it is produced at ground level. Ozone is produced around electric utilities plants where sparking is common, and in polluted areas exposed to intense sunshine (Los Angeles and Phoenix, for example). It enters into free-radical reactions (see page 427) that mimic the effects of ionizing radiation on the genetic material.

Our discussion has covered only a small sample of the atmospheric chemicals that are mutagenic. There are many others. The most dangerous are those that are weakly mutagenic and widespread, rather than those that are potent and more restricted in their distribution. The former affect many people over a long period of time, and are more difficult to detect. The latter affect fewer people more seriously. And the latter are probably easier to control.

For many aspects of our environment, we have no data at all to guide us. The drugs we take or the food additives we consume today may prove to be carcinogenic or mutagenic tomorrow. How should we respond to these potential threats to our well-being and the well-being of our children?

The future

In our society today, there prevails a noticeably antiscience attitude that does not augur well for the future. Public support of basic scientific research is diminishing to the point where important ideas cannot be explored because of lack of funds. This is in marked contrast to the 1960s, when the public generously supported a wide spectrum of research projects. One of the reasons for the support of scientific research during the 1960s was the international competition for scientific advancement that grew largely out of the Sputnik era. But why have things changed so drastically today? What is it about science that has antagonized the general public?

Much antiscience sentiment has its roots in the widespread opinion that scientific research has been grossly misused. The energy of the atom has been perverted into arsenals of atomic weaponry that threaten the existence of all human life. Chemical and biological research has produced a frightening array of products for use in chemical and biological warfare. Genetic research has developed to the point at which we can imagine the possibility of geneticists having unrestricted control over, and ability to design, human society. Scientific technology has poisoned our water, fouled our air, and ravaged the natural beauty of our landscapes.

But antiscience sentiment may also be due in part to the more benevolent applications of science. We have, after all, used science to vastly improve the quality of life for human beings all over the world. Diseases have been challenged and conquered. New techniques have enabled us to develop more productive varieties of food crops. Transportation and communication networks have opened the world up for all to see. Science has developed to a point where we, as a world society, dare to dream realistically of a life free of the maladies and turmoils that have plagued human existence so far. These benevolent applications have involved efforts to dominate nature and subjugate natural phenomena to suit our own needs. In our Western culture, we have not really tried to live harmoniously with nature, and there are many who feel that we have gone too far. Perhaps we ought to slow down and reassess where we have been and where we are going. This attitude may be important in the antiscience view that pervades our society today.

Genetic research has certainly been a focal point for the debate on science and society and the future of the human species. Genetic research in recent years has produced some truly amazing discoveries that many fear will be pressed into the service of a technology for controlling and designing the human population. Of all technologies, this is the most threatening and frightening. It is not at all surprising that people are expressing apprehension at the pace and type of research being done. We have already discussed recombinant DNA research, but let us examine some other areas of genetic research that have strong social implications.

One area of genetic research that engenders active criticism is cloning. *Cloning* refers to processes that can produce exact replicas of an organism. One such process entails the transplantation of nuclei into enucleated eggs—a process similar in concept to what we described in Chapter 13. Theoretically a single donor could provide nuclei to hundreds or thousands of enucleated eggs. Each egg with its new nucleus would be genetically identical to the other eggs, and to the donor. By this technique, we could generate copies of great musicians, Nobel Prize winning scholars, assembly-line workers, or soldiers.

But cloning need not depend on nuclear transplantation. Early embryos could be disaggregated in such a way that each separate cell could develop into a fully formed human. The techniques of cell fusion have also opened up new possibilities along these lines. Using radiation-killed Sendhai viruses, scientists could fuse an enucleated egg with a somatic cell, and the newly nucleated egg would develop into a fully formed human (review page 80 for cell-fusion techniques).

We must also consider the possibility that the entire process of development by cloning could be accomplished outside the living body. Amazing progress has been made in the development of artificial wombs. It will probably be possible to go from a specially constructed egg to a fully formed human infant with no human involvement other than donation of the original egg and nucleus, and the maintaining of the equipment. This is not a very attractive scenario, but cloning need not have such negative connotations. When some of the problems of cell differentiation are worked out, cloning could be used to grow healthy tissue to replace diseased or damaged tissue, without the worry of tissue rejection. Thus organ and tissue transplants would be more technically sound as well as more frequently possible.

There are less controversial techniques than cloning that still offer the prospect of altering the human genotype. We know, for example, that λ bacteriophage carrying gal^+ genes from *E. coli* can infect human cells growing in culture. The gal^+ genes are active in human cells and correct a gal^- genetic defect (a disease called galactosemia). The possibility of introducing nonmutant genes into cells carrying defective genes opens up exciting prospects for controlling some genetic diseases. Methods of introducing new genes into cells are being developed. They include the transduction approach just mentioned, transformation, and the *in vitro* synthesis of genes. Indeed, virus particles are being constructed *in vitro* to serve as special transducing particles.

These and other programs of genetic engineering offer tremendous potential for all human beings to live out their lives free of crippling diseases and the general infirmities of old age. But they also involve serious risks. Like anything else, this science can be misused. If it is, who is to blame? I would suggest that science should not be made a scapegoat if its discoveries are misused by society. Do we blame the discoverer of the wheel for automobile pollution? Should we blame the discoverer of specific restriction enzymes, used in recombinant DNA research, if an accident happens in the course of this research, or if a country uses the technique to develop a lethal biological-warfare weapon?

Our future as a species will depend on how we respond to the complexities brought about by developments in science and technology. As citizens informed of the ways of science, we have the responsibility of making public the issues involving science and society so that reasonable decisions can be made. What we accomplish will be evaluated by our children and our children's children. What will they see when they look back at us? Robert Sinsheimer has perhaps best stated the optimistic view.

> Perhaps when we've mutated the genes and integrated the neurons and refined the biochemistry, our descendants will come to see us as we see Pooh: frail and slow in logic, weak in memory and pale in abstraction, but usually warm-hearted, generally compassionate, and on occasion possessed of innate common sense and uncommon perception.

Issues for thought and discussion

16.1 A book was recently published claiming that a millionaire has had himself cloned. The child would be about two years old now. It was claimed that the child was produced by the process of cell fusion. Needless to say, many people were outraged by the alleged event. What are the pros and cons of human cloning, and how does it relate to individual freedom?

16.2 Some geneticists have dropped out of genetics research because they say that their discoveries will be misused. Do you agree with their position?

16.3 To what extent should the government control research and define its basic objectives?

16.4 Do you agree that public funds should be used only to support research that has clear and immediate practical application to our well-being?

16.5 Should research into the heritability of intelligence be continued? If so, what should its objectives be?

16.6 Do you think that research in recombinant DNA should continue?

16.7 Suppose that a hair dye is found to contain mutagenic chemicals, and a person argues that he (or she) will continue to use it because it is his (or her) personal choice, not the government's. Would you agree or disagree with this position? Is it consistent to remove hair dyes from the store shelves and at the same time allow cigarettes to be sold?

16.8 Soon we shall be able to initiate the development of an embryo by *in vitro* fertilization or cloning, and nurture that embryo to "birth" in an artificial womb. We shall be able to correct any genetic defects and even define the genotype of the individual. Should this kind of technology be allowed to develop? How is your answer related to the issue of free scientific inquiry?

16.9 How would you evaluate scientists, specifically geneticists, during the rise of National Socialism in Germany? Would your evaluation differ if you were looking at the passage of the Immigration Restriction laws?

References

Allen, G. E. 1975. Genetics, eugenics, and class struggle. *Genetics* 79:29–45.

Baer, A. 1977. *The Genetic Perspective.* Philadelphia, Saunders.

Baer, A. 1977. *Heredity and Society.* New York, Macmillan.

Bresler, J. B. 1973. *Genetics and Society.* Reading, Mass., Addison-Wesley.

Committee 17. 1975. Environmental mutagenic hazards. *Science* 187:503–514.

Dunn, L. C. 1967. Cross-currents in the history of human genetics. *Amer. J. Hum. Genetics* 14:1–13.

Gasman, D. 1971. *The Scientific Origins of National Socialism.* New York, American Elsevier.

Goodfield, J. 1977. *Playing God: Genetic Engineering and the Manipulation of Life.* New York, Random House.

Handler, P. H. 1970. *Biology and the Future of Man.* New York, Oxford.

Hollaender, A. (ed.). 1971–1976. *Chemical Mutagens: Principles and Methods for their Detection.* 4 vols. New York, Plenum.

Jensen, A. 1973. *Educability and Group Differences.* New York, Harper and Row.

Ludmerer, K. M. 1972. *Genetics and American Society.* Baltimore, Md., Johns Hopkins University Press.

Medvedev, Z. A. 1969. *The Rise and Fall of T. D. Lysenko.* New York, Columbia University Press.

Montagu, A. 1975. *Race and IQ.* New York, Oxford.

Provine, W. B. 1973. Geneticists and the biology of race crossing. *Science* 182:790–796.

Rogers, M. 1977. Biohazard. New York, Knopf.

Scarr-Salapatek, S. 1974. Some myths about heritability and IQ. *Nature* 251:463–464.

Smith, A. 1975. *The Human Pedigree.* New York, McGraw-Hill.

Stine, J. S. 1977. *Biosocial Genetics: Human Heredity and Social Issues.* New York, Macmillan.

Wade, N. 1977. *The Ultimate Experiment: Man Made Evolution.* New York, Walker.

Wertz, R. W. 1973. *Readings on Ethical and Social Issues in Biomedicine.* Englewood Cliffs, N.J., Prentice-Hall.

Glossary

Abortive transduction Transduction in which the transducing fragment is not incorporated into the host genome, nor is it replicated. It is, however, passed on to one of the two daughter cells, producing a unilinear mode of inheritance.

Acentric chromosome A chromosome or chromatid lacking a centromere.

Acquired characteristics A theory that argues that traits which an organism acquires by accommodating to the environment are assimilated into the genome and transmitted to the next generation.

Acrocentric chromosome A chromosome or chromatid with a centromere near the tip.

Adaptation Any characteristic of an organism that improves its chances for survival in a specific environment and thus makes it more likely that the organism will leave more offspring than other organisms.

Adaptive enzyme An enzyme that is produced in response to a specific environmental cue. We also call this an *inducible enzyme.*

Adaptor hypothesis The idea (advanced by F. H. C. Crick in 1958) that there is a molecule that serves as an adaptor to match up amino acids on an mRNA template. This adaptor has one end that matches the mRNA and another end that is attached to the amino acid. This adaptor molecule is tRNA.

Adenine A purine base found in DNA and RNA.

Aging The irreversible deterioration of bodily functions, and hence of adaptability, that normally occurs during the later stages of an organism's life span.

Albinism A genetically based condition characterized by the inability to form melanin.

Allele One of two or more forms of a given gene.

Allopatric speciation The formation of new species by the differentiation of populations that are geographically isolated.

Allopatric species Species that occupy regions that are geographically disjunctive (geographically separate).

Allopolyploidy Polyploidy that arises as a consequence of the union of chromosome sets of different species or perhaps subspecies.

Allosteric effect The consequence of the interaction of a small molecule with a protein, causing the protein to alter its shape and thus its capacity to interact with a second small molecule.

Amino acid The basic building block of polypeptides.

Aneuploidy A condition in which the chromosome number is not an exact multiple of the haploid number; for example, trisomy is $2n + 1$ and monosomy is $2n - 1$.

Anticodon The sequence of three bases on the tRNA that pair up with the three-base codon on the mRNA.

Antimutator gene A gene that reduces the normal spontaneous mutation rate.

Apomixis A form of asexual reproduction in plants.

Asexual reproduction Any type of reproduction that does not involve the union of gametes from two sexes or mating types.

Assortative mating A type of sexual reproduction in which the formation of mating pairs is not random. Males of one type tend to mate with females of another type.

Autogamy A process of self-fertilization in *Paramecium* that results in homozygosis.

Autopolyploidy Polyploidy that results from the proliferation of the basic haploid chromosome set within a species.

Autoradiography A technique in which low-energy radioactivity, usually from tritium, is recorded on photographic film. Dark, exposed spots on the film, caused by radioactive decay, indicate where tritium has been incorporated.

Autosome A chromosome other than the sex chromosome.

Auxotroph A mutant microorganism that can grow only on a medium that has been supplemented with nutrient(s) that the wild-type variety is able to synthesize itself.

Backcross Crossing an F_1 heterozygote back to one of the parental types.

Bacteriophage A bacterial virus.

Balbiani ring See *Chromosome puff.*

Barr body The condensed **X** chromosome seen in the nucleus of mammalian somatic cells.

Base analog A DNA or RNA base that resembles the normal bases, but is different, so that it can be incorporated into the nucleic acid molecule in place of the normal base. It is usually mutagenic.

Base-pair substitution A mutational event in which one base pair is exchanged for another base pair. See also *Transition* and *Transversion.*

Bidirectional replication DNA-replication scheme in which two replication forks move in opposite directions from a single origin point. See also *Unidirectional replication.*

Binding site The RNA-polymerase–binding site at the initiation of RNA transcription.

Binomial expansion $(a + b)^n$, where a and b are the probabilities of occurrence and nonoccurrence of an event and must total 1, and n is the number of times the event is tried.

Bivalent A pair of synapsed homologous chromosomes that may or may not have been replicated.

Blending inheritance An inheritance pattern in which the parental characteristics appear to blend into an intermediate form. There is no segregation apparent, even in later generations.

Breakage–fusion A recombinational mechanism in which chromosomes break and exchange segments.

cAMP Cyclic adenosine monophosphate, a compound that is important in gene regulation, among other things.

Cap A unique set of modified nucleotides at the 5' end of eukaryotic mRNA.

Carcinogen Any chemical capable of causing cancer.

Carrier An individual harboring a recessive allele in the heterozygous condition.

Catabolite repression A decrease in the synthesis of certain enzymes in bacteria grown on glucose. The basis for this decrease is a glucose catabolite that controls the level of cAMP, which is required for the function of all operons involved in glucose metabolism.

Catastrophism The idea that cataclysmic changes in the progression of life on Earth occurred as a consequence of geological catastrophes such as floods. This theory contrasts with the uniformitarianism of Lyell's geology.

Cell fusion The formation of a hybrid cell with nuclei and cytoplasm from different somatic cells.

Cell theory The theory that all living matter is composed of basic units, called cells.

Central dogma DNA can serve as a template for its own replication or as a template for RNA transcription. RNA serves as the template for protein translation. RNA can also serve as the template for DNA synthesis (reverse transcription).

Centromere The chromomere to which the spindle fiber is attached during cell division.

Chemical mutagenesis The induction of mutations by the use of chemical agents.

Chi-square test A way to statistically test whether an observed set of data matches the data expected from a specific hypothesis.

Chiasma The cross-shaped areas seen in early prophase of meiosis I between nonsister chromatids. Chiasma are believed to be manifestations of crossing-over.

Chromatid One of the two daughter strands of a replicated chromosome, joined by the centromere to the other daughter chromatid.

Chromatin The DNA–protein complex of chromosomes.

Chromomere The bands in polytene chromosomes, seen as serially aligned granules in chromosomes undergoing division.

Chromosome In eukaryotes, a DNA–protein complex that is a linear array of genes. During division, a chromosome is composed of two chromatids. In prokaryotes and viruses, the chromosome is a molecule of DNA or RNA.

Chromosome puff   A puff visible at a specific band or cluster of bands along a polytene chromosome. It is an indication of RNA or DNA synthesis.

Chromosome theory of inheritance   The established theory (first stated as such by Sutton and Boveri in 1902–1903) that chromosomes are the carriers of the genetic material, and that genes and chromosomes are linked.

Circular permutation   In a population of phage chromosomes, the chromosomes are redundant at the ends, and the redundancies are different for each chromosome. This causes physically linear chromosomes to yield a circular genetic map.

Cistron   A genetic unit of function, usually equated with the term *gene*. Commonly a cistron specifies a single polypeptide.

Codominance   In a heterozygote, full expression of both alleles. The A and B blood antigens in an AB heterozygote are an example.

Codon   Three bases in a mRNA molecule that specify an amino acid.

Codon-restriction theory of aging   The idea that with the passage of time, different sets of codons come into use and others are permanently repressed. This could lead to a more limited set of proteins being synthesized and thus to a more restricted capacity to adapt to changing environmental conditions.

Coefficient of Coincidence   Number of observed double crossovers divided by number of expected double crossovers. See also *Interference*.

Colinearity   Condition in which the sequence of mutant sites in a cistron can be directly correlated with the sequence of amino acid alterations in the polypeptide coded for by that cistron.

Comma-free code   The idea that there are two groups of codons: those specifying amino acids and those not specifying amino acids. Any one codon base is part of only one codon. It was hypothesized that there were 20 codons specifying amino acids and 44 codons specifying no amino acids.

Competence   The physiological state of a bacterial cell that enables it to be transformed.

Complementary bases   Those bases that normally form hydrogen bonds with each other, such as A with T and G with C.

Complementation   The production of a normal or wild-type phenotype in an organism carrying two different mutant alleles. This is a type of genetic analysis that examines the functional interaction between two alleles.

Complementation map   A map developed from the complementation relationships between alleles, normally in a small segment of the chromosome.

Complete linkage   The condition in which genes are linked to the same chromosome and do not exhibit any recombination.

Complex locus   A genetic locus comprised of functionally related alleles separable by recombination.

Concatenation   A single, large DNA molecule composed of two or more complete phage chromosomes.

Concordance   In twin studies, if both exhibit the trait in question, they are said to be concordant. See also *Discordance*.

Conjugation   A form of sexual reproduction in some unicellular organisms such as *Paramecium*. A parasexual form of reproduction in *Escherichia coli*, characterized by the unidirectional transfer of a chromosome through a conjugation tube.

Conservative replication   A replication scheme in which both parental DNA strands are conserved in the same daughter DNA molecule.

Constitutive enzyme   Enzyme synthesized at a regular rate regardless of the environment. See also *Inducible enzyme*.

Continuous variation   Exhibited by a trait that, when measured in a population, can assume a wide and continuous spectrum of values from one extreme to another. An example is body height.

Copy-choice model   A model of genetic recombination that suggests that the synthesis of a new DNA strand switches from one template to another, and that this strand-switching generates recombinants.

Corepressor   A repressing metabolite or effector molecule in repressible enzyme systems.

Correlation coefficient   A value from 0 to -1 or 0 to $+1$ that indicates the degree to which statistical variables vary together.

Cost of selection   The proportion of adaptively inferior genotypes that are produced in a population in the course of producing adaptively superior genotypes.

Crossing-over   The exchange of genetic material between homologous chromosomes.

Cytokinesis   Cytoplasmic division.

Cytoplasmic inheritance   The inheritance of traits through the cytoplasm instead of through nuclear chromosomes. Cytoplasmic inheritance usually involves organelles such as mitochondria or chloroplasts or intracellular symbionts such as viruses.

Cytosine A pyrimidine base found in DNA and RNA.

Dark repair The repair of DNA damage by DNA-repairing enzymes that do not require light for their activation.

Daughter chromosomes The two chromosomes produced by the replication of a single parental chromosome.

Deficiency See *Deletion*.

Degeneracy In the genetic code, the property that one amino acid is coded for by more than one codon.

Deletion The loss of a segment of the genetic material from a chromosome.

Deme A small group of interbreeding organisms.

Denature To alter the natural three-dimensional shape of a molecule through treatment by heat, chemicals, or other agents.

Density gradient centrifugation A method of separating molecules by subjecting them to centrifugal force in a medium containing a gradient of densities.

Deoxyribose The sugar component of DNA.

Determination The process by which a cell line becomes genetically fixed to differentiate into a specific type of tissue or organ.

Diamond code An early view of the genetic code in which diamond-shaped pockets in the DNA molecule served as sites for specific alignment of amino acids.

Dicentric chromosome A chromosome with two centromeres.

Differentiation The expression of the determined state in a line of cells. The changes involved in the progressive diversification and specialization of cell types.

Dihybrid cross A cross between individuals differing with respect to two pairs of genes.

Dimer In mutagenesis induced by ultraviolet irradiation, the covalent linking of two bases, usually pyrimidines, that occupy adjacent positions on a DNA strand.

Diploidy The condition in which the autosomes are each present twice; that is, each autosome has a homologous partner.

Directional selection Selection regime in which variants to the left or right of the mean are selected as parents for the next generation, the result being a shift in the mean in the direction of selection.

Discontinuous variation Variation in a population that falls into two or more nonoverlapping classes.

Discordance In twin studies, if both do not exhibit the same form of the trait in question, they are said to be discordant. See also *Concordance*.

Disjunct populations Populations whose ranges do not overlap.

Dispersive replication A model of DNA replication in which parental DNA is found to be distributed among all four progeny DNA strands.

Disruptive selection Selection against the mean, causing the extremes to be favored.

Ditype Two types of ascospores in the same ascus sac.

Dizygotic twins Twins formed from two eggs fertilized at the same time.

DNA Deoxyribose nucleic acid, the genetic material.

DNA strand One of the two DNA strands in a double-helical DNA molecule.

Dominance hypothesis The theory that heterosis is caused by the masking of harmful recessive alleles by dominant alleles. See also *Overdominance hypothesis*.

Dominance A relationship between two alleles in which one allele masks the effects of the other allele.

Dosage compensation A genetic mechanism that compensates for genes that are present in two doses in the homogametic sex and are present in just one dose in the heterogametic sex.

Duplication The occurrence of a chromosomal segment more than once in the genome.

Elongation The process of adding amino acids to a growing polypeptide chain.

Epigenesis The idea that development involves the appearance of new structures and functions by a process of differentiation from an undifferentiated zygote to the complex and fully differentiated adult.

Episomes DNA molecules, such as certain phage chromosomes, found in bacterial cells in addition to the normal cell chromosome. These molecules may replicate independently of the host cell's chromosome or in synchrony with it.

Epistasis The masking by one gene of the expression of another, nonallelic gene.

Error-catastrophe model Model of aging that suggests that mutations in the protein-synthesizing apparatus have a cascading effect on a widening series of cell products, causing cell malfunction.

Essentialism See *Typological thinking*.

Euchromatin The chromosomal regions that carry genes and show characteristically light staining properties. The banded segments in polytene chromosomes contain euchromatin. See also *Heterochromatin*.

Eugenics A program of improvement of the human genome that can involve the encouragement of individuals with desirable traits to have more offspring (positive eugenics) and/or the discouragement of individuals with undesirable traits from having offspring (negative eugenics).

Eukaryote Organism with cells that contain true nuclei and undergo meiosis.

Euploidy Polyploidy in which the chromosome number is an exact multiple of the haploid chromosome number of the species in question.

Evolution Descent with modification.

Exogenote The chromosome fragment donated to a host cell, making it partially diploid. The fragment is donated by processes such as conjugation, transformation, and transduction.

Expressivity The strength of the phenotypic expression of a mutant gene.

Extinction In evolutionary terms, the disappearance of a species.

Extranuclear inheritance See *Cytoplasmic inheritance*.

F₁ First filial generation.

F₂ Second filial generation.

F⁺ An *Escherichia coli* cell carrying a nonintegrated F factor.

F⁻ An *Escherichia coli* cell without an F factor.

F′ An F factor that is carrying some bacterial chromosome genes.

F factor A bacterial episome that functions as a fertility factor. Cells that have it are F⁺ and serve as donors of genetic information.

Factor A term Mendel used to refer to what we now call the *gene*.

Feedback inhibition Mechanism by which the end product of a metabolic pathway acts to decrease the production of enzymes active at the beginning of that pathway.

Fertilization The union of two gametes to produce a zygote.

Fitness A quantitative measure of an organism's ability to survive and transmit its genes to the next generation.

Fixity-of-species doctrine An old doctrine that argues that species were specially created and do not undergo any changes over time.

Founder effect A type of genetic drift, in which a very small number of individuals from a larger population break off and start a new population. These individuals are not representative of the gene frequencies expressed by the larger parent population.

Frameshift mutation A shift in the reading frame during translation, caused by the addition or deletion of bases in a DNA molecule.

Freemartin A female twin masculinized by its male sibling as male hormones are carried through a shared circulatory system.

Gamete A haploid germ cell.

Gaussian distribution See *Normal distribution*.

Gemmule An old and no-longer-appropriate term used to describe hypothetical units in the body that carried the essence of specific body traits to the gonads, from which they were transmitted to the offspring.

Gene The basic unit of inheritance that occupies a specific locus on the chromosome and has a specific function.

Gene amplification Selective DNA replication in genes that are functioning at specific points in development. Gene amplification serves to increase the amount of specific gene products required in large amounts at a particular developmental stage.

Gene conversion The case in which the meiotic products from an *Aa* individual are 3 *A* to 1 *a* or 1 *A* to 3 *a*, instead of 2 *A* to 2 *a*. One of the alleles appears to have been converted to the other type.

Gene frequency The number of times a specific allele appears in a population, divided by the total number of alleles that map to that specific locus in the given population.

Gene pool The total of all genetic information possessed by a sexually reproducing population.

Generalized transduction Transduction in which any bacterial gene(s) can be transduced. See also *Specialized transduction*.

Genetic code The nucleic acid sequence (base triplets) that specifies an amino acid sequence.

Genetic drift The random fluctuation of gene frequencies due to sampling errors. Genetic drift is usually associated with small samples.

Genetic equilibrium Condition that exists when the gene frequencies in a population remain the same generation after generation.

Genetic fine structure Detailed genetic mapping of a limited gene region, often down to the nucleotide level.

Genetic load The number of lethal or detrimental genes maintained in a population.

Genetic polymorphism The stable maintenance of different genotypes for the same locus in a population over a long period of time.

Genome All of the genes carried by a haploid gamete. This term can also refer to all the genes carried by a prokaryote.

Genotype The genetic constitution of an organism.

Germ plasm The cell line that gives rise to gametes.

Gray crescent A special region in the cytoplasm of an amphibian zygote that is crucial to future development.

Guanine A nucleic acid base

Gynandromorph An individual composed of both male and female tissue.

Haploid number The number of chromosomes found in a normal gamete.

Hardy–Weinberg equilibrium Condition that exists when the genotype and gene frequencies remain constant in a large population that experiences no drift, selection, migration, or mutation, and in which mating is random.

Hayflick limit The observation that cells have a specific number of cell divisions that they can go through.

Helper phage A virus that, when it infects a cell with a defective virus, allows the defective virus to multiply and complete its life cycle.

Hemizygous Genes that are sex-linked in the heterogametic sex. Also, genes present in only one dose, as in a haploid organism.

Heritability A measure of the degree to which the total phenotypic variance is the result of genetic factors and thus can be influenced by selection.

Hermaphrodite An individual with both male and female reproductive organs.

Heterochromatin That part of the chromosomal material that stains very deeply. It appears that heterochromatin is genetically inactivated chromosomal regions.

Heteroduplex A DNA molecule in which all or part of the molecule is composed of DNA strands from different sources.

Heterokaryon Two different nuclei residing in a common cytoplasm.

Heterosis The increased vigor expressed by a hybrid from two highly inbred lines. Heterosis is usually associated with increased heterozygosis.

Heterozygous In a diploid organism, characterized by two different alleles at a specific locus (for example, *Aa*). See also *Homozygous*.

Hfr High-frequency recombinant in bacteria, caused by an integrated F factor, which initiates transfer of chromosomal genes to a recipient cell. This may lead to genetic recombination.

HFT High-frequency transducing lysate, caused by the replication of a transducing phage.

Histones A class of six low-molecular-weight chromosomal proteins with basic properties.

hnRNA Heterogeneous nuclear RNA, usually a precursor of eukaryotic mRNA molecules.

Holandric inheritance The inheritance of genes linked to the Y chromosome.

Holism The idea that only the entire organism should be studied, because to separate the parts from the whole does not reveal how the separated parts function in the intact organism.

Homologous chromosomes Chromosomes that pair during meiosis.

Homozygous In a diploid organism, characterized by both alleles being the same at a specific locus (for example, *AA*). See also *Heterozygous*.

Hybrid The offspring produced by genetically dissimilar parents.

Hybrid vigor See *Heterosis*.

Hybridization The process of forming genetic hybrids. It can also refer to the formation of DNA–RNA hybrids through the annealing of different nucleic acid strands (or DNA–DNA hybrids involving strands from different sources).

Hydrogen bond The bond that forms between a hydrogen that is covalently bound to an oxygen or a nitrogen, and another atom that contains an unshared pair of electrons.

Inborn error of metabolism Genetically caused metabolic dysfunction.

Inbreeding The crossing of closely related organisms.

Inbreeding depression The decrease in vigor that normally accompanies a program of extensive inbreeding.

Incomplete dominance Condition that exists when the phenotype of the heterozygote is intermediate between the two parental extremes.

Incomplete linkage Linked genes that reassociate through genetic crossing-over.

Independent assortment Occurs when a pair of alleles located on different chromosome pairs line up and segregate in an independent fashion during meiosis. A pair of alleles may also assort independently if they are on the same chromosome and separated by 50 or more map units.

Inducer Small molecule that causes a cell to produce larger amounts of the enzyme(s) needed to metabolize that molecule.

Inducible enzyme Enzyme synthesized only in the presence of an inducer molecule.

Induction In viruses, the release from the repressed state of a lysogenic phage causing it to enter a lytic phase. In the operon theory, the stimulation of the synthesis of specific enzymes in response to a specific inducer molecule.

Informosome A complex of mRNA and protein that protects mRNA from degradation.

Initiation The stages involved in the beginning of polypeptide synthesis.

Initiation site The site on the DNA molecule where the synthesis of RNA begins.

Insertion elements Sequences of DNA that specify the location of insertion of different episomes into the genome.

Integrator gene One of the components of the Britten–Davidson model of gene regulation in eukaryotes.

Interference The degree to which crossing-over in one genetic region inhibits or enhances crossing-over in the adjacent region: 1 − Coefficient of coincidence = Interference. See also *Coefficient of coincidence*.

Intergenic suppressor A gene product produced at one locus that suppresses a mutant gene product at a second locus, causing the phenotype to be normal.

Inversion Changes in the arrangement of gene loci on a chromosome.

Ionizing radiation Radiation that generates charged molecules that are usually highly mutagenic.

Law of ancestral inheritance The idea that an individual contains, in equal amounts, genetic material from each of the four grandparents, each of the eight great-grandparents, and so on. Each generation is a blend of the previous generation, and each ancestor contributes a specific proportion of an individual's genome.

Leaky mutant A mutant gene that produces a protein with partially normal activity.

Linkage group Those genes located on a specific chromosome.

Locus The specific region or spot on a chromosome to which a gene maps.

Lyon hypothesis The hypothesis that the Barr body is an inactivated **X** chromosome and that the number of such bodies is equal to the number of **X** chromosomes less 1. See also *Barr body*.

Lysis The breaking open of a phage-infected bacterium caused by the phage-coded enzyme lysozyme.

Lysogeny The relationship between a virus and a host cell in which the viral genome is incorporated into the host cell's genome and replicated along with it.

Macroevolution See *Transspecific evolution*.

Macromutation Genetic changes leading to large, sudden changes in an organism or population. Darwin considered macromutation evolutionarily unimportant.

Map unit In genetic mapping, the distance that corresponds to a 1-percent recombination frequency.

Masked mRNA mRNA complexed with protein so that it is not translated or enzymatically degraded.

Master–slave hypothesis An idea that attempts to interpret mutation in terms of repetitive genes. A master gene, when mutant, dictates mutant sequences in all the repetitive genes. When normal, or nonmutant, the master gene dictates nonmutant sequences in the repetitive genes.

Maternal effect A phenotype determined by the maternal genes and usually transmitted through the ooplasm.

Mean The sum of all quantities, divided by the number of quantities.

Median The middle value in a group of numbers arranged in order of size.

Meiosis The process by which chromosomes replicate, form homologous pairs, and then segregate into different nuclei to produce the haploid condition.

Meiotic drive Any mechanism that results in one of the meiotic products being recovered in greater frequency than the other products.

Mendelian population A sexually reproducing population sharing a common gene pool.

Metacentric chromosome A chromosome with a centromere located in the middle.

Microevolution Evolutionary changes that occur in a population over a relatively short period of time. See also *Macroevolution*.

Migration Movement of individuals from one population to another, resulting in the possible alteration of gene frequencies for the two populations concerned.

Missense mutation A codon change that results in the substitution of one amino acid for another.

Mitosis A process of cell division in which the chromosomes replicate and divide equally so that identical daughter cells are produced, each with the same genetic constitution as the parent cell.

Mode The class of numbers that contains the largest number of representatives in a statistical sample.

Modifier gene A gene that modifies the expression of another, nonallelic gene.

Monohybrid cross A cross between individuals differing in one pair of genes.

Monoploid The basic number of chromosomes in a polyploid series.

Monozygotic twins Twins formed from a single fertilized egg that at some time in early cleavage divided into two embryos.

Mosaic An individual composed of cells of different genotypes.

mRNA Messenger RNA; translated into polypeptide.

Multiple allele One of three or more alternative forms of an allelic series that all map to a specific genetic locus.

Multiple-gene hypothesis The hypothesis that for some quantitative traits, several pairs of alleles all have an additive effect.

Mutant site The point within a gene at which a mutation has occurred.

Mutation The process by which the primary structure of a gene (its base sequence) or group of genes is altered.

Mutational equilibrium The equilibrium point at which the forward mutations balance the reverse mutations so that the allelic frequencies do not change as a consequence of mutations.

Mutation frequency Proportion of mutant alleles in a population.

Mutation rate The probability of a mutation occurring per round of DNA replication.

Mutator gene A gene whose product increases the spontaneous mutation rate throughout the entire genome.

Muton The basic unit of mutation (a base pair).

Natural selection A process of differential fertility, in which some genotypes under the influence of a defined set of environmental parameters are more successful in leaving progeny than others.

Naturphilosophie The eighteenth- and nineteenth-century idea that viewed nature in a holistic way. See also *Holism*.

Nearest-neighbor analysis The technique whereby a specific base's nearest neighbors are determined. Each base can have one of four bases lying next to it: A, T, G, or C.

Negative control In an operon, the regulatory gene has a repressing influence on the structural genes. See also *Positive control*.

Neutral mutation The idea that some gene mutations confer no selective advantage or disadvantage on an organism.

Nondisjunction The failure of chromosomes to segregate during cell division, producing cases of aneuploidy such as monosomy and trisomy.

Non-Mendelian inheritance See *Cytoplasmic inheritance*.

Nonparental ditype A tetrad in which only the two nonparental classes of meiotic products are produced.

Nonreciprocal recombination Recombination in which homologous chromosomes do not undergo reciprocal exchange of genetic material. Only one chromosome may be recombinant, the other never forming.

Nonsense codon A codon that does not specify an amino acid.

Nonsense mutation A mutation that changes a codon specifying an amino acid into one that specifies no amino acid.

Normal distribution The most commonly used probability distribution statistic. The normal distribution follows the formula

$$Y = \frac{1}{\sqrt{2\pi}\sigma}[e^{-(X-m)^2/2\sigma^2}]$$

where m is the mean, σ is the standard deviation, e is the base of the natural logarithm, π is 3.1416, and Y is the height of the ordinate for a given value of X. This formula produces a bell-shaped curve.

Norm of reaction The range of different phenotypes produced by a genotype under different environmental conditions.

Nuclear transplantation The technique by which a nucleus from one cell is inserted into another cell.

Nucleic acid A macromolecule composed of either ribose or deoxyribose, inorganic phosphate, and a series of bases.

Nuclein The nucleoprotein complex isolated by Meischer.

Nucleolar organizer The specific chromosomal segment that synthesizes rRNA and is the site of association between the nucleolus and the chromosome.

Nucleolus The rRNA-rich spherical body associated with the nucleolar organizer.

Nucleoside A nucleic acid base covalently bonded to a sugar.

Nucleosome A unit of structural organization involving the DNA and protein of eukaryotic chromosomes.

Nucleotide A nucleic acid base covalently bonded to a sugar and an inorganic phosphate.

Nullisomic Pertaining to a cell or organism in which one of the chromosome pairs is missing.

Okazaki fragments Small fragments of DNA synthesized during the discontinuous replication of DNA and ultimately linked by DNA ligase to form a newly synthesized DNA strand.

Oncogene Cancer gene, according to the theory that genes from tumor-causing viruses are part of the genome of animals and that they may cause cancer when the environment induces them to do so.

One-gene–one-enzyme theory The theory that one gene produces one enzyme.

Operator gene The gene that serves as the point of interaction between the proteinaceous repressor molecule and the DNA in an operon.

Operon A unit of coordinately controlled structural genes under the control of an operator and regulatory genes.

Orthogenesis The old and discredited idea that mystical forces are guiding the continued evolution of life.

Overdominance Condition in which, in a cross of $AA \times aa$, the F_1 is more extreme than either parent type.

Overdominance hypothesis The theory that heterosis is caused by the Aa genotype being superior to either the AA or the aa parent. See also *Dominance hypothesis*.

Overlapping code The discredited idea that the genetic code consists of base pairs that participate in more than one codon.

P The parental generation.

Pangenesis A theory, proposed by Darwin, in which gemmules, modifiable by the environment, filter through the body and are packaged into the gametes. See also *Gemmule*.

Panmixis Random mating. See also *Assortative mating*.

Parameter A value based on an entire population as opposed to a sample.

Parasexual reproduction Any reproductive process that generates recombinants without the formation of gametes.

Parental ditype In a tetrad, two of the meiotic products are like one parental type, and the other two are like the other parent.

Parthenogenesis The development of an individual from an egg without fertilization taking place.

Pascal's triangle A technique for generating coefficients in a binomial distribution. Each term in the triangle is obtained by adding the numbers to the immediate left and right on the line above that term.

$$
\begin{array}{ccccccccc}
 & & & & 1 & & & & \\
 & & & 1 & & 1 & & & \\
 & & 1 & & 2 & & 1 & & \\
 & 1 & & 3 & & 3 & & 1 & \\
1 & & 4 & & 6 & & 4 & & 1 \\
\end{array}
$$

Pedigree A diagram of the genetic history of an individual or family.

Penetrance The percentage of individuals of a specific genotype that exhibit an expected phenotype under a specific set of environmental conditions.

Peptide bond The covalent bond that forms between two amino acids. The NH_2 group of one amino acid is bonded to the COOH group of the second amino acid, with H_2O eliminated.

Paracentric inversion An inversion that excludes the centromere.

Pericentric inversion An inversion that includes the centromere.

Phage See *Bacteriophage*.

Phage "ghost" The phage protein shell left behind after the phage has injected its DNA.

Phenocopy A phenotype produced by environmental factors that mimics the phenotype produced by a specific gene.

Phenotype The observable properties of an organism, produced by the interaction of the genotype with the environment.

Phenotypic mixing A virus produced by the combination of the nucleic acid of one virus with the protein coat of another.

Photoreactivation The repair of genetic damage by enzymes that are activated by visible light.

Phyletic speciation The process of speciation caused by the gradual change in the genetic constitution of a population without the population splitting into demes and without any increase in the number of species produced by that population at any one time.

Plaque A clear spot in a lawn of bacteria, caused by the phage-induced lysis of the sensitive cells.

Plasmid An extrachromosomal segment of DNA frequently found in bacteria.

Pleiotropy A condition in which a single gene has a wide range of phenotypic effects.

Polycistronic mRNA An RNA molecule that codes for two or more polypeptides.

Polygene One of several genes that have a cumulative effect on a particular phenotype.

Polygenic inheritance The pattern of inheritance produced by polygenes. Polygenic inheritance commonly results in a phenotype with a normal distributional range.

Polymorphism The existence of two or more different genotypes for the same trait in a population.

Polypeptide A chain of covalently linked amino acids (joined by peptide bonds).

Polyploidy The condition in which the chromosome number of an individual is three or more times the haploid chromosome number.

Polyribosome An mRNA molecule with several ribosomes attached.

Population A group of individuals related in time and space.

Position effect A change in the expression of a gene caused by a change in its position relative to its neighbors.

Position-effect variegation An unstable phenotype caused by a position effect. A red and white mottled eye in *Drosophila* can be caused by changing the position of the w^+ allele on the chromosome.

Positive control In operons, coordinately controlled sequences of genes that require a regulator gene product in order to be transcribed. See also *Negative control*.

Postadaptive mutation A mutation that occurs as a direct response to an environmental stress and that enables the organism to adapt to that stress. A discredited idea.

Postzygotic isolation mechanism Any one of several mechanisms that keep populations reproductively isolated from each other even though fertilization and hybrid zygotes may form. These hybrids are either sterile, nonviable, or so weak that they do not survive.

Preadaptive mutation A mutation that occurs before a specific environmental stress is encountered. When the stress occurs, the mutation, which confers a selective advantage on the organism, is selected for.

Preformation The discredited idea that, during development, the zygote has miniature copies of all the adult structures. See also *Epigenesis*.

Presence and absence hypothesis The discredited idea that the normal phenotype is due to the presence of a factor and the mutant phenotype to its absence.

Prezygotic isolation mechanism Any one of several mechanisms that keep populations reproductively isolated from each other by preventing fertilization and zygote formation.

Primary protein structure The amino acid sequence in a protein.

Primer DNA strand The DNA strand to which nucleotides are added at the time of replication or synthesis.

Probability The ratio of a specific event to the total number of possible events.

Prokaryote Organisms that lack well-defined nuclei and do not undergo meiosis.

Promoter   A genetic region next to the operator to which RNA polymerase binds. The operator lies between the promoter and the structural genes.

Protoplast   A bacterial cell minus the cell wall.

Prototroph   An organism capable of synthesizing all of its metabolites from inorganic material and a carbon source such as glucose. A prototroph usually has no nutritional requirements.

Prophage   In lysogenic bacteria, the phage DNA.

Provirus   The viral DNA integrated into the host cell's DNA.

Pseudoalleles   Alleles that are functionally related and separable by recombination.

Pseudodominance   The expression of a recessive allele because of the absence of the dominant allele by deletion.

Pure line   A homozygous strain produced by inbreeding.

Purines   A class of organic bases found in DNA and RNA.

Pyrimidines   A class of organic bases found in DNA and RNA.

Quantitative inheritance   The inheritance pattern for a trait that is determined by the cumulative action of polygenes, each polygene producing a small additive effect.

Quantum speciation   The budding off of a new and very different daughter species from a semi-isolated peripheral population of the ancestral species in a cross-fertilizing organism.

Quarternary protein structure   The way in which two or more chains of amino acids that comprise a protein interact with each other.

Race   A genetically distinct, and usually geographically distinct, interbreeding division of a species.

Radiomimetic chemicals   Chemicals that mimic the effects of ionizing radiation on the genetic material.

Random genetic drift   See *Genetic drift*.

Recessive   In a diploid organism, the allele that is expressed only when it is homozygous and masked when it is heterozygous with a dominant allele.

Reciprocal cross   A genetic cross that can be symbolized as

$$A\male \times B\female \quad \text{and} \quad B\male \times A\female$$

where *A* and *B* represent different genotypes.

Reciprocal translocation   The reciprocal exchange of chromosome segments from nonhomologous chromosomes.

Recognition site   A sequence of bases within the promoter region that serve to recognize the RNA polymerase molecule. See also *Promoter, Binding site,* and *Initiation site.*

Recombination   Any one of several processes that result in the production of progeny with gene combinations that differ from those of the parent(s).

Recon   The smallest unit of DNA within which recombination can occur.

Reduction division   The first half of the meiotic process, in which the paired homologues segregate to different nuclei, thus reducing the chromosome number from $2n$ to n.

Redundancy   The repetition of genes or base sequences.

Reduplication hypothesis   The discredited idea that gametes with parental gene combinations undergo a selective mitosis so that their frequencies are greater than the nonparental gene combinations. The reduplication hypothesis was meant to counter the idea of gene combinations generated by crossing-over.

Regressive variation   Variation in which the offspring tend to approach the mean of the population rather than to exceed the parental extremes.

Regulatory gene   A gene the function of which is to control the rate of synthesis of distant gene products.

Relaxed response   In bacteria, under conditions of amino acid starvation, the cell's continued production of tRNA and rRNA. See also *Stringent response.*

Replication fork   The Y-shaped regions of a replicating DNA molecule where DNA replication is occurring.

Repressible enzyme system   A coordinately controlled gene system in which the regulator gene product blocks the transcription of the operon by combining with an effector molecule called a corepressor. See also *Inducible enzyme.*

Repression   The blocking of the transcription of a gene or genes.

Repressor   In an inducible operon, the protein product of the regulator gene. The repressor binds to the operator and prevents transcription of the structural genes.

Reproductive isolation   The inability of individuals in two different interbreeding populations to interbreed with each other.

Restriction enzyme One of a number of enzymes that break DNA molecules down by causing cleavage at specific points in the molecule; the points are determined by the base sequence.

Reverse mutation A mutation in a mutant gene that restores its ability to produce a functional gene product.

Reverse transcription The synthesis of DNA from an RNA template.

Ribose The sugar component of RNA.

Ribosome Subcellular structure composed of rRNA and protein, made up of two subunits, and functioning as the active site for protein synthesis.

RNA Ribose nucleic acid. RNA is transcribed from DNA—except in some viruses, in which it is the genetic material. Its primary function is in the process of protein synthesis.

Rolling-circle model of replication A replication scheme in which circular DNA can replicate by attaching to a cell membrane and unrolling as it replicates.

rRNA Ribosomal RNA. Along with ribosomal proteins, rRNA forms the structural components of the ribosome.

Saltation Evolution by stepwise changes in a population.

Secondary protein structure The type of helical structure exhibited by a polypeptide and caused by the formation of hydrogen bonds between amino acids.

Segregation Mendel's first law, calling for the separation of the members of a homologous pair of chromosomes into different gametes through the process of meiosis.

Segregation distortion Any mechanism that results in a significant deviation from the expected 1:1 segregation ratio of an *Aa* heterozygote.

Self-fertilization The formation of a zygote by fertilization involving gametes produced by only one parent.

Semiconservative replication The DNA-replication scheme in which daughter DNA molecules contain one parental and one newly synthesized strand.

Sensor region In the Britten–Davidson model of eukaryote gene regulation, the sensor region responds to a signal in the environment (perhaps a hormone) and activates integrator genes, which activate receptor genes, which activate structural genes.

Sex chromosome A chromosome, usually **X** or **Y**, that plays a major role in the determination of sex.

Sex factor A bacterial episome that enables its carrier to donate genetic material to a donor cell through a conjugation tube. See also *F factor* and *Episome.*

Sex linkage Linkage of genes to the sex chromosome.

Sex-limited trait A trait expressed in only one sex.

Sexduction Process by which bacterial chromosomal genes are incorporated into an episome and transmitted to a recipient cell via a conjugation tube.

Sexual reproduction Reproduction that involves the fusion of haploid gametes produced by meiosis.

sfa Sex factor affinity. A region on the bacterial chromosome that has a specific affinity for the sex factor. It is a region consisting of sex factor DNA left behind when the sex factor recombined out of the bacterial chromosome.

Simple translocation The translocation of a chromosome segment from one chromosome to a nonhomologous chromosome.

Social Darwinism The application of Darwinian principles to the operation of human society.

Somatic Pertaining to all cells and tissues that do not form gametes. See also *Germ plasm.*

Somatic mutation theory The theory that aging is the result of the accumulation of mutations in the somatic tissues.

SOS repair system A special series of enzymes that recognize and repair damage to DNA. It is error-prone.

Spacer DNA DNA sequences between genes that do not appear to be transcribed.

Special creation The discredited idea that all living things occupying the earth were specially created by God.

Specialized transduction Transduction in which a virus can transduce only a specific gene from one bacterium to another. See also *Generalized transduction.*

Speciation The process of forming a new species.

Species A population or populations of individuals that share in a common gene pool, are phenotypically similar, and are reproductively isolated from other species.

Species swarm A large number of closely related species. A species swarm can be the result of a population splitting into a number of founder populations that form new species.

744

Spindle fibers Cytoplasmic microtubules that attach to centromeres and move the chromosomes during division.

Spontaneous mutation A mutation that occurs in the absence of specific chemical or physical mutagenic agents. Spontaneous mutations occur under "natural" conditions, usually during replication.

Stabilizing selection Selection against individuals that deviate from the mean. The result is retention of the mean but reduction of the variance.

Stable mRNA An mRNA species that persists in time instead of being rapidly degraded. The stabilization is probably the result of complexing with a protein.

Standard deviation A measure of the variability in a population of N individuals, given by the formula

$$\sigma = \sqrt{\frac{\Sigma(x - \bar{x})^2}{N - 1}}$$

where N is the number of individuals in the sample and $\Sigma(x - \bar{x})^2$ is the sum of the squared deviations from the mean.

Statistics The application of probability theory to data collected from a sample of a population and then extended to the whole population.

Strand analysis Genetic analysis based on a sample of meiotic products.

Stringent response In bacteria, under conditions of amino acid starvation, repression of the cell's synthesis of tRNA and rRNA. See also *Relaxed response.*

Structural gene A gene that produces an mRNA, which is then translated into a polypeptide.

Subspecies See *Race.*

Supergene A sequence of genes preserved by selection and prevented from crossing-over by inversions or translocations.

Suppressor mutation A mutation that reverses the effects of a mutation at a distant locus or site.

Sympatric speciation Speciation in populations of organisms that overlap geographically.

Sympatric species Species that are reproductively isolated but occupy a common geographical area.

Synapsis The intimate pairing of homologous chromosomes during meiosis.

Synaptonemal complex Observed under the electron microscope, a tripartite structure between paired homologues during the pachytene stage of meiosis I. There are three dense lateral bands or ribbons connected by fine transverse elements.

Synkaryon The nucleus that results from the fusion of two nuclei; occurs in zygote formation and cell fusion

Target theory A theory that was used to estimate the size of the gene; based on the effects of ionizing radiation on a small volume within the cell. One or more ionizing events within this small volume are postulated as necessary to cause a mutation.

Tautology A circular argument.

Tautomeric shift In nucleic acid bases, the shifting back and forth between the keto and enol forms, or between the amino and imino forms.

Telocentric chromosome A chromosome with a terminal centromere.

Temperate phage A virus that can enter into a lysogenic relationship with a host cell.

Template DNA strand The DNA strand that dictates the sequence of bases in the newly synthesized strand.

Terminalization The movement of the chiasma toward the ends of the bivalent during meiosis.

Terminal redundancy Condition in which the ends of a linear DNA molecule have the same sequence of nucleotides.

Termination The processes involved in the completion of the synthesis of a polypeptide and its release from the ribosome.

Tertiary protein structure The manner in which a polypeptide chain folds back upon itself.

Testcross A cross involving a heterozygote crossed back to a homozygous recessive organism.

Tetrad The four meiotic products of a single meiosis.

Tetrad analysis A genetic analysis in which all four products of a single meiosis are analyzed.

Tetranucleotide theory The discredited idea that DNA is just a repeating sequence of the four bases (that is, ATGCATGC).

Tetratype A tetrad containing four different gene combinations: one type of each of the two parent types and two recombinant types.

Thymine One of the DNA bases.

Transcription The synthesis of RNA from a DNA template or the synthesis of DNA from an RNA template (reverse transcription).

Transcriptional control Control mechanisms that operate at the level of transcription.

Transduction The transfer of bacterial genes from one cell to another by means of a virus used as a vector.

Transformation The genetic alteration of a cell by the transfer of free DNA in the medium across the membrane and into the cell where it recombines with the host cell's DNA.

Transforming principle Identified by Avery, MacLeod, and McCarty as DNA.

Transgressive variation The production of F_2 phenotypes that exceed the parental extremes.

Transition mutation A base-pair substitution in which the orientation of the purine and pyrimidine bases on each strand remain the same (that is, AT to GC).

Translation The process of converting the information contained in a sequence of RNA bases into a sequence of amino acids.

Translational control Control mechanisms that operate at the level of translation.

Translocation A chromosomal aberration involving the interchange of nonhomologous chromosome segments.

Translocation shift Transfer of an interior segment of a chromosome to the interior of a nonhomologous chromosome.

Transspecific evolution The evolution of groups above the species level.

Transversion mutation A base-pair substitution in which the purine and pyrimidine orientation on each strand is reversed (that is, AT to CG).

Triangle code The discredited idea that RNA organizes itself into triangular shapes into which amino acids fit.

Trihybrid cross A cross that involves parents differing in three pairs of alleles.

Trisomy The diploid condition plus one extra chromosome.

tRNA Transfer RNA, which binds to mRNA and to an amino acid.

Typological thinking Thinking in terms of types or major categories. Important in evolutionary history.

Unidirectional replication DNA-replication scheme in which a single replication fork in a DNA molecule moves in one direction. See also *Bidirectional replication*.

Univalent A single chromosome seen segregating in meiosis.

Uracil A base in RNA.

Use and disuse doctrine The discredited idea that the more a structure is used, the more prominent it becomes in future generations, and that the less it is used, the less prominence it assumes in later generations. Also known as the doctrine of acquired characteristics.

Variance The mean of the squared deviations from the population mean.

Virulent phage Bacterial virus that always goes through a lytic cycle.

Wild-type phenotype The phenotype most frequently observed in nature and the one arbitrarily designated as normal.

Wobble hypothesis The idea that the third base in an anticodon has a certain amount of "wobble" so that it can pair with more than one base in the third position of the codon. This means that one tRNA can line up at more than one codon.

X-ray diffraction The technique of getting a diffraction pattern by scattering x rays through a crystal. X-ray diffraction is used to determine the three-dimensional structure of a molecule.

Zygotic induction The induction of prophage replication when a lysogenic Hfr bacterium transfers the prophage to a nonlysogenic F⁻ cell during conjugation.

Problems for review

These short answer problems are designed to give you an opportunity to review most of the concepts presented in the text. I emphasize *short answer*, because, with the exception of the last two problems, you can answer them with a number, a few letters, or a few words. The last couple of problems require only a sentence or two. There are various levels of difficulty represented in these problems. In the event that you encounter obstacles, you should go back and review the appropriate chapter. Answers to these problems are found on pages 753–754. [*Note:* In most answers, I have used the symbols *A* and *a*, or *B* and *b*, to represent dominant or recessive alleles not specifically defined in a problem.]

1. A carrier of the allele for sickle-cell anemia, an autosomal trait, marries another carrier. What proportion of the offspring can we expect to express hemolytic anemia?

2. What proportion of the above offspring can we expect to be carriers of the mutant alleles?

3. Huntington's chorea is a severe and rare human disease that is expressed relatively late in a person's life. People who have the disease almost always have a parent who died of it. Two normal people usually do not have offspring that express the trait. Is this a dominant or recessive trait?

4. If a trait usually appears in offspring of parents who do not express the trait, is the trait likely to be dominant or recessive?

5. A man who is color-blind marries a women who is not. They produce a color-blind daughter. Assuming that the trait is sex-linked and recessive, what is the genotype of the mother?

6. A woman whose blood type is A and M is married to a man whose blood type is B and M. She gives birth to dizygotic twins with blood types O-M and AB-MN; would you support the husband's contention that he is not the father of both children?

7. What are the A-B-O genotypes of the above two parents?

8. How can a person with Turner's syndrome be color-blind if both of her parents are normal in their vision?

9. Can a person with Klinefelter's syndrome be color-blind if both parents are normal in their vision?

10. A testcross for an individual heterozygous for three independently assorting pairs of alleles produces how many genotypic classes?

11. Suppose that one of the alleles at each of the above three loci is recessive. How many phenotypic classes will result from the above testcross?

12. How many kinds of gametes can be formed by an organism heterozygous at six independently assorting gene loci?

13. Two flies heterozygous for *dp* and *e* are crossed. The offspring produced are as follows: wild type = 461; *dp* = 142; *e* = 153; *dp*, *e* = 46. Use a chi-square test to see if these data fit a 9:3:3:1 ratio.

14. How would you explain the most likely cause of color-blindness of a woman who is color-blind in only one eye?

15. How would you explain the most likely cause of color-blindness of a man who is color-blind in only one eye?

16. A population is composed of 550 AA, 300 Aa, and 150 aa individuals. Is this population in Hardy-Weinberg equilibrium?

17. The ability to taste a particular chemical is due to the presence of a dominant allele. In a population of 1000 people, 16 are unable to taste the chemical. Estimate the frequency of the recessive allele.

18. A particular autosomal recessive allele is responsible for a genetic disease. The disease is found to occur with a frequency of 1 in 10,000. What is the expected frequency of carriers in the population?

19. Color-blindness is a sex-linked recessive trait. In a human population at equilibrium, the frequency of color-blind females is 36 per 1000. What is the frequency of color-blind males?

20. What proportion of the females in the above population would be heterozygous for color-blindness?

21. In a randomly mating population of 500 persons, 8 are found to be of blood type N. This group now intermarries at random with a second randomly mating population of 500 persons in which 25 are of blood group N. What is the N frequency in the combined population?

22. What proportion of the *offspring* of the combined population will be blood type N?

23. At a local blood bank, 10,000 pints of blood are collected; 1500 are type B, 400 are AB, 4900 are O, and 3200 are type A. Estimate the I^A, I^B, and I^O frequencies in this population.

24. Pelger anomaly is a blood disease in rabbits. In matings involving two Pelger rabbits, 30 offspring have the Pelger anomaly and 16 are normal. Similar patterns are observed in other matings between two Pelger rabbits. What is the genotype of the Pelger rabbits?

25. What ratio of offspring would be expected from a Pelger × normal mating?

26. Describe the phenotypes of all the genotypic classes produced in Problems 24 and 25.

27. A true-breeding strain of black-feathered chickens without head crests is crossed to a true-breeding strain with red feathers and a prominant head crest. Testcrossing the F_2 black, crested progeny of this cross produces an overall ratio of 4 black, crested:2 black, no crests:2 red, crested:1 red, no crests. Which traits are the result of dominant alleles?

28. What were the parental genotypes in the above problems?

29. What genotypic and phenotypic ratios would be expected from an F_1 × F_1 cross in Problem 27?

30. Yellow mice with crooked tails are crossed to each other. $\frac{6}{12}$ of the offspring are yellow, crooked; $\frac{2}{12}$ are yellow with normal tails; $\frac{3}{12}$ are agouti, crooked; and $\frac{1}{12}$ are agouti with normal tails. Which of the two traits is associated with a lethal allele?

31. Is crooked tail determined by a dominant or recessive gene?

32. What is the parental genotype of the yellow mice with crooked tails?

33. What are the genotypes of the 4 phenotypic classes described in Problem 30?

34. A white dog crossed to a brown dog produces an all-white F_1; $F_1 \times F_1$ crosses yield an F_2 composed of 12 white, 3 black, and 1 brown.
   (a) Is this a monohybrid, dihybrid, or trihybrid cross?
   (b) Describe the gene interaction.
   (c) What are the parental genotypes?
   (d) What are the offspring genotypes?

35. Crossing two white-flowered strains of sweet peas produces a purple F_1; $F_1 \times F_1$ matings produce 55 purple-flowered plants and 45 white-flowered plants. What type of phenotypic ratio is this?

36. What are the genotypes of the parents, F_1, and F_2 plants in the above cross?

37. A female *Drosophila* with yellow body color, when crossed to a wild-type male, produces all yellow females and normal males. Explain this.

38. Two normal parents have four sons, two of whom are hemophiliacs. These parents later divorce and each remarries a normal person. From the mother's new marriage come six more children. The four daughters are normal, but one of the two sons is a hemophiliac. From the father's new marriage come eight normal children, four male and four female. Is hemophilia a disease that is caused by a dominant or a recessive gene?

39. Is the type of hemophilia described in Problem 38 due to a sex-linked or autosomal gene?

40. What are the parental genotypes in Problem 38?

41. A normal female mates with a mutant male and produces two mutant males and ten normal females. A mutant male mates with his nonmutant female sib and they produce six mutant males and six nonmutant females.
   (a) Is this **X**-linked, **Y**-linked, or autosomal inheritance?
   (b) What are the genotypes of all individuals?

42. If a chiasma is observed between loci A and B in 36% of the tetrads, approximately what percentage of the meiotic products will be recombinant?

43. When a testcross is performed, the following results are obtained:

   AaBb × *aabb*—42% *Aabb*, 42% *aaBb*, 8% *AaBb*, and 8% *aabb*.

   (a) What are the crossover and noncrossover gametes?
   (b) What percentage of the tetrads had chiasma between the two loci?
   (c) What is the approximate genetic distance between the two points?

44. If a crossover occurs between A and B in 30% of the tetrads and between B and C in 20% of the tetrads, then what is the expected frequency of double crossover?

45. Suppose that the observed frequency of double crossover in the above cross is .005. What is the interference and coefficient of coincidence?

46. Suppose that the A–B distance is 30 map units and the B–C distance 20 map units. What are the expected gamete classes and frequencies produced by an *ABC//abc* individual?

47. On the basis of the following cross, determine the middle gene and construct a genetic map.   *AaBbCc* × *aabbcc*

Aabbcc	*aaBbCc*	*aabbCc*	*AaBbcc*
.36	.36	.09	.09
AabbCc	*aaBbcc*	*AaBbCc*	*aabbcc*
.04	.04	.01	.01

48. We wish to sequence five loci from three series of three-point crosses. From cross 1, we get $A\underset{}{^4}B\underset{}{^5}C$; from cross 2, we get $C\underset{}{^5}B\underset{}{^{11}}D$; from cross 3, we get $C\underset{}{^{15}}E\underset{}{^1}D$. What is the sequence of these five points and the distances between them?

49. Suppose that we have the following genetic map: $A\underset{}{^{10}}B\underset{}{^{20}}C$, and an interference of 20%. What is the coefficient of coincidence and the expected frequency of double crossovers?

50. Suppose that we have the same map as in Problem 49 and an interference of 40%. Give the frequencies of all gametes produced by an $ABC//abc$ parent.

51. A *Neurospora* strain carrying gene a is crossed to a strain carrying gene $b(a+ \times +b)$. The following results are obtained.

Sequence
(1) $\underline{a+ \ a+ \ +b \ +b}$ = 78%
(2) $\underline{a+ \ ++ \ ab \ +b}$ = 15%
(3) $\underline{a+ \ ab \ ++ \ +b}$ = 6%
(4) $\underline{a+ \ +b \ a+ \ +b}$ = 1%

(a) What is the sequence of the two loci and the centromere?
(b) What are the distances between the loci and the centromere?

52. A *Neurospora* carrying gene a is crossed to a strain carrying gene $b(a+ \times +b)$. The following results are obtained.

Sequence
(1) $a+ \ a+ \ +b \ +b$ = 48
(2) $ab \ ab \ ++ \ ++$ = 44
(3) $+b \ +b \ a+ \ a+$ = 43
(4) $++ \ ++ \ ab \ ab$ = 47

What can you conclude about the a and b loci?

53. A nucleic acid has the following base composition: $A = 20\%$, $C = 30\%$, $U = 20\%$, and $G = 30\%$. How would you characterize this nucleic acid?

54. If an $A + G/T + C = 1.5$, can we argue that that nucleic acid is double-stranded DNA?

55. How can a single bacteriophage produce both b and b^+ progeny phage?

56. Several mutations are induced in T4; then the mutations are tested for their ability to complement each other. The following results are obtained.

	A	B	C	D	E
A	0	+	0	0	+
B		0	+	+	+
C			0	0	+
D				0	+
E					0

Which of these mutants are members of the same cistron, and how many cistrons are there?

57. The following two-point crosses are carried out in the λ bacteriophage: $a^+b^+ \times ab$; $c^+b^+ \times cb$; $a^+c^+ \times ac$. The results obtained are as follows.

$a^+b^+ \times ab$	$c^+b^+ \times cb$	$a^+c^+ \times ac$
↓	↓	↓
a^+b^+ : 450	c^+b^+ : 1170	a^+c^+ : 800
ab : 470	cb : 1182	ac : 820
a^+b : 42	c^+b : 19	a^+c : 90
ab^+ : 38	cb^+ : 19	ac^+ : 90

What is the gene sequence and what are the distances separating the three points?

58. In a transformation experiment, a^+b^+ DNA is used to transform ab host cells. The data show 140 a^+b transformants, 120 ab^+ transformants, and 416 a^+b^+ transformants. Can one reasonably assume that the two genes are linked? If so, what is the approximate map distance between a and b?

59. Consider the following gene sequences as determined by conjugation in four different *E. coli* Hfr strains.

(a) 1234567
(b) 4321765
(c) 7123456
(d) 7654321

Are these data consistent with a single *E. coli* map?

60. We use a generalized transducing phage to map three genes in a bacterium. The host cell is a^+bc^+. Transducing phage infect this cell, and, after release, infect ab^+c cells. The following genotypes are recovered.

a^+b^+c	3%
a^+bc^+	46%
$a^+b^+c^+$	27%
abc^+	1%
a^+bc	23%

What is the gene order and what are the map distances? (Assume that abc are always cotransduced.)

61. A polypeptide chain has 173 amino acids. What is the minimal number of mRNA bases needed to code for it?

62. Proflavin is used to study mutations in the lac operon. A mutation is induced that results in abnormal gene products from genes z and y. Gene a product is normal and the operon remains inducible. What happened?

63. In a transformation experiment, the recipient cell is $ABCDE$, but the gene order is unknown. A donor strain provides $abcde$ DNA and the following double transformants are found: $AbcDe$, $abCdE$, $aBcDe$, and $AbcdE$. Single transformants for each of the five genes are also found. Determine the order of the five genes.

64. Consider four metabolites: u, q, p, and s and a series of mutations in genes that affect the metabolism of these metabolites. The effect of one mutation can be overcome by adding u, s, or q to the medium, but p will not reverse the effects. A second mutation's effects can be overcome only by adding q, and a third mutation's effects can be overcome only by adding u or q. What is the sequence of metabolite formation? Indicate the points blocked by the mutations.

65. Six mutations are found that all result in the formation of a nonfunctional gene product. In a complementation study using these six mutants, the following results were obtained.

1	2	3	4	5	6	
0	+	+	0	+	+	1
	0	+	0	0	+	2
		0	+	+	0	3
			0	0	+	4
				0	+	5
					0	6

+ = Complement
0 = No complement

(a) These data indicate that there are how many cistrons involved?

(b) One mutation involves more than one cistron. Which mutation is it?

66. What is the minimum number of mutations required to go from

(a) *leu* to *ser*

(b) *leu* to *arg*

(c) *met* to *cys*

67. An amino acid site is known to mutate in a single step to *val, lys,* or *ileu.* What amino acid is the most likely to occupy this site?

68. An amino acid site normally occupied by glycine can also be occupied by *val, arg,* or *met.* What codon is the most likely for glycine at this site?

69. Two mutants are in the same cistron. Will the phenotype be mutant or normal in the following genotype: $m^1m^2//++$?

70. What would the answer be if the arrangement in Problem 69 were $m^1 +//+ m^2$?

71. Can mutations induced by nitrous acid be reversed by nitrous acid?

72. Would you expect most x-ray-induced mutations to be revertible?

73. Are transversions revertible by nitrous acid?

74. Red hair in humans is believed to be caused by a recessive gene. In one case, however, two red-haired parents had a brown-haired child. Assuming that illegitimacy is not an issue, how would you explain this?

75. A female, known to be a carrier of the sex-linked gene for hemophilia, marries a normal male. They have a hemophiliac daughter. How would you explain it?

1. $\frac{1}{4}$
2. $\frac{1}{2}$
3. Dominant
4. Recessive
5. Aa
6. Yes; no MN possible
7. AO and BO
8. Mother was a carrier; male **X** was lost
9. Yes; mother was a carrier; nondisjunction at meiosis II
10. 8
11. 8
12. 64
13. $\chi^2 = 0.99$; $p = 0.8$
14. Lyon hypothesis
15. Mosaic from a somatic mutation
16. No
17. 0.126
18. 0.0198
19. 0.19
20. 0.31
21. 0.17
22. 0.03
23. $I^A = 0.2$; $I^B = 0.1$; $I^O = 0.7$
24. Aa
25. 1 Pelger:1 normal
26. AA = normal; Aa = Pelger; aa = lethal
27. Black, crested
28. $AAbb$ and $aaBB$
29. 9 black, crested
    3 black, noncrested
    3 red, crested
    1 red, noncrested
30. Coat color
    Lethal = AA; yellow = Aa; agouti = aa
31. Dominant
32. $AaCc$
33. 6 $AaC-$; 2 $Aacc$; 3 $aaC-$; 1 $aacc$
34. (a) Dihybrid
    (b) $I-$: inhibits color
        ii: color
        $B-$: black
        bb: brown
    (c) $IIBB \times iibb$
    (d) 9 $I-b-$ } white
        3 $I-bb$ }
        3 $iiB-$ black
        1 $iibb$ brown
35. 9:7

36. $aaBB \times AAbb$
    ↓
    $AaBb$
    ↓
    9 $A-B-$ purple
    3 $A-bb$ ⎫
    3 $aaB-$ ⎬ white
    1 $aabb$ ⎭
37. Attached **X**
38. Recessive
39. Sex-linked
40. $Aa \times A$
41. (a) **Y**-linked
    (b) $XY^m \times XX$
        ↓
        XY^m and XX
        ↓
        XX and XY^m
42. 18%
43. (a) $CO = ab, AB$
        $NCO = Ab, aB$
    (b) 32%
    (c) 16 map units
44. 0.015
45. $CC = 0.33$; Int. = 0.67
46. $ABC = 0.27$; $abc = 0.27$
    $Abc = 0.12$; $aBC = 0.12$
    $ABc = 0.07$; $abC = 0.07$
    $AbC = 0.03$; $aBc = 0.03$
47. $C-10-A-20-B$
48. $D-1-E-6-A-4-B-5-C$
49. $CC = 0.8$; expected $DCO = 0.016$
50. ABC
    abc ⟩71.2% Abc
                      aBC ⟩8.8%
    ABc
    abC ⟩18.8% AbC
                      aBc ⟩1.2%
51. $a-8-\cdot-3.5-b$
52. a and b on different chromosomes; a and b close to the centromere
53. Double-stranded RNA
54. No; single-stranded DNA
55. Heteroduplex
56. Cistron 1: ACD
    Cistron 2: B
    Cistron 3: E
57. $a-8-b-2-c$
    $-10-$
58. Yes; 38 map units
59. Yes; different points of F insertion and different direction of transfer

60. b—31—a—27—c

61. 519

62. Frame-shift mutation through z and into y

63. *CEADB*

64. $p \xrightarrow{1} s \xrightarrow{3} u \xrightarrow{2} q$

65. (a) 3

(b) Number 4

66. (a) 1

(b) 1

(c) 3

67. *met*

68. *GGG*

69. Normal

70. Mutant

71. Yes

72. No; most are breaks

73. No

74. One parent was *Aa,* but the normal allele was impenetrant (*a* was dominant).

75. Daughter is *Aa*; normal **X** is inactive via Lyon hypothesis.

Answers to selected problems

Chapter 1   The establishment of Mendelian principles of inheritance

1.4 This reconstruction should include a discussion of the significance of mitosis and meiosis. It should also include a discussion of the parallel between meiosis and Mendel's principles, especially as seen in haploid organisms like *Neurospora* and *Chlamydomonas*.

1.6 (a) The same (b) Some tetrad types would be NPD (A, $+$; a, $-$); some tetrad types would be the same as shown in the figure; and some tetrad types would be PD (a, $+$; A, $-$). It all depends on how the chromosome pairs line up in metaphase.

1.8 (a) Anaphase I (b) Prophase I (c) Metaphase I (d) Telophase II (e) Anaphase II

1.10 This is not a genetically determined trait

1.12 There is more than one possibility. The purple parent could be *AaBB* or *AABb*, for example, and the white parent *aabb*. It is a cross with a single gene difference. It could also work as a 5:3 ratio, *AaBb* × *Aabb* with both *A* and *B* required for purple.

1.14 9 white disc, 3 white spheroid, 3 yellow disc, 1 yellow spheroid

1.16 Most likely *MMBB*

1.18 1 red narrow, *RRNN*; 2 red medium, *RRNn*; 1 red broad, *RRnn*; 2 pink narrow, *RrNN*; 4 pink medium, *RrNn*; 2 pink broad, *Rrnn*; 1 white narrow, *rrNN*; 2 white medium, *rrNn*; 1 white broad, *rrnn*

1.20 (a) Recessive (b) Both are *AaBb*

Chapter 2   Mendelism and the chromosome theory of inheritance

2.2 Sex-linked inheritance refers to genes located on the sex chromosomes (**X** or **Y**); sex-limited inheritance refers to traits expressed in only one sex. **Y**-linked genes in humans are both sex-linked and sex-limited.

2.6 *Neurospora* is haploid and thus not subject to the complicating factor of dominance seen in diploid organisms like *Drosophila*; also in *Neurospora*, all the products of a single meiosis are contained in sequence in an ascus sac. This contrasts with *Drosophila* where one can look at only a sample of meiotic products.

2.12 th vg cv   799.5 (0.40)
     + + +   799.5 (0.40)
     th + +   115.5 (0.06)
     + vg cv   115.5 (0.06)
     th vg +   83.5 (0.04)
     + + cv   83.5 (0.04)
     th + cv   1.5 —
     + vg +   1.5 —

Expected DCO is 19.89; coefficient of coincidence (from Figure 2.17) is 0.15; therefore observed DCO is 3.

2.14 (a) As in Problem 2.13, the gene for pattern baldness is influenced by the sex of the carrier. Male heterozygotes express A, but female heterozygotes do not. Only *AA* females are bald. (b) The mother was nonbald, and the father was bald.

2.16 centromere $\dfrac{5.01}{}$ *nic* $\dfrac{5.23}{}$ *ad*
        $\vdash\!\!-\!\!-10.24\!\!-\!\!-\!\!\dashv$

Chapter 3   The expansion of Mendelian principles, I

3.6 Epistasis is the interaction between nonallelic genes; dominance is the interaction between alleles.

3.10 Multiple allelism played an instrumental role in overcoming the presence and absence hypothesis, and it led to the concept of pseudoalleles. Both of these notions should be discussed in this analysis.

3.12 The hypothesis is a 9:6:1 ratio.

o	e	o − e	$(o - e)^2$	$\dfrac{(o - e)^2}{e}$
89	91	−2	4	0.04
62	61	1	1	0.02
11	10	1	1	0.10
162	162	0	—	0.16

$$\chi^2 = \frac{(o - e)^2}{e} = 0.16 \qquad\qquad p \cong 93 \text{ percent}$$

The data fit a 9:6:1 hypothesis.

There is complete dominance at both loci. Interaction between both dominants produces a new disc phenotype.

         A = Spheroid   *a* = Long
         B = Spheroid   *b* = Long

     (spheroid) *AAbb* × *aaBB* (spheroid)
               ↓
         F₁: *AaBb* (disc)
         F₂: ↓
         9 *A−B−* (disc)
    6 ⎰ 3 *A−bb* (spheroid)
       ⎱ 3 *aaB−* (spheroid)
         1 *aabb*   (long)

3.14 Montezuma = *Aa*

Aa	×	*Aa*
	↓	
AA	*Aa*	*aa*
1	2	1
Wild type	Montezuma	Lethal

3.16 Try to go to other sources here. In this book, think about retinoblastoma, *Cy* wing, *Pm* eyes, *Sb* bristles, and many others.

3.18 Dihybrid cross; 9:3:4 ratio; *A* = black, *a* = brown, *B* = color present, *b* = no color.

3.20 *A*− solid   *BB* = Yellow
   aa striped   *Bb* = Green
                 bb = Lethal

Chapter 4   The expansion of Mendelian principles, II

4.5 (a)   63.5   (c)   65.9
    (b)   68.1   (d) $\sigma^2 = 145.6$, $\sigma = 12.1$

4.7 (a) This is a trihybrid cross with three independently assorting loci.

(b) The parental genotypes were *AABBCC* × *aabbcc*, and the F_1 was *AaBbCc*. We would expect 1/64 of the F_2 to be white, but in a sample of 78, it is so low in frequency that it may not have been produced.

4.9 (a) The *i* trait (white or striped) shows maternal inheritance once it is in an *ii* strain. Once established, the trait is independent of the *i* gene and is thus permanent.

(b) 1.   Cross-3 white plants die.
     2.   Cross-3 white plants die.
     3.   All *Ii* or *II* and green
     4.   All *Ii* and green, white, and striped

4.11 All XX flies born to *aa* females die due to a maternal effect in the ooplasm.

4.13 (a) 3 killers:1 sensitive; all sensitive. From the sensitive exconjugant will come all sensitive; from the killer exconjugant will come 3 killers:1 sensitive.

(b) All progeny of *KK* sensitive will be sensitive; all progeny of *Kk* killer will be killer.

4.15 10 dominants   (1)
      9 dominants   (10)
      8 dominants   (45)
      7 dominants   (120)
      6 dominants   (210)
      5 dominants   (252)
      4 dominants   (210)
      3 dominants   (120)
      2 dominants   (45)
      1 dominants   (10)
      0 dominants   (1)

Chapter 5   Variation in chromosome number and structure

5.5 The addition or deletion of chromosomes leads to aberrant Mendelian ratios, and a cytological basis for these ratios can be established (see the answer to Problem 5.2 and pp. 60–64).

5.7 35 $B-$ to 1 bb

5.9 5 wild-type to 1 eyeless

5.11 Nondisjunction occurred in meiosis II in the female, who is a carrier.

5.13 The experiment indicates that v^+ in a single dose could not overcome vv:

5.17 C′: a b c d e f g h i j k l m n

      a b m l k j i h g f e d c n

  C: a b m l k j f g h i e d c n

Chapter 6   The identification and structure of the genetic material

6.1 No, because the sequences can be different:

$$A\ A\ T\ A\ A\ T \atop T\ T\ A\ T\ T\ A \quad \text{and} \quad A\ T\ A\ A\ T\ A \atop T\ A\ T\ T\ A\ T$$

6.3 RNA is found mainly in the cytoplasm, whereas the nucleus is the carrier of the genetic material. There is no constancy in the amount of cellular RNA as a function of cell division, but DNA concentration per cell and cell replication are well correlated. Most viruses have no RNA, but they do have DNA. These and other points all argue that DNA, not RNA, is the genetic material.

6.5 Two techniques that would work are UV absorption and DNA-specific enzymes.

6.7 RNA is single-stranded, has U instead of T, and has ribose. DNA is double-stranded, has T instead of U, and has deoxyribose.

6.9 DNA was thought to be a repeating polymer. Chargaff showed that this was not possible, because $(A + T)/(G + C) \neq 1$.

Chapter 7   The replication of the genetic material

7.1 To discriminate between these two DNA molecules, I would suggest a nearest-neighbor analysis. For the left molecule, there will be all G → C and C →

G label transferrals. For the right molecule, there will be only G → G and C → C label transferrals.

7.5 RF: A = T = 0.245
           G = C = 0.255
        Complementary strand: A = 0.22
                          C = 0.31
                          G = 0.20
                          T = 0.27

7.7 Yes, because complementary sequences carry different informational contents. They would code for different proteins.

7.9 (a) 2x (b) x (c) 2x (d) x (e) x (f) 0.5x

7.11 Cairns discovered a mutant that lacked DNA polymerase 1 yet still replicated normally. The mutant was, however, defective in its ability to repair DNA.

7.15 It initiates DNA replication by polymerizing some RNA bases in order to provide 3' OH group for the DNA polymerase III to start from.

Chapter 8   Gene function

8.2 Thymine is required for DNA synthesis but not RNA synthesis. Thus RNA is synthesized and polypeptides are assembled in the absence of DNA synthesis.

8.10 (a) *DCBA*   Direction of transcription
           3'→5'

   (b) *D C B A*

8.12 30S   50S
         L   L
         H   H
         H   L
         L   H

8.14 See the Glossary for a definition of colinearity. It relates sequence of mutant sites to amino acid sequence, thus establishing a link between gene and protein.

Chapter 9   The genetic code

9.4 The wobble hypothesis addresses this problem. The I at the third anticodon position has pairing flexibility with U, C, and A in the codon (see Table 9.8).

9.6 There are two main approaches to this problem. One approach assumes the reversion occurs at a different site within the same codon; for example: UAC (*tyr*) → UCC (*ser*) → UCU (*ser* → UAU (*tyr*). The mutant sites in this case are different. The other approach invokes the notion of a suppressor tRNA mutation, an event occurring in another gene.

9.12 (a) 1: AUU → AGU   (c) 1: AGU → GGU
         (b) 1: ACU → GCU   (d) 2: ACU → UAU

$$\boxed{UGG}$$

AGG	UAG	UGA
(Arg)	x	x
GGG	UUG	UGC
(Gly)	(Leu)	(Cys)
CGG	UCG	UGU
(Arg)	(Ser)	(Cys)

Chapter 10   Mutation and the gene concept

10.2 Mosaicism in the F_1 organisms could produce two classes of F_2 Cy//?. In one class, ? = lethal; in the other class, ? = nonlethal. This mosaicism would be missed in anything other than a single-pair mating situation.

10.4 The approach to this problem should focus on reversion analysis. A base-pair substitution mutation is usually revertible with base analogs; deletions are not. A deletion mutation may be revertible by frameshift mutagens but not by base analogs.

10.6 The *cis-trans* complementation test is designed to examine the functional relatedness of mutant alleles.

10.8 The mutation may go undetected if it is a mosaic and the mutant tissue is not involved in the mutant expression. Or the environment may modify the expression of the mutant genotype so that it appears normal. Another possibility for being unable to detect a mutation is when a mutation occurs that leads to no loss of function of the gene product. (Chapter 8 elaborates on this point.)

10.10 It is located to the right of the left end of the C deletion and to the left of the left end of the D deletion.

10.12 Two cistrons:

```
    Mutants          Mutants
   1, 3, 5, 6          2, 4
 ├──────────────┼───────────────┤
        A                B
     Cistron          Cistron
```

10.14 (a) Primarily CG → AT; some AT → GC   (b) Frameshift   (c) Deletions

Chapter 11   The genetics of viruses

11.1 Cross (a) must be the double recombinant cross, therefore *m* must be the middle gene. In crosses (b) and (c), wild-type recombinants can be generated by single break events.

11.3 Viral genetics has shown DNA to be the genetic material (Hershey-Chase). Studies on the mechanism of recombination have correlated genetic recombination with the physical exchange of DNA molecule segments.

11.9 The results show that $rIIB_2$ and $rIIB_3$ are point mutations 0.9 unit apart. The $rIIB_1$ mutation looks to be a deletion covering the $rIIB_2$ and $rIIB_3$ sites.

11.11 The mutants do not complement each other, so they do not grow on K cells. They do grow on B cells. Some progeny from this cross are recombinant (double-mutant and wild-type), and the wild-type recombinants grow on K.

11.13 It can do so when it is terminally redundant and circularly permutated.

11.15 Lambda is not terminally redundant. When the circular chromosome produces a linear form, it breaks at a specific point so that genes on each side of the break are always separated and do not appear next to each other at any time.

Chapter 12 The transmission of the genetic material in bacteria

12.9 We can, but there are other possibilities. The phage may not be able to infect the cells because of a phage defect or a resistance mutation in the *E. coli* cells. Other possibilities may also be put forth.

12.11 (a) The genes are indeed linked. (b) They are located 48 map units apart.

12.13 (a) The strain (a^+ b^+ c^- d^+) was the donor strain. (b) Most of the recombinants come from F$^-$ cells, and only some genes are transferred from Hfr → F$^-$.

12.15 Data support the model shown in Figure 12.20(b). In Figure 12.20(a), we see that infective and transducing phage have roughly the same density. In Figure 12.20(b), the transducing particles have the density of BU transducing particles in Figure 12.20(c), and the infective phage have the density of the infective phage in Figure 12.20(a). If the model shown in Figure 12.20(a) were valid, transducing phage would have intermediate densities (between T and BU).

Chapter 13 Developmental genetics

13.3 Our developing gene concept has a strong functional component. That is to say, we have characterized genes by their function and structure. Nuclear transplantation experiments add a new dimension to the gene concept by emphasizing differential gene activity in development.

13.11 It might have an I^s or P mutation; it could also be Z^-.

13.13 An O^c mutation

13.15 (a) Constitutive *A* synthesis; C^c is dominant over C^-.

(b) No constitutive *A* synthesis; C^+ is dominant over C^c.

(c) and (d) Normal *A* levels; inducible; C^+ is dominant over C^- in both the *cis* and the *trans* configuration.

13.17 It is like the *lac* operon.

13.19 Pre-existing RNA in the embryo is used up to gastrulation; then new RNA synthesis is blocked and development stops.

14.2 A mutant recessive gene that is homozygous lethal may be maintained in the heterozygous state if it confers on the heterozygote a selective advantage. Second, selection operates against phenotypes, not individual genes.

14.4
$$2pq = .50$$
$$pq = .25$$
$$q = 1 - p$$
$$p(1 - p) = .25$$
$$p^2 - p + 0.25 = 0$$
$$(p - 0.5)^2 = 0$$
$$p = 0.5$$
$$q = 0.5$$

14.6 No. There is some selection against Aa as well as $S_{aa} = 1.0$. Since S cannot be greater than 1.0, the loss of the lethal gene must also occur in heterozygous carriers.

14.8 q_0 $(a) = 0.997$ and p_0 $(A) = 0.003$ after three generations. The allele changes are -0.397 for A and $+0.397$ for a.

14.10 Here *ele* was effectively lethal, OR *e* was eliminated by selection on both *ele* and *e/+*.

14.12 $A = a = 0.5$.

14.14 $\dfrac{27}{108} \div \dfrac{582}{457} = 0.197$

14.16 $T = 0.25$ $t = 0.75$ $TT = 0.06 = $ 1.6
$Tt = 0.38 = $ 9.4 carriers
$tt = 0.56 = 14.0$

Chapter 15   Genes in populations, II

15.1 Examine their morphology, and then try to interbreed them to see if they will (1) mate and (2) produce viable fertile offspring.

15.3 Yes. Clones diverge into different environments and independently accumulate their own set of mutations. Environmental heterogeneity and genetic polymorphism set the stage for natural selection.

15.5 Race is a population phenomenon, involving gene frequency differences in populations that inhabit parts of the distribution area of a polytypic species. A single individual does not represent a gene frequency distribution.

15.7 Not necessarily. The two species may be isolated in nature by their habitat and so remain effectively isolated from each other reproductively. Furthermore, the F_1 hybrids may experience hybrid breakdown, in which the F_2 and F_3 rapidly lose vigor.

15.11 All people can interbreed.

15.13 We share a common ancestor, but we have each undergone extensive evolutionary divergence since the split.

Index

Grant, Madison, 717
Gratia, A., 464
Gray crescent, 574
Green, M.M., 446
Grell, Rhoda, 93
Griffith, F., 242–244, 506
Gross, A., 314
Gurdon, J.B., 570, 576
Gynandromorph, 208

Hadorn, E., 610
Haeckel, Ernst, 718–719
Hammerling, J. 574–575
Haploidy, 30, 185–186, 415
  crossing-over and, 90
  gene mapping in, 84–90
  Mendelian principles and, 33–37
Hardy, G.H., 641
Hayflick, Leonard, 622
Hayflick limit, 621–622
Head, of T-phage, 469
Hegel, G.W.F., 3
Hela cells, 582
Hemoglobin
  synthesis of, 578
  unequal crossing-over and, 215–216
Hemophilia, 58
  treatment of, 121
Henking, H., 47
Henschel, A.W., 3
Heredity, see Inheritance
d'Hérelle, Felix, 464
Heritability (H), 165–166
Hermaphroditic reproduction, 186
Hershey, A.D., 248–253, 466, 472,
  478–480, 489, 491
Hertwig, O., 25, 26, 239
Heteroduplex, 490
Heterogametic females, 48
Heterogeneous nuclear RNA (hnRNA),
  614–615, 617
Heterokaryon, 81, 447
  complementation tests on, 447, 449
  formation of, 34
Heterosis, 652–653
  mechanism of, 653–654
Heterozygosity, 14, 15
  in bacteriophages, 489–490
  heterosis and, 652
  inversion and, 223–224
  translocation and, 227, 229–231
Heterozygote superiority, 666–667
Hiesey, W.M., 685
High-frequency recombinant (Hfr), 521
  sexduction and, 534–536
Himmler, Heinrich, 719
Histidine, 384
Histones, gene activity and, 611
Hitler, Adolf, 718, 719
Holley, R.W., 346
Holliday, R., 304
Homogametic males, 48
Homolog, 30

Homozygosity, 14–15
Hooke, Robert, 24
Hormones
  as regulators of gene activity, 578–581
  sex determination and, 52–54
Horse, evolution of, 703–704
Hot spots, 421
Humans
  aneuploidy in, 196–211
  diplotene in, 30–31
  gene mapping in, 80–83
  genetic analysis in, 32
  segregation principle seen in, 17–19
  sex chromosomes in, 49, 51
Hutchison, C.A., III, 455
Hybridization
  Darwin's rejection of, 4
  doctrine of fixed species and, 3
  pre-Mendelian, 7–10
  study of RNA transcription by,
    338–339
Hybrid vigor, 652
Hydrocarbons, 728
Hydrogen bonds, 258–259
Hydroxylamine, mutaton induced by,
  435
Hyphae, 34

Imaginal discs, 610
Immigration Restriction Act, 716, 717
Inbreeding, 651
  depression caused by, 651–652
Incorporation, error of, 435
Independent assortment, 14, 20–23
  in haploid organisms, 36–37
  during meiosis, 32
Indole, in protein synthesis, 321
Inducer, enzyme synthesis and, 588,
  598–590
Induction, 476
  in bacteriophages, 546
  of enzymes, 589–590
  zygotic, 548
Informosomes, 584
Inheritance
  chemical basis of, 238–255
  chromosome theory of, 32, 46
  cytological basis of, 24–38
  extranuclear, 166–179
  holandric, 58
  polygenic, 157
  quantitative, 152–166
  of sex, 46–54
Initial recognition site, 343
Initiation site, 343
Inman, R.B., 292
Insecticides, resistance to, 661–662
Insect pigments, genetic control of,
  316–318
Insertion elements, 551–553
Insulin, 121
Integration protein, 551

Intelligence
  autosomal aneuploidy and, 201
  fluid and crystallized, 125
  genotypic and environmental con-
    tributions to, 125–127, 722–723
  sex-chromosome aneuploidy and,
    198, 201
Intelligence tests, 125
Intercalation, 441
Interference, 77
  in bacteriophages, 489
  negative, 77–78, 489
Interphase
  meiosis, 31
  mitosis, 29
Interrupted-mating technique, 521–522
Interstitial region, 230
Intervening sequences, 617
Intracellular organelles, 169. See also
  Mitochondria
Intracellular symbionts, 168–171
Inversion, 220–226, 409
  paracentric, 220, 222–223
  pericentric, 220, 221–222
Inversion loop, 221
Ionizing radiation, mutations induced
  by, 426–430
IQ, see Intelligence
Isolating mechanisms, 693–696

Jacob, F., 353, 354, 521–523, 524, 535,
  590, 591, 593
Janssens, F.A., 64–65
Jayaraman, R., 340
Jenkins, J.B., 414
Jensen, Arthur, 127, 722
Johannsen, W., 155
Johnson Act, 716
Joseph, Anton, 19
Joseph Family Disease, 19
Josse, J. 279
Judd, B.H., 145
Jukes, T.H., 403

Kaiser, A.D., 279, 485–487
Kappa factor, 169–170
Keck, D.D., 685
Kettlewell, H.B.D., 662
Khorana, H.G., 384
Khrushchev, Nikita, 721
Kimura, Motoo, 674–675
King, J.T., 585
King, M.C., 707, 708
King, T.J., 567, 568
Klinefelter's syndrome, 198, 201
Knight, T.A., 8–9
Kölliker, R.A., 26
Kölreuter, Josef Gottlieb, 3, 8, 25–26,
  152
Kornberg, A., 275–278
Kornberg, Arthur, 279, 285
Kornberg, T., 289

pre- and postadaptive, 423–426
rate of, 420–421
recessive sex-linked lethal, 409–411,
427, 428–429
recombination and, 450–451, 453
somatic theory of, 623–624
spontaneous, 420–421
suppressor, 348, 394–399
temperature-sensitive, 121
termination, 325–327
transition, 434
transversion, 434
ultraviolet-induced, 430–431
Mutational equilibrium, 656
Mutational events, 420
Mutational hot spots, 421
Mycelium, 34

Nägeli, Karl von, 23–24, 26
National Institutes of Health (NIH),
724, 726
Natural selection
basic concepts of, 657–661
cost of, 667
directional, 664
disruptive, 665–667
evolution by, 635–639
examples of, 661–664
mutation balanced with, 667–669
stabilizing, 665
Nature–nurture controversy
intelligence and, 125–127, 722–723
twin studies and, 121–124
Nature-philosophers, 3, 9
Naudin, Charles, 9
Nazism, eugenics and, 718–719
Nearest-neighbor analysis, 279–283,
337
Needle, of T-phage, 469
Nester, E.W., 514
Neurospora, 318–319
crosses in, 37–38
gene mapping in, 84–90
life cycle of, 33–35
mutation study in, 415
negative interference in, 78
Neutralist hypothesis, 674–676
Newcombe spreading technique, 424–
426
Nilsson-Ehle, H., 155–157, 640
Nirenberg, M.W., 379, 381, 384, 386
Nitrous acid, mutation induced by, 435
Nondisjunction, 60–64
cause of, 211
during meiosis, 194
primary, 60, 62
secondary, 60, 62
Nonparental ditype, 37
Normal distribution, 164
Norm of reaction, 119, 155
Novick, A., 499
Novitski, E., 144

Nuclease activity, of DNA polymerase,
286–288
Nucleic acids, genetic code and, 375
Nuclein, 26
function of, 239
isolation of, 238–242
Nucleocytoplasmic interaction, 567
equivalence of nuclei and, 568–570
regional differentiation and, 570–575
Nucleolar organizer (NO) region, 576
cell differentiation and, 584
gene amplification and, 612–613
Nucleolus, 576
Nucleosomes, 611–612
Nucleus. *See also* Meiosis; Mitosis
discovery of role of, 25
equivalence during early development, 568–570
during fertilization, 25–26
RNA in, 614–615
Nullisomy, 194

Ochoa, S., 384, 387
Ohta, T., 676
Okazaki, R., 296
Okazaki fragments, 296, 298, 299
Olbrycht, T.M., 72
Oliver, C.P., 110
Oncogene, 497–498
One gene–one enzyme theory, 318–
319
"One gene–one ribosome–one pro-
tein" hypothesis, 353, 355
One-step growth experiment, 466, 471
Oocytes
during diplotene, 30–31
primary and secondary, 36
Oogonia, 36
Operon theory, 591–593
arabinose operon, 601–603
cyclic AMP and, 599–600
genetic proof of, 593–597
histidine operon, 600–601
lysogeny and, 603–606
promoter and, 597–599
Orgel, L.E., 624
Orthogenesis, 701, 704–706
Ovary, 36
Overdominance, 112, 653
Ovum, 36
Ozone, mutagenic hazard of, 728

Pachytene, 30
Pair formation, specific and effective,
537
Pangenesis, 635
theory of, 5–6
Panmixis, 639, 651
Parameters, 162
Parasexual processes, 47, 506, 516
Pardee, A.B., 590
Parental ditype, 37

Parthenogenetic reproduction, 186
Pascal's triangle, 7
Pastan, I., 599
Pauling, Linus, 256, 258, 260, 262
Pea, *see* Garden pea
Pearson, Karl, 152
Penetrance, 118
incomplete, 118
Peptide bonds, 327, 330
Peptidyl site, 363
Peptidyl transferase, 364
Perithecium, 35
Perlman, R., 599
Petite mutation, 172–173, 177
neutral, 173, 175–176
segregational, 173
suppressive, 176–177
Phage, *see* Bacteriophage
Phenocopy, 119–120
Phenotype, 15, 16
acquired characters and, 676–678
components of, 155
cytoplasmic influence on, 574–575
genotype differing from, 51
intersexual, 47
natural selection and, 660–661
sex chromosomes and, 49, 51
Phenotypic mixing, 499
Phenotypic ratio, 15
Phenylalanine
code word for, 381–382
metabolism of, 314–316
Phenylketonuria (PKU), 315–316
Phenylthiocarbamide (PTC), ability to
taste, 160–161, 646
Philadelphia chromosome, 219
Phocomelia, thalidomide-induced,
119–120
Photoreactivation, 302
Physiological isolation, 695
Pigments, genetic control of, 316–318
Pili, 537
Pistil, 13
Plaque, 418, 471
size and shpe of, 476–477
Plasmids, 556–557
recombinant DNA and, 557–558
Pleiotropy, 660–661
Polar bodies, 31, 36
Pollin, W., 127–128
Pollination, 13
Pollister, A.W., 247
Pollution, mutagenic hazard of, 727–
729
Polymorphism, 639–640
of enzymes, 674–675
Polynuclear aromatic hydrocarbons,
728
Polynucleotide phosphorylase, 381
Polypeptides, 330
DNA coding for, 456–457
elongation of, 363–364